Ivan Zelinka, Sergej Celikovsky, Hendrik Richter, and Guanrong Chen (Eds.)

Evolutionary Algorithms and Chaotic Systems

Studies in Computational Intelligence, Volume 267

Editor-in-Chief

Prof. Janusz Kacprzyk
Systems Research Institute
Polish Academy of Sciences
ul. Newelska 6
01-447 Warsaw
Poland
E-mail: kacprzyk@ibspan.waw.pl

Further volumes of this series can be found on our homepage: springer.com

Vol. 246. Thanasis Daradoumis, Santi Caballé, Joan Manuel Marquès, and Fatos Xhafa (Eds.)
Intelligent Collaborative e-Learning Systems and Applications, 2009
ISBN 978-3-642-04000-9

Vol. 247. Monica Bianchini, Marco Maggini, Franco Scarselli, and Lakhmi C. Jain (Eds.)
Innovations in Neural Information Paradigms and Applications, 2009
ISBN 978-3-642-04002-3

Vol. 248. Chee Peng Lim, Lakhmi C. Jain, and Satchidananda Dehuri (Eds.)
Innovations in Swarm Intelligence, 2009
ISBN 978-3-642-04224-9

Vol. 249. Wesam Ashour Barbakh, Ying Wu, and Colin Fyfe
Non-Standard Parameter Adaptation for Exploratory Data Analysis, 2009
ISBN 978-3-642-04004-7

Vol. 250. Raymond Chiong and Sandeep Dhakal (Eds.)
Natural Intelligence for Scheduling, Planning and Packing Problems, 2009
ISBN 978-3-642-04038-2

Vol. 251. Zbigniew W. Ras and William Ribarsky (Eds.)
Advances in Information and Intelligent Systems, 2009
ISBN 978-3-642-04140-2

Vol. 252. Ngoc Thanh Nguyen and Edward Szczerbicki (Eds.)
Intelligent Systems for Knowledge Management, 2009
ISBN 978-3-642-04169-3

Vol. 253. Roger Lee and Naohiro Ishii (Eds.)
Software Engineering Research, Management and Applications 2009, 2009
ISBN 978-3-642-05440-2

Vol. 254. Kyandoghere Kyamakya, Wolfgang A. Halang, Herwig Unger, Jean Chamberlain Chedjou, Nikolai F. Rulkov, and Zhong Li (Eds.)
Recent Advances in Nonlinear Dynamics and Synchronization, 2009
ISBN 978-3-642-04226-3

Vol. 255. Catarina Silva and Bernardete Ribeiro
Inductive Inference for Large Scale Text Classification, 2009
ISBN 978-3-642-04532-5

Vol. 256. Patricia Melin, Janusz Kacprzyk, and Witold Pedrycz (Eds.)
Bio-inspired Hybrid Intelligent Systems for Image Analysis and Pattern Recognition, 2009
ISBN 978-3-642-04515-8

Vol. 257. Oscar Castillo, Witold Pedrycz, and Janusz Kacprzyk (Eds.)
Evolutionary Design of Intelligent Systems in Modeling, Simulation and Control, 2009
ISBN 978-3-642-04513-4

Vol. 258. Leonardo Franco, David A. Elizondo, and José M. Jerez (Eds.)
Constructive Neural Networks, 2009
ISBN 978-3-642-04511-0

Vol. 259. Kasthurirangan Gopalakrishnan, Halil Ceylan, and Nii O. Attoh-Okine (Eds.)
Intelligent and Soft Computing in Infrastructure Systems Engineering, 2009
ISBN 978-3-642-04585-1

Vol. 260. Edward Szczerbicki and Ngoc Thanh Nguyen (Eds.)
Smart Information and Knowledge Management, 2009
ISBN 978-3-642-04583-7

Vol. 261. Nadia Nedjah, Leandro dos Santos Coelho, and Luiza de Macedo de Mourelle (Eds.)
Multi-Objective Swarm Intelligent Systems, 2009
ISBN 978-3-642-05164-7

Vol. 262. Jacek Koronacki, Zbigniew W. Ras, Slawomir T. Wierzchon, and Janusz Kacprzyk (Eds.)
Advances in Machine Learning I, 2009
ISBN 978-3-642-05176-0

Vol. 263. Jacek Koronacki, Zbigniew W. Ras, Slawomir T. Wierzchon, and Janusz Kacprzyk (Eds.)
Advances in Machine Learning II, 2009
ISBN 978-3-642-05178-4

Vol. 264. Olivier Sigaud and Jan Peters (Eds.)
From Motor Learning to Interaction Learning in Robots, 2009
ISBN 978-3-642-05180-7

Vol. 265. Zbigniew W. Ras and Li-Shiang Tsay (Eds.)
Advances in Intelligent Information Systems, 2009
ISBN 978-3-642-05182-1

Vol. 266. Akitoshi Hanazawa, Tsutom Miki, and Keiichi Horio (Eds.)
Brain-Inspired Information Technology, 2009
ISBN 978-3-642-04024-5

Vol. 267. Ivan Zelinka, Sergej Celikovsky, Hendrik Richter, and Guanrong Chen (Eds.)
Evolutionary Algorithms and Chaotic Systems, 2010
ISBN 978-3-642-10706-1

Ivan Zelinka, Sergej Celikovsky, Hendrik Richter,
and Guanrong Chen (Eds.)

Evolutionary Algorithms and Chaotic Systems

Springer

Prof. Ivan Zelinka
Department of Applied Informatics
Faculty of Applied Informatics
Tomas Bata Univerzity in Zlin
Nad Stranemi 4511
Zlin 76001
Czech Republic
E-mail: zelinka@fai.utb.cz

Prof. Hendrik Richter
HTWK Leipzig
Faculty of Electrical Engineering &
Information Technology
04251 Leipzig
Germany
E-mail: richter@fbeit.htwk-leipzig.de

Prof. Sergej Celikovsky
Department of Control Engineering
Faculty of Electrical Engineering
Czech Technical University in Prague
UTIA AV CR
Pod vodarenskou vezi 4
182 08 Prague 8
Czech Republic

Prof. Guanrong Chen
Department of Electronic Engineering
City University of Hong Kong
83 Tat Chee Avenue, Kowloon
Hong Kong SAR
P. R. China
E-mail: gchen@ee.cityu.edu.hk

ISBN 978-3-642-10706-1 e-ISBN 978-3-642-10707-8

DOI 10.1007/978-3-642-10707-8

Studies in Computational Intelligence ISSN 1860-949X

Library of Congress Control Number: 2009943373

Typeset & Cover Design: Scientific Publishing Services Pvt. Ltd., Chennai, India.

Printed in acid-free paper

9 8 7 6 5 4 3 2 1

springer.com

Ivan Zelinka dedicates this book to his wonderful wife Martina, his beautiful daughters Marketa and Katerina, and to his parents.

Sergej Celikovsky dedicates this book to his parents Karel and Galina, his wife Avgustina, daughter Klara and son Viktor.

Guanrong Chen dedicates this book to the memory of his mentor Professor Mingjun Chen (1934-2008).

Foreword

Ever since the historical discovery of the now-famous Lorenz system in 1963, a large number of nonlinear systems that can produce chaos have been observed, constructed and analyzed. In fact, chaos theory has become indispensable for science and engineering at all levels of research today. The most active recent research includes chaos control and chaos synchronization, among others, with a visible trend toward real-world applications.

The book titled "Evolutionary Algorithms and Chaotic Systems", edited by Ivan Zelinka, Sergej Celikovsky, Hendrik Richter and Guanrong Chen, is a timely volume to be welcome by the chaos community as well as computational intelligence community and beyond. This book is devoted to the studies of common and related subjects in two intensive research fields of chaos theory and evolutionary computation. It was not typical that evolutionary computing techniques are used for effective chaos control, chaos synchronization, chaos identification, and in particular for chaos analysis and synthesis, therefore this edition of collective state-of-the-art articles on such interdisciplinary subjects is especially valuable for the scientific and engineering communities. For these reasons, I enthusiastically recommend this book to our scientists and engineers working in the fields of nonlinear dynamics, evolutionary algorithms, control theory, circuits and systems, and scientific computing alike.

University of California at Berkeley, September 2009 Leon O. Chua

Preface

Deterministic chaos is a fairly active area of research in the last few decades. Well known chaotic attractors can even be produced by some simple three-dimensional autonomous systems of ordinary differential equations, for example the Lorenz system, which originates from modelling of atmospheric dynamics. For discrete chaos, there is another famous chaotic system, called logistic equation, which was found based on a predator-prey model showing complex dynamical behaviors. These simple models are widely used in the study of chaos today, while other similar models exist (e.g., canonical logistic equation and 1D or 2D coupled map lattices). To date, a large set of nonlinear systems that can produce chaotic behaviors have been observed and analyzed. Chaotic systems thus have become a vitally important part of science and engineering at the theoretical as well as the practical level of research. The most interesting and applicable notions are, for example, chaos control and chaos synchronization related to secure communications, among others. Recently, the study of chaos is focused not only along the traditional trends but also on the understanding and analyzing principles, with the new intention of controlling and utilizing chaos toward real-world applications.

This book discusses the mutual intersection of two interesting fields of research, i.e. deterministic chaos and evolutionary computation. Evolutionary techniques are discussed in this book, which are able to handle tasks such as control of various chaotic systems and synthesis of their structures (i.e., handling symbolic objects to create more complex structures). In this way, evolutionary techniques are capable of synthesizing chaotic behavior in the sense that mathematical descriptions of chaotic systems are generated symbolically. Another capability of evolutionary computation - identification of chaotic system structure-is also discussed in this book. Part of the book is focused on how chaos can be observed in the dynamics of evolutionary algorithms and used to improve performance of selected evolutionary techniques.

Chapter authors background: Chapter authors are to the best of our knowledge the originators or closely related to the originators of the above mentioned applications of evolutionary computation as well as the applications of chaos principles in selected evolutionary algorithms. Hence, this book will be one of the few books discussing the benefit from intersection of two modern and fruitful scientific fields of research.

Organization of the Chapters and Book Structure: The book consist of three parts. The first part is presented by Zelinka and Chen as a motivation for the application of evolutionary computation on chaotic systems (Chapter 1). It is followed by a brief introduction of evolutionary algorithms for chaos researchers (Zelinka and Richter, Chapter 2). The next chapter is a complementary and serves as an introduction of chaos theory for evolutionary algorithms researchers (Celikovsky and Zelinka, Chapter 3). The last chapter of this first part discusses the appearance of the so-called edge of chaos in evolutionary algorithms (Davendra, Chapter 4).

The second part discusses the use of evolutionary algorithms on chaotic dynamics. A reader can find here an approach of evolutionary algorithms to 1D chaos control (Senkerik et. al., Chapter 5), spatiotemporal chaos control (Zelinka, Chapter 6) or chaos reconstruction by means of standard methods (Chapter 7, Chadli), which is followed by Chapter 8 (Zelinka, Raidl) in which evolutionary algorithms are used to reconstruct chaotic systems from measured data. Chapter 9 (Ping Li et. al.) is focused on the use of chaos in encryption, Chapter 10 (Zelinka, Jasek) demonstrates possible benefit and drawback of the usage of evolution on decryption of chaotically encrypted information. In Chapter 11 (Zelinka, Chen, Celikovsky) synthesis of chaotic structure by means of genetic programming like techniques is discussed. Furthermore, Chapter 12 is centered on the application of evolutionary algorithms on chaos synchronization (Zelinka, Raidl). Finally, the application of evolutionary optimization in chaotic CML-based fitness landscapes by Richter (Chapter 13) is discussed.

The third part discuss the appearance and use of deterministic chaos in evolutionary techniques. Chapter 14 (Davendra, Zelinka) describes the impact of various chaotic systems use on mutation of individuals and Chapter 15 shows the appearance of chaos in selected evolutionary techniques (Davendra, Zelinka and Onwubolu) and discusses the impact of chaos on permutative optimization.

The book is based on original research and contains all important results including more than 589 pictures.

Audience: The book will be an instructional material for senior undergraduate and entry-level graduate students in computer science, physics, applied mathematics and engineering, who are working in the area of deterministic chaos and evolutionary algorithms. Researchers who want to investigate how evolutionary algorithms can be used for chaos control as well as researchers interested in the appearance of chaos in evolutionary algorithms will find this book a very useful handbook and starting step-stone. The book will also be a resource and material for practitioners who want to apply these methods to solve real-life problems in their challenging applications.

Appendix: The appendix contains description of Mathematica software and user manual for 40 notebooks. Their actual versions can also be downloaded from www.fai.utb.cz/people/zelinka/evolutionarychaos or www.ivanzelinka.eu/evolutionarychaos. It consist Mathematica notebooks of different evolutionary (or random-like) algorithms and test functions, interactive notebooks allowing manipulation of different chaotic systems and notebooks supporting selected case studies reported in this book.

Motivation: The decision as why to write this book was based on a few facts. The main one is that the research field on evolutionary algorithms and deterministic chaos is an interesting area, which is under intensive research from many other branches of science today. Evolutionary algorithms with its applications can be found in biology, physics, economy, chemical technologies, air industry, job scheduling, space research (i.e. antena design for space mission), amongst others. The same can be stated for deterministic chaos. This kind of behavior can be observed in physical as well as biological, economical systems etc. Due to the fact that evolutionary algorithms are capable of solving many problems including problems containing imprecise information or uncertanities, it is obvious that it can also be used on chaotic systems to control, synchronize or/and synthesize them. On the other hand, chaotic systems and their behavior are very important in engineering, because such behavior can be used to encrypt important information or, for example, cause damage if not expected or desired in designed device. Together with "classical" techniques, evolutionary algorithms can be used to solve various tasks based on deterministic chaos. This book was written to contain simplified versions of our experiments with the aim to show how, in principle, evolutionary algorithms can be used on chaotic systems, and vice versa.

It is obvious that this book does not encompass all aspects of these two fields of research due to limited space. Only the main ideas and results are reported here. The authors and editors hope that the readers will be inspired to do their own experiments and simulations, based on information reported in this book, thereby moving beyond the scope of the book.

September 2009

Czech Republic	Ivan Zelinka
Czech Republic	Sergej Celikovsky
Germany	Hendrik Richter
Hong Kong	Guanrong Chen

Acknowledgements

We are grateful to all the contributors of this book, for their willingness to work on this interdisciplinary book project which focuses on intersection of two interesting fields of research: deterministic chaos and evolutionary algorithms. To write such a book was a little bit complicated due to the fact that both areas are quite special and researchers are usually experts only of one of the two areas. Fortunately, we were able to establish a very good team of contributors, who were interested in participation and were working professionaly and quickly. Also, the help and support of other colleagues and friends are indispensable to this book. Thus, we would like to thank all of them.

Firstly, we would like to thank Prof. Leon O. Chua (University of California at Berkeley) for his very kind, nice and inspirative Foreword.

To Springer-Verlag in Heidelberg, Germany, our special thank goes to Dr. Thomas Ditzinger for his enthusiasm and editorial guidance throughout the period of our preparing the book, and for his patience.

Our special thanks are dedicated to our families for their patience with our time spent on book writing as well as supporting software preparation. Especially, to Martina, Marketa and Katerina Zelinkova for their patience and support to the first editor, who spent almost all holidays on this book instead of spending wonderful time with his little children.

Very special thanks are dedicated to Donald Davendra for his enormous help with final book collection and TeX error correction. Without his help, the time needed to finish this book would be significantly longer.

Contents

List of Contributors

Ivan Zelinka
Tomas Bata University in Zlin,
Faculty of Applied Informatics,
Nad Stranemi 4511,
Zlin 76001, Czech Republic
VSB-TUO, Faculty of Electrical
Engineering and Computer Science,
17. listopadu 15, 708 33
Ostrava-Poruba, Czech Republic,
zelinka@fai.utb.cz

Guanrong Chen
Department of Electronic
Engineering,
City University of Hong Kong,
Kowloon, Hong Kong SAR,
P.R. China
eegchen@cityu.edu.hk

Hendrik Richter
HTWK Leipzig, Fakultät
Elektrotechnik und
Informationstechnik,
D–04251 Leipzig, Germany
richter@fbeit.htwk-leipzig.de

Sergej Celikovsky
Department of Control Engineering
Faculty of Electrical Engineering
Czech Technical University in Prague
Control Theory Department
and
Institute of Information
Theory and Automation,
Academy of Sciences of
Czech Republic,
Pod Vodarenskou vezi 4,
182 08, Praha 8, Czech Republic
celikovs@utia.cas.cz

Donald Davendra
Tomas Bata University in Zlin,
Faculty of Applied Informatics,
Nad Stranemi 4511, Zlin 76001,
Czech Republic
davendra@fai.utb.cz

Roman Senkerik
Tomas Bata University in Zlin,
Faculty of Applied Informatics,
Nad Stranemi 4511, Zlin 76001,
Czech Republic
senkerik@fai.utb.cz

Zuzana Oplatkova
Tomas Bata University in Zlin,
Faculty of Applied Informatics,
Nad Stranemi 4511, Zlin 76001,
Czech Republic
oplatkova@fai.utb.cz

Mohammed Chadli
University of Picardie Jules Verne, Laboratory of Modeling Information & Systems. 7, Rue du Moulin Neuf, 80000, Amiens, France
Tel.: +33(0)3 82227680
mohammed.chadli@u-picardie.fr

Ales Raidl
Charles University, Faculty of Mathematics and Physics, V Holesovickach 2, 180 00 Prague 8, Czech Republic
ales.raidl@mff.cuni.cz

Ping Li
Department of Electronic Engineering, Shunde Polytechnic, Kanton, P.R. China
Kikiliping@hotmail.com

Zhong Li
Faculty of Electrical and Computer Engineering, FernUniversität in Hagen, 58084 Hagen, Germany
zhong.li@fern-hagen.de

Wolfgang A. Halang
Faculty of Electrical and Computer Engineering, FernUniversität in Hagen, 58084 Hagen, Germany

Roman Jasek
Tomas Bata University in Zlin, Faculty of Applied Informatics, Nad Stranemi 4511, Zlin 76001, Czech Republic
jasek@fai.utb.cz

Godfrey Onwubolu
School of Applied Technology, Humber Institute of Technology and Advanced Learning, Toronto, ON, Canada M9W 5L7
godfrey.onwubolu@humber.ca

Part I
Theory

Chapter 1
Motivation for Application of Evolutionary Computation to Chaotic Systems

Ivan Zelinka and Guanrong Chen

Abstract. This chapter focuses on motivating an application of evolutionary computation to complex problems especially with respect to chaotic systems. In this chapter, general evolutionary techniques are first reviewed, including the so-called evolvable hardware, with some selected examples of their applications. Then, motivation of studying chaotic systems as an interesting application domain for evolutionary algorithms is provided with brief discussions.

1.1 Introduction

Evolutionary algorithms (better known as "evolutionary computational techniques" or simply "evolutionary techniques") are powerful tools that can be used to solve various very complex engineering problems. Generally speaking, evolutionary techniques can be divided into two main categories: evolutionary algorithms (such as genetic algorithms, particle swarming, ant colony optimization, ...) and evolvable hardware (application of evolutionary techniques to hardware design, adaptation, self-repairment, ...). In principle, evolutionary techniques solve selected problems in the same way as human, which in general can be used successfully to a large set of engineering problems like the design of different devices and complex systems identification, control and modeling, etc.

Ivan Zelinka
Tomas Bata University in Zlin, Faculty of Applied Informatics, Nad Stranemi 4511, Zlin 76001, Czech Republic
and
VSB-TUO, Faculty of Electrical Engineering and Computer Science, 17. listopadu 15, 708 33 Ostrava-Poruba, Czech Republic
e-mail: zelinka@fai.utb.cz

Guanrong Chen
Department of Electronic Engineering, City University of Hong Kong, Kowloon, Hong Kong SAR, P.R. China
e-mail: eegchen@cityu.edu.hk

I. Zelinka et al. (Eds.): Evolutionary Algorithms and Chaotic Systems, SCI 267, pp. 3–36.
springerlink.com

To describe the application capability of evolutionary techniques, some selected representative examples are first reviewed, followed by a discussion of evolvable hardware, which is a major part of evolutionary techniques and can be used to solve complex technical problems in various engineering designs.

1.2 Evolutionary Computation and Selected Examples

The application of evolutionary techniques to hardware design (such as electronic devices, chaotic circuits, ...) is termed *evolvable hardware*, aiming to replace some traditional design methods with evolutionary techniques for specified applications that either are not achievable by using traditional methods or can benefit from the evolutionary approach.

The development of evolutionary hardware has been inspired by several other fields as illustrated by Fig. 1.1.

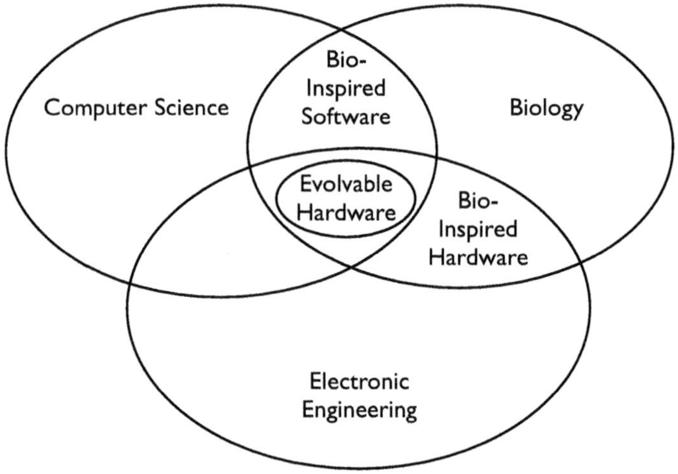

Fig. 1.1 Evolvable hardware and sciences of computer, biology and electrical engineering

Recently, the field of bio-inspired hardware has emerged, based on ideas from biology, to explore methods of fault tolerance and reconfigurability in modern hardware designs. There are many interchangeable ideas between the fields of evolvable hardware and bio-inspired hardware. However the main focus in this paper is on the field of evolvable hardware, which lies in the overlapping zone among basic sciences of computer, biology and electronic engineering.

Evolutionary techniques has been in existence for quite a long time, successfully solved many complex problems, showing its powerful applications in engineering practice and theory (see Section 1.2.8). The following examples demonstrate the successful applications of evolutionary techniques:

- **Real-time plasma reactor control.** Selected evolutionary algorithms have been used to control a plasma reactor. No mathematical model was needed. The reactor was running in real time during experiments. Evolutionary algorithm was used to estimate 14 parameters so as to eliminate noise from measured signals (see Section 1.2.10, [68] and [47]). The most important aspect of this contribution is the demonstrated ability of evolutionary techniques for controlling black-box problems in real-time. Another very important contribution is the use of laboratory hardware equipment to calculate the fitness of just-in-time synthesized solutions without knowing a cost function.
- **Fingerprint identification.** Evolutionary algorithms have been used by computer scientists (e.g., Grasemann and Miikkulainen, Neural Networks Research Group from the University of Texas at Austin, USA). Genetic algorithm has been used to develop a program, which can better digitally improve the quality of fingerprint images than programs created by human programmers. During evolution, the best solutions of each generation in the algorithm were recorded and discovered. After 50 generations, the algorithm outperforms the comparable human program. The algorithm was then synthesized. In a typical FBI application, taking into consideration that the FBI has more than 50 million sets of fingerprints in its archives, so any algorithm needs to perform about 60,000 digital fingerprint image transactions every day, it is clear that evolutionary algorithms might soon help speed up such a time-consuming identification process. This kind of tasks were investigated by e.g., Ammar and Tao from the West Virginia University, USA [21]. In their application, they used genetic algorithm to optimize the alignment of a pair of fingerprint images. To test the performance, other two algorithms were compared with the genetic algorithm. All simulations were carried out on 250 pairs of fingerprint images. Their results showed that the genetic algorithm was about 13% more accurate than the well-known 2D algorithm within the same running time.
- **Airplane optimization.** Evolutionary techniques are also widely used in airplane engineering [24]. There are numerous examples of wing optimization and optimal design of various mechanical parts of an airplane under investigation. An interesting approach is described in [33], where minimization of sonic boom on a supersonic aircraft was based on an evolutionary algorithm. Evolutionary techniques are also used for optimization of flight dynamics [60], where calculation of the aircraft trimming process was presented using an efficient combination of global optimization theory and the direct hand-on computer simulation. In this approach, the methods of artificial intelligence are combined with heuristic algorithms to deal with the complicated equations of airplane motion. The method may be practically utilized in constructing airplanes as well as flight simulators design.
- **Antenna design.** In this application, evolutionary hardware has been used to design special antenna for NASA space mission (e.g., the Space Technology 5 Project (ST5), http://nmp.jpl.nasa.gov/st5/). The ST5 space program is focused on the use of identical satellites to test new space technologies. It is part of the New Millennium Program (NMP). The NMP was created to identify, develop,

build, and test innovative technologies and concepts for infusion into future space missions. Despite numerous engineering designs, this particular design was deemed the best.

- **Flow shop scheduling - permutation-based combinatorial optimization.** A very important application in industry is scheduling, to which a number of manufacturing problems are associated. There is many such problems that cannot be solved by using conventional methods or cannot be solved in reasonable time (see Chapter 2). Such problems can be successfully solved by evolutionary algorithms, however. A typical problem is the traveling salesman problem. Today, there exist the so-called ACO algorithm (see Chapter 2), which is able to solve this hard problem with cardinality 10000! in a reasonable time yielding good results. The class of permutation-based combinatorial optimization problem is one of the famous optimization problems just like traveling salesman problem and vehicle routing problem. The most realistic and interesting are the shop scheduling problems for flow shop and job shop. What makes the permutation-based problem complex is that the solution representation is very concise, since it must have a discrete number of values and each variable in the solution is unique. Given a problem of size n, a representation can be described as $x = \{x_1, x_2, x_3, \dots, x_n\}$, where each value x_i in the solution is unique and the entire set of solutions is an integer representation from 1 to n. From an optimization point of view, this represents a number of problems. Firstly, the search space is discrete and a number of validations inevitably have to be conducted in order to have a viable solution. Secondly, the search space is very large, to $n!$. Consequently, these problems are generally NP or NP-hard [22].
- **Chemical reactor design.** In chemical engineering, evolutionary optimization has been applied to system identification [50], [51], where a model of a process is built and then its parameters are identified by error minimization against experimental data. Evolutionary optimization has been widely applied to the evolution of neural network models for use in control applications (e.g., [35]). There has been increasing awareness of textbook knowledge and heuristics [34], which were commonly employed in the development of chemical reactors, were deemed responsible for the lack of innovation, quality, and efficiency that characterizes many industrial designs. In such examples, among many others, it has been proved that evolutionary techniques are highly effective for the application in chemical engineering.
- **Bioinformatics.** Bioinformatics applies information technology to the field of molecular biology. The term bioinformatics was coined by Hogeweg in 1978 for the study of informatics processes in biotic systems. Today, it entails the creation and advancement of databases, algorithms, computational and statistical techniques, as well as the theory to solve formal and practical problems arising from management and analysis of biological data. Evolutionary algorithms have been successfully applied in, for example, multi-objective optimization in modeling of protein structure prediction [31], evolutionary optimization of metabolic pathways in [9], and so on (see [10])

There are many other examples worthy of mentioning, where evolutionary algorithms find successful applications. Standard evolutionary algorithms "exist" since the first famous seminal paper [23], while evolvable hardware is merely a "product" of later years. Both algorithms and hardware are complementarily joined together to solve various complex and difficult problems. The inter-relationships among various areas of hardware design, synthesis, and evolutionary computation are shown below in Fig. 1.2.

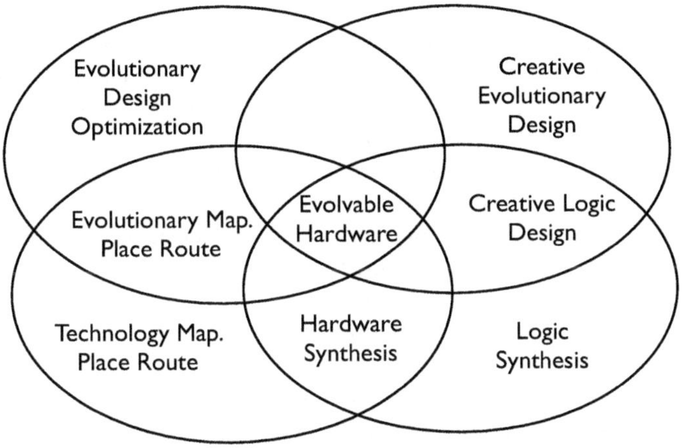

Fig. 1.2 Inter-relational aspects of evolvable hardware

The field of evolvable hardware is still in its infancy, and there are many problems that must be tackled before one can see a large-scale industrial applications of the technology. Two important parts of evolvable hardware design are:

Extrinsic evolution uses software simulation of the underlying hardware to evaluate the fitness value of each individual in the algorithm. This may be an advantage if one does not wish to be too technology-specific in the sense that the model of the hardware can be rather general and even abstract. On the other hand, if technology itself is the goal then more accurate fitness values may be obtained from a real implementation other than from numerical simulation. Since fitness steers the selection process, and thus the evolution, abstraction from technology can lead to a less-optimal solution.

Intrinsic evolution is based on hardware implementation, where each individual in the algorithm is implemented and evaluated based on the target technology. This approach can be used to explore properties of the technology, which otherwise cannot be utilized by traditional design methods. The evolution process runs on a host computer responsible for selection and performance of genetic operators. Each individual in the algorithm is down-loaded to the chip as a configuration data-design description. Fitness evaluation of a given individual is achieved by applying test vectors to the implemented individual and then calculating the fitness value from its response.

After all, the main objective is then to analyze the design phase and all its attributes. The following section will give a brief overview of the design phase. Then, the third section will introduce some applications of the evolvable hardware. The fourth section will discuss some research aspects of evolvable hardware, and the final section will provide the criteria for successful evolutionary platforms.

1.2.1 Evolutionary Design

For evolvable hardware design problems, assuming design characteristics, the evolutionary algorithm determines some of the structure and/or parameters of a reconfigurable object. This object may exist in software, though it could be a simulation of the hardware of a final implementation.

The reconfigurable object might alternatively be physically changeable hardware. Typically, the object is embedded in some sort of environment, where it responds, influences, and behaves. The evolutionary algorithm designer devises a fitness evaluation procedure that monitors and possibly manipulates the environment and object, returning objective function values.

Figure 1.3 shows how such a situation appears to the evolutionary algorithm. It generates structural/parametric variations of the object, by applying variation operators (such as mutation and crossover) to some representation of the object's configuration. All it gets back are the measured objective values: one may think of the entire evaluation/environment/object complex process as a black-box system.

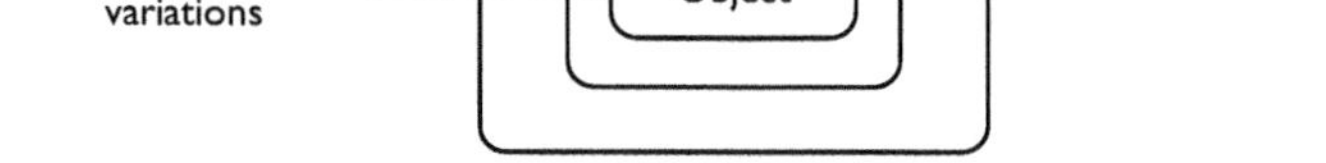

Fig. 1.3 Evolutionary algorithm for a black-box system

Defining this black-box system allows to consider three separate cases that distinguish the differences between evolutionary and conventional designs:

1. **Inverse model is tractable:** If there is a tractable "inverse model" of the black-box system, then there is a way of working out in advance a sequence of variations that brings about a desired set of objective values. "Conventional" methods can be applied: the blind generate-and-test nature of evolution is not essential, although evolutionary methods are vary competitive.

2. **Inverse model is not tractable, but forward model is:** In this case, one can predict the influence of variations upon the objective values, but the black-box system is not tractably invertible, so one cannot derive in advance a sequence of variations to bring about a desired set of objective values. This implies an iterative approach, where variations are carefully selected according to the forward model and are applied in sequence. This kind of iterative design-and-test is a common component of traditional approaches. Search techniques, including evolutionary algorithms, can be competitive or in some cases the only viable choice [7].
3. **Neither forward nor inverse models are tractable:** There is neither a way of discerning which variations will give improvements in the objective values, nor a way of predicting what will be the effects of variations on the objective values. Without evolution, all will be lost. By a tentatively agreed definition, evolutionary methods are those that proceed by incrementally applying variations which are essentially blind. Selection can lead to an improvement in objective values with neither a forward nor an inverse model. Whether evolutionary methods actually succeed in finding a satisfactory design is another issue, but these are the only way to go in general.

Thus, there indeed is an entire class of design problems that can only be tackled by evolutionary methods.

1.2.2 Application of Evolvable Hardware

Evolutionary computation is a field of solving problems using learning algorithms inspired by biological evolution. These kind of algorithms are collectively known as evolutionary algorithms. They model the cycle of selection, recombination and reproduction that biological organisms undergo. Typically, they work on a population of prospective solutions at every time step. Each individual of the population is evaluated according to a problem-specific fitness function, which tests how well the trial solution performs the required task. A selection operator then probabilistically chooses the solutions within the population that the algorithm will subsequently focus on, on the basis of the fitness function evaluation. The selected solutions are recombined and mutated in order to search new but related areas in the problem space, and then the process iterates.

The evolutionary algorithm most commonly used to evolve hardware design today is the genetic algorithm [15], where each trial circuit design is encoded as a bit-string. Recombination, or crossover, is achieved by the probabilistic exchange of bits between individuals, and then mutated by the probabilistic toggling of bits in each individual, normally according to predefined rates. Thus, the algorithm explicitly separates the genetic information that is recombined and mutated (the genotype) from the actual circuit that is evaluated (the phenotype). Another evolutionary algorithm commonly used is genetic programming [3], where an individual solution is a computer program typically represented by a tree, without explicit mappings between genotype and phenotype.

Using evolution to design, for example, circuits brings a number of important benefits to electronics, allowing design automation and innovation for an growing range of applications. Some important areas where evolvable hardware finds good applications include:

- Automatic design of low-cost hardware;
- Coping with poorly specified problems;
- Creation of adaptive systems;
- Creation of fault tolerant systems;
- Innovation in poorly understood design spaces.

The remainder of this section will explore these benefits in a little more detail.

1.2.3 Automatic Design of Low-Cost Hardware

The ideas of design automation can be of significant benefit to hardware that requires a low cost per unit. One example is low-volume hardware. Low-cost reconfigurable hardware can be used to embody evolved designs. For low-volume designs, this reduces cost by avoiding the need for a VLSI fabrication process. The use of reconfigurable hardware also allows changes in specification to be applied not only to new applications of a design but also to hardware already in use, thus avoiding replacement costs. Risk, and its associated cost, may also be reduced, as design faults could be corrected either manually or through further evolution.

Evolutionary automation can even make the prospect of evolving hardware design realistic to suit a specific application. Many medical applications have not been suitable for hardware solutions owing to the expenses of personalization. Evolvable hardware, on the contrary, allows cheap-and-fast solutions to such medical applications. For example, a system has been developed to control a prosthetic hand by recognizing patterns of myoelectric signals in a user's arm [32].

1.2.4 Poorly Specified Problems

It is difficult to specify the functionality of some technical problems. In these cases, design automation may allow feasible solutions be generated from a behavioral description of the problem. Evolution is one of a class of soft computing techniques that can be used to handle the situation with poor specifications. For instance, artificial neural networks (ANNs) have been applied to such problems as noisy pattern recognition [54]. Evolvable hardware techniques are similar but with some particular advantages over ANNs, as noted by Yao and Higuchi [65]. Both of them can be feed-forward networks, and both can learn nonlinear functions successfully. However, hardware is by nature a fast medium and in many cases such as when restricted to feed-forward networks, evolvable hardware designs are more easily understood and implemented than ANNs. Therefore, evolvable hardware is often suited to those problems commonly tackled by ANNs, with fast operation and good solution tractability. Evolvable hardware suitable for these purposes has already been developed for industrial applications [42].

1.2.5 Adaptive Systems

With sufficient automation (for example, real-time synthesis provided by PLDs), evolvable hardware has good potential to autonomously adapt to changes in its environment. This can be very useful for situations where real-time control over systems by humans is impossible, such as deep-space missions. In addition, it could be particularly useful when harsh or unexpected conditions are encountered.

1.2.6 Fault Tolerant Systems

Another practical class of adaptive systems is one that can adapt to faults in its own hardware, thereby implementing a level of fault-tolerance. Higuchi et al. [26] developed an adaptive hardware system that learned the behavior of an expert robot controller by examples using a genetic algorithm. It could then be used as a backup controller if the expert controller failed.

On-line autonomous hardware fault detection and repair mechanisms have been developed [14, 48]. Although these architectures are examples of bio-inspired hardware and have been proposed as a platform for evolutionary experiments, they do not use evolution as an adaptive repair mechanism. Off-line systems can also be evolved to provide fault tolerance, as first shown by Thompson [56]. Thompson also showed that evolution may generate fault tolerant solutions implicitly through the incremental nature of the evolutionary design process. Fault tolerance can exhibit itself at the population level as well as at the individual level.

1.2.7 Design Innovation in Poorly Understood Design Spaces

The design space of all circuits contains infinitely many components that can be wired together in an infinite number of ways. In order to find useful circuits, human designers need to reduce this search space to a manageable size.

To do so, one works in a space of lower dimensionality, in which one needs skills for searching. The evolutionary approach may allow to search the space with a lower or different abstraction. This means that exploration of designs from a much larger and often richer solution space beyond the realm of the traditional hardware search spaces is possible, resulting in novel designs.

Such innovative solutions are needed when one does not have a good understanding of the design space. A great deal of work center around the optimization of parameters for design, which fits more to the field of evolutionary optimization rather than evolutionary design. Work in less abstract search spaces has also been carried out recently.

Handling the increase in search space size, when moving from optimization to design, may require additional evolutionary techniques.

1.2.8 Hummies Competition

Other than simulations, which use evolutionary algorithms to solve various problems of different levels of complexity, "Hummies" were developed, which aims to show that solutions reached by evolutionary computation are fully comparable with solutions designed by "human designers." Rules of this competition were defined by Koza, which states that if a solution (result, design, etc.) are fully comparable with a design by human, then such solution win the competition. These rules of hummies competition are:

- A reached result by evolutionary algorithms should be patented in the past, or is an improvement of an existing version, or can be classified as a patentable solution;
- A result is of the same quality or better than a result published in a recognized scientific journal;
- A result is better then or the same as a result recorded in database or archive by widely recognized scientists;
- A result is ready to be published despite the fact that it was "mechanically" created;
- A result is of the same quality or better than most real results created by human, based on a long-term known problem and its "classical" solution;
- A result is better than a solution which was accepted in the time of its discovery;
- A result solves a problem, which is very hard and complex;
- A solution created by evolutionary algorithms wins human or wins programs created by human.

During the course of this competition, a large set of results were obtained from various fields of science. Table 1.1 summarizes some selected and accepted results. A large part of accepted results has been reached in analog circuit design; and results in other fields like physics and chemistry were also obtained.

Table 1.1 An overview of Hummies awards

Applications	Number of Awards
Analog circuits	25
Quantum circuits	8
Physics - optical systems	7
Logical circuits	5
Optimization problems	5
Game strategies	4
Chemistry - molecular design	3
Antenna design	2
Applied mathematics	1

1.2.9 Problems Solvable by Evolutionary Computation

Generally speaking, any optimization problem in the real world of engineering may be solved by evolutionary algorithms. A concerned issue is the effectiveness of such numerical algorithm as evolutionary algorithms in applications. Various artificial problems have been defined to test the effectiveness and speeds of such algorithms, in particular how many cost-function evaluations are needed for an algorithm to reach an optimal solution.

A few selected samples of such test functions are depicted in Fig. 1.4 (Eq. 1.1) – Fig. 1.15 (Eq. 1.6). Landscapes depicted on those figures are "suffered" by multimodality (more than one global extreme along with many local extremes), nonlinearity (nonlinear dependance), noise (additive noise), and the so-called pathologies. For example, the cost function in Fig. 1.1 (1.1) is very simple and can be solved very fast by any numerical scheme. In contrast, the cost functions in Fig. 1.5 (Eq. 1.2), Fig. 1.6 (Eq. 1.3), Fig. 1.13 (Eq. 1.5) and Fig. 1.15 (Eq. 1.6) are hardly solvable by conventional methods such as the gradient-based methods.

Despite the fact that the aforementioned examples are artificial, many problems of the same complexity can be found in real life. For instance, consider a very simple problem defined by Eq. 1.4. This is based on a simple problem from tee seller, which tries to increase profit by right packing of three kinds of mixes of tee (x_1, x_2, x_3) with weight limits (2850g and 1380g) on the final amount of tee. This problem, especially its mathematical definition and description looks quite simple, but location of the global extreme is surprisingly hard even for evolutionary algorithms. A closer look at the geometry (i.e., the cost function landscape) shows that the global extreme is in a very small spot, surrounded by a wide flat area. Fig. 1.7 – Fig. 1.12 depict three views on the global extreme. In fact, the cost-function landscape is in the 4D space (with three variables $x = x_1$, $y = x_2$, $z = x_3$). So, the easiest way to see the global extreme is to fix one of the three variables to better view the position of the global extreme and then to draw the landscape in terms of the other two variables. In this way, Fig. 1.7 – Fig. 1.12 are obtained.

$$\sum_{i=1}^{D} x_i^2 \tag{1.1}$$

$$\sum_{i=1}^{D-1} \left(\sqrt[4]{(x_i^2 + x_{i+1}^2)} \sin(50 \sqrt[10]{(x_i^2 + x_{i+1}^2)})^2 + 1 \right) \tag{1.2}$$

$$\sum_{i=1}^{D-1} \left(0,5 + \frac{\sin(\sqrt{100x_i^2 - x_{i+1}^2})^2 - 0,5}{(1 + 0.001(x_i^2 - 2x_i x_{i+1} + x_{i+1}^2)^2)} \right) \tag{1.3}$$

$$-(2x_1 + 3x_2 + 2x_3)g, \quad \text{with } g = \left\{ \begin{array}{ll} 1, & \text{if} \\ -100, & \text{if} \end{array} \right\} \begin{array}{l} 10x_1 + 6x_2 + 5x_3 \le 2850 \\ \text{and} \quad 4x_2 + 5x_3 \le 1380 \end{array} \tag{1.4}$$

$$-1\sum_{j=1}^{30}\frac{1}{c_j+\sum_{i=1}^{D}(x_i-a_{j,i})^2} \tag{1.5}$$

$$\prod_{i=1}^{D}\frac{\sqrt{\frac{s}{\pi}}}{e^{s(x_i-o_i)^2}} \tag{1.6}$$

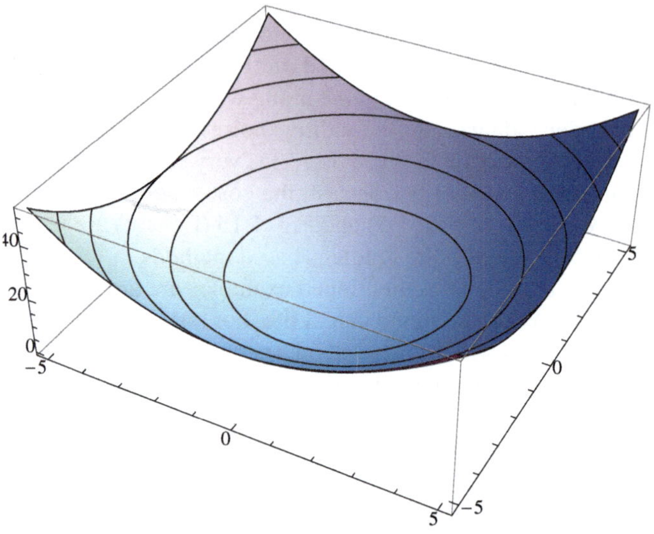

Fig. 1.4 Artificial cost function, see Eq. 1.1

The performance of an algorithm depends not only on the algorithmic structure, but also on the problem complexity. Complexity [61] is a theory that describes how "fast" can a problem be solved in dependance on input size (i.e., the number of parameters such as the number of cities in the traveling salesman problem, see Chapter 2). Fig. 1.16 – Fig. 1.19 show some selected dependance on input size. One can see that the "speed" of dependance on input size is stabilized to far from the origin, and the input size is the biggest for exponential and factorial growths [61]. There are in fact two classes of problems (for more precise description, see [61] and [13]), i.e., P and NP classes of problems [12]. According to complexity theory, the class of P problems consists of all those decision-making problems that can be solved by a deterministic sequential machine (e.g., by a deterministic algorithm) in an amount of time that is polynomial in input size. On the contrary, the class of NP problems consists of all those decision-making problems whose positive solutions can be verified in polynomial time given the right information, or equivalently, whose solutions can be found in polynomial time by a non-deterministic machine (e.g., by nondeterministic algorithm).

In a deterministic Turing machine, the set of rules prescribes at most one action to be performed for any given situation, i.e., each step is strictly determined, no

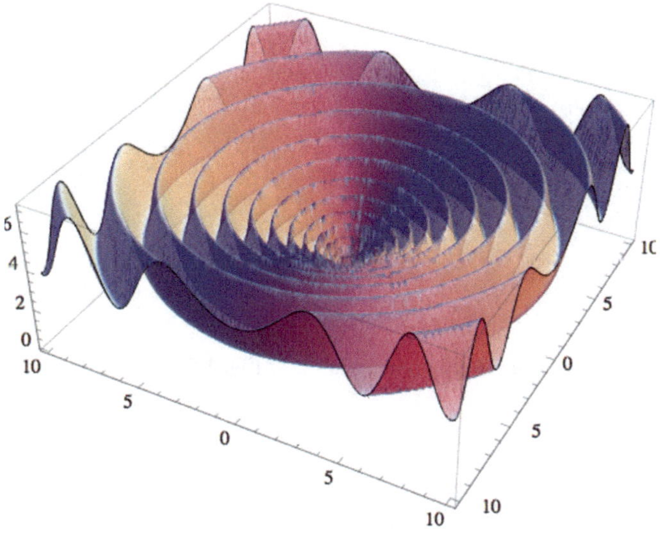

Fig. 1.5 Artificial cost function, see Eq. 1.2

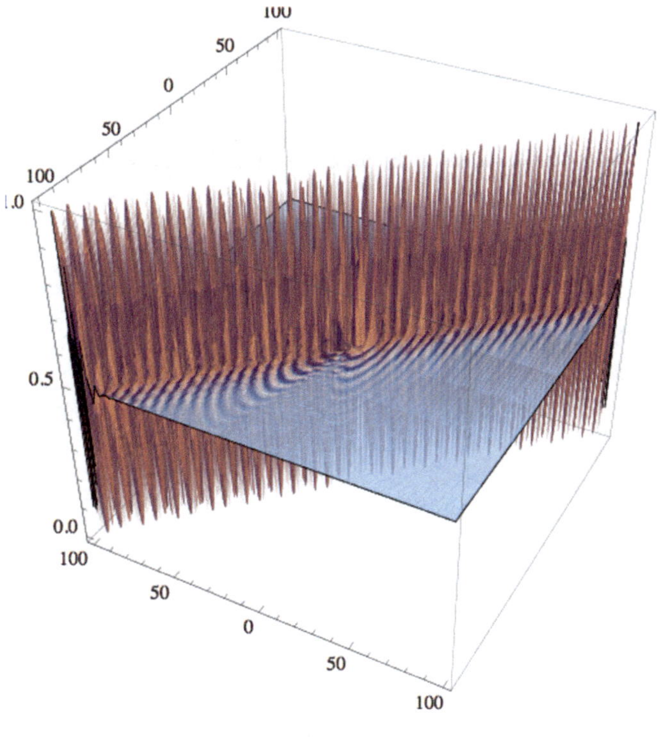

Fig. 1.6 Artificial cost function, see Eq. 1.3

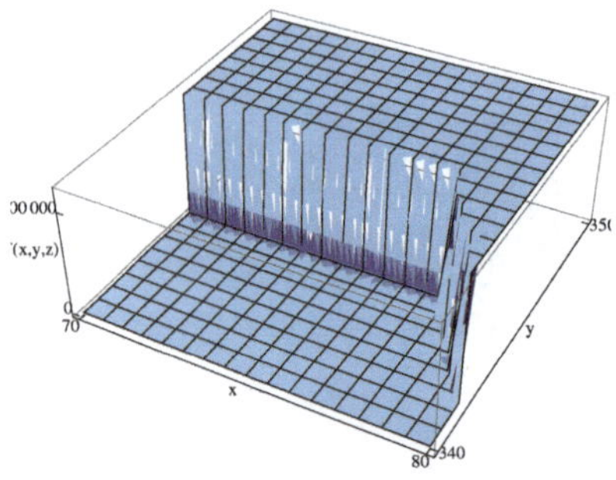

Fig. 1.7 Cost function of tea seller problem, a complete view

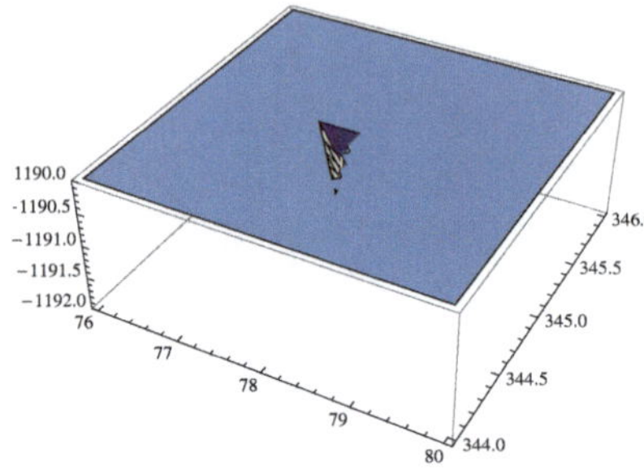

Fig. 1.8 A detailed view, see Eq. 1.4

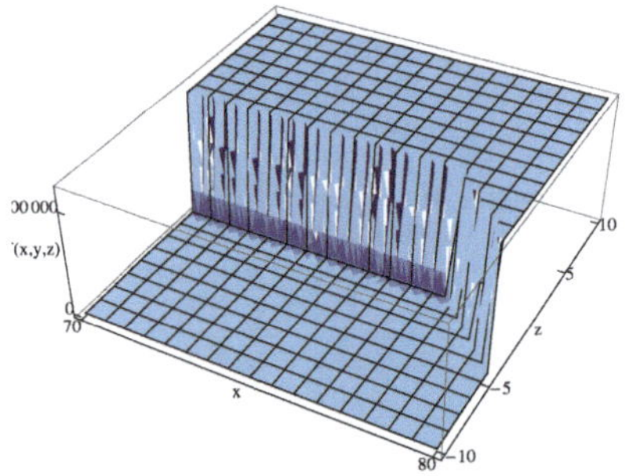

Fig. 1.9 Cost function of tea seller problem, a complete view

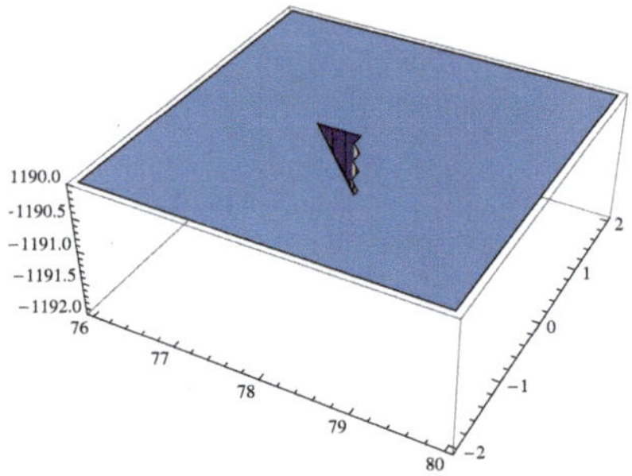

Fig. 1.10 A detailed view, see Eq. 1.4

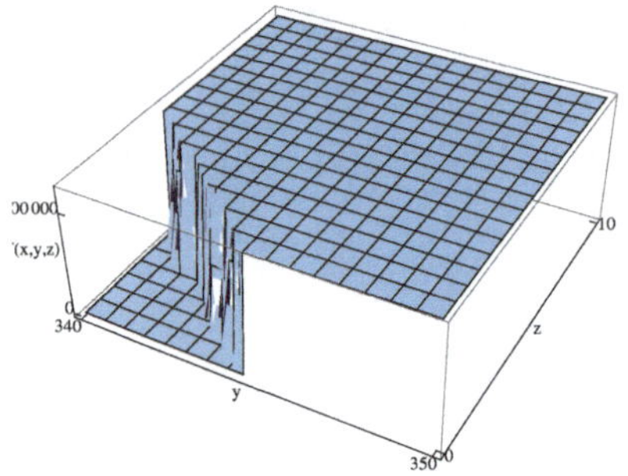

Fig. 1.11 Cost function of tea seller problem, a complete view

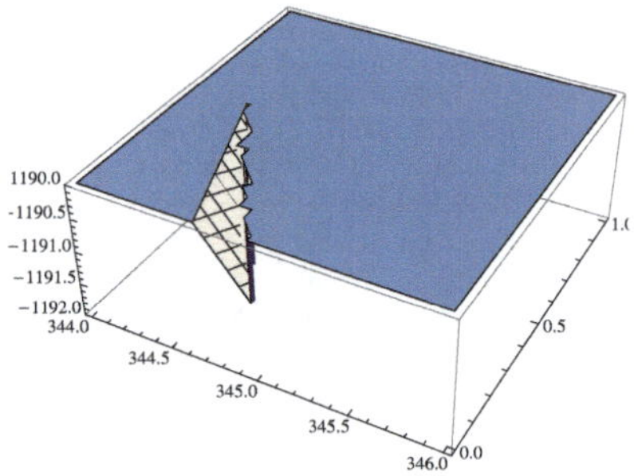

Fig. 1.12 A detailed view, see Eq. 1.4

randomness play any role here. Because such a theoretical machine can be described algorithmically, one may say that deterministic algorithms are solving this class of problems. A nondeterministic Turing machine, in contrast, may have a set of rules

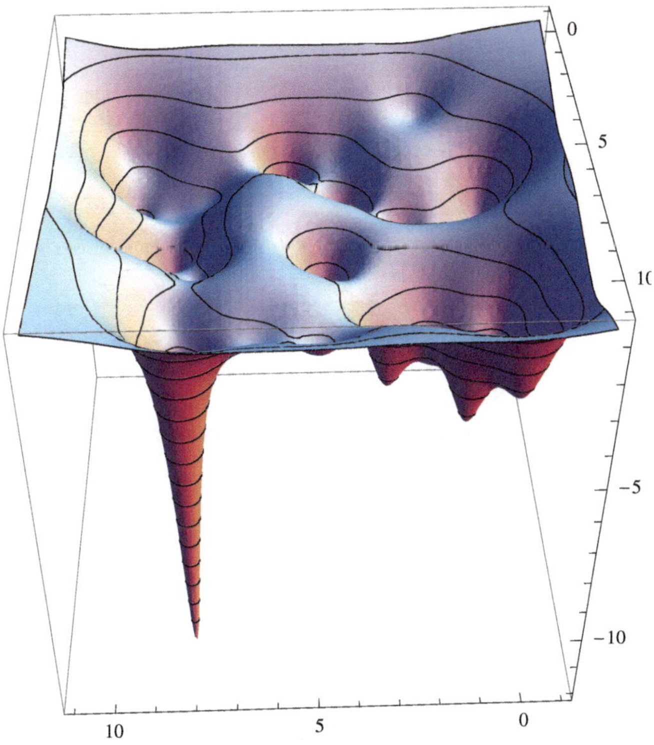

Fig. 1.13 Artificial cost function, see Eq. 1.5

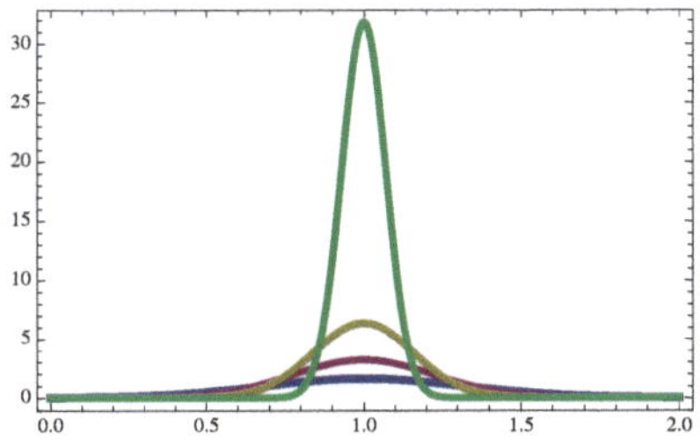

Fig. 1.14 Artificial cost function, see Eq. 1.6

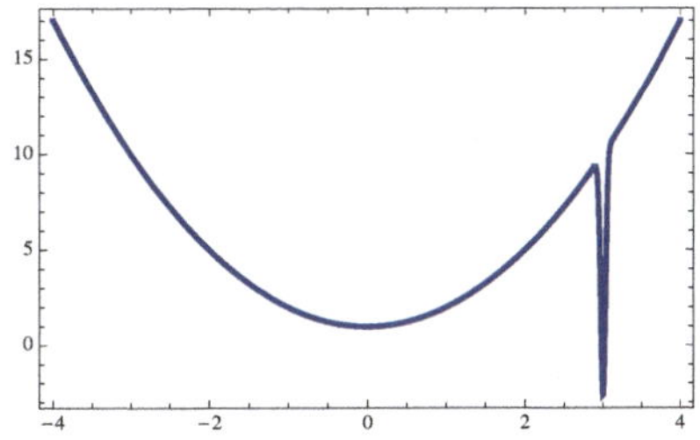

Fig. 1.15 Appended to Eq. 1.1

that prescribes more than one action for a given situation, i.e., some steps are (or can be) influenced by randomness. This is basically the fundamental principle of evolutionary techniques, and in some sense is the behavior of evolutionary techniques influenced by randomness in the form of mutation, selection, etc. (see Chapter 2). A lot of engineering problems belong to these classes of problems, and the next subsection shows a real one in a laboratory plasma reactor control.

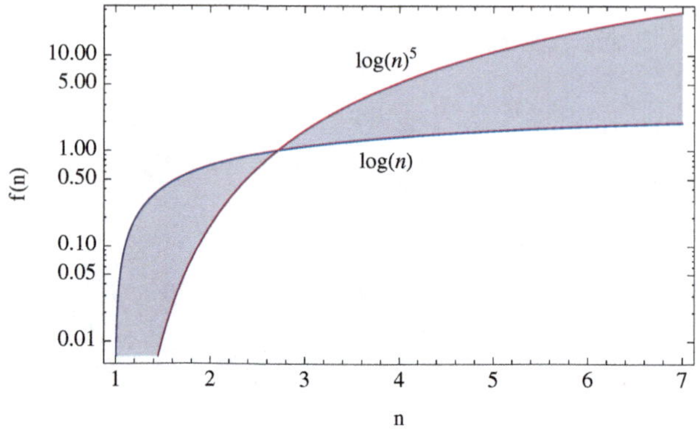

Fig. 1.16 Growth of a complexity problem represented by $\log(n)$ and $\log^5(n)$

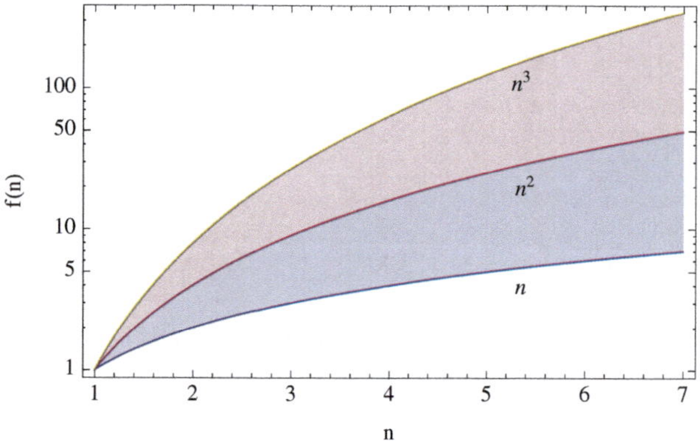

Fig. 1.17 Growth of a complexity problem represented by n, n^2 and n^3

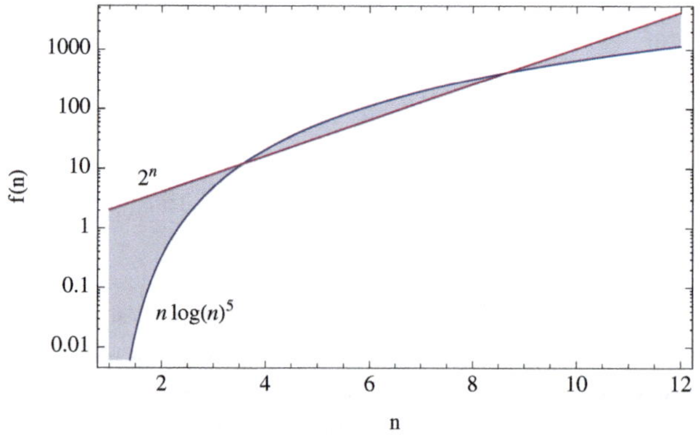

Fig. 1.18 Growth of a complexity problem represented by $n\log^5(n)$ and 2^n

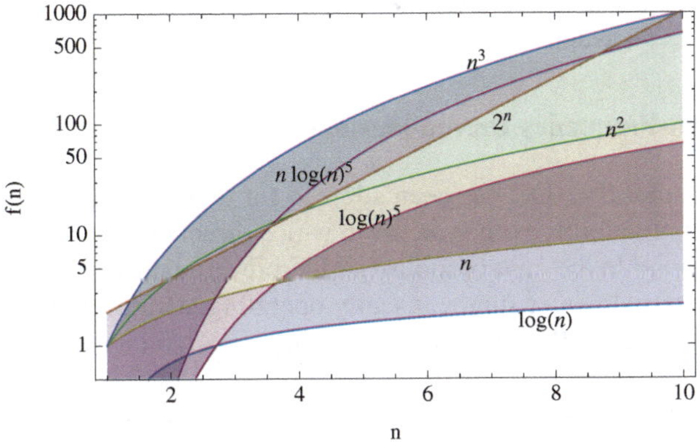

Fig. 1.19 Growth of a complexity problem represented by $\log(n)$, $\log^5(n)$, n, n^2, n^3, $n\log^5(n)$ and 2^n

1.2.10 Example: Real-Time Compensation of Plasma Reactor

The performance of a self-organizing migration algorithm (SOMA) is examined here, where a new stochastic optimization algorithm is compared with simulated annealing (SA) and differential evolution (DE), for a typical engineering application (see Chapter 2 and Chapter 6; for a more detailed study of this example, see [46], [67]).

More precisely, this application is the automated deduction of fourteen Fourier terms in a radio-frequency (RF) waveform to tune a Langmuir probe. Langmuir probes are diagnostic tools used to determine the ion density and the electron energy distribution in plasma processes. RF plasmas are inherently nonlinear, and many harmonics of the driving fundamental can be generated in the plasma. RF components across the ion sheath formed around the probe distort the measurements. To improve the quality of the measurements, these RF components are removed by an active-compensation method. This can be achieved by applying an RF signal to the probe tip that matches both the phase and the amplitude of the RF signal generated from the plasma. Here, seven harmonics are used to generate the waveform applied to the probe tip. In so doing, fourteen mutually interacting parameters (seven phases plus seven amplitudes) had to be tuned on-line. SA and DE were applied to this problem and compared with the performance of SOMA.

1.2.10.1 A Low-Temperature Plasma System

Artificially produced plasmas are typically generated through an application of electrical energy to a certain gas. Under normal conditions, gases do not conduct electrically with almost all electrons being bound to atoms or molecules. However, if electrons are introduced and given enough energy by an external power source, then they have a potential to collide with gas atoms or surfaces thereby releasing more electrons, which may then release other electrons. The resulting electrical breakdown is known as an avalanche effect.

1.2.10.2 Radio-Frequency Driven Plasmas

The use of RF rather than DC has been adopted for a number of reasons including efficiency and compatibility with systems in which direct electrical contact with the plasma is infeasible. In the case of industrial RF-powered plasmas, an RF generator is used as the external power source, usually operating at 13.56MHz or a harmonic of this frequency. Fig. 1.20 shows a typical configuration for a capacitively coupled system using electrodes.

A main application of RF-powered plasmas is to produce a flux of energetic positive ions, which impinge continuously over a large area of work piece, e.g. for etching or deposing. By nature the plasma medium is quasi-neutral, to ensure which a plasma sheath forms between the plasma and the bounding surfaces. This results in a plasma potential that tends to be positive relative to the surfaces. The plasma sheath prevents electrons from leaving the plasma at a greater rate than the ions.

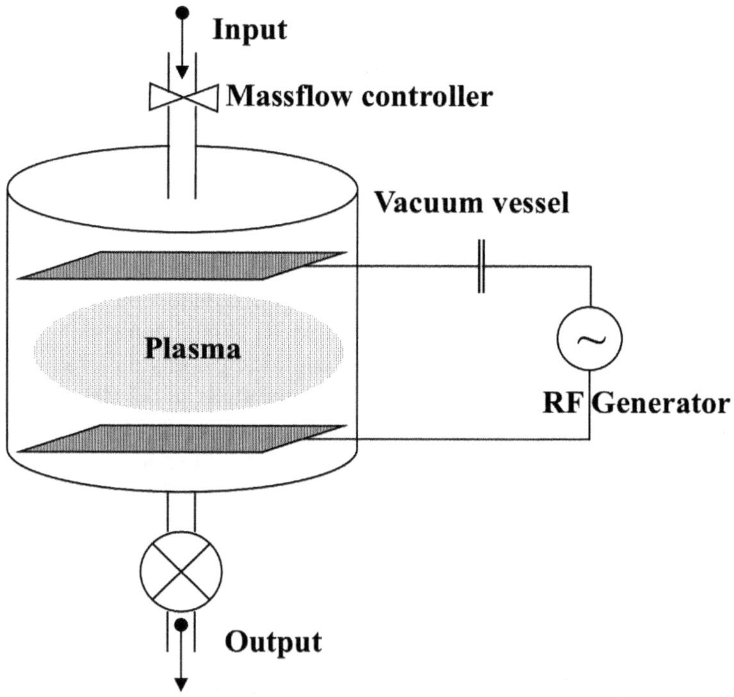

Fig. 1.20 Schematic of an RF-driven plasma system

1.2.10.3 Langmuir Probes

Langmuir probes [55], developed in 1924 by Langmuir, are one of the oldest probes used to obtain information about low-pressure plasma properties. By applying a positive or negative DC potential to the probe relative to the plasma, either an ion or an electron current can be drawn from the plasma.

The region of space-charge (the sheath), which forms around a probe immersed in a plasma, has a highly nonlinear electrical characteristic. In RF-generated plasmas, this is a major issue as the excitation process leads to the space potential in the plasma having RF components. As a result, harmonic components of the RF potential across this layer give rise to serious distortion of the probe's measured DC signal. In order to achieve accurate measurements, the harmonic components across the probe sheath have to be eliminated.

1.2.10.4 Active Compensation in RF-Driven Plasmas

To eliminate the time variation of RF potential difference between the probe and the plasma, the probe potential is made to follow that of the plasma [4]. This can be achieved by superimposing a synchronous waveform of appropriate amplitude and phase onto the probe tip. Because plasmas are inherently nonlinear, they can generate many harmonics of the exciting fundamental. As a consequence, the RF signal necessary for satisfactory compensation not only has to match the amplitude and the phase of the exciting RF fundamental, but also has to match the complex waveform of the harmonics generated in the plasma.

Conveniently, the electrostatic probe generates a useful control signal that indicates the degree of uncompensated RF voltage across the probe sheath. In the presence of a plasma, and without any deliberate biasing of the probe, the isolated electrostatic probe tip adopts a floating potential, at which it draws zero net current. This floating potential of the probe is also referred to as its DC bias. This DC bias was used in an automated control system as a feedback parameter for compensation of the harmonics at the Langmuir probe tip.

1.2.10.5 Automated Control System

An additive synthesizer (harmonic box) with seven harmonics has been developed in [44], to generate the appropriate waveforms for compensation of a Langmuir probe system attached to a Gaseous Electronics Conference (GEC) standard reference plasma reactor [27]. Fig. 1.21 shows the schematic of the control system for waveform tuning.

The control software selects set-points for the harmonic generator and sets the parameters, i.e., seven amplitudes and seven phases, using 14 D/A converters. The harmonic generator, which is synchronized with the main RF power generator, generates the required waveform which is applied to the Langmuir probe. The probe's floating potential (DC bias) is used as a fitness-measure: the higher the DC bias the

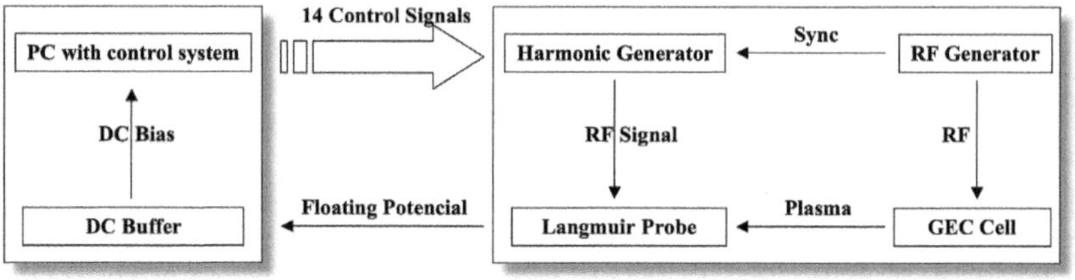

Fig. 1.21 Closed control loop for waveform tuning

better the compensation. The DC bias is read on-line via a DC buffer and a A/D converter by the computer system. Depending on the optimization algorithm used in the system, the software then calculates a new set-point based on the actual measure of the fitness. It can be seen that all the fitness evaluations are actually measurements rather than simulation results. This implies time restrictions on the search process.

The fourteen input parameters interact, to some degree, due to the technical realization of the synthesizer hardware and the nature of the problem. Small variations in the 14 parameters caused by these interactions could lead to a large deviation from optimal tuning and, hence, the probe measurement itself. As a consequence, the number of points in the discrete search space has to be calculated, as follows:

$$n = (2^b)^p \tag{1.7}$$

Here,

- **n**: number of points in search space
- **b**: resolution per channel in bits
- **p**: number of parameters to be optimized

The D/A and A/D converters used had a resolution of 12 bits and the dimensionality of the search space was 14. Hence, the search space consists of n $\approx 3.7 \times 10^{50}$ search points. Due to the system time constant, mapping out the entire search space would take approximately 10^{41} years with the plasma system used. Hence, mapping out the entire search space was not a practical option at all.

SA [44] and DE [67] were used successfully to tune the Langmuir probe. The results were improved further by introducing step width adaptation to SA [43]. The performance was compared with the performance of SOMA. All experiments were carried out at the Open University, Oxford Research Unit, UK.

Fig. 1.22 – 1.23 show the experimental set-up. Three different optimization algorithms were used for the automated waveform tuning experiments. A digital oscilloscope was used to measure the actual waveforms found by the three algorithms. The control software ran on a PC under the Linux operating system. The algorithms used for these experiments were written in C++ and integrated with the existing Langmuir probe control software. The plasma system used was a standard GEC reference cell.

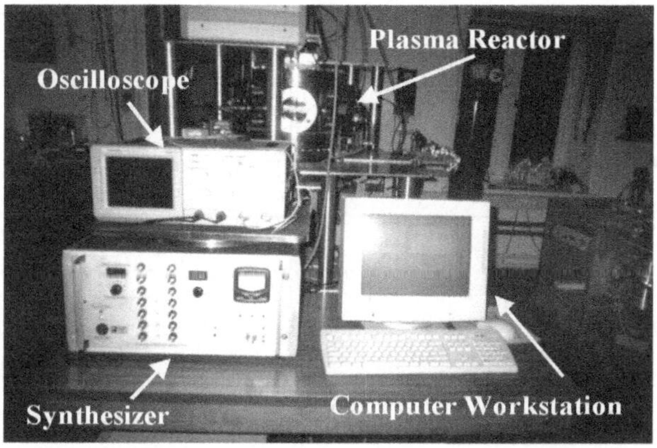

Fig. 1.22 Experimental set-up: computer with control software (right), wave synthesizer (bottom left), and oscilloscope (top left)

Fig. 1.23 Experimental set-up: plasma reactor with Langmuir probe

1.2.10.6 Experimental Results

Each algorithm was applied 20 times in total. In order to compensate for drifts of the plasma over time, the three algorithms were applied alternatively, i.e., an algorithm was applied once followed by the other two algorithms before the first one was applied again.

The experimental results can be seen in Fig. 1.24 – 1.26 (only selected figures are shown here). These figures show a typical search run over time: the average fitness of the population in A/D Converter units, the best individual in the current generation, and the standard deviation. Fig. 1.25 shows the average values of the D/A Converters, and the deviation for all 14 parameters found by the evolutionary algorithm.

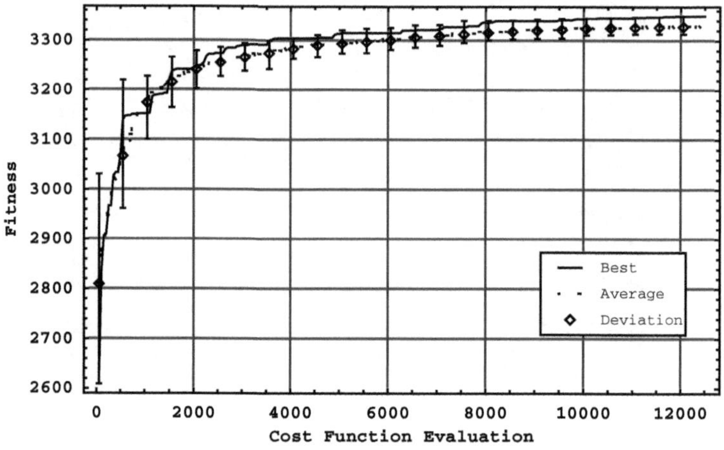

Fig. 1.24 Graphical visualization of all 20 repeated simulations with DE. Every curve is the history of the best solution from each generation during simulated evolution. Evolutionary algorithms are searching for the maximal value of the cost function.

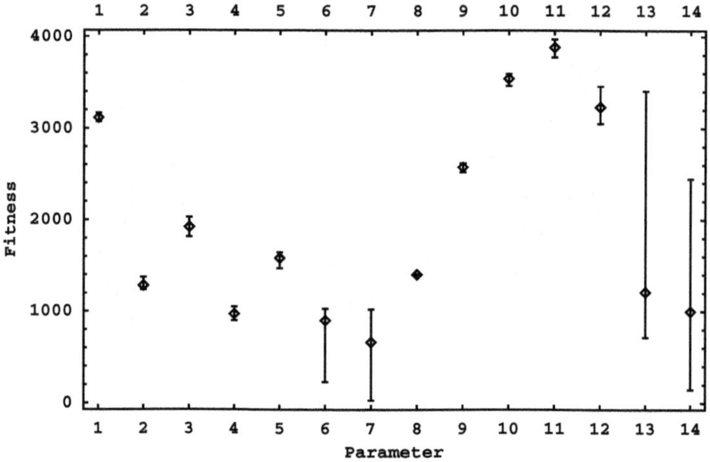

Fig. 1.25 Optimal-value setting of all 14 parameters by SOMA. Small rectangles represent average values from all 20 experiments. Minimal and maximal values are also depicted.

Fig. 1.26 shows the average waveforms found by the evolutionary algorithm. In Fig. 1.24, the example of evolution runs is depicted. Also other results show a linear drift of plasma over time, which was observed during all experiments [46].

From these results, it can be seen that basically all the three optimization algorithms can find good solutions in a very short time (about 4 minutes, for the limit was set to 12000 times of cost-function evaluations for each algorithm). Also, based

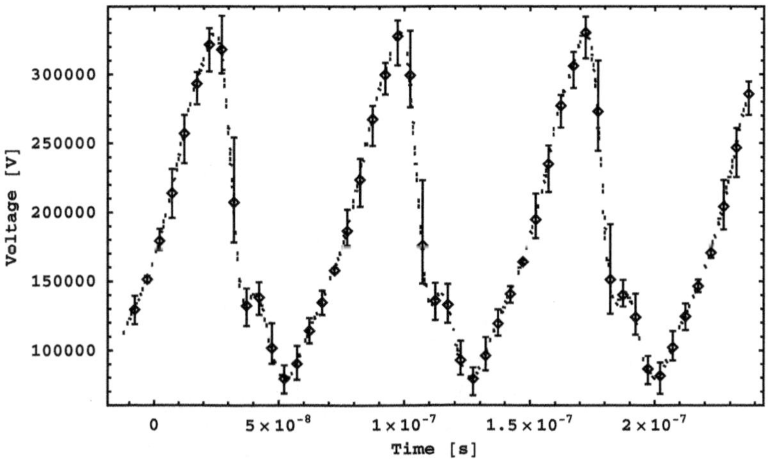

Fig. 1.26 Final waveform based on 14 estimated parameters by DE. Small rectangles represent average values from all 20 experiments. Minimal and maximal values are also depicted. Dotted line shows average waveform synthesized by evolutionary algorithm.

on [43] and [44], it can be concluded that the tested algorithms significantly outperformed the parameter settings by human operator in precision as well as in time.

This is only one of the many existing examples that can be used to demonstrate the capability of evolutionary algorithms from various fields of research and applications. It is clear that the use of evolutionary algorithms should be possible also in the field of chaotic dynamics, which, in fact, has been preliminary investigated, with focus on chaos control and synthesis, as further discussed in the next section.

1.3 Chaotic Systems

Mutual interaction of evolutionary algorithms and chaotic systems may be studied from two points of view. The first is how to use evolutionary algorithms as a tool for analysis, understanding, and control of chaotic systems. The second is how chaos exists and behaves throughout the algorithmic evolution.

One of main advantages of evolutionary algorithms in chaos control is the fact that such algorithms are able to solve complex black-box problems without requiring auxiliary information about the problem itself. In chaos control, some widely accepted artificial cost functions are demonstrated in Fig. 1.4 (Eq. 1.1) – Fig. 1.15 (Eq. 1.6). The first example comes from chaos synchronization (see Chapter 12), from very simple case of two chaotic systems coupled via one scalar variable.

Fig. 1.27 depicts an example of the cost-function surface, which shows the dependance of cost value on the coupling parameter c and the control parameter a. The version with only parameter c is depicted in Fig. 1.28. This cost function is very simple – it is based on a simple sum of the differences $(x-x_1, y-y_1, z-z_1)$ of

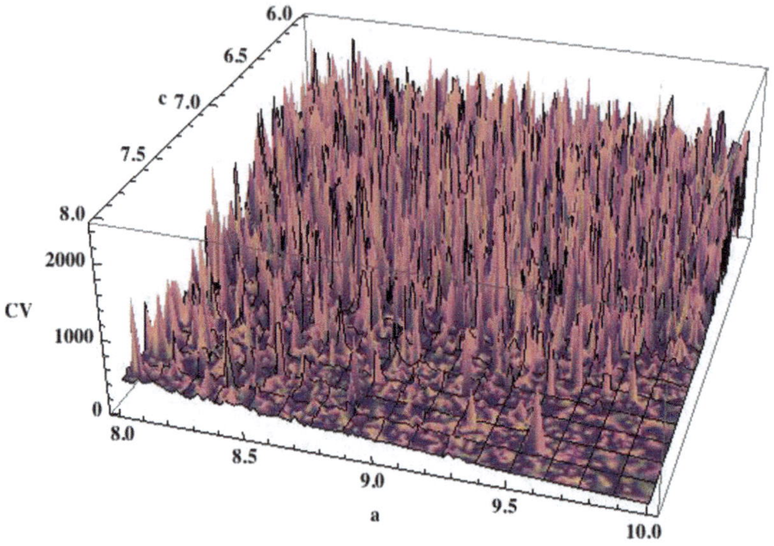

Fig. 1.27 Cost-function surface representing problem of synchronization, see Chapter 12

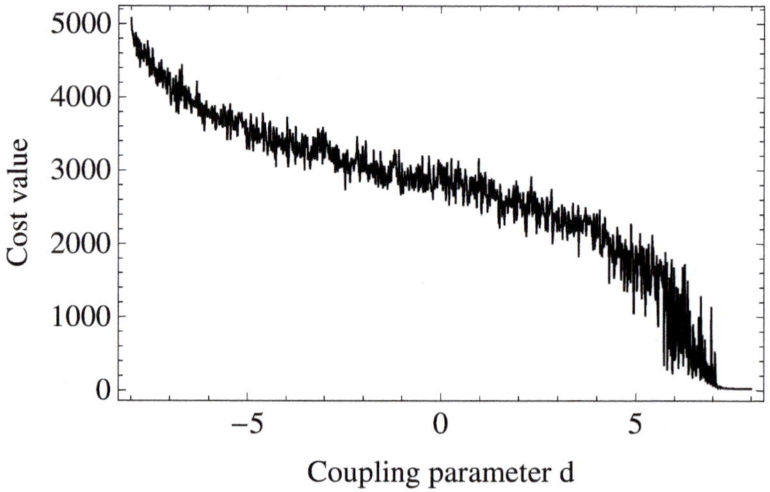

Fig. 1.28 Cost-function curve representing problem of synchronization, see Chapter 12

all the three variables. For state variable x, it is depicted in Fig. 1.29. The difference is shown by the light-gray surface between two curves.

Other examples are shown in Fig. 1.30 – Fig. 1.32. Cost-function surfaces were obtained in a similar way as in the previous example (see Chapter 5). Again, despite

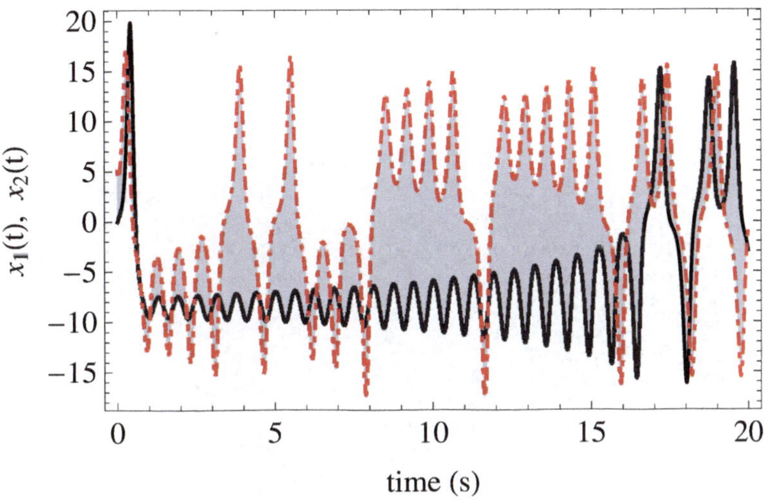

Fig. 1.29 Demonstration of cost function calculation - difference between two different kinds of behavior (light-gray surface), see Chapter 12

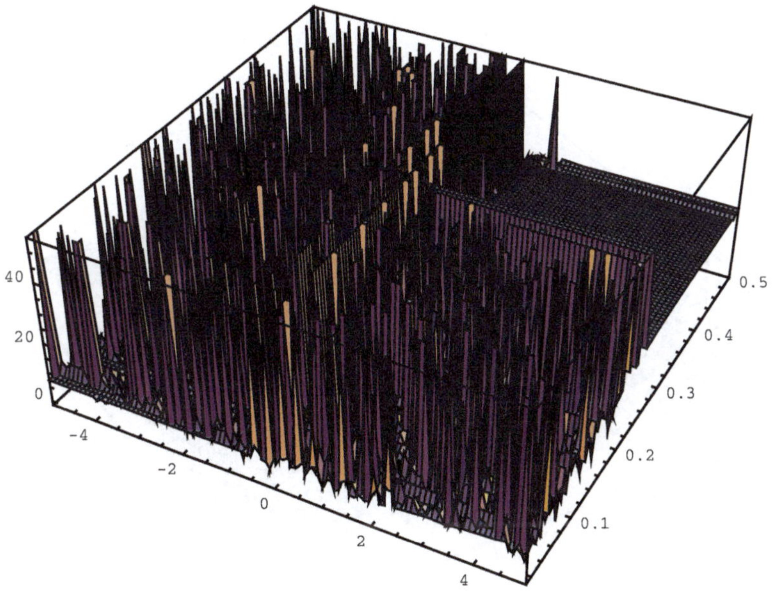

Fig. 1.30 Cost-function surface representing problem of chaos control, see Chapter 5

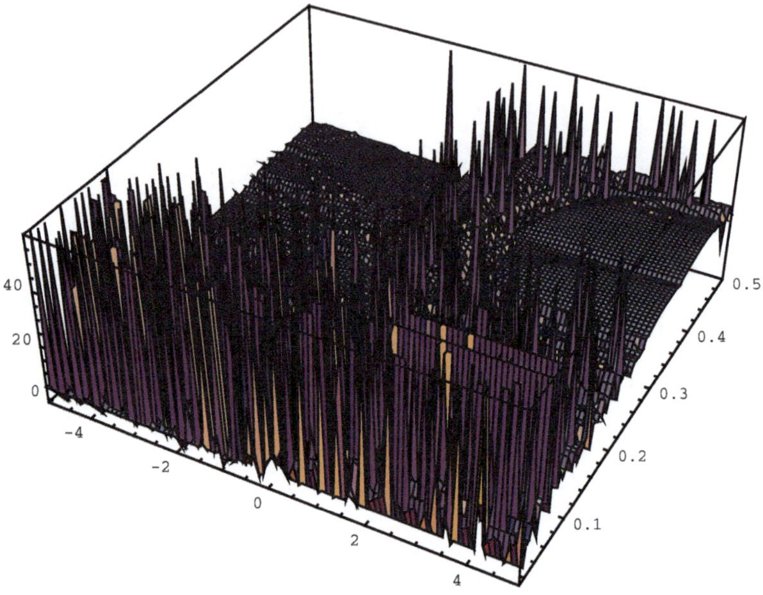

Fig. 1.31 Cost-function surface representing problem of chaos control, see Chapter 5

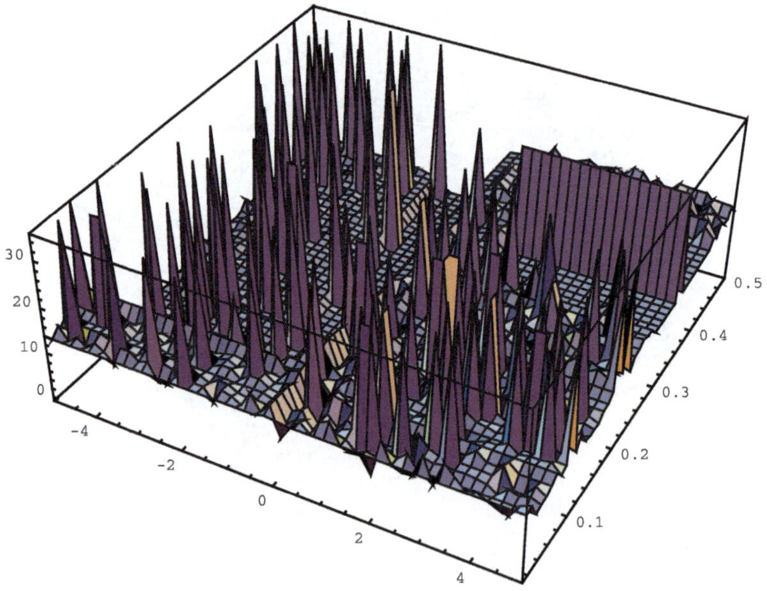

Fig. 1.32 Cost-function surface representing problem of chaos control, see Chapter 5

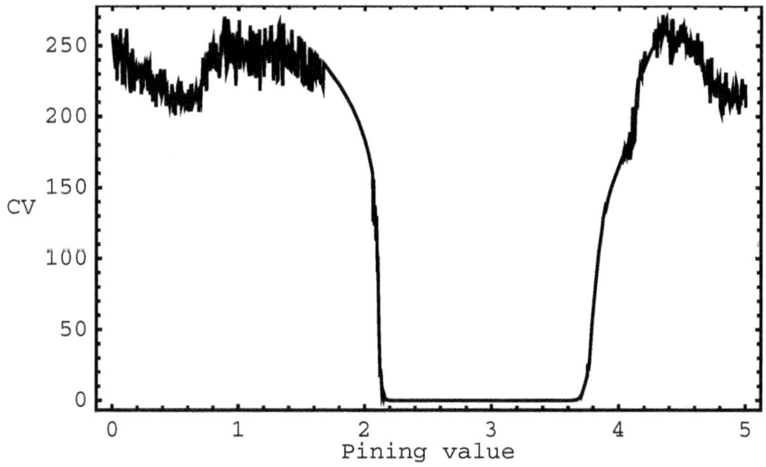

Fig. 1.33 Cost-function curve representing problem of 1D coupled-map-lattice control, see Chapter 6

the simple structure of all the cost functions, visualization shows very erratic-like surfaces with high degrees of multi-modality and nonlinearity.

The last example is demonstrated by Fig. 1.33 and Fig. 1.34. This problem is discussed in Chapter 6. It is clearly visible that any of the above-discussed cost functions (problems) is hardly solvable by classical numerical methods.

Many methods for chaos control have been developed [11] and some are based on the original OGY control method [49]. The main principle lies in waiting for a natural passage of the chaotic orbit close to the desired periodic behavior and then applying a small perturbation to it, so as to stabilize the system. This relies on linearization of the corresponding Poincaré Map [5] - [2]. Moreover, there are variants using for example the pole placement principle [20], [18], and many others (see Chapter 5). Together with all those "classical" control methods, evolutionary techniques have been applied [53], [70], [69].

The second issue of interest is about chaos throughout the algorithmic evolution. Chaos can be observed from inside evolutionary algorithms, as reported in [32]. In this work, models of genetic algorithms were introduced, which exhibit under certain conditions cycling and chaotic behavior. Chaos in a genetic algorithm can potentially be used as a source of diversity. In [64], with the logistic map, a simple equation involving chaos was proposed as the basis of a special mutation operator. Similar research articles include [64], where dynamic clone and chaos mutation methods were sued for optimization; [63], where chaos was used to keep individuals of subgenerations ergodically distributed in a defined space and to circumvent premature individuals of the subgenerations; [37], where the impact of ergodicity induced by chaos on the population diversibility was investigated.

Chaos can be also observed in evolutionary dynamics, as studied in [45]. Chaotic patterns are discovered from simulations on games with the prisoners' dilemma.

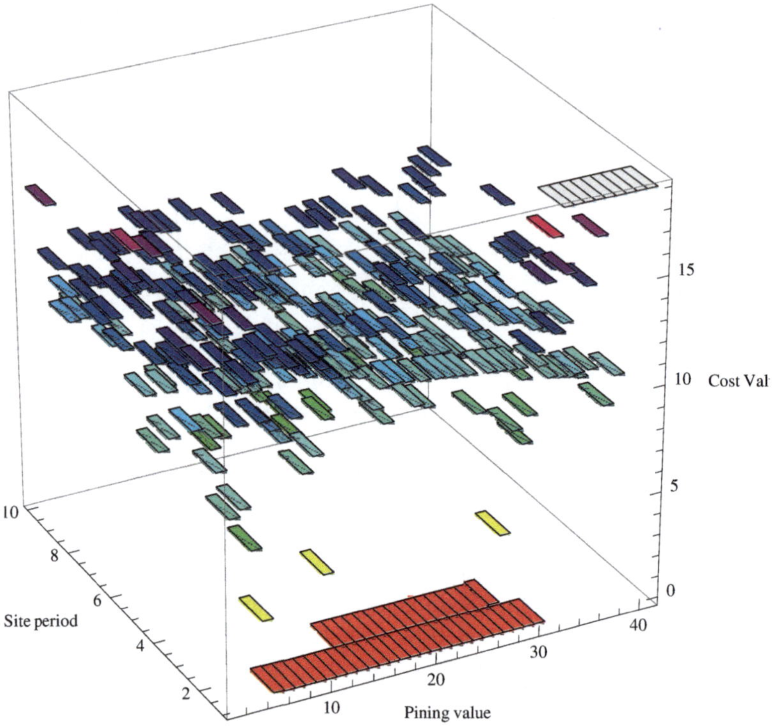

Fig. 1.34 Cost-function surface representing problem of 2D coupled-map-lattice control, see Chapter 6

More comprehensive and compact studies about chaos existence in evolutionary dynamics are reported in the book [45], which contains rich information about chaotic behavior generated by evolution and has many interesting examples.

It is also interesting to study the use of chaotic dynamics embedded inside evolutionary algorithms for driving the evolutionary process. The background idea is that evolutionary algorithms require some randomness to proceed. The behavior of chaotic systems appears to be random, therefore chaotic systems can be used as for the needed randomness by evolutionary algorithms. The main advantage is that there is a fairly good understanding of chaotic systems, which in turn leads to a good understanding of the evolutionary process. An application of such an approach is discussed in Chapter 14, where chaotic Lozi and delayed logistic maps are embedded inside the Differential Evolution algorithm for the optimization of PID control.

The aforementioned examples are merely a few of the many. There are lots of cases where the model of a system is unknown and, thus, one has no idea about its complexity landscape beforehand (see Chapter 11). In case studies, evolutionary algorithms (genetic programming-like techniques) are commonly used to synthesize some new discrete chaotic systems (designed by computer). Cost functions were basically algorithms, i.e., programs, but not standard mathematical formulas. The objects of evolution were typically symbolic objects, which together with vaguely defined cost functions do not allow to draw searched landscapes.

Based on the above observations and personal experiences, the following areas of mutual interaction of evolutionary techniques and deterministic chaos will be further studied and discussed throughout the present book:

- **Chaos control.** In this application (Chapter 5), selected evolutionary algorithms will be used on chaotic dynamics control on, e.g., logistic equation and Hennon map. Results were compared with standard control methods.
- **CML chaos control.** Coupled map lattice (CML) systems based on mutually coupled logistic equations are controlled by different evolutionary algorithms. Simulations reported in this case study (Chapter 6) will be designed to be non-realtime, i.e., each repetition of a simulation is started with the same CML from the beginning under randomly generated initial conditions. Control of a CML to two stabilized patterns will be defined and simulated.
- **Chaotic systems reconstruction.** In this case study (Chapters 7 and 8), chaotic systems will be reconstructed using “classical” methods. In the end, a set of experiments based on evolutionary algorithms will be used for system identification and reconstruction, using observed data.
- **Chaos-based encryption and its evolutionary decryption.** In this application (Chapters 9 and 10), conventional methodologies and principles of chaos-based encryption will be studied, and evolutionary decryption of chaos-based encrypted information will be discussed.
- **Evolutionary chaos synthesis.** This study (Chapter 11) is a continuation of the investigation in [71]. It will be shown that evolutionary techniques can be used for chaotic system synthesis, based on some predefined conditions. Some preliminary results on synthesis of continuous chaotic systems will be briefly discussed.
- **Synchronization of chaotic systems.** As an interesting application (Chapter 12), a few selected evolutionary algorithms will be used to synchronize two kinds of coupled systems: Lorenz-Lorenz system and Rössler-Lorenz system.
- **Evolutionary optimization in chaotic CML-based fitness landscapes.** In this study (Chapter 13), fitness landscapes will be analyzed and quantified by using topological and dynamical landscape measures, such as modality, ruggedness, information content, dynamic severity, and two types of dynamic complexity measures. The main focus is on dynamic fitness landscapes that exhibit spatiotemporal chaotic behavior.
- **Controller parameters optimization on a representative set of systems using deterministic-chaotic-mutation evolutionary algorithms.** In this investigation (Chapter 14), selected chaotic maps will be used for evolutionary operators like mutation and selection. Statistically massive simulations will be carried out to show the impact of chaotic maps on the performance of selected evolutionary techniques.

- **Chaotic attributes and permutative optimization.** Population dynamics and its relation to chaotic systems will be analyzed (Chapter 15). Using basic chaotic principles of attractors and edges, a dynamic population is developed, which is then used to induce and retain diversity in a metaheuristic population. Simulation will be performed with genetic algorithm, differential evolution, and self-organizing migrating algorithm, on the combinatorial problem of quadratic assignment, with promising results.

It should be noted that all these selected areas are still under intensive research today, and the role of evolutionary algorithms has been and will continue to be important and interesting.

1.4 Conclusions

This chapter has outlined and summarized, in a preliminary and brief manner, the ongoing research about the capability of some evolutionary algorithmic methods, which will be used and analyzed in the present book. For this purposes, contemporary methods from evolutionary techniques, their selected applications, some classes of problems solvable by evolutionary algorithms, as well as graphical visualization of selected cost functions, and so on, have been the focal topics. Some related key references have also been referred to, as information for the readers' further studies.

Acknowledgements. This work was supported by Grant No. MSM 7088352101 of the Ministry of Education, the Czech Republic, and by Grant GACR 102/09/1680 of the Grant Agency of the Czech Republic.

References

1. Andersen, P.: Evolvable Hardware: Artificial Evolution of Hardware Circuits in Simulation and Reality, M.Sc. Thesis, University of Aarhus, Denmark (1998)
2. Andrievski, B., Fradkov, A.: Control of Chaos: Methods and Applications. Automation and Remote Control 64(5), 679–719 (2003)
3. Banzhaff, W., Nordin, P., Keller, E., Francone, F.: Genetic Programming. Morgan-Kaufmann, San Francisco (1998)
4. Benjamin, N., Braithwaite, N., Allen, J.: Self bias of an r.f. driven probe in an r.f. plasma. Proc. Mat. Res. Soc. Symp. 117, 275–280 (1998)
5. Chen, G., Dong, X.: From Chaos to Order: Methodologies. Perspectives and Applications. World Scientific, Singapore (1998)
6. de Garis, H.: An Artificial Brain: ATR's CAM-Brain Project Aims to Build/Evolve an Artificial Brain with a Million Neural Net Modules Inside a Trillion Cell Cellular Automata Machine. New Generation Computing J. 12(2), 215–221 (1994)
7. de Oliveira, A., Ramos, F., Gatto, R.: A research agenda for iterative approaches to inverse problems using evolutionary computation. In: Proc. 3rd IEEE Int. Conf. on Evolutionary Computation, pp. 55–60 (1996)

8. Determan, J., Foster, J.: Using chaos in genetic algorithms. In: Evolutionary Computation, CEC 1999 (1999), doi:10.1109/CEC.1999.785533
9. Ebenhöh, O., Heinrich, R.: Evolutionary optimization of metabolic pathways. Theoretical reconstruction of the stoichiometry of ATP and NADH producing systems, Bull. Math. Biol. 63(1), 21–55 (2001)
10. Fogel, David, W., Corne (eds.): Evolutionary Computation in Bioinformatics. Morgan Kaufmann, San Francisco (2002)
11. Fradkov, A., Evans, R.: Control of Chaos: Survey 1997 - 2000. In: Preprints of 15th Triennial World Congress IFAC, Plenary Papers, Survey Papers, Milestones, Barcelona, pp. 143–154 (2002)
12. Garey, M., Johnson, D.: Computers and Intractability: A Guide to the Theory of NP-Completeness. Freeman, New York (1979)
13. Garey, M., Johnson, D.: Computers and Intractability: A Guide to the Theory of NP-Completeness. W.H. Freeman, New York (1979)
14. Girau, B., Marchal, P., Nussbaum, P., Tisserand, A.: Evolvable Platform for Array Processing: A One-Chip Approach. In: Proc. of the 7th Int. Conf. on Microelectronics for Neural, Fuzzy and Bio-inspired Systems, Granada, Spain, pp. 187–193 (1999)
15. Goldberg, D.: Genetic Algorithms in Search, Optimization and Machine Learning. Addison-Wesley, Reading (1989)
16. Gonzales-Miranda, J.: Perturbing Chaotic Systems to Control Chaos. In: Synchronization and Control of Chaos - An Introduction for Scientists and Engineers. Imperial College Press, London (2004)
17. Gordon, D., des Jardins, M. (eds.): Mach. Learn. 20, 1–17 (1995)
18. Grebogi, C., Lai, Y.: Controlling Chaotic Dynamical System. Phys. Rep. 31, 307–312 (1997)
19. Grebogi, C., Lai, Y.: Controling Chaos. In: Schuster, H. (ed.) Handbook of Chaos Control. Wiley-VCH, Weinheim (1999)
20. Grebogi, C., Lai, Y.: Pole placement Method of Controling Chaos in high dimensions. In: Schuster, H. (ed.) Handbook of Chaos Control. Wiley-VCH, Weinheim (1999)
21. Hany, H., Yongyi, T.: FingerPrint Registration Using Genetic Algorithms. In: 3rd IEEE Symposium on Application-Specific Systems and Software Engineering Technology (ASSET 2000), p. 148 (2000)
22. Hochbam, D.: Approximation Algorithms for NP - Hard Problems. PWS Publishing Company, USA (1997)
23. Holland, J.: Adaptation in natural and artificial systems. Univ. of Michigan Press, Ann Arbor (1975)
24. Houghton, E., Carpenter, P.: Aerodynamics for Engineering Students, 5th edn. Elsevier, Butterworth-Heinemann, Oxford (2003)
25. Harvey, I., Thompson, A.: Through the Labyrinth Evolution Finds a Way: A Silicon Ridge. In: Proc. of the 1st Int. Conf. on Evolvable Systems, Tsukuba, Japan, pp. 406–422 (1996)
26. Higuchi, T., Iwata, M., Kajitani, I., Iba, H., Hirao, Y., Manderick, B., Furuya, T.: Evolvable Hardware and its Applications to Pattern Recognition and Fault- tolerant Systems. In: Sanchez, E., Tomassini, M. (eds.) Towards Evolvable Hardware: The Evolutionary Engineering Approach, pp. 118–135. Springer, Berlin (1996)
27. Hargis, P.: The Gaseous Electronics Conference Radiofrequency Reference Cell - A Defined Parallel-Plate Radiofrequency System For Experimental And Theoretical-Studies of Plasma-Processing Discharges. Rev. Sci. Instrum. 65(1), 140–154 (1994)

28. Hirst, A.: Notes on the Evolution of Adaptive Hardware. In: Proc. of Adaptive Computing in Engineering Design and Control, Plymouth, U.K., pp. 212–219 (1996)
29. Hofbauer, J., Sigmund, K.: Evolutionary Games and Population Dynamics. Cambridge University Press, Cambridge (1998)
30. Huynen, M., Stadler, P., Fontana, W.: Smoothness within ruggedness: The role of neutrality in adaptation. Proc. of the National Academy of Science 93, 397–401 (1996)
31. Judy, M., Ravichandran, K., Murugesan, K.: A multi-objective evolutionary algorithm for protein structure prediction with immune operators. Comput. Meth. Biomech. Biomed. Eng. 12(4), 407–413 (2009)
32. Kajitani, I., Hoshino, T., Nishikawa, D., Yokoi, H., Nakaya, S., Yamauchi, T., Inuo, T., Kajihara, N., Iwata, M., Keymeulen, D., Higuchi, T.: A Gate-Level EHW Chip: Implementing GA Operations and Reconfigurable Hardware on a Single LSI. In: Proc. of the 2nd Int. Conf. on Evolvable Systems, Lausanne, Switzerland, pp. 1–12 (1998)
33. Karr, C., Bowersox, R., Singh, V.: Minimization of Sonic Boom on Supersonic Aircraft Using an Evolutionary Algorithm. In: Cantú-Paz, E., Foster, J.A., Deb, K., Davis, L., Roy, R., O'Reilly, U.-M., Beyer, H.-G., Kendall, G., Wilson, S.W., Harman, M., Wegener, J., Dasgupta, D., Potter, M.A., Schultz, A., Dowsland, K.A., Jonoska, N., Miller, J., Standish, R.K. (eds.) GECCO 2003. LNCS, vol. 2724. Springer, Heidelberg (2003)
34. Levenspiel, O.: Chemical reaction engineering. John Wiley and Sons, New York (1962)
35. Li, Y., Haubler, A.: Artificial evolution of neural networks and its application to feedback control. Artif. Intell. Eng. 10, 143–152 (1996)
36. Liu, W., Murakawa, M., Higuchi, T.: ATM Cell Scheduling by Function Level Evolvable Hardware. In: Proc. of the 1st Int. Conf. on Evolvable Systems, Tsukuba, Japan, pp. 180–192 (1996)
37. Xingwei, L., Yongxiang, P., Gao, H.: Using chaos-parallel evolutionary programming to solve the flow-shop scheduling problem. In: Intelligent Control and Automation, WCICA, vol. 3, pp. 2001–2003 (2000)
38. Miller, J., Thomson, P.: Aspects of Digital Evolution: Geometry and Learning. In: Proc. of the 2nd Int. Conf. on Evolvable Systems, Lausanne, Switzerland, pp. 25–35 (1998)
39. Miller, J., Job, D., Vassilev, V.: Principles in the Evolutionary Design of Digital Circuits - Part II. Genetic Programming and Evolvable Machines 1(3), 259–288 (2000)
40. Mitchell, M.: An Introduction to Genetic Algorithms. MIT Press, Cambridge (1998)
41. Murakawa, M., Yoshizawa, S., Kajitani, I., Furuya, T., Iwata, M., Higuchi, T.: Hardware Evolution at Function Level. In: Proc. of the 4th Conf. on Parallel Problem Solving from Nature, Berlin, Germany, pp. 62–71 (1996)
42. Murakawa, M., Yoshizawa, S., Kajitani, I., Yao, X., Kajihara, N., Iwata, M., Higuchi, T.: The GRD Chip: Genetic reconfiguration of DSPs for Neural Network Processing. IEEE Trans. on Computers 48(6), 628–639 (1999)
43. Nolle, L., Goodyear, A., Hopgood, A., Picton, P., Braithwaite, N.: On Step Width Adaptation in Simulated Annealing for Continuous Parameter Optimisation. In: Reusch, B. (ed.) Fuzzy Days 2001. LNCS, vol. 2206, pp. 589–598. Springer, Heidelberg (2001)
44. Nolle, L., Goodyear, A., Hopgood, A., Piction, D., Braithwaite, N.: Automated control of an actively compensated Langmuir probe system using simulated annealing. Knowl. Base Syst. 15(5-6), 349–354 (2002)
45. Nowak, M., May, R.: Evolutionary games and spatial chaos. Nature 359, 826–829
46. Nolle, L., Zelinka, I., Hopgood, A., Goodyear, A.: Comparison of an self-organizing migration algorithm with simulated annealing and differential evolution for automated waveform tuning. Adv. Eng. Software 36(10), 645–653 (2005)

47. Nolle, L., Zelinka, I., Hopgood, A., Goodyear, A.: Comparison of an self organizing migration algorithm with simulated annealing and differential evolution for automated waveform tuning. Adv. Eng. Software 36(10), 645–653 (2005)
48. Ortega, C., Tyrrell, A.: Biologically Inspired Fault-tolerant Architectures for Real-time Control Applications. Contr. Eng. Pract. 7(5), 673–678 (1999)
49. Ott, E., Grebogi, C., Yorke, A.: Controlling Chaos. Phys. Rev. Lett. 64, 1196–1199 (1990)
50. Pham, Q.: Dynamic optimization of chemical engineering processes by evolutionary method (2005)
51. Pham, Q., Coulter, S.: Modelling the chilling of pig carcasses using an evolutionary method. Proc. Int. Congress of Refrig. 3a, 676–683 (1995)
52. Rendell, L.: Similarity-based Learning and its Extensions. Comput. Intell. 3, 241–266 (1987)
53. Richter, H., Reinschke, K.: Optimization of local control of chaos by an evolutionary algorithm. Physica D 144, 309–334 (2000)
54. Rumelhart, D., Widrow, B., Lehr, M.: The Basic Ideas in Neural Networks. Comm. ACM 37(3), 87–92 (1994)
55. Swift, J., Schwar, M.: Electrical Probes for Plasma Diagnostics, Ilitte, London (1970)
56. Thompson, A.: Evolving Inherently Fault-Tolerant Systems. Proc. IME J. 211(1), 365–371 (1997)
57. Thompson, A.: Hardware Evolution. Springer, London (1998)
58. Thompson, A., Harvey, I., Husbands, P.: Unconstrained Evolution and Hard Consequences. In: Sanchez, E., Tomassini, M. (eds.) Towards Evolvable Hardware 1995. LNCS, vol. 1062, pp. 136–165. Springer, Heidelberg (1996)
59. Torresen, J.: Possibilities and Limitations of Applying Evolvable Hardware to Real-World Applications. In: Proc. of the 10th Int. Conf. on Field Programmable Logic and Applications, Villach, Austria, pp. 230–239 (2000)
60. Tupy, J., Zelinka, I.: Database and Expert Systems Application. In: 19th International Conference on DEXA, September 1-5, pp. 524–530 (2008)
61. Wegener, I.: Complexity Theory: Exploring the Limits of Efficient Algorithms. Springer, Heidelberg (2005)
62. Wright, A., Agapie, A.: Cyclic and Chaotic Behavior in Genetic Algorithms. In: Proc. of Genetic and Evolutionary Computation Conference (GECCO), San Francisco, July 7-11 (2001)
63. Yan, X., Chen, D., Hu, S.: Chaos-genetic algorithms for optimizing the operation conditions based on RBF-PLS model. Comput. Chem. Eng. 28(4), 579 (2004)
64. Yang, M., Guan, J.: Dynamic Clonal and Chaos-Mutation Evolutionary Algorithm for Function Optimization. In: Kang, L., Cai, Z., Yan, X., Liu, Y. (eds.) ISICA 2008. LNCS, vol. 5370, pp. 19–27. Springer, Heidelberg (2008)
65. Yao, X., Higuchi, T.: Promises and Challenges of Evolvable Hardware. In: Proc. of the 1st Int. Conf. on Evolvable Systems, Tsukuba, Japan, pp. 55–78 (1996)
66. Zebulum, R., Aurélio Pacheo, M., Vellasco, M.: Evolvable Systems in Hardware Design: Taxonomy, Survey and Applications. In: Proc. of the 1st Int. Conf. on Evolvable Systems, Tsukuba, Japan, pp. 344–358 (1996)
67. Zelinka, I., Nolle, L.: Plasma Reactor Optimizing Using Differential Evolution. In: Price, K., Storn, R., Lampinen, J. (eds.) Differential Evolution: Global Optimization for Scientists and Engineers. Springer, Heidelberg (2005)

68. Zelinka, I., Nolle, L.: Plasma reactor optimizing using differential evolution. In: Price, K., Lampinen, J., Storn, R. (eds.) Differential Evolution: A Practical Approach to Global Optimization, pp. 499–512. Springer, New York (2006)
69. Zelinka, I., Senkerik, R., Navratil, E.: Investigation on Real Time Deterministic Chaos Control by Means of Evolutionary Algorithms, CHAOS 2006. In: Proc. 1st IFAC Conference on Analysis and Control of Chaotic Systems, Reims, France, June 28–30, pp. 211–217 (2006)
70. Zelinka, I., Senkerik, R., Navratil, E.: Investigation on Evolutionary Optimitazion of Chaos Control, Chaos, Solitons & Fractals (2007), doi:10.1016/j.chaos.2007.07.045
71. Zelinka, I., Guanrong, C., Celikovsky, S.: Chaos Synthesis by Means of Evolutionary algorithms. Int. J. Bifurcat. Chaos Appl. Sci. Eng. 18(4), 911–942 (2008)

Chapter 2
Evolutionary Algorithms for Chaos Researchers

Ivan Zelinka and Hendrik Richter

Abstract. Evolutionary algorithms are search methods that can be used for solving optimization problems. They mimic working principles from natural evolution by employing a population–based approach, labeling each individual of the population with a fitness and including elements of random, albeit the random is directed through a selection process. In this chapter, we review the basic principles of evolutionary algorithms and discuss their purpose, structure and behavior. In doing so, it is particularly shown how the fundamental understanding of natural evolution processes has cleared the ground for the origin of evolutionary algorithms. Major implementation variants and their structural as well as functional elements are discussed. We also give a brief overview on usability areas of the algorithm and end with some general remarks of the limits of computing.

2.1 Historical Facts from a Slightly Different Point of View

Evolutionary algorithms, or better evolutionary computational techniques (ECT), are based on principles of evolution which have been observed in nature long time before they were applied to and transformed into algorithms to be executed on computers. When next reviewing some historical facts that led to evolutionary computation as we know it now, we will mainly focus on the basic ideas, but will also

Ivan Zelinka
Tomas Bata University in Zlin, Faculty of Applied Informatics, Nad Stranemi 4511, Zlin 76001, Czech Republic
and
VSB-TUO, Faculty of Electrical Engineering and Computer Science, 17. listopadu 15, 708 33 Ostrava-Poruba, Czech Republic
e-mail: zelinka@fai.utb.cz

Hendrik Richter
HTWK Leipzig, Fakultät Elektrotechnik und Informationstechnik, D–04251 Leipzig, Germany
e-mail: richter@fbeit.htwk-leipzig.de

I. Zelinka et al. (Eds.): Evolutionary Algorithms and Chaotic Systems, SCI 267, pp. 37–88.
springerlink.com

allow to glimpse at the people who did the pioneering work and established the field. Maybe the two most significant persons whose research on evolution and genetics had the biggest impact on modern understanding of evolution and its use for computational purposes are Gregor Johann Mendel and Charles Darwin.

Gregor Johann Mendel (Fig. 2.1, July 20, 1822 - January 6, 1884) was an Augustinian priest and scientist, and is often called the father of genetics for his study of the inheritance of certain traits in pea plants. He was born in the family of farmers in Hyncice (Heinzendorf bei Odrau) in Bohemia (that time part of Austrian - Hungary empire, today Czech Republic). The most significant contribution of Mendel for science was his discovery of genetic laws which showed that the inheritance of these traits follows particular laws (published in [52]), which were later named after him. All his discoveries were done in Abbey of St. Thomas in Brno (Bohemia). Mendel published his research at two meetings of the Natural History Society of Brünn in Moravia (east part of Bohemia) in 1865 [52]. When Mendel's paper was published in 1866 in Proceedings of the Natural History Society of Brünn, it had little impact and was cited only about three times over the next thirty-five years. His paper was criticized at the time, but is now considered a seminal work. The significance of Mendel's work was not recognized until the turn of the 20th century. Its rediscovery (thanks to Hugo de Vries, Carl Correns and Erich von Tschermak) prompted the foundation of the discipline of genetics. Very peculiar historical fact about Mendel's research is also that his letters about his discovery, sent to many of scientific societies, had been found after many years in their libraries unopened. Mendel died on January 6, 1884, at age 61, soon after his death the succeeding abbot burned all papers in Mendel's collection, to mark an end to the disputes over taxation [10].

The other important (and much more well–known and therefore here only briefly introduced) researcher whose discoveries founded the theory of evolution was the British scientists Charles Darwin. Darwin (Fig. 2.2) published in his work [17] the

Fig. 2.1 Gregor Johann Mendel (July 20, 1822 - January 6, 1884)

Fig. 2.2 Gregor Charles Darwin 12 February 1809 - 19 April 1882

main ideas of the evolutionary theory. The full and original title was *"On the Origin of Species by Means of Natural Selection, or the Preservation of Favored Races in the Struggle for Life"*. Word *"races"* refers here to biological varieties. The title has been changed to [17] for the 6th edition of 1872. In Darwin's book On the Origin of Species (1859) established evolutionary descent with modification as the dominant scientific explanation of diversification in the nature.

The above mentioned ideas of genetics and evolution have been formulated long before the first computer experiments with evolutionary principles had been done. The beginning of the ECT is officially dated to the 70s of the 20th century, when famous genetic algorithms were introduced by J. Holland [37, 38] or to the late 60s with evolutionary strategies, introduced by Schwefel [64] and Rechenberg [60] and evolutionary programming by L.J. Fogel [29]. However, when certain historical facts are taken into consideration, then one can see that the main principles and ideas of ECT as well as its computer simulations had been done earlier than mentioned above. Conceptionally, ECT can be traced back to the famous A.M. Turing, first numerical experiments to the (far less famous) N.A. Barricelli and others. Their understanding and formulation of basic ideas of ECT was remarkably clear, see e.g. Turing in his essay "Intelligent machinery" (1948) [67] where he say:

"...if the untrained infant's mind is to become an intelligent one, it must acquire both discipline and initiative... discipline is certainly not enough in itself to produce intelligence. That which is required in addition we call initiative...our task is to discover the nature of this residue as it occurs in man, and to try and copy it in machines...".

In other words he speaks about simulation of an intelligent creature. Turing continues in his text by

"...further research into intelligence of machinery will probably be very greatly concerned with 'searches'...",

and suggested that *'searches'* will be done probably in the space of numbers and basically he describes the central idea of ECT by

"...there is the genetic or evolutionary search by which a combination of genes is looked for, the criterion being the survival value".

Turing further improved this idea in his article "Computing Machinery and Intelligence" (1950) [67]:

"We cannot expect to find a good child-machine at the first attempt. One must experiment with teaching one such machine and see how well it learns. One can then try another and see if it is better or worse. There is an obvious connection between this process and evolution, by the identifications"

- *structure of child machine = hereditary material*
- *changes of the child machine = mutations*
- *natural selection = judgement of the experimenter".*

One of the first to transform Turing's ideas into real computer numerical experiments was N. Barricelli (1954) [4, 5]. Results were published in the journal *"Methods"* with the title *"Esempi Numerici di processi di evoluzione"* and consequently repeated and improved in 1962 [6] when ECT numerical experiment with 500 of 8 bits strings had been successfully done. Based on this numerical simulations Barricelli reported that:

"we have created a class of numbers which are able to reproduce and to undergo hereditary changes...the constitution for an evolutionary process according to the principle of Darwins theory would appear to be present. The numbers which have the greatest survival in their environment will survive. The other numbers will be eliminated little by little. A process of adaptation to the environmental conditions, that is, a process of Darwinian evolution, will take place".

Barricelli's early computer-assisted experiments, which were focused on symbiogenesis and evolution (based on Darwin's ideas), can be accepted like pioneering experiments in artificial life research. Barricelli has been working in Institute for Advanced Study in Princeton, New Jersey in 1953, 1954 and 1956. Later he worked at the University of California, Los Angeles, at Vanderbilt University, in the Department of Genetics of the University of Washington, Seattle and then at the Mathematics Institute of the University of Oslo. He is also author of a variety of articles in fields as different as theoretical physics and mathematical language, virus genetics, DNA, theoretical biology, space flight, etc. Barricelli's experiments are probably some of the first historically recorded numerical ECT experiments. Another interesting tread of ECT's pre-history are the works of Box (1957) [7] and Friedberg (1958) [30]. Although the original papers are meanwhile hardly accessible, it is in particular thanks to D.B. Fogel (son of evolutionary programming pioneer L.J. Fogel), who edited some of these works [27] and recollected some technical details and implications [26, 28], that this early history of ECT can now be rediscovered. However, in some respect all these works were slightly ahead of time, as the results have clearly shown the potential of ECT methods, but the lack of computing power at that time prevented to solve "real problems" and hence to widespread the

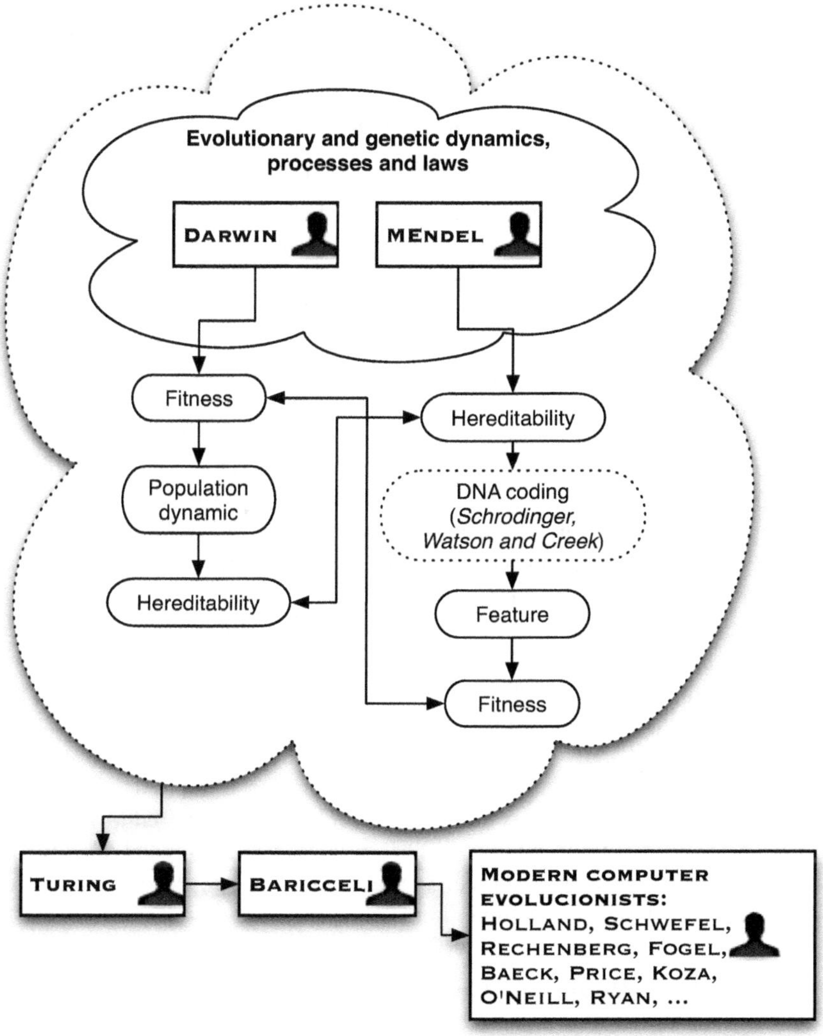

Fig. 2.3 Evolutionary theory and its historical relations

methods. So, the "golden era" of ECT began, when genetic algorithms by J. Holland [37], evolutionary strategies, by Schwefel [64] and Rechenberg [60] and evolutionary programming by Fogel [29] had been introduced. All these designs were favored by the upcoming of more powerful and more easily programmable computers so that for the first time interesting problems could be tackled and ECT started to compete with and became a serious alternative to other optimization methods. Since that time other successful algorithms using ECT ideas have been developed, for instance scatter search, particle swarm, memetic algorithms, differential evolution, ant colony optimization, and many others. Before their brief description it is important to outline the main principle (lets call it *central dogma*) of evolutionary computation in general. History of ECT is of course more rich and complex, than described here. Main ideas and relations can be more clearly visible from Fig. 2.3.

2.2 Evolutionary Algorithms – Outline

In recent years, a broad class of algorithms has been developed for stochastic optimization, i.e. for optimizing systems where the functional relationship between the independent input variables and the output (objective function) of a system is not explicitly known. Using stochastic optimization algorithms such as Genetic Algorithms, Simulated Annealing and Differential Evolution, a system is confronted with a random input vector and its response is measured. This response is then used by the algorithm to tune the input vector in such a way that the system produces the desired output or target value in an iterative process. Most engineering problems can be defined as optimization problems, e.g. the finding of an optimal trajectory for a robot arm, the optimal thickness of steel in pressure vessels, the optimal set of parameters for controllers, optimal relations or fuzzy sets in fuzzy models, etc. Solutions to such problems are usually difficult to find as their parameters usually include variables of different types, such as floating point or integer variables. Evolutionary algorithms, such as the Genetic Algorithms, Particle Swarm, Ant Colony Optimization, Scatter Search, Differential Evolution etc., have been successfully used in the past for these engineering problems, because they can offer solutions to almost any problem in a simplified manner: they are able to handle optimizing tasks with mixed variables, including the appropriate constraints, and they do not rely on the existence of derivatives or auxiliary information about the system, e.g. its transfer function. This chapter is concerned with a brief introduction on so-called evolutionary computational techniques (ECT). Although the editors of this book assume that most of the readers will have at least basic knowledge of ECT, there might be the wish for clarification and broadening. For this reason, this chapter was also included to describe in simple terms what ECT actually means.

2.2.1 Central Dogma of Evolutionary Computational Techniques

The evolutionary computational techniques are numerical algorithms that are based on the basic principles of Darwin's theory of evolution and Mendel's foundation of genetics. The main idea is that every individual of a species can be characterized by its features and abilities that help it to cope with its environment in terms of survival and reproduction. These features and abilities can be termed its fitness and are inheritable via its genome. In the genome the features/abilities are encoded. The code in the genome can be viewed as a kind of "blue–print" that allows to store, process and transmit the information needed to build the individual. So, the fitness coded in the parent's genome can be handed over to new descendants and support the descendants in performing in the environment. Darwinian participation to this basic idea is the connection between fitness, population dynamics and inheritability while the Mendelian input is the relationship between inheritability, feature/ability and fitness. Both views have been brought together in molecular Darwinism with the idea of a genetic code (first uttered in its full information–theoretical meaning by Erwin Schrödinger in his 1944 book *What is life?*) and the discovery of the structure of the DNA and its implications to genetic coding by James Watson and Francis Crick in 1953.

By these principles and in connection with the occurrence of mutations that modify the genome and hence may produce inheritable traits that enhance fitness, a development of individuals and species towards best adaption to the environment takes place. Here, the multitude of individuals in a species serve two connected evolutionary aspects, (i.) provide the opportunity to "collect" mutations and pass traits that are or are not fitness enhancing to descendants on an individual level and (ii.) allow fitness enhancing genomes to spread in the species from one generation to the next if the traits bring advantages in survival and reproduction.

However, it should be noted that Darwin or, as the case may be, Mendel, were not the first. Already in the ancient era, there were thinkers who came with the same idea as Darwin and Mendel. An outstanding thinker, who supported the idea of evolution before Darwin, was Anaximander, a citizen from Miletus, an Ionian city of Asia Minor. Anaximander's philosophical ideas are summarized in his philosophical tract *"On nature"*, however, this name is of a later date, because this book was not preserved. According to Anaximander, the original principle of the world and the cause of all being is "without a limit" (*"apeiron"* in Greek), from which cold and warm and dry and wet is separated - essentially, one can imagine this principle in the sense of unlimited and undifferentiated wetness, from which all other natural substances and individual species of living creatures arise.

By his idea that the Earth, which he imagined as freely floating in space, was initially in a liquid state and later, when it was drying, gradually gave rise to animals, who at first lived in water and later migrated onto land, Anaximander in part anticipates the modern theory of evolution.

The ECT technology stands or falls with the existence of the so-called evolutionary algorithms (EA) that in principle form the majority of ECT. Besides evolutionary algorithms, there still exist other extensions, such as genetic programming, evolutionary hardware, etc. With respect to the fact that evolutionary algorithms are the backbone of ECT, attention will be paid in this chapter just to these algorithms, whose understanding is **absolutely necessary** for understanding the rest of this publication.

From the above mentioned main ideas of Darwin and Mendel theory of evolution, ECT uses some building blocks which the diagram in Fig. 2.4 illustrates. The evolutionary principles are transferred into computational methods in a simplified form that will be outlined now.

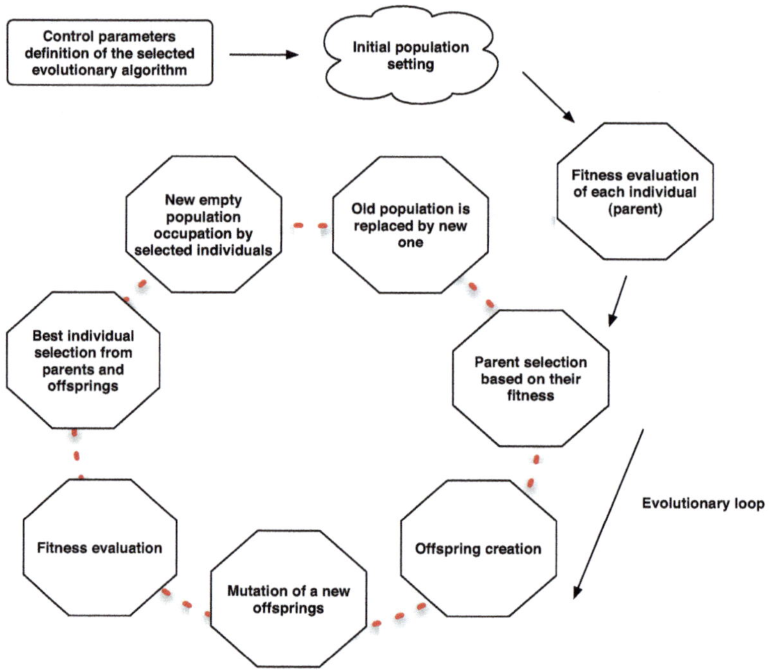

Fig. 2.4 General cycle of the evolutionary algorithm. The termination of the evolution after *n* generations and the selection of the best individual are not indicated in this figure - solution from the last population.

If the evolutionary principles are used for the purposes of complicated calculations (in accordance with Fig. 2.4), the following procedure is used:

1. Specification of the evolutionary parameters: For each algorithm, parameters must be defined that control the run of the algorithm or terminate it regularly, if the termination criterions defined in advance are fulfilled (for example, the number of cycles - generations). Part of this point is the definition of the cost function (objective function) or, as the case may be, what is called fitness - a modified return value of the objective function). The objective function is usually a mathematical model of the problem, whose minimization or maximization (generally therefore extremization) leads to the solution of the problem.

This function with possible limiting conditions is some kind of "environmental equivalent" in which the quality of current individuals is assessed.

2. Generation of the initial population (generally $N \times M$ matrix, where N is the number of parameters of an individual - D is used hereinafter in this publication - and M is the number of individuals in the population): Depending on the number of optimized arguments of the objective function and the user's criterions, the initial population of individuals is generated. An individual is a vector of numbers having such a number of components as the number of optimized parameters of the objective function. These components are set randomly and each individual thus represents one possible specific solution of the problem. The set of individuals is called population.
3. All the individuals are evaluated through a defined objective function and to each of them is assigned a) Either a direct value of the return objective function, or b) A fitness value, which is a modified (usually normalized) value of the objective function.
4. Now parents are selected according to their quality (fitness, value of the objective function) or, as the case may be, also according to other criterions.
5. Descendants are created by crossbreeding the parents. The process of crossbreeding is different for each algorithm. Parts of parents are changed in classic genetic algorithms, in a differential evolution, crossbreeding is a certain vector operation, etc.
6. Every descendant is mutated. In other words, a new individual is changed by means of a suitable random process. This step is equivalent to the biological mutation of the genes of an individual.
7. Every new individual is evaluated in the same manner as in step 3.
8. The best individuals are selected.
9. The selected individuals fill a new population.
10. The old population is forgotten (eliminated, deleted, dies,..) and is replaced by a new population; step 4 represents further continuation.

Steps 4 - 10 are repeated until the number of evolution cycles specified before by the user is reached or if the required quality of the solution is not achieved. The principle of the evolutionary algorithm outlined above is general and may more or less differ in specific cases. So, methods that work by an algorithmic structure as outlined in the steps 1-10 share the following main evolutionary principles:

- **Biological inspiration:** The algorithms mimic and use in an abstracted way working mechanisms of biological systems.
- **Population–based calculations:** By structuring data in the algorithm by the individual–and–species model, individual search is coordinated to other individuals and so to the whole population. This has the effect of parallelism in the search which is assumed to be the main reason for success of evolutionary search.
- **Repeated calculation of fitness for all individuals:** This principles provides a spectrum of fitness to the population from which search can be guided by noticing and discriminating individuals of different fitness.

- **Generational search:** Repeated generational search guided by the fitness spectra allows to accumulate individuals with high fitness.
- **Stochastic and deterministic driving forces:** Random influences, for instance in form of mutations are balanced by the deterministic elements in the flow of the algorithm.
- **Coordination between individuals:** Some kind of communication on (or even in–between) the individuals of the population (e.g. in the selection (crossover) or recombination process) allows to recognize and exploit individual differences in fitness.

There are also exemptions that do not adhere to steps 1 - 10; in such a case, the corresponding algorithms are not denoted as evolutionary algorithms, but usually as algorithms that belong to ECT. Some evolutionists exclude them completely from the ECT class. The ACO algorithm (*Ant Colony Optimization*), see [21] and [56] may be an example - it simulates the behavior of an ant colony and can solve extremely complicated combinatory problems. It is based on the principles of cooperation of several individuals belonging to the same colony - in this case ants.

The evolution diagrams are not only popular because they are modern and differ from classical algorithms, but mainly because of the fact that they are able to replace a man in the event of a suitable application. This is illustrated in Fig. 2.5. There are two methods of the problem solution illustrated in this figure. The first one represents steps of a human investigator, the second one represents the procedure if ECT is used.

This publication thus deals with ECT's that in most cases adhere to the above indicated evolutionary scheme; nevertheless, exemptions are also indicated.

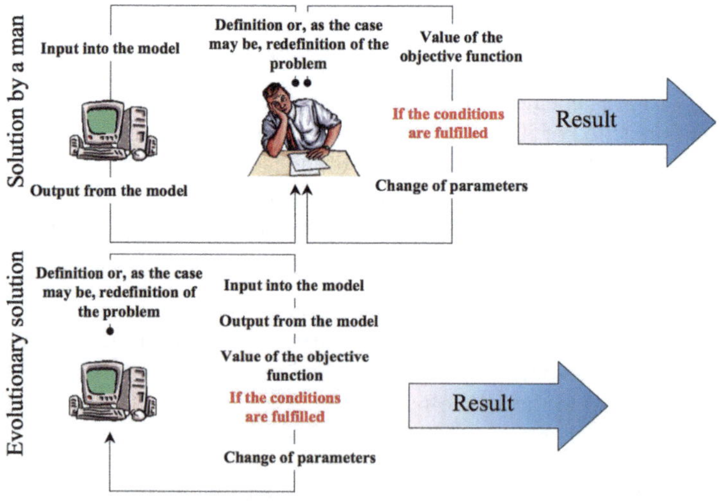

Fig. 2.5 Comparison of the problem solution by means of ECT and a man. Simplified illustration.

2.2.2 Evolutionary Algorithms and Importance of Their Use

Comparing to standard optimization techniques, evolutionary algorithms can be used on *almost* arbitrary optimization problem, however, it is important to remember that with different performance. As mentioned in the section 2.5.2, there are problems with different level of complexity, from the simplest (solvable by standard techniques) to the most complex, whose solution would take much more longer time, than our universe exist. Thus, some simple problems, that can be very easily and quickly solved by gradient based techniques, should not be solved by heuristic methods, because its use would be expensive, i.e. user would "pay" by big number of cost function evaluations. Another important fact, having impact on EA use is so called No Free Lunch Theorem (NFLT), see [70]. Main idea of this theorem is that there is no ideal algorithm which would be able to solve **any** problem. Simply, if there are for example two algorithms **A** and **B**, then for certain subset of possible problems is more suitable algorithms **A** and for another subset algorithm **B**. All those subsets can be of course totally disconnected, or/and overlapped.

Based on those facts it is important to remember that evolutionary algorithms are suitable for problems which are more complex rather simple, and also that their selection and setting depend on user experiences, expertise etc. More exact classification is mentioned in the section 2.4.

2.3 Selected Evolutionary Techniques

2.3.1 Overview

Optimization algorithms are a powerful tool for solving many problems of engineering applications. They are usually used where the solution of a given problem analytically is unsuitable or unrealistic. If implemented in a suitable manner, there is no need for frequent user intervention into the actions of equipment in which they are used.

The majority of the problems of engineering applications can be defined as optimization problems, for example, finding the optimum trajectory of a robot or the optimum thickness of the wall of a pressure tank or the optimum setting of the regulator's parameters. In other words, the problem solved can be transformed into a mathematical problem defined by a suitable prescription, whose optimization leads to finding the arguments of the objective function, which is its goal.

Countless examples can be found illustrating this problem. The solution of such problems usually requires working with the arguments of optimized functions, where the definition ranges of these arguments may be of a heterogeneous character, such as, for example, the range of integers, real or complex numbers, etc. Moreover, it may happen (depending on the case) that for certain subintervals from the permitted interval of values, the corresponding argument of the optimized function may assume values of various types (integers, real, complex,..). Besides this, various penalizations and restrictions can play a role within optimization, not only for given arguments, but also for the functional value of the optimized function. In many cases, the analytical solution of such an optimization problem is possible, nevertheless, considerably complicated and tedious.

A class of very efficient algorithms has been developed for the successful solution of such problems in the past two decades that make it possible to solve very complicated problems efficiently. The algorithms of this class have their specific name, namely "evolutionary algorithms". They solve problems in such an elegant manner that they became very popular and are used in many engineering fields.

From the point of view of the most general classification, the evolutionary algorithms belong to heuristic algorithms. Heuristic algorithms are either *deterministic* or *stochastic*. The algorithms of the second group differ in that their certain steps use random operations, which means that the results of the solutions obtained with their use may differ in the individual runs of the program. It is therefore meaningful to run the program several times and select the best solution obtained.

Stochastic heuristic methods are sometimes called *metaheuristics*, because they only provide a general framework and the algorithms of the operation itself must be chosen (for example, by the operation of crossbreeding and mutation in genetic algorithms, operation of neighborhood in simulated annealing, "tabu search", etc.) in dependence on the problem investigated. Because these methods are frequently inspired by natural processes, they are also called evolutionary algorithms. Depending on their strategy, they can be divided in two classes:

1. Methods based on point-based strategy such as, for example, *simulated annealing* ([41], [13], [66]), *hill-climbing algorithm* [63] and *tabu-search* [31]. These algorithms are based on the *neighborhood* operation of the current solution, in which we are looking for a better solution.
2. *Population-based strategy. Genetic algorithms* [19], [33], [53], [14] are based on *population strategy*.

These methods differ from classic gradient methods by admitting (with a certain probability) a worse solution into the next iteration; in this manner, they try to avoid local minima. For more details, see, for example, books [31], [33],[53],[54] and [61].

2.3.2 Current State

Evolutionary algorithms serve for finding the minima (or maxima) of a given objective function by looking for the optimum numerical combination of its arguments. These algorithms can be divided according to the principles of their action, complexity of the algorithm, etc. Of course, this classification is not the only possibility, nevertheless, because it fits the current state rather well, it can be considered as one of the possible views on the classical and modern optimization methods. There are slight differences in opinions on their classification. One can encounter statements that, for example, simulated annealing does not belong to evolutionary techniques, which is true to some extent. On the other side, other "evolutionists" state that simulated annealing does belong to evolutionary techniques, at least as their direct predecessor. It is true that if simulated annealing is taken into account with elitism, then one could consider this alternative as an evolutionary algorithm.

2.3.2.1 Classes Optimization Approaches

Figs. 2.6 - 2.8 illustrate various views on the classification of evolutionary algorithms that exhibit certain differences although they have a visible common line. These differences may be caused not only by the classification of the algorithm according to the principles by which it is controlled, but, for example, according to the classes of problems for which it is "predestined". The individual classes of algorithms represent generally solutions of a given problem by the methods of various degrees of efficiency and complexity. Depending on their properties, we classify algorithms into the following categories:

Enumerative: The algorithm calculates all possible combinations of a given problem. This approach is suitable for problems where the arguments of the objective function have a discrete character and assume a small number of values. Should it be applied generally, it might need more time for its successful termination than is the time of the existence of the universe.

Deterministic: This group of algorithms is based only on the rigorous methods of classical mathematics. The algorithms of this character usually require limiting assumptions that enable these methods to provide efficient results. Usually, these assumptions are as follows:

- The problem is linear.
- The problem is convex.
- The space of the possible solutions is small.
- The space of the possible solutions is continuous.
- The objective function is unimodal (it has only one extreme).
- There are no nonlinear interactions between the parameters of the objective function.
- Information on the gradient is available, etc.
- The problem is defined in an analytical form.

The result of the deterministic algorithm is then only one solution.

Stochastic: The algorithms of this type are based on the use of chance. This is essentially a purely random search of the values of the objective function; the result is always the best solution that was found during the entire random search. The algorithms of this type are usually;

- slow.
- suitable only for small spaces of possible solutions (small range of arguments of the objective function),
- suitable for a rough estimation.

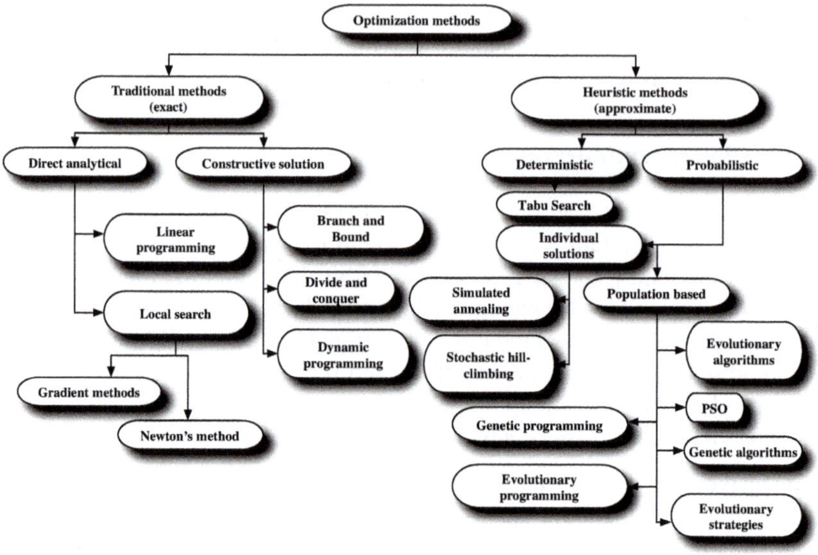

Fig. 2.6 Classification of the optimization methods according to [68]

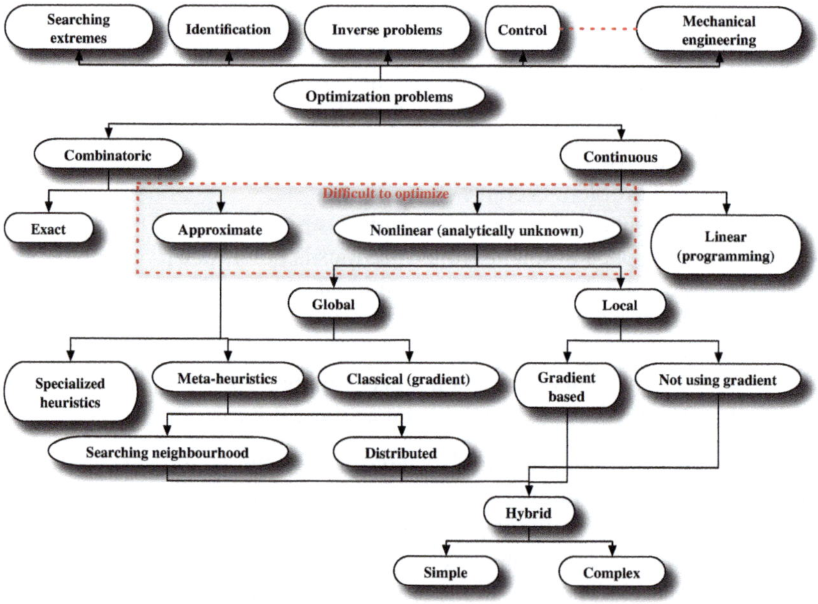

Fig. 2.7 Classification of the optimization methods according to [22]

Mixed: The algorithms of this class represent a "sophisticated" mixture of deterministic and stochastic methods that achieve surprisingly good results in mutual cooperation. The evolutionary algorithms mentioned above are a

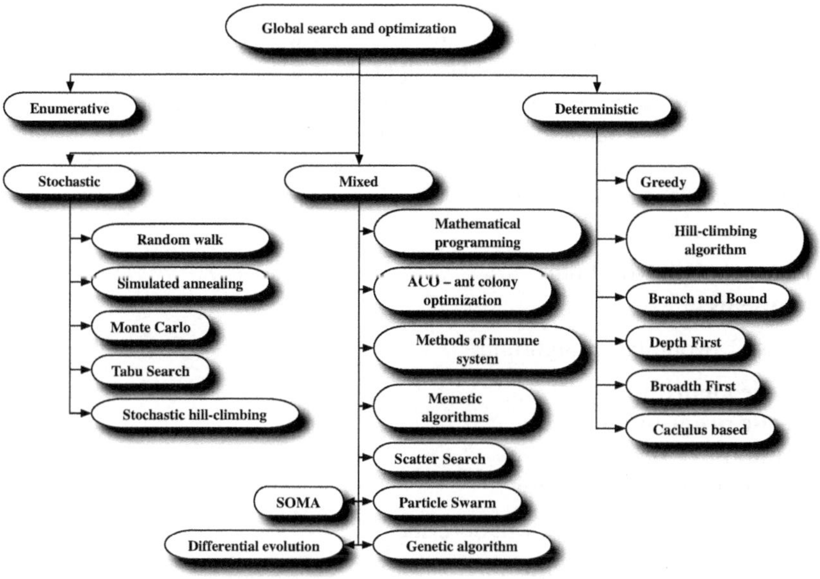

Fig. 2.8 Other possible organization of optimization algorithms [72]

relatively strong sub-set of these algorithms. The algorithms of the mixed character are:

- Robust, which means that they very frequently find a quality solution independently of the initial conditions; this solution is usually represented by one or several global extremes.
- Efficient and powerful. The terms "efficient and powerful" here mean that they are able to find a quality solution during a relatively small number of evaluations of the objective function.
- Differ from purely stochastic methods (thanks to the presence of deterministic approaches).
- Have minimum or no requirements for preliminary information.
- They are able to work with problems of the "black box" type, i.e., they do not need an analytical description of the problem for their activity.
- Are able to find several solutions during one run.

We can briefly summarize these features as follows:

- The enumerative and stochastic optimization is not suitable for problems where an extensive space of possible solutions must be searched.
- The deterministic optimization works well with problems where the space of possible solutions is not too extensive.
- Mixed optimization is suitable for problems without limitations to the size of the space of possible solutions.

In this book, selected algorithms are described with emphasis put on their explanation and testing. The evolutionary algorithms are now very popular thanks to properties that are characteristic for the entire class. However, before we start to discuss their details, it is suitable to mention both the generally known algorithms and the newer algorithms in this field, which will be developed in more detail later in this publication. From the well known algorithms, we have selected a few random–driven algorithms:

- **Stochastic Hill Climbing**
- **Tabu search**

the "classic" evolutionary algorithm

- **Genetic algorithms**
- **Genetic programming**
- **Evolutionary strategy**
- **Evolutionary programming**

and from the newest ones, we will describe

- **Learning classifier systems**
- **Population-based incremental learning**
- **Ant Colony Optimization**
- **Immunology System Method**
- **Memetic Algorithms**
- **Scatter Search**
- **Particle Swarm**
- **Differential Evolution**
- **SOMA**

2.3.2.2 The Outline of the Principles of Action of Random–Driven Search Algorithms

At the present time, there is a broad spectrum of publications dealing with optimization algorithms, for example, [2]. The purpose of this chapter is to outline only the principles of some selected algorithms for better information for the reader. The discussed algorithms are:

Stochastic Hill-Climbing: (SHC) is a version of the hill-climbing mechanism enriched by the stochastic component [40], [63]. It belongs among the gradient methods, which means that it searches the space of possible solutions in the direction of the steepest gradient. Thanks to its gradient nature, it frequently gets stuck in a local extreme. The basic version functions so that it always starts from the random point in the space of possible solutions. For the momentary proposed solution, the certain neighborhood is proposed by means of a final set of transformations (Figure 2.9) and the given function is minimized only in this neighborhood. The local solution obtained is then used as a start point for the calculation of the new neighborhood. The entire process is then repeated iteratively. The best found

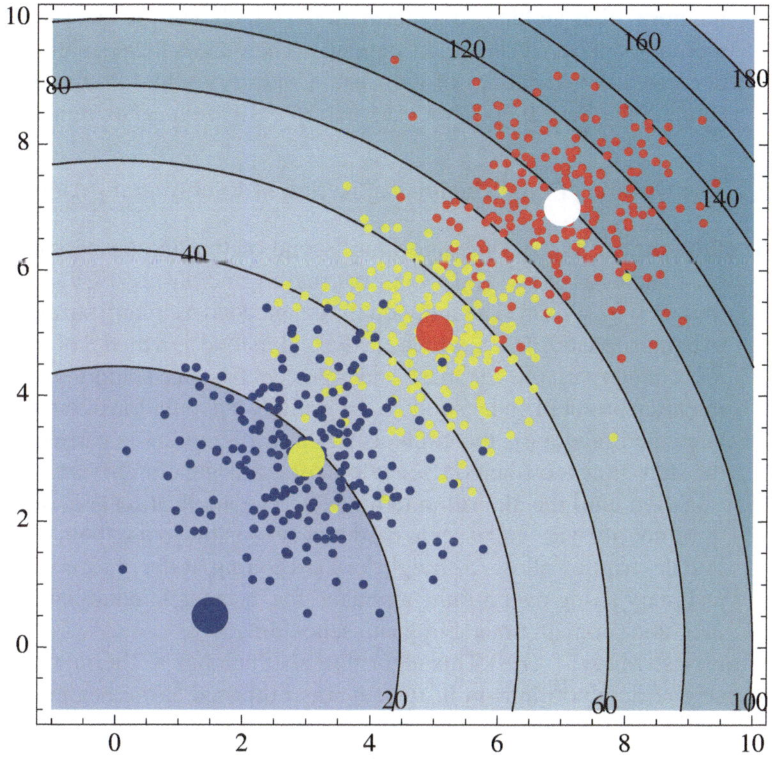

Fig. 2.9 Principle of the hill-climbing algorithm. The white dot is the starting point of the algorithm. The set of red solutions is generated for this dot. The best solution is denoted by the bold red dot that serves also as a new position for the generation of a new (yellow) set of solutions. The best solution (bold yellow dot) is then used for the generation of the following set of solutions (blue).

solution is recorded during the process that serves as the optimum found after termination. The stochastic hill-climbing algorithm is basically only the multiple repetition of the standard hill-climbing algorithm, each time from another randomly selected position. The disadvantage of this algorithm is that runaway may occur in it under certain conditions and the solution gets stuck in a local extreme [40].

Tabu Search: (TS) (prohibited search,[31]) is the improved version of the hill-climbing algorithm. This algorithm was created by Professor Fred Glover from the University of Colorado. The improvement consists in introducing a "short-term memory" into the hill-climbing algorithm, whose task is to remember those transformations by means of which the current median was calculated. The final consequence is that runaway cannot occur due to the forbidden use of these transformations. The name "Tabu Search" originates from this feature. This method was improved by a "long term memory" that includes transformations, which are not short term in the memory, but have been used frequently. Their use is then penalized, which decreases the frequency of their use. Contrary to the hill-climbing algorithm, Tabu Search does not get stuck so easily in local extremes.

2.3.2.3 The Outline of the Principles of Action of Evolutionary Algorithms

Genetic algorithm: (GA) This algorithm is one of the first successful applied ECT methods [37, 33]. In GAs the main principles of ECT are applied in their purest form. The individuals are encoded as binary strings (mostly over the alphabet $[0,1]$), which can be understood as a model of the biological counterpart, the genome,[1] and represent possible solutions to the optimization problem under study. After initially a population of binary strings is created randomly, the circle as given in Figure 2.4 is carried out with the steps fitness evaluation, selection, offspring generation (crossover) and mutation until the algorithm terminates. The application area of these algorithms are wide and it seem particularly sensible to use them if the problem description allows a straightforward coding of the objects to optimize as binary string over a finite alphabet, for instance in combinatorial optimization problem timetabling and scheduling.

Evolutionary strategy: (ES) This algorithm also belongs to the first successful stochastic algorithms in history. It was proposed at the beginning of the sixties by Rechenberg [60] and Schwefel [64]. It is based on the principles of natural selection similarly as the genetic algorithms. Contrary to genetic algorithms, the evolutionary strategy works directly with individuals described by vectors of real values. Its core is to use candidate solutions in the form of vectors of real numbers, which are recombined and then mutated with the help of a vector of random numbers. The problem of accepting a new solution is strictly deterministic. Another distinctive feature is that ES use self-adaptation, that is the mutation strength for each individual is variable over the generational run and subject to an own evolutionary adaption and optimization process.

Evolutionary programming: (EP) EP algorithms [29] have much similarity to ES in using vectors of real numbers as representation. The main operator in the generational circle is mutation, in most (particularly early) implementations no recombination is carried out. In recent years, by adopting elements of their algorithmic structure EP more and more tends to become similar to ES.

Learning classifier systems: (LCS) LCS [9] are machine learning algorithms which are based on GAs and reinforcement learning techniques. Interestingly, LCS were introduced by Holland[2] [37] and for a certain time

[1] The genome is coded over the alphabet $[A,C,G,T]$, which stand for the amino acids adenine A, cytosine C, guanine G, thymine T.

[2] Holland is also know as the father of GAs.

regarded as a generalization of GAs. LCS optimize over a set of rules that are intended to best–fit inputs to outputs. The rules are coded binary and undergo an adaption using GA–like optimization that modifies and selects the best rules. The fitting of the rules is determined by reinforcement learning methods.

Population-based incremental learning: (PBIL) PBIL was proposed by Baluja [3] and combines ideas from evolutionary computation with methods from statistical learning [49]. It uses a real valued representation that is usually restricted to the interval $[0,1]$ and can be interpreted as the probability to have a "1" - bit at a certain place in a binary string. From these probabilities, a collection of binary strings is created. These strings are subjected to a standard evolutionary circle with fitness evaluation, selection and discarding of inferior samples. In addition, based on the evaluation of the fitness, a deterministic statistical-learning-like updating of the probability vector takes place, which afterwards is also altered by random mutation.

Ant Colony Optimization: (ACO), [21] This is an algorithm whose action simulates the behavior of ants in a colony. It is based on the following principle. Let there be a source of ants (colony) and the goal of their activity (food), see Fig. 2.10. When they are released, all the ants move after some time along the shorter (optimum) route between the source and goal. The effect of finding the optimum route is given by the fact that the ants mark the route with pheromones. If an ant arrives to the crossroads of two routes that lead to the same goal, his decision along which route to go is random. Those ants that found food start marking the route and when returning, their decision is influenced thanks to these marks in favor of this route. When returning, they mark it for the second time, which increases the probability of the decision of further ants in its favor. These principles are used in the ACO algorithm. Pheromone is here represented by the weight that is assigned to a given route leading to the goal. This weight is additive, which makes it possible to add further "pheromones" from other ants. The evaporation of pheromones is also taken into account in the ACO algorithm in such a way that the weights fade away with time at individual joints. This increases the robust character of the algorithm from the point of view of finding the global extreme. ACO was successfully used to solve optimization problems such as the travelling salesman problem or the design of telecommunication networks, see [56].

Immunology System Method: (ISM) This algorithm is unusual by its algorithm based on the principles of functioning of the immunology system in living organisms. As indicated in [56], there are several principles based on this model. In this work, the immunology system is considered as a multivalent system, where individual agents have their specific tasks. These agents have various competencies and ability to communicate with other agents. On the basis of this communication and a certain "freedom" in making decisions of individual agents, a hierarchic structure is formed able to solve complicated problems. As an example of using this method,

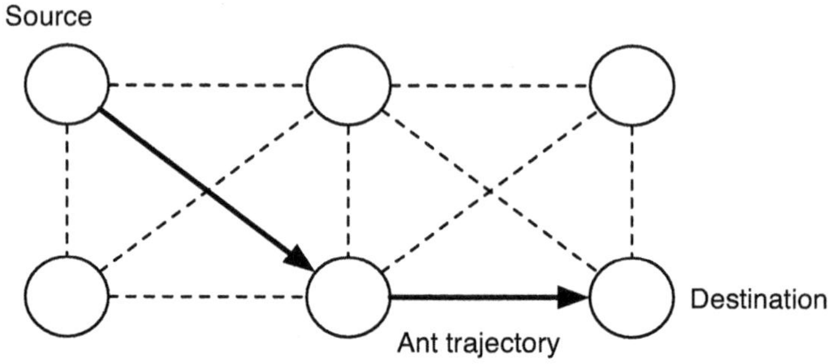

Fig. 2.10 Principle of the ACO algorithm

antivirus protection can be mentioned in large and extensive computer systems [18], [12].

Memetic Algorithms: (MA) This term represents a broad class of metaheuristic algorithms [56], [35], [32], [65]. The key characteristics of these algorithms are the use of various approximation algorithms, local search techniques, special recombination operators, etc. These metaheuristic algorithms can be basically characterized as competitive-cooperative strategies featuring attributes of synergy. As an example of memetic algorithms, hybrid combinations of genetic algorithms and simulated annealing or a parallel local search can be indicated. Memetic algorithms were successfully used for solving such problems as the traveling salesman problem, learning of a neural multilayer network, maintenance planning, nonlinear integer number programming and others (references see [56].

Scatter Search: (SS) This optimization algorithm differs by its nature from the standard evolutionary diagrams. It is a vector oriented algorithm that generates new vectors (solutions) on the basis of auxiliary heuristic techniques. It starts from the solutions obtained by means of a suitable heuristic technique. New solutions are then generated on the basis of a subset of the best solutions obtained from the start. A set of the best solutions is then selected from these newly found solutions and the entire process is repeated. This algorithm was used for the solution of traffic problems, such as traffic control, learning neural network, optimization without limits and many other problems [56], [45].

Particle Swarm: (PSO) The "particle swarm" algorithm is based on work with the population of individuals, whose position in the space of possible solutions is changed by means of the so-called velocity vector. According to the description in [56], [71] and [15], there is no mutual interaction between individuals in the basic version. This is removed in the version with the so-called neighborhood. In the framework of this neighborhood, mutual interaction occurs in such a manner that individuals belonging to one neighborhood migrate to the deepest extreme that was found in this neighborhood.

Differential Evolution: (DE) Differential Evolution [57] is a population-based optimization method that works on real-number coded individuals. For each individual $\vec{x}_{i,G}$ in the current generation G, DE generates a new trial individual $\vec{x'}_{i,G}$ by adding the weighted difference between two randomly selected individuals $\vec{x}_{r1,G}$ and $\vec{x}_{r2,G}$ to a third randomly selected individual $\vec{x}_{r3,G}$. The resulting individual $\vec{x'}_{i,G}$ is crossed-over with the original individual $\vec{x}_{i,G}$. The fitness of the resulting individual, referred to as perturbated vector $\vec{u}_{i,G+1}$, is then compared with the fitness of $\vec{x}_{i,G}$. If the fitness of $\vec{u}_{i,G+1}$ is greater than the fitness of $\vec{x}_{i,G}$, $\vec{x}_{i,G}$ is replaced with $\vec{u}_{i,G+1}$, otherwise $\vec{x}_{i,G}$ remains in the population as $\vec{x}_{i,G+1}$. Differential Evolution is robust, fast, and effective with global optimization ability. It does not require that the objective function is differentiable , and it works with noisy, epistatic and time-dependent objective functions.

SOMA: (Self-Organizing Migrating Algorithm) is a stochastic optimization algorithm that is modeled on the social behavior of cooperating individuals [73]. It was chosen because it has been proven that the algorithm has the ability to converge towards the global optimum [73]. SOMA works on a population of candidate solutions in loops called *migration loops*. The population is initialized randomly distributed over the search space at the beginning of the search. In each loop, the population is evaluated and the solution with the highest fitness becomes the leader L. Apart from the leader, in one migration loop, all individuals will traverse the input space in the direction of the leader. Mutation, the random perturbation of individuals, is an important operation for evolutionary strategies (ES). It ensures the diversity amongst the individuals and it also provides the means to restore lost information in a population. Mutation is different in SOMA compared with other ES strategies. SOMA uses a parameter called PRT to achieve perturbation. This parameter has the same effect for SOMA as mutation has for GA. The novelty of this approach is that the PRT Vector is created before an individual starts its journey over the search space. The PRT Vector defines the final movement of an active individual in search space. The randomly generated binary perturbation vector controls the allowed dimensions for an individual. If an element of the perturbation vector is set to zero, then the individual is not allowed to change its position in the corresponding dimension. An individual will travel a certain distance (called the path length) towards the leader in n steps of defined length. If the path length is chosen to be greater than one, then the individual will overshoot the leader. This path is perturbed randomly.

The evolutionary algorithms can be essentially used for the solution of very heterogeneous problems. Of course, for the solution of the optimization problems, there are many more algorithms than were indicated here. Because their description would exceed the framework of this text, we can only refer to the corresponding literature, where the algorithms indicated above are described in more details.

2.4 Selected Basic Terms from the Evolutionary Algorithms

For work with evolutionary diagrams (Figure 2.4), it is necessary to know the meaning of certain terms that occur in the terminology of evolutionary algorithms and optimization. Some of them will be explained in this section.

2.4.1 The Usability Areas of Evolutionary Algorithms

Until the present date, there are many algorithms that belong to the class of evolutionary diagrams or can be included into this class under certain conditions. Typical examples are the already mentioned Ant Colony Optimization algorithms, Immunology System Method, Scatter Search or Particle Swarm. These algorithms, like many others, are not universal, but from the principle of their action, they are always suitable for solving certain classes of problems. The class of problem may be of various "size" for each algorithm. Genetic algorithms, for example, can be used for a wide class of problems, while the ACO algorithm, acting on the principle of the behavior of ants, is essentially predetermined for combinatoric problems of the type of a traveling salesman, where its performance is excellent.

It is therefore obvious that it is not only sufficient to have a good algorithm, but it is frequently of vital importance to know with what class of problems a given algorithm can work. This means that it is therefore necessary to determine the usability range of a given algorithm. In the case of evolution algorithms, we will understand by this term the class of problems for which a given algorithm provides at least satisfactory results.

Most optimization problem can be viewed as a geometrical problem, whose goal is to the find the lowest (minimum) or highest point (maximum) on the N dimensional surface. Such surfaces, defined usually by some functional prescription, may suffer from various pathologies from the mathematical point of view. With respect to the tests carried out on the functions tested, whose algorithms are described below, one can state that the evolutionary algorithms are very efficient and usually suitable for global optimization (see Fig. 2.11 and Fig.2.12). This set of test function can be viewed as the usability range of the evolutionary diagrams. It holds for the test functions mentioned above:

1. The graph of the function does not have a fractal character;
2. They are defined on real, integer or discrete arguments;
3. They are multimodal (one or several extremes);
4. They have various limits (imposed on the arguments or the value of the objective function);
5. They are strongly nonlinear;
6. They represent problems of the type "needle in a haystack";
7. Finding the global extreme with evolutionary algorithms is less or more complicated;

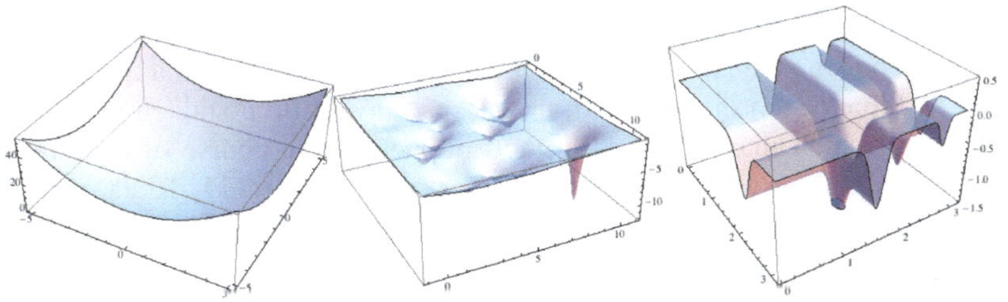

Fig. 2.11 Examples of functions that exhibit certain combinations of properties 1-7, a-c

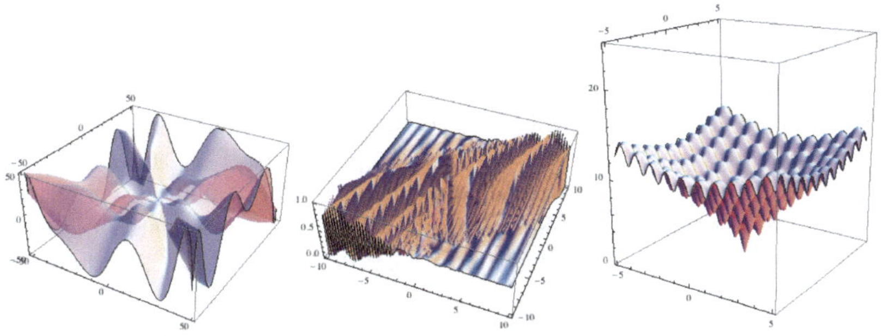

Fig. 2.12 Examples of functions that exhibit certain combinations of properties 1-7, a-c

moreover, it may hold true that:

(a) The function is separable (non-separable), which means that it can be (cannot be) decomposed into several simpler functions that can be optimized separately;
(b) The number of variables is high;
(c) The space of possible solutions may be large and discontinuous.

Generally speaking, evolutionary algorithms can be used to find the optima of functions from a very large class. Other information such as gradient, etc., are usually not necessary.

2.4.2 Common Features

Evolutionary algorithms have certain common features.

1. **Simplicity**, because algorithms can usually be programmed in a simple manner.
2. **Hybridity** of numbers with which the algorithms work. Numbers of the *integer, real* type, or, as the case may be, only selected sets of numbers (usually denoted as discrete), such as, for example, -5, 2, 8, 55, 3, 100, can be combined without any problem.

3. **Use of decimal numbers** - The individual need not be converted into the binary code that is commonly used in genetic algorithms. By the conversion into a binary code, a given number is distorted (the binary string has a limited length). When binary recording is used, mutations may cause a sudden change of the number, which may not have a good impact on the course of evolution. For example, numbers 15, 16 and 17 are represented as 01111, 10000 and 10001. The transition from 15 to 16 means the inversion of all five bits, i.e. a 100% mutation. However, transition from 16 to 17 requires the mutation of only one bit. Although this "unevenness" can be removed with what is called Gray coding (see Section 4.5), work with real numbers is still more convenient.
4. **Speed** - Thanks to its relative simplicity, particularly in comparison with classical methods, one can say that the required solutions are found much faster.
5. **The ability to find an extreme also for functions that are flat** from the graphical point of view and the extreme is just a "hole" in this plane. With some exaggeration, searching the extreme in such a function can be denoted as "looking for a needle in a haystack". Unfortunately, the efficiency of any algorithms, including evolutionary, is very low in these problems. If the surface around the extreme is a plane, than finding the extreme is usually a matter of chance.
6. **Ability to provide manifold solutions** - The best individual is the result of evolution - one solution. However, if, for example, the three best individuals are selected from the last population, they represent three different solutions of the problem. Of course, with graduated quality. If there are more global extremes in a given problem, one can expect that they will also be found by the evolutionary process. Therefore, there will be several solutions of the same quality available. As an example, the design of the optimum geared transmission can be used [47], [73], for which the method of differential evolution obtained four alternative solutions with the same value of objective function or a test function that has two global extremes at different points. The majority of good evolutionary techniques are able to localize these extremes.

In other words, the evolutionary algorithms are suitable for looking for extremes of functions suffering from such pathologies, such as, for example, noise, a high number of dimensions, "multi-modality" (several local extremes).

2.4.3 Population

A typical feature of the evolutionary algorithms is that they are based on work with the population of individuals. Population can be represented as matrix NxM (Fig. 2.13), where the columns represent individuals. Each individual represents the current solution of the problem. Essentially this is a set of arguments of the objective function, whose optimum numerical combination is searched for gradually. Moreover, a value of the objective function is connected with each individual (sometimes "fitness") that tells how the individual is suitable for further evolution of the population. This value does not participate in the evolution process itself. It only carries information on the quality of the corresponding individual.

	I_1	I_2	I_3	I_4	..	..	..	I_M
Fitness	**55.2**	**68.3**	**5.36**	**9.5**	..	..	..	**0.89**
P_1	2.55	549.3	-55.36	896.5	..	..	..	1.89
P_2	0.25	66.2	2	-10	..	..	..	-2.2
P_3	-66.3	56	4	15.001	..	..	..	-83.66
..	..	..	..	..	..	..	..	..
P_N	259.3	-10	22.22	536.22	..	..	..	-42.22

Fig. 2.13 Population (of the *NM* size), J_x is the x-th individual, P_i is the y-th individual Fitness - individual quality measured by means of the objective function

For the generation of population, it is necessary to define a specimen, see (2.1), according to which the entire initial population is generated. This sample individual is also used for correcting the parameters of individuals, who exceed the boundaries of the space searched.

$$\text{Specimen} = \{\{\text{Real}, \{Lo, Hi\}\}, \{\text{Integer}, \{Lo, Hi\}\}, \ldots\ , \{\text{Real}, \{Lo, Hi\}\}\} \quad (2.1)$$

In the sample, three constants are defined for each parameter of a specific individual from the population: The type of variable (i.e. integer, real, discrete, etc.) and the boundaries of the interval in which the value of the parameter may be. For example, {Integer, {*Lo*, *Hi*}} defines an integer parameter with the bottom limit *Lo* and upper limit *Hi*. The choice of limits is a very important step, because if they are chosen in an unsuitable manner, it may happen that solutions will be found that are not physically real (for example, a negative thickness of the pressure vessel), or will not be substantiated (for example, an airplane without wings, a pressure vessel whose wall thickness equals its radius, etc.).

Another equally important meaning of the boundaries is related to the evolutionary process itself. It may happen that a given optimization problem will be represented by a surface, on which the local extremes will assume greater values with the increasing distance from the origin (Fig. 2.14). This will cause that the evolution will be finding new solutions until infinity. Of course, if termination is not specified in dependence on the number of evolutionary cycles (generations, migrations, annealing,). This is caused by the fact that the evolutionary process always proceeds to deeper and more distant extremes on this (Schwefel's) function.

Population is generated on the basis of the sample individual by means of 2.2, see also [47]. $P^{(0)}$ represents the initial population, x_{ij} is the j-th parameter of the i-th individual.

$$\begin{array}{l} P_{i,j}^{(0)} = x_{i,j}^{(0)} = rnd[0.1] \cdot (x_{i,j}^{(Hi)} - x_{i,j}^{(Lo)}) + x_{i,j}^{(Lo)} \\ i = 1,\ \ldots\ ,\ M \quad , \quad j = 1,\ \ldots\ ,\ N \end{array} \quad (2.2)$$

In Fig. 2.14 and Fig. 2.15, two randomly generated populations of ten individuals are represented. The individuals were two-dimensional in this case. It is obvious

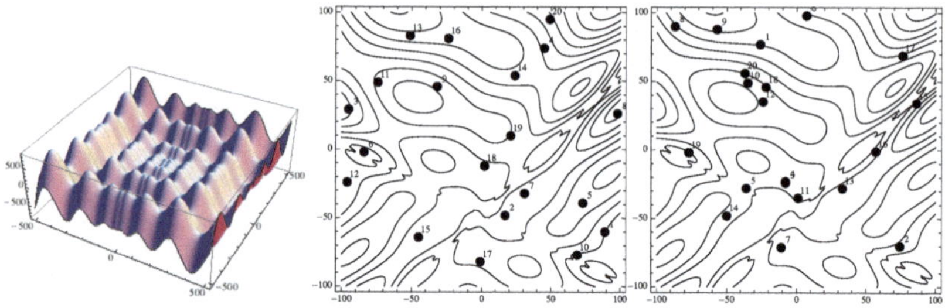

Fig. 2.14 Schwefel's function [64] with the growing extremes in the direction from the origin (a) and two randomly generated populations (b-c) on another function

I_1	I_2	I_3	I_4	..	..	..	I_{10}
424	104	53.3	942.9	..	..	..	178.008
-1.8	-1	0.7	-1.25	..	..	..	-1.19
1.2	2	1.2	-1.5	..	..	..	0.1

Fig. 2.15 Numerical representation of two randomly generated populations from Fig. 2.14 b)

from both figures that the generation of population is essentially a random distribution of individuals in the space of possible solutions. It occurs during the run of a given evolutionary algorithm that the individuals gather around one (usually global) or more extremes, which is graphically illustrated in Fig. 2.17. Mapping of information on how the evolution proceeded qualitatively is carried out by means of the evolution history of the objective function value in form of a simple graph. The dependence of the evolution of the value of the objective function on the evolution cycle is illustrated in this figure (Fig. 2.16). This is the sequence of the worst (upper curve) and best (bottom curve) solutions from individual populations. A mapping more convenient than that described just now is plotting the dependence of the value of the objective function on the current number of objective function evaluations. This approach is suitable because during evolutionary cycles (generation, annealing cycles, migration cycles, etc.), various numbers of evaluations of the profit function are carried out in individual algorithms. In the first method of graphical mapping, the slower convergence of the values of the objective function may be displayed as the faster one and vice versa. The true information on the quality of evolution may then be distorted. However, if we use the second method, it is then possible to compare various types of algorithms irrespective of their inner structure.

Besides the evolution of the best individual (or the best individuals when repeating the simulation), it is also suitable to display the evolution of the worst individual from the population in one graph. This reveals the overall convergence of the population as such. In a case where the courses of the best and worst individuals meet soon in the same extreme, it is possible that this is the case of a local extreme. If the

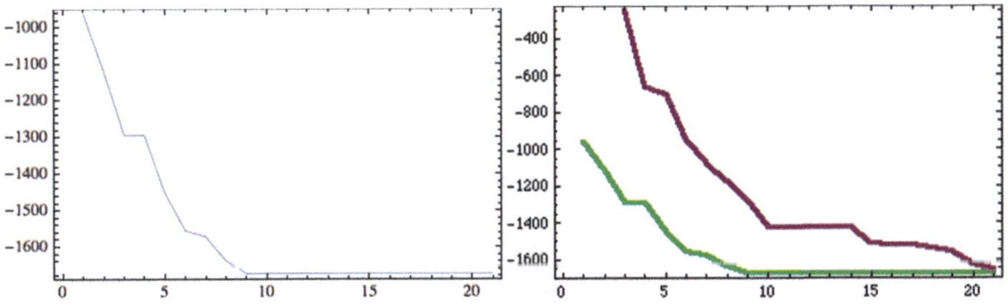

Fig. 2.16 Evolution of the value of the objective functions during evolution. This is a sequence of the best solutions from individual populations in dependence on the evolution cycle (generation, migration cycle, etc.).

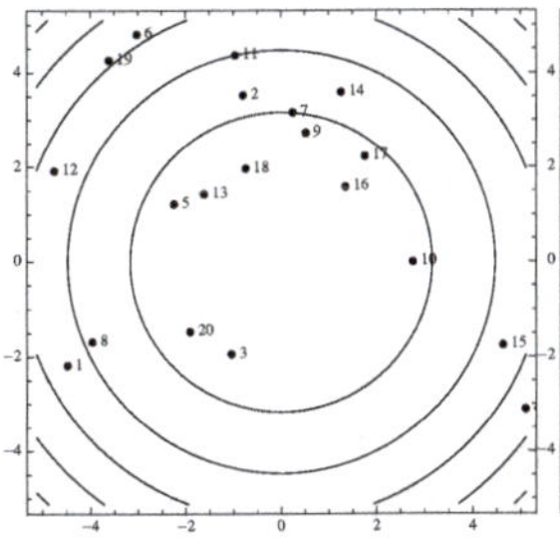

(a) Generation 5

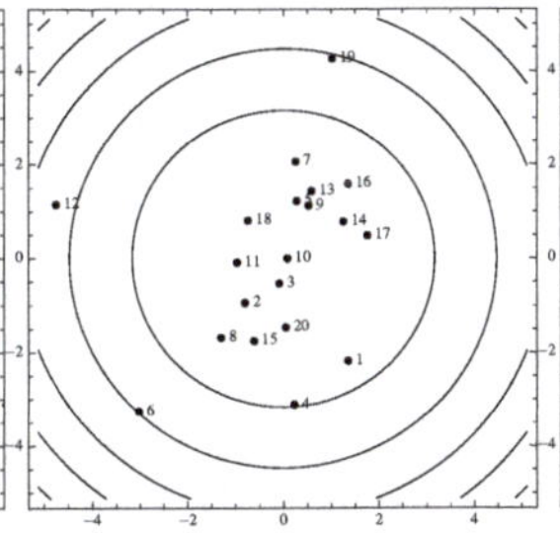

(b) Generation 10

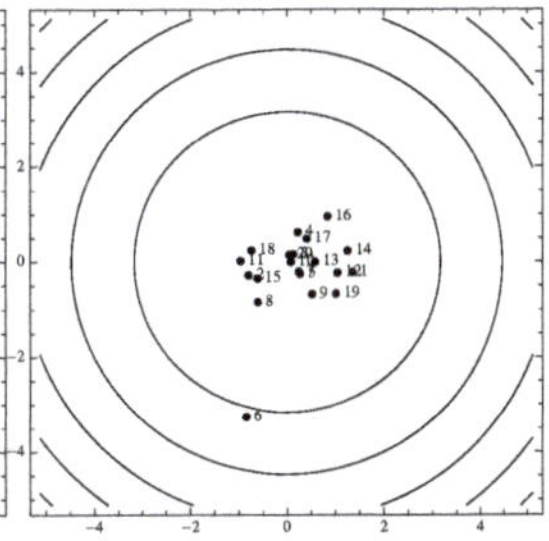

(c) Generation 15

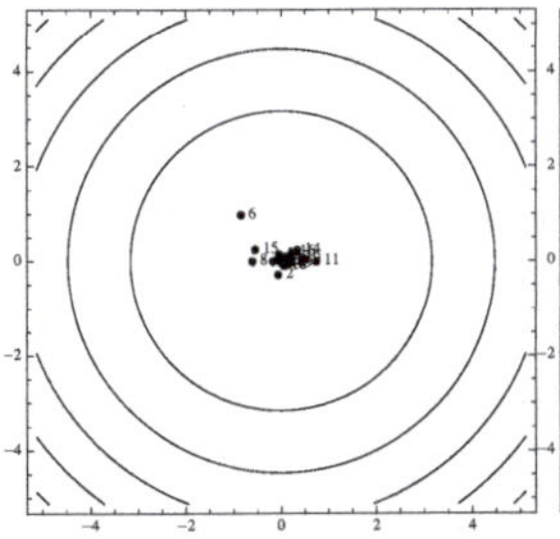

(d) Generation 20

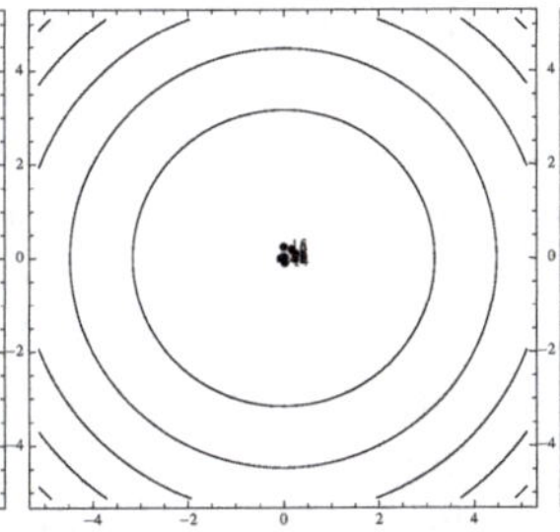

(e) Generation 25

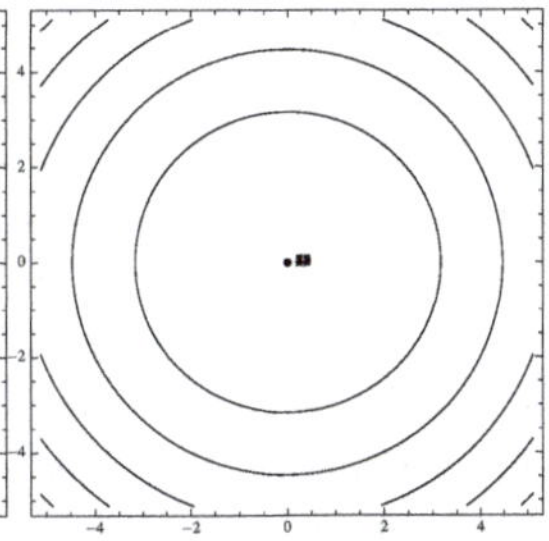

(f) Generation 30

Fig. 2.17 Convergence of the population to the global extreme in individual evolutionary cycles. The start display was omitted; individuals are uniformly distributed on the entire surface during the start.

best and worst individuals have the same value of the objective function, then only two explanations are possible:

1. The population is distributed in several extremes with the same value of the objective function or
2. The entire population is in one extreme, which is more probable, because a lot of problems are represented by a function with one global extreme.

In case No. 1, there is a chance that the evolution will proceed further, while in case No. 2, the same value of both individuals shows that further evolution is useless. The evolution of the population **must** always converge to better values, which means that it may never show divergence. If a minimum (maximum) is sought, the evolution must converge to lower (higher) values. If this is not the case, then in a given algorithm, "elitism" is somehow disrupted (elitism serves as some kind of a one way filter, which transmits only *such solutions that are better or equally good as those from the old population*). In the event of its dysfunction, the given algorithm would degrade to merely a random search.

2.4.4 Individuals and Their Representation

Several methods are being used when representing individuals in the evolutionary algorithms. The binary representation is historically the oldest one. In this case, the individual is formed by a 0 and 1 sequence called a chromosome [53], [2]. This representation of individuals has its historic roots and is being used in genetic algorithms up to date. In spite of its extension, it has its disadvantages. The basic disadvantage is the step change of the structure of chromosomes, the corresponding real values during a continuous change. In order to prevent these undesirable changes in the behavior of the binary code, the so-called Gray code is used. This is again a binary code, however, completely without the step changes mentioned above. The transformation of the standard binary code into the Gray binary code and the inverse transformation are illustrated in Fig. 2.18. The accuracy with which individuals are able to occur in the space of possible solutions is related to the binary representation of individuals. Basically, if the binary individual is short and represents a number with a few digits behind the decimal point, then it will occur only in certain positions of the real space of possible solutions. If its length and thus also accuracy grow, then the density of the positions of possible occurrences will grow. Nevertheless, one should note that the growing length of the individual causes considerable problems during further operations in the corresponding evolutionary algorithm (mutations, crossbreeding ...).

Furthermore, individuals can be represented in the form of real or integer numbers [57] or, as the case may be, their combinations, depending on the type of the algorithm [73]. The individual that also contains non-numerical values is a special representation. With the use of special techniques, it is possible to work numerically also with this individual. The last special form of representation is the so-called "tree". This form of representation makes it possible to visualize the tree structure, nevertheless, in the computer sense, the string of suitable symbols of a certain character is still the individual. This kind of visualization is is usually used in so called genetic programming [48], which is advanced evolutionary technique, used to manipulate with symbolic structures and create in this way more complex structures. Another similar approach is grammatical evolution [55], [20].

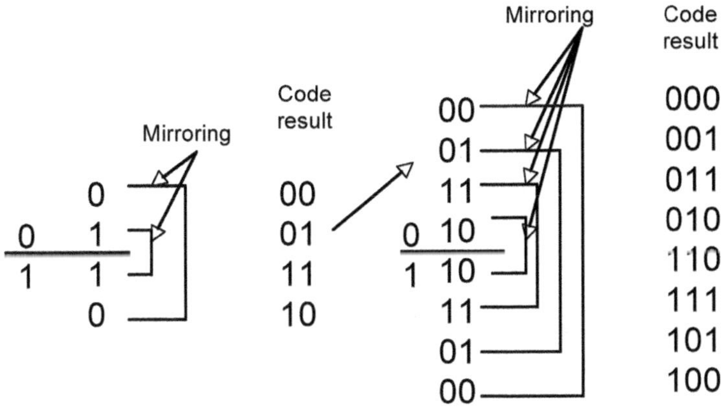

Fig. 2.18 Generation of the Gray code - reflection method

2.4.4.1 Binary and Gray Code Representation

The binary code is formed over the alphabet $[0,1]$ as a number representation system with base–2. Any (integer) number n can be expressed by the binary string $(b_N b_{N-1} \ldots b_i \ldots b_2 b_1 b_0)$ with all the $b_i \in [0,1]$. The number n converts to the binary string by

$$n = b_N 2^N + b_{N-1} 2^{N-1} + \ldots + b_2 2^2 + b_1 2^1 + b_0. \tag{2.3}$$

Note that in the binary string every digit (bit) b_i counts for a distinct numerical value, 2^{b_i}. So if this bit changes, it depends on the exact bit position how big the change in the represented numerical value is. Clearly, this change in represented numerical value can be considerable if the bit is far left in the bit string. Next to the binary code, the Gray code is frequently used for genetic algorithms. It is also called the constant change code since when mutating a Gray–coded individual, the real number that corresponds to the corresponding binary sequence does not change much.

The Gray code was patented in 1947, when Frank Gray asked to register it under the name of reflected binary code. However, users started using the name Gray code according to its founder [34]. One of the alternatives of the construction of the Gray code is described in Fig. 2.18. It is the so-called "reflection" method. It consists in taking the n bit code, for example, for n = 2 code 00, 01, 11, 10. This is then extended by its mirror copy 00, 01, 11, 10, **10, 11, 01, 00** and 0 is added to the original part, while to the reflected 1: **0**00, **0**01, **0**11, **0**10, **1**10, **1**11, **1**01, **1**00. This is repeated until we have the required m bit string of the Gray code.

Another method that can be easily implemented by a computer is using the XOR operation. The principle of transformation of the Gray code into the binary code is prescribed by relation (2.4) and illustrated in Fig. 2.19a. The inverse transformation, i.e. from the binary into the Gray code is given by relation (2.5) and illustrated in

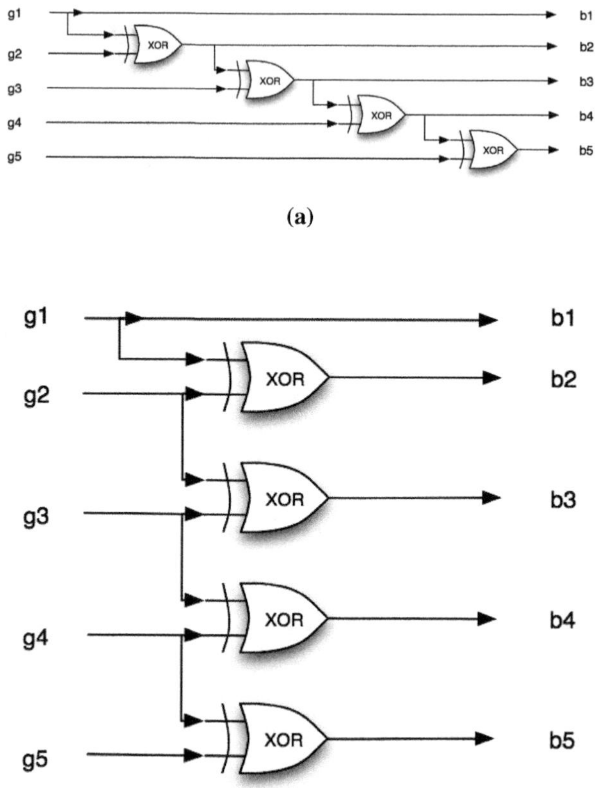

(b)

Fig. 2.19 Generation of the Gray code - XOR method, a) Conversion from the Gray code into the binary code, b) Conversion from the binary code into the Gray code

Fig. 2.19b. Symbol *bk* and *gk* represents the *k*-th bit of the standard binary code or, as the case may be, Gray code.

As mentioned above, the binary strings in the Gray code always differ only in one bit when changing their decimal equivalent by one. The distances between individual numbers have therefore the Hamming distance equal to one. The Hamming distance is defined as the number of bites in which two binary strings differ. The difference between the binary and Gray coding for numbers from zero to seven can be seen in Table 2.1.

$$\begin{aligned}
b1 &= g1 \\
b2 &= g1 \oplus g2 = b1 \oplus g2 \\
b3 &= g1 \oplus g2 \oplus g3 = b2 \oplus g3 \\
b4 &= g1 \oplus g2 \oplus g3 \oplus g4 = b3 \oplus g4 \\
b5 &= g1 \oplus g2 \oplus g3 \oplus g4 \oplus g5 = b4 \oplus g5
\end{aligned} \tag{2.4}$$

$$\begin{aligned} g1 &= b1 \\ g2 &= b1 \oplus b2 \\ g3 &= b2 \oplus b3 \\ g4 &= b3 \oplus b4 \\ g5 &= b4 \oplus b5 \end{aligned} \tag{2.5}$$

Table 2.1 Difference between the Gray and standard binary code

Decimal value	Gray code	Binary code
0	0	0
1	1	1
2	11	10
3	10	11
4	110	100
5	111	101
6	101	110
7	100	111

The advantage of the Gray code appears in the more uniform mutation of individuals and generally in faster convergence to the global optimum. However, there are differences in opinions. Some authors insist that the Gray code slows down the genetic algorithm (GA) due to the process of conversion [34]. On the contrary, other authors prefer the use of the Gray code ([11], [36]).

When using genetic algorithms during mutation or crossbreeding of standard binary individuals, the argument (gen) may change considerably, see Fig. 2.20. Such big changes do not occur in individuals in the Gray coding, Fig. 2.21.

Despite fact that evolutionary algorithms has very good performance, it is important to remember, that there are still problems, whose solution obtaining is still impossible and also, that there are some limits given to the computation by quantum physics. This is discussed in the following (last) section of this chapter.

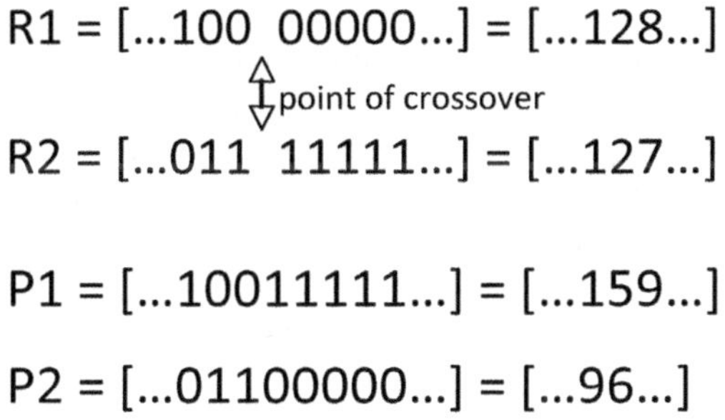

Fig. 2.20 Crossbreeding of individuals in standard binary coding (R1, R2 - parents; P1, P2 - descendants)

R1 = [...110 00000...] = [...128...]

⇕ point of crossover

R2 = [...010 00000...] = [...127...]

P1 = [...11000000...] = [...128...]

P2 = [...01000000...] = [...127...]

Fig. 2.21 Crossbreeding of individuals in Gray coding (R1, R2 - parents; P1, P2 - descendants)

2.4.4.2 Real, Integer and Discrete

Another way of individual representation is, when individuals are represented not only by by binary strings, as written above, but also in strings of real and integer numbers. Special case of integer representation are so called discrete sets, as explained later.

In the real representation is individual represented by string of real numbers like for example 2.3, 22.56, -569.2, ... Process of mutation, crossover, etc are then governed by used evolutionary algorithm.

In its canonical form, EAs are usually only capable of handling continuous variables. However, extending it for optimization of integer variables is rather easy. Only a couple of simple modifications are required. First, for evaluation of the cost-function, integer values should be used. Despite this, the EAs itself may still work internally with continuous floating-point values. Thus,

$$\begin{aligned} & f_{\text{cost}}(y_i) \quad i = 1,..,n_{param} \\ & where: \\ & \quad y_i = \begin{cases} x_i & \text{for continuous variables} \\ INT(x_i) & \text{for integer variables} \end{cases} \\ & x_i \in X \end{aligned} \tag{2.6}$$

INT() is a function for converting a real value to an integer value by truncation. Truncation is performed here only for purposes of cost function value evaluation. Truncated values are not assigned elsewhere. Thus, EA works with a population of continuous variables regardless of the corresponding object variable type. This is essential for maintaining the diversity of the population and the robustness of the algorithm.

Secondly, in case of integer variables, the population should be initialized as follows:

$$\begin{aligned} & P^{(0)} = x_{i,j}^{(0)} = r_{i,j}\left(x_j^{(High)} - x_j^{(Low)} + 1\right) + x_j^{(Low)} \\ & i = 1,\ldots,n_{pop}, \quad j = 1,\ldots,n_{param} \end{aligned} \tag{2.7}$$

Additionally, the boundary constraint handling for integer variables should be performed as follows:

$$x_{i,j}^{(ML+1)} = \begin{cases} r_{i,j}\left(x_j^{(High)} - x_j^{(Low)} + 1\right) + x_j^{(Low)} \\ \quad if \quad INT\left(x_{i,j}^{(ML+1)}\right) < x_j^{(Low)} \vee INT\left(x_{i,j}^{(ML+1)}\right) > x_j^{(High)} \\ x_{i,j}^{(ML+1)} \quad \text{otherwise} \end{cases} \tag{2.8}$$

where,
$i = 1,...,n_{pop}, \quad j = 1,...,n_{param}$

Discrete values can also be handled in a straight forward manner. Suppose that the subset of discrete variables, *X(d)*, contains *i* elements that can be assigned to variable *x*:

$$X^{(d)} = x_i^{(d)} \qquad i = 1,...,l \quad \text{where} \quad x_i^{(d)} < x_{i+1}^{(d)} \tag{2.9}$$

Instead of the discrete value x_i itself, its index, *i*, can be assigned to *x*. Now the discrete variable can be handled as an integer variable that is boundary constrained to range $\{1,2,3,..,N\}$. In order to evaluate the objective function, the discrete value, x_i , is used instead of its index *i*. In other words, instead of optimizing the value of the discrete variable directly, the value of its index *i* is optimized. Only during evaluation is the indicated discrete value used. Once the discrete problem has been converted into an integer one, the previously described methods for handling integer variables can be applied. The principle of discrete parameter handling is depicted in Fig 2.22. This technique is called *discrete set handling* (DSH), see [46].

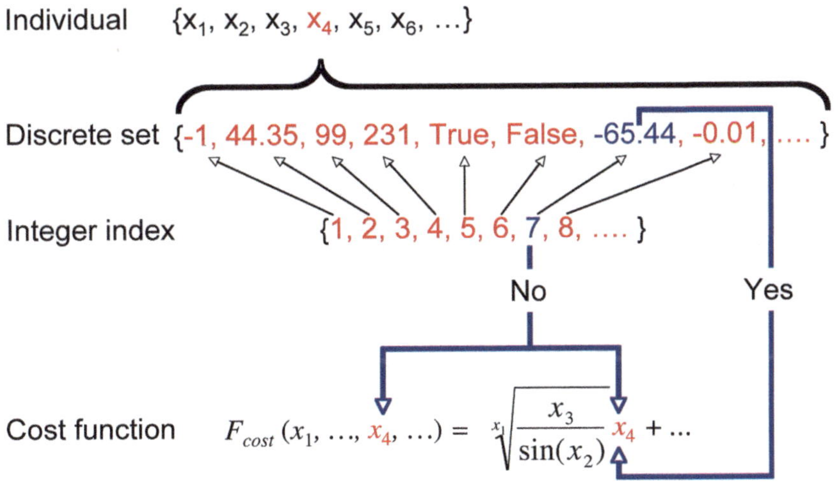

Fig. 2.22 Discrete parameter handling

2.4.4.3 Tree Representation

The tree representation used by EAs, come from initial idea of so called symbolic regression by means of a computer program. It was proposed in Genetic Programming (GP), [43], [42]. Genetic programming was the first tool for symbolic regression carried out by means of computers instead of humans. The main idea comes from genetic algorithms (GA), which was used in GP [43], [42]. Its ability to solve very difficult problems is well proved; for example, GP performs so well that it can be applied to synthesize highly sophisticated electronic circuits [44].

The main principle of GP is based on GA, which is working with populations of individuals represented in LISP programming language. Individuals in a canonical form of GP are not binary strings, different from GA, but consist of LISP symbolic objects like *sin*, +, *Exp*, etc. These objects come from LISP, or they are simply user-defined functions. Individuals in genetic programming, in its canonical form, are thus commands of Lisp language. There are also another techniques like Read's linear coding [59] and DSH technique. DSH technique mentioned above, allow EAs to manipulate with such a structure also via integer index. Individual is in fact integer string, which is converted into symbolic expression. Individuals can be then visualized in the form of so called trees, which are graphical unfolding of nonnumerical expressions, see Fig. 2.23. Individuals in this form can represent not only classical mathematical expressions, but also logical functions (Fig. 2.24 and Fig. 2.25), or elements of electronics circuits (Fig. 2.26). Read's linear code is representation based on string of integer numbers, as shown on Fig. 2.27 and Fig. 2.28. In the case of Read's representation each vertex has associated number according to number of

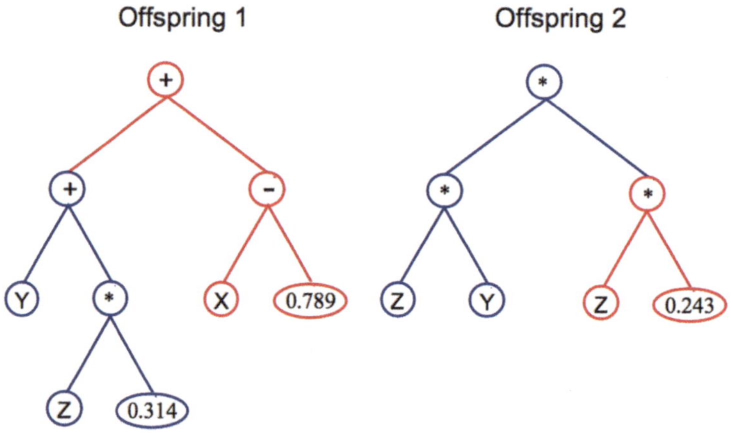

$(Y + Z * 0.314) + X - 0.789 \qquad Z * Y * (Z * 0.243)$

Fig. 2.23 An example of tree representation. Individuals are represented graphically by trees. Crossover is nothing more than cutting and exchange of randomly selected sub-trees.

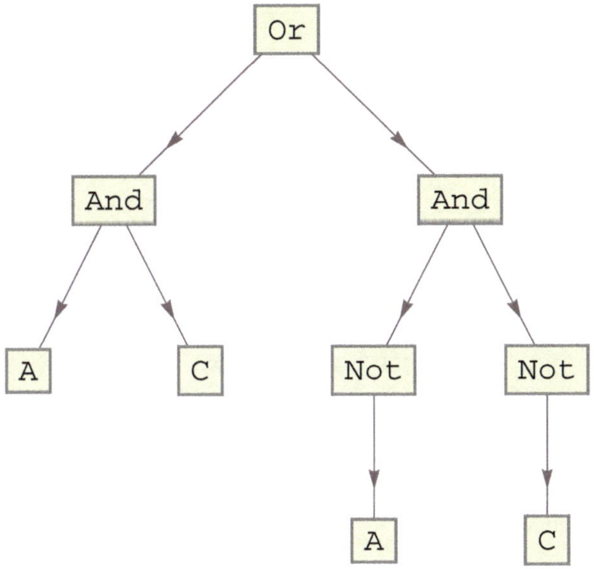

Fig. 2.24 Tree representation of logical function

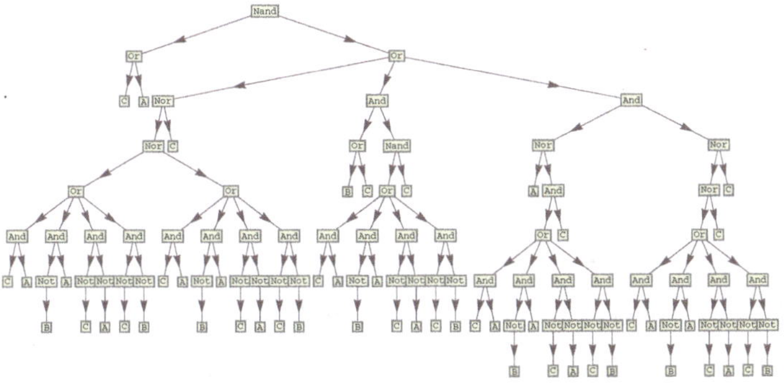

Fig. 2.25 Tree representation - more complicated example

outcomming vertexes. Code of arbitrary tree is obtained so that labels of vertexes are "joined" (according to dotted line with arrows, see Fig. 2.28) into integer string with a such condition that used vertex label is further ignored, to keep unicity of description. There is more techniques of how to represent individuals for genetic programming techniques, however above mentioned techniques (Lisp, Read's linear code) are well known. DSH technique can be regarded like experimental novelty technique, which is mostly used in this book.

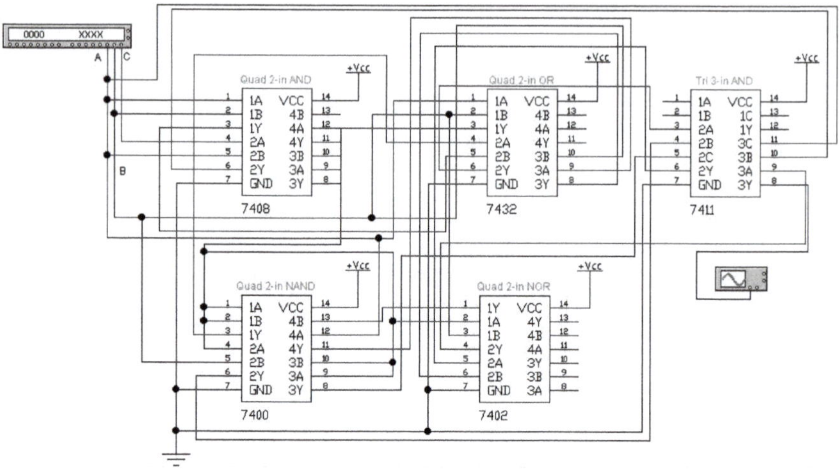

Fig. 2.26 Electronic realization of evolutionary designed circuit via evolution with individuals in DSH representation

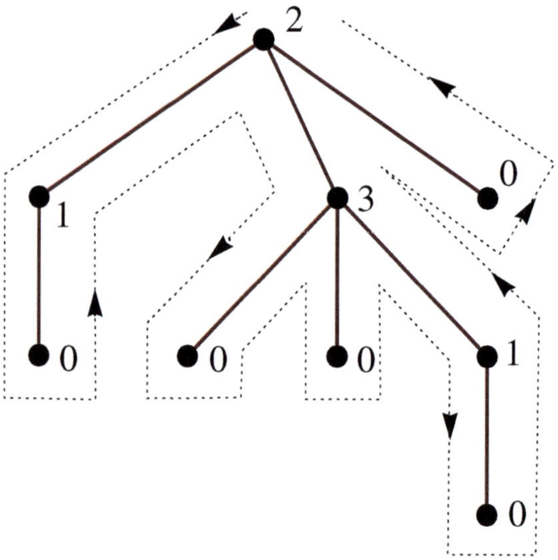

Fig. 2.27 Read's tree. Each vertex has associated number according to edges coming out of vertex. Unique code of tree is constructed according to dashed arrows.

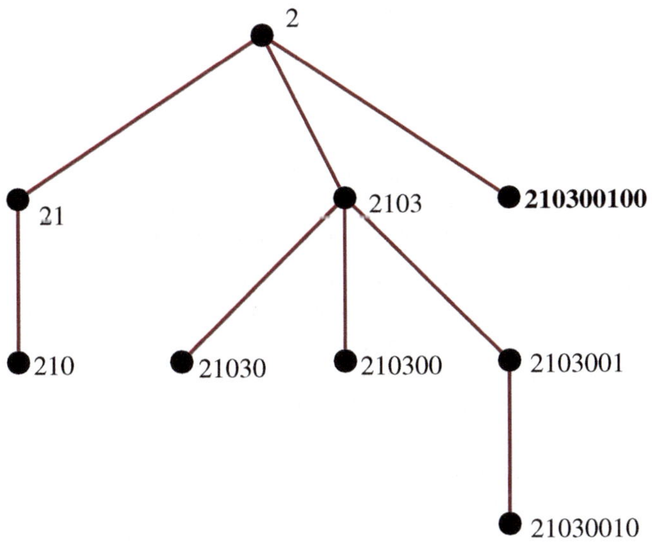

Fig. 2.28 Code of Read's Tree - 210300100

2.4.5 Evolutionary Operators: Selection, Recombination, Mutation

The algorithmic cycle of evolutionary algorithms relies upon the working of three main operators, selection, recombination and mutation. All three operators play a distinctive role in solving the posed optimization problem. In general, during the evolutionary run, we can separate two phases, exploring and exploiting. In the exploring the individuals should cover large (ideally all) parts of the search space in order to find regions where optimal values are likely to be found. Hence, individuals should be different, the diversity of the population should be high. After that exploration phase, a promising region in the search space should be searched more detailed, so exploiting the knowledge about the distribution of fitness in the search space should set in. In this phase the individuals of the population should be pushed towards the actual optima. Clearly, both phases are necessary for successful problem solving. If the exploration phase is missing or too short, the individuals might settle for local optima, missing the global optima in this process. Without exploitation, the exact location of the best solution might not be found. Overlaying both phases is the intention to exclude clearly inferior solutions from hindering the search process. Against this background, the working of the evolutionary operators can be understood.

Selection: The selection mechanism organizes that individuals with higher fitness become the material from which the next generation is produced. In doing so, it is important (particularly in the exploration phase) that not only the very best individuals are kept, but also "promising second-bests". A good selection mechanism should balance the selective pressure (that is only the best are to survive) with maintenance of residual diversity in the population. For achieving this balance, different kinds of selection schemes have been proposed. They can be roughly distinguished in purely deterministic selection, where either based on ranking or by setting a fitness threshold a certain percentage of the population is kept, and guided stochastic selection, where samples of the population are randomly picked and based on the comparison of their fitness values a decision is made to discard or keep them.

Recombination: In recombination (also called crossover, in particular for GAs) new individuals are created based on those previously selected for their superior fitness. In a first step, two (or sometimes even more) individuals are appointed to be parents. This appointment can be either deterministic by working off the whole selected population, or stochastic where the individuals are determined randomly. Then, the schemes proposed for performing the step differ largely for different kinds of representation. For evolutionary algorithms that use binary representation, we find that both parents swap or shuffle (sometimes at more than one place) subsections of their binary string. For real valued representation, a (sometimes weighted) arithmetic or geometric mean between both parents yields the offspring.

Mutation: Following the recombination step, the produced offspring are altered randomly by mutation, mostly in a marginal manner only. Therefore, mutation rate (which defines the probability that the offspring are subjected to mutation) and the mutation strength (which fixed the magnitude of the changes in the offspring) must be set. Again, we find a difference between binary and real valued representation. For binary stings, we have a flipping of a bit ($0 \rightarrow 1$ or $1 \rightarrow 0$) at one or more places in the binary string. For vectors of real numbers, realizations of a (mostly normally distributed) random variable are added to the offspring.

The evolutionary operators just considered can be regarded as the backbone for the majority of ECT methods, although not all of them must be part in a specific implementation and also their respected role and importance might be largely different. In general, mutation is next to a sensible choice of the initial population the source of random and diversity enhancement in the algorithm. It helps to explore the search space. Selection, on the other hand, acts mainly as a filtering for superior solution candidates. In this, it exercises selection pressure on the population. However, for a given type of ECT the same operator might have a different flavor. So, in GAs mutation is a minor operator, while in ESs it is the main component. Such differences depend frequently on the type of representation. For a binary representation, as GAs use, a single bit change in the right place of the binary string, as induced by mutation, can cause a very dramatic change of the encoded solution. The very same bit flip can alter the string largely or a little, depending on where in the string it happens. This is not the case for real value representation, where the magnitude of the change caused by mutation can be closely controlled but mutation rate and strength.

2.5 Limits to Computation

Unfortunately, many people believe that everything can be computed if we have a sufficiently powerful computer and elegant algorithm. The goal of this chapter is to show that some problems cannot be solved algorithmically due to their nature. Popularly speaking, there is not, has not been and will not be enough time for their solution.

Part of these restrictions are also physical limits that follow from the material nature of the universe, which restricts the output of every computer and algorithm by its space-time and quantum-mechanical properties. These limits, of course, are based on the contemporary state of our knowledge in physical sciences, which means that they might be re-evaluated in the case of new experimentally confirmed theories (strings, etc.). At this moment, however, this is only a speculation and we must adhere to the generally accepted and confirmed facts from which these limits follow.

2.5.1 Searched Space and Its Complexity

The complexity of the optimization problems can be demonstrated by many examples. Let us follow examples from [54]. A typical representative is the so-called SAT problem (boolean satisfiability problem). This is a problem from the field of logic that is represented by a complex logical function with a great number of logical variables. Relation 2.10 is an example from [54].

$$F(x) = (x_{17} \vee \bar{x}_{37} \vee x_{73}) \wedge (\bar{x}_{11} \vee \bar{x}_{56}) \wedge \ldots \wedge (x_2 \vee x_{43} \vee \bar{x}_{77} \vee \bar{x}_{89} \vee \bar{x}_{97}), \tag{2.10}$$

that contains 100 variables and the objective is to find such values of individual arguments of this function for which the resulting value of relation 2.10 is TRUE. At first sight, this problem looks very trivial; nevertheless, it is a problem that cannot be solved by classical methods. If we take into account that the expression contains 100 unknown variables that can assume two values (0,1), then the number of all possible combinations is 2^{100}, which is approximately 10^{30}. In order to get a better impression on the monstrous size of this number, it is sufficient to imagine how long it would take to evaluate all the combinations, if 10^{13} of these combinations are evaluated within one second (which is of course impossible on single processor). The correct answer is 10^9 years. This essentially means that the solution of this problem would take approximately the time of the existence of the universe.

Another complication related to this problem is the fact that function 2.10 as defined does not make it possible to evaluate the quality of the current solution. This is a substantial drawback, particularly if the evolutionary techniques are used, because there is no possibility how to determine whether the qualities of two subsequently found solutions are close or not. As will be shown further, when using the evolutionary algorithms, it is of vital importance that the information on the quality of the solution is available for the determination in which "direction" the optimum solution lays. This is not possible in the case of the SAT problem, because the function

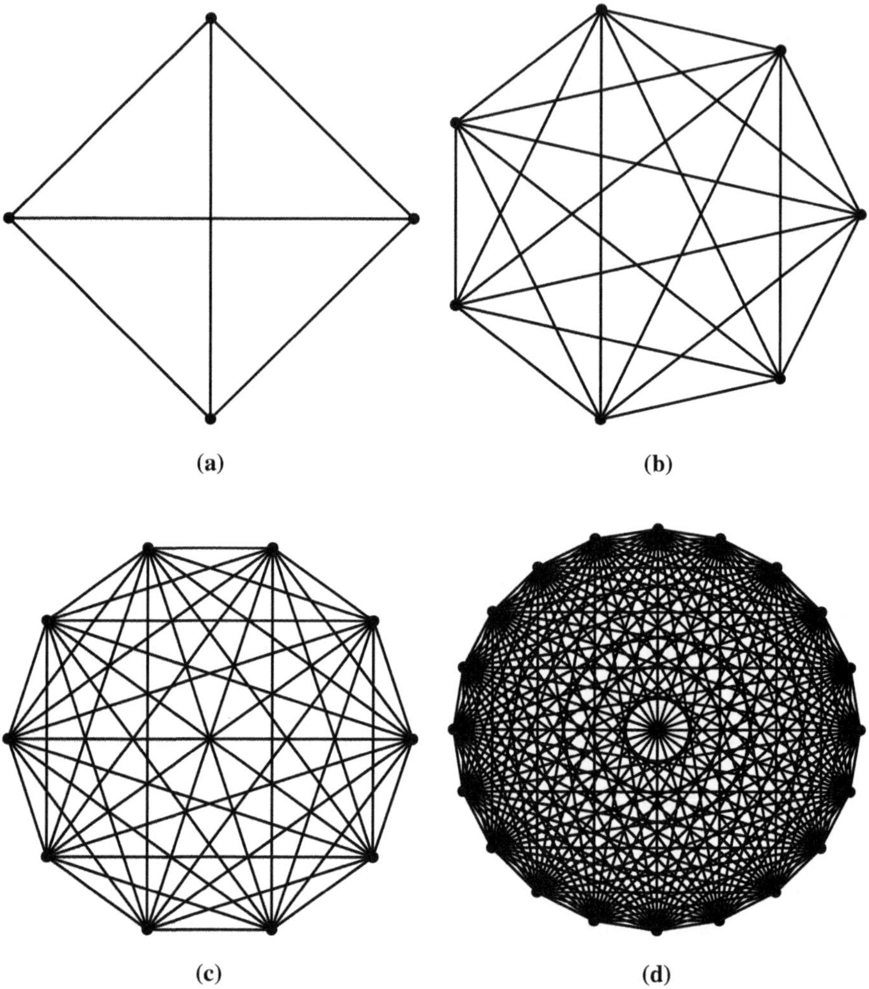

Fig. 2.29 Connections in the traveling salesman problem that form n! possible trajectories (see Fig. 2.32). We indicate the number of cities / number of connections between the cities a) 4/6, b) 7/21, c) 10/45, d) 20/190.

only returns TRUE or FALSE, i.e. "good" or "bad". It does not return how good or bad a given solution is.

The SAT problem is more or less a scholastic problem. As a more practical problem from real life, one can use the well known traveling salesman problem. This is a problem, in which a traveling salesman must visit a set of N cities in the shortest possible time or with the smallest fuel consumption or, as the case may be, fulfill other criteria. The traveling salesman problem can be visualized by means of graphs, as demonstrated in Fig. 2.29 - 2.31.

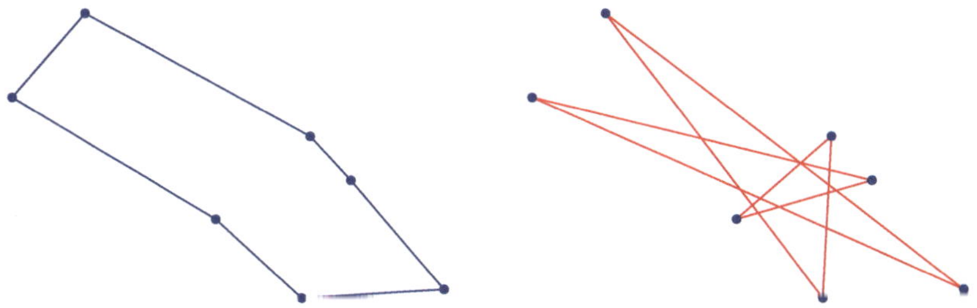

Fig. 2.30 Traveling salesman visiting seven cities: The best route is on the left and the worst route is on the right

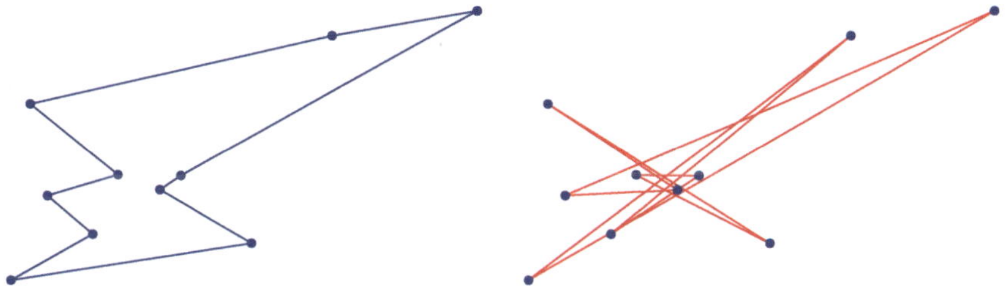

Fig. 2.31 Traveling salesman visiting ten cities: The best route is on the left and the worst route is on the right

The condition is that each route must start and end in the same city and each city should be visited only once. This is therefore a purely practical problem. The trajectory of the traveling agent represents a sequence of dots, such as, for example, "2 - 3 - ... - 7 - 26 ... ". The number of all possible combinations is $n!$. In the case of a symmetrical problem of a traveling salesman (the distance from city A to B is the same as from city B to A), $2n$ routes repeats. In this case the final number of all possible combinations is $(n\text{-}1)!/2$. However, this number is still large. As shown in Fig. 2.33, the number of all possible combinations very quickly grows with the number of cities. Already for $n > 6$, there are more combinations in the traveling agent problem than in the SAT problem. Fig. 2.33 shows the growth of the number of solutions of the SAT problem in comparison with the growth of the complexity of the traveling salesman problem.

Let us look further. The traveling salesman problem has 181,440 possible solutions for 10 cities. There are 10^{16} possible solutions for 20 cities and 10^{62} for 50 cities. If 60 cities is used, then there is 10^{79} of possible solutions. This number is equal to the estimated number of protons in our universe, i.e. if one proton is used like memory to store one possible solution, then all protons in universe can store only TSP with size 60 cities. No more. It is worth mentioning that there is approximately 10^{21} liters of water on our Planet [54]. It is a trivial task to calculate how many globes could be covered with this volume of water had we used a reservoir

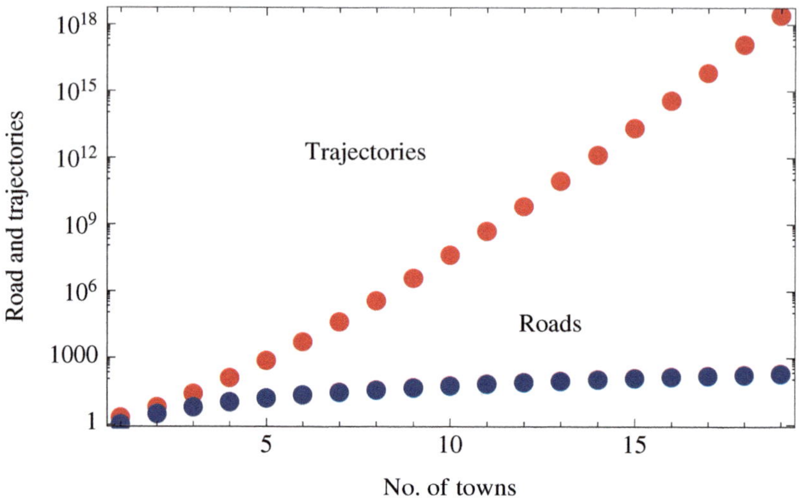

Fig. 2.32 Visualization of the travelling salesman complexity. The difference is illustrated between the number of roads (blue dots) and possible trajectories (red dots).

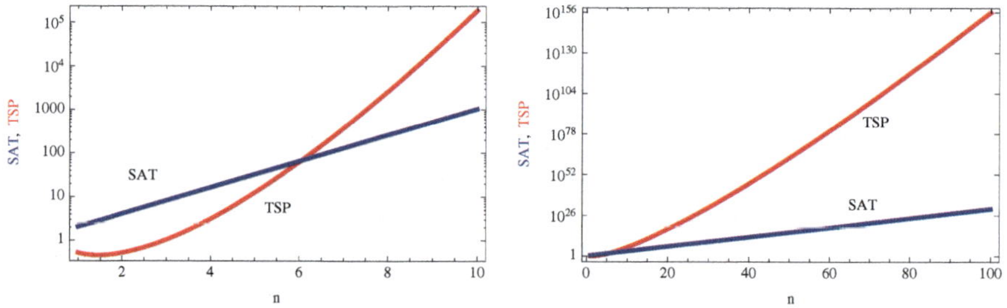

Fig. 2.33 Growth of the problem complexity for SAT (blue curve) and traveling salesman (red curve). Starting with seven cities (or variables in SAT), the traveling salesman problem is more time consuming.

with a volume of 10^{62} liters water. It is therefore obvious that even from such a trivial example as the optimum distribution of parcels, a problem may arise, whose optimum solution is not known. It is worth mentioning that at the present time, there are special types of evolutionary algorithms (ACO - Ant Colony Optimization) that manage up to 10,000 cities satisfactorily. We leave it to the kind reader to calculate what is the number of combinations (hint: 2.846259680910^{35659}).

The third and last sample problem is the artificial testing function, depicted in Fig. 2.34 that is used as a testing function for various types of evolutionary techniques;

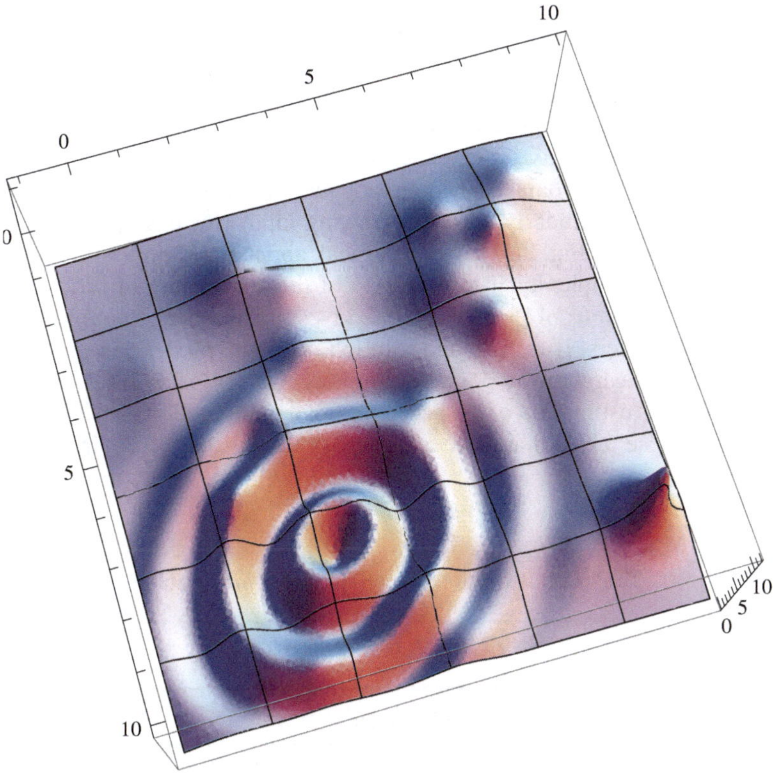

Fig. 2.34 Graph of test fuction

(for another example see [54]). This function is strongly nonlinear (for another similar functions see Fig. 2.11 and Fig.2.12) and it is complicated. Although the function in this example is artificial, one can encounter even "wilder" functions that represent real physical problems. This type of function looks innocent, however, it is the contrary in this case. It is necessary to realize that everything is running in computers, thus also the optimization of such a function, is digitized. If this would not be so, then it would necessary to calculate the value of the function in an infinite amount of points. Due to digitization, this infinity reduces to a set of values of the function, whose cardinality is finite, even though it is still immense. Let us assume that the computational accuracy of the computer used is 6 decimals. In this case every variable in a given function assumes real values. Through digitization, the infinity mentioned above reduces to a set of possible solutions, however the cardinality of this is still immense. Let us assume that variable in a given function Fig. 2.34 may assume up to 10^7 different values. In general terms, this function will assume 10^{7^n} values (n is a number of variables here). This number is many times greater than the number of solutions of the traveling salesman problem for $n \leq 10^7$. For $n = 50$, there are 10^{350} solutions. It is necessary to realize that the accuracy of present computers is much higher and the problem therefore generates a gigantic number of possible solutions.

Let us mention that the complexity of problems is not measured in theoretical informatics by the time demand factor (even though it is so de facto in the result), but primarily by the complexity or dependence of the capacity of the algorithm on the growing number of input data. As was already mentioned, there are problems whose complexity grows nonlinearly with the growing input (for example, the traveling salesman problem, see Fig. 2.33). We then speak about algorithms with polynomial, exponential, etc., complexity. The examples of the complexity of problems are in Table 2.2 - 2.4 (taken from [51]). Table 2.2 gives the number of possible solutions for n input parameters. If testing one solution takes the predefined time, the time demand factor for searching all possible solutions is in Table 2.3. If faster computers are used, the gross estimation of the acceleration of computation is in Table 2.4. It is obvious from these tables that there are many problems that no computer can help to solve.

Table 2.2 Estimation of the values of some functions

n Function	10	50	100	300	1,000
			Polynomial		
5n	50	250	500	1,500	5,000
$n \log_2 n$	33	282	665	2,469	9,966
n^2	100	2,500	10,000	90,000	1 million (7 digits)
n^3	1,000	125,000	1 million (7 digits)	27 million (8 digits)	1 billion (10 digits)
			Exponential		
2"	1,024	16 digit number	31 digit number	91 digit number	302 digit number
$n!$	3.6 million (7 digits)	65 digit number	161 digit number	623 digit number	giant number
n^n	10 billion (11 digits)	85 digit number	201digit number	744 digit number	giant number

For comparison: The number of protons in the visible Universe has approximately 79 digits The number of microseconds from the "big bang" has 24 digits.

Table 2.3 Estimation of the time of *f(n)* operations if 1 operation takes 1 μs

n / Function	10	20	50	100	300
			Polynomial		
n^2	1/10,000 s	1/2,500 s	1/400 s	1/100 s	9/100 s
n^5	1/10 s	3.2 s	5.2 s	2.8 hours	28.1 days
			Exponential		
2^n	1/1,000 s	1 s	35.7 years	400 trillion centuries	75 digit # of centuries
n^n	2.8 days	3.3 trillion years	70 digit # of centuries	185 digit # of centuries	728 digit # of centuries

Table 2.4 Estimation of the time of *f(n)* operations if 1 operation takes 1 μs

	Maximum dimension of the input manageable in a reasonable time		
Function	Current computers	100 times faster computers	1,000 times faster computers
n	N_1	100 N_1	1,000 N_1
n^2	N_2	10 N_2	31.6 N_2
2^n	N_3	N_3 + 6.64	N_3 + 9.97
$n!$	N_4	N_4 + 1	N_4 + 2

2.5.2 *Physical Limits of Computation*

As was already mentioned, there are limits restricting the output of any computer that follow from the quantum-mechanical nature of mass. These limits restrict both the output of the computer and its memory. It is obvious from these restrictions that there are many problems that no computer can help to solve.

Basic restriction in this direction is the so-called Bremermann's limit [8], according to which it is not possible to process more than 10^{51} bites per second in every kilogram of matter. In the original work of this author [8], the value of $2\text{x}10^{47}$ bites per second in one gram of matter is indicated. At first sight, this limit does not look frightening, but only until we take "elementary" real examples for comparison. Let us consider chess-mate for illustration. For this game, the estimated number of combinations is 10^{120}. As another example, let us consider the lattice of cellular automata [39] of 100 x 100 cells that can only assume black and white values that represents $2^{10,000}$, which is approximately $10^{3,000}$ combinations - images. The current TV sets with a LCD monitor have approximately 1,300x700 pixels, which can assume various colors and degrees of brightness. It is clear that the number of combinations is much higher on an LCD monitor.

This limit can be derived in the following relatively simple manner: For making it possible to measure, process and transfer information, it is **necessary** to store it on some physical carrier. This information may be electromagnetic radiation, paper tape, laser beam, etc., therefore always something material. Information alone, i.e., without a physical carrier, cannot exist. Because elementary particles and their energy states can also be used as a carrier of information, it is obvious that the limit of how much information the matter can carry follows from the restriction that was discovered at this physical level.

In order to make it possible to measure this information, it must be modulated on the corresponding carrier to resolve the individual carrier's states that represent the value of the information. Von Neumann[69] called the resolvable states "markers". The lowest resolvable energy states are the quantum states of matter, whose resolvability from the bottom is limited by Heisenberg's uncertainty relation. When deriving the already mentioned limits, it does not matter whether mass or energy types of carriers are considered. Both types are physically interchangeable. Therefore, if quantum states are considered as the smallest resolvable energy states, which will be considered as bits in this case, then the "energy-bit" resolution is given by Heisenberg's uncertainty relation. Generally, one can say that according to the Heisenberg principle of uncertainty it is possible to always identify the final number of states. Because nobody can say which state will be observed, probability has to be used. It is common to say that variable X will have n different values with probability $p_1, p_2, .., p_n$ Based on information theory is clear that we can get

$$H(p_1, p_2, .., p_n) = -\sum_{i=1}^{n} p_i \log_2 p_i \qquad (2.11)$$

bits of information. This function has one global extreme only if it hold $p_1 = p_2 = \ldots = p_n = 1/n$ true. Then

$$H(1/n, ..., 1/n) = -\sum_{i=1}^{n} (1/n)\log_2(1/n) = n(1/n)\log_2 n = \log_2 n. \qquad (2.12)$$

Such marker can carry maximally $\log_2 n$ bits of information. Based on quantum nature of our world it is clear that there is no better marker than marker represented by n states (i.e. energy levels) of selected quantum system. All levels have to be in interval $[0, E_{\max}]$ where E_{max} is maximum of energy. If one can measure energy with precision ΔE, then in the marker, can be distinguished maximally $n+1 = (E_{\max}/\Delta E) + 1$ energy levels. When one marker with *n+1* energy levels will be taken into consideration, then by this marker can be represented maximally $\log_2(n+1)$ of bits. On the contrary, when two markers will be used with energy levels in $[0, 1/2E_{\max}]$ it can represent $2\log_2(n/2+1) = \log_2(n/2+1)^2$ bits whereas $n+1 << (n/2+1)^2 = (n^2/4) + n + 1$ and so on. Based on this, it is clear that for representation of the maximal information carried by marker is optimal, when n different markers with energy levels in $[0, \Delta E]$ is used, i.e. with two energy levels which represents 0 and 1. In total it is possible to represent maximally $n \log_2 (n/n+1) = n \log_2 (n/n+1) = n \log_2 2$ i.e. n bits of information because clearly $\log_2 2 = 1$ hold.

Carrier with mass m is according to Einstein equation equal to $E_{\max} = mc^2$. It is obvious that in such a carrier it is possible to maximally have

$$n = \frac{E_{\max}}{\Delta E} = \frac{mc^2}{\Delta E} \tag{2.13}$$

bits of information. To calculate the exact amount of information stored by 2.13, then we need to use Heisenberg principle of uncertainty

$$\Delta E \Delta t \geq \frac{\hbar}{2} \tag{2.14}$$

In which $\hbar = h/2\pi$ (h is Planck constant, $\hbar$ is Dirac constant). If in 2.14 the equality is taken into consideration, then one obtains for the upper estimation

$$n = \frac{mc^2}{\frac{\hbar}{2\Delta t}} = 4\pi \frac{mc^2}{h} \Delta t \tag{2.15}$$

During time interval Δt it is possible to process maximally $4\pi \frac{mc^2}{h} \Delta t$ bits of information. When $\Delta t = 1s$ one can get maximal number of bits which can be processed or stored in mass per 1s. For $m = 1kg$ this number (lets call it BL) is

$$BL = 4\pi \frac{c^2}{h} \tag{2.16}$$

where $[BL] = 1kg^{-1}s^{-1}$

In this moment it is only a matter of simple calculation to get exact numerical value of BL, lets: speed of light and Planck constant $h = 6,62607 \cdot 10^{-34} J.s$. Finally we get

$$BL \approx 1,7045 \cdot 10^{51} kg^{-1} \cdot s^{-1} \tag{2.17}$$

This number, which we call BL here, is the so called Bremermann limit. It is the definite limit which gives maximal number of bits which can be processed or / and stored by an arbitrary matter. In the original paper Bremermann suggested 10^{47} which is caused by the use of nonstandard units (*cm* instead of *m* and *grams* instead of *kg*, as already mentioned before).

Based on this, it is visible that in our universe the computational power is limited by matter and basically there is no computer (existing or theoretical) which would be able to solve arbitrary problems.

If the mass of the Earth (5.9742×10^{24} kg) is taken into account, then a computer of such a mass might store (and subsequently also process) approximately 10^{76} bits every second. During the life of Earth (10^9 years), a computer of its mass might process maximally 10^{92} bits. If the output of a fictive computer is plotted against its mass, it is obvious (Fig. 2.35) that its "computational capacity" is exceeded already during the solution of the traveling salesman problem for a small number of cities/computer mass.

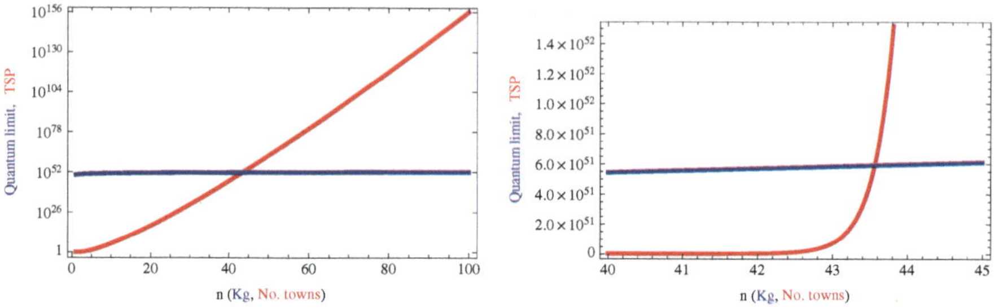

(a) **(b)** Detailed view

Fig. 2.35 Simultaneously plotted dependence of the number of possible solutions of the traveling salesman problem on the number of cities n (red) and the number of bits processed in a computer of mass m (blue). Let us add for more attentive readers that there is a logarithmic scale in the left figure, while a "normal" in the right figure. This is the reason why the plots appear considerably different in both figures.

It is clearly obvious from Fig. 2.35 that the break between the number of cities and the computer mass occurs somewhere between 43 and 44. Perhaps it is not necessary to mention that the output of our computer is illustrated in bits, which is a little bit misleading, because one bit is not sufficient for storing information on one possible solution of the traveling salesman problem. Had this been taken into consideration during the computation, then the result would have been different, nevertheless approximately the same as for the order of magnitude.

If we take into account the ACO (Ant Colony Optimization) algorithm that satisfactorily solves the traveling salesman problem up to approximately 10,000 cities, then we would need a computer of the mass of 10^{35608} kg for storing and processing the information on all possible trajectories. In other words, 10^{35566} computers of the mass of the Earth, should the computation be finished during the life of the universe (10^{17} s). In a similar way [50], we would derive the shortest possible time during which it is possible to process the stored information. This value is $t = 10^{-12}$ s; the current computers work in a region of 10^{-9} s.

In the publication [50], these considerations have been worked out in more details and applied to the transfer of information through an information channel (computation can also be considered as a transfer of information through a special channel). Beside other things, it was found that if a certain mass (or energy) of the transfer medium is reached, further information cannot be transferred through the channel, because the channel collapses into an astrophysical object called black hole. According to[50], the transfer of information is efficient (optimum, maximally usable), if the information channel is on the brink of collapsing into a black hole.

Independently of whether these calculations are accurate or only approximate, it is obvious that physical limits restrict the possibilities of any computer and also of the mathematical computational methods.

2.6 Conclusion

The principles of evolutionary techniques are described in many publications focused both on evolutionary diagrams and in publications "outside" the field, where it is necessary to inform the corresponding community of experts about these techniques. A representative example is [29]. In these and similar monographs, the problems of evolutionary algorithms are introduced at a very vague level. However, there are more suitable sources of literature intended for the needs of the technical community. Here we mention several publications that are suitable for the possible extension of knowledge on ECT. We can recommend the book [2], which is very comprehensive. The book [53] is in principle sufficient for understanding the basic principles; the paper [1] can also be recommended. Among many book monographs, it is possible to mention [23] and also [54]. Both are written in a very understandable manner and the reader will not get lost in theorems, definitions and proofs that do not bring much information for practitioners and beginners. Description of specialized algorithms can be found in [53] and [38] (GA), [15] and [71] (PSO), [57], [56], [58] and [24] (DE), [73] (SOMA), [35], [32] and [65] (MA), [41] and [13] (SA), [45] (SS).

Popularly written books from the field of computers are much less represented than specialized expert books. Much information can be found on the Internet, however, one should mention that this source of information is not always reliable and one can also encounter untrue and misleading information.

There are many publications on the limits of computational technologies based on quantum physics. However, these publications are relatively very demanding on the knowledge from the field of quantum mechanics and mathematics. For extending the information indicated in this chapter, we recommend the already mentioned publications [8] and [50]. The substantial part of limits imposed by mass on processing and storing data is described in the first part [8]. The explanation is so understandable that even a reader at a high school level will understand it. In the paper [50], the relation between transfer channels and black holes is discussed.

You can also read in [2] on the representation of individuals, basic concepts of ETV and the properties of the test functions. Of course, there are other monographs and Internet sources providing this information, but we consider publications mentioned above as sufficiently representative.

References

1. Babu, B.: Evolutionary Computation - At a Glance. NEXUS, Annual Magazine of Engineering Technology Association, BITS, Pilani, 3–7 (2001)
2. Back, T., Fogel, B., Michalewicz, Z.: Handbook of Evolutionary Computation, Institute of Physics, London (1997)

3. Baluja, S.: Population-based incremental learning: A method for integrating genetic search based function optimization and competitive learning. Technical Report CMU-CS-94-163, Carnegie Mellon University, USA (1994)
4. Barricelli, N.A.: Esempi Numerici di processi di evoluzione. Methodos, 45–68 (1954)
5. Barricelli, N.A.: Symbiogenetic evolution processes realized by artificial methods. Methodos 9(35-36), 143–182 (1957)
6. Barricelli, N.A.: Numerical testing of evolution theories: Part I: Theoretical introduction and basic tests. Acta Biotheor. 16(1-2), 69–98 (1962)
7. Box, G.E.P.: Evolutionary Operation: A Method for Increasing Industrial Productivity. Appl. Stat. 6(2), 81–101 (1957)
8. Bremermann, H.: Optimization through evolution and recombination Self- Organizing Systems. In: Yovits, M., Jacobi, G., Goldstine, G. (eds.), pp. 93–106. Spartan Book, Washington (1962)
9. Bull, L., Kovacs, T.: Foundations of Learning Classifier Systems. Springer, Heidelberg (2005)
10. Carlson, E.: Doubts about Mendel's integrity are exaggerated. In: Mendel's Legacy, pp. 48–49. Cold Spring Harbor Laboratory Press, Cold Spring Harbor (2004)
11. Caruana, R., Schaffer, J.: Representation and hidden bias: Gray vs. binary coding for genetic algorithms. In: Proc. 5th Int. Conf. on Machine Learning, Los Altos, pp. 153–161. Morgan Kaufmann, San Francisco (1988)
12. Castro, L., Timmis, J.: Artificial Immune Systems: A New Computational Intelligence Approach. Springer, Heidelberg (2002)
13. Cerny, V.: Thermodynamical approach to the traveling salesman problem: An efficient simulation algorithm. J. Opt. Theory Appl. 45(1), 41–51 (1985)
14. Chu, P.: A Genetic Algorithm Approach for Combinatorial Optimisation Problems. Ph.D. Thesis. The Management School Imperial College of Science, Technology and Medicine, London, p. 181 (1997)
15. Clerc, M.: Particle Swarm Optimization. ISTE Publishing Company (2009)
16. Coveney, P., Highfield, R.: Mezi chaosem a radem, Mlada fronta (2003)
17. Darwin, C.: On the origin of species by means of natural selection, or the preservation of favoured races in the struggle for life, 1st edn. John Murray, London (1859)
18. Dasgupta, D.: Artificial Immune Systems and Their Applications. Springer, Berlin (1999)
19. Davis, L.: Handbook of Genetic Algorithms. Van Nostrand Reinhold, Berlin (1996)
20. Dempsey, I., O'Neill, M., Brabazon, A.: Foundations in Grammatical Evolution for Dynamic Environments. Springer, Heidelberg (2009)
21. Dorigo, M., Stützle, T.: Ant Colony Optimization. MIT Press, Cambridge (2004)
22. Dreo, J., Petrowski, A., Siarry, P., Tailard, E.: Metaheuristic for Hard Optimization: Methods and Case Studies. Springer, Heidelberg (2005)
23. Eiben, A., Smith, J.: Introduction to Evolutionary Computing. Springer, Heidelberg (2007)
24. Feoktistov, V.: Differential Evolution - In Search of Solutions. Springer, Heidelberg (2006)
25. Fogel, B., Corne, W.: Evolutionary Computation in Bioinformatics. Morgan Kaufmann, San Francisco (2002)
26. Fogel, D.B.: Unearthing a Fossil from the History of Evolutionary Computation. Fundamenta Informaticae 35(1-4), 1–16 (1998)
27. Fogel, D.B.: Evolutionary computation: the fossil record. IEEE Press, Piscataway (1998)
28. Fogel, D.B.: Nils Barricelli - Artificial Life, Coevolution, Self-Adaptation. IEEE Comput. Intell. Mag. 1(1), 41–45 (2006)

29. Fogel, L., Owens, J., Walsh, J.: Artificial Intelligence through Simulated Evolution. John Wiley, Chichester (1966)
30. Friedberg, R.M.: A learning machine: Part I. IBM Journal Research and Development 2, 2–13 (1958)
31. Glover, F., Laguna, M.: Tabu Search. Springer, Heidelberg (1997)
32. Goh, C., Ong, Y., Tan, K.: Multi-Objective Memetic Algorithms. Springer, Heidelberg (2009)
33. Goldberg, D.: Genetic Algorithms in Search, Optimization, and Machine Learning. Addison-Wesley Publishing Company Inc., Reading (1989)
34. Haupt, R., Haupt, S.: Practical genetic algorithms, 2nd edn. John Wiley & Sons, USA (2004)
35. Hart, W., Krasnogor, N., Smith, J.: Recent Advances in Memetic Algorithms. Springer, Heidelberg (2005)
36. Hinterding, R., Gielewski, H., Peachey, T.: The nature of mutation in genetic algorithms. In: Eshelman, L. (ed.) Proc. 6th Int. Conf. on Genetic Algorithms, Los Altos, pp. 70–79. Morgan Kaufmann, San Francisco (1989)
37. Holland, J.: Adaptation in natural and artificial systems. Univ. of Michigan Press, Ann Arbor (1975)
38. Holland, J.: Genetic Algorithms. Sci. Am., 44–50 (1992)
39. Ilachinski, A.: Cellular Automata: A Discrete Universe. World Scientific Publishing Company, Singapore (2001)
40. Jones, T.: Evolutionary Algorithms, Fitness Landscapes and Search, Ph.D. Thesis, University of New Mexico, Alburquerque (1995)
41. Kirkpatrick, S., Gelatt Jr., C., Vecchi, M.: Optimization by Simulated Annealing. Science 220(4598), 671–680 (1983)
42. Koza, J.: Genetic Programming. MIT Press, Cambridge (1998)
43. Koza, J.: Genetic Programming: A paradigm for genetically breeding populations of computer programs to solve problems. Stanford University, Computer Science Department, Technical Report STAN-CS-90-1314 (1990)
44. Koza, J., Keane, M., Streeter, M.: Evolving inventions, pp. 40–47. Scientific American (2003)
45. Laguna, M., Martí, R.: Scatter Search - Methodology and Implementations in C. Springer, Heidelberg (2003)
46. Lampinen, J., Zelinka, I.: Mechanical Engineering Design Optimization by Differential Evolution. In: Corne, D., Dorigo, M., Glover, F. (eds.) New Ideas in Optimization, pp. 127–146. McGraw-Hill, London (1999)
47. Lampinen, J., Zelinka, I.: Mechanical Engineering Design Optimization by Differential Evolution. In: Corne, D., Dorigo, M., Glover, F. (eds.) New Ideas in Optimization. McGraw-Hill, London (1999)
48. Langdon, W.: Genetic Programming and Data Structures. Springer, Heidelberg (1998)
49. Larrañaga, P., Lozano, J.A.: Estimation of Distribution Algorithms: A New Tool for Evolutionary Computation. Kluwer Academic Publishers, Dordrecht (2002)
50. Lloyd, S., Giovannetti, V., Maccone, L.: Physical limits to communication. Phys. Rev. Lett. 93, 100501 (2004)
51. Marik, V., Stepankova, O., Lazansky, J.: Artificial Intelligence III. Czech (ed.) Artificial Intelligence III. Academia, Praha (2001)
52. Mendel, J.: Versuche über Plflanzenhybriden Verhandlungen des naturforschenden Vereines in Brünn, Bd. IV für das Jahr. Abhandlungen, 3–47 (1865); For the English translation, see: Druery, C.T., Bateson, W.: Experiments in plant hybridization. Journal of the Royal Horticultural Society 26, 1–32 (1901), `http://www.esp.org/foundations/genetics/classical/gm-65.pdf`

53. Michalewicz, Z.: Genetic Algorithms + Data Structures = Evolution Programs. Springer, Berlin (1996)
54. Michalewicz, Z., Fogel, D.: How to Solve It: Modern Heuristics. Springer, Berlin (2000)
55. O'Neill, M., Ryan, C.: Grammatical Evolution - Evolutionary Automatic Programming in an Arbitrary Language. Springer, Heidelberg (2003)
56. Onwubolu, G., Babu, B.: New Optimization Techniques in Engineering. Springer, New York (2004)
57. Price, K.: An introduction to differential evolution. In: Corne, D., Dorigo, M., Glover, F. (eds.) New Ideas in Optimisation, pp. 79–108. McGraw Hill, International, UK (1999)
58. Price, K., Storn, R., et al.: Differential Evolution - A Practical Approach to Global Optimization. Springer, Heidelberg (2005)
59. Read, R.C.: Coding of Unlabeled Trees. In: Read, R. (ed.) Graph Theory and Computing. Academic Press, London (1972)
60. Rechenberg, I.: (1971) Evolutionsstrategie - Optimierung technischer Systeme nach Prinzipien der biologischen Evolution (PhD thesis), Printed in Fromman-Holzboog (1973)
61. Reeves, C.: Modern Heuristic Techniques for Combinatorial Problems. Blackwell Scientific Publications, Oxford (1993)
62. Rego, C., Alidaee, B.: Metaheuristic Optimization via Memory and Evolution: Tabu Search and Scatter Search. Springer, Heidelberg (2005)
63. Russell, Norvig, S.J., Peter: Artificial Intelligence: A Modern Approach, 2nd edn., pp. 111–114. Prentice Hall, Upper Saddle River (2003)
64. Schwefel, H.: Numerische Optimierung von Computer-Modellen, PhD thesis (1974); Reprinted by Birkhäuser (1977)
65. Schönberger, J.: Operational Freight Carrier Planning, Basic Concepts. In: Optimization Models and Advanced Memetic Algorithms. Springer, Heidelberg (2005)
66. Telfar, G.: Acceleration Techniques for Simulated Annealing. MSc Thesis. Victoria University of Wellington, New Zealand (1996)
67. Turing, A.: Intelligent machinery, unpublished report for National Physical Laboratory. In: Michie, D. (ed.) Machine Intelligence, vol. 7 (1969); Turing, A.M. (ed.): The Collected Works, vol. 3, Ince D. North-Holland, Amsterdam (1992)
68. Vesterstrom, J., Riget, J.: Particle Swarms (May 2002), Dostupny z `www.evalife.dk/publications/JSV_JR_thesis_2002.pdf` (cit.10.2.2007)
69. Von Neumann, J.: The computer and the brain. Yale University Press, New Haven (1958)
70. Wolpert, D., Macready, W.: No Free Lunch Theorems for Search, Technical Report SFI-TR-95-02-010, Santa Fe Institute (1995)
71. Li, X.: Particle Swarm Optimization - An introduction and its recent developments (2006), `www.nical.ustc.edu.cn/seal06/doc/tutorial_pso.pdf` (4.10.2006) (cit. 20. 2. 2007)
72. Zelinka, I.: Artificial Intelligence in problems of global optimization. Czech (ed.) BEN, Praha (2002) ISBN 80-7300-069-5
73. Zelinka, I.: SOMA - Self Organizing Migrating Algorithm. In: Onwubolu, Babu, B. (eds.) New Optimization Techniques in Engineering. Springer, New York (2004)
74. Zvelebil, M., Jeremy, B.: Understanding Bioinformatics. Garland Science (2007)

Chapter 3
Chaos Theory for Evolutionary Algorithms Researchers

Sergej Celikovsky and Ivan Zelinka

Abstract. This chapter deals with chaotic systems. Based on the characterization of deterministic chaos, universal features of that kind of behavior are explained. It is shown that despite the deterministic nature of chaos, long term behavior is unpredictable. This is called sensitivity to initial conditions. We further give a concept of quantifying chaotic dynamics: the Lyapunov exponent. Moreover, we explain how chaos can originate from order by period doubling, intermittence, chaotic transients and crises. In the second part of the chapter we discuss different examples of systems showing chaos, for instance mechanical, electronic, biological, meteorological, algorithmical and astronomical systems.

3.1 Introduction

The discovery of the phenomenon of deterministic chaos brought about the need to verify manifestations of this phenomenon also in experimental data. Deterministically chaotic systems are necessarily nonlinear, and conventional statistical procedures, which are mostly linear, are insufficient for their analysis. If the output of a deterministically chaotic system is subjected to linear methods, such signal will appear as the result of a random process. Examples include Fourier spectral analysis, which will disclose nonzero amplitudes at all frequencies in a chaotic system,

Sergej Celikovsky
Institute of Information Theory and Automation,
Academy of Sciences of the Czech Republic,
Faculty of Electrical Engineering, Czech Technical University in Prague
e-mail: celikovs@utia.cas.cz

Ivan Zelinka
Tomas Bata University in Zlin, Faculty of Applied Informatics, Nad Stranemi 4511,
Zlin 76001, Czech Republic
and
VSB-TUO, Faculty of Electrical Engineering and Computer Science, 17. listopadu 15,
708 33 Ostrava-Poruba, Czech Republic
e-mail: zelinka@fai.utb.cz

I. Zelinka et al. (Eds.): Evolutionary Algorithms and Chaotic Systems, SCI 267, pp. 89–143.
springerlink.com

and so chaos can be easily mistaken for random noise. Apart from the (now mature) signal analysis in both the time and frequency domains, methods operating in the phase space are gaining in importance. Within such methods, the trajectory of a dynamic system in the phase space is first reconstructed from the (usually scalar) time series, and the chaos descriptors are subsequently estimated or modelling is applied. This is a quite recent field of research, just going back to the discovery of the immersion theorem [19], [26] in the early 1980s. Despite lack of rigorous mathematical explanation of some issues, well interpretable results can be obtained with some caution. This is so, in particular, for low-dimensional systems, for the analysis of which such procedures have been primarily developed. Nevertheless, some rather naive applications and interpretations of results were attempted in the past. Examples of such simplified interpretations which contradict physical intuition have been cited by Drazin and King [11]. According to those authors, the early successes of nonlinear analysis of time series raised hopes that the day will come when we will be able, from periodical air temperature measurements behind the window, to identify the dynamics of the whole atmosphere, based on which the future climatic situation should be predictable.

One of the goals of this chapter is to explain why such a mechanistic interpretation of determinism is not correct. The main reason for that naive idea to fail is exactly the existence of the deterministic chaos.

This chapter discusses the most common topics of chaos theory, especially from the practical application aspects point of view. Common approaches to the reconstruction of the system trajectory in the phase space are summarized and procedures are outlined for estimating the correlation dimension, entropy and the largest Lyapunov exponent. Thereby, it a priori assumed that the sources of the time series examined are nonlinear chaotic systems. This is why, for example, nonlinearity tests are not described here. For the same reason, as well as due to the limited extent of this book some important components of nonlinear modelling of time series, such as nonlinear methods for noise prediction and reduction, are omitted. The interested reader may find more detailed information on that topic in monographs [2], [17], [10], [12].

3.2 Characterization of Deterministic Chaos

When hearing the word "chaos", people who are not experts in this field may imagine a process which is of a purely random nature and lacks any internal rules. Just a few people realize that "being chaotic" actually means complying well defined and strictly deterministic rules". As indicated in the historical outline, chaos is a discipline which obtained its name only in the 20th century but whose roots date back to the 18th and 19th centuries, associated with the finding that even simple problems may generate very complex and unpredictable solutions. For historical reasons, Hamiltonian systems were the first systems to be studied, represented then by celestial mechanics problems. Many rules valid for a wide class of Hamiltonian systems generating chaotic behavior were discovered. Later on, these rules were extended to apply to some dissipative chaotic systems as well. Although it deals basically with dissipative systems, this publication will include a short excursion to other chaos generating systems as well.

3.2.1 Roots of Deterministic Chaos

The term "chaos" covers a rather broad class of phenomena whose behavior may seem erratic and unpredictable at the first glance. Often, this term is used to denote phenomena which are of a purely stochastic nature, such as the motion of molecules in a vessel with gas etc. This publication focusses on the deterministic chaos, a phenomenon which - as its name suggests - is not based on the presence of a random, stochastic effects. On the contrary, it is based on the absence of such effects what may seem surprising at the first glance. Broadly used, the term "chaos" can denote anything that cannot be predicted deterministically (e.g. motion of an individual molecule, numbers in a lottery, ...). If, however, the word "chaotic" is combined with an attribute such as "stochastic" or "deterministic", then a specific type of chaotic phenomena is involved, having their specific laws, mathematical apparatus and a physical origin. Stochastic system (not stochastic chaos) is the appropriate term for a system such as plasma, gas, liquid, which should be studied by using a suitable apparatus of plasma physics, statistical mechanics or hydrodynamics. On the contrary, if a double pendulum, billiard or the similar objects are the subject of examination, a mathematical apparatus which is based on classical mathematics and does not exhibit "stigmata" of statistics is employed. The mathematical apparatus for the description and study of the systems was not chosen at random; in fact, it is related with the physical nature of the system being studied. Considering the class of systems of deterministic chaos as mentioned above, signs of chaotic behavior are usually conditional on the presence of nonlinearities, either in the system itself (i.e. the system is a nonlinear system) or in links between linear systems [14]. Usually, such nonlinearities are only visible after making up a mathematical model of the system or after analysis of observed data. Simple systems exhibiting deterministic chaos include, for instance, double pendulum, magnetic pendulum, electronic circuit or a set of bars (Fig. 3.1) over which balls are poured from "the same" starting position. Since the individual examples are discussed in this monograph below, the focus will now be on the last-mentioned example. The example involves a very simple mechanical system which is an analogy of the well-known billiard problem. As Fig. 3.1 demonstrates, the entire mechanical system consists of a set of bars which are held by a vertical board and over which balls are poured from "the same" position. Although released from the same position, each ball follows a different pathway. This is so because the starting conditions are not absolutely identical; instead, they differ very slightly, even negligibly at first glance. It is those differences that are responsible for the fact that the trajectories differ appreciably. In other words, the system is sensitive to the initial conditions.

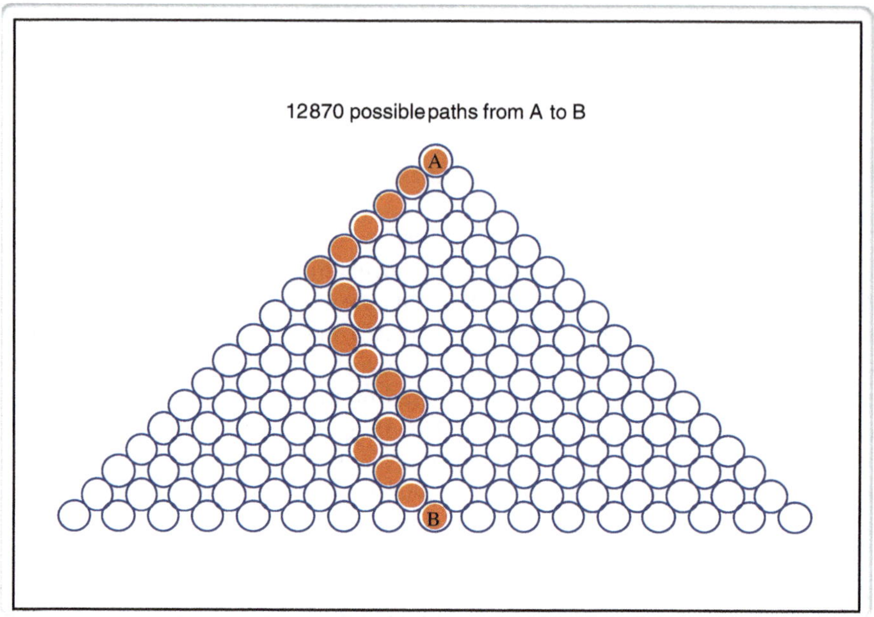

Fig. 3.1 One of the possible realizations of the billiard problem

This sensitivity to the initial conditions is a phenomenon which is related to the billiard problem. Basically, the cause is in the fact that the mechanical objects hitting each other do not possess ideally smooth surfaces. Due to this, even the slightest differences in the initial conditions are "amplified", ultimately giving rise to different trajectories. The nonlinear model shown in Fig. 3.1 can serve as a next model of the billiard problem. Two types of trajectory are involved: periodic (Fig. 3.2) and chaotic (Fig. 3.3). The axes of incidence and recoil of the hypothetical ball are shown. Fig. 3.4 demonstrates the creation of chaos. The ball was started from a nearly identical position with a difference of 1×10^{-12} in this simulation. Different trajectories (red and green) can be discriminated after 25 iterations.

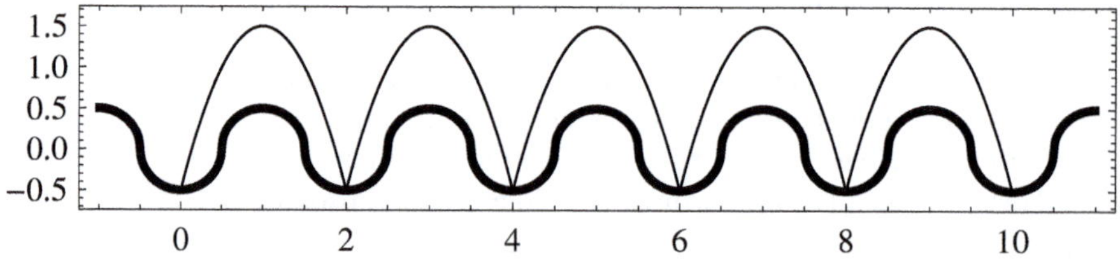

Fig. 3.2 Deterministic behavior at billiard

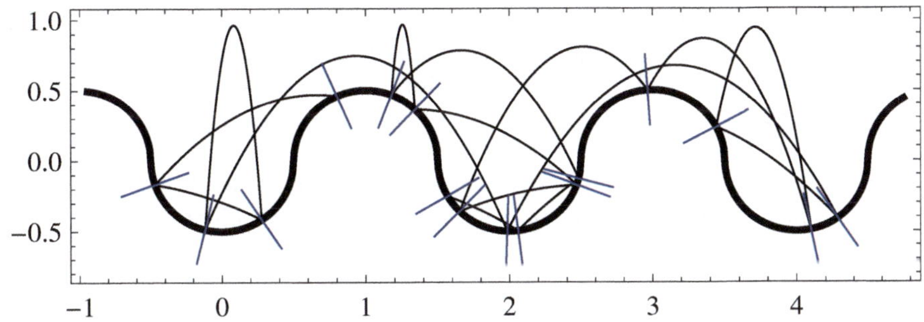

Fig. 3.3 Chaotic behavior at billiard

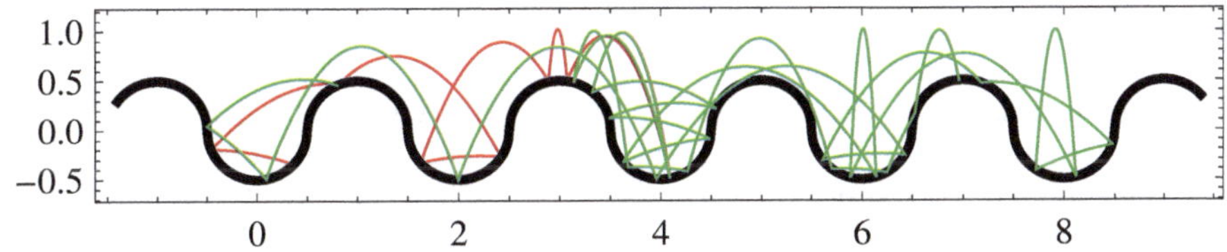

Fig. 3.4 Chaotic behavior at billiard, difference in the initial conditions was 1×10^{-12}. After 25 iterations trajectories has diverged.

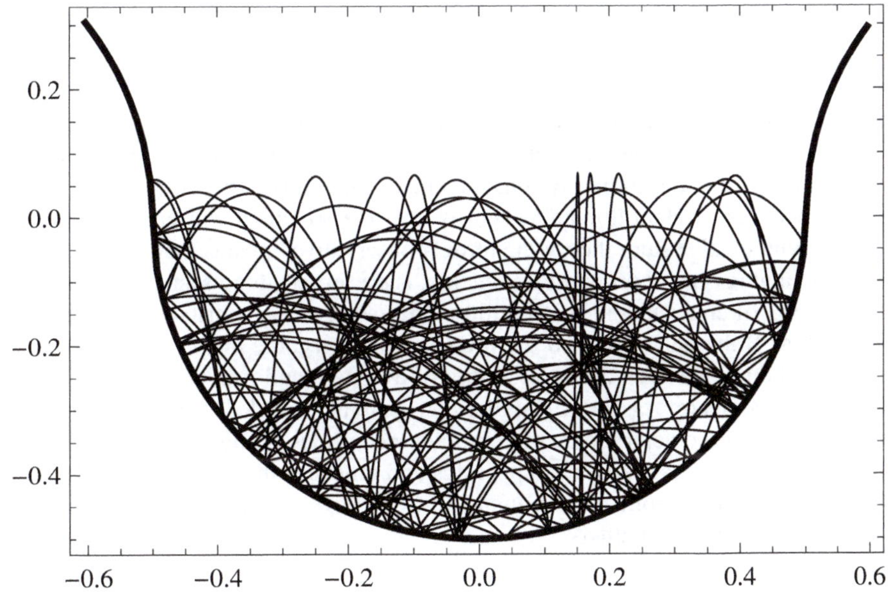

Fig. 3.5 Chaotic behavior at simple billiard, no periodical behavior is visible. 500 iterations from 10 000 is depicted at this picture.

Chaos can be visualized not only in the manner shown in the figures but also by means of interdependences of quantities of state. A trajectory having a total length of 10 000 iterations (Fig. 3.5, only 500 iterations are shown) was generated for this purpose A rather wealthy set of types of behavior can be encountered in the real world. One of the possible categorizations is included in Table 3.3. The table encompasses both purely stochastic types of behavior (coin toss, thermal noise, ...) and deterministic types of behavior (celestial mechanics), including chaos (intermittence, chaotic attractors, ...).

Table 3.1 Possible types of behavior of dynamic systems [14]

Behavior	Example
Predictable	Planets
Unpredictable	Coin toss
Chaotic transitions	Billiard problem
Intermittence	Logistic equation (for A = 3.8284)
Narrow-band chaos	Rossler attractor
Low-dimensional broadband chaos	Lorenz attractor
High-dimensional broadband chaos	Neural networks
Correlated (colored) noise	Random walk
Pseudorandomness	Computer-generated randomness
Randomness	Thermal noise, radioactivity
Combination of the above types of behavior	Real data

3.2.1.1 Hamiltonian Systems

The study of Hamiltonian systems has its roots in the 19th century when it was introduced by Irish mathematician William Hamilton. For mechanical systems, a typical feature of Hamiltonian systems that no dissipation of energy occurs in them, so that mechanical Hamiltonian system is also the so-called conservative one. In general dynamical system theory the term "conservative" means that certain scalar function, having typical properties of energy, is preserved along system trajectories. The creation of chaos theory for Hamiltonian systems was contributed to by scientists such as Boltzman (who laid the foundations of ergodic theory and discovered the contradiction between the reversibility of a system and irreversibility of its behavior) and Poincare. Assets of Hamiltonian systems included their amenability to solution without the deployment of computer techniques, something we can hardly imagine today. The mathematical apparatus and thus also the philosophy of Hamiltonian systems find application in many areas of physics, such as plasma physics, quantum mechanics and others.

3.2.1.2 Dissipative Systems

Dissipative dynamic systems are systems where energy escapes into the surroundings and state space volume is reduced. Typical examples include weight on spring (dissipation being caused by friction between the body and air and energy losses inside the material), motion on a wheel, electronic resonance circuits. Since the topics of dissipative dynamic systems is the subject of a whole monograph, demonstration of a concrete real system, see the classical oscillating cell, will be given here. This well-known classical example of a dynamic system is defined by the Lorenz system (3.2).

3.3 Universal Features of Chaos

Deterministic chaos possesses many features that are common to chaotic behavior irrespective of the physical system which is the cause of this behavior. This common nature is expressed by the term universality so as to stress the universal nature of the phenomena. The quantity and properties of the features as well as the complexity of links between them are so extensive that they could make up a topic for a separate publication, such as [9]. It is not the aim of this part of the publication to make a detailed analysis - this would be like carrying coals to Newcastle; instead, only the best-known features, to be used in the subsequent sections of this book, will be highlighted. These include, in particular, Feigenbaum's constants α and δ, the U-sequence, Lyapunov exponents, self-similarity and processes by which a system usually passes from deterministic behavior to chaotic behavior: intermittence, period doubling, metastable chaos and crises. Another property which is, curiously, not included in the pantheon of universalities will be mentioned at the beginning: the deterministic nature and non-predictability of deterministic chaos.

3.3.1 Determinism and Unpredictability of the Behavior of Deterministic Chaos – Sensitivity to Initial Conditions

The deterministic structure of systems which generate chaos and their unpredictability constitute another typical feature of the universal properties of deterministic chaos. It is actually irrelevant what type the chaotic system is (chemical, biological, electronic, economic, ...): it holds invariably that their mathematical models are fully deterministic (there is no room for randomness as such in them) and they are long-term unpredictable in their behaviour. The Rössler (3.1) and Lorenz (3.2) attractors are the typical examples:

$$\begin{aligned} \dot{x}_1(t) &= -x_2(t) - x_3(t) \\ \dot{x}_2(t) &= -x_1(t) - \tfrac{x_2(t)}{5} \\ \dot{x}_3(t) &= (x_1(t) - 5.7)x_3(t) + 0.2 \end{aligned} \tag{3.1}$$

$$\begin{aligned} \dot{x}_1(t) &= -a(x_1(t) - x_2(t)) \\ \dot{x}_2(t) &= -x_1(t)x_3(t) + bx_1(t) + x_3(t) \\ \dot{x}_3(t) &= x_1(t)x_2(t) - x_3(t). \end{aligned} \tag{3.2}$$

It is clear from the structure of the equations that no mathematical term expressing randomness is present. That apparent randomness that can be seen in deterministic chaos at first glance is not purely fortuitous; in fact, it is related to the sensitivity to initial conditions. This sensitivity can be demonstrated well on the example of a smooth hill from whose top a ball is let run down. The ball will take a different trajectory in each experiment, which is due to two factors: the first is the non-ideality of the hill surface, the other, impossibility of setting the starting position absolutely identically when repeating the experiment. The inaccuracies are due to the ubiquitous error of measurement (in manufacturing the hill, in setting the position, in manufacturing the ball, ...), and even if all the errors could be eliminated, the uncertainty of the quantum world (i.e. Heisenberg uncertainty principle) would ultimately take effect and act in the macro-world as well (which it actually does). Hence, fluctuations cannot be avoided, and so "declaring total war" on fluctuations is a waste of time and akin to Don Quixote's tilting at windmills. A normal PC with appropriate software will do for experiments with sensitivity to initial conditions. Fig. 3.6 demonstrates sensitivity to initial conditions for the Lorenz attractor. Two time developments of the variable of state x (Fig. 3.6) are shown for a difference between the initial conditions $\Delta y(0) = 0.001$, which appears as a negligible error at first glance. However, in a time as short as 24 seconds the two state trajectories diverge, as emphasized by the grey area between them. (Fig. 3.7) shows the same for $\Delta y(0) = 10^{-9}$. Sensitivity to initial conditions is thus one of the characteristic features of deterministic chaos and can be used as an indicator when classifying a dynamic system.

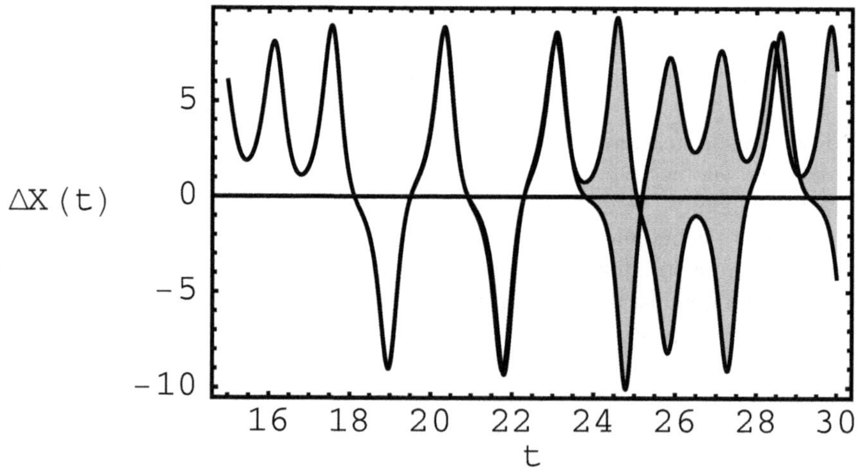

Fig. 3.6 Sensitivity of the variable x of the Lorenz attractor for $\delta y(0) = 0.001$

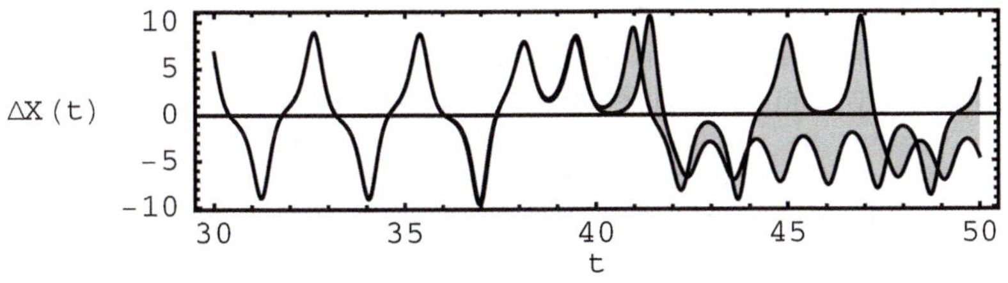

Fig. 3.7 Sensitivity of the variable x of the Lorenz attractor for $\delta y(0) = 10^{-9}$

3.3.2 Lyapunov Exponents

Lyapunov exponents are another member of the family of universal features of deterministic chaos. They are numbers which basically express the divergence (or also convergence) of the state trajectories of a dynamic system. The exponents can be calculated relatively simply, both for discrete-time systems and for continuous-time systems. As will be explained later, Lyapunov exponents are closely related to the structure of the state space, which (in dynamic systems theory) is represented by an array of arrows determining the time development of the system at each point of the space. The development of the system in this space is then represented by a (usually) continuous curve [24].

The effect of Lyapunov exponents on the behavior of the dynamic system is apparent from Fig. 3.8 and 3.9. Figure 3.8 shows the state space of a simple dynamic system along with two different time developments starting from two different initial conditions, which only differ by $\Delta x = 0.01$ in the x-axis. The behavior in the

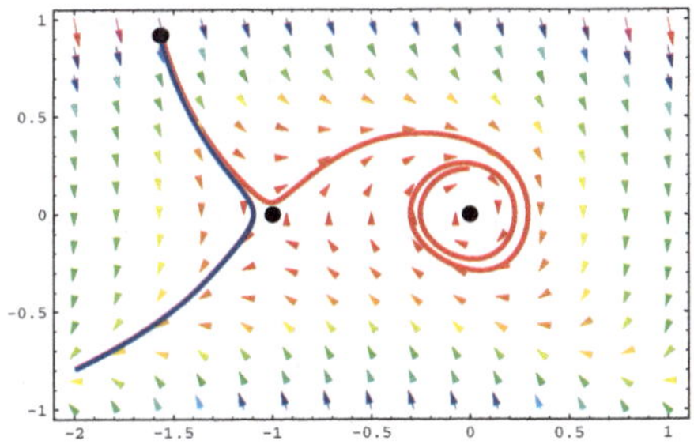

Fig. 3.8 State space trajectory for a dynamic system with 2 singular points s_1 and s_2. On the position $s_1 = \{0,0\}$ is repeller and at the position $s_2 = \{-1,0\}$ saddle. Start points of both trajectories diverge despite fact that this coordinates (x_1 = {-1.56, 0.92} and x_2 = {-1.57, 0.92}) are very close.

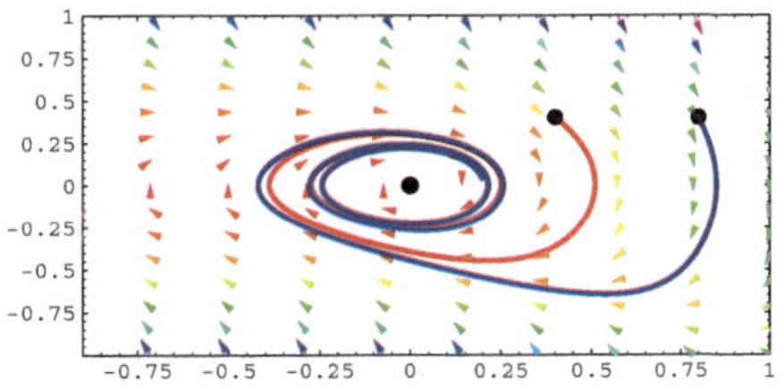

Fig. 3.9 Different behavior can be observed when both trajectories will start in different part of state space. Despite its bigger difference in starting position (x_1 = {0.4, 0.4} and x_2 = {0.8, 0.4}) trajectories merge together after certain time.

two cases is entirely different. Figure 3.9 shows different behavior. Hence, the behavior of a dynamical system is determined by its physical structure, which in the mathematical description is represented by the state space whose quantifiers can be Lyapunov exponents. If one is to follow colored arrows in Fig. 3.8, it can be noticed that they are separating with increasing time. On the other hand in Fig. 3.9 they after certain time occupy the same set of points in the state space, in this case called limit cycle. This observation can be described in a mathematical way by the Lyapunov exponent λ, see eq. 3.3. The structure of the exponents can help assess whether chaotic behavior is present in the system or not.

Consider a situation where at time $t_0 = 0$, a hypersphere whose radius is $l(0)$ exists in the m-dimensional phase space. Let different points of the hypersphere surface represent different initial conditions of the dynamical system. Hence, starting from each point of the hypothetical hypersphere, construct a trajectory through the phase space. After time t the hypersphere transforms into a new object. In the general case, this object can have a very complicated shape, especially if the dynamics are chaotic. However, if we restrict ourselves to very short time segments $[0,t]$ and if the initial radius $l(0)$ is also very small, one can assume for simplicity that the initial hypersphere is transformed, in the ideal case, into a hyper ellipsoid. Denote l_i(t) the length of the semi-major axes of the ellipsoid formed at time t. The i_{th} Lyapunov exponent

$$\lambda_i = \lim_{t\to\infty} \frac{1}{t} \ln \frac{l_i(t)}{l(0)} \tag{3.3}$$

is a measure of the extension or contraction of the i_{th} semi-major axis of the ellipsoid. For graphic reasons, Lyapunov exponents are arranged by magnitude, i.e. $\lambda_1 \geq \lambda_2 \geq ... \geq \lambda_m$, where m is the dimension of the phase space; this is referred to as the Lyapunov exponents spectrum. For a chaotic trajectory, at least one Lyapunov exponent must be positive, although, in addition, the existence of any asymptotic periodicity must be ruled out to confirm the chaotic nature - see, e.g., [3]. In other words, the possibility that the trajectory converges to some periodic orbit with $t \to \infty$ must be eliminated. But it is just this requirement that can pose a problem in practice if the dynamical system is investigated during a finite time interval only. Chaotic systems with more than one Lyapunov exponent are referred to as hyperchaotic [22].

Owing to the limit $t \to \infty$, Lyapunov exponents introduced by eq. (3.3) are global quantities describing the system dynamics on average. Nevertheless, relating Lyapunov exponents to a certain part of the trajectory for a relatively short time segment t also proved to be useful. This leads to the concept of a local Lyapunov exponent [2], [30]. It will be clear from the above text that Lyapunov exponents represent the rate of divergence (or convergence) of near trajectories in the phase space, thus providing a measure of predictability. Hence, this warrants the question as to how Lyapunov exponents relate to Kolmogorov entropy. The relation can be expressed as follows [22]:

$$K \leq \sum_{i,\lambda_i>0} \lambda_i \tag{3.4}$$

where equality occurs for the Sinai-Ruelle-Bowen measure, i.e. the measure which is smooth along an unstable manifold. Equality between Kolmogorov entropy and the sum of positive Lyapunov exponents is referred to as Pesin identity [23]. Now, examine the relationship between Lyapunovs exponents and fractal dimension. If Lyapunov exponents are negative for all i's ($\lambda_i < 0$), each attractor of such a system must be a fixed point and thus have a zero dimension. And on the contrary, if $\lambda_i > 0$ for all i's, the trajectories in the phase space go apart constantly in all directions and the dimension converges to that of the phase space [30]. Hence, one will ask what the relation between Lyapunov exponents and the fractal dimension is. Using the spectrum of Lyapunov exponents $\lambda_1 \geq \lambda_2 \geq ... \geq \lambda_m$, J. L. Kaplan and J. A. Yorke introduced the concept of Lyapunov dimension, sometimes referred to as the Kaplan-York dimension. If k is the largest non-negative integer for which

$$\sum_{i=1}^{k} \lambda_i \geq 0 \tag{3.5}$$

then Lyapunov dimension is defined as follows [15]:

$$d_L = \begin{cases} 0 & if \text{ no such } k \text{ exists} \\ k + \frac{\sum_{i=1}^{k} \lambda_i}{|\lambda_{k+1}|} & if \quad k < m \\ m & if \quad k = m \end{cases} \tag{3.6}$$

In this definition, m has the meaning of the phase space dimension. General equality between Lyapunov dimension and some of the other fractal dimensions has not been proved so far. Many numerical experiments lead to the approximate equality $d_L \approx d_1$, d_L is given by eq. (3.6), d_1 is so called informational dimension, see eq. (3.7)

$$d_1 = \lim_{r \to 0} \frac{-S(r)}{\log_2 r} = \lim_{r \to 0} \frac{\sum_i p_i \log_2 p_i}{\log_2 r}, \tag{3.7}$$

equality between these two dimensions being found for two dimensional mappings [18]. It is generally believed that Lyapunov dimension and information dimension are equal for "typical" attractors [20], [15]. A general rule holds [8] that Lyapunov dimension is the upper limit of Hausdorff dimension. The fact that knowledge of Lyapunov exponents gives us an idea of fractal dimension can be used when testing procedures for estimating attractor dimension from time series. With the knowledge of the system control equations in the form of difference equations or ordinary differential equations the calculation of Lyapunov exponents is "merely" a technical - although not necessarily easy - task. Having calculated Lyapunov dimension from the Lyapunov exponents spectrum and adopting the hypothesis of its closeness to other fractal dimensions, the value of the dimension so obtained can be compared with the estimate based on the time series generated by the system control equations and, tentatively at least, assess the adequacy of some algorithms for nonlinear analysis of time series. The same approach can be used to examine procedures for the calculation of Kolmogorov entropy from experimental data, assuming validity of Pesinov identity. Now, pay some attention to the time of predictability of the system behavior, as mentioned earlier. Imagine a dynamic system with one positive Lyapunov exponent λ. The initial state of the system is known with accuracy ε. After time T, the position of the system in the phase space is known with accuracy L. Taking into the account eq. (3.3) we have [5]

$$\lambda \approx \frac{1}{T} \ln \frac{L}{\varepsilon}, \tag{3.8}$$

which implies that

$$T \approx \frac{1}{\lambda} \ln \frac{L}{\varepsilon} \approx \frac{1}{K} \ln \frac{L}{\varepsilon}, \tag{3.9}$$

where K is Kolmogorov entropy. Time T expresses the time in which inaccuracy ε in the determination of the initial conditions increases exponentially. This time is usually referred to as the system behavior predictability time. However, the relation above indicates that this time is not only dependent on the dynamics of the system (Lyapunov exponents); in fact, the magnitude of the initial error also plays a role: time T increases logarithmically with increasing initial accuracy. Time T can be only crudely identified with the predictability time and only within the context of the accuracy considered, which should be chosen reasonably. If you forecast, for instance, that the next winter will be colder than the past summer, you will probably be right but such prediction is actually useless for the vast majority of purposes.

3.3.3 The U-Sequence

The universal sequence, or the U-sequence, is another universal feature of deterministic chaos. The U-sequence is frequently demonstrated on iterated maps, whose typical representative is the well-known logistic equation. The U-sequence can be observed in the behavior of a number of dynamic systems whose mathematical model contains unimodal mapping (with one extremum). The logistic equation, formulated as eq. (3.11), is a typical example. The term unimodal mapping denotes the dependence of a next value on the preceding values when the control parameter is varied. For instance, if eq. (3.11) is considered and the control parameter A is varied within the interval of [0, 4], the functional dependence shown in Fig. 3.10 emerges. The value at which this dependence attains its maximum is usually referred to as the critical point [16]. This value is 0.5 in Fig. 3.10, as indicated by a vertical ordinate. When the initial conditions are set, the development of the system is shown graphically as a sequence of points (Fig. 3.11) and 3.11 on the unimodal curve. The points of this sequence are assigned the letter L or R according to whether they lie to the left or to the right of the critical point.

U-sequences listed in Table 3.3 can be observed for the logistic equation. This and other sequences are also observable with other mathematical models of dynamic systems.

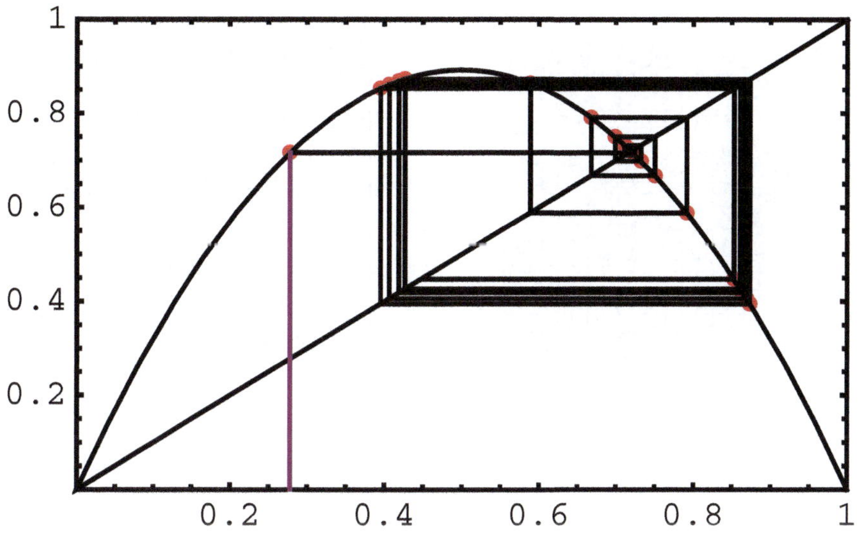

Fig. 3.10 Unimodal sequence of the logistic equation

Table 3.2 U-sequence according to [0]

Perioda	U-sequence	Parameter A value
2	R	3.2361
4	RLR	3.4986
6	RLRRR	3.6275
5	RLRR	3.7389
3	RL	3.8319
6	RLLRL	3.8446
5	RLLR	3.9057
6	RLLRR	3.9375
4	RLL	3.9603
6	RLLLR	3.9778
5	RLLL	3.9903

A graphic presentation of such sequences is also possible in 2D graphs by assigning white color to the R-positions and black color to the L-positions. Therefore, one can easily see when U-sequences agree with one another. Figs 3.11 and 3.12 depict the U-sequences for the logistic equation and for the following equation

$$x_{n+1} = 1 - Cx_n^2 \tag{3.10}$$

called as the quadratic one. The sequences are the same for $A = 3.3$ in the former equation and C = 1.1 in the latter equation.

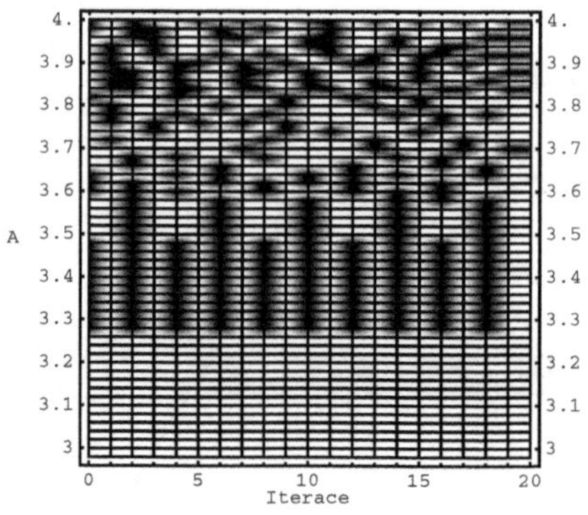

Fig. 3.11 Graphical representation of the U-sequence for the logistic equation with parameter A

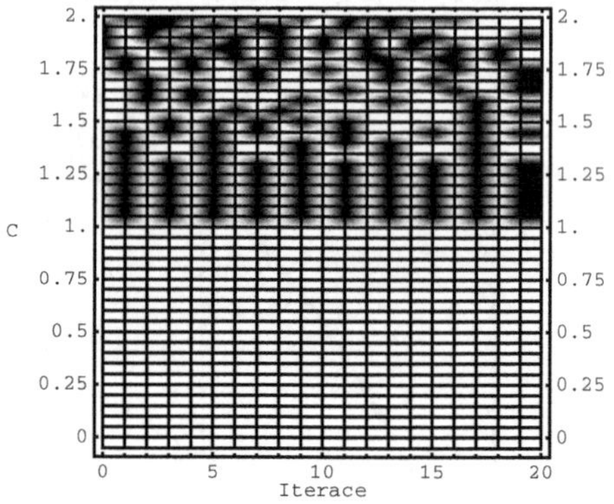

Fig. 3.12 Graphical representation of the U-sequence for the quadratic equation with parameter C

3.3.4 Intermittence, Period Doubling, Metastable Chaos and Crises

The emergence of chaos is not a phenomenon that can be described as a purely discrete event; instead, it has a "transient phase" during which the system behavior changes from predictable to chaotic, both by a deterministic pathway and by a random pathway. The two processes are often intertwined, representing thus a kind of "universal" pathway to chaos. Period doubling is a typical example of a deterministic transition [16]. This is a phenomenon where the period of the system behavior doubles and at some control parameter levels transforms into chaotic behavior. This is demonstrated for the logistic equation in Fig. 3.13 and 3.14, where the left part displays the period doubling mode and the right part displays intermittence. It is of interest to note that the geometric objects which are seen on the right in Fig. 3.14, having a triangular shape (iterations 30 - 40, 50 - 60), are known from stock exchange developments and are employed for near-future time series behavior estimates.

The emergence of intermittence [16] is associated with very fine changes in the control parameter, which can be due to noise or, for instance, to numerical instability. Due to such fine changes the system behavior changes dramatically, being transferred from one type of behavior to the other. The emergence of intermittence from the logistic equation is shown in Figs 3.15 and 3.16 by means of the WEB diagram [16], [6]. A web diagram, also sometimes called a cobweb plot, is a graph that can be used to visualize successive iterations of a function $x_{n+1} = f(x_n)$. The diagram is called WEB because its straight line segments "anchored" to the functions and can resemble a spider web - thus WEB diagram.

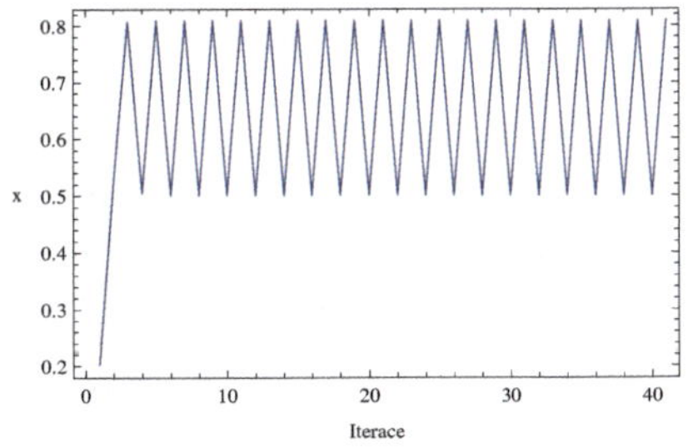

Fig. 3.13 Period doubling for the logistic equation...

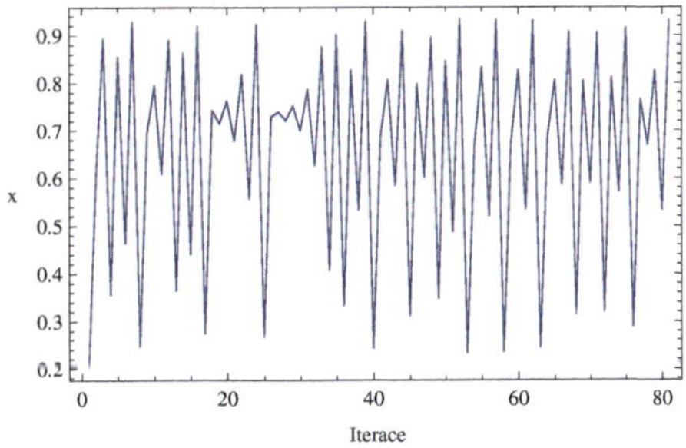

Fig. 3.14 ... and intermittence.

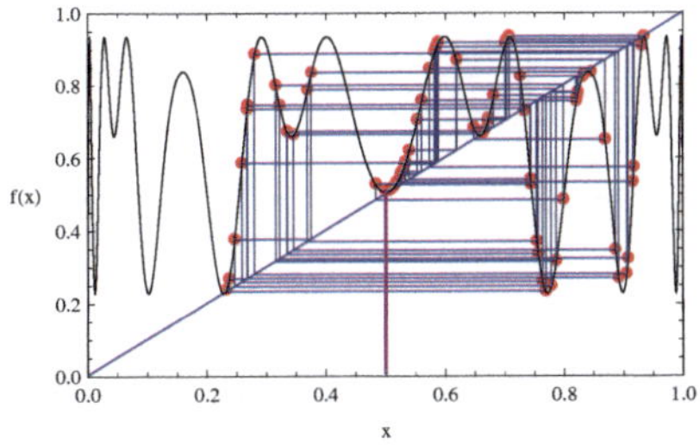

Fig. 3.15 WEB diagram of the logistic equation for A = 3.7375 and 70 iterations

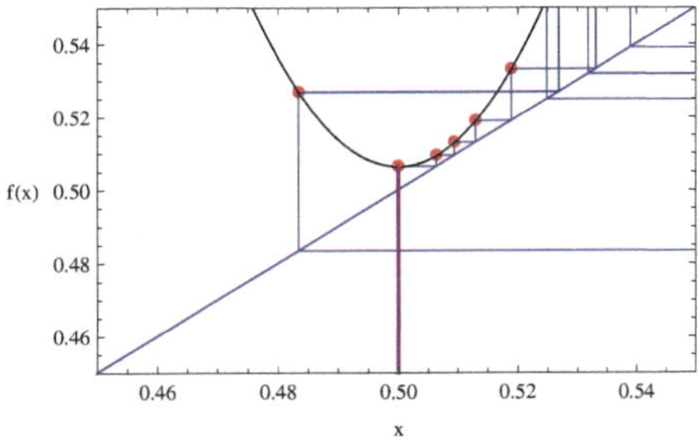

Fig. 3.16 WEB diagram (detail) of the logistic equation for A = 3.7375 and 70 iterations

In Fig. 3.16 a small change causes the behavior to "switch" to the chaotic mode whose overall appearance is shown in Fig. 3.15. If this change is due to a continuous change in the control parameter, a crisis (see later) can take place if promoted by the configuration of the system. This means that the entire chaotic attractor can vanish or be replaced by another attractor [16]. A little bit more detailed analysis of the various pathways leading to chaos will be presented later in this Chapter.

3.3.5 *Feigenbaum Constants*

As mentioned in the section highlighting the history of theories dealing with deterministic chaos, the theoretical physicist Mitchell Feigenbaum devised two constants which certainly belong to the set of universalities of deterministic chaos. Their nature and application can be best explained using examples which include graphical visualization of the development of a chaotic system, specifically bifurcation diagrams (Fig. 3.17).

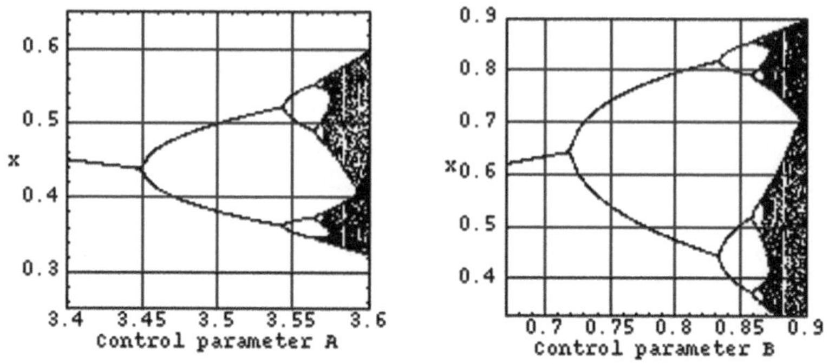

Fig. 3.17 Source of Feigenbaum's constants - self-similarity of bifurcation diagrams. Left: diagram for the logistic equation (3.11); right: section for the equation containing the trigonometric function, eq. (3.12).

The diagrams were generated by using (3.11) and (3.12):

$$x_{n+1} = Ax_n(1 - x_n) \tag{3.11}$$

$$x_{n+1} = Bsin(\pi x_n) \tag{3.12}$$

They differ in a comprehensive representation (Fig. 3.17) but a detailed view shows that different systems can produce virtually identical behavior: Two Feigenbaum's constants α and δ follow from Fig. 3.17. Basically, they are numbers (constants) representing geometric convergence of bifurcation diagrams. Both diagrams exhibit branch splitting which proceeds in a very similar manner, as regards both the x-axis and the y-axis. This can be seen in detail in Fig. 3.17.

Fig. 3.17 demonstrates that when the control parameter is changed, the system behavior changes so that the branches in the bifurcation diagram are divided into two additional branches each while a distance from the most recent division is progressively diminishing. If the branching is projected into the x-axis and the ordinates in which the branching has taken place are denoted sequentially, a sequence of numbers is obtained expressing the geometric convergence of the bifurcation diagram with respect to the x-axis. This set of numbers also expresses the second Feigenbaum's constant, δ, given by relations (3.13) and (3.14).

$$\delta_n = \frac{A_n - A_{n-1}}{A_{n+1} - A_n} \tag{3.13}$$

$$\delta = \lim_{n \to \infty} \delta_n = 4.66920161... \tag{3.14}$$

Constant δ is the limit of numbers which can be understood, with some exaggeration, as "local Feigenbaum's constants". The first Feigenbaum's constant is α (which also precedes δ in the Greek alphabet). This constant is derived by a similar procedure. The branching process is accompanied by changes in the distance between the points of branching, denoted d_n. Once again, constant α is given by the limit of the ratio of the current distance to the previous distance. The mathematical formula is given by eq. (3.15). The limiting sequences leading to the above constants can be calculated even from simple mathematical models, such as the logistic equation.

$$\alpha = \lim_{n\to\infty} \frac{d_n}{d_{n+1}} = 2.5029... \tag{3.15}$$

Feigenbaum's constants are physical parameters which are common to a wide class of systems. From how the constants are derived (as also indicated in [16]) also follows how they can be used, specifically, how δ can be used to predict additional bifurcations in the system. Realizing that δ describes the measure of subsequent bifurcations, the prediction principle is quite clear. Starting from eq. (3.13) and (3.14) and rearranging, one arrives at eq. (3.16), which can be used to calculate the control parameter value at which the next bifurcation will take place.

$$A_{n+1} = \frac{A_n - A_{n-1}}{\delta} + A_n \tag{3.16}$$

In this manner the values can be obtained, or as shown by relations (3.17) and (3.18).

$$A_3 = \frac{A_2 - A_1}{\delta} + A_2 \tag{3.17}$$

$$A_4 = \frac{A_3 - A_2}{\delta} + A_3 \tag{3.18}$$

Hence, the result is fully determined by the two preceding bifurcations. Substitution of eq. (3.17) in (3.18) gives eq. (3.19), which enables us to calculate from two values of the control parameter A_n at which bifurcation takes place. In this manner one can proceed up to the value (3.20) at which chaos appears. In fact, this prediction is approximate only; nevertheless, as proved by various experiments [16], the predictions fit the reality quite well.

$$A_4 = \frac{A_2 - A_1}{\delta^2 + \delta} + A_2 \tag{3.19}$$

$$A_\infty = \frac{A_2 - A_1}{\delta - 1} + A_2 \tag{3.20}$$

3.3.6 Self-similarity

Another common feature of chaos is self-similarity [6], a phenomenon which can be seen quite well on bifurcation diagrams. Self-similarity is best demonstrable in fractal geometry. Basically, self-similarity is the property of a geometric object that contains a component part which is identical with or very similar to the geometric structure of the whole object. In other words, a sub-set of the parent object is similar to the parent object. This property is actually only a geometric-linguistic expression of rather complex mathematical structures and the associated mathematical apparatus which is used in fractal geometry. Self-similarity can be demonstrated graphically on two classic fractal objects - snowflake and fern (Fig. 3.18 - 3.19). Take any part of the object: its structure will resemble that of the basic object.

Fig. 3.18 Self-similarity in snowflake ...

Fig. 3.19 ... and in fern.

The same applies, for instance, to bifurcation diagrams. Since their structure is determined by Feigenbaum's constants, which are universal for chaos as such, some graphical visualizations of chaos can be expected to exhibit self-similarity, viz. within a single visualization (a single bifurcation diagram) or among several bifurcations diagrams of different systems. This is well illustrated by the demonstration of self-similarity using bifurcation diagrams (Fig. 3.17). The diagrams clearly display self-similarity. The result will be the same with other bifurcation diagrams also. Self-similarity and other fractal properties can also be found in other visualizations of course (chaotic attractors), but bifurcation diagrams are apparently most graphic for this purpose.

3.4 From Order to Chaos

Deterministic chaos as such does not exist on its own. In fact, it is a type of behavior that can be observed in some nonlinear systems and which can be tackled from various sides. Usually, two methods to get to chaotic behavior are described in the literature: through local bifurcations and through global bifurcations. The two categories are then classed further into special subgroups of transition to chaos. For local bifurcations these include period doubling, quasi-periodicity, and intermittence, the last-mentioned being further granulated into Type I (tangent bifurcation), Type II (Hopf's bifurcation), and Type III (period doubling). For global bifurcations, these include chaotic transients and crises. An overview of the transitions is shown in Table 3.3.

Table 3.3 Ways to chaos

Way to chaos	Note
Local bifurcation	Period doubling, quasi-periodicity, intermittence (type I, II a III)
Global bifurcation	Transients, crisis

Transient to chaotic behavior is very often combination of transients mentioned in the Table 3.3. Complexity of final transient depend on dynamical system structure, but also on the set of signals which influent behavior of given dynamical system.

3.4.1 Period Doubling

Period doubling is another way to reach chaos domain and is joined with so called limit cycles. Term "period doubling" means that under certain conditions is behavior of dynamical system doubling its periodical behavior (from period 2 to period 4, etc...) which is remoted by certain control parameter of observed system. Period doubling is easily observable on so called Poincare section, which is in fact, $N-1$ dimensional plane through which trajectory is going. All intersections with plane are recorded and are observable like points on Poincare plane, as is depicted at Fig. 3.21, Fig. 3.23 or Fig. 3.25. Under changes of control parameter, system's trajectory is doubling (number of intersection increase) till chaotic behavior is reached. Period doubling is observable in systems which containing "internal" frequency and are controlled by external signal. In the case that there is no external control input and period doubling is observable, system must contain both signals (frequencies) generated under suitable conditions.

Both frequencies, or better their mutual combination, determine resulting behavior of dynamical system, which is determined by mutual ratios of both frequencies (lets call them for now ω_R and ω_r) which can be rational or irrational. In the case of rational ratio, is resulting trajectory periodical, in the case of irrational ratio one can observe quasi-periodical trajectories. The influence of both frequencies can be easily generated by (3.21). Equations parametrically describe dynamics of trajectory in 3D on a torus, with radius R and r. Frequencies ω_R and ω_r are of rotation around main torus radius R and radius of its body r. On figure 3.20 and 3.21 is depicted trajectory for $\omega_R = 3$ and $\omega_r = 2$ including Poincare's surface with three points. Trajectory is periodical. For $\omega_r = 2.1$ is trajectory more complicated, see Fig. 3.22 and 3.23. If the raion of both frequencies become to be more irrational, then torus surface is more densely covered and at Poincare section is cutting trajectory creating a circle, see Fig. 3.24 and 3.25.

$$
\begin{aligned}
x_1(t) &= \cos(t\omega_R)(r\sin(t\omega_r)+R) \\
x_2(t) &= r\cos(t\omega_r) \\
x_3(t) &= \sin(t\omega_R)(r\sin(t\omega_r)+R)
\end{aligned}
\tag{3.21}
$$

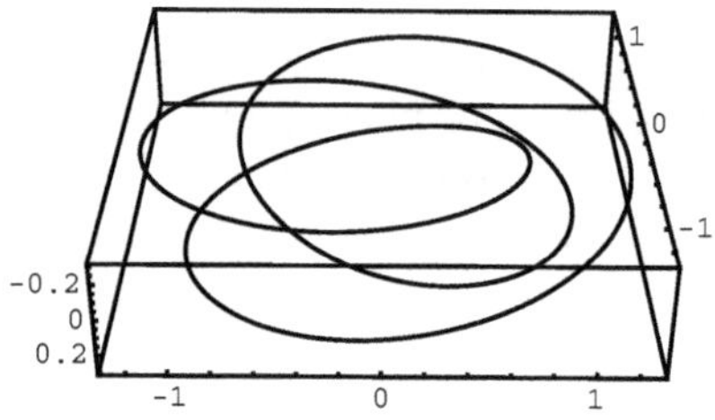

Fig. 3.20 Trajectory and its Poincare section for $\omega_R = 3$

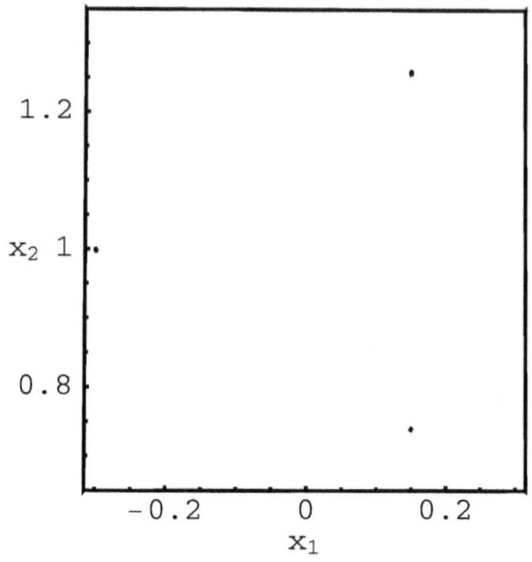

Fig. 3.21 and $\omega_r = 2$

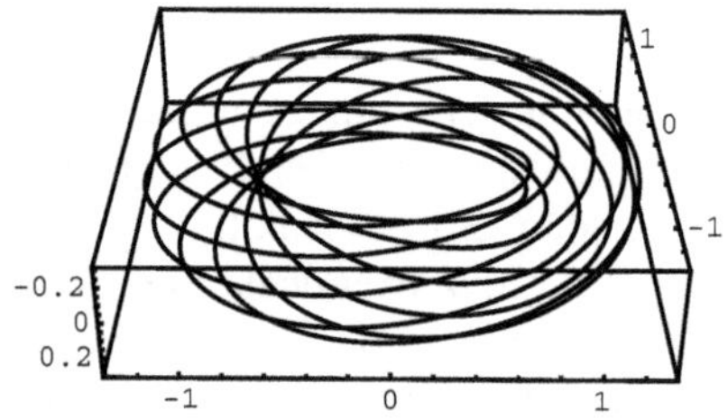

Fig. 3.22 Trajectory and its Poincare section for $\omega_R = 3$

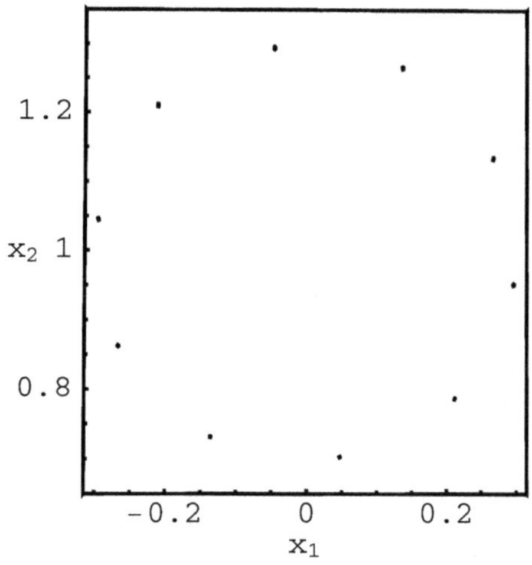

Fig. 3.23 and $\omega_r = 2.1$

If any of the two frequencies is changed, the resulting trajectory need not necessarily be more chaotic; on the contrary, if the two frequencies are in suitable ("more rational") ratios, "deterministic windows" can appear in the trajectory behavior, i.e. the trajectory does not exhibit chaotic motion. This is demonstrated in Fig. 3.26 - 3.29, where more or less chaotic behavior of the resulting trajectory can be observed

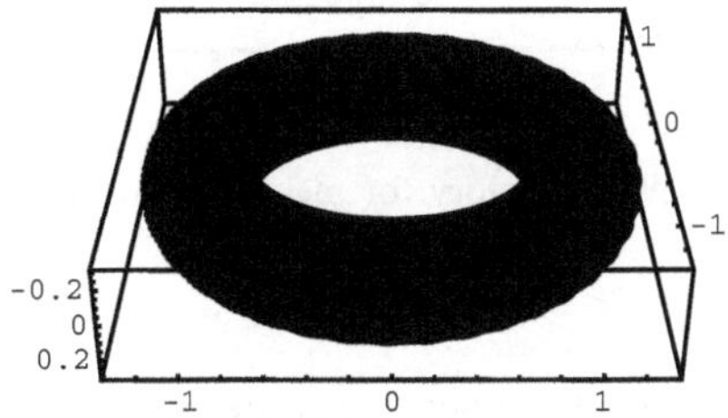

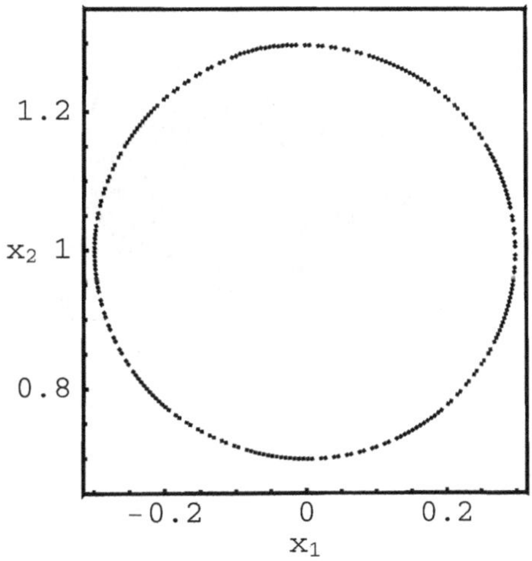

Fig. 3.24 Trajectory and its Poincare section for $\omega_R = 3$

Fig. 3.25 and $\omega_r = 2.33$

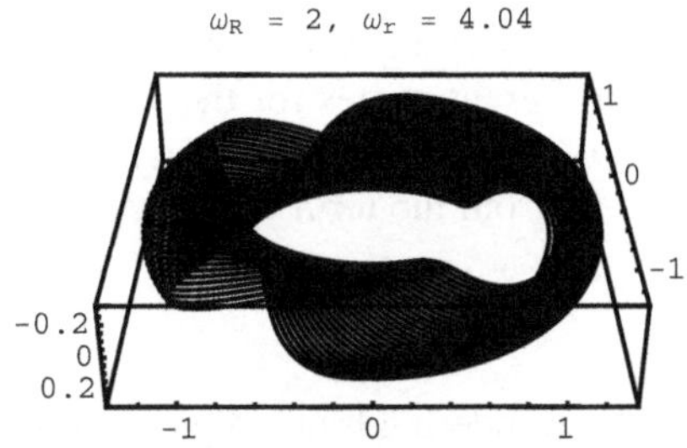

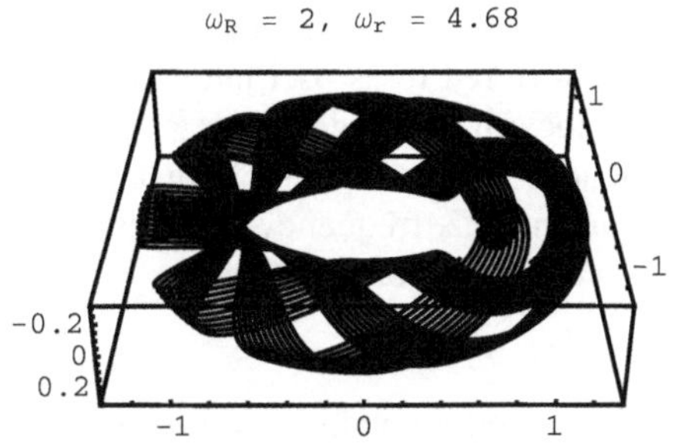

Fig. 3.26 Trajectory for $\omega_R = 2$ and different ω_r

Fig. 3.27 Trajectory for ω_R= 2 and different ω_r

for different. If the behavior becomes chaotic, the trajectory forms a ring, called a drift ring, on the Poincare plane.

The pathway leading to chaos and containing period doubling has the following structure: singular point $\rightarrow$ limiting cycle $\rightarrow$ period doubling $\rightarrow$ quasi-periodicity $\rightarrow$ chaos. Period doubling and quasi-periodicity play the parts of chain links only. Apart from special cases, transition from quasi-periodicity to chaos is only possible if a new, third frequency appears in the system with a constant change in the control parameter. Three dimensions as a minimum are needed for chaos to emerge. If (except for special cases as mentioned) chaos could emerge for less that 3 dimensions, this would be in violation of the Poincare-Bendixon theorem, according to which

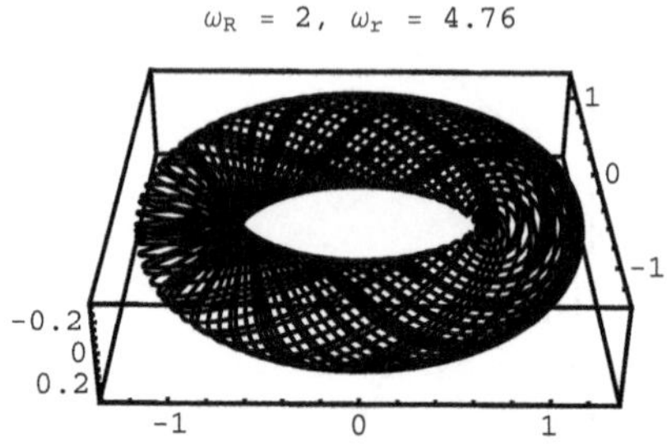

Fig. 3.28 Trajectory for ω_R= 2 and different ω_r

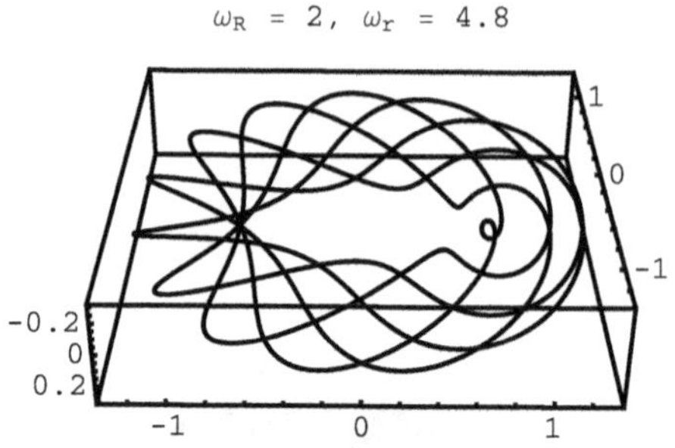

Fig. 3.29 Trajectory for ω_R= 2 and different ω_r

chaos cannot emerge in 2D. Period doubling with subsequent quasi-periodicity is a universal phenomenon which can be observed in a wide range of dynamic systems. The only condition that must be met is that a suitable number of frequencies exist in the system, while the physical structure of the system does not matter. When studying the phenomenon of period doubling, the system can be looked upon as pair of systems where one system is superior to (affects - controls) the other system. This is also referred to as oscillator locking (coupling), specifically frequency locking, phase locking or mode locking [16], which are different names for the same phenomenon. The extent of locking is given by the ω_R and ω_r, or more generally by ω_1 and ω_2 in 3.22, frequency ratio, and is denoted w, from the term winding number (also called rotation number).

$$w = \frac{\omega_2}{\omega_1} \tag{3.22}$$

If w is determined by a rational ratio, then the wandering trajectory only forms a finite set of points on Poincare section, and vice versa. It is noteworthy that if the winding number w is plotted in dependence on a suitably chosen system parameter, a fractal called "devil's staircase" [16], [6] appears. Devil's staircase for the "circular sine" (eq. 3.23) discrete dynamic system is shown in Fig. 3.30. Meaning of ϕ in eq. (3.23) is such that it is based on general description $\phi_{n+1} = f(\phi_n)$ in which $f(\phi)$ is periodic in angle ϕ, see [16], page 263-265. This staircase is a monotonically increasing curve whose horizontal segments correspond to the winding number (calculated as the ω_R and ω_r frequency ratio) at which frequency locking takes place.

$$\phi_{n+1} = \left\lceil -\frac{K\sin(2\pi\phi_n)}{2\pi} + \phi_n + \Omega \right\rceil \tag{3.23}$$

Period doubling can also be observed in systems whose mathematical model does not directly include any frequency (which does not imply that such a model cannot be set up for the system). Typical examples include the above logistic equation, as demonstrated in Fig. 3.31.

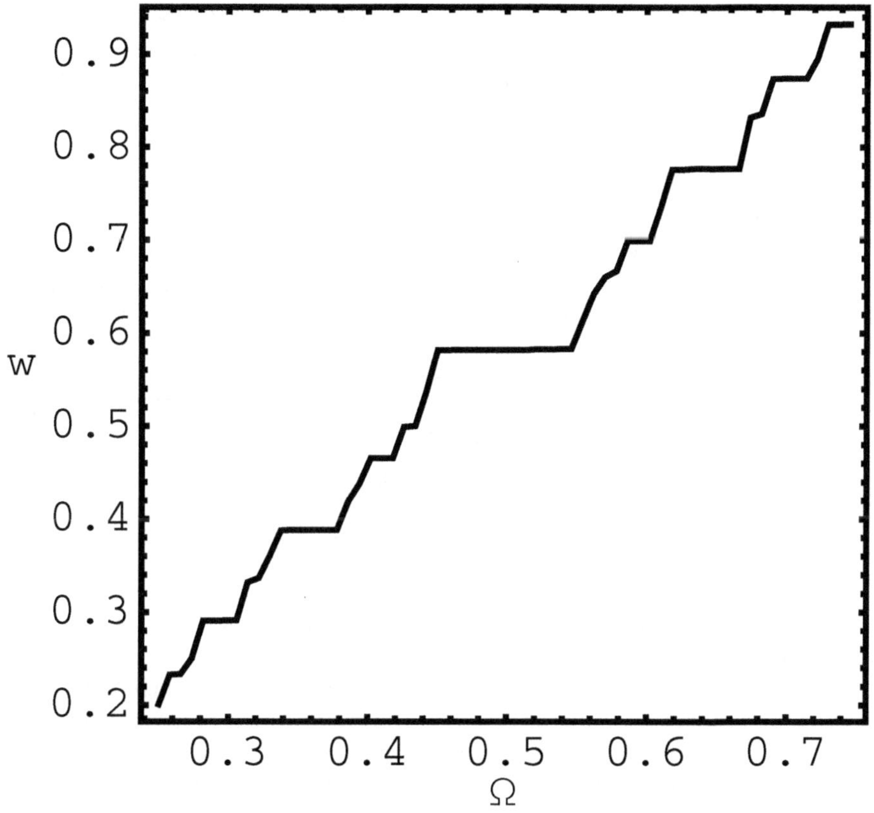

Fig. 3.30 Devil's staircase for the "circular sine" discrete dynamic system with $K = 1.2$ and $\Theta = 0.3$ and $\Omega \in [0.25, 0.75]$

3.4.2 Intermittence

Intermittence is a next pathway to chaos. During this transition to chaos, irregularly appearing regions of chaotic behavior whose length and frequency of occurrence depend on the appropriate system control parameters can be observed in the time development of the system. As the parameters are gradually changed, the chaotic segments can be more and more frequent and ultimately become the only observable behavior of the system (or vice versa). Once again, the behavior of the logistic equation (3.11) can be used to demonstrate intermittence (Fig. 3.14). Intermittence, of both types in which it is usually classed, is seen in both graphs. First type intermittence is a phenomenon where deterministic (periodic) behaviour alternates with chaotic behavior (Fig. 3.14). Second type intermittence is characterized by changes in behavior between chaos and quasi-periodicity (Fig.3.13). It is noteworthy that in both cases, an object whose apexes fill an imaginary triangle appears in the development roughly at the $30th$ iteration. This object has its name and is amply used

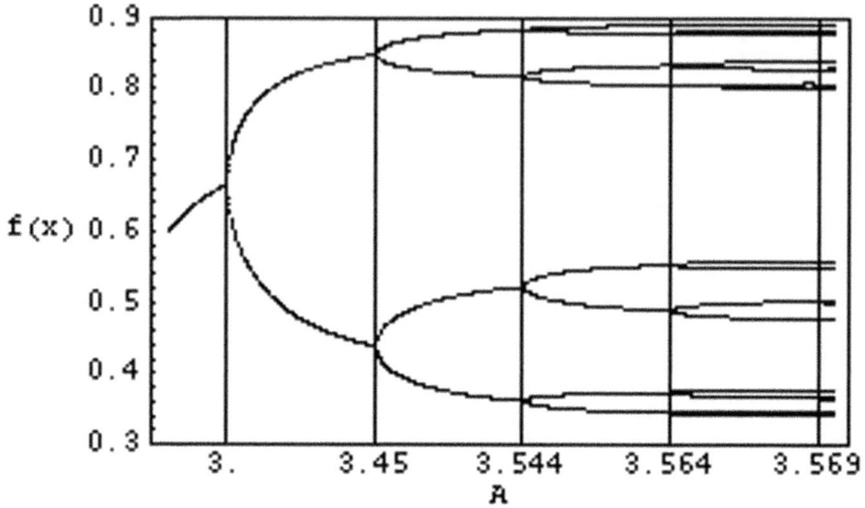

Fig. 3.31 Period doubling in the logistic equation

in stock exchange speculations to predict the near-future behavior of stocks. One of the technical indicators, its name is Triangle.

Generally speaking, intermittence is based on the existence of singular points in the dynamic system's state space. The abrupt dramatic change in the system behavior is due to the fact that some singular points vanish when the control parameter is changed slightly and are not replaced by other singular points. As some singular points vanish, the remaining singular points and their attractivity basins undergo overall position reconfiguration, and as a consequence, a trajectory which was periodic becomes chaotic and vice versa. The reverse phenomenon is also feasible of course, singular points can "be formed", whereupon the state space is reconfigured and the system behavior changes.

The dependence of the existence of singular points on an external control parameter can be well demonstrated on iterative mappings, e.g. on the logistic equation. Fig. 3.32 shows the logistic equation in five-fold iteration for different values of the control parameter A. If $A = 3.74$, this "system" includes some singular points of the sink type, by which trajectories are attracted, and some source type points, by which trajectories are repulsed. In the steady state the behavior can then be deterministic. If the A-values start to change towards 3.72, singular points vanish (no point of intersection with the logistic equation curve with a slope of 45° exists). If the system development reaches this area, it starts to exhibit deterministic behavior, because it cannot do otherwise in the limited space between the slope and logistic equation curve (see Fig. 3.34). Since the intersection of the curve and 45° straight line emerges or vanishes here, this phenomenon is called tangent bifurcation or also saddle-node bifurcation.

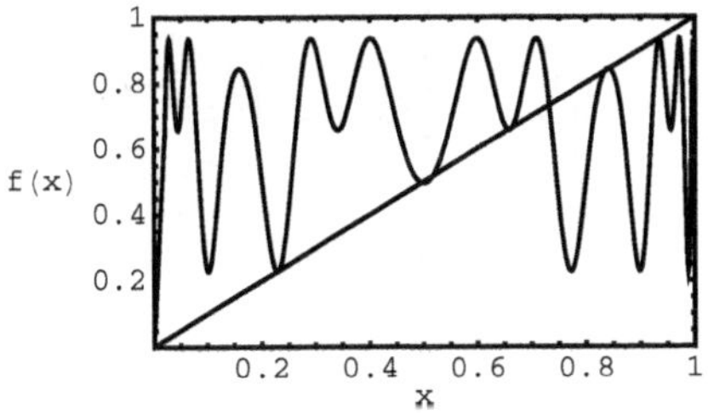

Fig. 3.32 Logistic equation intermittence for A = 3.7375

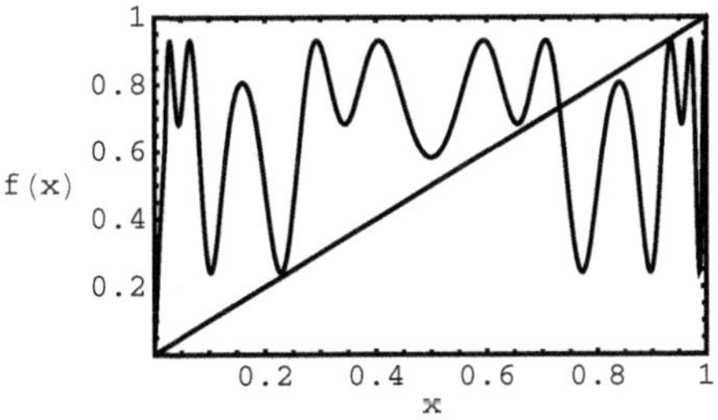

Fig. 3.33 and 3.7427

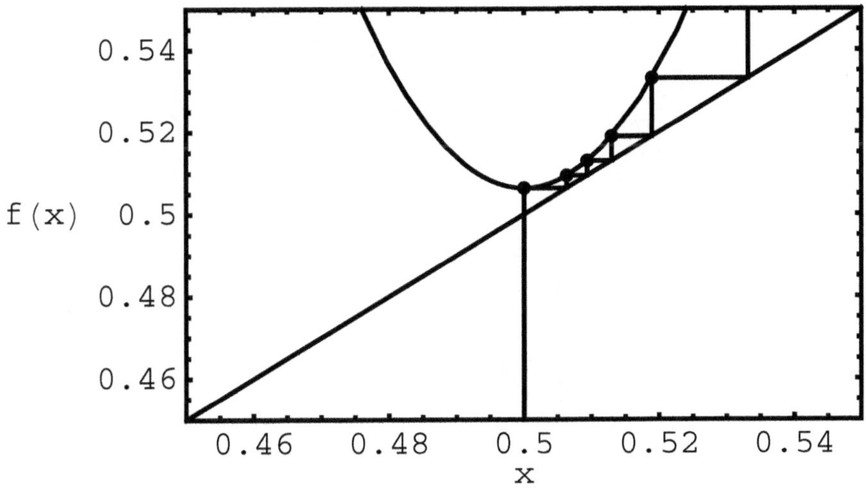

Fig. 3.34 Occurrence of intermittence for the logistic equation with A = 3.7375, detail from Fig. 3.33

In this case the logistic equation tends to chaos (Fig. 3.36). If the parameter varied from 3.72 to 3.74, deterministic sequences would be more and more frequent in the chaotic behavior and ultimately the behavior would be purely deterministic. So far it was tacitly assumed that the intermittence was induced by purely deterministic *A*-parameter setting. In the real world, however, virtually everything is affected by noise, which can superpose control signals as well as other signals. This implies that noise can also affect the *A*-parameter, which otherwise can also be constant, approaching tangent bifurcation. It will be clear that with a suitable noise intensity and nature, the *A*-parameter can take values at which singular points vanish, and furthermore, that due to the properties of noise, this value will be transient rather than permanent and that the *A*-parameter will eventually return to its initial value. The frequency of occurrence of intermittence so induced can be quite different from that obtained by deterministic "excitation". The effect of noise on the existence or

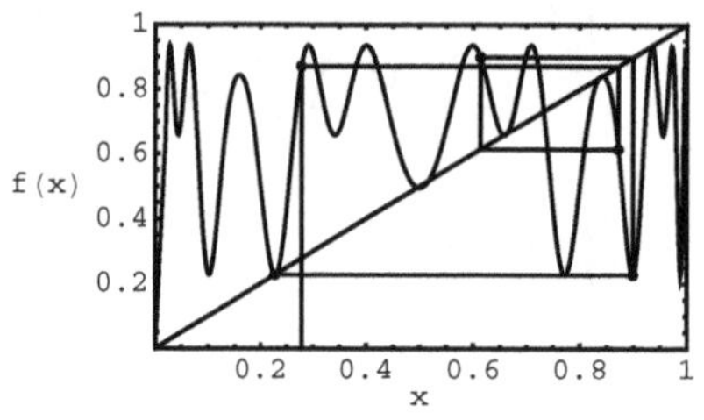

Fig. 3.35 Behaviour of the logistic equation with A = 3.74

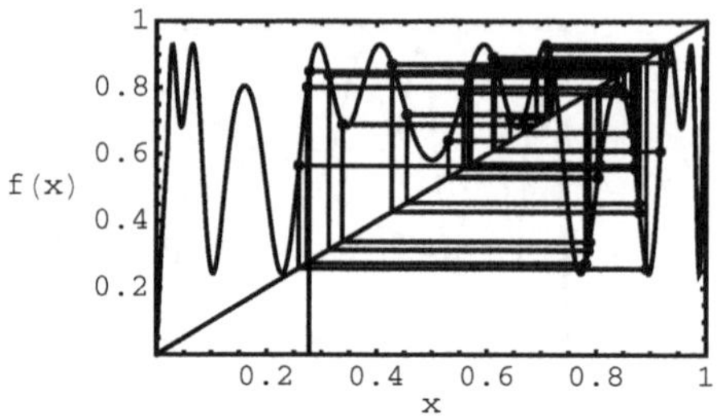

Fig. 3.36 and 3.72, 60 iterations

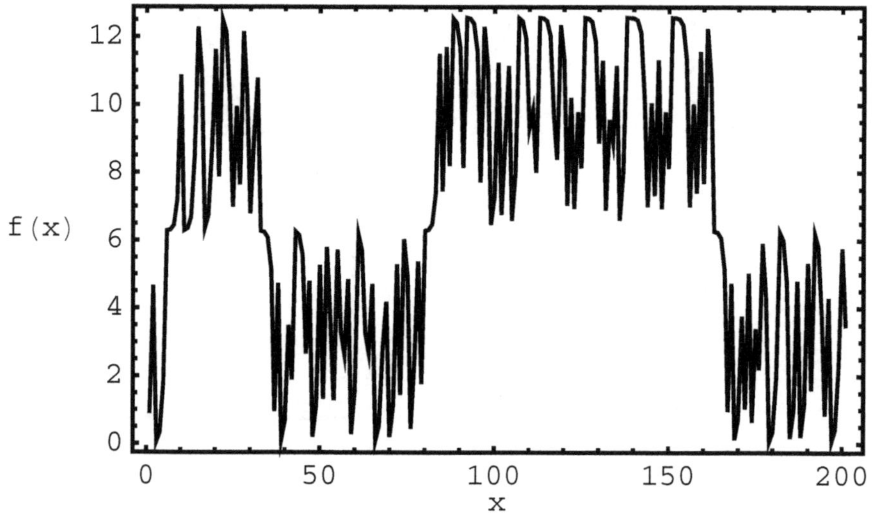

Fig. 3.37 Behaviour of the relation $x_n + 1 = Asin(x_n) + x_n$ at A = 4.61 and $x_0 = 0.91$

non-existence of intermittence is a problem which is too complex to be discussed in this publication.

3.4.3 *Chaotic Transients*

Chaotic transients are a typical phenomenon accompanying models that are based on differential equations. The state space of such a system-model generally includes n singular points lying on the intersections of separatrices dividing the state space into regions with different types of behavior. A state space can generally have N dimensions and so a separatrix may not be a mere curve; instead, it constitutes a smooth, differently wavy plane referred to as manifold. Such manifolds can get, without any collision, as far as the state space boundary or else they can intersect.

The source of deterministic chaos in dynamic systems modeled by differential equations is the manner in which the manifolds intersect and thus separate the state space regions from one another. Two types of manifold intersections exist: homoclinic and heteroclinic. The principle can be best explained on manifolds in 3D with a Poincare plane. An artificial example of manifolds is shown in Fig. 3.38 and 3.39, exhibiting also their Poincare plane. Manifolds are classed into stable manifolds (in-set) and unstable manifolds (out-set). If a state trajectory starts its path on a stable manifold, it is attracted directly into a singular point, whereas repulsion occurs if the manifold is unstable. This also holds for reasonably near manifold neighborhoods. Generally, the nearer a state trajectory is to a manifold, the more its behavior will be affected by that manifold.

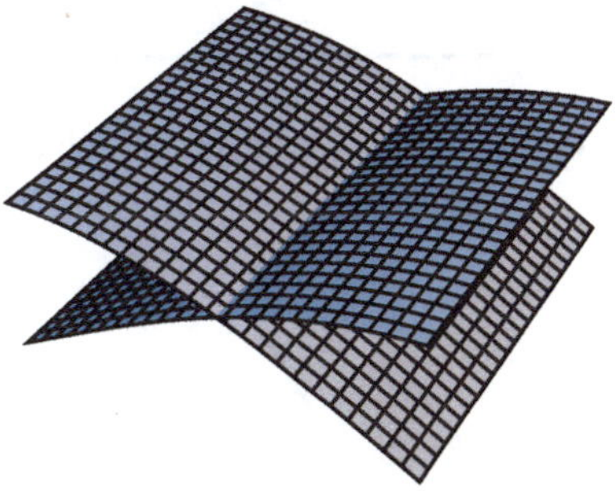

Fig. 3.38 Manifolds in 3D (left) and their Poincare plane

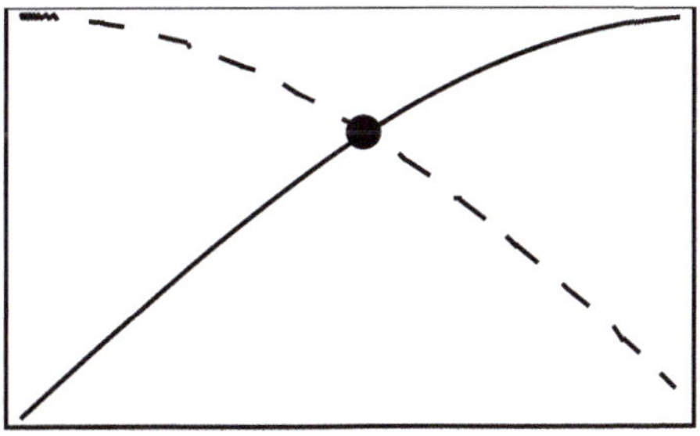

Fig. 3.39 The dashed area represents an unstable manifold

Homoclinic intersection is an intersection of manifolds originating from the same singular point. This is demonstrated in Fig. 3.40 and 3.41 showing a special case of intersection of manifolds which is more interlinking than intersection. Homoclinic intersections, demonstrated in Fig. 3.40 and 3.41, are less common. A classic example is the intersection of manifolds shown in Fig. 3.42, exhibiting what will happen in such case. If two manifolds intersect in this manner, the intersecting manifold will create a set of intersections of which there are infinitely many and whose "density" increases towards the singular point. It will be clear that 3D representation of such intersection creates a much more complex structure - the old state space is broken down and trapped in any regions from which the trajectory cannot escape and an attractor emerges.

The appearance of an attractor is thus determined by the formation of a kind of "pocket" whose boundaries are formed by two manifolds of opposite nature. As explained above, manifolds affect the behavior in their neighborhood. If a trajectory starts its path anywhere within such a region, then it is necessarily attracted by one of the manifolds and repulsed by the other manifold. The moment the trajectory is attracted near to a set of points that appear as a singular point on the Poincare

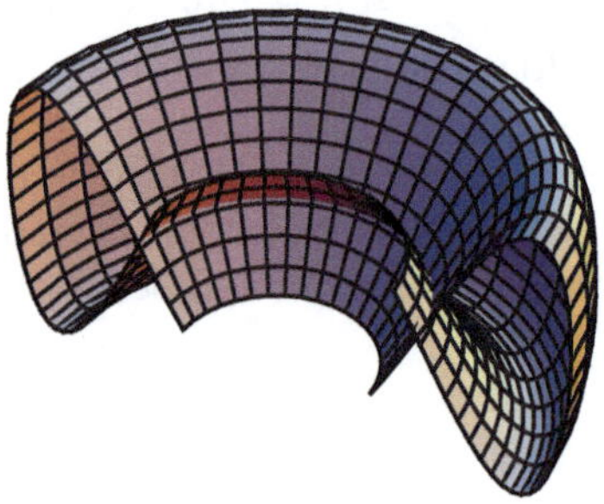

Fig. 3.40 Homoclinic intersection in 3D representation...

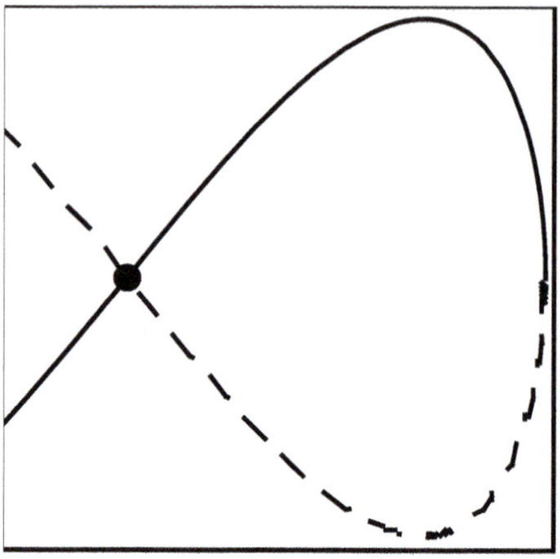

Fig. 3.41 and its representation on Poincare plane

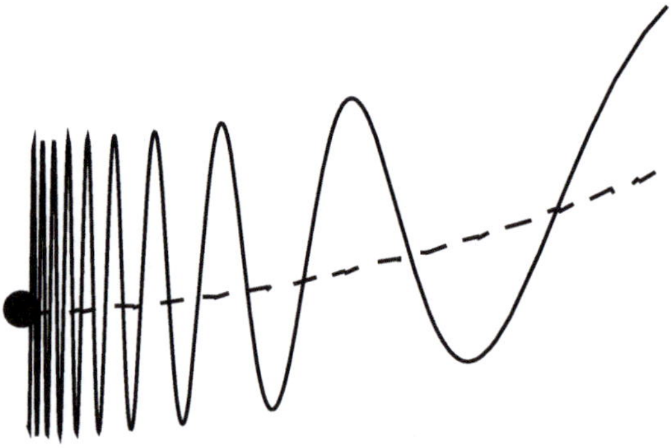

Fig. 3.42 Homoclinic intersection of manifolds - a more common case

plane and forms a homoclinic trajectory in 3D, it is hurled off due to the presence of an unstable manifold. Such a trajectory moves constantly on trajectories which do not repeat. An artificial case of such development is shown in Fig. 3.43 and 3.44. Singular points (and manifolds rising from them) are saddle type, and the chance that the trajectory will starts its path precisely in the position of a homoclinic point set is nearly certainly nil. This is contributed to by the ubiquitous noise, inaccuracy of measurement, etc., including quantum uncertainty which is actually transformed as far as to the macroworld.

The Lorenz attractor is a clear example of the emergence of chaos based on intersecting manifolds, we recommend to read for more [13].

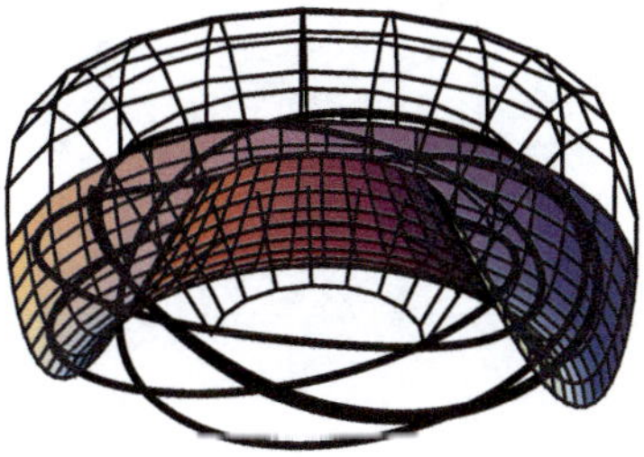

Fig. 3.43 Trajectory in a region bounded by manifolds and by their intersections

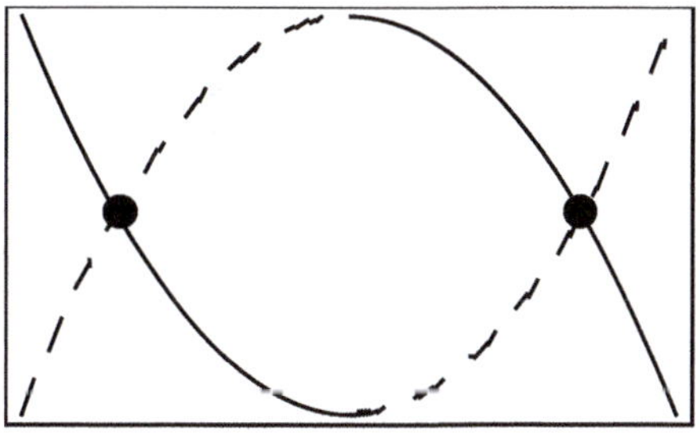

Fig. 3.44 Cut through region from Fig. 3.43

3.4.4 Crises

Crises are a phenomenon where chaotic behavior usually changes dramatically. Such changes can be of various nature. Deterministic behavior can vanish altogether, to be replaced by pure chaos, or conversely, the magnitude of the attractor changes, as does the size of its basin of attraction. Such changes have a common denominator, namely, the quality and configuration of singular points in the state space, including their change in dependence on the control parameters. Crises are categorized into 2 classes: boundary crises and interior crises. Boundary crises occur on imaginary boundaries of attractors, which are determined by a suitable control parameter value. For the logistic equation, the boundary is $A = 4$. Beyond this boundary the chaotic attractor, represented by "snowing" in the bifurcation diagram, vanishes. This is due to the divergence of the trajectory away from the region in which the chaotic attractor was initially present. For the logistic equation with $A = 4$ and $x_0 \in [0, 1]$ the trajectory is confined in the chaotic attractor, because any calculated value of it again lies within the interval of [0, 1], and since it serves as the logistic equation argument in the next iteration, it is clear that such a number would also belong to that interval. However, if A is changed, say, to $A = 4.1$, then levels in excess 1 can be attained in the area of the apex of the parabola generated by the logistic equation. The time needed to attain that area is relatively short. If a trajectory "strays" into that area, it starts running away from the area where the chaotic attractor was initially present at $A = 4$. In other words, if the value is changed to $A > 4$, a "creep-hole" in the chaotic attractor opens up, enabling the trajectory to escape. Such change can be caused by deterministic influences (control, ...) or by random effects (noise). Boundary crisis is demonstrated for the logistic equation in the form of the WEB diagram in Fig. 3.45. When the number of iterations exceeds 11, the trajectory reaches the apex of the parabola and escapes to infinity in this case. Something similar can also be observed on the "circular sine" bifurcation diagram (Fig. 3.46), where chaos vanishes abruptly at $K = 3.8$ and purely deterministic behavior establishes in a different region of the state space (up to a value of approximately 4.27). The same effect can be observed in Henon bifurcation diagram at $C = 1.8$.

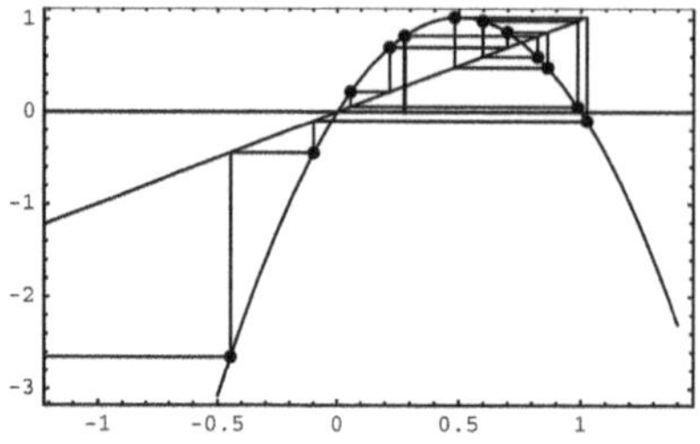

Fig. 3.45 WEB diagram of the logistic equation for A = 4.1. Example of boundary crisis

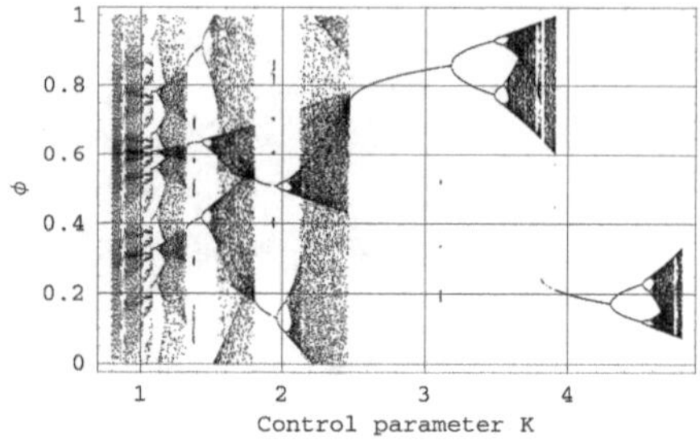

Fig. 3.46 "Circular sine" bifurcation diagram

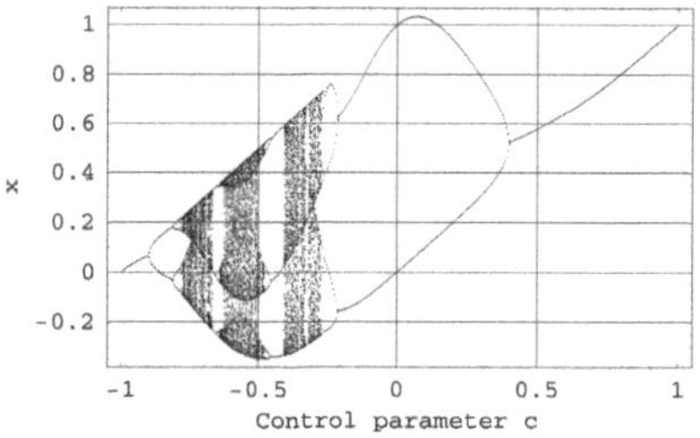

Fig. 3.47 Bifurcation diagram of "Gausian map" $x_{n+1} = e^{-bx_n^2} + c$

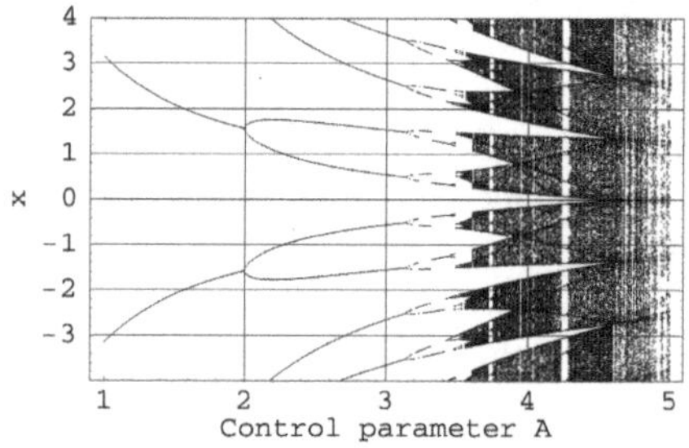

Fig. 3.48 Bifurcation diagram of $x_{n+1} = A \sin x_n + x_n$

Interior crises are changes in behavior during which the chaotic attractor undergoes dramatic changes but does not vanish. The bifurcation diagram of the Gaussian map (Fig. 3.47) is a graphic example showing how the chaotic attractor structure changes in dependence on the control parameter c. The expansion of the chaotic attractor is usually due to collision of a trajectory with a source type or unstable limiting cycle type singular point. In such case the trajectory is "hurled off" to regions where it normally would not get or would get in an extremely long time. Like in intermittences, noise plays an important role in crises.

Due to crises, attractors can be linked up into a single one, or conversely, can decompose into several attractors [25]. Fig. 3.48 shows the behavior of equation $x_{n+1} = A \sin x_n + x_n$ in dependence on A-parameter and different initial conditions. Observing what happens when this parameter is increased, one finds that all attractors are combined into a single one starting from $A \approx 4.603$. Before this level, the trajectory develops in one of the attractors shown only, in dependence on A and on the starting value. The trajectories only merge at $A > 4.603$. Decrease in A is accompanied by the reverse effect - decomposition of the bound attractor into a number of disjoint attractors at $A < 4.603$.

3.5 Selected Examples

Deterministic chaos can be observed in many dynamic systems of different nature. Included are electronic systems (Chua's circuit, circuits with diodes, circuits with digital filters,...), mechanical systems (double pendulum, magnetic pendulum, billiard problem, ...), biological systems (logistic equation, evolutionary dynamics systems, ...), physical systems (physical plasma, the three-body problem, hydrodynamics, ...) and others. Some can be simply materialized on the bench, whereas others can only be observed within a natural process. The objective of this chapter is to demonstrate deterministic chaos on selected examples, specifically from the domains of mechanics, electronics, biology, meteorology and numbers theory.

3.5.1 Mechanical System – Billiard

There are countless examples of deterministic chaos in classical mechanics. A very didactic example is the experiment with small balls falling through a system of bars fixed in a wall. This problem concerns the reflection of two bodies with curved surfaces - balls in this case - or of a radius (beam) from a spherical surface. Taking into account the curvature of the surfaces it will be clear that even the slightest change in the initial conditions will bring about differences in the repeated trajectory. Sensitivity to initial conditions in the billiard problem can be clearly seen on the simulation of falling of a ball through a system of bars with 20 rows (Fig. 3.49). Here the simulation was repeated four times with differences in the initial conditions (x-axis) of 0, 0.00001, 0.00002, and 0.00003, respectively. The difference in the initial conditions

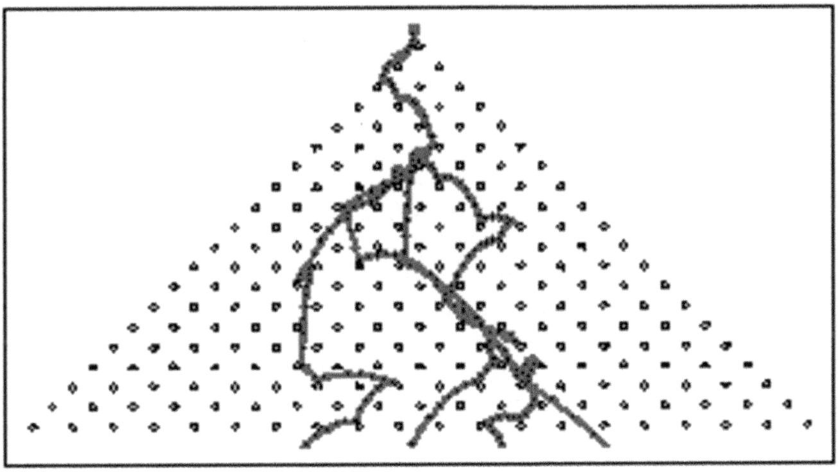

Fig. 3.49 Variant of trajectories in the billiard problem

was thus in the order 10^{-5}. Despite the small number of bar rows (exactly 20) the trajectories are apparently different starting from the seventh row.

The billiard problem can be demonstrated not only on classic balls but also on many other types of "billiard", which are basically curved surfaces forming together closed objects in which the divergence of colinear radii can be well observed. Another example is at Fig. 3.50. It is clearly visible that trajectories diverge after a few iterations. Starting positions were $x_1 = 0.936578, y_1 = 1.31709$ and $x_2 = 0.936578, y_2 = 1.3063$.

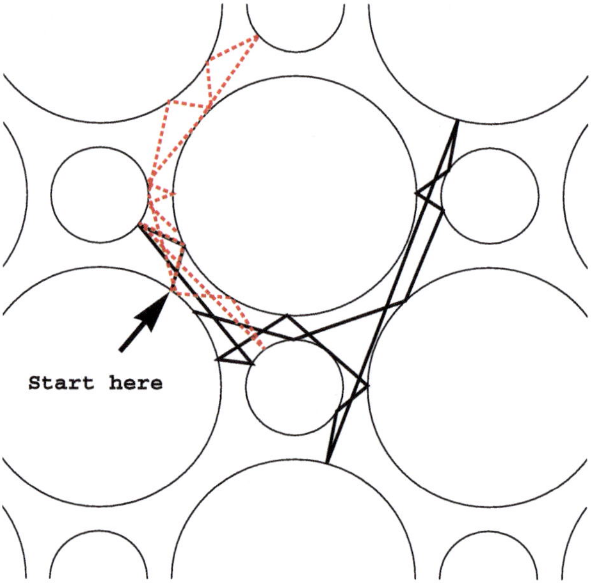

Fig. 3.50 Another variant of the billiard - trajectories diverge after a few iterations. Starting positions were $x_1 = 0.936578, y_1 = 1.31709$ and $x_2 = 0.936578, y_2 = 1.3063$

3.5.2 *Mechanical System – Duffing's Equation*

Duffing's equation describes Duffing's oscillator, designed in 1918. Duffing's oscillator consists of a metallic strip with an ac electromagnet located near the centre of the strip. The electromagnetic field which is formed by the magnet displaces the strip sideways. Duffing's oscillator is modeled by (3.24) which, however, describes the ideal case where no energy is lost. In *a* real Duffing's oscillator, energy losses must be taken into account, as in eq. (3.25). This equation transforms into eq. (3.26) for the external excitation setup.

$$\ddot{q}(t) - aq(t) + bq(t)^3 = 0 \tag{3.24}$$

$$\ddot{q}(t) - aq(t) + bq(t)^3 + cq'(t) = 0 \tag{3.25}$$

$$\ddot{q}(t) - aq(t) + bq(t)^3 = f_0 \cos(t\omega_d) \tag{3.26}$$

Equation (3.24) is a starting point for understanding the origin of chaos in this system. The model contains 3 components: acceleration $\ddot{q}(t)$, linear force effect $aq(t)$, and nonlinear force effect $bq(t)^3$. Various types of the steady state can be achieved in the oscillator by varying parameters a and b. The states can be determined by means of the first integral (3.27) of the system, describing total energy of the oscil lator. The total energy consists of 2 components: kinetic energy and potential energy, described by the last term and by the remaining terms in (3.28), respectively.

$$\int \dot{q}(t)\left(-aq(t) + bq(t)^3 + \ddot{q}(t)\right)\, dt \tag{3.27}$$

$$-\frac{1}{2}aq(t)^2 + \frac{1}{4}bq(t)^4 + \frac{1}{2}\dot{q}(t)^2 \tag{3.28}$$

The first two terms in (3.28) can be used to set up the potential (Fig. 3.51 and 3.52) describing its dependence on parameter a. If $a > 0$, the oscillator has three equilibrium states - two stable states (minima) and one unstable state (maximum between the two minima). If $a < 0$, the oscillator possesses one stable state only. The minima and maxima in the potential shown represent states to which the oscillator behavior is attracted or from which it is repulsed. If the entire equation (3.28) is considered, the basin of attraction of Duffing's oscillator can be depicted as shown in Fig. 3.53. In the picture, the variables are interchanged according to scheme . Figs 3.53 and 3.54 display both the basins of attraction and the energy equipotentials - points in which the oscillator possesses the same energy.

The plots in Fig. 3.51 - 3.54 differ in that only the components of the potential energy of the first integral were used in Fig. 3.51. The components contained $q(t)$ only and the graph was generated as the $q(t)$ vs a plot. In Fig. 3.53 and 3.54, kinetic

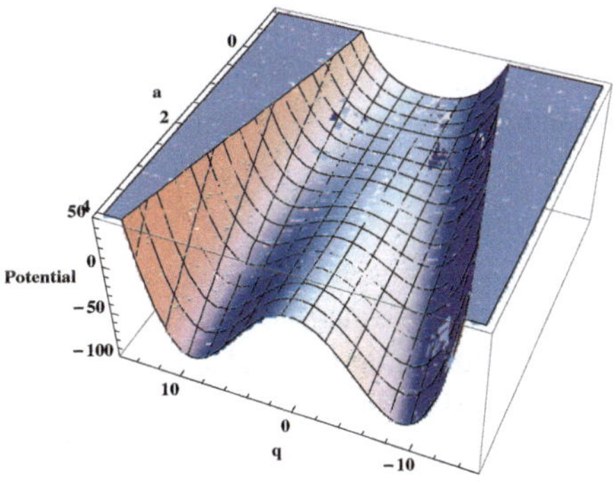

Fig. 3.51 Duffing's equation potential at b = 0.05...

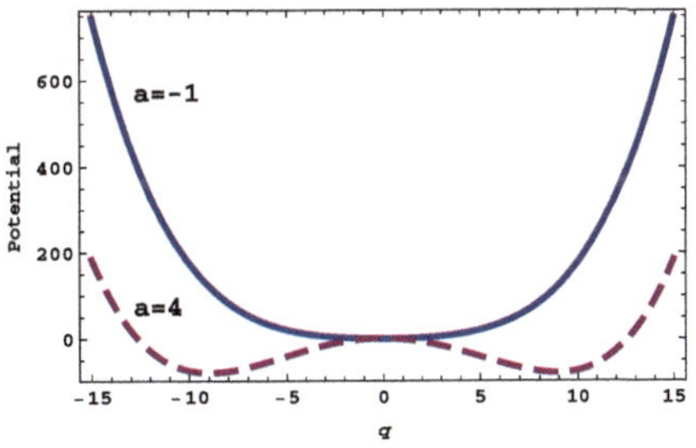

Fig. 3.52 ... and 2D view.

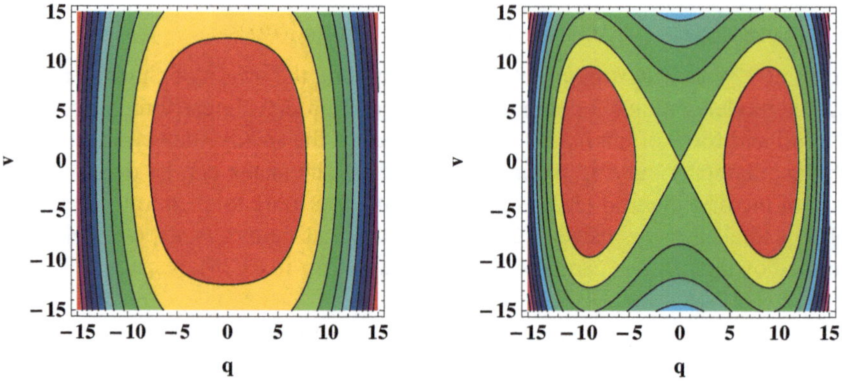

Fig. 3.53 Duffing's equation basins of attraction, a = -1 (left) and a = 4 (right); b = 0.05

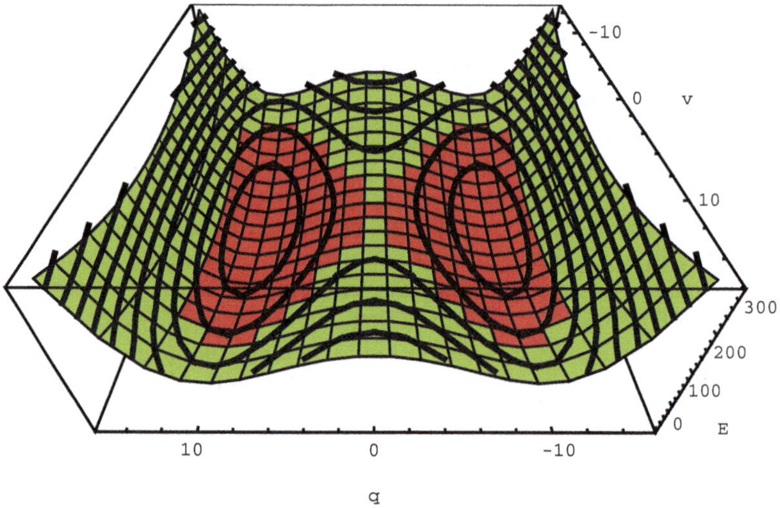

Fig. 3.54 Duffing's equation basins of attraction and equipotentials in 3D for a = 4 and b = 0.05

energy was also included, enabling non-parametric representation to be applied to the system total energy. The potential in Fig. 3.51 can be imagined as a wire with a ball on it. If the ball is positioned at the local maximum, any impulse can displace the ball from this position. The ball then travels further to some of the sinks, and since friction is not considered in this model, the ball will oscillate about the local minimum infinitely long. If the ball were released from a higher-energy position (level), it would travel cyclically from one local minimum to another through a local energy maximum. If energy dissipation is considered, (3.24) takes the form of (3.25) where the term $c\dot{q}(t)$ represents dissipation. In this modification the ball motion on the wire will slow down (energy is irreversibly lost) and ultimately stop. Behavior of this type is better represented in terms of the state space and state trajectories. For this purpose, eq. (3.24) is modified to the form (3.29).

$$\begin{aligned} &\dot{p}(t) - aq(t) + bq(t)^3 = 0 \\ &\dot{q}(t) = p(t) \end{aligned} \tag{3.29}$$

In this manner the n_{th} order differential equation is transformed into n first-order equations. The corresponding variables then represent state variables. This system of differential equations can serve to simply draw a "state portrait" (Fig. 3.55) in which the arrows show the direction of the state trajectory (corresponding, in fact, to the equipotential lines in Fig. 3.53 and 3.54). Different types of behaviour of Duffing's equation with dissipation can be obtained by solving (3.25), in dependence on the extent of dissipation and on initial energy.

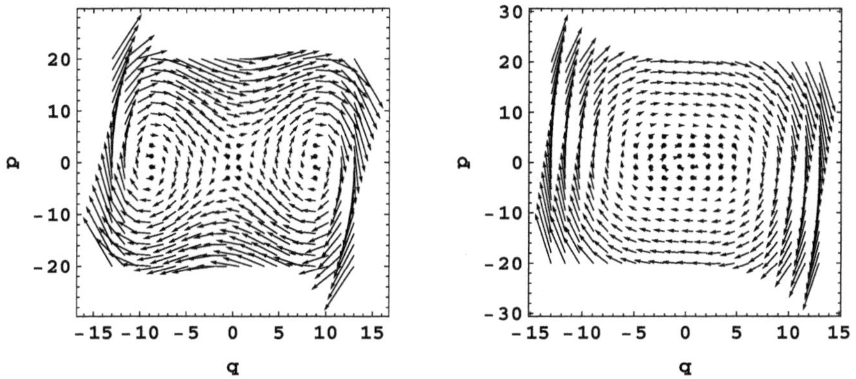

Fig. 3.55 State portrait of Duffing's equation, a = -1 (left) and a = 4 (right); b = 0.05

It is clear from Fig. 3.56 and 3.57 that the oscillator's ultimate steady state depends both on initial energy and on initial position, in other words, on initial conditions. The trajectories of the system behaviour are attracted to one of the basins of the system's state space (Fig. 3.58). Fig. 3.59 - 3.61 shows both the state portrait and the system behavior of eq. (3.25). The gradual energy loss causes the trajectory to "sink" slowly to one of the attractors. In this manner the state space is divided into basins in which the state trajectory gets into one or another attractor lobe. Geometric appearance of such basins can be very complex. See for example system Fig. 3.62 and 3.63 where are depicted basins of attraction with clear fractal border. Black area is the domain of attraction, i.e. if arbitrary trajectory start in it, then will end in white attractor depicted inside black area, otherwise it goes out of the basin. Other color layers represent trajectory "speed" of escaping.

Chaotic behavior of Duffing's oscillator by can be obtained by choosing suitable excitation conditions. This is described by (3.26). The right-hand side excitation

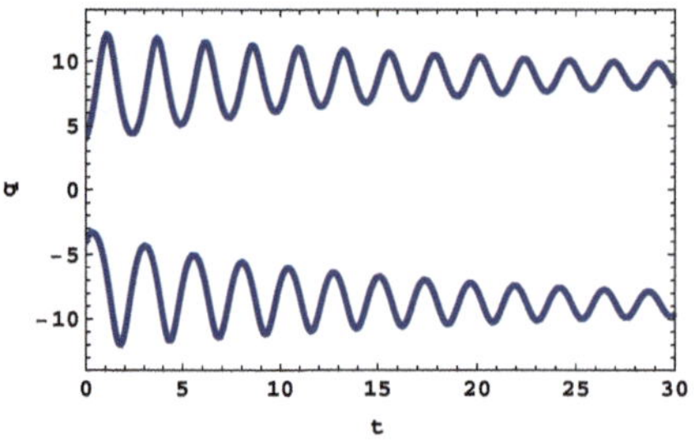

Fig. 3.56 Behavior of Duffing's equations with dissipation...

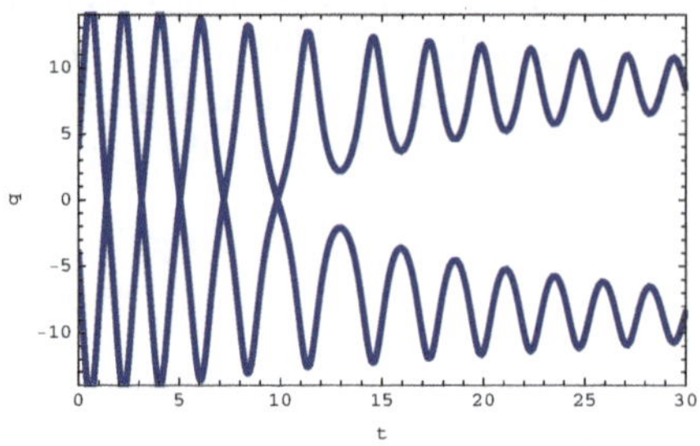

Fig. 3.57 ... another level of dissipation.

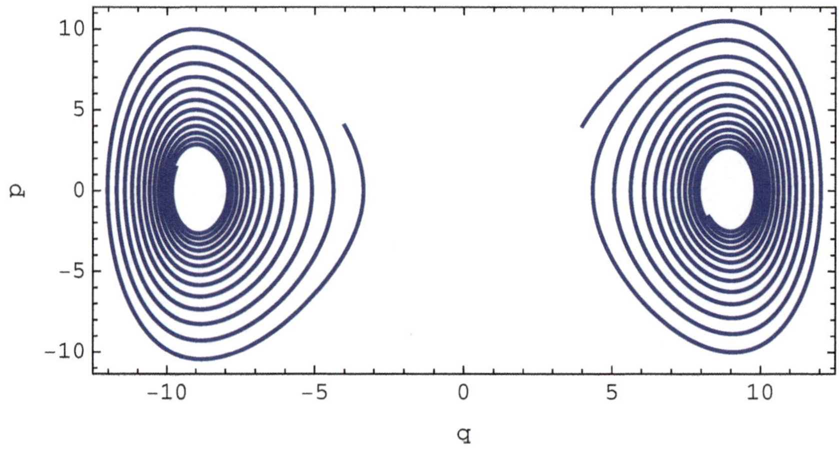

Fig. 3.58 State trajectories of Duffing's equation with dissipation

term consists of the term $f_0 \cos(t\omega_d)$. Both deterministic and chaotic behavior can be observed for Duffing's equation for certain values of the two terms. A typical example of chaos is shown in Fig. 3.64 - 3.65. If the above setup of a ball on a wire is "transformed" into the setup of a ball rolling on a plane, then the appearance of chaos can be understood so that external excitation by the element $f_0 \cos(t\omega_d)$ provides sufficient energy not only to cover dissipation losses but also for chaotic motion of the ball.

3.5.3 Electronic System – Chua's Circuit, Circuit with a Diode

Electronic circuits are among the most popular systems used to demonstrate deterministic chaos. Their popularity stems from the fact that electronic circuits are easy

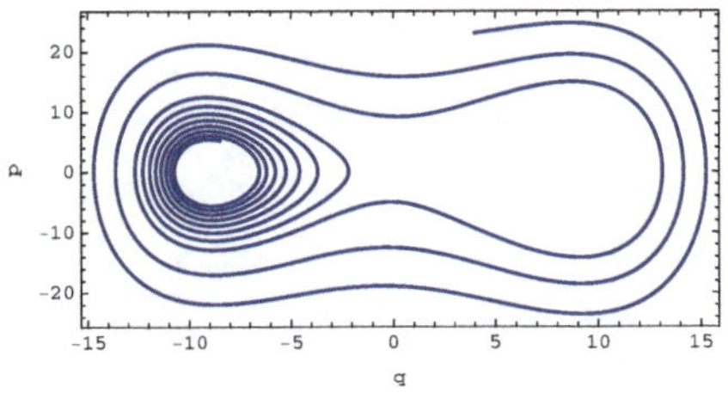

Fig. 3.59 State trajectories of Duffing's equation with dissipation.

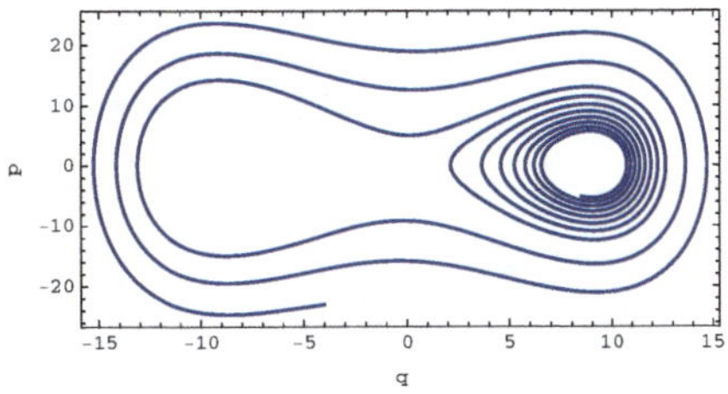

Fig. 3.60 ... another level of dissipation.

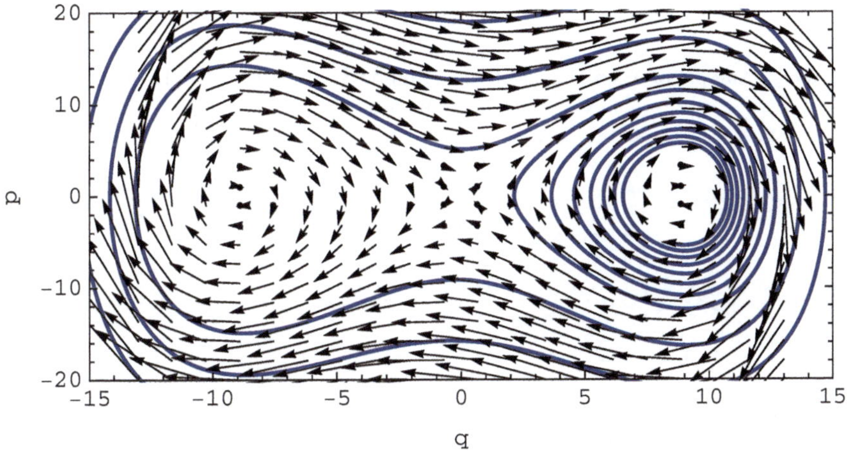

Fig. 3.61 State trajectories of Duffing's equation with dissipation

to set up and provide fast response to impulse. Typical representatives of electronic circuits with deterministic chaos include Chua's circuit, whose hardware design and behaviour are shown in Fig. 3.66-3.67 and Fig. 3.68-3.69, respectively. The core of Chua's circuit is a nonlinear resistor, eq. 3.32, sometimes called Chua's diode [33].

On Fig. 3.69 Chua's attractor visualized by the program Mathematica (left) and on the oscilloscope connected to its hardware implementation shown in Fig. 3.67 (left) Chua's circuit can be described mathematically by eq. (3.30), which can be used to simulate the behavior of the circuit:

$$
\begin{aligned}
C_1 v\dot{c}_1(t) &= G(vc_2(t) - vc_1(t)) - g(vc_1(t)) \\
C_2 v\dot{c}_2(t) &= G(vc_1(t) - vc_2(t)) + i_L(t) \\
L\dot{i}_L(t) &= -vc_2(t)
\end{aligned}
\tag{3.30}
$$

$$vc_1(0) = 0.15264,\ vc_2(0) = -0.02281,\ i_l(0) = 0.38127 \tag{3.31}$$

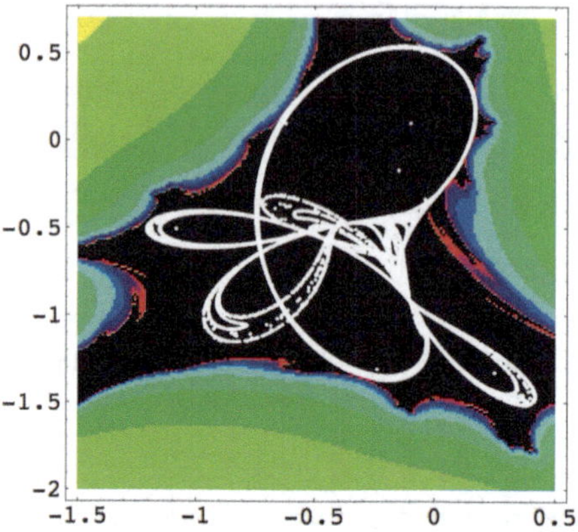

Fig. 3.62 Example of basin of attraction

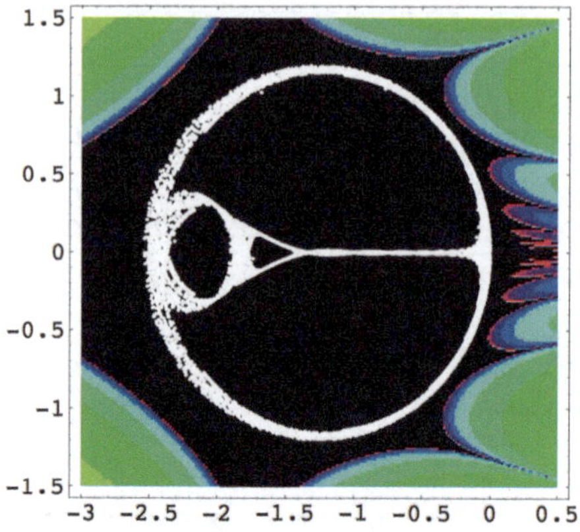

Fig. 3.63 Another example of basin of attraction

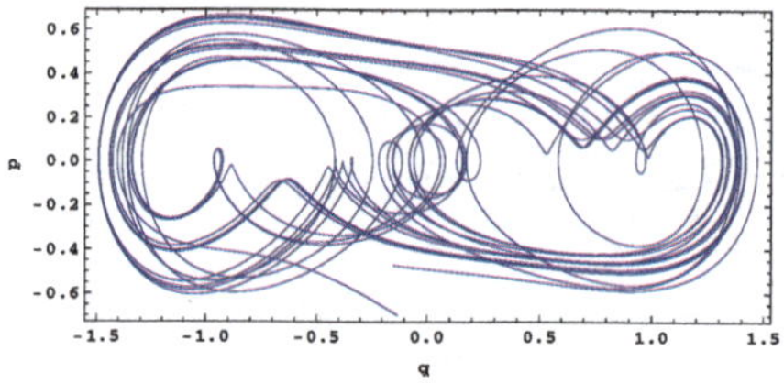

Fig. 3.64 Duffing's equation chaos for $f_0 = 0.29$

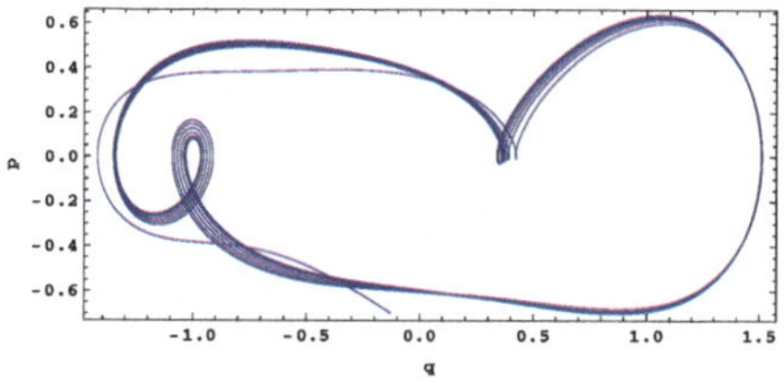

Fig. 3.65 ... and for $f_0 = 0.32$; $\omega_d = 1$.

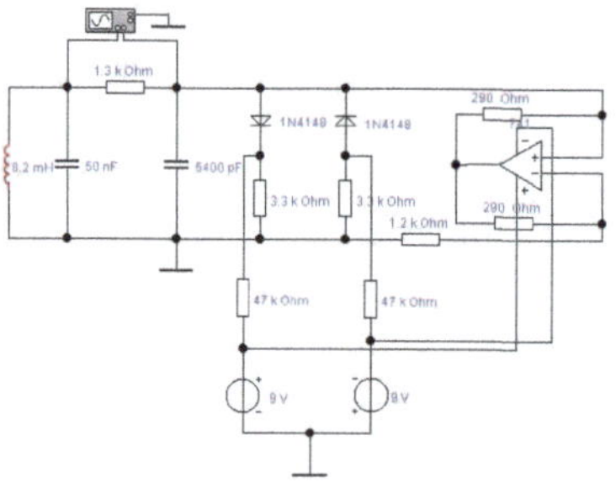

Fig. 3.66 Scheme of the Chua's circuit ...

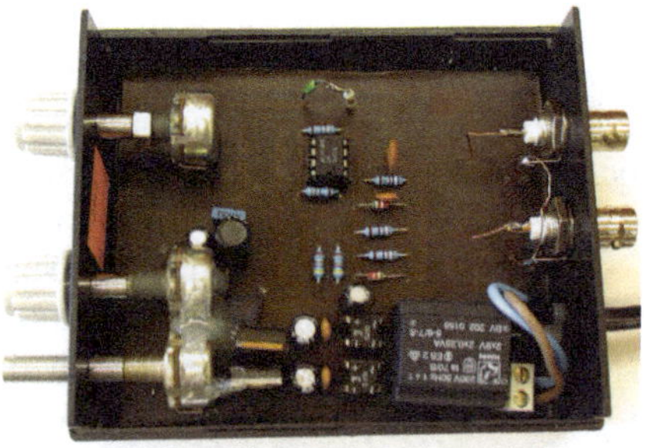

Fig. 3.67 ... and hardware design of Chua's circuit.

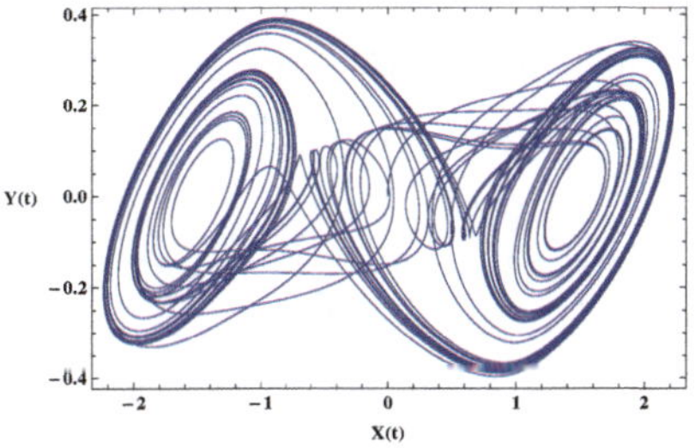

Fig. 3.68 Simulation of the Chua's circuit ...

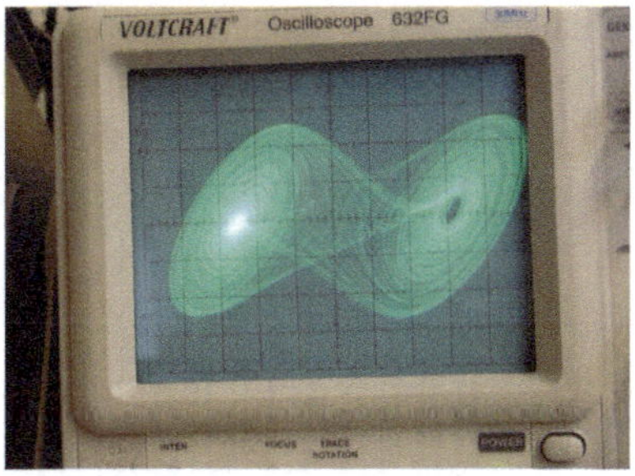

Fig. 3.69 ... and the real behavior.

where the nonlinear resistor $g(x)$ is represented by (3.32),

$$g(x) = m_0 x + \frac{m_1 - m_0}{2}(|x+b_1| - |x-b_1|) + \frac{m_2 - m_1}{2}(|x+b_2| - |x-b_2|) \quad (3.32)$$

If suitable initial conditions are set as described by (3.31), a chaotic attractor can be found in the system (Fig. 3.68).

A simple electronic circuit (Fig. 3.70) where an excitation source, resistor, coil and diode are connected in series can serve as a next example. The diode provides nonlinearity which is the cause of chaotic behavior in this circuit.

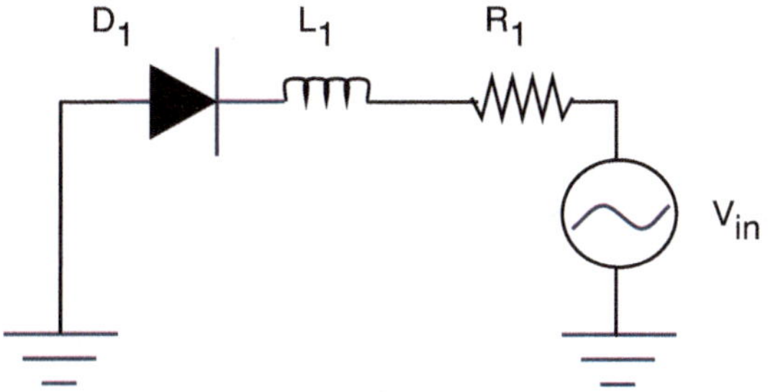

Fig. 3.70 Layout of the circuit with a diode

The mathematical model of this physical system consists of a system of equations and initial conditions (3.33) where the diode is modeled by means of a piecewise linear capacitance, namely:

$$\begin{aligned} \dot{q}(t) &= i(t) \\ L_1 \dot{i}(t) &= v\sin(2\pi f t) - \left(\tfrac{|q(t)|(C_2 - C_1)}{2C_2C_1} + \tfrac{|q(t)|(C_2 + C_1)}{2C_2C_1} + e_0\right) + i(t)(-R_1) \\ q(0) &= 0 \\ i(0) &= 0 \end{aligned} \quad (3.33)$$

Chaotic behavior can be observed when analyzing the dependence of charge q on control voltage V. Numerical simulations of this circuit are shown in Fig. 3.71 and 3.72, displaying the time development of the behavior of the circuit, and in Fig. 3.73 and 3.74 displaying the behavior of the dependence of current i on $q(t)$.

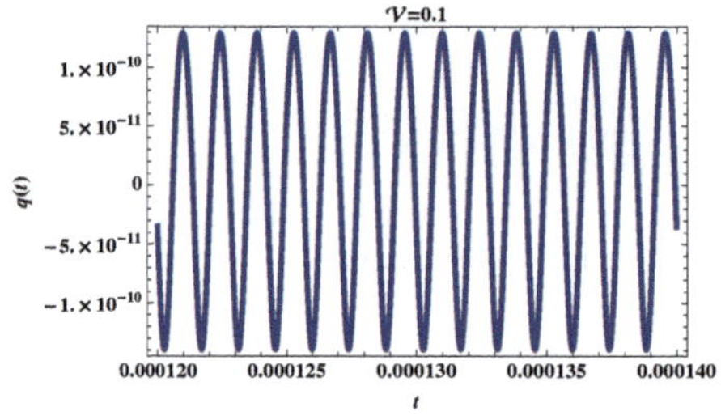

Fig. 3.71 Simulation of the diode circuit ...

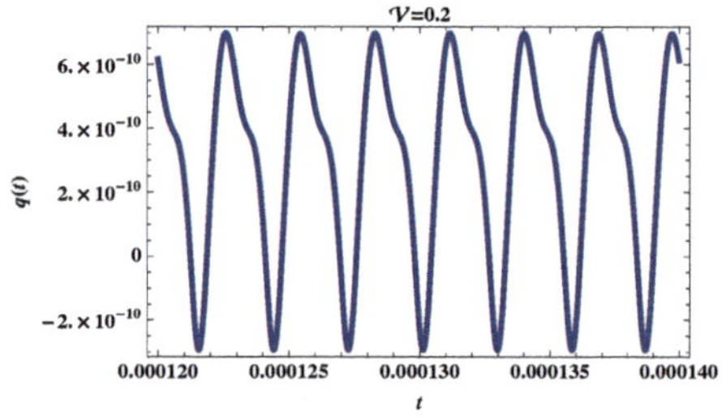

Fig. 3.72 ... for different values of v.

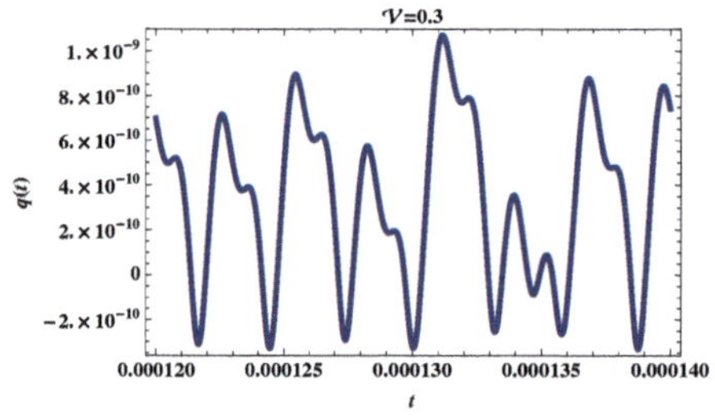

Fig. 3.73 Simulation of the diode circuit ...

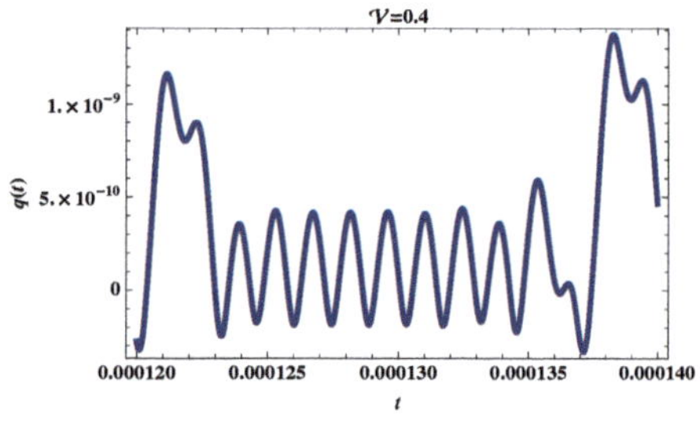

Fig. 3.74 ... for different values of v.

The bifurcation diagram of the circuit with a diode is shown in Fig. 3.79. The diagram clearly displays transition to chaotic behavior with increasing parameter V. From the structure of the bifurcation diagram one can not only see structure repetition (self-similarity) but also the fact that all three parts are visually very similar to bifurcation diagrams of the logistic equation (Fig. 3.81), which is just another evidence in a series of experimental evidences of universality of chaos as such.

3.5.4 Biological System – Logistic Equation

The logistic equation is the most typical example in the domain of biological systems. This equation models the evolution of dynamic co-evolutionary systems of the predator-prey type in which all the relevant behavior types are present. The logistic equation is modeled by relation eq. (3.34). An important element in this equation is the control parameter A, whose gradual change in the equation gives rise to behavior

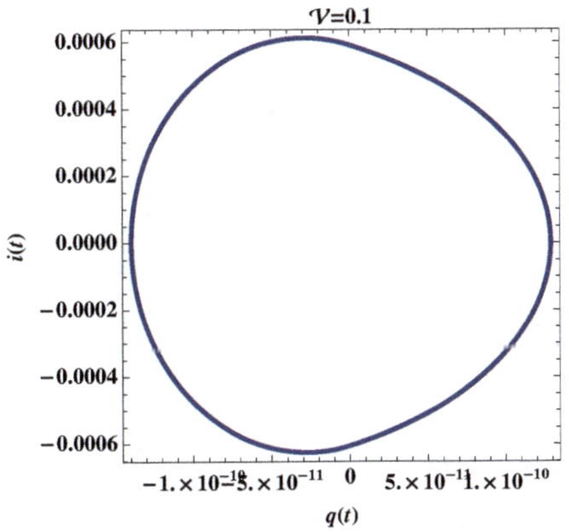

Fig. 3.75 Dependance of i on $q(t)$ ...

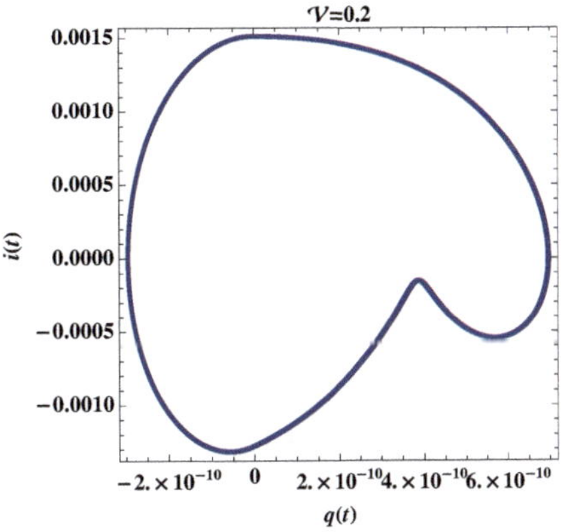

Fig. 3.76 ... for different values of ν.

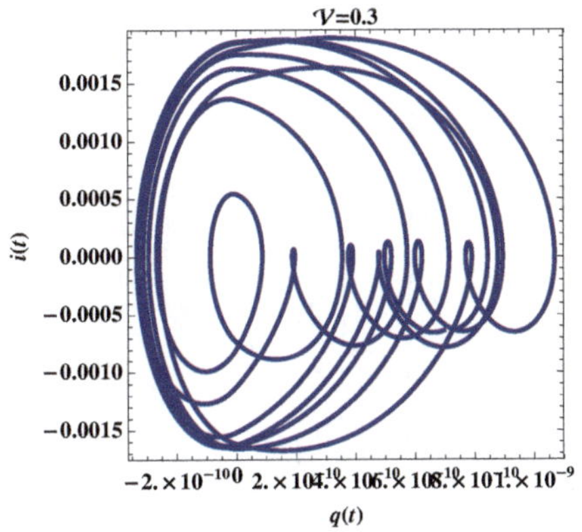

Fig. 3.77 Dependance of i on $q(t)$ ...

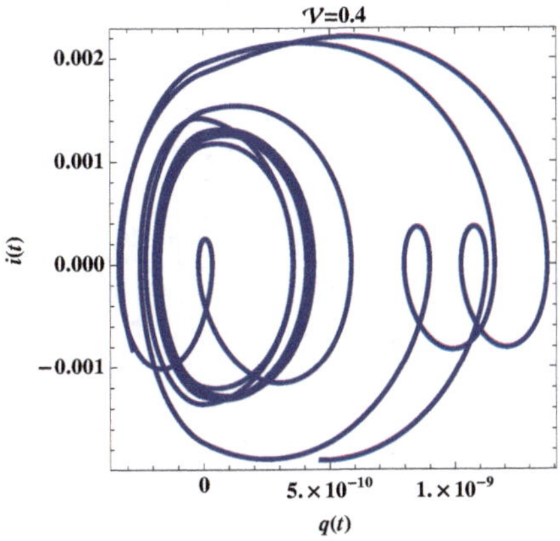

Fig. 3.78 ... for different values of ν.

which can be visualized conventionally (Fig. 3.80) or by means of the bifurcation diagram (Fig. 3.81). Logistic equation is a suitable tool for studying the transition from deterministic behavior to chaotic behavior as well as phenomena accompanying that transition, such as intermittence and period doubling. Recall that logistic equation takes the form

$$x_{n+1} = Ax_n(1 - x_n) \tag{3.34}$$

Fig. 3.80 shows chaotic behavior of the logistic equation for precisely defined initial conditions and control parameter A. The behavior depends both on the initial conditions and on the control parameter, as the two bifurcation diagrams in Fig. 3.81 clearly demonstrate.

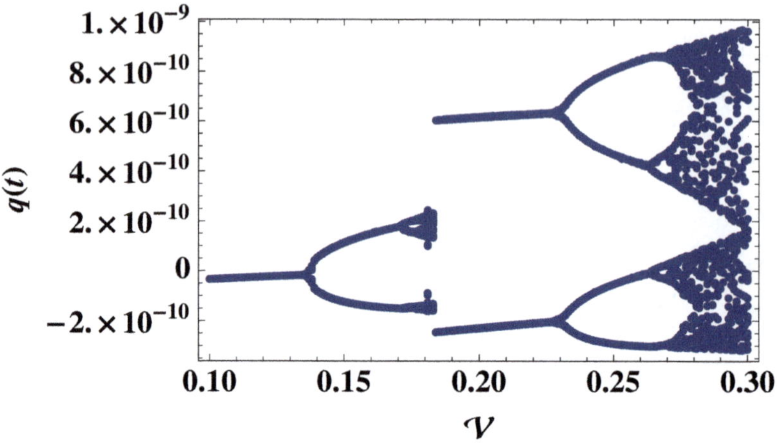

Fig. 3.79 Circuit with a diode - bifurcation diagram

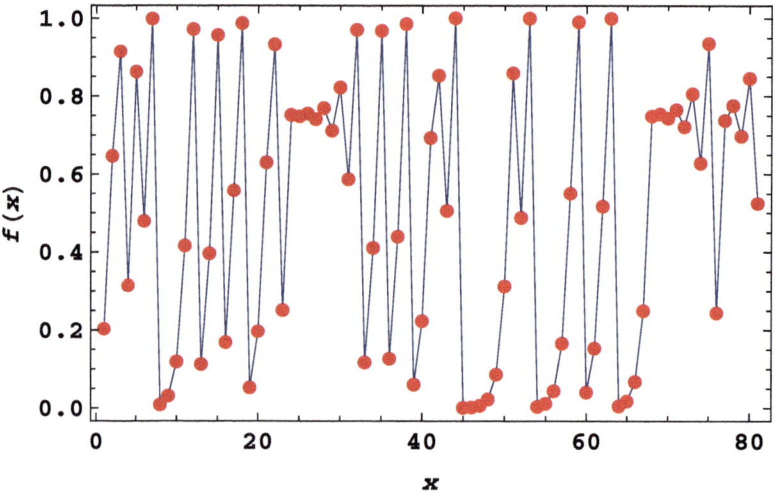

Fig. 3.80 Behavior of the logistic equation in time for $A = 4$, $x_0 = 0.2027$

The diagrams show the chaotic patterns of the system behavior in dependence on the control parameter. We would like also to note that the bifurcation diagram (and bifurcation in general) is related to abrupt changes in the system behavior, referred to as catastrophes, in dependence on the control parameter (Thom's catastrophe theory, see also [13], [4], [21]).

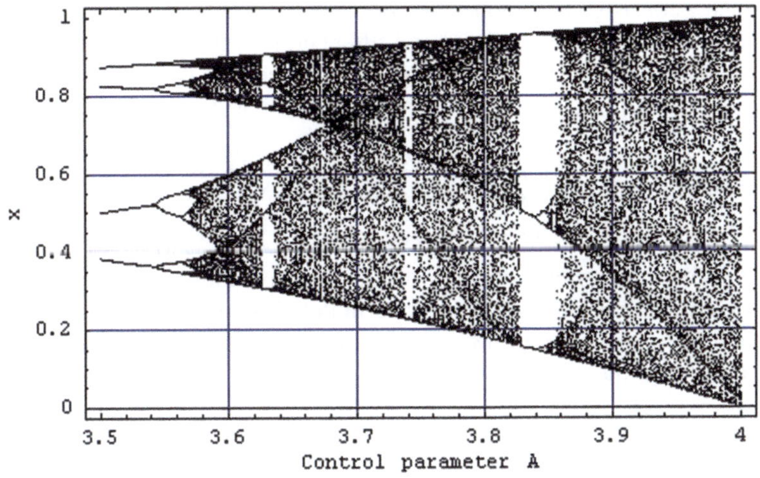

Fig. 3.81 Bifurcation diagram of the logistic equation

3.5.5 Meteorological System – Lorenz Weather Model

A typical representative of deterministic chaos is a very simple model of the behavior of weather expressed by a system of equations devised by Edward Lorenz at MIT in 1963. Lorenz is generally regarded as the discoverer of deterministic chaos. The equations, including the initial conditions, are given by (3.35). They represent a hydrodynamic model of the behaviour of a gas or liquid during external heating [16]. A simulation of (3.35) provides the chaotic attractor that is shown in Fig. 3.82-3.83.

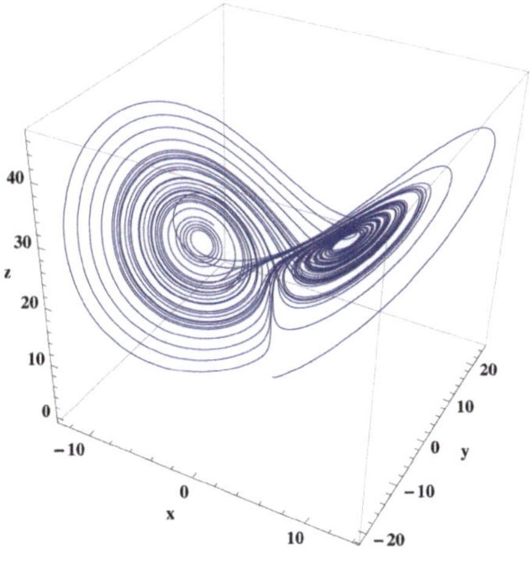

Fig. 3.82 The Lorenz attractor in 3D ...

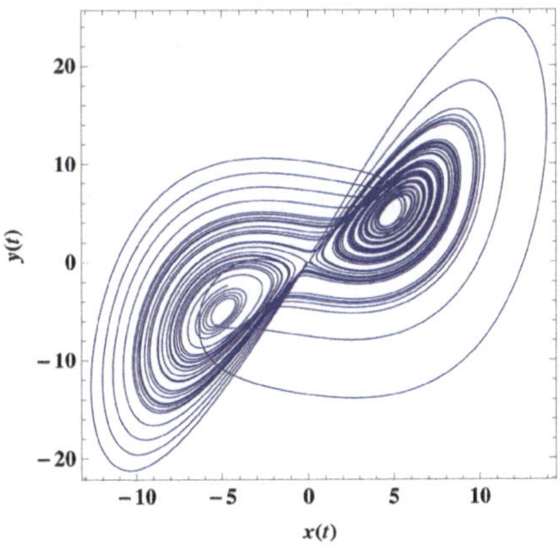

Fig. 3.83 ... and 2D representation.

The attractor consists of two lobes in whose centers are singular points that attract trajectories from their neighborhood and, after certain attraction, repulse them away. The arrangement of the two singular points is such that the repulsed trajectories get into the attraction domain of the opposite singular point, where the process is repeated.

$$\begin{aligned}\dot{x}_1(t) &= -a\,(x_1(t)-x_2(t))\\ \dot{x}_2(t) &= -x_1(t)x_3(t)+bx_1(t)+x_3(t)\\ \dot{x}_3(t) &= x_1(t)x_2(t)-x_3(t)\end{aligned} \tag{3.35}$$

The origin of the Lorenz attractor, including modifications in the nature and positions of the singular points, is described in detail in [13]. It should be noted that the accuracy of calculation of the behavior of a chaotic system also depends on the software and method used.

3.5.6 Spatiotemporal Chaos

The systems discussed so far demonstrated deterministic chaos in the time domain, i.e. where chaotic behavior can be observed in the system behavior developing in time. In addition to this type of chaotic behavior, another type exists, see spatiotemporal behavior ([16], [24]), occurring in systems that are described, e.g., by partial differential equations. Hence, they are systems with distributed parameters. This type of behavior can be nicely and simply demonstrated on the logistic equation discussed above (other iteration equations can also be used, of course) in parallel connection, referred to as Coupled Map Lattices (CML). This is a spatiotemporally coupled system with the development of n equations that affect each other via a coupling constant, usually denoted ε. CML can be regarded as a field of kind of "oscillators" which affect each other. Mathematical description of a CML using an iteration equation for its activity consists in (3.36) where the function which is denoted $f(...)$ represents the iteration equation.

$$x_{n+1}(i) = (1-\varepsilon)f(x_n(i)) + \frac{\varepsilon}{2}(f(x_n(i-1)) + f(x_n(i+1))) \tag{3.36}$$

Equation (3.36) is referred to as a symmetric CML because the kth equation acts on its neighbors (through the coupling constant ε) equally on both sides. Asymmetric CMLs whose description is, naturally, slightly modified, also exist. Such types of relatively simple spatiotemporal chaotic systems provide a very wide scale of behavior, which is used for modelling this type of chaos as well as for the study of its control and use in information transmission and encoding. Figs 3.84 to 3.85 show the behavior of a CML according to eq. (3.36) where term $f(...)$ is replaced by the logistic equation, or more precisely by 100 logistic equations that affected each other during 100 iterations. In Fig. 3.84, black points denote values exceeding the level of 0.88 (according to [24]). The other points remain white, due to which information regarding the actual diversity of the spatiotemporal chaos is lost. This is demonstrated by Fig. 3.85, where a gray-scale picture is depicted. Fig. 3.86 shows

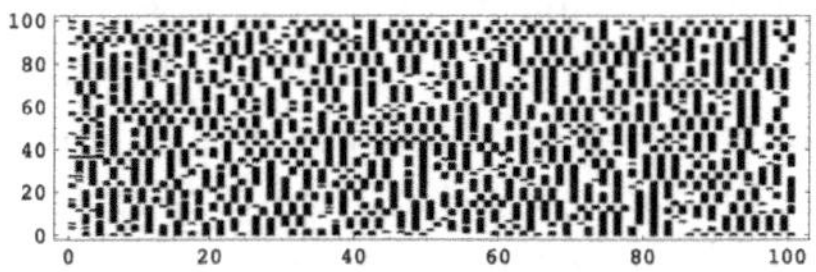

Fig. 3.84 CML in black-white visualization...

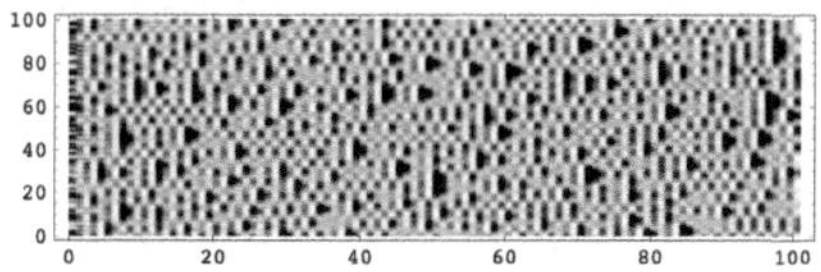

Fig. 3.85 ... and its gray-scale version.

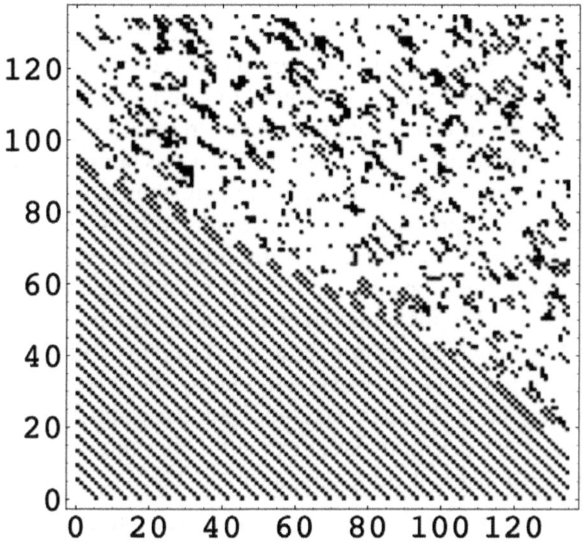

Fig. 3.86 2D CML

another version of CML: 2D version, i.e. both axes x and y are logistic equations joined together. Time line is axe z, which is not visible in Fig. 3.86, this figure is basically only slice cut of 2D CML in iteration 200.

Naturally, CML is not the only method to simulate spatiotemporal chaos. Considerably more complex descriptions (as regards mathematical formalism and solution) exist and will be discussed in the Chapter 6, dealing with the control of chaos.

3.5.7 Cellular Automata – Game of Life

Cellular automata [31] represent a tool that can be employed to simulate extensive or complex systems. The history of cellular automata can be traced back to ancient China, specifically to the year 1303. This is the era of origin of the Chinese arithmetic triangle, better known as Pascal's triangle (after the French mathematician Blaise Pascal, 1623 - 1662) published in 1527, which indirectly led to the later development of probability theory. Cellular automata only enjoyed boom with the development of PCs, which enabled their use in virtually any branch of human activity. Among applications of cellular automata are, for instance, simulation of forest fires, differentiation of cells in human body (Kuffman's model), the human body's immune failure, hydrodynamic phenomena (e.g. motion of particles of a fluid, was used to simulate the behavior of 4 million molecules) and passage of a liquid through unordered geometric structures such as sand. Given the computational capacity of currently available hardware, cellular automata appeared to be so to say predestined for technically demanding "parallel" simulations of systems such as the flow of molecules of a liquid or gas, etc. In such "huge" simulations cellular automata feature simplicity as well as a high speed as compared to conventional calculations. Cellular automata can also be used to simulate tessellations, i.e. mosaics, which find application in investigations into the creation of mosaics in various materials, the shape of boundaries of territories of various predators or the propagation of epidemics. A cellular automaton can be imagined as a grid/matrix, where each square/matrix element represents a cell. In simple automata all cells are subject to a single law, owing to which the most bizarre images can emerge. Apart from their geometrical meaning, such images can provide information about the dynamics of the process involved. If a phenomenon is simulated which is not homogeneous or isotropic (which means identical properties in all points and directions), then this fact must be taken into account when formulating the rule governing the cellular automaton. Among the best known and most popular cellular automata is Game of Life, governed by a very primitive rule and still exhibiting very complex behavior. The rules are very simple and are identical for all cells:

- Any live cell with more than three live neighbors dies, as if by overcrowding.
- Any live cell with fewer than two live neighbors dies, as if caused by underpopulation.
- Any dead cell with exactly three live neighbors becomes a live cell.
- Dead cells are shown in white, live cells shown in black.

This simple set of rules gives rise to incredibly complex behavior (Fig. 3.88) forming groups of cells that die and become live cells again (blinkers), travel along the cellular automaton (gliders), shoot down gliders (guns) or travel leaving blinkers in their traces (star ships). Cellular automata generate both chaotic behavior and deterministic behavior (Fig. 3.88). The above CML simulation can also be considered a cellular automaton based on eq. (3.36 which is the single rule for all cells here.

3.5.8 *Artificial Intelligence – Neural Networks*

Neuron networks - biological or artificial - represent another chaos-generating system. The presence of chaos in biological networks is associated with diseases such as epilepsy, in artificial networks, with the phases of learning and recollection. A neuron network [7] can be represented by an oriented graph whose nodes are neurons, i.e. simple computational units performing primitive mathematical operations

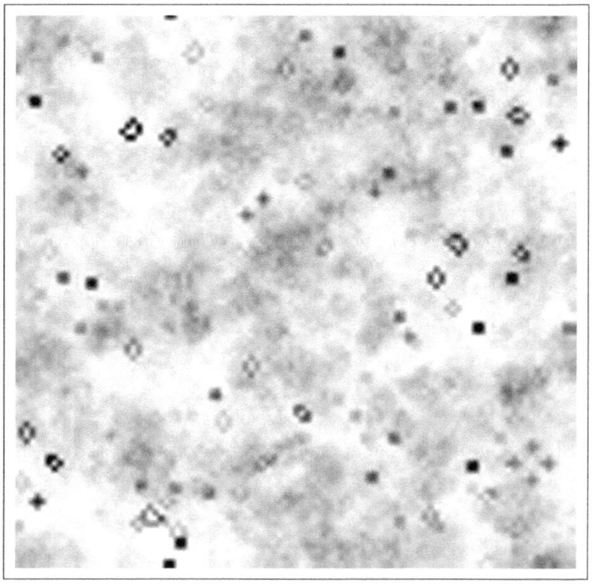

Fig. 3.87 Game of Life containing both deterministic and chaotic structures

Fig. 3.88 Order in a cellular automaton

such as summation, multiplication, etc. Since a network is formed by discrete objects, it can be looked upon as a special type of cellular automaton, with a special set of cells (input and output neurons). Hence, it is reasonable to expect information

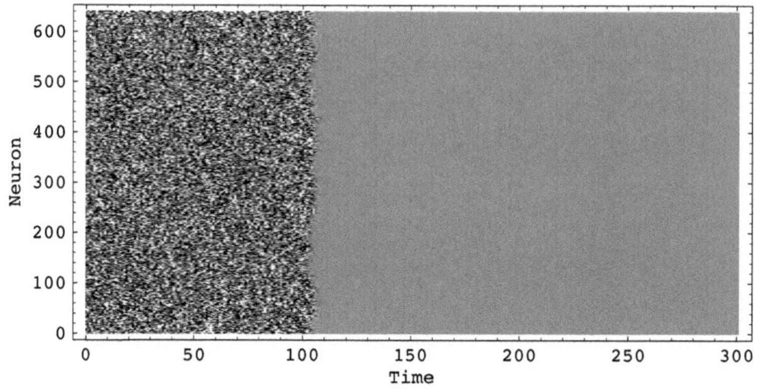

Fig. 3.89 Steadying of chaos generated by a neuron network ($w \in [-0.3, 0.3]$)

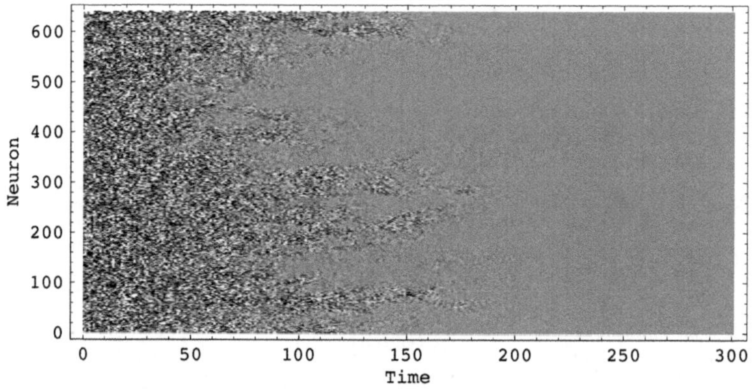

Fig. 3.90 "Intermittence" in the behavior of a neuron network ($w \in [-0.6, 0.6]$)

processing by neuron networks to be accompanied by chaotic behavior. This was confirmed both experimentally (association with epilepsy found) and by simulations (numerical studies on various models). By way of example, consider a simple network [25] which is defined by (3.37). This is a single-layer network where outputs from neurons not farther than r enter the i_{th} neuron. Hyperbolic tangent is the transfer function [7] and w is weight, which is generated at random. Fig. 3.89 to 3.91 display the network's behavior for identical initial conditions with differently large intervals at which weights w were generated. The color of each point represents the state of the neuron, of which they are 640. Fig. 3.89 clearly demonstrates that starting from an initial chaotic state, all neurons will ultimately assume the same value. It is clear from Fig. 3.90 and 3.91 that even a slight change in the weight generating

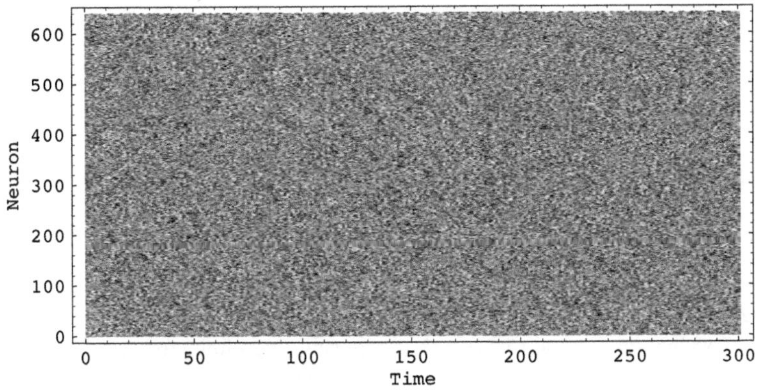

Fig. 3.91 Chaos generated by a neuron network ($w \in [-0.9, 0.9]$)

interval brings about non-uniform stabilization, and regions can be observed where neurons pass from a chaotic regime to a steady-state regime and back (see the group of neurons around neuron 100-150 in Fig. 3.91) - a situation called intermittence. When the interval for weight generation is extended again, the network's behavior is free from any determinism and the networks is in the chaotic regime:

$$x_{n+1}(i) = \tan\left(\sum_{j=1}^{r} w_i\left(x_n(i-j) + x_n(i+j)\right)\right) \tag{3.37}$$

3.5.9 *Artificial Intelligence – Evolutionary Algorithms*

Optimization algorithms are powerful tools in solving many problems in practical engineering. They are typically used where solving a problem by an analytical method is inappropriate or infeasible. Suitably implemented, optimization algorithms can be used without frequent user interventions into the performance of the facility where they are used. The majority of problems in engineering practice can be defined as optimization problems, such as finding the optimal trajectory for a robot, optimal pressure vessel wall thickness, optimal controller parameter setting, optimal relation between fuzzy sets, etc. In other words, the problem to be solved can be transformed into a mathematical problem defined by a functional prescription whose optimization leads to the finding of arguments of the objective function, which is the goal of the optimization exercise. A number of highly efficient algorithms were developed during the past two decades, enabling highly complex problems to be solved very efficiently and effectively. This class of algorithms has a specific name of evolutionary algorithms. Such algorithms are capable of solving highly complex problems quite well, owing to which they are widespread and popular in many fields of technology. A typical feature of evolutionary algorithms is that

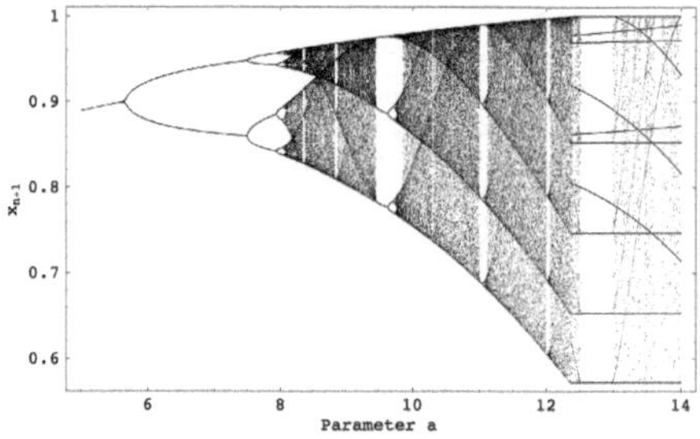

Fig. 3.92 Bifurcation diagram of simple genetic algorithm for $a \in [4,15]$, $b = 1$, $T = 7/8$

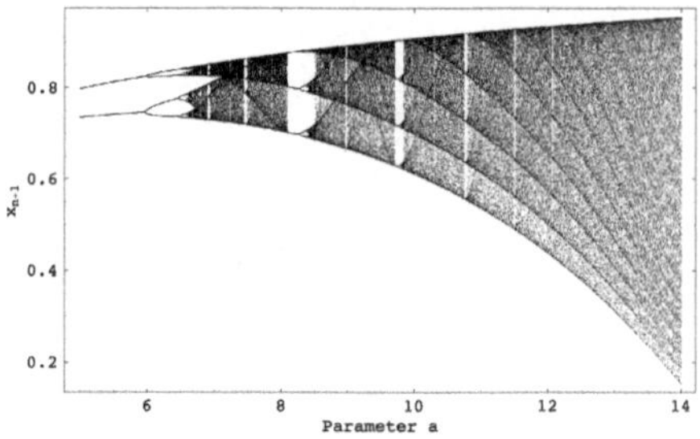

Fig. 3.93 Bifurcation diagram of simple genetic algorithm for $a \in [4,15]$, $b = 7$, $T = 7/8$

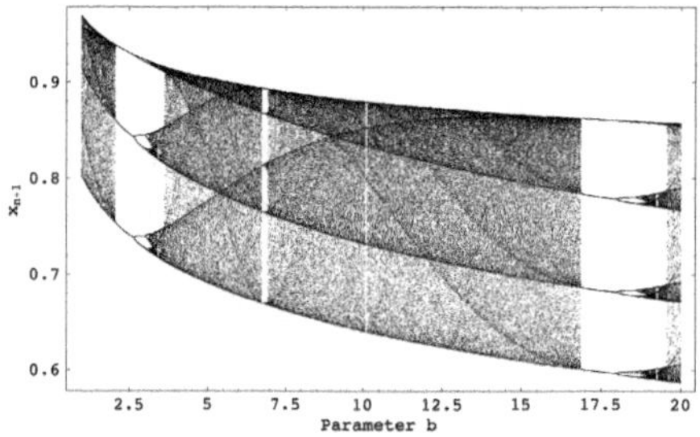

Fig. 3.94 Bifurcation diagram of simple genetic algorithm for $a = 9$, $b \in [1,20]$, $T = 7/8$

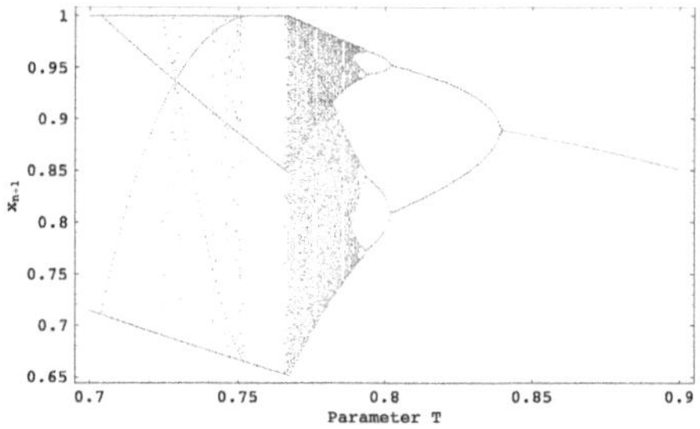

Fig. 3.95 Bifurcation diagram of simple genetic algorithm for $a = 4$, $b = 1$, $T \in [0.7, 0.9]$

they work on populations of possible solutions, called individuals. Such individuals affect each other's quality based on certain evolutionary principles in cycles, usually bearing the name "Generation".

Deterministic chaos has been also observed, mathematically proven and numerically demonstrated in evolutionary algorithms, especially in genetic algorithms as reported in [32].

In that research, dynamical system models of genetic algorithms were considered with the expected behavior of the algorithm analyzed as the population size goes to infinity. Their work is based on the research of [28], [29] and [27]. An elegant theory of simple genetic algorithms is based on random heuristic search on the idea of a heuristic map G. An important point of the research in [32] is that the map G includes all of the dynamics of the simple genetic algorithm, based on eq. 3.38 (truncation selection) and eq. 3.39 (mutation heuristic function). It is defined by $G_{a,b,T} = F_T \circ U_{a,b}$. In both equations, p represents population and $T = t/r$ is an ratio of t most fitted individuals selected from population of size r for reproduction.

Sample bifurcation diagrams are depicted in Figs. 3.92 - 3.95. Ideas about chaos in simple genetic algorithm are explained in detail in [32]. In this chapter, it has been proven that chaos in heuristic algorithms can be observed. This observation is certainly not valid only for simple genetic algorithms.

$$F_T(p) = \begin{cases} 1 & if\, T < p \\ \frac{p}{T} & if\, T > p \end{cases} \tag{3.38}$$

$$U_{a,b}(p) = p - \frac{b}{2}\,|2p-1|^a\,(2p-1) \tag{3.39}$$

3.5.10 Astronomy – The Three-Body Problem

Quite a number of systems are encountered in astrophysics exhibiting chaotic behavior. As a typical example, let us discuss the three-body problem. This is a celestial mechanics problem describing the motion of three (or more) bodies affecting one another by gravitational forces. Mathematically, the three-body problem is formulated by a system of equations of motion, see (3.40).

$$m_j\ddot{q}_j = \gamma \sum_{k\neq j}^{n} \frac{m_k m_j (q_j - q_k)}{\left|q_j - q_k\right|^3}, \; j = 1, \; ..., \; n \tag{3.40}$$

In this system of equations, m is the mass of the mutually affecting bodies and q is a vectorial function of time defining the positions of the bodies. The problem of n bodies involves $6n$ variables (because each body has 3 position components and 3 velocity components). The motion of a system of n bodies is practically analytically unsolvable starting from $n = 3$, and simulations of the behavior are performed numerically on computers. This problem attracted interest of such mathematicians as Euler (1767, discovery of colinear periodic trajectories), Lagrange (1772, central configuration of a system of n bodies), Charles-Eugene Delaunay (1860-1867, a study 900 pages volume dealing with the Earth-Moon-Sun system).

A simplified version of the three-body problem, called the restricted three-body problem, has been formulated in this context. In this simplification, the mass of one of the bodies is disregarded or the trajectories of the bodies are reduced to some shapes such as circular or elliptical. Fig. 3.96 - 3.99 shows the behavior of three bodies for different initial conditions. Chaotic behavior, or more precisely chaotic orbits of the three bodies are clearly seen.

The n-body problem (or more precisely its restricted version) can also be simulated by means of a relatively simple device called a mad pendulum. This pendulum consists of N magnets located in the apexes of an N-angle, above which hangs a steel ball on a thin string (see Fig. 3.100). The mathematical model describing the behavior of the pendulum is given by (3.41).

$$
\begin{aligned}
&-\sum_{i=1}^{6}\frac{X_i-x(t)}{\sqrt{d^2(X_i-x(t))^2+(Y_i-y(t))^2}}+sc*x(t)+R*\dot{x}(t)+\ddot{x}(t)=0\\
&-\sum_{i=1}^{6}\frac{Y_i-y(t)}{\sqrt{d^2(X_i-x(t))^2+(Y_i-y(t))^2}}+sc*y(t)+R*\dot{y}(t)+\ddot{y}(t)=0
\end{aligned}
\tag{3.41}
$$

Each magnet attracts in some way the ball suspended from the starting position, and a chaotic trajectory results. For example, Fig. 3.100 - 3.101 shows two trajectories which are entirely different although the starting conditions only differ by one-hundredth in the velocity.

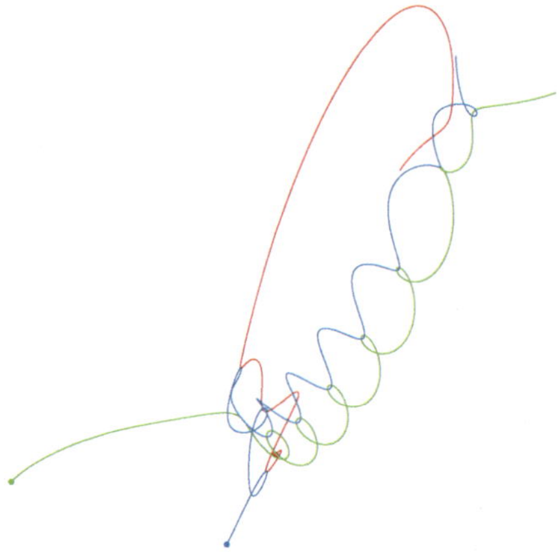

Fig. 3.96 Three body problem - random initial conditions

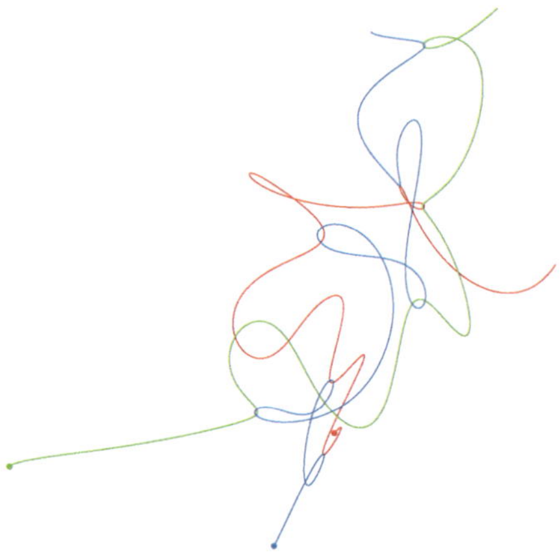

Fig. 3.97 Three body problem - different initial conditions

Fig. 3.98 Three body problem - different initial conditions

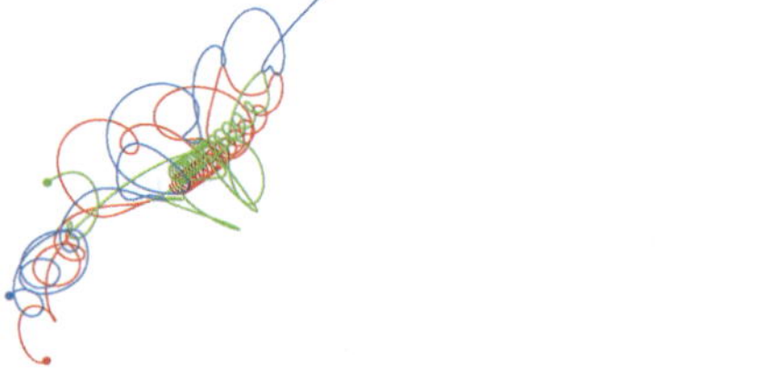

Fig. 3.99 Three body problem - different initial conditions

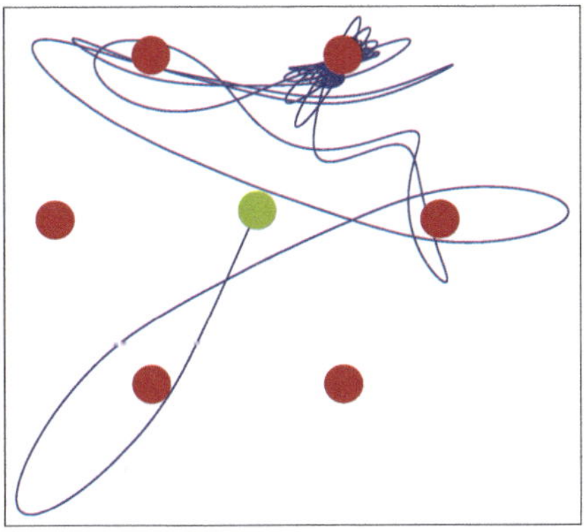

Fig. 3.100 Trajectories of a mad pendulum for $v_{x,y}$ = -1.05, -2.31...

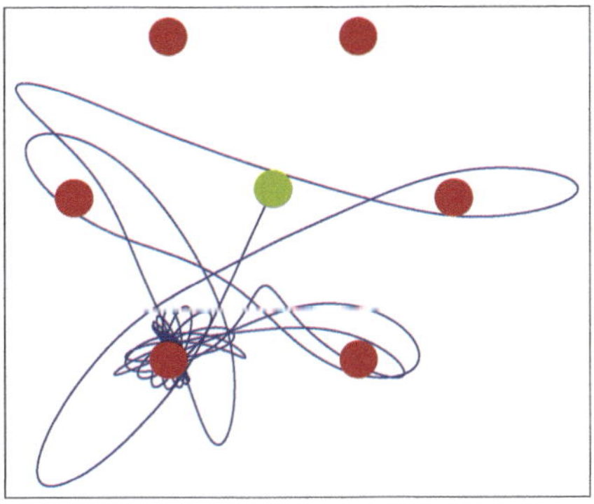

Fig. 3.101 ... and another trajectories for $v_{x,y}$ = -1.05, -2.3.

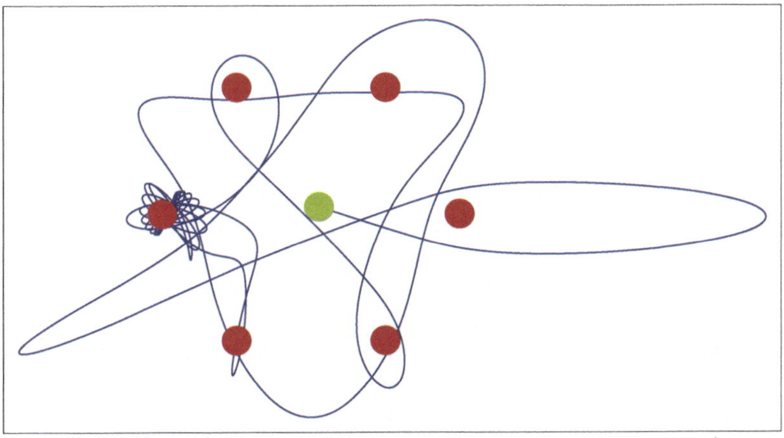

Fig. 3.102 Another trajectory for $v_{x,y}$ = 3, -1

It is clear from the pictures and from the physical nature of the problem that the chaotic mode can only be observed during a certain time interval. Due to energy dissipation the pendulum will eventually stay in the resting position at one of the magnets or in the origin of the N-angle. Fig. 3.103 shows the development of the x-component of the pendulum motion. Chaotic behavior can be observed during the first 20 seconds of development. Subsequently, chaos vanishes due to energy dissipation, quasi-periodic oscillation follows, and ultimately the pendulum remains at rest.

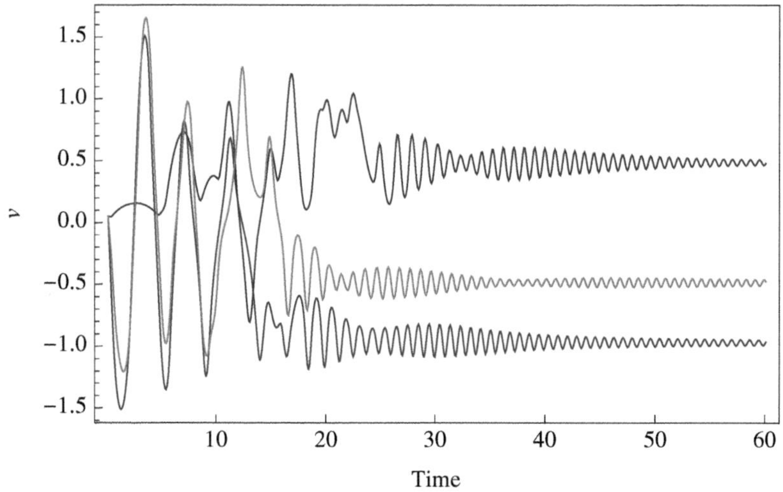

Fig. 3.103 Time development of the behavior of the pendulum for different starting values of $v_{x,y}$

3.6 Conclusion

This chapter presents a very simple introduction to deterministic chaos theory. Main and well known icons of chaos, like Lyapunov exponent, Feigenbaum's constant, U-sequence, self-similarity etc has been introduced. The way how deterministic behavior can be changed into a chaotic one is also discussed like intermittence, period doubling, crises as well as chaotic transients. At the end of this chapter, selected examples from mechanics, astrophysics, computer sciences, electronics amongst others are described. Main attention has been paid to demonstration of deterministic chaos behavior. For more detailed explanation and description of deterministic chaos it is recommended to study literature in the references.

Acknowledgements. This work was supported by grant No. MSM7088352101 of the Ministry of Education of the Czech Republic and by grants of the Grant Agency of the Czech Republic GACR 102/09/1680 and 102/08/0186.

References

1. Abarbanel, H.: Analysis of observed chaotic data. Springer, New York (1996)
2. Abarbanel, H., Brown, R., Kennel, M.: Variation of Lyapunov exponents on a strange attractor. J. Nonlinear Sci. 1, 175 (1991)
3. Alligood, K., Sauer, T., Yorke, J.: Chaos - an introduction to dynamical systems. Springer, New York (1997)
4. Arnold, V.: The Theory of Singularities and Its Applications, Accademia Nazionale Dei Lincei, Pisa, Italy (1991)

5. Baker, G., Gollub, J.: Chaotic dynamics: an introduction. Cambridge University Press, Cambridge (1996)
6. Barnsley, M.: Fractals Everywhere. Academic Press Professional, London (1993)
7. Bose, N., Liang, P.: Neural Network Fundamentals with Graphs, Algorithms, and Applications. McGraw-Hill Series in Electrical and Computer Engineering (1996)
8. Constantin, P., Foias, C.: Global Lyapunov exponents, Kaplan-Yorke formulas and the dimension of attractors for 2D Navier-Stokes equations. Commun. Pure Appl. Math. 38, 1 (1985)
9. Cvitanovic, P.: Universality in Chaos. Taylor and Francis, Abington (1989)
10. Diks, C.: Nonlinear time series analysis, Methods and applications. World Scientific, Singapore (1999)
11. Drazin, P., Kind, G.(eds.): Interpretation of time series from nonlinear Systems. Special issue of Physica D, 58 (1992)
12. Galka, A.: Topics in nonlinear time series analysis with implications for EEG analysis. World Scientific, Singapore (2000)
13. Gilmore, R.: Catastrophe Theory for Scientists and Engineers. John Wiley and Sons, Chichester (1993)
14. Haken, H.: Synergetics: Introduction and Advanced Topics. Springer, Heidelberg (2004)
15. Kaplan, J., Yorke, J.: Chaotic behavior of multidimensional difference equations. In: Walter, H., Peitgen, H. (eds.) Functional differential equations and approximation of fixed points. Lect. Notes Math., vol. 730, p. 204. Springer, Berlin (1979)
16. Hilborn, R.: Chaos and Nonlinear Dynamics. Oxford University Press, Oxford (1994)
17. Kantz, H., Schreiber, T.: Nonlinear time series analysis. Cambridge University Press, Cambridge (1997)
18. Ledrappier, F., Young, L.: The metric entropy of diffeomorphisms, Parts I and II. Ann. Math. 122, 509 (1985)
19. Packard, N., Crutchfield, J., Farmer, D., Shaw, R.: Geometry from a time series. Phys. Rev. Lett. 45, 712 (1980)
20. Pesin, Y.: Characteristic Lyapunov exponents and smooth ergodic theory. Russ. Math. Surv. 32, 55 (1977)
21. Poston, T., Stewart, I.: Catastrophe Theory and its Applications, Pitman, pp. 842–844. IEEE Press, New York (1977)
22. Rössler, O.: An equation for hyperchaos. Phys. Lett. A 71, 155 (1979)
23. Ruelle, D.: Thermodynamics Formalism. Addison-Wesley, Reading (1978)
24. Schuster, H.: Handbook of Chaos Control. Wiley-VCH, New York (1999)
25. Sprott, J.: Chaos and Time-Series Analysis. Oxford University Press, Oxford (2003)
26. Takens, F.: Detecting strange attractors in turbulence. Lecture Notes in Math., vol. 898 (1981)
27. Vose, M.: The Simple Genetic Algorithm: Foundations and Theory. MIT Press, Cambridge (1999)
28. Vose, M., Liepins, G.: Punctuated equilibria in genetic search. Complex Systems 5, 31–44 (1991)
29. Vose, M., Wright, A.: Simple genetic algorithms with linear fitness. Evol. Comput. 4(2), 347–368 (1994)
30. Wolff, R.: Local Lyapunov exponents: Looking closely at chaos. J. R. Statist. Soc. B 54, 301 (1992)
31. Wolfram, S.: A New Kind of Science, Wolfram Media (2002)
32. Wright, A., Agapie, A.: Cyclic and Chaotic Behavior in Genetic Algorithms. In: Proc. of Genetic and Evolutionary Computation Conference (GECCO), San Francisco, July 7-11 (2001)
33. Wyk, M.: Chaos in Electronics. Springer, Heidelberg (1997)

Chapter 4
Evolutionary Algorithms and the Edge of Chaos

Donald Davendra

Abstract. An unconventional approach of the edge of chaos and its application to discrete systems is described in this chapter. Langton's approach to cellular automata and its unique ordered and chaotic behavior is discussed. The expansion of this approach to genetics and random networks by Kauffman is described with a brief analogy provided of chaos in evolutionary algorithms in terms of stagnation.

4.1 Introduction

The linkage between chaos and its manifestation in a number of systems of discrete states is described in this chapter. The most famous example of systems with discrete states are cellular automata [15]. Whereas the traditional view of chaos as non-linear dynamical system is well entrenched, a number of discrete systems, some in nature are also shown to have chaotic attributes. Chaos in this respect is seen as the course toward *unnatural* and *destructive* behavior. Langton [7] and Kauffman [5] have shown that the boundary between these two phases, *edge of chaos* is where the most interesting features can be found.

Edge of chaos has more recently been discovered in living systems. [1] and [6] have discovered neural activity in the brain which they believe shows that the brain operates on the edge of chaos, which is the boundary between stable, orderly behavior - such as a swinging pendulum - and the unpredictable world of chaos, as exemplified by turbulence. It is believed that near-chaotic states may be crucial to memory, and could explain why some people are smarter than others.

Donald Davendra
Tomas Bata University in Zlin, Faculty of Applied Informatics, Nad Stranemi 4511, Zlin 76001, Czech Republic
e-mail: davendra@fai.utb.cz

I. Zelinka et al. (Eds.): Evolutionary Algorithms and Chaotic Systems, SCI 267, pp. 145–161.
springerlink.com

This research looks at Kauffman's approach to chaos in genetics as the backdrop to the exploration of the dynamic nature of evolutionary algorithms and its convergence. An analytical viewpoint in permutative optimization demonstrates how stagnation can arise and how it can adversely affect the stability of the system.

A practical application as to how to develop efficient anti-bias rules to overcome stagnation is given in Chapter 15.

4.2 Edge of Chaos

The phrase **edge of chaos** was first coined by Christopher Langton in 1990 [7]. It simply refers to the region of space in cellular automata (CA) which exhibits transitory behavior between **deterministic** and **chaotic** while a parameter lambda, λ is varied. This region is believed to provide a region capable to *universal computation* in CA.

A brief description of CA is now presented. CA are discrete spatially-extended dynamical systems that have been studied extensively as models of physical processes and as computational devices [8]. In its simplest form, a CA consists of a spatial lattice of cells of size N, each of which, at time t, can be in one of k states. A CA has a single fixed rule, which is used to update each cell. The lattice starts out with some initial configuration of local states and, at each t, the states of all cells in the lattice are synchronously updated based on the rules.

Assume a simple CA, which is one-dimensional and has two possible states per cell (binary 1 or 0). The neighborhood of this cell is defined by a radius **r**, which is the allowable interaction of this cell. In order to update the cell, a set of rules have to be applied to each and every cell. These rules are kept in a rules table. As an example, a single 3 grid lattice CA is represented in Figure 4.1. Since $k = 2$, the binary state (1,0) is used. State *1* is represented as dark while state *0* is presented in light sequence.

Fig. 4.1 Single 3 grid CA representation

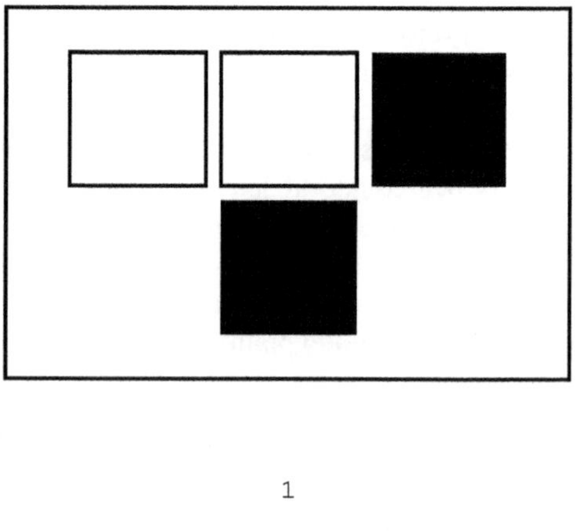

One of the most common rules table in CA is **Rule 30** table, which is given in Figure 4.2. Each possible grid combination is presented and the next state is given underneath the cell.

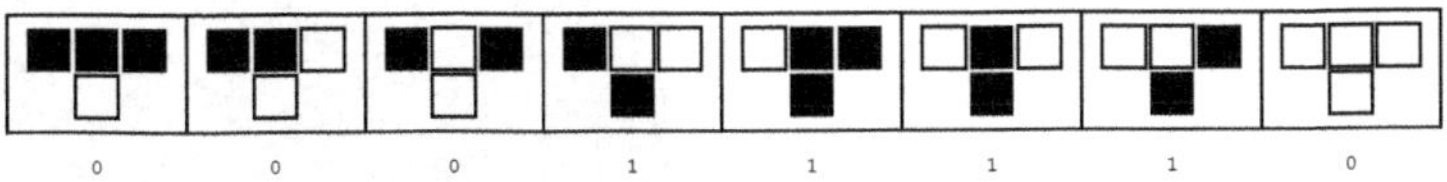

Fig. 4.2 Rule 30 table.

The rule representation for *Rule 30* is given in Figure 4.3.

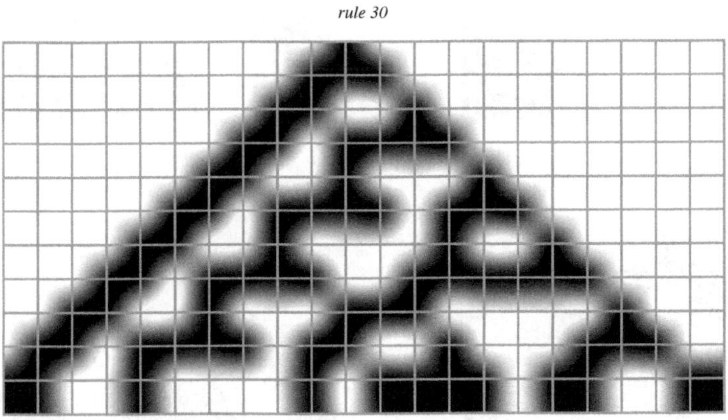

Fig. 4.3 Rule 30 representation

Using this rule table as an update sequencer, an initial random configuration can be iterated, producing more and more complex CA configurations.

The following Figures 4.4 - 4.13 represent the iteration of a single CA lattice using Rule 30. Figure 4.4 is the initial cell, with single cell of state 1 and 0. Figures 4.5 - 4.9 show the iteration from $t = 1$ to $t = 5$. At each level the CA grows by an additional row, where the preceding cells use Rule 30 to calculate the new states. Figure 4.10 shows the CA at $t = 10$, whereas Figures 4.11, 4.12 and 4.13 shows the growth of the CA lattice at times $t = 20$, $t = 50$ and $t = 100$ respectively.

CA are of interest as models of physical processes because, like many physical systems, they consist of a large number of simple components (cells) which are modified only by local interactions, but which acting together can produce global complex behavior. Even simple CA exhibit two unique states of behavior; limit cycles (periodic behavior) to unpredictable ("chaotic") behavior.

Fig. 4.4 Initial Cell

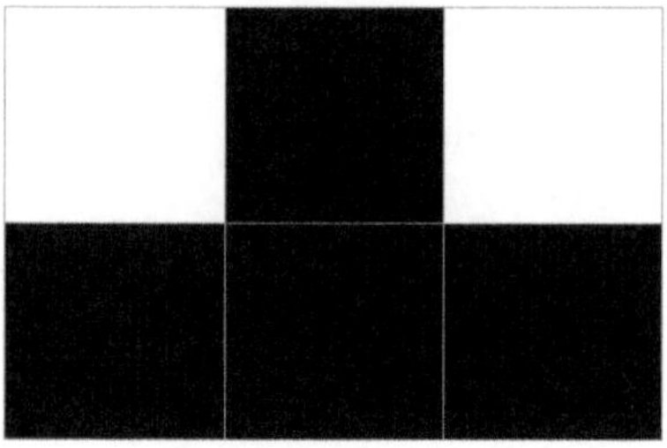

Fig. 4.5 1 Iteration

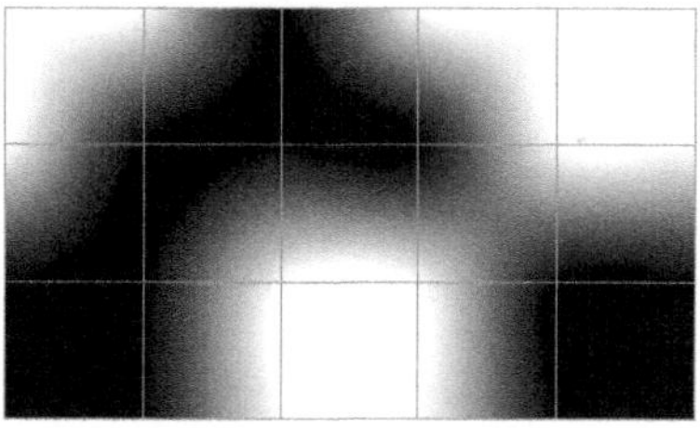

Fig. 4.6 2 Iteration

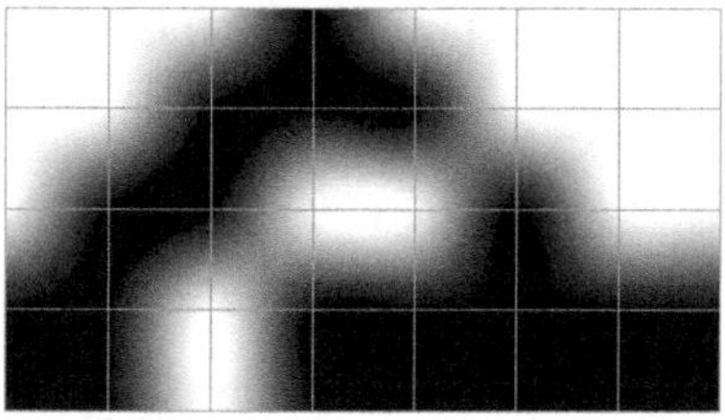

Fig. 4.7 3 Iteration

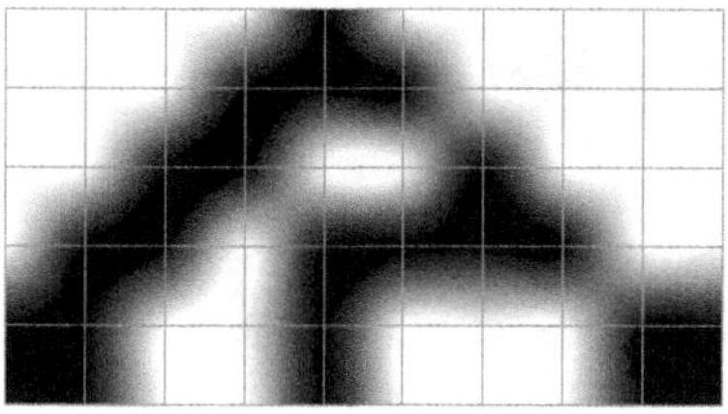

Fig. 4.8 4 Iteration

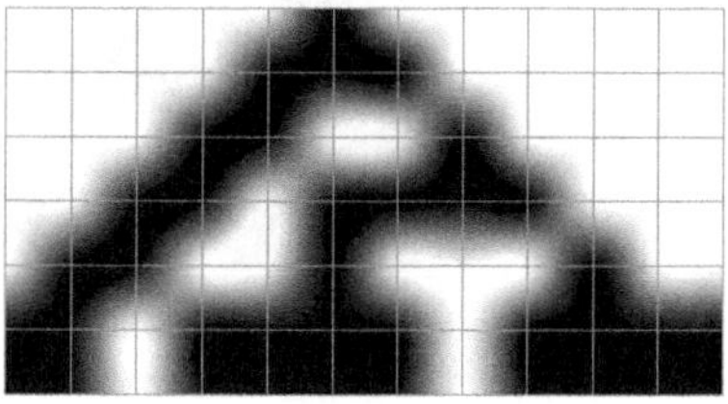

Fig. 4.9 5 Iteration

One of the core proponents and researchers of CA, Stephen Wolfram considered a coarse classification of CA behavior in terms of these categories. He proposed the following four classes with the intention of capturing all possible CA behavior [15]:

Class 1 Almost all initial configurations relax after a transient period to the same fixed configuration (e.g., all 1's).

Class 2 Almost all initial configurations relax after a transient period to some fixed point or some temporally periodic cycle of configurations, but which one depends on the initial configuration.

Class 3 Almost all initial configurations relax after a transient period to chaotic behavior.

Class 4 Some initial configurations result in complex localized structures, sometimes long-lived.

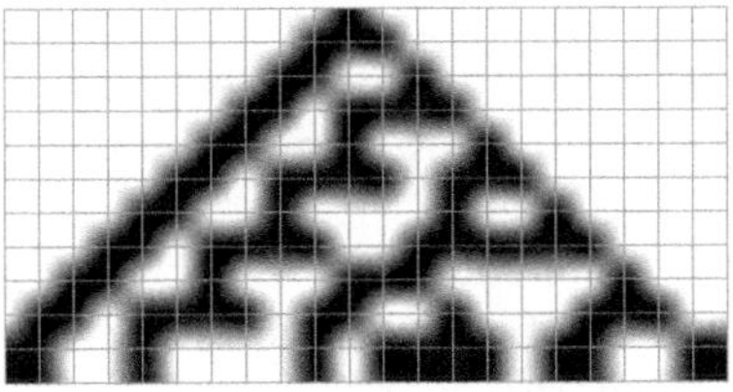

Fig. 4.10 10 Iteration

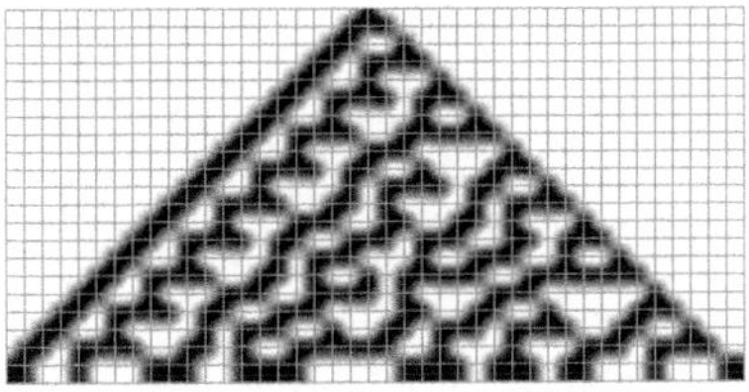

Fig. 4.11 20 Iteration

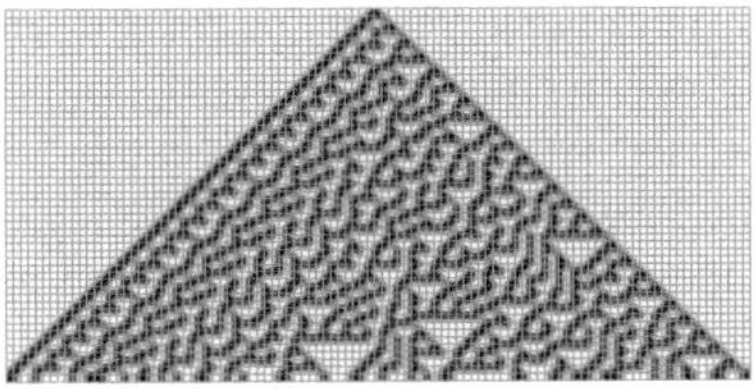

Fig. 4.12 50 Iteration

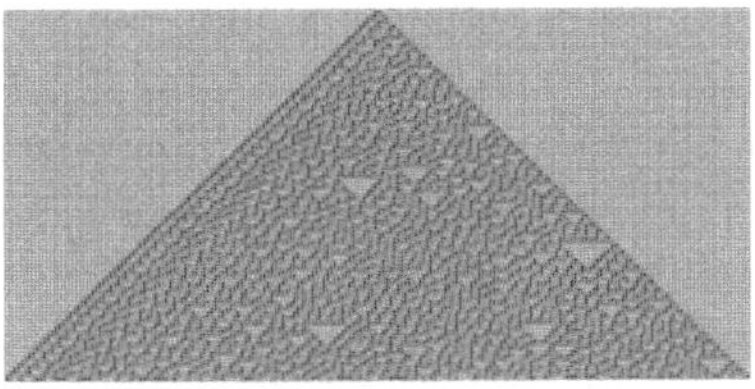

Fig. 4.13 100 Iteration

However, prior to Wolfram's work, Langton studied the relationship between the "average" dynamical behavior of cellular automata and a particular statistic (λ) of a CA rule table [7].

The drawback of CA and its interpretation as a chaotic system is its discrete nature. Chaos and its identification is ruled by *non-linear differential systems*, where the *Lyapunov Exponent* can easily differentiate a chaotic system, whereas a CA is a discrete system. Therefore. Langton defined a parameter lambda λ that varies incrementally as single output bits are turned on or off in a given rule table.

In order to calculate λ, the following parameters are required:

- Designate one state (0,1) to be **quiescent** state
- Let $k \rightarrow$ number of states
- Let $N = 2r + 1 \rightarrow$ area of neighborhood
- Let $T = K^N \rightarrow$ number of entries in the rule table
- Let $n_q \rightarrow$ number mapping to quiescent state

The formulation of λ is given in Equation 4.1.

$$\lambda = \frac{T - n_q}{T} \tag{4.1}$$

A "quiescent" state can be any of the available states in a system, therefore in a binary system, it can be either 1 or 0. For a binary-state CA, if 0 is chosen to be the quiescent state, then λ is simply the fraction of output 1 bits in the rule table.

Langton performed experimentation on a number of Monte Carlo samples of two-dimensional nature, starting with $\lambda = 0$ and gradually increasing λ to 1, in order to have a shift from the most homogeneous to the most heterogeneous rule tables.

In terms of analysis. Langton used various statistics methods such as single-site entropy, two-site mutual information, and transient length to classify CA "average" behavior at each λ value. Shannon's work on the Theory of Communication [11] has formed a major part of the entropy work done by Langton.

The notion of "average behavior" is intended to capture the most likely behavior observed with a randomly chosen initial configuration for CAs in a fixed-λ subspace. These studies revealed correlation between the various statistics and λ. The correlation is quite good for very low and very high λ values. However, for intermediate λ values in finite-state CAs, there is a large degree of variation in behavior.

Langton claimed on the basis of these statistics that as λ is incremented from 0 to 1, the average behavior of CAs undergoes a "phase transition" from ordered (fixed point or limit cycle after some short transient period) to chaotic (apparently unpredictable after some short transient period). As λ reaches a "critical value" λ_c, the claim is that rules tend to have longer and longer transient phases. Also CAs close to λ_c tend to exhibit long-lived, "complex"-nonperiodic, but nonrandom-patterns. Langton proposed that the λ_c regime roughly corresponds to Wolfram's Class 4 CAs [15], and he then hypothesized that "computationally capable" CAs and, in particular, CAs capable of universal computation will have "critical" λ values corresponding to a phase transition between ordered and chaotic behavior. Packard experimentally tested this hypothesis by using a genetic algorithm (GA) to evolve CAs to perform a particular complex computation [9]. He interpreted the results as showing that the GA tends to select CAs close to "critical" λ regions, hence the "edge of chaos" [8].

Langton specified four regions according to the rule of λ as given in Figure 4.14. The following describes the range of λ parameter.

1. If *all* configurations map to quiescent state:
   $\lambda = 0$

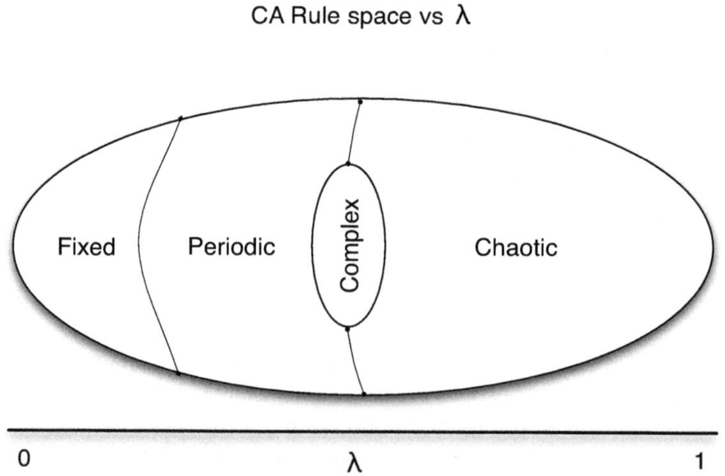

Fig. 4.14 Schematic of CA Rule space vs λ

2. If *no* configurations map to quiescent state:
   $\lambda = 1$
   a. If every state is represented *equally*:
      $\lambda = 1 - 1/k$

4.3 Antichaos and Self-organization

A refreshing look at Langton's work was made by Stuart Kauffman in his article on *AntiChaos and Adaption* [4]. Kauffman is a proponent of self-organization ability of a complex system, where he believes that biological order reflects in part a spontaneous order on which selection has acted, and that the capacity to evolve and adapt is directly linked to evolution rules [4].

Kauffman deduced that chaos is a part of behavior of complex system. A very important observation made by him is that any network or dynamical system, which is finite and oscillatory, will have a finite number of states, and must eventually reenter a state it has previously encountered. It will consequently cycle repeatedly through the same states. These states can be labelled as **dynamic attractors** of a system which once a system's trajectory carries it onto a state cycle, it stays there. The set of states that flow into a cycle or that lie on it constitutes the **basin of attraction** of the state cycle.

Left to itself, a network will eventually settle into one of its state cycle attractors and remain there. Yet if the network is perturbed in some way, its trajectory may change. Two types of perturbation are possible: *minimal* perturbations and *structural* perturbations [4].

Minimal perturbation is a change of a single or few *states* of the system; equivalent to the changing of bits in a boolean system. Two outcomes are possible; the first is that the system will stay in the same basin of attraction, or the second, that the change will be enough to push it out of the basin of attraction and change its trajectory.

Structural perturbation is a permanent mutation in the system, equivalent to a change of system rule. This will cause a permanent change in the system and its underlying stability.

4.3.1 A Butterfly Sleeps

Kauffman studied the effects of self organization on the random $K = N$ boolean networks where N is the number of boolean logic elements and K is the number of inputs into each element. These networks are of interest since the number of inputs to each element equals the total number of elements - in other words, everything is connected to everything else [3] . The state S of each element at a given time t is given as $S_i(t) \in \{0,1\}\,(i = 1,..,N)$.

Each state S_i is dynamically (randomly) updated by means of a boolean function A_i. This dynamical system now can be defined as Equation 4.2.

$$S_i(t+1) = A_i\left[S_1(t), S_2(t), \ldots .S_K(t)\right] \tag{4.2}$$

During his experimentations, he observed the following points of interest:

- As the number of elements N increase in a network, the length of the state cycles grows exponentially.
- The average length of a state cycle is around the square root of the number of different states.
- Due to random succession of states, there is maximal sensitivity to initial conditions.
- Number of state cycles (basins of attraction) is very small.
- The expected number of state cycles equals the number of elements divided by the logarithmic constant e.
- The stability of an attractor is proportional to its basin size, which is the number of states of trajectories that drain into the attractor.

A strange occurrence is when the number of inputs per element, K decreases to 2; $K=2$. This system is very stable, in almost all perturbations, with only structural perturbation having slight effect; hence the butterfly sleeps [4].

Consequently, in random networks with only two inputs per element, each attractor is stable to most minimal perturbations. Similarly, most mutations in such networks alter the attractors only slightly. The ordered network regime is therefore characterized by a homeostatic quality: networks typically return to their original attractors after perturbations [4].

4.3.2 Chaos and Antichaos

Kauffman picked up Langton's approach for CA to describe the application of network interactivity. Network connectivity and bias is considered the most important aspect of self organization. The influence a section of network has to another has lasting effect on the final trajectory of the network. Some network behavior is ordered while others are chaotic.

The transcendence between the two regions are controlled by the bias of the network with K acting as the key parameter [13]. Langton proposed the following analogy to describe a system; ordered networks are solid, chaotic networks are gaseous and networks in an intermediate state are liquid. In order to move from one state to another require the lowering of the bias to a critical value. At the point where the network is in "liquid" state, the **edge of chaos** of the network is reached and this is where a number of interesting dynamic behavior emerges [4]. The bias of the random boolean network K_c is set as 2, where the chaotic regime is in $K > K_c$ and the ordered regime is in $K < K_c$. This critical point $K_c = 2$ was analytically determined by [2].

Systems poised in the liquid transition state may also have special relevance to evolution because they seem to have the optimal capacity for evolving. A *antichaotic* system is one which changes from a chaotic to an ordered system with the freezing of its network, while a *chaotic* system is one which changes from a ordered to a chaotic system with the melting of its network.

4.4 Edge of Chaos in Evolutionary Algorithms

The major impact of Langton and Kauffman works on chaos is that they have demonstrated that chaos is not just in the preview of non-linear dynamical systems, but their impact can be rightly felt in discrete systems. Kauffman most importantly showed that his interpretation of chaos is not in the low dimensional deterministic chaos but in a phase where *damage spreading* takes place.

Evolutionary algorithms have a common base with evolution, since they are based on the fundamental's of natural selection.The major advent of these algorithms has been in the proliferation on complex engineering and mathematical problems, The problems usually require robust techniques for resolutions, and evolutionary algorithms are at its forefront.

Most evolutionary algorithms share a common framework. A population P is randomly generated consisting a number of solutions S, each of which is of size x. This x is defined by the particular problem being solved $F(x)$. Each solution is vetted for its fitness by $F(x)$. Each solution is then systematically combined with another randomly selected solution using *selection rules* and the new solution is then vetted for its fitness. If this new solution S_{i+1} is better then the older solution S_i, it then replaces the older solution in the population. Iteratively, all solutions are subjected to this routine. The routine continues for a number of generations G.

4.4.1 Stagnation

The major drawback of evolutionary algorithms is that occasionally the solutions converge to a particular position in the solution space. What is most undesirable is that these solutions are then unable to move away to another region, and thus the entire evolution process stagnates. This phenomena is usually referred to as *local optima convergence* or *stagnation*.

Stagnation is of a special interest in evolutionary algorithms. Under what conditions do random systems converge or as Kauffman stipulated - get frozen?

Firstly, a linkage has to be found between evolutionary algorithms and boolean systems. Evolutionary algorithms share a common background with the Kauffman's Random Boolean Networks (RBN). Both are randomly generated. The selection criteria for each new solution is synchronous and random as in RBN.

Secondly, the main attribute of evolutionary algorithms is the *bias* in the system as in RBN. These can be viewed as the rules or selection criteria of an evolutionary algorithms. Using these bias, it is feasible for solutions to converge and form basins of attraction.

A solution like a gene, is of finite length, and thus has a fixed number of *states* which it cycles through evolution. The premise is that evolution would somehow take each variable in a solution through a different trajectory on each iteration, which would then assist in the mapping of the solution space and allow for the discovery of the global optimal solution.

However, the selection criteria is *biased* with the other solutions in the population, from which the new solution obtains new genetic material for evolution. If both selected solutions for mating have the same *state* in the population, then the new solution will replicate this information and therefore increase the basin of attraction. During evolution, these basin of attractions would encompass a majority of solutions and based on the probability of random selection, the opportunity for solutions to discover new regions in solution space decreases. In effect the population stagnates.

4.4.2 Anti-stagnation

Stagnation is usually observed form the standpoint of the fitness of the solution in respect of the problem being solved. However, using the Kauffman analogy, stagnation can be viewed from the perspective of the solution, and the *state* it occupies in respect to the other solutions in the population. Using the *damage spreading* viewpoint, the question then arises as to what can exactly be done to have *anti-chaos*, or the free movement of the solutions in a evolutionary algorithms?

An approach can be the introduction of **reverse bias** of a population. This is equivalent to the *structural perturbation* of the system. A number of *bias rules* will enforce the phase shift between ordered and chaotic behavior. However, these rules cannot encroach on the underlying heuristic since they will permanently shift its paradigm.

4.5 Analytical Observation

A brief analytical observation is presented in this section in order to describe the stagnation process in an evolutionary algorithm. A simplified generic Genetic Algorithm (GA) is selected as the evolutionary algorithm. The operating parameters are given in Table 4.1.

Table 4.1 GA operating parameters

Parameters	Value
Population size	30
Solution size	20
Crossover	2 point
Mutation rate	Single
Generation	100

The problem to be solved is a *strict sense permutative* scheduling problem of flowshop (FSS). This problem is selected since a permutative problem is, first and foremost, *discrete*, and secondly since it is permutative, its *state* cycle can easily be observed.

The minimization of completion time (makespan) for a flow shop schedule is equivalent to minimizing the objective function $\Im$:

$$\Im - \sum_{j=1}^{n} C_{m,j} \tag{4.3}$$

s.t.

$$C_{i,j} = \max\left(C_{i-1,j}, C_{i,j-1}\right) + P_{i,j} \tag{4.4}$$

where, $C_{m,j}$ = the completion time of job j, $C_{i,j} = k$ (any given value), $C_{i,j} = \sum_{k=1}^{j} C_{1,k}$; $C_{i,j} = \sum_{k=1}^{j} C_{k,1}$ machine number, j job in sequence, $P_{i,j}$ processing time of job j on machine i [10] .

The problem instance to be solved is the first problem instance of the Taillard flowshop benchmark set [14].

4.5.1 Diversity Measure

In order to measure the trajectory of each solution, it becomes imperative to define some analytical measure of the solution. In terms of permutative strings, where each variable in a solution is unique and discrete, the deviation of the solution can be assigned as one of the diversity measures. The formulation for its calculation is given in Equation 4.5, where n is the size of the solution, and x represents each variable in the solution.

$$\delta = \left(\frac{\sum_{i=1}^{n-1} |x_i - x_{i+1}|}{n}\right); \; x_i \in \{x_1, x_2, \ldots x_n\} \tag{4.5}$$

The second measure is the frequency of each solution in the population. This shows the level and general depth of the basin of attraction in the solution space. This is easily given through the use of histograms.

The third and final measure is the representation of the solutions in 3D space. This can be done through catenation of each solution string into 3 parts. What this does, is give a clear representation as to how many basins of attraction exist, and most importantly, their distance to each other.

4.5.2 Population Representation

The population was randomly generated and iterated for 100 generations. The population is represented at ten generation/iteration intervals with the three diversity measures as given in Figures 4.15 to 4.25.

The initial population is given in Figure 4.15. The deviation plot in Figure 4.15a displays the population as scattered in the deviation space, with the frequency plot in Figure 4.15b displaying the same trait. The 3D representation in Figure 4.15c shows a wide range of solution displacement in the population. This is typical for random population generation.

From iteration 10 to 30, shown in Figures 4.16 - 4.18, it can be seen that the solutions are drifting in the solution space. There is a small level of attraction between the solutions as shown in the histograms, where the number of unique solutions decreases, and the basins of attraction becomes larger. The 3D representation also displays the similar traits with the displacement between the different basins reducing in size.

From iteration 40 to 70, shown in Figures 4.19 - 4.22, a clear indication of the emergence of the basins of attraction is shown. The deviation plot shows the solutions collating in uniques regions, and the histogram shows in increasing depth of these basins. The 3D representation clearly shows that the distance between these basins is decreasing in size.

Stagnation of the population is seen from iteration 80 onwards in Figures 4.23 - 4.25. At this point only 3 main basin of attractions exist in the population as shown in the 3D plots, which occupy 2 distinct deviation points in the deviation plots. At time increases, the solutions simply drift from one basin to another without being able to find new regions of space. At this point there is simply *genetic drift* of the population with no new solutions being produced. Evolution can be said to have stagnated.

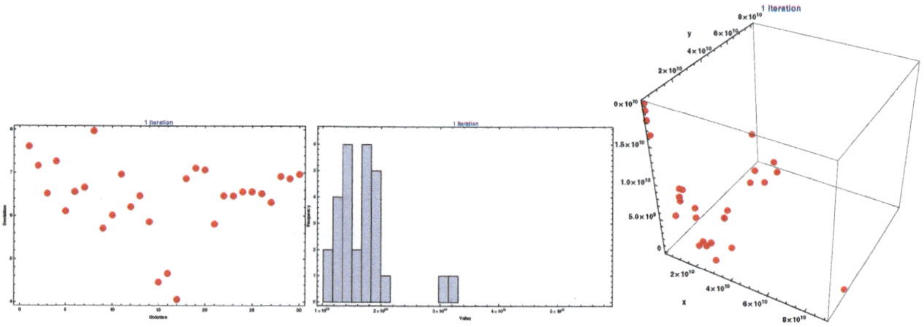

(a) Deviation **(b)** Histogram **(c)** 3D representation

Fig. 4.15 Initial state

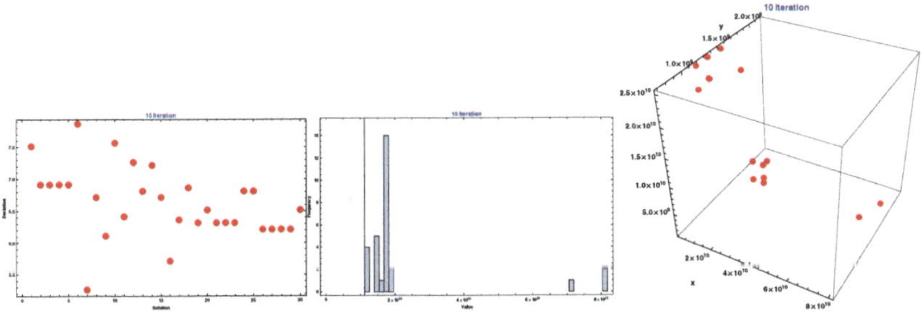

(**a**) Deviation (**b**) Histogram (**c**) 3D representation

Fig. 4.16 Iteration 10

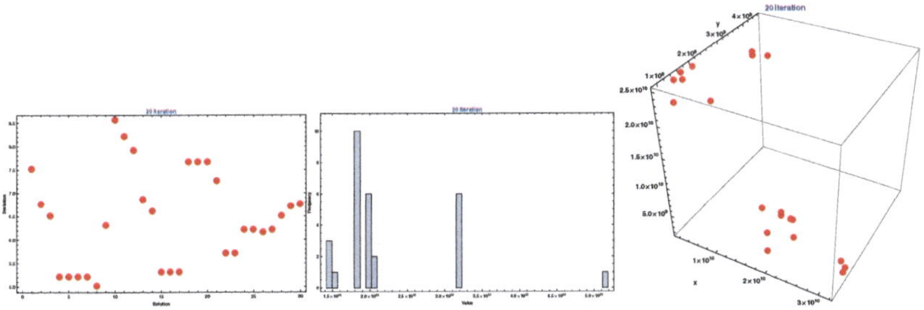

(**a**) Deviation (**b**) Histogram (**c**) 3D representation

Fig. 4.17 Iteration 20

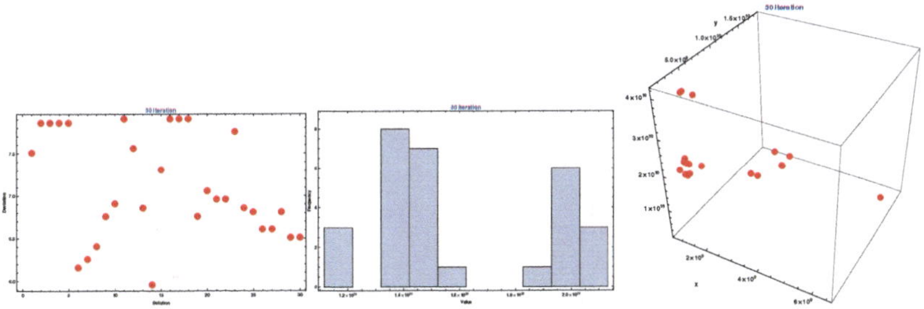

(**a**) Deviation (**b**) Histogram (**c**) 3D representation

Fig. 4.18 Iteration 30

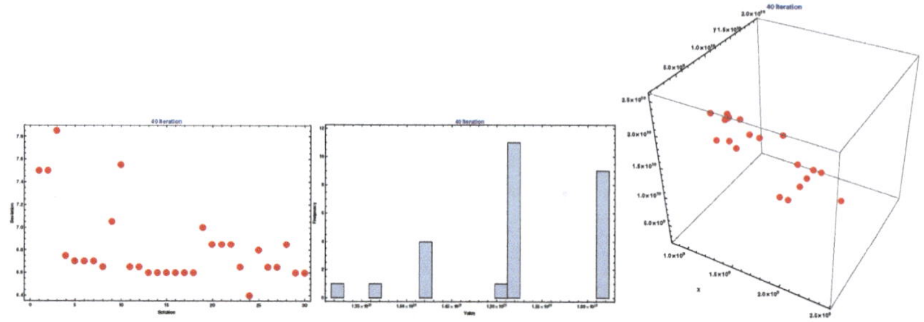

(**a**) Deviation (**b**) Histogram (**c**) 3D representation

Fig. 4.19 Iteration 40

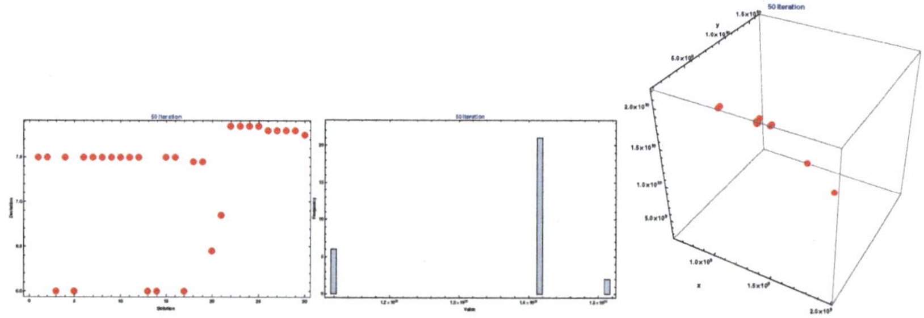

(**a**) Deviation (**b**) Histogram (**c**) 3D representation

Fig. 4.20 Iteration 50

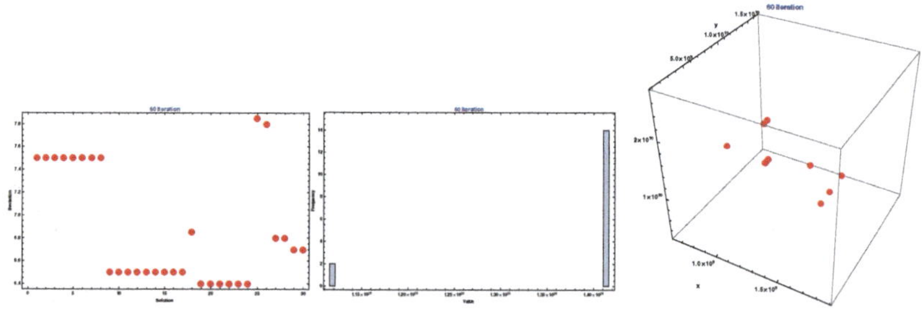

(**a**) Deviation (**b**) Histogram (**c**) 3D representation

Fig. 4.21 Iteration 60

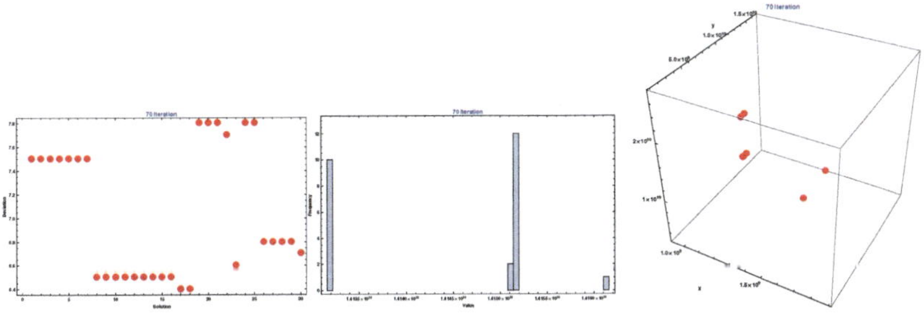

(a) Deviation (b) Histogram (c) 3D representation

Fig. 4.22 Iteration 70

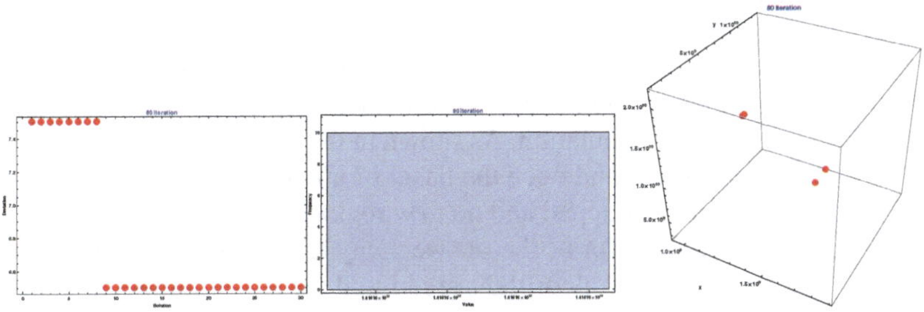

(a) Deviation (b) Histogram (c) 3D representation

Fig. 4.23 Iteration 80

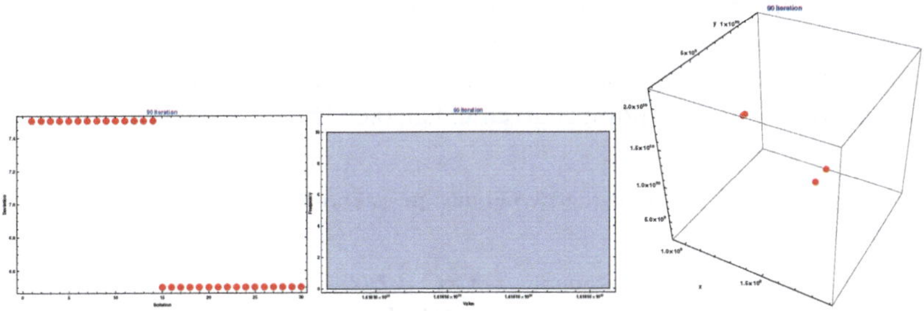

(a) Deviation (b) Histogram (c) 3D representation

Fig. 4.24 Iteration 90

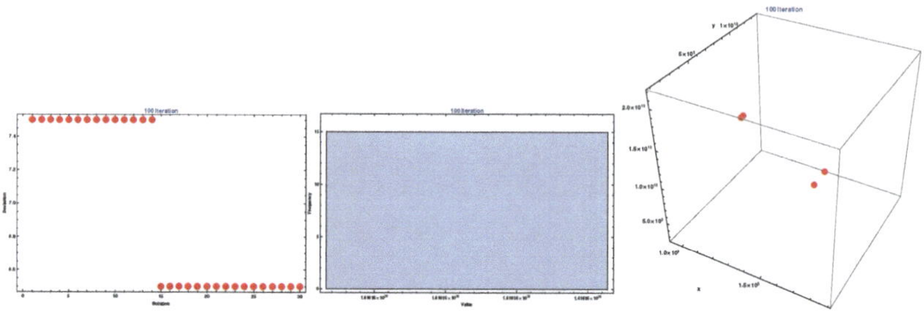

(**a**) Deviation (**b**) Histogram (**c**) 3D representation

Fig. 4.25 Iteration 100

4.6 Conclusion

Stagnation as described in the chapter is a major concern to researchers. These regions first and foremost decrease the evolutionary process and secondly destroy genetic information in the population. As shown in the analytical example, these regions form during evolution and once the basin of attraction become large enough, all the solutions are simply trapped and no new regions are then explored.

As to how these regions form, the answer may lie in the use of the *selection rules* or *bias*. All evolutionary algorithms are biased towards finding regions of best fitness, however the engine that drives this process is the uniqueness of the solutions in the population. Stagnation has the opposite effect, where all solutions are simply cloned. This is the drift from *ordered* to *destructive* behavior, the transition from ordered to chaotic system. The solutions are frozen and simply drift in the attractors which have formed between the basin of solution.

From an engineering point of view, this is highly undesirable. *Reverse bias* rules can provide an answer to this predicament which can keep the trajectories of the solutions separated. An approach concerning the creation and development of such rules based on this presumption is given in Chapter 15.

Acknowledgements. The following two grants are acknowledged for the financial support for this research.

1. Grant Agency of the Czech Republic **GARC 102/09/1680**
2. Grant of the Czech Ministry of Education **MSM 7088352102**

References

1. Bassett, D., Meyer-Lindenberg, A., Achard, S., Duke, T., Bullmore, E.: Adaptive reconfiguration of fractal small-world human brain functional networks. The National Academy of Sciences of the USA 103(51), 19518–19523 (2006)
2. Derrida, B., Pomeau, Y.: Europhysics Letters 1, 45 (1986)

3. Fronczak, P., Fronczak, A.: Critical line in undirected Kauffman Boolean networks - the role of percolation. J. Phys. A: Math. Theor. 41, 224009 (2008)
4. Kauffman, S.: Antichaos and Adaptation. Scientific America, 78–84 (August 1991)
5. Kauffman, S.: At Home in the Universe. Oxford University Press, Oxford (1995)
6. Kitzbichler, M., Smith, M., Christensen, S., Bullmore, E.: Broadband Criticality of Human Brain Network Synchronization. PLoS Comput. Biol. 5(3), e1000314 (2009)
7. Langton, C.: Computation at the edge of chaos: Phase transitions and emergent computation. Physica D 42, 12–37 (1990)
8. Mitchell, M., Hraber, P., Crutchfield, J.: Revisiting the Edge of Chaos: Evolving Cellular Automata to Perform Computations. Complex Systems 7, 89–130 (1993)
9. Packard, N.: Adaptation toward the edge of chaos. In: Kelso, J., Mandell, A., Shlesinger, M. (eds.) Dynamic Patterns in Complex Systems, pp. 293–301. World Scientific, Singapore (1988)
10. Pinedo, M.: Scheduling: theory, algorithms and systems. Prentice Hall, Inc., New Jersey (1995)
11. Shannon, C.: A Mathematical Theory of Communication. Bell System Technical Journal, 623–656 (October 27, 1948)
12. Shermer, M.: Exorcising Laplace's Demon: Chaos and Antichaos, History and Metahistory. History and Theory 34(1), 59–83 (1995)
13. Sole, R., Luque, B.: Phase transitions and antichaos in generalized Kauffman networks. Phys. Lett. 196, 331–334 (1995)
14. Taillard, E.: Benchmarks for basic scheduling problems. Eur. J. Oper. Res. 64, 278–285 (1993)
15. Wolfram, S.: A New Kind of Science. Wolfram Media, Champaign (2002)

Part II
Applications

Chapter 5
Evolutionary Design of Chaos Control in 1D

Roman Senkerik, Ivan Zelinka, Donald Davendra, and Zuzana Oplatkova

Abstract. The main aim of this work is to show that powerful optimizing tools like evolutionary algorithms can be in reality used for the optimization of deterministic chaos control. This work is aimed on explanation of how to use evolutionary algorithms (EAs) and how to properly define the cost function (CF). It is also focused on selection of control method and, the explanation of all possible problems with optimization which comes together in such a difficult task, which is chaos control.

5.1 Introduction

In general, methods for control of chaos deal with a process wherein a tiny perturbation is applied to a chaotic system in order to realize a desirable chaotic, periodic or stationary behavior. The problems of control of chaos have attracted the attention of researchers and engineers since the early 1990's.

Many methods for control of chaos [8] have been developed and based on the original OGY control method [26]. The main principle consisted in waiting for a natural passage of the chaotic orbit close to the desired periodic behavior and then applying a small perturbation, in order to stabilize the system. This is the main principle of OGY method - Linearization of Poincare Map [10] - [1]. Moreover there exist also special versions which use the pole placement principle [13], [11].

Roman Senkerik · Donald Davendra · Zuzana Oplatkova
Tomas Bata University in Zlin, Faculty of Applied Informatics, Nad Stranemi 4511, Zlin 76001, Czech Republic
e-mail: {senkerik,davendra,oplatkova}@fai.utb.cz

Ivan Zelinka
Tomas Bata University in Zlin, Faculty of Applied Informatics, Nad Stranemi 4511, Zlin 76001, Czech Republic
and
VSB-TUO, Faculty of Electrical Engineering and Computer Science, 17. listopadu 15, 708 33 Ostrava-Poruba, Czech Republic
e-mail: zelinka@fai.utb.cz

I. Zelinka et al. (Eds.): Evolutionary Algorithms and Chaotic Systems, SCI 267, pp. 165–190.
springerlink.com

But there is generally one big disadvantage of OGY and it is the long initial chaotic transient before trajectories are stabilized. Consequently many targeting algorithms were introduced to shorten the time of stabilization. The question of targeting with application to chaos control has attracted the researchers. The several first approaches for targeting have used special versions of OGY control scheme [5], [21] or collecting of information about trajectories, which fall close to desired state [10]. Also the Pyragas control method is mentioned in this paper as the tool for successful targeting [22]. Later, lots of methods were developed based on adaptive approach [30], optimal feedback controller [31], center manifold targeting [38] or neural networks [18], [19].

Unlike the OGY the Pyragas's delayed feedback control technique [10], [20] can be simply considered as targeting and stabilizing algorithm together in one package. From the point of view of soft computing and optimizations another big advantage of Pyragas method is the amount of accessible control parameters, which are set up by using a priori knowledge or mathematical analysis. This is very advantageous for successful use of optimization of parameters set up by means of EA, leading to avoidance of any kind of mathematical analysis and mainly to improvement of system behavior and better and faster stabilization to the desired periodic orbits.

During recent years, a lot of other control techniques have been developed. Some of them are based on classical linear control law - Open loop and Open plus loop [18], [7], or they represent new approaches in linear/nonlinear deterministic chaos control such as adaptive nonlinear control [49] - [17], an indirect adaptive control algorithm [14], Lyapunov-Bellman technique [2], impulsive control [39] - [41], sliding modes technique [6], using of neural network controller [40], fuzzy control [3] or back-stepping design [23]-[24].

5.2 Evolutionary Techniques in Chaos Control

These days the evolutionary algorithms (EA) are known as powerful tool for almost any difficult and complex optimization problem. But the quality of optimization process results mostly depends on proper design of used cost function, especially when the EAs are used for optimization of chaos control. The results of numerous simulations lends weight to the argument that deterministic chaos in general and also any technique to control of chaos are sensitive to proper parameter set up, initial conditions and in the case of optimization they are also extremely sensitive to the construction of used cost function.

The main aim of this work is focused on the examples of EA implementation to methods for chaos control for the purpose of obtaining better results, which means faster reaching of desired stable state and superior stabilization, which could be robust and effective to optimize difficult problems in the world. In other words this work deals with an investigation on the optimization of the control of chaos by means of EA and constructing of the cost function securing the improvement of system behavior and faster stabilization to desired periodic orbits. The control law is based on two Pyragas methods: Delay feedback control - TDAS and Extended delay feedback control - ETDAS. As models of deterministic chaotic systems, one dimensional Logistic equation and two dimensional Henon map were used. The evolutionary algorithm SOMA (Self-Organizing Migrating Algorithm) and Differential Evolution (DE) were used. Also the comparison with classical control technique - OGY is presented.

Some research in this field has been recently done using the evolutionary algorithms for optimization of local control of chaos based on a Lyapunov approach [33], [32]. But the approach described here is unique and novel and up to date were not used or mentioned anywhere. We use EA to search for optimal setting of adjustable parameters of arbitrary control method to reach desired state or behavior of chaotic system.

In general we would like to show how evolutionary algorithms can be used in the challenging task of optimization of deterministic chaos control. The main principle of our approach and the role of EA are depicted in Fig. 5.1.

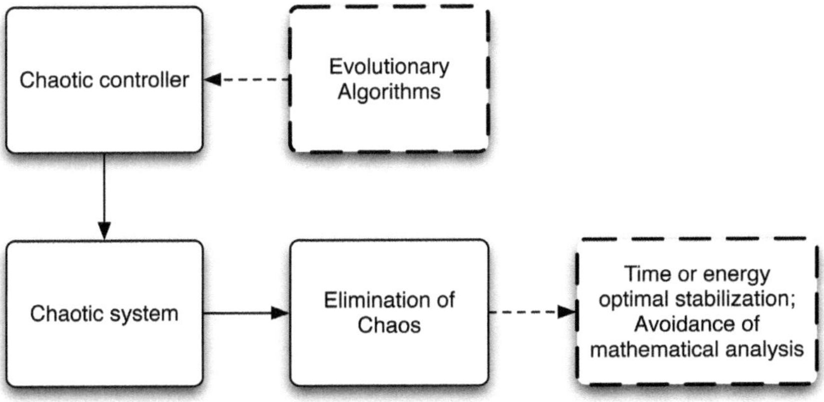

Fig. 5.1 The scheme of evolutionary chaos control optimization

5.3 Chaotic Systems

5.3.1 Logistic Equation

The logistic equation (logistic map) (LQ) is a one-dimensional discrete-time example of how complex chaotic behavior can arise from very simple non-linear dynamical equation. This chaotic system was introduced and popularized by the biologist Robert May [25]. It was originally introduced as a demographic model by Pierre Francois Verhulst as a typical predator - prey relationship. Mathematical notation is given by Equation 5.1 [15]:

$$x_{n+1} = rx_n(1 - x_n) \tag{5.1}$$

Where (in case of biological meaning) x_n is the population at year n, and r is a positive number, which represents a special parameter - combination of rate for reproduction and starvation.

The chaotic behavior can be observed by varying the parameter r. At $r = 3.57$ is the beginning of chaos, at the end of the period-doubling behavior. At $r > 3.57$ the system exhibit chaotic behavior. All of this behavior can be clearly seen from bifurcation diagram (Fig. 5.1).

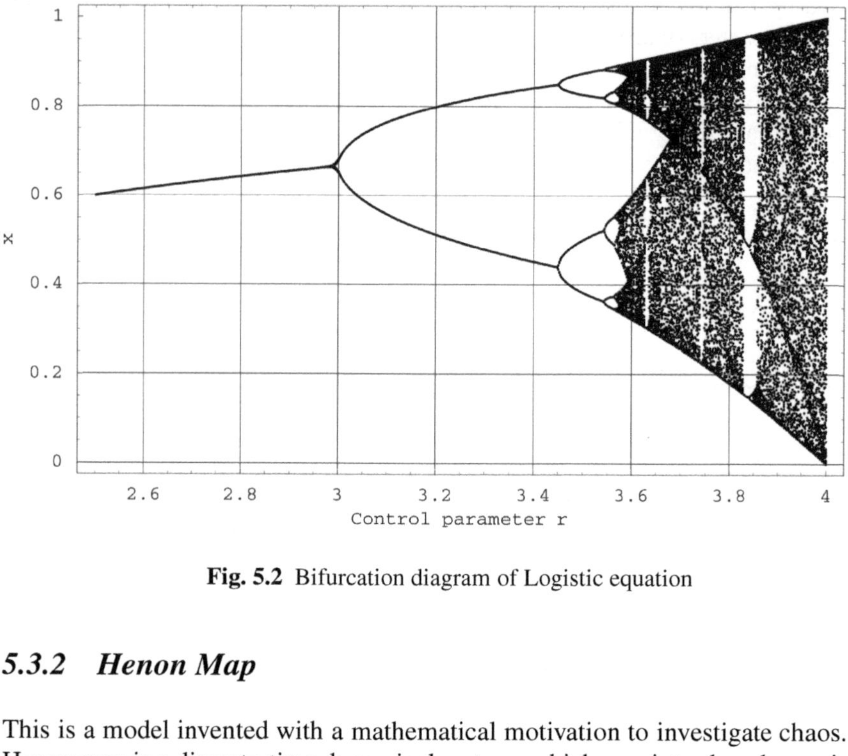

Fig. 5.2 Bifurcation diagram of Logistic equation

5.3.2 *Henon Map*

This is a model invented with a mathematical motivation to investigate chaos. The Henon map is a discrete-time dynamical system, which was introduced as a simplified model of the Poincare map for the Lorenz system. It is one of the most studied examples of dynamical systems that exhibit chaotic behavior and in fact it is also a two-dimensional extension of the one-dimensional quadratic map.

Mathematical notation is given by Equation 5.2 [15]:

$$\begin{aligned} x_{n+1} &= 1 + y_n - a x_n^2 \\ y_{n+1} &= b x_n \end{aligned} \tag{5.2}$$

The map depends on two parameters, a and b. For the values of a = 1.4 and b = 0.3 the Henon map is chaotic. For other values of a and b the map may be chaotic, intermittent, or converge to a periodic orbit.

Fig. 5.3 shows the bifurcation diagram for the Henon map created by plotting of variable x as a function of the one control parameter for the fixed second parameter.

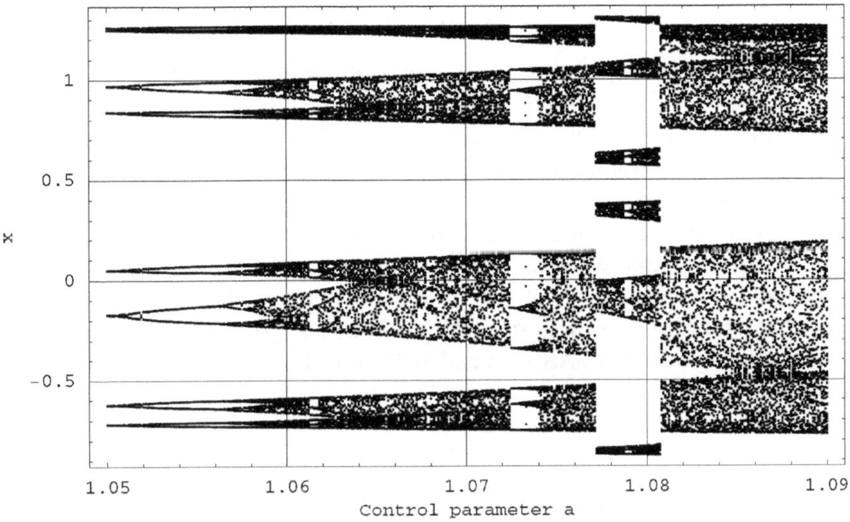

Fig. 5.3 Bifurcation diagram of Henon map

5.4 Selected Method for the Controlling of Chaos

Subsequently is given a description of method used for all optimizations and simulations.

5.4.1 Delayed Feedback Control (Pyragas Method)

This is a method developed to stabilize UPO by means of applying small time-continuous control (perturbation) to a system parameter in continuous time. This is the main difference from OGY method which is suitable for a discrete control. It is also known as the Time Delayed Auto Synchronization (TDAS method) and it was proven that it is very easy to implement and is effective for the less order UPOs, i.e. orbits with smaller periods. This is one of the most important limitations for this technique. There also exists the discrete version suitable for control the chaos within chaotic maps [27].

It is assumed the system P is described by variables x with F as an external controllable parameter, which has numerical value $F = 0$ in the absence of control (external perturbation) Equation 5.3.

$$\frac{dx}{dt} = P(x) + F(t) \tag{5.3}$$

Desired UPO of period τ which fulfills the following logical condition $x(t+\tau) = x(t)$ can be stabilized by means of delayed feedback control by calculating and applying control parameter F to the system based on following control law 5.3.

$$F(t) = K[x(t-\tau) - x(t)] \tag{5.4}$$

where: the parameter K represents the strength of the perturbation. By proper choice of the value of K, the desired UPO may be stabilized. The big advantage of this method lies in the fact, that there is no need of additional information about UPO except its period τ or only its order in case of discrete-time control.

Once the control is achieved, the size of the perturbation is very small, although during the previous chaotic transient passage it may be very large and of have to be limited. But this kind of absence of perturbation can lead to either low quality stabilization or none, especially in case of higher periodic orbits. Due to this problem, the extended version of delayed feedback control method was developed to solve it. (Also called ETDAS - Extended Time Delayed Auto Synchronization) Equation 5.6 [29].

$$\begin{aligned} \frac{dx}{dt} &= P(x) + F(t) \\ F(t) &= K[(1-R)S(t-\tau) - x(t)] \\ S(t) &= x(t) + RS(t-\tau) \end{aligned} \tag{5.5}$$

where: R is adjustable constant and S is given by a delay equation utilizing previous states of the system.

This modification particularly solved the problems with stabilization of higher order UPOs in discrete or continuous time systems.

This method is very simple and can be applicable to a wide variety of systems; of course it is possible to use it for discrete-time systems. There are only small changes in the form of equations 5.3, 5.4 and 5.6. The discrete-time version of TDAS method has the following form 5.6:

$$\begin{aligned} x_{n+1} &= P(x_n) + F_n \\ F_n &= K[x_{n-m} - x_n] \end{aligned} \tag{5.6}$$

The discrete-time version of ETDAS method has form 5.8:

$$\begin{aligned} x_{n+1} &= P(x_n) + F_n \\ F_n &= K[(1-R)S_{n-m} - x_n] \\ S_n &= x_n + RS_{n-m} \end{aligned} \tag{5.7}$$

Where the symbol m represents the order of desired UPO.

5.5 Evolutionary Algorithms

Four versions of SOMA (AllToOne (ATO), AllToOneRand (ATR), AllToAll (ATA), AllToAllAdaptive (ATAA)) [45] and six versions of DE (DERand1Bin, DERand2Bin, DEBest2Bin, DELocalToBest, DERand1DIter, DEBest1JIter) [28] were used for all simulations. See Table 5.1 and Table 5.2 for parameter setting. These parameters for optimizing algorithms were set up in this "common" way in order to reach the same value of maximal CF evaluations. This fact is very important due to further possibility of creating the complete statistical overview of EAs performance. This statistical summary is significant not only for comparison of both used evolutionary algorithms and its versions but for example in the task of the decision, as to which algorithm gives better results for all runs when final CF value of the best individual solution is the same as the CF Value of other best individual solution given by different versions or algorithms.

Table 5.1 Parameter settings for SOMA

Parameter / Version	ATO/ATR	ATA/ATAA
PathLength	3	3
Step	0.33	0.33
PRT	0.1	0.1
PopSize	25	10
Migrations	25	7
Max. CF Evaluations (CFE)	5400	5670

Table 5.2 Parameter settings for DE

Parameter	Value
F	0.9
Cr	0.2
PopSize	25
Generations	215
Max. CF Evaluations (CFE)	5375

5.6 Optimization of Chaos Control

5.6.1 Problem Design

This section primarily consists of five case studies. All of them are focused on estimation of accessible control parameters for TDAS or EDTAS method for five proposed Cost Functions used in optimizations to stabilize selected UPOs, which are the following: p-1 (a fixed point), p-2 and p-4 (examples of higher periodic orbits). The chosen examples of chaotic systems were one dimensional Logistic equation in the form 5.1 [15] and two dimensional Henon map in the form 5.2 [15]. Here is the list of desired UPOs:

Logistic Equation with $r = 3.8$:
p-1 (fixed point): $x_F = 0.73842$
p-2 orbit: $x_1 = 0.3737, x_2 = 0.8894$
p-4 orbit: $x_1 = 0.3038, x_2 = 0.8037, x_3 = -0.5995, x_4 = 0.9124$

Henon Map with $a = 1.2$ and $b = 0.3$:
p-1 (fixed point): $x_F = 0.8$
p-2 orbit: $x_1 = -0.562414, x_2 = 1.26241$
p-4 orbit: $x_1 = 0.139, x_2 = 1.4495, x_3 = -0.8595, x_4 = 0.8962$

All simulations were mostly repeated 50 times for each EA version, in order to find the actual optimum and to show and check robustness of used method. The control method - original TDAS has the form 5.3, 5.4 [27] and ETDAS has the form 5.6 [29].

In the case of Logistic Equation (LQ), optimization proceeded with these parameters: K and F_{max} for TDAS control method, which is obtained in the form 5.8 after modification into discrete form suitable for logistic equation.

$$\begin{aligned} x_{n+1} &= rx_n(1-x_n)+F_n \\ F_n &= K[x_{n-m}-x_n] \end{aligned} \tag{5.8}$$

The question as to why the TDAS was chosen and used in selected cases of p-1 orbit, although it has proven lower stabilizing performance, has this simple answer. To avoid any long simulations and evolutionary computations it is better to search in lower dimensional space and to work with simpler control algorithm, which is not so demanding for computational time. Furthermore numerous simulations proved that performance of this control technique in case of p-1 orbit is very satisfactory.

Due to problems with stabilization of higher periodic orbits, it was necessary to try the optimization by EA for another control method - ETDAS in the form 5.9 suitable for the logistic equation. Thereafter optimization proceeded with these parameters: K, F_{max} and R.

$$\begin{aligned} x_{n+1} &= rx_n(1-x_n)+F_n \\ F_n &= K[(1-R)S_{n-m}-x_n] \\ S_n &= x_n+RS_{n-m} \end{aligned} \tag{5.9}$$

In case of the Henon map, the ETDAS control method was used for all simulations in the form 5.10 after modification into discrete form suitable for the used system.

$$\begin{aligned} x_{n+1} &= a-x_n^2+by_n+F_n \\ F_n &= K[(1-R)S_{n-m}-x_n] \\ S_n &= x_n+RS_{n-m} \end{aligned} \tag{5.10}$$

All results are shown only for variable x of two dimensional Henon map because of its form 5.4-5.8, where the variable y has the same values as variable x but it is only phase shifted.

The perturbation F_n in equations 5.8-5.10 may have arbitrarily large value, which can cause diverging of the system outside the interval $\{0, 1\}$ for logistic equation or $\{-1.5, 1.5\}$ in the case of Henon map. Therefore, F_n should have a value between - F_{max} , F_{max} and EA should find an appropriate value of this limitation to avoid diverging of the system.

The ranges of all estimated parameters were in general these:

$$-2 \leq K \leq 2, 0 \leq F_{\max} \leq 0.5 \text{and} 0 \leq R \leq 0.5$$

The optimization interval for p-1 orbit was $\tau_i = 100$ iterations, for higher periodic orbits it was mostly $\tau_i = 150$ iterations.

5.6.2 The Cost Function

In this work several types of cost function (CF) were developed and tested for stabilization of p-1 orbit (fixed point) and higher periodic orbits (p-2 and p-4). The CF has been calculated in general from the distance between desired state and actual system output. The minimal value of this cost function revealing the best solution is zero. The aim of all the simulations was to find the best solution that returns the cost function value as close as possible to zero.

5.6.2.1 Basic CF – Case Study 1

This proposal of the basic cost function is in general based on the simplest CF, which could be used only for the stabilization of p-1 orbit. The idea was to minimize the area created by the difference between the required state and the real system output on the whole simulation interval - τ_i.

But another cost function (CF) had to be used for stabilizing of higher periodic orbit. It was synthesized from the simple CF and other terms were added. In this case, it is not possible to use the simple rule of minimizing the area created by the difference between the required and actual state on the whole simulation interval - τ_i, due to the many serious reasons, for example: degrading of the possible best solution by phase shift of periodic orbit.

This CF, is in general based on searching for desired stabilized periodic orbit and thereafter calculation of the difference between desired and found actual periodic orbit on the short time interval - τ_s (approx. 20 - 50 iterations) from the point, where the first min. value of difference between desired and actual system output is found. Such a design of CF should secure the successful stabilization of higher periodic orbit anywise phase shifted.

Furthermore, because of CF values being very close to zero, this CF also allows using of decision rule avoiding very time demanding simulations. This rule stops

EA immediately, when the first individual with good parameter structure is reached, thus the value of CF is lower then the acceptable (CF_{acc}) one. Typically CF_{acc} = 0.001 at time interval τ_s = 20 iterations, thus difference between desired and actual output has value 0.0005 per iteration - i.e. successful stabilization for used control technique. This CF was also used for p-1 orbit. The CFBasic has the form 5.11.

$$CF_{Basic} = penalization_1 + \sum_{t=\tau 1}^{\tau 2} |TS_t - AS_t| \tag{5.11}$$

where: TS - target state, AS - actual state
τ_1 - the first min. value of difference between TS and AS
τ_2 -the end of optimizing interval (τ_1 + τ_s)
$penalization_1$ = 0 if τ_i - $\tau_2 \geq \tau_s$;
$penalization_1$= 10* (τ_i - τ_2) if τ_i - $\tau_2 < \tau_s$ (i.e. late stabilization)

5.6.2.2 Targeting CF Simple – Case Study 2

In this case study the simplest CF proposal outlined above was used. It is based on minimizing the area created by the difference between the required state (stabilized fixed point) and the real system output on the whole simulation interval - τ, thus this proposal of CF should secure fast targeting into the close neighborhood of p-1 orbit and its stabilization. The CF_{Simple} is given by 5.12.

$$CF_{Simple} = \sum_{t=0}^{\tau_i} |TS_t - AS_t| \tag{5.12}$$

5.6.2.3 Targeting CF NA – Case Study 3

It was necessary to modify the definition of CF in order to decrease the average number of iteration required for successful stabilization and avoidance of any associated problem. The CF_{simple} is not suitable for adding any term of penalization for slowly stabilizing solutions, thus the CF_{basic} was modified to use for all required UPOs. The CF value is multiplied by the number of iterations *(NI)* of the first found minimal value of difference between desired and actual system output (i.e. the beginning of fully stabilized UPO). To avoid problems associated with CF returning value 0 and to put the penalization to similar level as the non-penalized CF value, the small constant *(SC)* is added to CF value before penalization (multiplying by *NI*). The modified CFNA has the form 5.13.

$$CF_{NA} = NI \left(SC + penalization1 + \sum_{t=\tau 1}^{\tau 2} |TS_t - AS_t| \right) \tag{5.13}$$

where: $SC = 10^{-16}$ for p-1 orbit, $SC = 10^{-8}$ for p-2 orbit.

5.6.2.4 Targeting CF Targ1 – Case Study 4

The next proposal of CF design is based on the previous one with small change, which should avoid any problems with defining the value of small constant SC in advance (especially for stabilization of higher periodic orbit). The SC value (5.16) here is computed with the aid of power of non-penalized basic part of CF (5.15).

In general, there exists two possible ways for applying the multiplication by number of iterations required for stabilization *(NI)*. The first version of final design of targeting CF (CF_{TARG1}) has the form (5.14). Here the sum of basic part of CF and automatically computed *SC* is multiplied by *NI*. Consequently, the EA should find the solutions securing the fast targeting into desired behavior of system.

$$CF_{TARG1} = NI\left(SC + penalization1 + \sum_{t=\tau 1}^{\tau 2} |TS_t - AS_t|\right) \tag{5.14}$$

where

$$EXPCF = \log_{10}\left(\sum_{t=\tau 1}^{\tau 2} |TS_t - AS_t| + 10^{-15}\right) \tag{5.15}$$

$$SC = 10^{EXPCF} \tag{5.16}$$

5.6.2.5 Targeting CF Targ2 – Case Study 5

In the second version of targeting CF (CF_{TARG2}), there is only slight change in comparison with the previous proposal. Here the number of steps for stabilization *(NI)* multiplies only the small constant *(SC)* which is counted in the same way as in the previous case (5.14). This version of targeting CF (CF_{TARG2}) has the form (5.17)

$$CF_{TARG2} = (NI \cdot SC) + penalization1 + \sum_{t=\tau 1}^{\tau 2} |TS_t - AS_t| \tag{5.17}$$

5.6.2.6 Graphical CF Overview

The difference between proposed CFs can be clearly seen in Fig. 5.4 - 5.6, which shows the dependence of CF values on the adjustable parameters K and F_{max} (left part of image - 3D diagram) and dependence of CF values on the adjustable parameter K (right part of image 2D - diagram). Possible remaining parameters were set at the best values reached in optimizations; consequently the two-dimensional diagram always shows the section of global minimum. From these figures, it is obvious that together with growing complexity of the used CF, the nonlinearity and unpredictability of CF surface also increases. Thus this is the answer for the question as to why EA were used. The illustrative examples related to CF_{TARG1} and CF_{TARG2} are not presented here due to the close graphical similarity with CF_{NA} (Fig. 5.6).

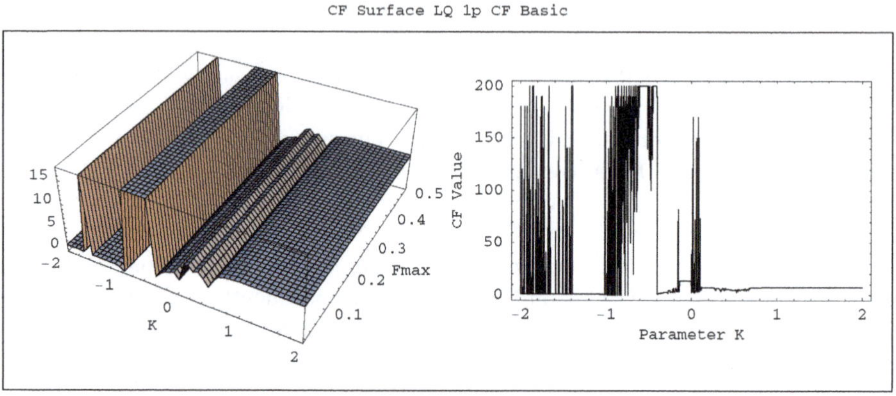

Fig. 5.4 Dependence of CF value on parameters K and F_{max}; R = 0 (left); and parameter K; F_{max} = 0.1944, R = 0 (right); CF Basic, p-1 orbit, Logistic equation, $x_{initial}$ = 0.8

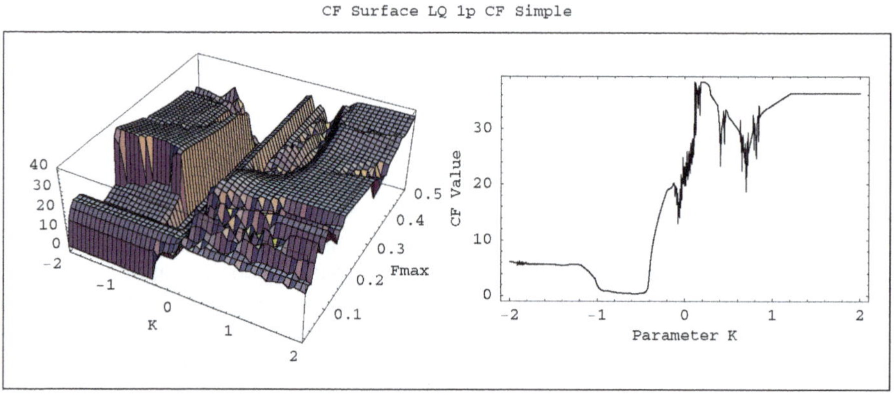

Fig. 5.5 Dependence of CF value on parameters K and F_{max}; R = 0.0180 (left); and parameter K; F_{max} = 0.1030, R = 0.0180 (right); CF Simple, p-1 orbit, Logistic equation, $x_{initial}$ = 0.8

5.6.3 Experimental Results

This section presents an accumulation of research [48] and also collates and elaborates the experiences with application of EA to chaos control [46] - [34]. It contains the brief overview of results given by optimizations by means of CF Targ2 (case study 5) developed on the basis of successful CF NA design. This new CF is able to firstly, successfully resolve the issue of fast stabilization and secondly, adds more robustness to the execution of the heuristic. The presented data lends weight to the argument, that this CF design is a serious consideration in the robust stabilization of chaotic systems for wide range of initial conditions and seem to be the best choice

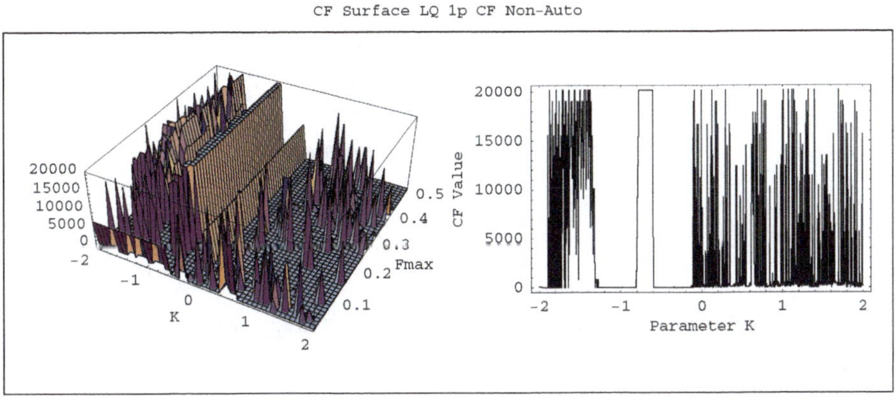

Fig. 5.6 Dependence of CF value on parameters K and F_{max}; R = 0.4977 (left); and parameter K; F_{max} = 0.3195, R = 0.4977 (right); CF NA, p-1 orbit, Logistic equation, $x_{initial}$ = 0.8

for the task of finding of "universal and robust solution". The most of the problems, which arose with previous CF designs during numerous repetitive simulations (case studies 1 - 4) were here either successfully suppressed or their negative influence to the simulations results were reduced. The only disadvantage of this proposal is relatively big computational-time demands.

The figures shows the simulation of the best individual solutions (with the lowest final CF value) given by SOMA and DE under identical initial conditions used in optimization (left part) and for the uniformly distributed initial conditions in the range $0 < x_{initial} < 1$, 100 samples were used in this kind of simulation (right part).

For the comparison of average number of iterations required for successful stabilization (all 50 repeated simulations), see Table 5.3 and Table 5.4. The value in braces represents corrected one, which shows the average IStab value (iterations required for stabilization) only for solutions, which leads to successful stabilization.

5.6.3.1 One Dimensional Example

For the excellent results of optimization in the case of one-dimensional Logistic equation, please refer to Fig. 5.7 (p-1 orbit), Fig. 5.8 (p-2 orbit) and Fig. 5.9 (p-4 orbit). From these presented results, it follows that the control method reached very good performance from the point of view of quickness and quality of the stabilization for both types of simulations. Only in the case of p-4 orbit there occurs intensifying of the unpleasant fact that this CF Targ2 allowed finding of faster stabilizing solutions, nevertheless this solutions are not suitable for complex simulation with uniformly distributed initial conditions. Thus the system was not stabilized on p-4 orbit for all 100 samples and this p-4 orbit seems to be a hard task for EA to find optimal setting up of control method.

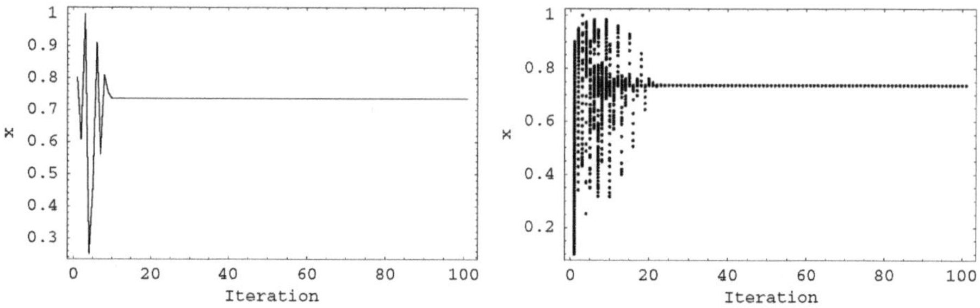

Fig. 5.7 Best individual solution, CF Targ2, p-1 orbit, DERand1DIter

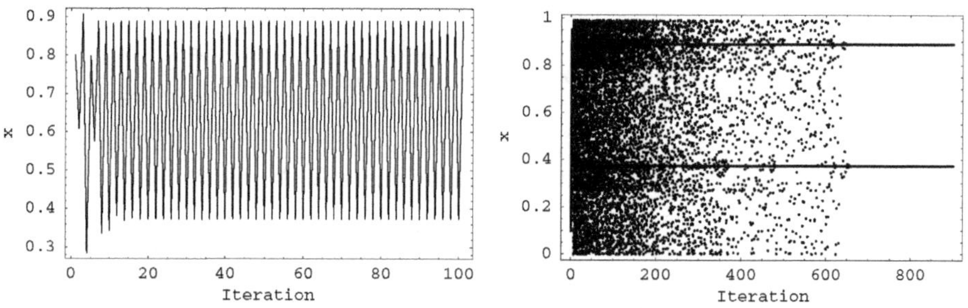

Fig. 5.8 Best individual solution; CF Targ2, p-2 orbit, DERand1DIter

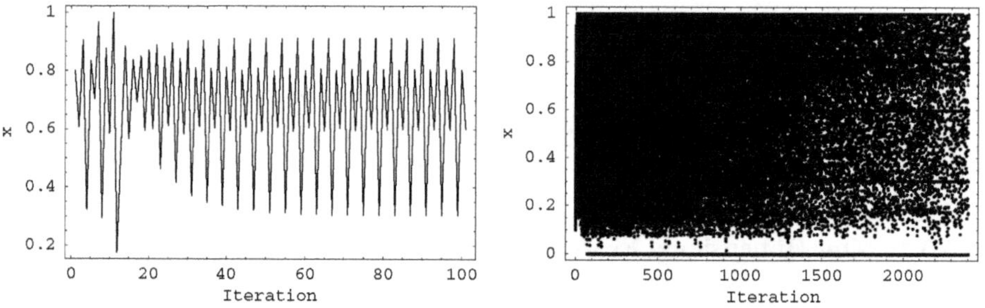

Fig. 5.9 Best individual solution, CF Targ2, p-4 orbit, DEBest2Bin

5.6.3.2 Two Dimensional Example

The results of optimization in the case of two-dimensional Henon map are depicted in Fig. 5.10 (p-1 orbit), Fig. 5.11 (p-2 orbit) and Fig. 5.12 (p-4 orbit). As can be seen from the simulation results, this CF design gives similar excellent performance from the point of view of quickness and quality of stabilization in as in the case of Logistic equation. But on the other hand in case of usually not problematic p-1 orbit this CF give several solutions, where the final CF value is not divisible by the *NI*

Table 5.3 Comparison of Average IStab values - LQ - Case studies 1-5

UPO	p-1		p-2		p-4	
EA Version	**SOMA**	**DE**	**SOMA**	**DE**	**SOMA**	**DE**
CF Basic	97	97	123	123	197	218
CF Simple	98	99	-	-	-	-
CF NA	33	31	73 (109)	102 (116)	-	-
CF Targ1	33	31	78 (112)	104 (115)	77 (195)	121 (215)
CF Targ2	31	30	95 (112)	113 (113)	151 (184)	182 (196)

value (or IStab) without remainder Thus it seems, that every subsequent simulation affirms the fact, that this design of CF secures very fast reaching of desired state but with slightly lower quality of stabilization (basic part of CF > 0). This confirms the phenomenon that endeavors for fast stabilization is at the cost of arising problems with quality of stabilization and in the proper CF design there should be quickness and quality balanced.

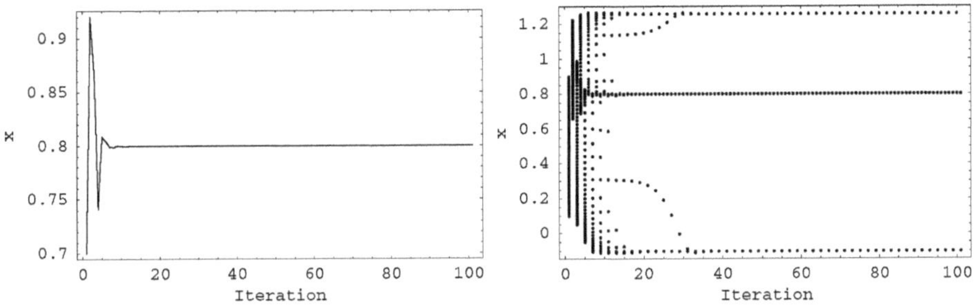

Fig. 5.10 Best individual solution, CF Targ2, p-1 orbit, DELocalToBest

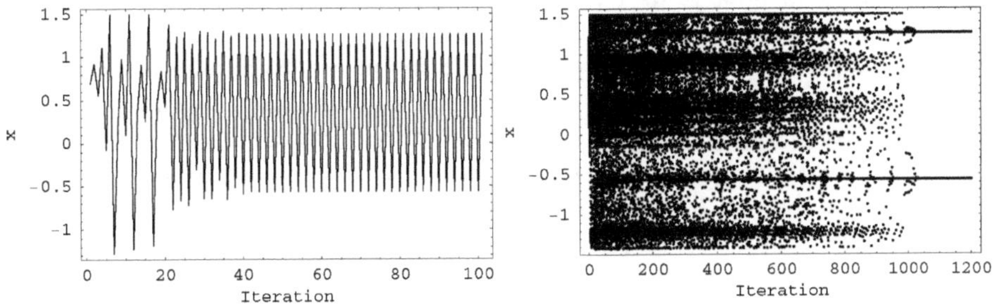

Fig. 5.11 Best individual solution, CF Targ2, p-2 orbit, SOMA ATO

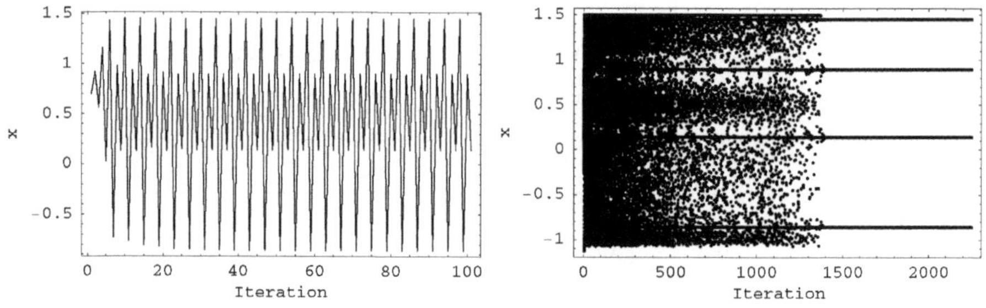

Fig. 5.12 Best individual solution, CF Targ2, p-4 orbit, DEBest1JIter

Table 5.4 Comparison of Average IStab values - HENON - Case studies 1-5

UPO	p-1		p-2		p-4	
EA Version	**SOMA**	**DE**	**SOMA**	**DE**	**SOMA**	**DE**
CF Basic	77	75	124	125	122	122
CF Simple	65	70	-	-	-	-
CF NA	49	47	84 (114)	110 (118)	-	-
CF Targ1	49	47	72 (113)	109 (118)	110 (121)	121 (123)
CF Targ2	39	38	91 (108)	113 (117)	115 (118)	123 (123)

5.6.4 Analysis of All Results

This work covers five case studies with different used CF as presented. The results of numerous simulations and previous statistical comparisons of all case studies in Tables 5.3 and 5.4 give the following piece of knowledge.

- The first proposed CF Basic gives satisfactory results and can be used wherever the good quality of stabilization is expected and the speed of stabilization and "universality of this solution" for wider range of initial conditions are not decisive. This CF does not require any special experiences and knowledge about the system.
- The second CF simple represents the simplest example of targeting CF suitable only for stabilization of fixed point. In comparison with CF Basic it gives similar results in the case of LQ and slightly better results in the case of Henon map.
- The next proposal of CF NA represents the progressive targeting CF suitable for p-1 and p-2 orbit, which gives very good results in the task of shortening of the initial chaotic stage. The results for p-1 orbit are significantly better than in the previous two CFs, on the other hand the slightly better results for p-2 orbit were achieved at the cost of arising of problem with worse performance of EAs and obtaining of solutions with only temporary stabilization or none at all. Moreover this CF requires some knowledge about results achieved in the case of CF Basic due to proper setting of *SC* value.
- The fourth CF Targ1 brings the advantage of automatically computed SC value, thus it is suitable for any desired UPO. The obtained results were similar as in case of CF NA.

- In the last proposal of CF Targ2 there were only slight changes in CF design, but from the presented results it can be seen, how such a small change can influence the performance of a controlled system, especially when it is an extremely sensitive chaotic system. Here, another improvement from the point of view of quickness of stabilization was achieved and furthermore the performance of EAs was increased, thus the proportion of non-stabilizing and stabilizing securing solutions. This seems to be the best choice, when very good solution from the close neighborhood of initial conditions is expected.

An apt comparison of all case studies is depicted in the following Figures 5.13 - 5.18. These figures show selected complex simulations from the best solutions

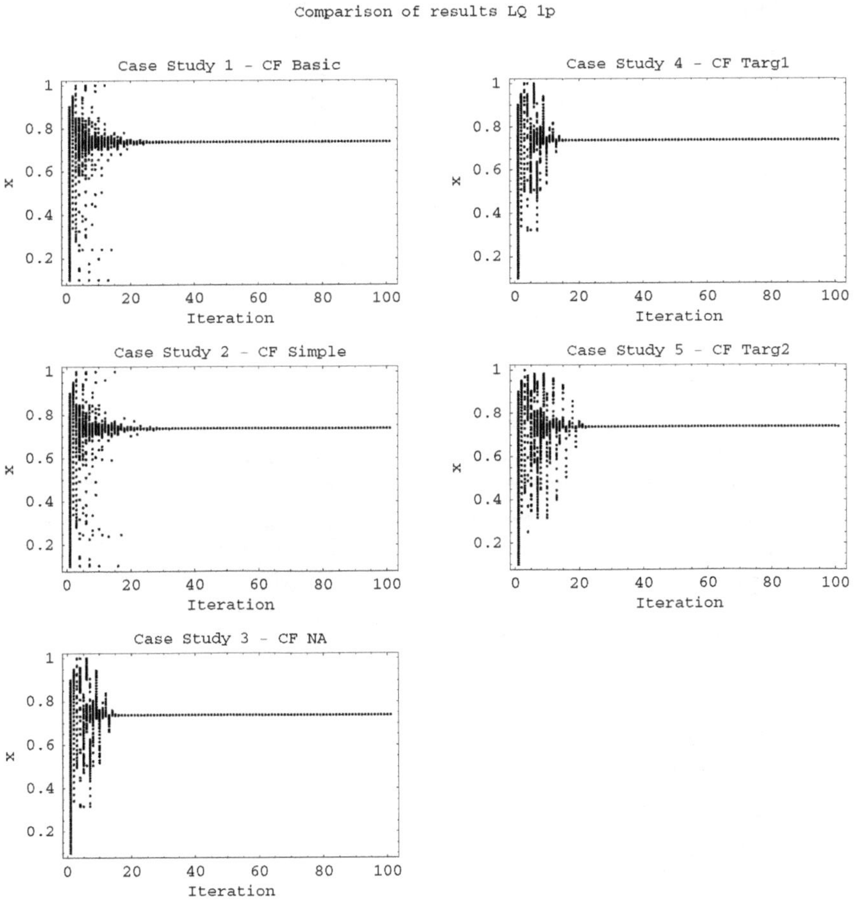

Fig. 5.13 Comparison of results for LQ - 1p, simulations with distributed initial conditions; $0 < x_{initial} < 1$, 100 samples; 1-SOMA ATA, 2-DELocalToBest, 3-DEBest1JIter, 4-DERand1Bin, 5-DERand1DIter

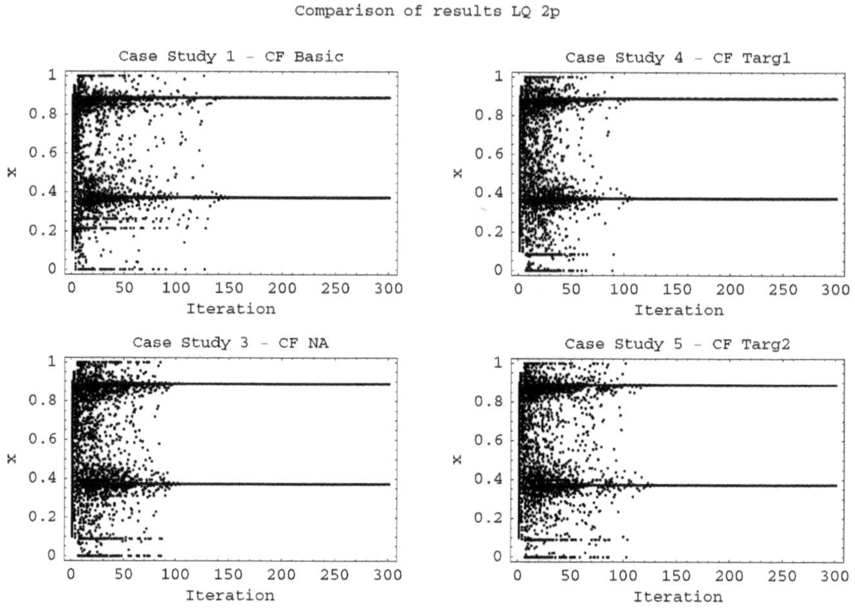

Fig. 5.14 Comparison of results for LQ - 2p, simulations with distributed initial conditions: $0 < x_{initial} < 1$, 100 samples; 1-SOMA ATR, 3- DERand1Bin, 4-SOMA ATAA, 5-DERand2Bin

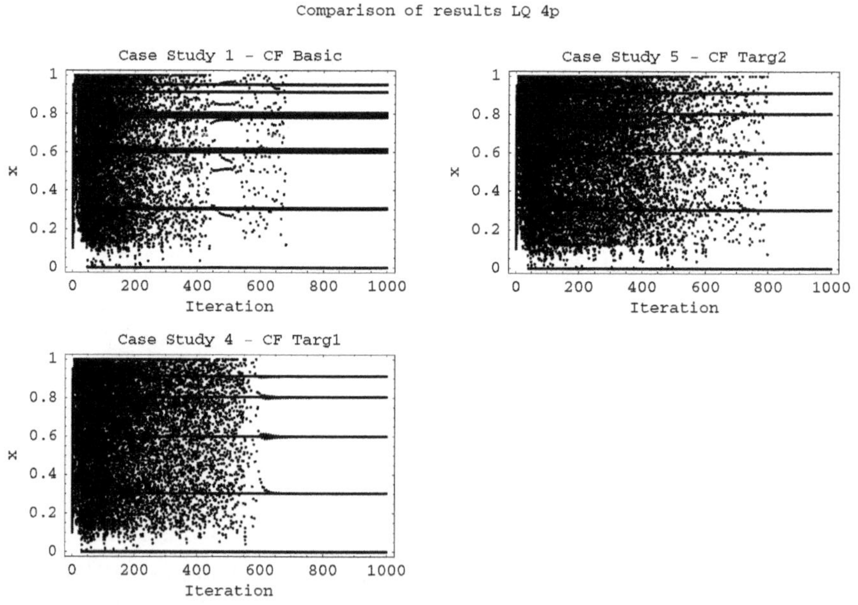

Fig. 5.15 Comparison of results for LQ - 4p, simulations with distributed initial conditions: $0 < x_{initial} < 1$, 100 samples; 1-SOMA ATA, 4-DEBest2Bin, 5- DELocalToBest

given by all 10 versions of evolutionary algorithms. The statistical comparison in Tables 5.3 and 5.4 and the average results of complex simulations of all 10 best solutions which were the source for following figures 5.13 - 5.18 lends weight to the argument, that targeting cost functions (CF NA, CF Targ1, and CF Targ2) allows to reach faster and mostly better quality of stabilization.

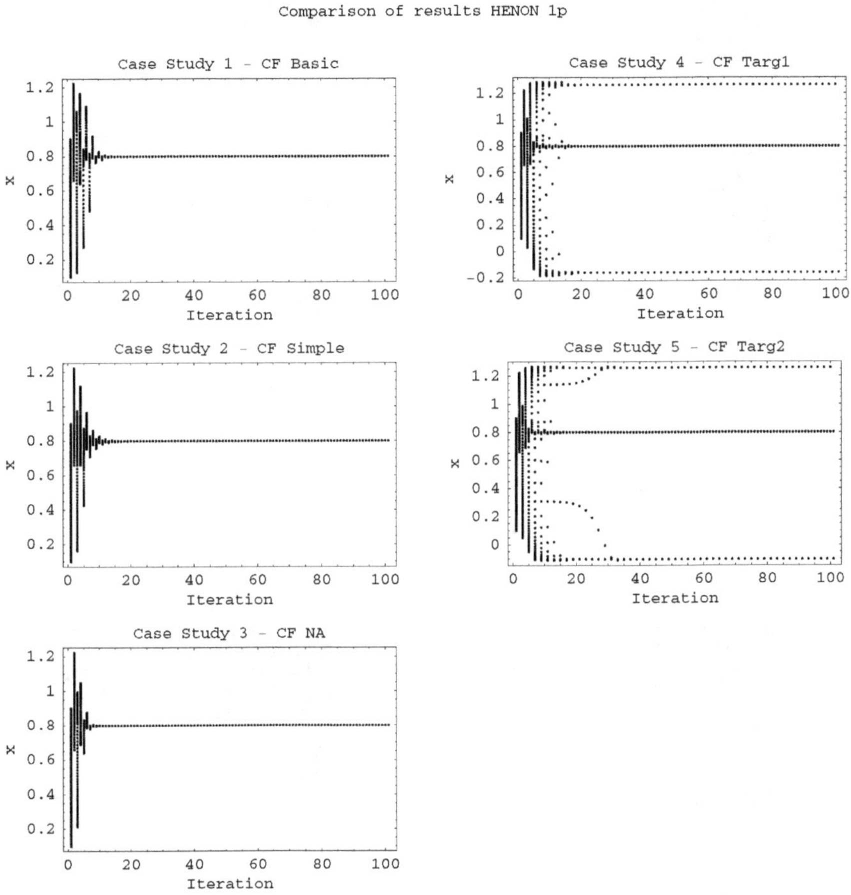

Fig. 5.16 Comparison of results for HENON - 1p, simulations with distributed initial conditions: $0 < x_{initial} < 1$, 100 samples; 1-DERand2Bin, 2-DERand1Bin, 3-DEBest2Bin, 4-DELocalToBest, 5-DELocalToBest

5.7 Comparison with OGY Method

The comparison was done for these two cases: p-1 orbit and p-2 orbit. These comparisons are derived from superimposition of 100 examples. The ETDAS control algorithm was not mostly set up identically as the best individual solution given by

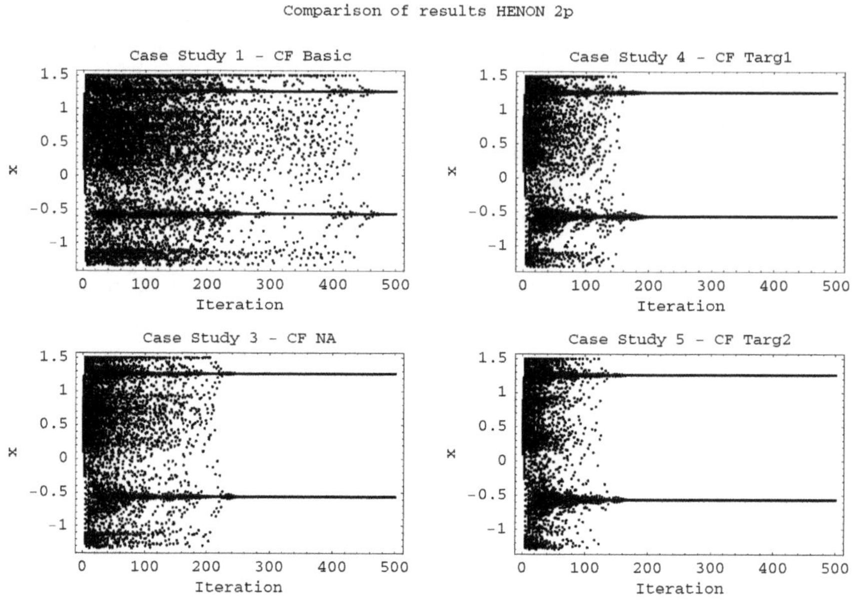

Fig. 5.17 Comparison of results for HENON - 2p, simulations with distributed initial conditions: $0 < x_{initial} < 1$, 100 samples; 1-DEBest2Bin, 3-SOMA ATO, 4-SOMA ATO, 5-SOMA ATR

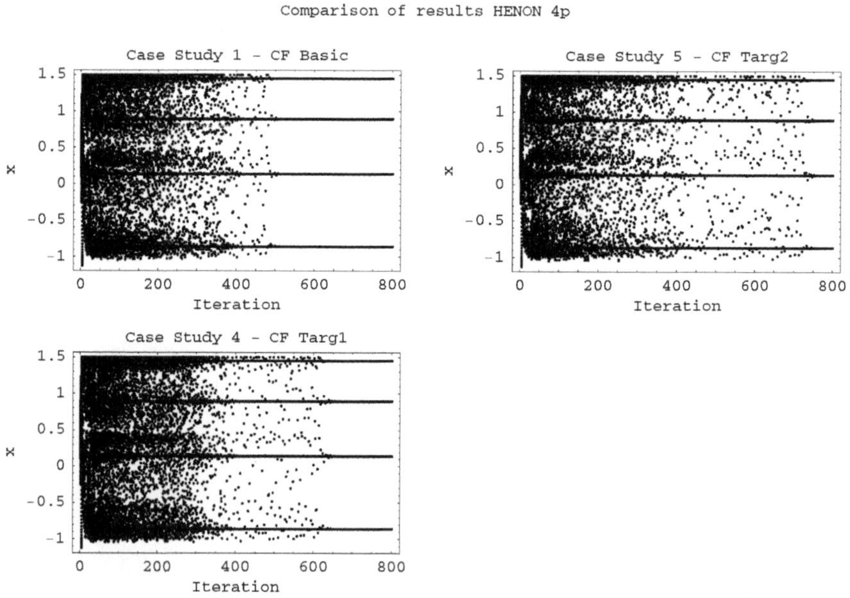

Fig. 5.18 Comparison of results for HENON - 4p, simulations with distributed initial conditions: $0 < x_{initial} < 1$, 100 samples; 1-SOMA ATAA, 4-DEBest1JIter, 5-DELocalToBest

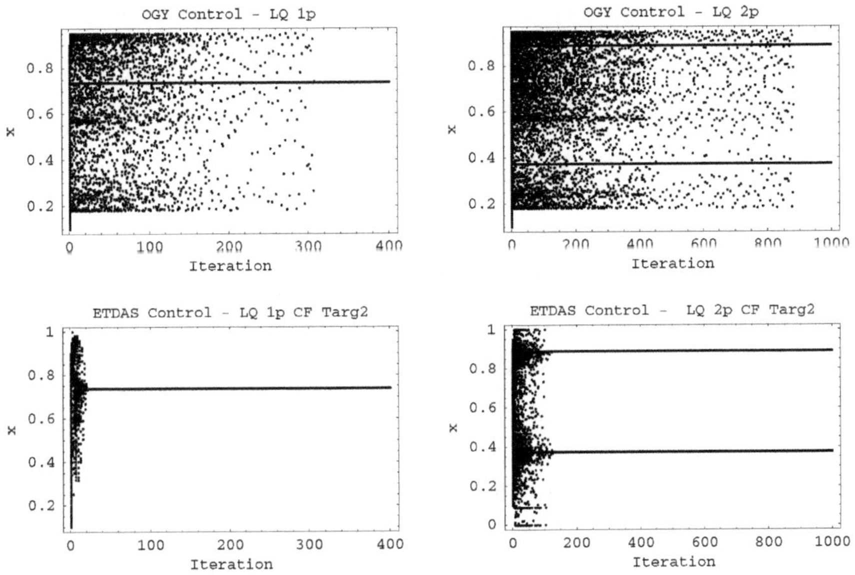

Fig. 5.19 Comparison of OGY, optimized ETDAS for p-1 orbit (left) and p-2 orbit (right), LQ, CF Targ 2, $0 < x_{initial} < 1$

SOMA or DE as ideally expected (only one exception - LQ, p-1 orbit), due to above discussed problems and negative phenomenon. Also it has to be considered that the performance of OGY method is very dependent on the size of tiny neighborhood of desired UPO where the linearization is applied. Consequently it is possible to reach either better result than presented here (larger neighborhood increases the chance, that the chaotic attractor will reach it, but at the cost of breaking the basic OGY idea, which is linearization in the very tiny neighborhood of UPO) or of course worse results in the sense of longer initial chaotic stage before stabilization is applied.

5.7.1 Logistic Equation

In the first case ETDAS was set up identically as the best solution given by DERand1DIter.

As can be seen from Fig. 5.19, ETDAS method steers the chaotic system very quickly to the desired state. The OGY stabilize 50% of examples in first 100 iterations, but ETDAS stabilize more than 50% of examples in first 20 iterations. Thus this supports the theory that ETDAS based control method can be simply considered as targeting and stabilizing algorithm.

In the second case, ETDAS was set up identically as the best solution given by DERand2Bin. From Fig. 5.19 it follows that ETDAS method steers the chaotic system very quickly to close neighborhood of p-2 orbit. To stabilize all of the examples around 150 iterations are required. However, more than 50% of examples oscillate in the close neighborhood after first 50 iterations. Then the stage of reducing of the neighborhood size, caused by progressive converging into p-2 orbit, ensues. The OGY method stabilizes about 50% of examples in first 400 iterations and to stabilize all of the examples about 900 iterations are required.

5.7.2 Henon Map

In the first case, ETDAS was set up identically as the best solution given by SOMA ATR.

From Fig. 5.20 it follows, that optimized ETDAS method steers the chaotic system very quickly to the stable state. The difference between two compared methods is very considerable because the OGY stabilize 50% of examples in first 200 iterations, whereas EDAS stabilize more than 50% of examples in first 10 iterations. From the similar comparison for logistic equation given in Fig. 5.19 it can be clearly seen that the difference between these two control methods increase together with higher dimension or complexity of controlled system. The performance of ETDAS is almost the same whereas OGY needs twice more iterations to achieve stabilization. In this case the performance of ETDAS with Henon map is even better then with simpler one dimensional equation.

In the second case, ETDAS was set up identically as the best solution given by DELocalToBest. From Fig 5.20 it follows that optimized ETDAS method needs approximately 100 iterations to stabilize 50% of chaotic samples system at desired

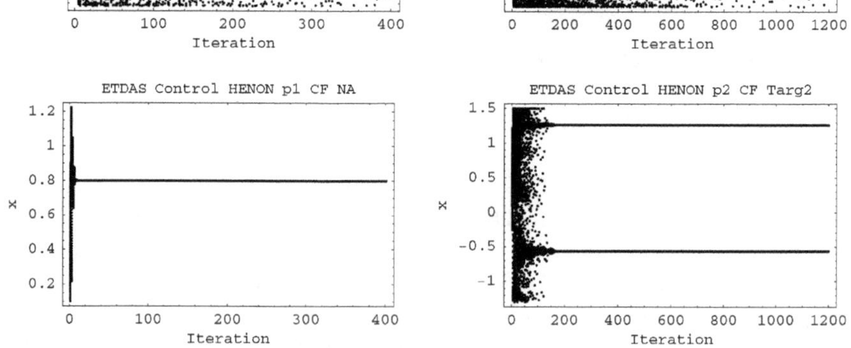

Fig. 5.20 Comparison of OGY, optimized ETDAS for p-1 orbit (left) and p-2 orbit (right), HENON, CF Targ2, $0 < x_{initial} < 1$

p-2 orbit. To stabilize the rest of the samples around 200 iterations are required. As in previous case the OGY stabilizes about 50% of examples in first 500 iterations and to stabilize all of the examples more than 1000 iterations are required.

5.8 Conclusion and Discussion

The optimization of chaos control described here is relatively simple and easy to implement. Based on obtained results, it may be claimed that all simulations give satisfactory results and thus EAs are capable of solving this class of difficult problems and the quality of results does not depend only on the problem being solved but also on the proper definition of the CF. The matter of selection of control method and seemingly simple design of cost functions has this solution. The effective usage of evolutionary computation (notably from the point of view of time demands) claims fast and elementary operations. Presented CFs and selected control algorithm are quite simple and their representation in the Wolfram Mathematica environment requires only a few rows of code and do not contain any time-consuming operations, for example matrix manipulations or analysis of system state. Furthermore as mentioned in the introduction, the Pyragas control method has lot of easy accessible parameters, which can be tuned by EA.

From the comparison with classical control technique - OGY follows that ETDAS based control method can be simply considered as targeting and stabilizing algorithm and their performance is much better than OGY.

As can be seen from the optimization results presented here, they are extremely sensitive to the construction of used CF. Any small change in the design of CF can cause radical improvement of system behavior, but of course on the other hand can cause worsening of observed parameters and behavior of chaotic as well. For example the problem with fast stabilization not only for initial conditions used in optimization process, but for the whole range of the initial conditions or other described problems. The sensitivity is confirmed by the phenomenon, that the best individual solution is not sometimes suitable for complex simulation at all, although the best solution secures fast and precise stabilization. The weight of this phenomenon grows together with endeavor for effective complex targeting CF, which can even bring the increase of the nonlinearity and uncertainness of the CF surface.

From the pair of tables 5.3 and 5.4 it is complex to answer as to which evolutionary algorithm is better or worse. For the first view SOMA seems to be the better choice, when the average IStab value is taken into account. But from the point of view of the behavior of each EA during optimizations, it is a difficult question. The SOMA rapidly heads towards global optimal solution, whereas DE slowly and "carefully" searches in the erratic CF surface. And this is the reason for the phenomenon, where DE gives slightly more "stabilization securing" solutions, whereas SOMA got stuck in one of huge amount of local minimums. Eventually both EAs give very satisfactory results.

From all presented results follows, that it is hard task to propose a CF, which gives excellent results, especially "universal results" suitable for simulation with wide range of initial conditions. The series of simulations demonstrates extremely sensitivity of chaotic systems to proper settings of control algorithm and of course the sensitivity to even very tiny change in any parameter. This extreme sensitivity is transferred into complexity of CF surface thus it is also hard task for EAs to find good solution. It is also difficult to determine the conditions for optimizations and subsequent simulations. For example the specification of correct length of optimization interval τ_i is very difficult and it can be stated that it is alike to balancing at the edge of knife when considering this fact, that the difference in final CF value of size 1.10^{-4} and subsequent very tiny change in combination of estimated parameters can cause improvement or worsening of system behavior. And this small difference can be caused by change in CF design or just only by adding 50 iterations (if the system is not absolutely stabilized, the difference between ideal target state and actual output of system can be relatively appreciable).

As a consequence of these facts it is possible to say that all presented CFs gives good results and each one is more or less suitable depending on concrete demands for quickness or quality of stabilization, computational-time demands, order of UPO, whether it should be the solution only for limited circle of initial conditions or for wider range etc.

Lastly, as a conclusion it seems that CF Targ2 (case study 5) is the best choice for optimizations and simulations with limited circle of initial conditions. Also it gives satisfactory results for wide range of initial solutions, thus gives the "universal" solutions, which contradict with words chaos or chaotic system.

There is no problem for the future research in defining much more complex CF comprising as subcriteria control of stability, costs, time-optimality, controllability, or any of their arbitrary combinations. Furthermore parameter settings for EA were based on heuristic approach; therefore there is also possibility for the future research.

The total amount of optimizations was 36 000. The optimization took from one minute in the case of p-1 orbit, CF Basic and TDAS control method to six minutes in case of CF Targ2, Logistic equation ETDAS control and p-4 orbit. All simulations were performed in Wolfram Mathematica environment. The total number of cost function evaluations (CFE) for all presented results was 150 millions.

Acknowledgements. This work was supported by grant No. MSM 7088352101 of the Ministry of Education of the Czech Republic and by grants of the Grant Agency of the Czech Republic GACR 102/09/1680.

References

1. Andrievski, B., Fradkov, A.: Control of Chaos: Methods and Applications. Autom. Rem. Contr. 64(5), 679–719 (2003)
2. Awad, E., Ammar, S.: Optimal control and synchronization of Lorenz system with complete unknown parameters. Chaos, Solitons & Fractals 30(5), 1122–1132 (2006)
3. Bing, C., Xiaoping, L., Shaocheng, T.: Adaptive fuzzy approach to control unified chaotic systems. Chaos, Solitons & Fractals 34(4), 1180–1187 (2007)
4. Bird, C., Aston, P.: Targeting in the Presence of Noise. Chaos, Solitons, & Fractals 9(1), 251–259 (1998)

5. Bollt, E., Kostelich, E.: Optimal Targeting of Chaos. Phys. Lett. 245, 399–406 (1998)
6. Cannas, B., Cincotti, S., Pisano, A., Usai, E.: Controlling Chaos via Second-Order Sliding Modes. In: Proc. International Symposium on Circuits and Systems, ISCAS 2003, pp. 156–159 (2003)
7. Chen, L.: The Open-plus-closed-loop Control of Chaotic Maps and its Robustness. Chaos, Solitons & Fractals 21, 113–118 (2004)
8. Fradkov, A., Evans, R.: Control of Chaos:Survey 1997 - 2000. In: Preprints of 15th Triennial World Congress IFAC, Plenary Papers, Survey Papers, Milestones, Barcelona, pp. 143–154 (2002)
9. Fradkov, A., Evans, R.: Control of Chaos: Methods and Applications in Engineering. Annu. Rev. Contr. 29(1), 33–56 (2005)
10. Gonzales-Miranda, J.: Perturbing Chaotic Systems to Control Chaos. In: Synchronization and Control of Chaos - An Introduction for Scientists and Engineers. Imperial College Press, London (2004)
11. Grebogi, C., Lai, Y.: Controlling Chaotic Dynamical System. Phys. Rep. 31, 307–312 (1997)
12. Grebogi, C., Lai, Y.: Controling Chaos. In: Schuster, H. (ed.) Handbook of Chaos Control. Wiley-VCH, Weinheim (1999)
13. Grebogi, C., Lai, Y.: Pole placement Method of Controling Chaos in high dimensions. In: Schuster, H. (ed.) Handbook of Chaos Control. Wiley-VCH, Weinheim (1999b)
14. Hassan, S., Mohammad, S.: Indirect adaptive control of discrete chaotic systems. Chaos, Solitons & Fractals 34(4), 1188–1201 (2007)
15. Hilborn, R.: Chaos and Nonlinear Dynamics: An Introduction for Scientists and Engineers. Oxford University Press, Oxford (2000)
16. Hua, Ch., Guan, X.: Adaptive Control for Chaotic systems. Chaos, Solitons & Fractals 22, 55–60 (2004)
17. Huang, W.: Stabilizing nonlinear dynamical systems by an adaptive adjustment mechanism. Phys. Rev. E 61, R1012–1015 (2000)
18. Iplikci, S., Denizhan, Y.: Control of chaotic systems using targeting by extended control regions method. Phys. Nonlinear Phenom. 150(3-4), 163–176 (2001)
19. Iplikci, S., Denizhan, Y.: An improved neural network based targeting method for chaotic dynamics. Chaos, Solitons & Fractals 17(2), 523–529 (2003)
20. Just, W.: Principles of Time Delayed Feedback Control. In: Schuster, H. (ed.) Handbook of Chaos Control. Wiley-VCH, Weinheim (1999)
21. Kostelich, E., Ott, E., Grebogi, C., Yorke, J.: Higher-dimensional Targeting. Phys. Rev. E 47(1), 305–310 (1993)
22. Kwon, J.: Targeting and Stabilizing Chaotic Trajectories in the Standard Map. Phys. Lett. 258, 229–236 (1999)
23. Mascolo, S.: Backstepping Design for Controlling Lorenz Chaos. In: Proc. 36th IEEE Conference on Decision and Control, San Diego, pp. 1500–15001 (1997)
24. Mascolo, S., Grassi, G.: Controlling Chaos via Backstepping Design. Phys. Rev. E 56(5), 6166–6169 (1997)
25. May, R.: Stability and Complexity in Model Ecosystems. Princeton University Press, Princeton (2001)
26. Ott, E., Grebogi, C., Yorke, A.: Controlling Chaos. Phys. Rev. Lett. 64, 1196–1199 (1990)
27. Pyragas, K.: Continuous control of chaos by self-controlling feedback. Phys. Lett. 170, 421–428 (1992)
28. Price, K., Storn, R., Lampinen, J.: Differential Evolution: A Practical Approach to Global Optimization. Natural Computing Series. Springer, Heidelberg (2005)

29. Pyragas, K.: Control of chaos via extended delay feedback. Phys. Lett. 206, 323–330 (1995)
30. Ramaswamy, R., Sinha, S., Gupte, N.: Targeting Chaos Through Adaptive Control. Phys. Rev. E 57(3), 2507–2510 (1998)
31. Paskota, M., Lee, J.: Targeting moving targets in chaotic dynamical systems. Chaos, Solitons & Fractals 8(9), 1533–1544 (1997)
32. Richter, H.: An Evolutionary Algorithm for Controlling Chaos:The Use of Multi - Objective Fitness Function. In: Guervós, J.J.M., Adamidis, P.A., Beyer, H.-G., Fernández-Villacañas, J.-L., Schwefel, H.-P. (eds.) PPSN 2002. LNCS, vol. 2439, pp. 308–320. Springer, Heidelberg (2002)
33. Richter, H., Reinschke, K.: Optimization of local control of chaos by an evolutionary algorithm. Physica D 144, 309–334 (2000)
34. Senkerik, R., Zelinka, I., Davendra, D.: Comparison of Evolutionary Algorithms in the Task of Chaos Control Optimization. In: Proc. IEEE Congres on Evolutionary Computation 2007, CEC 2007, Singapore, September 25-28, pp. 3952–3958 (2007)
35. Senkerik, R., Zelinka, I., Navratil, E.: Optimitazion of Feedback Control of Chaos by Evolutionary Algorithms. In: Proc. 1st IFAC Conference on Analysis and Control of Chaotic Systems, CHAOS 2006, Reims, France, pp. 97–102 (2006)
36. Senkerik, R., Zelinka, I., Navratil, E.: Design of Targeting Cost function for Evolutionary Optimization of Chaos Control. In: Proc. 21st European Conference on Modelling and Simulation 2007, ECMS 2007, Prague, Czech Republic, June 4-6, pp. 345–350 (2007)
37. Senkerik, R., Zelinka, I., Navratil, E.: Cost function Design for Evolutionary Optimization of Chaos Control. In: Proc. 9th European Control Conference 2007, ECC 2007, Kos, Greece, July 2-5, pp. 1682–1687 (2007)
38. Starrett, J.: Time-optimal Chaos Control by Center Manifold Targeting. Phys. Rev. E 66(4), 6206–6211 (2002)
39. Sun, J.: Impulsive Control of a New Chaotic System. Math. Comput. Simulat. 64, 669–677 (2004)
40. Sun, J., Zhang, Y.: Control of Chaotic Systems Using an on-line Trained Linear Neural Controller. Physica D 100, 423–438 (1997)
41. Sun, J., Zhang, Y.: Impulsive Control of Rossler System. Phys. Lett. 306, 306–312 (2003)
42. Tian, Y., Gao, F.: Adaptive Control of Chaotic Continuous-time systems with delay. Phys. Nonlinear Phenom. 117, 1–12 (1998)
43. Yang, T., Yang, L., Yang, C.: Impulsive Control of Lorenz System. Physica D 110, 18–24 (1997)
44. Yongai, Z.: Controlling chaos based on an adaptive adjustment mechanism. Chaos, Solitons & Fractals 30(5), 1069–1073 (2006)
45. Zelinka, I.: SOMA - Self Organizing Migrating Algorithm. In: Babu, B., Onwubolu, G. (eds.) New Optimization Techniques in Engineering. Springer, Heidelberg (2004)
46. Zelinka, I.: Investigation on Evolutionary Deterministic Chaos Control - Extended Study. In: ECMS 2005, Riga, Latvia (2005)
47. Zelinka, I., Senkerik, R., Navratil, E.: Investigation on Real Time Deterministic Chaos Control by Means of Evolutionary Algorithms. In: Proc. 1st IFAC Conference on Analysis and Control of Chaotic Systems, CHAOS 2006, Reims, France, June 28-30, pp. 211–217 (2006)
48. Zelinka, I., Senkerik, R., Navratil, E.: Investigation on Evolutionary Optimitazion of Chaos Control. Chaos, Solitons & Fractals (2007), doi:10.1016/j.chaos.2007.07.045
49. Zeng, Y., Singh, S.: Adaptive Control of Chaos in Lorenz System. Dynam. Contr. 7, 143–154 (1997)

Chapter 6
Evolutionary Control of CML Systems

Ivan Zelinka

Abstract. This chapter is a continuation of an investigation on deterministic spatiotemporal chaos real-time control by means of selected evolutionary techniques. Real-time like behavior is specially defined and simulated with spatiotemporal chaos model based on mutually nonlineary joined n equations, so called Coupled Map Lattices. In total five evolutionary algorithms has been used for chaos control: differential evolution, self-organizing migrating algorithm, genetic algorithm, simulated annealing and evolutionary strategies in a total of 15 versions. For modeling of spatiotemporal chaos behavior, the so called coupled map lattices were used based on logistic equation to generate chaos. The main aim of this investigation was to show that evolutionary algorithms, under certain conditions, are capable of controlling of CML deterministic chaos, when the cost function is properly defined alongside the parameters of selected evolutionary algorithms. Investigation consists of four different case studies with increasing simulation complexity. For all algorithms each simulation was evaluated 100 times in order to show and check robustness of used methods. All data were processed and used in order to get summarized results and graphs.

6.1 Introduction

Deterministic chaos, discovered by E. Lorenz [23] is a fairly active area of research in the last few decades. The Lorenz system produces one of the well-known canonical chaotic attractors in a simple three-dimensional autonomous system of ordinary

Ivan Zelinka
Tomas Bata University in Zlin, Faculty of Applied Informatics, Nad Stranemi 4511,
Zlin 76001, Czech Republic
and
VSB-TUO, Faculty of Electrical Engineering and Computer Science, 17. listopadu 15,
708 33 Ostrava-Poruba, Czech Republic
e-mail: zelinka@fai.utb.cz

I. Zelinka et al. (Eds.): Evolutionary Algorithms and Chaotic Systems, SCI 267, pp. 191–235.
springerlink.com

differential equations [23], [32]. For discrete chaos, there is another famous chaotic system, called logistic equation [24]. Logistic equation is based on a predator-prey model showing chaotic behavior. This simple model is widely used in the study of chaos, where other similar models exist (canonical logistic equation [12] and 1D or 2D coupled map lattices [31]). Since then, a large set of nonlinear systems that can produce chaotic behavior have been observed and analyzed. Chaotic systems thus have become a vitally important part of science and engineering in theoretical as well as in practical levels of research. The most interesting and applicable notions are, for example, that chaos control and chaos synchronization are related to secure communication, amongst others. Recently, the study of chaos is focused not only along the traditional trends but also on the understanding and analyzing principles, with the new intention of controlling and utilizing chaos as demonstrated in [4] and [33]. The term chaos control was first coined by Ott, Grebogi and Yorke in 1990. It represents a process in which a control law is derived and used so that the original chaotic behavior can be stabilized on a constant level of output value or a n-periodic cycle. Since the first experiment of chaos control, many control methods have been developed and some are based on the first approach [27], including pole placement [13], [42] and delay feedback [19], [20]. Another research has been done on CML control by [10], special feedback methods for controlling spatio-temporal on-off intermittency has been used there and [10]. This paper introduces a controller (based on discrete-time sliding mode and Lyapunov function) for controlling of spatiotemporal chaos system. Many methods were adapted for the so-called spatiotemporal chaos represented by coupled map lattices (CML). Control laws derived for CML are usually based on existing system structures [31], or by using an external observer [3]. Evolutionary approach for control was also successfully developed, for example in, [30], [29] and [38]. Many published methods of deterministic chaos control (DCC) were (originally developed for classic DCC) adapted for so called spatiotemporal chaos represented by CML, given by eq. (6.1). Models of this kind are based on a set of spatiotemporal (for 1D, Fig. 6.1) or spatial (for 2D, Fig. 6.2)) cells which represents appropriate state of system elements. Typical example is CML based on so called logistic equation, [24], [15], [3] which is used to simulate behavior of system which consists of n mutually joined cells via nonlinear coupling, usually noted like ε. Mathematical description of CML system is given by eq. (6.1). The function, which is represented by $f(x_n(i))$ is an "arbitrary" discrete system - in this case study logistic equations has been selected to substitute $f(x_n(i))$. CML description based on eq. (6.1) in *Mathematica* software is given in Fig. 6.3.

$$x_{n+1}(i) = (1-\varepsilon)f(x_n(i)) + \frac{\varepsilon}{2}(f(x_n(i-1)) + f(x_n(i+1))) \tag{6.1}$$

The main aim of this participation is to show that evolutionary algorithms (EA's) are capable of controlling (as was also shown for temporal DCC in [30], [29]), CML as well as deterministic methods without internal system knowledge operating with CML as like a black box system. The ability of EAs to successfully work with black box type of problems have been proven; see for example real-time control of plasma

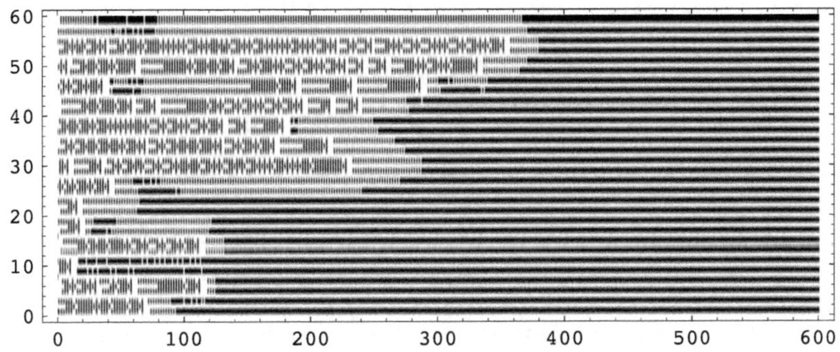

Fig. 6.1 1D CML with stabilized pattern T1S2

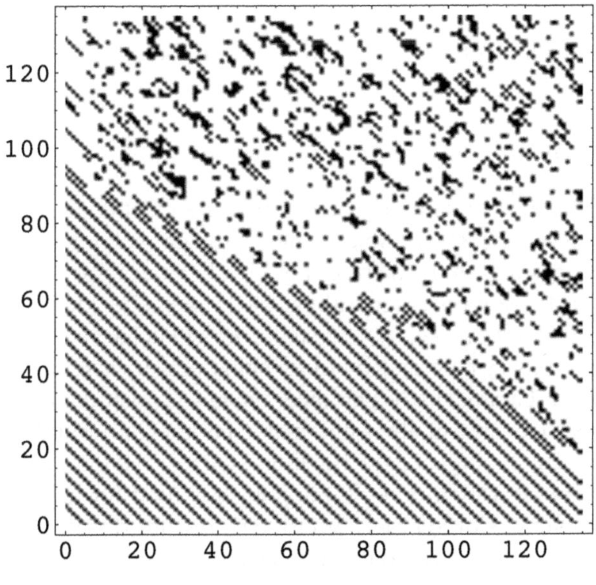

Fig. 6.2 2D CML with pinning imported through lattice on position (0,0). Resulting control pattern (left) is visible as well as spatiotemporal chaos (right)

reactor [25], [26] and [40] or CML non real-time control by evolutionary algorithms [36], [37] and [41].

This chapter is organized as follows. The first part outlines the motivation of this investigation. This is followed by a brief survey of evolutionary algorithms which follow, along with a brief description of the idea of CML chaos control. Evolutionary chaos control is then described, and finally experimental results are reported, followed by conclusion.

```
Logistic = Compile[{{x, _Real}, {A, _Real}}, A x (1 - x)]
SPL = Compile[{{x, _Real, 1}, {ε, _Real}, {A, _Real}, {L, _Integer}},
   MapIndexed[
    If[#2[[1]] == 1, (1 - ε) Logistic[x[[#2[[1]]]], A] +
      ε/2 (Logistic[x[[L]], A] + Logistic[x[[#2[[1]] + 1]], A]),
     If[#2[[1]] == L, (1 - ε) Logistic[x[[#2[[1]]]], A] +
       ε/2 (Logistic[x[[#2[[1]] - 1]], A] + Logistic[x[[1]], A]),
      (1 - ε) Logistic[x[[#2[[1]]]], A] + ε/2 (Logistic[x[[#2[[1]] - 1]], A] +
          Logistic[x[[#2[[1]] + 1]], A])
     ]
    ] &, x]
  ];
```

Fig. 6.3 Realization of eq. (6.1) in *Mathematica* code

6.2 Motivation

Motivation of this investigation is quite simple. As mentioned in the introduction, evolutionary algorithms are capable of hard problem solving. A lot of examples about evolutionary algorithms can be easily found. For example in [7] are developed statistically robust evolutionary algorithms, on the opposite side [18] used evolutionary algorithms for fuzzy power system stabilizer which has been applied on single-machine infinite bus system and multi-machine power system. Another research was done by [22]. Parameters of permanent magnet synchronous motors has been optimized by particle swarm algorithm and experimentally validated on servomotor. [6] used swarm intelligence for IIR filter synthesis and in [14] co-evolutionary particle swarm optimization (CoPSO) approach is used for design of constrained engineering problems. CoPSO was used for the pressure vessel, compression spring and welded beam problem with only 30 independent runs.

The main question in this research was if EAs are able to control and stabilize chaotic systems like CML, and if they are capable to control CML like a black box system, i.e. when the structure of the controlled system is unknown. All experiments here were designed to check this idea and either confirm or reject it. Comparison has been done with control based on analysis of CML system [17], [31] and analytic derivation of control law for CML. Behavior of controlled CML is as demonstrated on Fig. 6.4 - Fig. 6.6. Snapshot of frontwave stabilization of CML is depicted here. Fig. 6.4 shows the initial phase of the frontwave. It is clearly visible that it is fully random. On Fig. 6.4 CML is shown after 60 iterations – pattern-like structure is visible there. The last snapshot has been done after 344 iterations – CML has been successfully stabilized. Thus, the main aim was to stabilize CML with the quality to that of standard controlling techniques.

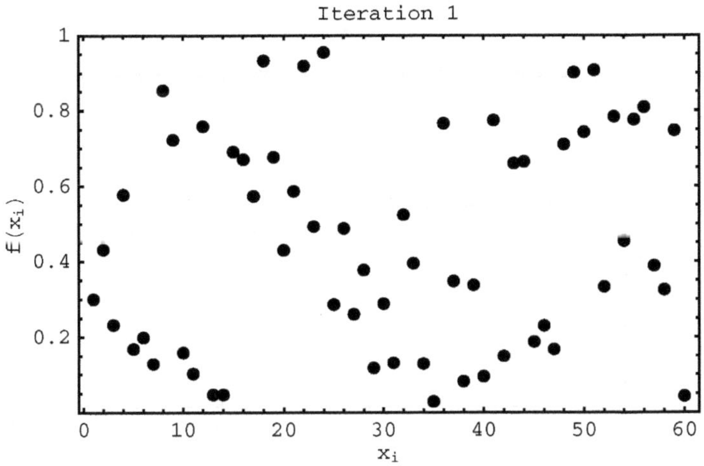

Fig. 6.4 Successful stabilization of CML in T1S3 pattern - start.

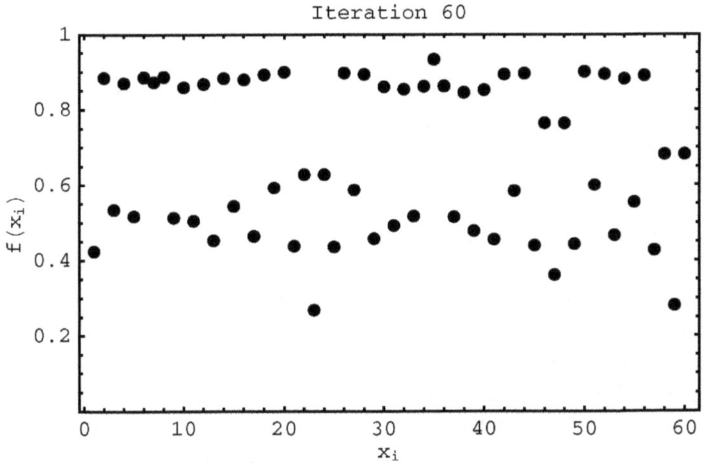

Fig. 6.5 Successful stabilization of CML in T1S3 pattern - iteration 60.

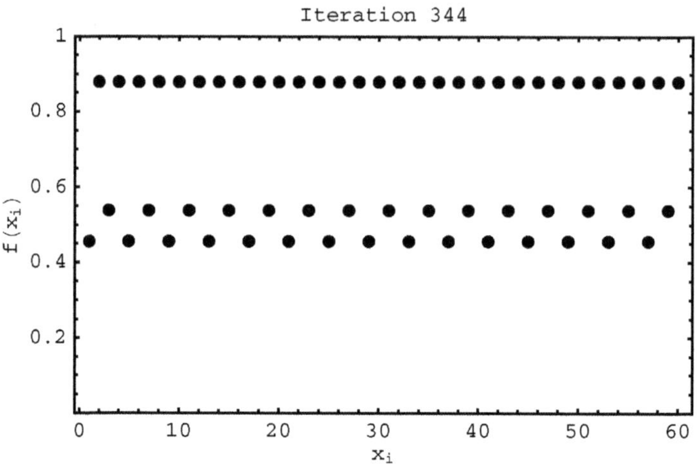

Fig. 6.6 Successful stabilization of CML in T1S3 pattern - pattern is stabilized.

6.3 Selected Evolutionary Algorithm – A Brief Introduction

For the numerical and symbolic experiments described here, stochastic optimization algorithms such as Differential Evolution (DE) [28], Self Organizing Migrating Algorithm (SOMA) [35], Genetic Algorithms (GA) [16], Simulated Annealing (SA) [21], [2] and Evolutionary Strategies (ES) [1] were selected.

6.3.1 Differential Evolution

Differential Evolution (Fig. 6.7) is a population-based optimization method that works on real-number coded individuals. For each individual $\mathbf{x}_{i,G}$ in the current generation G, differential evolution (DE) generates a new trial individual $\mathbf{x}'_{i,G}$ by adding the weighted difference between two randomly selected individuals $\mathbf{x}_{r1,G}$ and $\mathbf{x}_{r2,G}$ to a randomly selected third individual $\mathbf{x}_{r3,G}$. The resulting individual $\mathbf{x}'_{i,G}$ is crossed-over with the original individual $\mathbf{x}_{i,G}$. The fitness of the resulting individual, referred to as a perturbed vector $\mathbf{u}_{i,G+1}$, is then compared with the fitness of $\mathbf{x}_{i,G}$. If the fitness of $\mathbf{u}_{i,G+1}$ is greater than the fitness of $\mathbf{x}_{i,G}$, then $\mathbf{x}_{i,G}$ is replaced with $\mathbf{u}_{i,G+1}$; otherwise $\mathbf{x}_{i,G}$ remains in the population as $\mathbf{x}_{i,G+1}$. Differential Evolution is robust, fast, and effective with a global optimization ability. It does not require the objective function to be differentiable, and it works well even with noisy, epistatic and time-dependent objective functions. Pseudocode for DE, especially for DERand1Bin, is given in eq. (6.2).

$$
\begin{array}{l}
1.\text{Input}: D, G_{\max}, NP \geq 4, F \in (0, 1+), CR \in [0,1], \text{ and initial bounds }: \mathbf{x}^{(lo)}, \mathbf{x}^{(hi)}. \\
2.\text{Initialize}: \left\{ \begin{array}{l} \forall i \leq NP \wedge \forall j \leq D : x_{i,j,G=0} = x_j^{(lo)} + rand_j[0,1] \bullet \left(x_j^{(hi)} - x_j^{(lo)} \right) \\ i = \{1,2,...,NP\}, j = \{1,2,...,D\}, G = 0, rand_j[0,1] \in [0,1] \end{array} \right. \\
\left\{ \begin{array}{l} 3.\text{While } G < G_{\max} \\
\forall i \leq NP \left\{ \begin{array}{l} 4.\text{Mutate and recombine}: \\ 4.1 r_1, r_2, r_3 \in \{1,2,....,NP\}, \text{randomly selected, except }: r_1 \neq r_2 \neq r_3 \neq i \\ 4.2 j_{rand} \in \{1,2,...,D\}, \text{randomly selected once each } i \\ 4.3 \forall j \leq D, u_{j,i,G+1} = \left\{ \begin{array}{l} x_{j,r_3,G} + F \cdot (x_{j,r_1,G} - x_{j,r_2,G}) \\ \text{if}(rand_j[0,1] < CR \vee j = j_{rand}) \\ x_{j,i,G} \text{otherwise} \end{array} \right. \\ 5.\text{Select} \\ \mathbf{x}_{i,G+1} = \left\{ \begin{array}{l} \mathbf{u}_{i,G+1} \text{ if } f(\mathbf{u}_{i,G+1}) \leq f(\mathbf{x}_{i,G}) \\ \mathbf{x}_{i,G} \text{ otherwise} \end{array} \right. \end{array} \right. \\
G = G + 1 \end{array} \right.
\end{array}
\tag{6.2}
$$

An example of DE is demonstrated on Fig. 6.7.

6.3.2 SOMA

SOMA (Fig. 6.8) is a stochastic optimization algorithm that is modeled based on the social behavior of competitive - cooperating individuals [35]. It was chosen because it has been proved that this algorithm has the ability to converge towards the global optimum [35]. SOMA works on a population of candidate solutions in loops, called migration loops. The population is initialized by uniform random distribution over the search space at the beginning of the search. In each loop, the population is evaluated and the solution with the lowest cost value becomes the leader. Apart from the leader, in one migration loop, all individuals will traverse the searched space in the direction of the leader. Mutation, the random perturbation of individuals, is an important operation for evolutionary strategies (ES). It ensures diversity amongst all the individuals and it also provides a means to restore lost information in a population. Mutation is different in SOMA as compared with other ES. SOMA uses a parameter called PRT to achieve perturbations. This parameter has the same effect for SOMA as mutation for GA. The PRT vector defines the final movement of an active individual in the search space. The randomly generated binary perturbation vector controls the allowed dimensions for an individual. If an element of the perturbation vector is set to zero, then the individual is not allowed to change its position in the corresponding dimension. An individual will travel over a certain distance (called the PathLength) towards the leader in finite steps of the defined length. If the PathLength is chosen to be greater than one, then the individual will overshoot the leader. This path is perturbed randomly. Pseudocode for SOMA is eq. (6.3).

Input : $N, Migrations, PopSize \geq 2, PRT \in [0,1], Step \in (0,1], \text{MinDiv} \in (0,1]$,
PathLength $\in (0,5]$, *Specimen* with uper and lower bound $x_j^{(hi)}, x_j^{(lo)}$

$$\text{Initialization}: \begin{cases} \forall i \leq PopSize \wedge \forall j \leq N : x_{i,j,Migrations=0} = x_j^{(lo)} + rand_j[0,1] \bullet \left(x_j^{(hi)} - x_j^{(lo)}\right) \\ i = \{1,2,\ldots,Migrations\}, j = \{1,2,\ldots,N\}, Migrations = 0, rand_j[0,1] \in [0,1] \end{cases}$$

$$\begin{cases} \text{While } Migrations < Migrations_{\max} \\ \forall i \leq PopSize \begin{cases} \text{While } t \leq PathLength \\ if\ rnd_j < PRT\ pak\ PRTVector_j = 1\ else\ 0, \quad j = 1,\ldots,N \\ x_{i,j}^{ML+1} = x_{i,j,start}^{ML} + (x_{L,j}^{ML} - x_{i,j,start}^{ML})\ t\ PRTVector_j \\ f\left(x_{i,j}^{ML+1}\right) = \text{ if } f\left(x_{i,j}^{ML}\right) \leq f\left(x_{i,j,start}^{ML}\right) \text{ else } f\left(x_{i,j,start}^{ML}\right) \\ t = t + Step \end{cases} \\ Migrations = Migrations + 1 \end{cases} \tag{6.3}$$

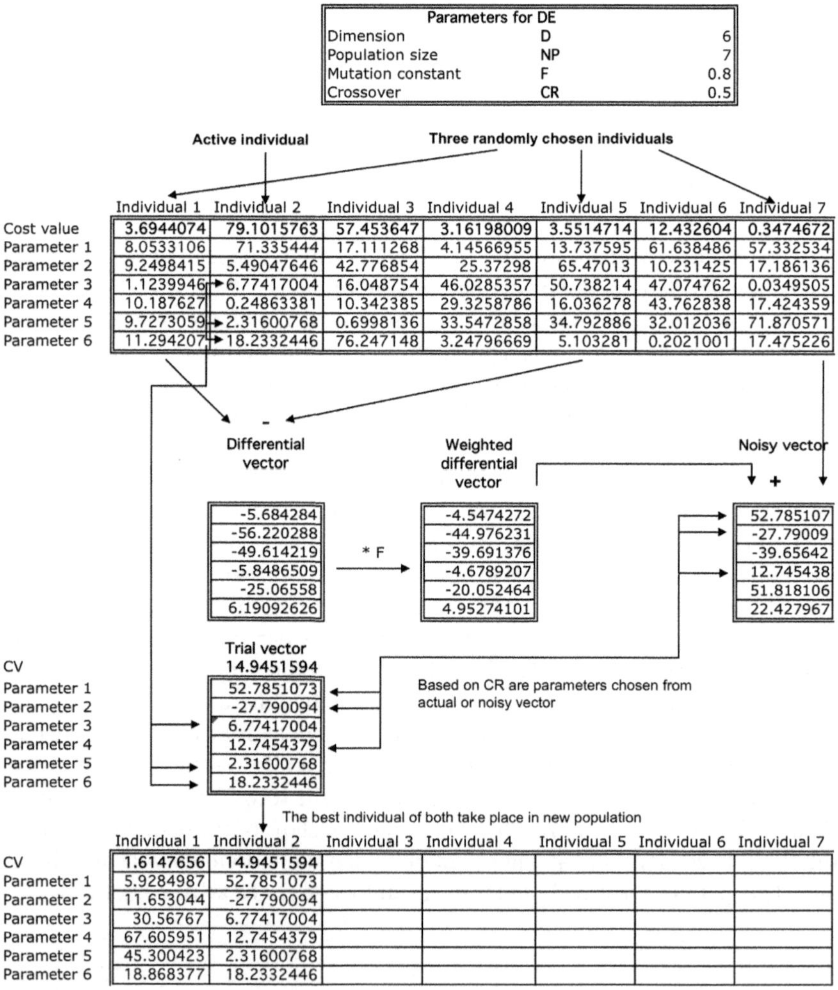

Fig. 6.7 Differential evolution, (http://www.icsi.berkeley.edu/ storn/code.html)

An example of SOMA is demonstrated on Fig. 6.8.

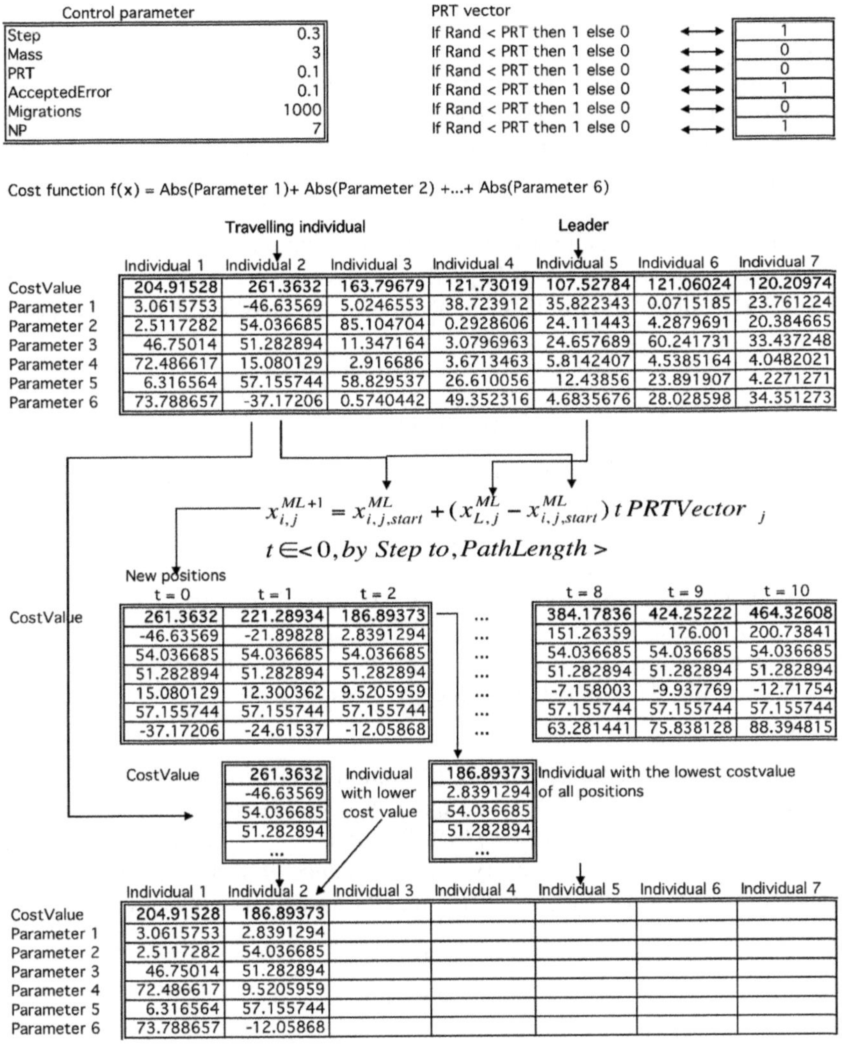

Fig. 6.8 SOMA, (http://www.fai.utb.cz/people/zelinka/soma)

SOMA can be also regarded as a member of the swarm intelligence class of algorithms. This class contains algorithms such as particle swarm, which is also based on a population of particles, which are mutually influenced amongst themselves. Some similarities as well as differences exist between SOMA and particle swarm, for details see [11], [5].

6.3.3 Simulated Annealing

Simulated annealing (SA, S. Kirkpatrick, C. D. Gelatt and M. P. Vecchi in 1983, and by V. Cerny in 1985) is a generic probabilistic meta-algorithm for the global optimization problem, namely locating a good approximation to the global optimum of a given function in a large search space. By analogy with metallurgical processes, each step of the SA algorithm replaces the actual solution by a randomly generated solution from the neighborhood, chosen with a probability depending on the difference between the corresponding function values and on a global parameter so called temperature - T. Temperature is decreasing during the process. Current solution changes almost randomly when T is large, but increasingly "downhill" as T goes to zero. The allowance for "uphill" moves saves the method from becoming stuck at local minimum. Simulated annealing is a stochastic algorithm defined by:

$$SA = (M, x_0, N, f, T_0, T_f, \alpha, n_T), \tag{6.4}$$

The meaning of parameters is as follow:

- M: space of possible solutions
- x_0: initial solution, randomly selected
- $N(x, \sigma)$: subset of possible solution $x \in M$
- T_0: initial temperature.
- T_f: stopping temperature (temperature of crystallization)
- n_T: number of iterations of Metropolis algorithm
- α: temperature reduction $\alpha : T \rightarrow T', T' < T$, usually is used function of reduction like $T' = \alpha \times T$. Parameter α is usually a single value between 0.8 - 0.99.

In a real world, all objects consist of a number of particles. Physical state can be described by vector $\mathbf{x} = (x_1, x_2, ..., x_n,)$ describing particle position for example. This state is related to energy $y = f(\mathbf{x})$. If such system is on the same temperature T long enough, then the probability of the existence of such states is given by the Boltzmann distribution. The probability that the system is in state $\mathbf{x}$ is then given by

$$\frac{e^{-f(\mathrm{x})/T}}{Q(T)} \tag{6.5}$$

with

$$Q(T) = \sum_i e^{-f(\mathrm{x})/T} \tag{6.6}$$

For sufficiently small T, the probability that the system will be in state x_{min} with minimal energy $f(\mathbf{x_{min}})$ is almost 1. It has been suggested that the simulation of annealing by means of Monte Carlo method can be accomplished with a new decision function as given in eq. 6.7.

$$P(x \rightarrow x_0) = \begin{cases} 1, & for f(x) < f(x_0) \\ e^{-(f(x)-f(x_0))/T} & for f(x) \geq f(x_0) \end{cases} \tag{6.7}$$

This decision function stipulates whether new state, say x_{new} (when for example one particle will change its position) is accepted or not. In the case that x_{new} is related to lower energy, then the old state is replaced by a new one. On the contrary, x_{new} is accepted with probability $0 < \mathrm{P}(\mathbf{x} \rightarrow \mathbf{x}_0) < 1$. If r is random number from [0, 1], then the new state is accepted only if $r < \mathrm{P}(\mathbf{x} \rightarrow \mathbf{x}_0)$ In eq. (6.7) temperature T has an important influence on probability $P(x \rightarrow x_0)$ when $f(x) \geq f(x_0)$; for big T is basically accepted for any new state (solution), however for low T states with higher energy are accepted only rarely. If this algorithm (Metropolis algorithm) is repeated for one state in a sufficient number of repetitions, then the observed distribution of generated states is basically the Boltzmann distribution. This allows the realization of simulated annealing on a PC. Algorithm of simulated annealing, repeating the Metropolis algorithm for decreasing temperature uses the final state x_n related to T_n as the initial state for the next iteration x_m with function $\alpha(T_n)$. This function decreases actual temperature and can be represented for example by $T_m = T_n - \alpha$ or $T_m = T_n \times \alpha$. Variable α is an arbitrarily small number.

Pseudocode for simulated annealing is given in eq. (6.8).

$$
\begin{array}{l}
1.\text{Input}: \text{ initial solution } \mathbf{x_0}, \text{ temperature } T_0, \text{ function of temperature decrement } \alpha(t), \\
\text{final temperature } T_f, \text{ number of iterations } n_T, \text{ and initial bounds}: \mathbf{x}^{(lo)}, \mathbf{x}^{(hi)}. \\
\left\{\begin{array}{l}
2.\text{While } T < T_f \\
\forall i \leq n_T \left\{\begin{array}{l}
3.\text{ Randomly select } \mathbf{x} \text{ from neighbor of } N(\mathbf{x_0}, \sigma) \text{and calculate } \Delta f := f(\mathbf{x}) - f(\mathbf{x_0}) \\
3.1\ if\ \Delta f < 0 \left\{\begin{array}{l}
\mathbf{x_0} := \mathbf{x} \text{ and } if f(\mathbf{x}) < f(\mathbf{x}*) \left\{\begin{array}{l} then\ \mathbf{x}* := \mathbf{x} \\ \{\text{ actualization of the best solution}\} \\ end \end{array}\right. \\
else \left\{\begin{array}{l} \text{randomly select } r \text{ from } [0,1] \text{ and} \\ if\ r < e^{-\Delta f/T} \left\{\begin{array}{l} then\ \mathbf{x}* := \mathbf{x}\{\text{ move to the worst solution}\} \\ end \end{array}\right. \\ end \end{array}\right.
\end{array}\right. \\
3.2\ T_m := \alpha(T_n)
\end{array}\right.
\end{array}\right.
\end{array}
\tag{6.8}
$$

6.3.4 Genetic Algorithms

Genetic algorithms [16] has been developed according to the ideas of biological evolution. GA in its canonical version consists of binary strings-individuals, but other encodings are also possible. Individuals are processed through evolutionary operators like selection, crossover, mutation etc. with the aim to get better individuals-solutions. The evolution usually starts from a population of randomly generated individuals and loops of generations. In each generation, the fitness of every individual in the population is evaluated, two or more individuals are selected from the current population, and modified to form a new population. There exists a whole class of genetics algorithms, which was used to solve very rich class of problems. Principles of genetic algorithms are the same as reported in Chapter 2, i.e. it is running in loops called generations G. In the beginning randomly created initial individuals are generated (binary strings in basic version of GA), which are then evaluation by the fitness function (usually cost function unified into interval [0, 1]). When individuals are evaluated, we can classify them as parents. Usually two parents (P_n, P_m) are selected for crossover operation Ω (exchange of parts of their binary "bodies"). Selection is usually the process when two parents are less or more selected by means of their fitness. New individuals, called offsprings, are created like a product of crossover. Each of them has to be mutated (Ω_m, random reverse of randomly selected bits of their binary bodies) and again evaluated by the fitness function. In the last step, before the actual generation is finished, the best individuals (from parents and offsprings) is selected for the new population. The pseudocode for genetic algorithm is given in eq. (6.9).

$$
\begin{array}{l}
1.\text{Input}: \text{ initial population } \mathbf{P} \text{ consist of individuals } p_i, \text{mutation } T_0, \\
\text{operators of parent selection } \Theta, \text{ mutation } \Omega_m, \text{ crossover } \Omega, \\
\text{and initial bounds}: \mathbf{x}^{(lo)}, \mathbf{x}^{(hi)}. \\
\left\{
\begin{array}{l}
2.\text{While } g < G_{\max} \\
\forall p_i \in \mathbf{P} \left\{
\begin{array}{l}
3. \{P1, P2\} := \Theta(P(t)); \{\text{ parents selection }\} \\
4.\ O := \Omega(P1, P2)); \{\text{ offspring creation by crossover}\} \\
5.\ O := \Omega_m(O, p_m)); \{\text{ mutation}\} \\
6. \text{ find } S_0 \in P(t) \text{ such, that } f(S_0) \geq f(S), \forall S \in P(t); \\
\{S_0 \text{ is the worst solution in } P(t)\} \\
7.\ S_0 := O; \{\text{ The worst solution is replaced by offspring}\} \\
8.\ iff(O) < f(S*) \left\{
\begin{array}{l}
then\ S* := O; \{\text{ updating of the best solution}\} \\
end
\end{array}
\right.
\end{array}
\right. \\
g := g + 1
\end{array}
\right.
\end{array}
\tag{6.9}
$$

6.3.5 Evolutionary Strategies

Evolutionary strategies [1] was originally developed by P. Bienert, I. Rechenberg and H. P. Schwefel. The first application was focused on mechanical engineering problems. The original version of ES was different from GA in two main points:

- ES did not used individuals like GA in binary strings,
- ES did not used operators of crossover, selection and mutation, as in GA.

EAs are usually defined like ES$((\mu + \lambda))$ and ES$((\mu, \lambda))$. Symbols μ and λ represents a set of parents and offsprings, symbols $+$ and , represents whether selection of the best individuals will be done from the set of parents (μ) or both $(\mu \cup \lambda)$. Both sets (populations) can have size from 1 individual to a number which is limited by the memory of the used computer. Size of both ES's consist of a few strategies like *two-membered* ES, *multi-membered* ES, *recombinative* ES and *self-adaptive* ES. ES is running in a simple loop as described by the pseudocode below; general ES$((\mu + \lambda))$ is described there. In the beginning, a population of parents (μ) is created. Then the selected parents are mutated by means of so called "Gaussian mutation operator" $N(0, \sigma)$ and thus, a set (population) of offsprings (λ) is created.

After their union into the population $P := \mu \cup \lambda = \left(\bigcup_{j=1}^{\lambda} y^j\right) \cup \left(\bigcup_{i=1}^{\mu} x^i\right)$ the best solutions from P are selected and replace the worst solutions in the population. The whole process is then repeated. Pseudocode for evolutionary strategies is described in eq. (6.10).

$$
\begin{array}{l}
1.\text{Input} : \mu - \text{ randomly generated parents } x_i,\\
\sigma - \text{ standard deviation of normal distribution},\\
f\text{: costfunction, } i_{\max}\text{: maximal number of iterations of ES},\\
FV\text{: fitness value (cost value)},\\
\text{used to stop ES, auxiliary variables} : \lambda - \text{ offsprings population},\\
y_i - i^{th} \text{ offspring};\\
\text{P} - \text{populationof } \lambda \text{ and } \mu.\\
\left\{\begin{array}{l}
2.\text{While } i < i_{\max}\\
\forall x_i \in \mu \left\{\begin{array}{l}
3.y_i := x_i + N(0,\sigma), \{\text{offspring creation}\}\\
4.P := \mu \cup \lambda = \left(\bigcup_{j=1}^{\lambda} y^j\right) \cup \left(\bigcup_{i=1}^{\mu} x^i\right),\\
\{\text{joining of both populations}\}\\
3.\mu := \text{ selection of the best solutions out of } P\\
5.if \text{ the best } f(\mu) < FV \left\{\begin{array}{l} then \text{ stop ES}\\ \text{end}\end{array}\right.
\end{array}\right.\\
i := i + 1
\end{array}\right.
\end{array}
\tag{6.10}
$$

For exact descriptions of the above mentioned algorithms, see [28] for DE, [35] for SOMA, [16], [8] for GA, [21], [2] for SA and [9] or [1] for ES.

6.4 CML Control

6.4.1 Used Hardware

CML control in this case study has been done on special grid computer, compared to a simple PC as in [39]. This grid computer, called Emanuel, consist of two special Apple servers (David and Goliath). The bigger one (Goliath, see Fig. 6.10) is based on 16 XServers, each 2x2 GHz Intel Xeon, 1 GB RAM, 80 GB HD i.e. 64 CPUs. The second one (David, see Fig. 6.9) is created from 7 Apple Minimacs CoreDuo i.e. number of accessible CPUs is 14. In total 78 CPUs are available for computation. Emanuel has been used for calculations in two ways. The first one was focused on use of each CPU as a single processor and thus a rich set of statistically repeated experiments were conducted were time was not a factor. The second way involved the use of Emanuel like a grid machine in order to increase speed of selected time demanding simulations, as reported in this chapter. However, this does not means that such class of problems can be solved only on special computers. All solved problems and reported case studies in this chapter can also be done on single PC but in a different time scale.

Fig. 6.9 Emanuel server: David (14 CPUs) ...

Fig. 6.10 ... and Goliath (64 CPUs).

6.4.2 Problem Selection and Case Studies

The class of CML problems chosen for this comparative study was based on case studies reported in [31]. In general, CML control means setting of such pinning sites (controlled CML sites) and their pinning values (control values) so that the system stabilizes itself on expected spatiotemporal pattern. CML as an object of study was chosen because it shows chaotic behavior and its level of complexity can be quite rich.

All simulations designed and reported here are based on previous simulations, like [36], [37] or [39]. In the previous simulations, EAs has been found to be capable of controlling CML chaos. Some of them were modified (cost functions was redefined) to increase speed (i.e. number of cost function evaluations) of simulations. To highlight the impact of proposed changes in this chapter, we have used all five evolutionary algorithms to control CML, size of 10 inputs, see Fig. 6.11 and 6.12.

Comparing to simulations described further, this simulation was defined so that 20 unknown parameters has been estimated. The reason of why exactly 20, is simple. CML size was 10 and EAs estimated which pinning site (10 inputs of CML) shall be used and what pinning value (10 control signals) will be applied to each input. Thus evolutionary search has run in the 20 dimensional solution space. Based on informations in [31] and previous experiences [36], [37] the cost function (6.12) has been used and it has been empirically discovered that the cost value ≤ 5 should guarantee stabilized CML (at least in our experimentation in *Mathematica* code).

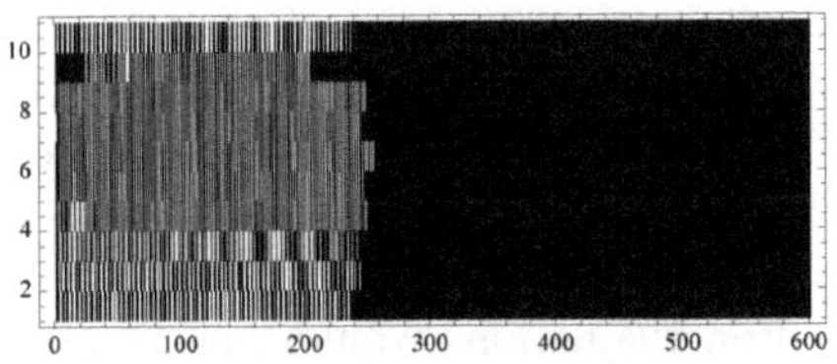

Fig. 6.11 An example from 1500 simulated CML behavior. Stabilization to T_1S_1 has been reached after 250 iterations

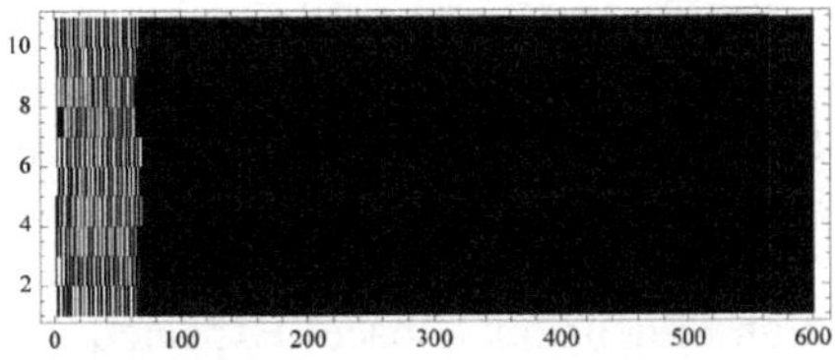

Fig. 6.12 Another example from 1500 simulated CML behavior. Stabilization to T_1S_1 has been reached before 100 iterations

We have found, that when EAs stop above cost value ≤ 5.1, then the CML is stabilized in almost all cases between 300 - 600 iterations. To safely stabilize CML before 100 iterations, it is enough when EAs stop below this level, like for example cost value ≤ 5.00001. Thus it is quite critical, what stopping cost value is selected. On Fig. 6.13 and Fig. 6.14 are two examples (of 15) which shows what pinning sites (black squares, white means not used) were used to control CML. Pinning values were estimated in the interval $[0,5]$ for each pinning site, and are depicted in Fig. 6.15 and 6.16. On both figures all pinning values are shown, however only those related used pinning sites (Fig. 6.13, Fig. 6.14), has been used.

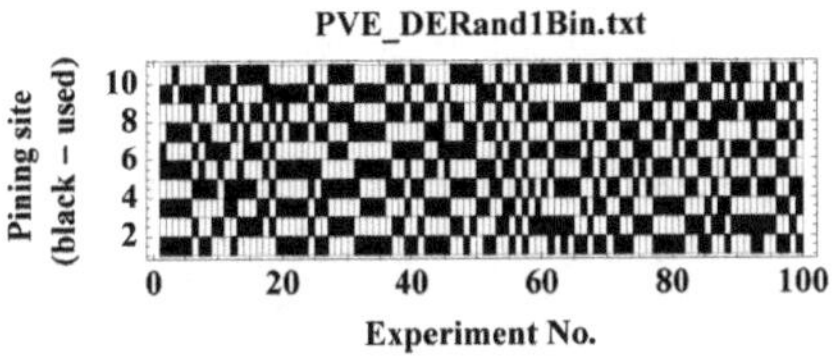

Fig. 6.13 Used pinning sites for 100 DE repeated simulations. Black sites were used for CML simulations.

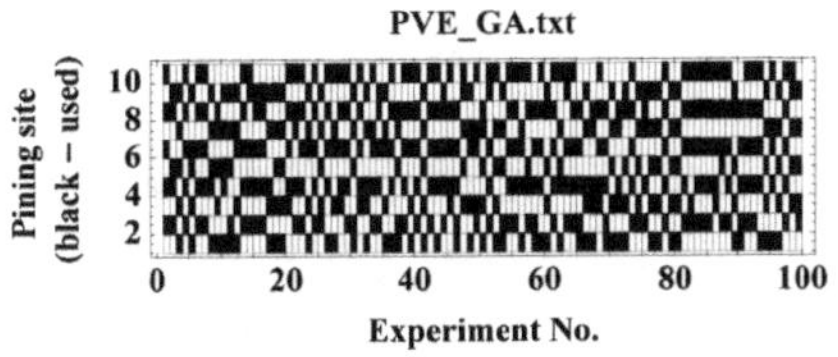

Fig. 6.14 Used pinning sites for 100 GA repeated simulations. Black sites were used for CML simulations.

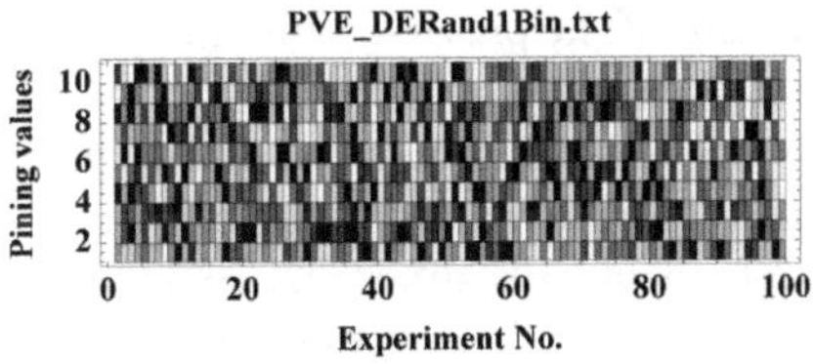

Fig. 6.15 Used pinning values for sites depicted in Fig. 6.13.

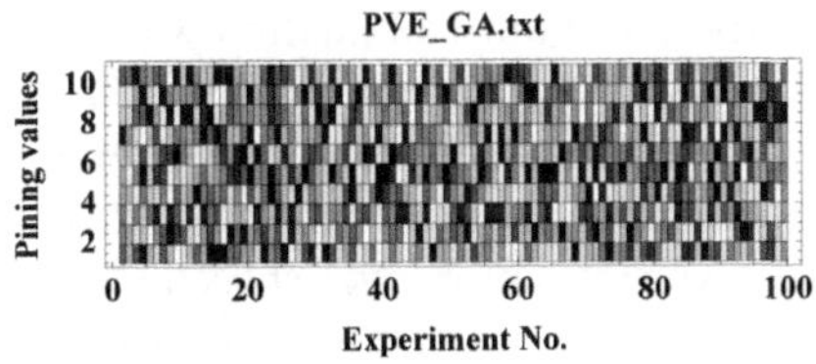

Fig. 6.16 Used pinning values for sites depicted in Fig. 6.14.

In Fig. 6.17 and 6.18 the summarized cost function needed to find stabilizing combination of all 20 parameters is given. It is visible, that the number of cost function evaluations needed to reach solution was 3964 on average. All those results are valid for CML of 10 inputs only, however, more often one can use CML with 50, 100 or more inputs and in such a case search algorithms would search in really high-dimensional space. Expected cost function evaluations would then be much more higher.

To improve the performance and speed of simulations, two modifications are suggested here. Number of used pinning sites is omitted, only period of used pinning site is estimated (i.e. **only one variable instead of *n* variables**) which means that only each n_{th} site is used to apply pinning value. Pinning value is estimated in the same manner. Only one value is estimated and then applied to each n_{th} pinning site. In such a case, problem of generally n dimensional problem (n can be 20, 50 , 100,.....) is reduced only to search in the 2D solution space.

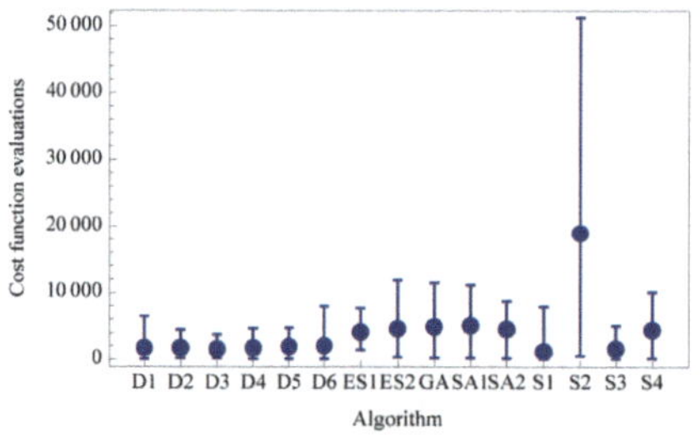

Fig. 6.17 Cost function evaluations, total view.

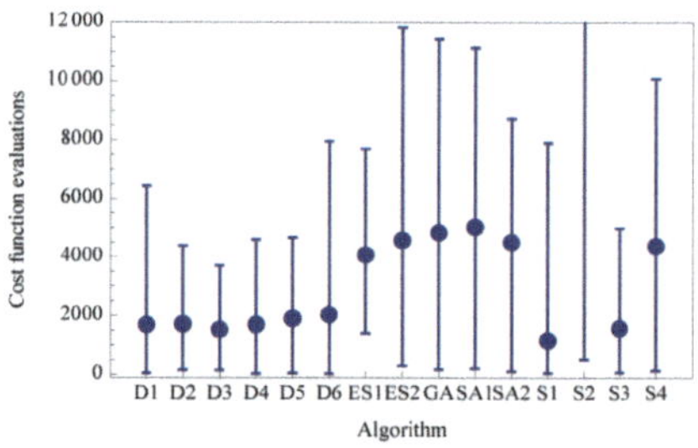

Fig. 6.18 Cost function evaluations, detail.

Selected modifications has improved the performance of selected algorithms, as reported in the following section. The investigation consists of four parts in increasing order from the calculation complexity point of view and was based on [17]. The first one is focused on pinning values estimation for *a priori* given pinning sites. In the second one, the pinning sites with *a priori* given pinning values were estimated by EAs. The third simulation was the enlargement of the previous simulation - EAs were used to find minimal number of pinning sites and the fourth simulation was focused on mutual estimation of pinning sites and values, i.e. EA was searching for the minimal number of pinning sites and optimal (as much as possible) pinning values. All simulations were based on the same model and 100 times repeated for each EA with new initial conditions for each simulation. Simulations were done for two basic CML stabilized configuration - T_1S_1 (CML is stabilized on period Time=1 and Space=1, i.e. after stabilization is CML as in Fig. 6.19) and T_1S_2 (CML is stabilized on period Time=1 and Space=2, i.e. after stabilization is CML as in Fig. 6.1). In

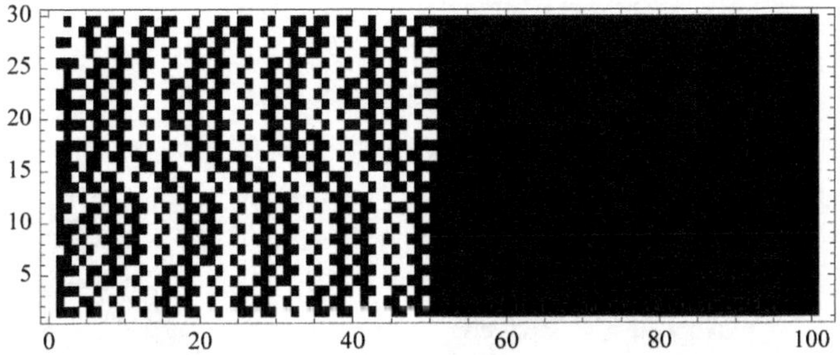

Fig. 6.19 Successful stabilization of CML (30×100 - 30 pinning sites, 100 iterations) in T_1S_1 pattern - stabilization after 52 iterations is visible.

total $4 \times 2 \times 1500 = 12000$ independent simulations ($4 \times T_1S_{1,2}$, 15 algorithms, each for 100 independent runs) of spatiotemporal CML were carried out.

6.4.3 Cost Function

The fitness (cost function) has been calculated according to using the distance between desired CML state and actual CML output, eq. (6.11). The minimal value of this cost function, which guarantees the best solution is 0. The aim of all simulations was to find the best solution, i.e. a solution that returns the cost value 0. This cost function was used for the first two case studies (pinning values setting, pinning sites setting). In the next (last) two case studies, the cost function (6.12) was used. It is synthesized from cost function (6.11) so that two penalty terms are added. The first one, p_1, represents the number of used pinning sites in CML. The second one, p_2, is added here to attract attention of evolutionary process on the main part of the cost function. If this would not be done, then mainly p_1 would be optimized and the results should not be acceptable (proved by simulations). Indexes i and j are coordinates of lattice element, i.e. $CML_{i,j}$ is i_{th} site (equation) in j_{th} iteration. For all simulations of T_1S_1 the stabilized state was set to $S_1 = 0.75$, and for T_1S_2 to period $S_2 = (0.880129, 0.536537)$, i.e. CML behavior was controlled to this state.

Knowledge (at least approximate) about complexity and variability of used cost function is very important. Such knowledge can be important when the class of optimizing algorithms is selected. Thus a few ideas and examples has been selected here to show complexity and its dependance on chaotic system parameter setting. The complexity of a cost function is clearly visible from Fig. 6.20 - 6.24 and Fig. 6.26 - 6.30. Different shape of geometrical visualization of (6.11) is given in Fig. 6.20 - 6.24 for different number of iterations and pinning sites. It is clearly visible, that the cost function is partly chaotic and for certain pinning value, the global minimum representing stabilization is accessible. Chaoticity of such a graphical

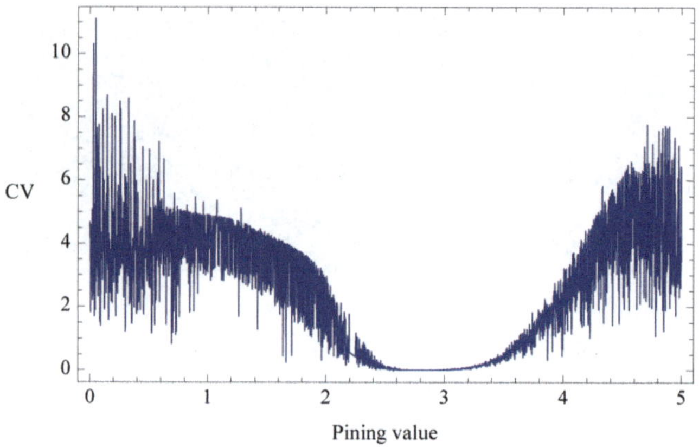

Fig. 6.20 Landscape of eq. (6.11) for T_1S_1 in configuration 10×10, compare with Fig. 6.22.

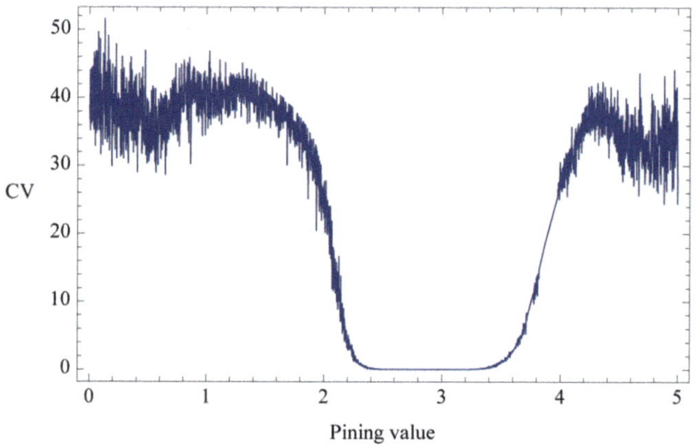

Fig. 6.21 Landscape of eq. (6.11) for T_1S_1 in configuration 10×20.

representation is caused due to the fact that calculations are based on the chaotic system. If an average value (over many of such runs) would be calculated, then we would get graphs like in Fig. 6.30 (case for T_1S_2). However, because our simulations were running on a single run, not over a number of them, are given in Fig. 6.20 - 6.24 and Fig. 6.26 - 6.30 shows the real representation of landscape of our cost function. It is also important to note, that for each simulation of CML, its exact shape and chaoticity can be slightly different from the previous ones, due to the sensitivity of initial conditions. Another more complex visualizations of landscape of eq. (6.11) is depicted in Fig. 6.31 - 6.35. In Fig. 6.31 is a 3D visualization of the cost function for T_1S_1, case study D (see below). Dependence of eq. (6.11) on pinning value

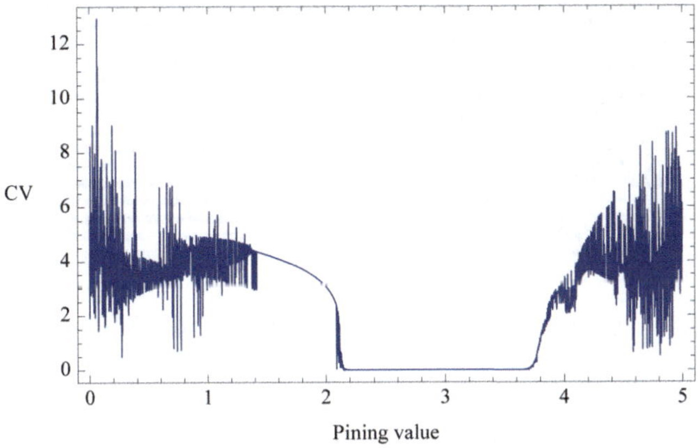

Fig. 6.22 Landscape of eq. (6.11) for $T_1 S_1$ in configuration 10×100.

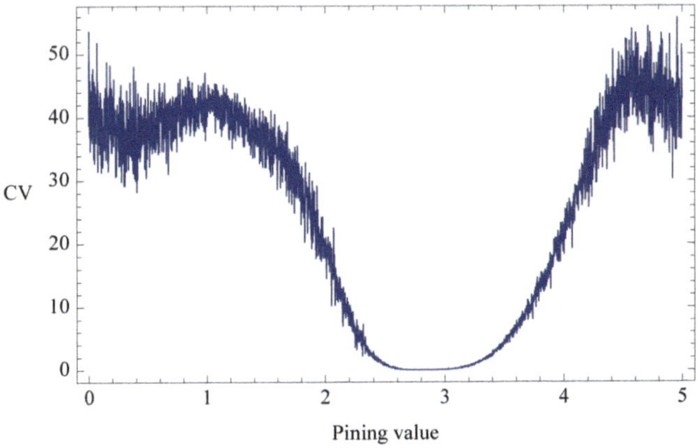

Fig. 6.23 Landscape of eq. (6.11) for $T_1 S_1$ in configuration 100×10.

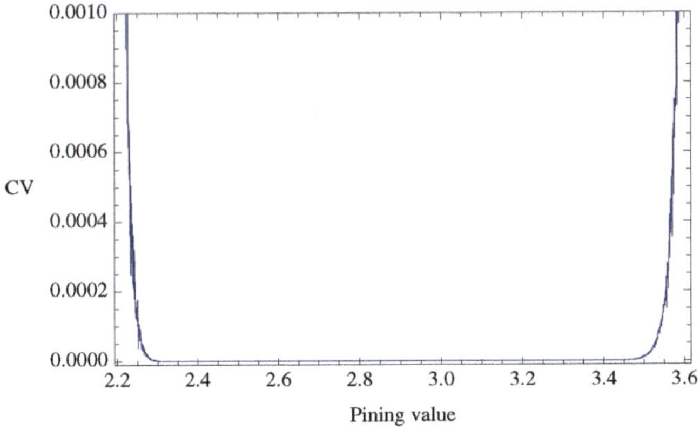

Fig. 6.24 Zoom of the landscape of Fig. 6.22 (eq. (6.11)) for $T_1 S_1$ in configuration 10×100.

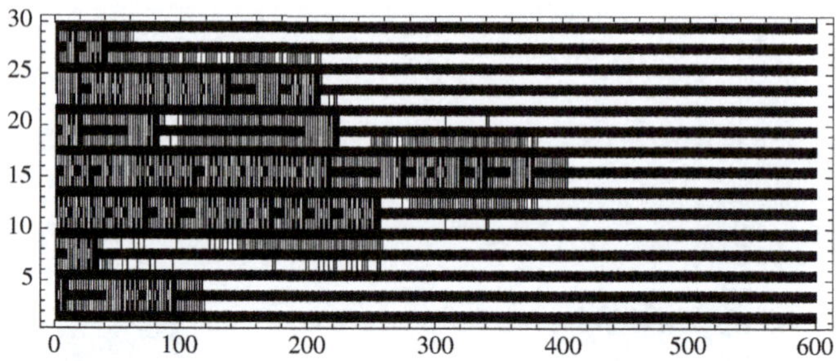

Fig. 6.25 CML T_1S_2 in configuration 30×600 – stabilization after 400 iterations is visible.

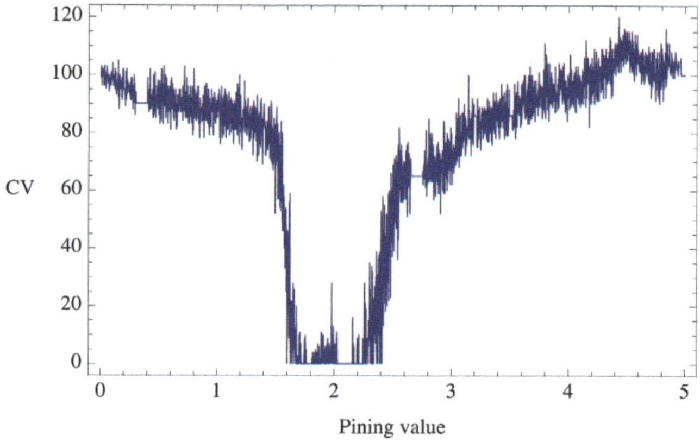

Fig. 6.26 Landscape of eq. (6.11) for T_1S_2 in configuration 30×600. Comparing with landscapes for T_1S_1 is this much more complex.

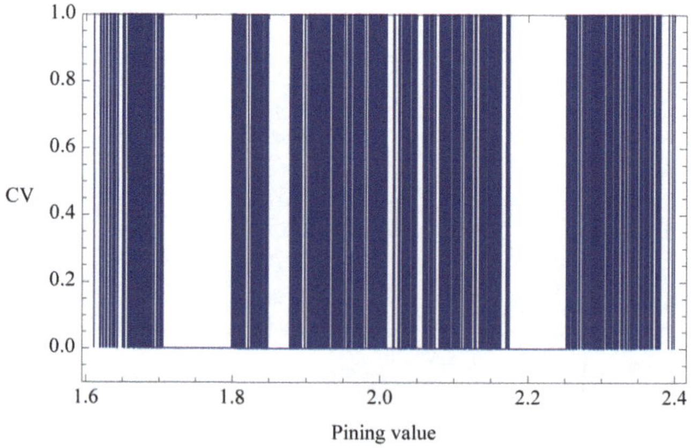

Fig. 6.27 Zoom into [1.6, 2.4] of the landscape of eq. 6.11 for T_1S_2 from Fig. 6.26. Fractal like structure is observable.

(continuous value) and used pinning site (discrete value) is depicted here. This figure has been calculated and depicted like "continuous" landscape, i.e. landscape shape between calculated points has been interpolated. Discrete (no interpolation is used) and more realistic visualization, is given in Fig. 6.32. Suitable parameter setting for CML stabilization is visible from this figure. Better readability of this information is in Fig. 6.33. It is clear that for pinning values belonging to [1.5, 4] and used site period (pinning sites) 1 or 2, CML is stabilized, after suitable number of iterations. Two remaining examples are given in Fig. 6.34 and Fig. 6.35. Both figures represent CML for T_1S_2 with the use of each 2nd and 4th pinning site with pinning value 2. Black color represent stabilizing combination of pinning value and used pinning sites. It is clear that complexity of the used cost function is big, despite its simple mathematical description. Also, suitable stabilizing combination of control parameters depend on a number of CML iterations (after certain number of iterations does a combination become permanent) and configuration ($T_1S_{1,2}$) of stabilized state.

$$f_{cost} = \left|\sum_{i=1}^{30}\sum_{j=a}^{b} \left|TS_{i,j} - CML_{i,j}\right|\right|^2 \tag{6.11}$$

$\mathrm{TS_{i,j}}$ – target state of CML
$\mathrm{CML_{i,j}}$ – actual state of controlled CML
$\{a,b\} = \{80,100\}$ for $\mathrm{T_1S_1}$ and $\{a,b\} = \{580,600\}$ for $\mathrm{T_1S_2}$

$$f_{cost} = p1 + \left(p2 + \sum_{i=1}^{30}\sum_{j=a}^{b} \left|TS_{i,j} - CML_{i,j}\right|\right)^2 \tag{6.12}$$

$\mathrm{TS_{i,j}}$ – target state of CML
$\mathrm{CML_{i,j}}$ – actual state of controlled CML
p1 – number of actually selected pinning sites
p2 – 100, heuristically set weight constant
$\{a,b\} = \{80,100\}$ for $\mathrm{T_1S_1}$ and $\{a,b\} = \{580,600\}$ for $\mathrm{T_1S_2}$

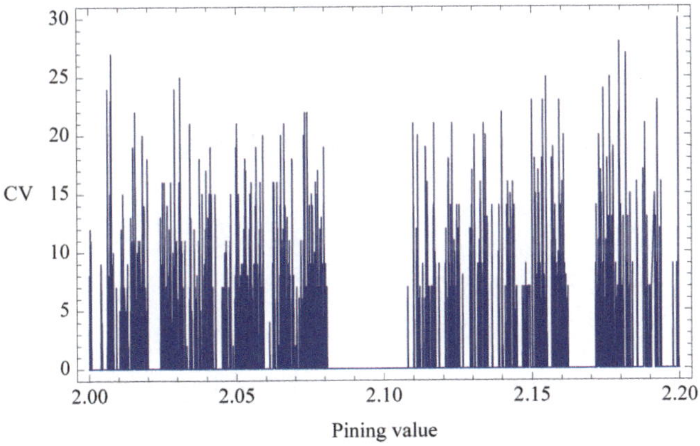

Fig. 6.28 Zoom into [2, 2.2] of the landscape of eq. 6.11 for T_1S_2 from Fig. 6.26. Fractal like structure is observable.

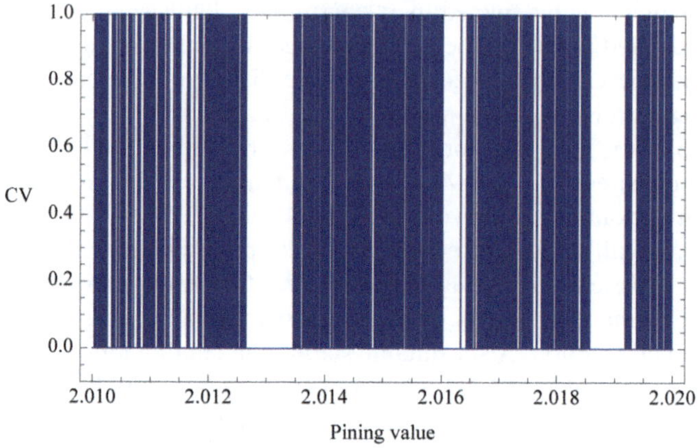

Fig. 6.29 Zoom into [2.01, 2.02] of the landscape of eq. 6.11 for T_1S_2 from Fig. 6.26. Fractal like structure is observable.

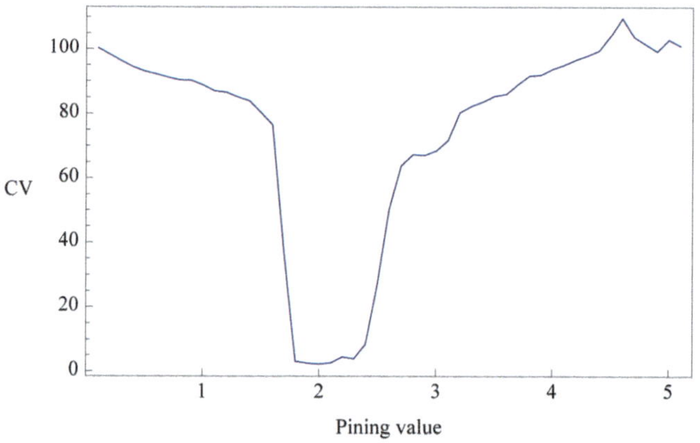

Fig. 6.30 Average landscape of eq. 6.11 for T_1S_2 from Fig. 6.26. Note that smooth landscape shape is caused by averaging of 100 independent CML runs, each from randomly chosen initial conditions.

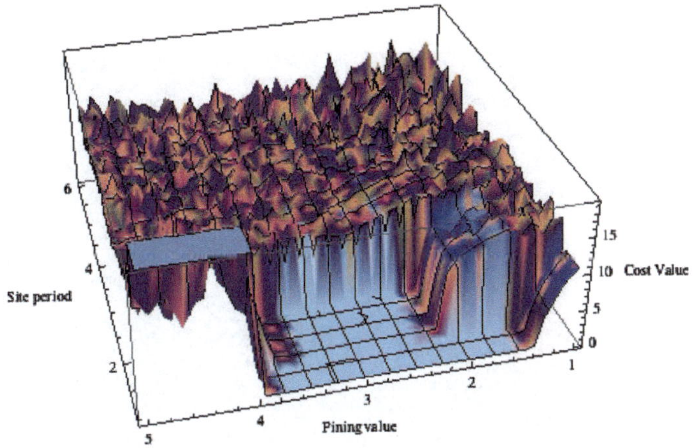

Fig. 6.31 3D cost function visualization for $T_1 S_1$, case study D, continuous landscape

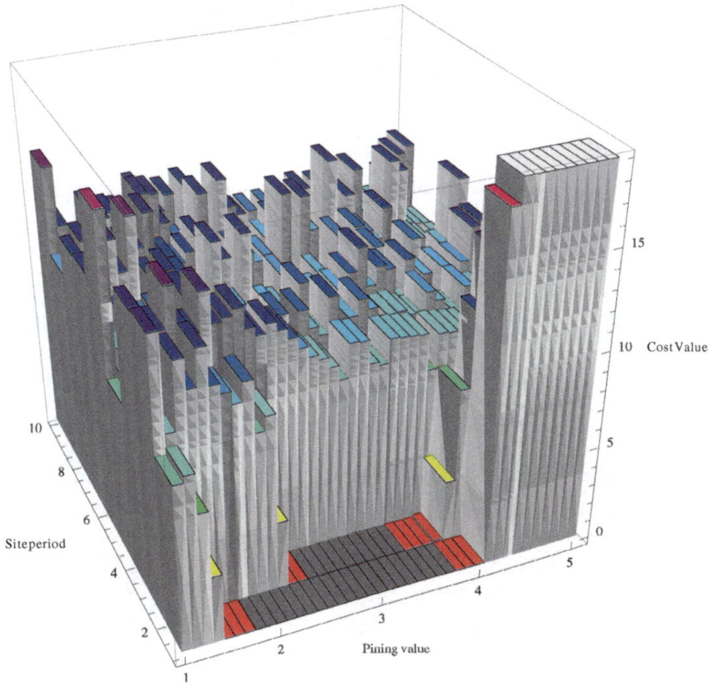

Fig. 6.32 3D cost function visualization for $T_1 S_1$, case study D, discrete landscape in 3D

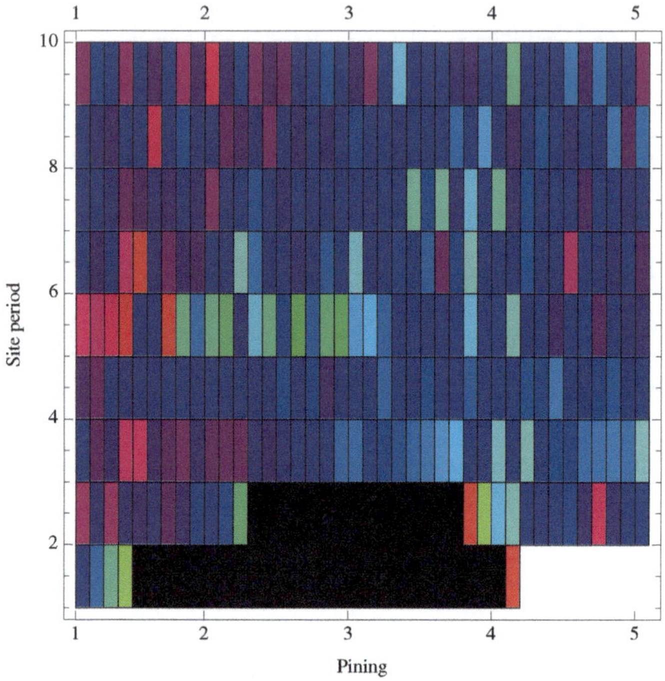

Fig. 6.33 2D cost function visualization for T_1S_1, case study D, discrete landscape in 2D

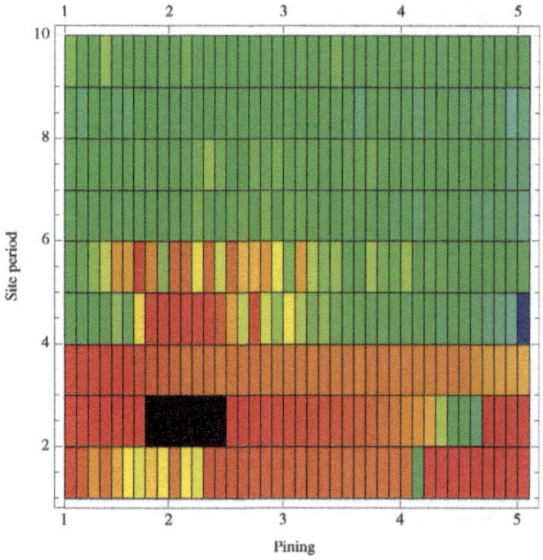

Fig. 6.34 2D cost function visualization for T_1S_2, case study D, discrete landscape in 2D. Original CML used each 2nd pinning site with pinning value 2.

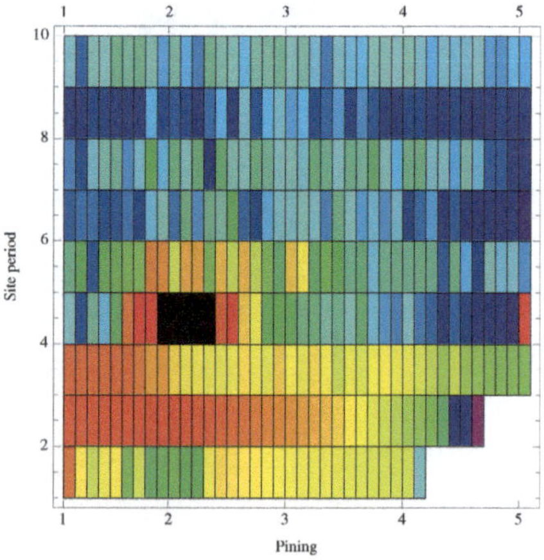

Fig. 6.35 2D cost function visualization for T_1S_2, case study D, discrete landscape in 2D. Original CML used each 4th pinning site with pinning value 2.

6.4.4 Parameter Setting

The control parameter settings have been found empirically and are given in Table 6.1 - 6.6. The main criterion for this setting was to keep the same setting of parameters as much as possible and of course the same number of cost function evaluations as well as the population size. Individual length represents the number of optimized parameters (number of pinning sites, values, ...). Comparing to previous [37] experiments, the length of an individual has been set to 1 or 2, according to the case study. In the [37] and [39], the individual length was equal to the number of pinning sites, which has increased the complexity of calculations. To simplify simulations here, a simple presumption has been made - instead of the exact number of pinning sites as in [37], their periodicity has been estimated in the evolution, i.e. if parameter for the pinning site was for example 4, then each 4th site has been used for pinning value application, etc.

All algorithms (SOMA, DE, SA, GA, ES) have been evaluated 100 times in order to find the optimal setting. The primary aim of this comparative study is not to show

Table 6.1 Algorithms abbreviation

Algorithm	Version	Abbreviation
Differential Evolution	DEBest1JIter	D1
	DEBest2Bin	D2
	DELocalToBest	D3
	DERand1Bin	D4
	DERand1DIter	D5
	DERand2Bin	D6
Evolutionary strategies	(μ,λ)	ES1
Evolutionary strategies	$(\mu+\lambda)$	ES2
Genetic Algorithm		G
Simulated annealing with elitism		SA1
Simulated annealing without elitism		SA2
SOMA	AllToAllAdaptive	S1
	AllToAll	S2
	AllToOne	S3
	AllToOneRandomly	S4

Table 6.2 DE setting for case studies A, B, C and D

Case Study	A	B	C	D
NP	10	10	10	10
F	0.9	0.9	0.9	0.9
CR	0.3	0.3	0.3	0.3
Generations	500	500	500	500
Individual Length	1	1	1	2

Table 6.3 ES setting for case studies A, B, C and D

Case Study	A	B	C	D
μ,λ	10	10	10	10
σ	0.8	0.8	0.8	0.8
Iterations	200	200	200	200
Individual Length	1	1	1	2

Table 6.4 GA setting for case studies A, B, C and D

Case Study	A	B	C	D
Population size	10	10	10	10
Mutation	0.4	0.4	0.4	0.4
Generations	500	500	500	500
Individual Length	1	1	1	2

Table 6.5 SA setting for case studies A, B, C and D

Case Study	A	B	C	D
No. of particles	10	10	10	10
σ	0.5	0.5	0.5	0.5
$k_{\max}$	66	66	66	66
T_{min}	0.0001	0.0001	0.0001	0.0001
T_{max}	1000	1000	1000	1000
α	0.95	0.95	0.95	0.95
Individual Length	1	1	1	2

Table 6.6 SOMA setting for case studies A, B, C and D

Case Study	A	B	C	D
PathLength	3	3	3	3
Step	.11	.11	.11	.11
PRT	0.3	0.3	0.3	0.3
PopSize	10	10	10	10
Migrations	10	10	10	10
MinDiv	-0.1	-0.1	-0.1	-0.1
Individual Length	1	1	1	2

which algorithm is better and worst, but to show that evolutionary deterministic chaos control (EDCC) can be really used for different problems of spatiotemporal chaos control based at least on CML. Outputs of all simulations are depicted in Fig. 6.36 - 6.61, which shows results of all 100 simulations for each case study. In each case study, the mutual comparison of obtained results between T_1S_1 and T_1S_2 stabilized state is done.

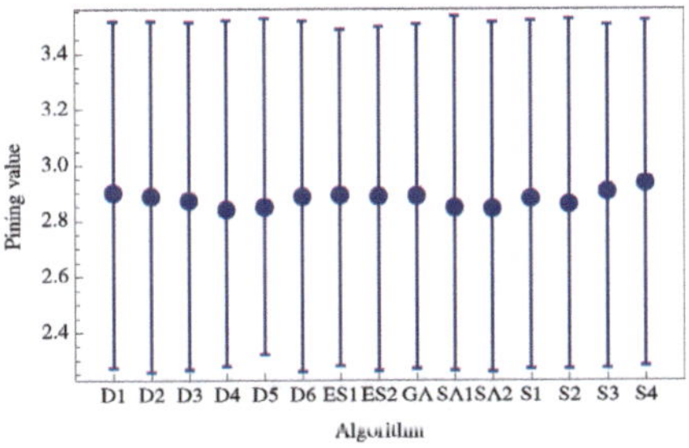

Fig. 6.36 Pinning value for CML/T1S1 ...

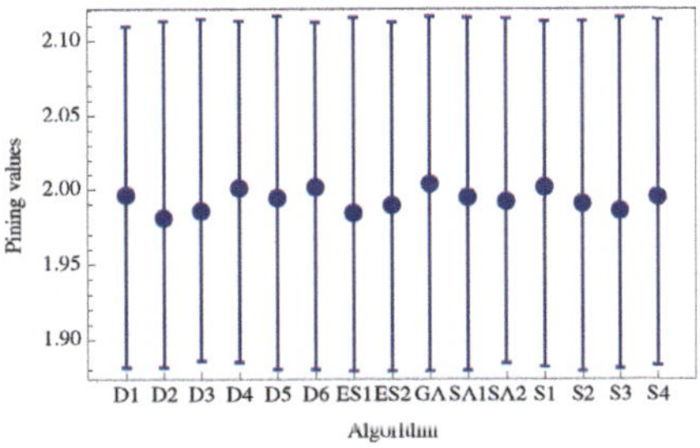

Fig. 6.37 ... and for CML/T1S2.

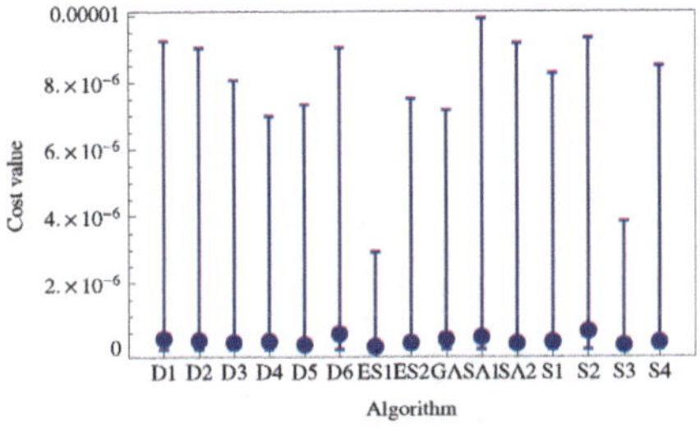

Fig. 6.38 Cost value for CML/T1S1 ...

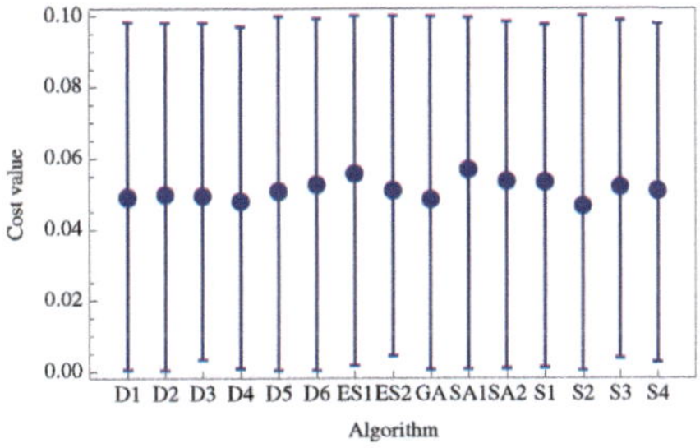

Fig. 6.39 ... and for CML/T1S2.

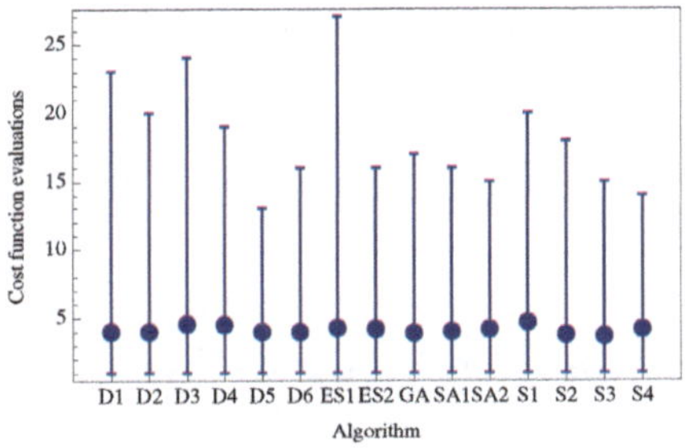

Fig. 6.40 Cost function evaluations for CML/T1S1 ...

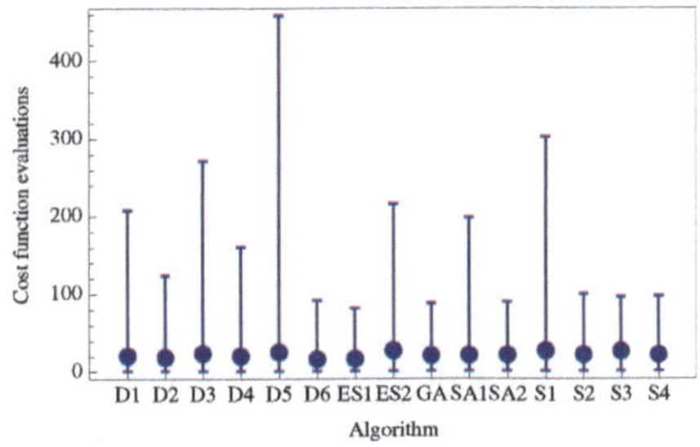

Fig. 6.41 ... and for CML/T1S2.

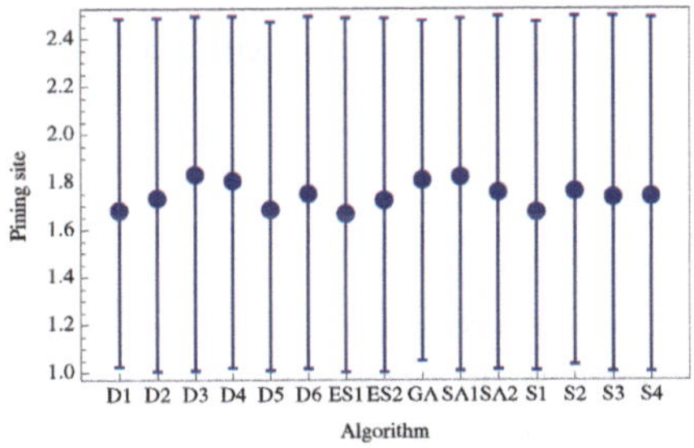

Fig. 6.42 Pinning sites for CML/T1S1 ...

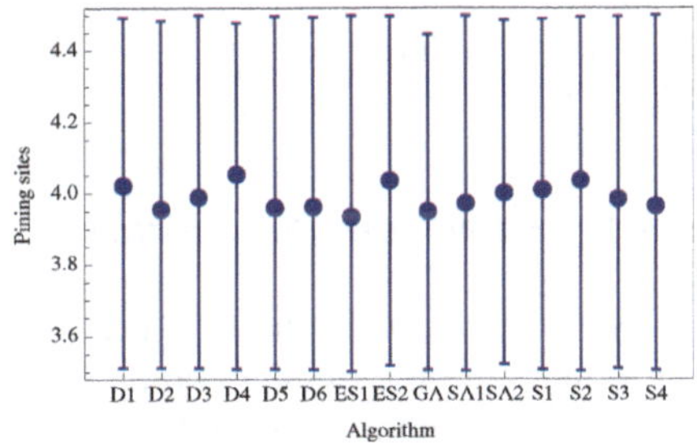

Fig. 6.43 ... and for CML/T1S2.

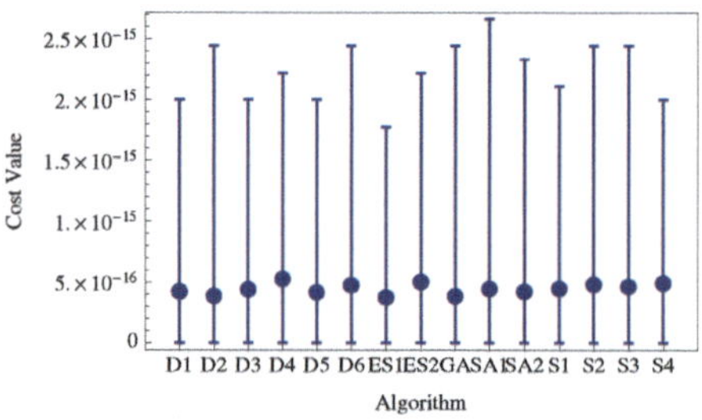

Fig. 6.44 Cost value for CML/T1S1 ...

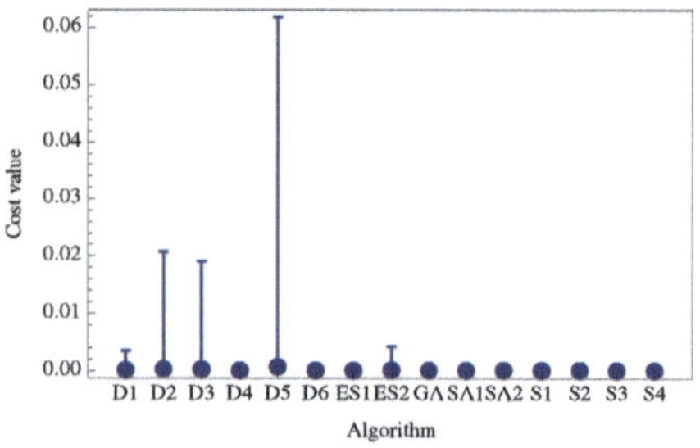

Fig. 6.45 ... and for CML/T1S2.

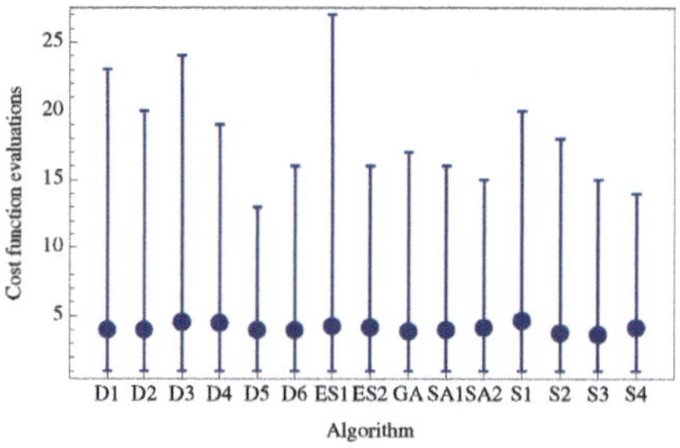

Fig. 6.46 Cost function evaluations for CML/T1S1 ...

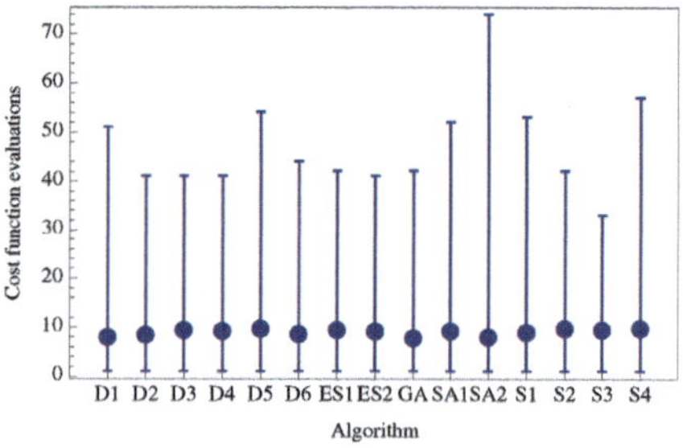

Fig. 6.47 ... and for CML/T1S2.

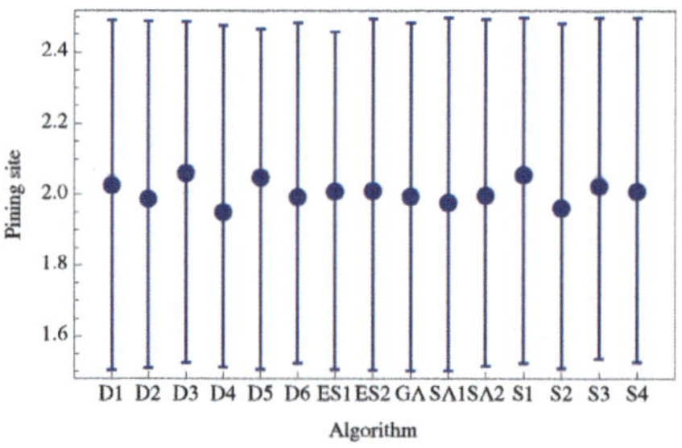

Fig. 6.48 Minimal pinning sites for CML/T1S1 ...

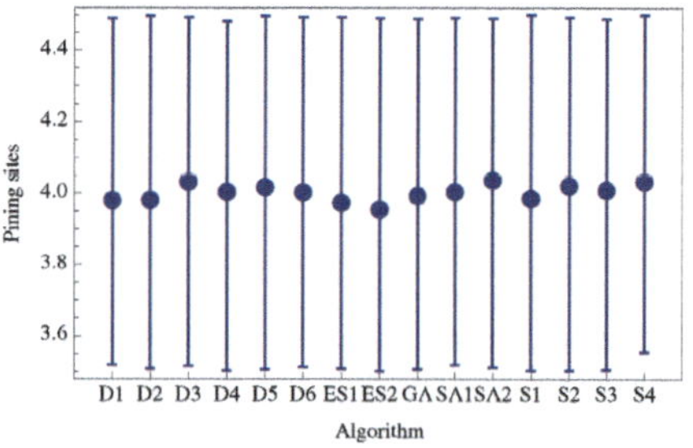

Fig. 6.49 ... and for CML/T1S2.

6.4.5 Experimental Results

All simulations were done in the frame of four case studies A, B, C and D. In all numerical case studies, both CML regimes T_1S_1 and T_1S_2 were compared to show how complexity of both regimes can influence the number of cost function evaluations etc.

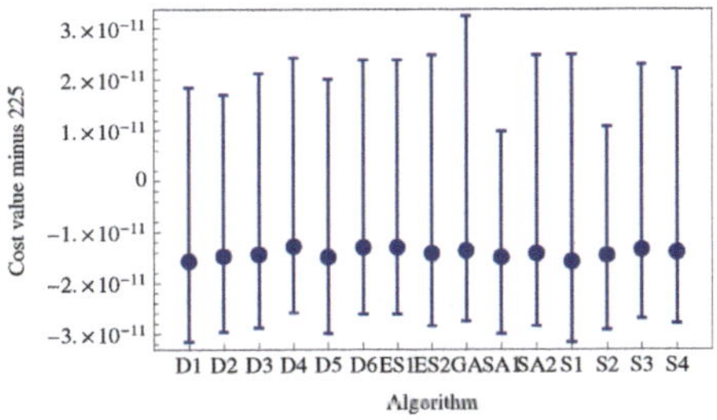

Fig. 6.50 Cost value for CML/T1S1 ...

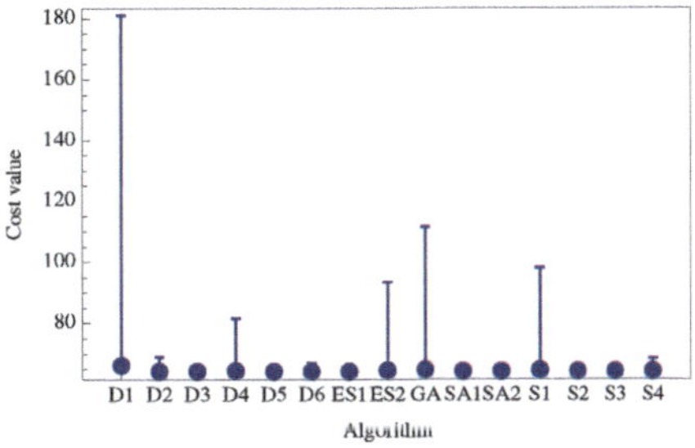

Fig. 6.51 ... and for CML/T1S2.

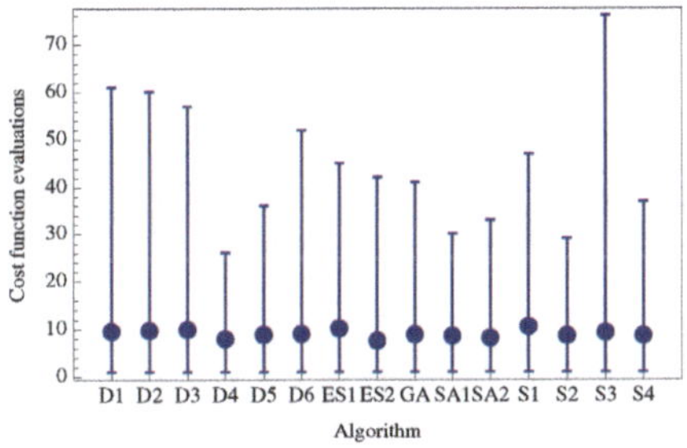

Fig. 6.52 Cost function evaluations for CML/T1S1 ...

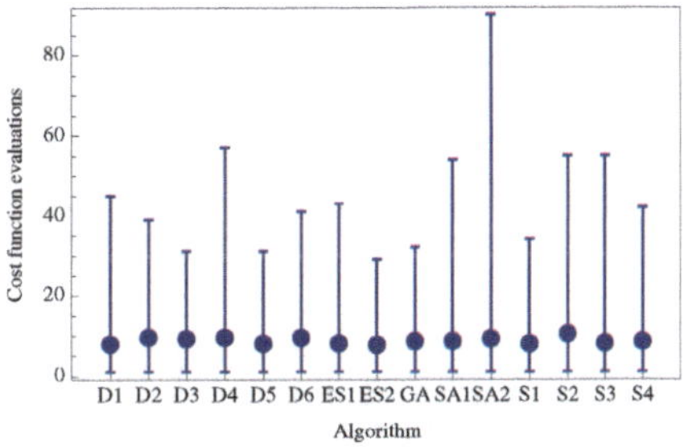

Fig. 6.53 ... and for CML/T1S2.

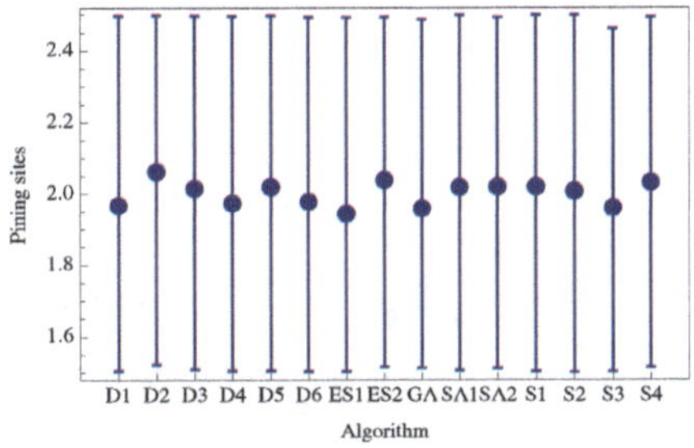

Fig. 6.54 Minimal pinning sites for CML/T1S1 ...

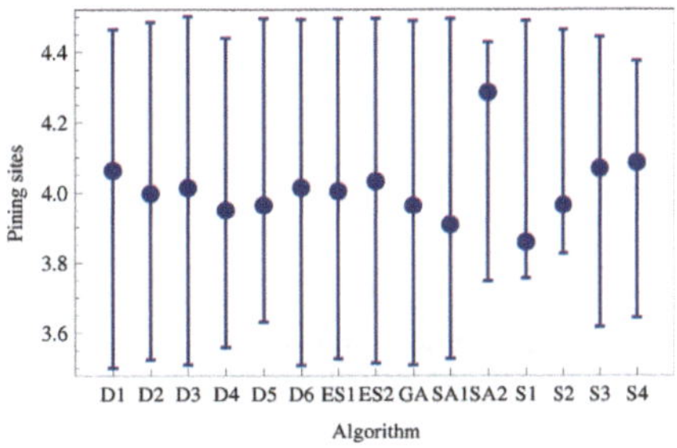

Fig. 6.55 ... and for CML/T1S2.

6.4.5.1 Case Study A – Pinning Value Estimation

In this case study, EAs were used to estimate the pinning value for CML. Pinning sites were a priori set according to [17]. Estimated pinning value was used for all a priori defined pinning sites (each odd). Calculation was 100 times repeated and

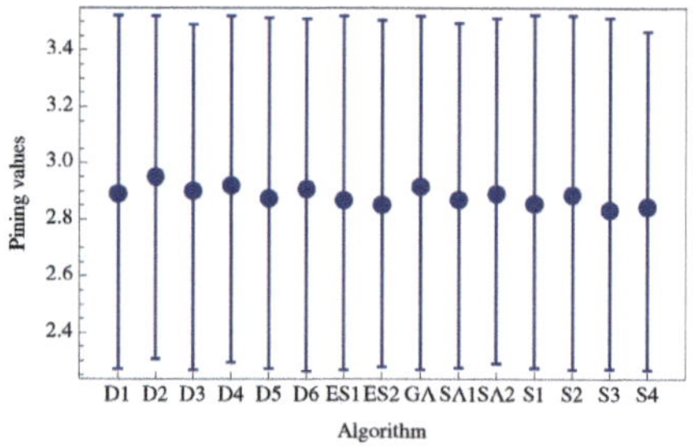

Fig. 6.56 Pinning values for CML/T1S1 ...

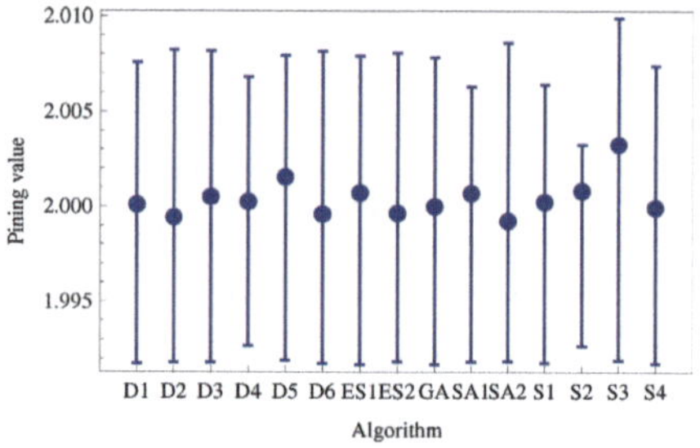

Fig. 6.57 ... and for CML/T1S2.

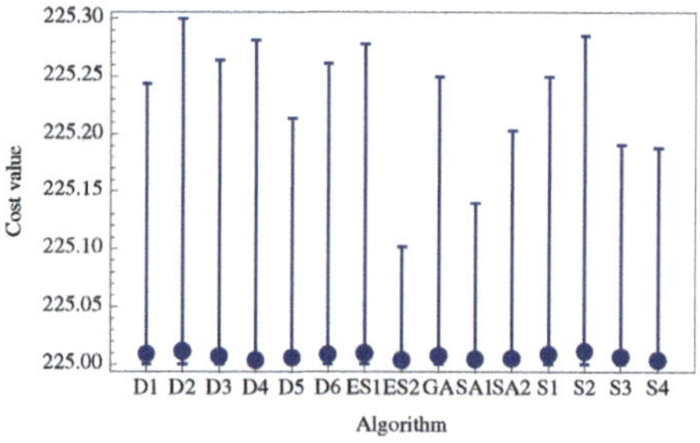

Fig. 6.58 Cost value for CML/T1S1 ...

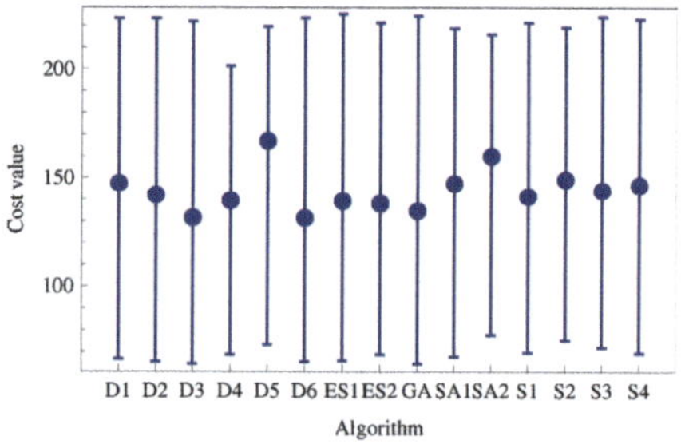

Fig. 6.59 ... and for CML/T1S2.

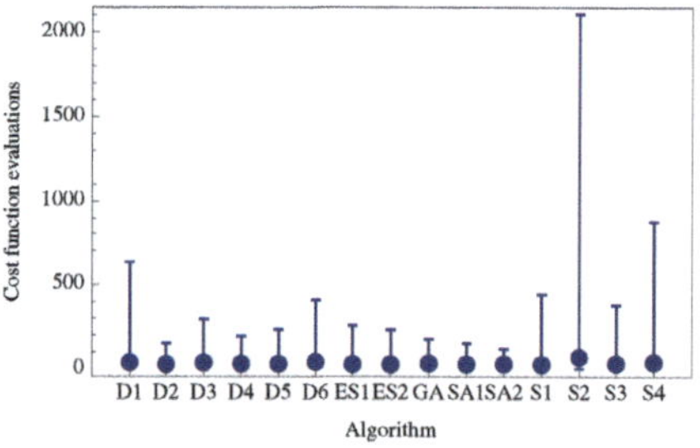

Fig. 6.60 Cost function evaluations for CML/T1S1 ...

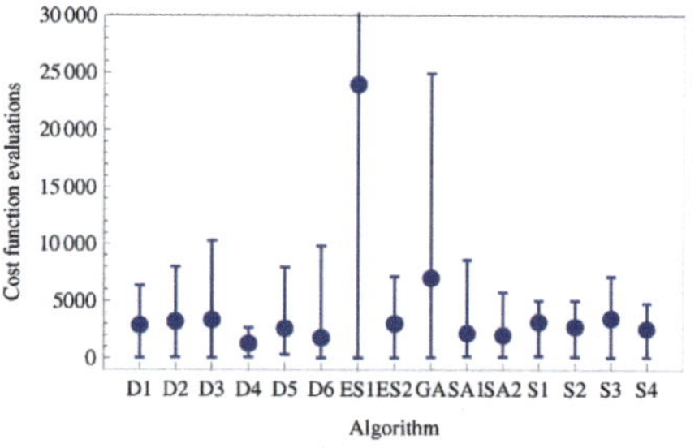

Fig. 6.61 ... and for CML/T1S2.

from the last population in each simulation the best solution was recorded. Consequently, the best, the worst and the average values were calculated. This has been done also for cost value and cost function evaluations. All 100 triplets (best, worst, average) for pinning value, cost value and cost function evaluations were used to create Fig. 6.36 to 6.37. For results verification, the dependance of cost value

Table 6.7 T1S1, case study A - experiment summarization, part 1.

Algorithm	D1	D2	D3	D4	D5	D6	ES1	ES2
Pinning values see. Fig 6.36								
Minimum	2.2705	2.2559	2.2644	2.2770	2.3190	2.2592	2.2772	2.2607
Average	2.8974	2.8848	2.8700	2.8379	2.8472	2.8838	2.8892	2.8849
Maximum	3.5146	3.5137	3.5109	3.5178	3.5239	3.5150	3.4842	3.4946
Cost Values see. Fig 6.38								
Minimum	0	0	0	0	0	0	0	0
Average $\times 10^{-}17$	3.2538	2.6701	2.0487	2.462	1.5003	4.6961	0.8441	1.9491
Maximum $\times 10^{-}16$	9.2493	9.0279	8.0505	6.9654	7.3313	9.047	2.9231	7.5095
Cost function evaluations see Fig. 6.40								
Minimum	1	1	1	1	1	1	1	1
Average	4	4	5	4	4	4	4	4
Maximum	23	20	24	19	13	16	27	16
Total for each algorithm	400	399	455	449	398	397	426	418

Table 6.8 T1S1, case study A - experiment summarization, part 2.

Algorithm	G	SA1	SA2	S1	S2	S3	S4
Pinning values see. Fig 6.36							
Minimum	2.2677	2.2630	2.2586	2.2697	2.2687	2.2712	2.2797
Average	2.8807	2.8444	2.8412	2.8776	2.8569	2.9022	2.9329
Maximum	3.5036	3.5328	3.5109	3.5165	3.5223	3.5021	3.5187
Cost Values see. Fig 6.38							
Minimum	0	0	0	0	0	0	0
Average $\times 10^{-}17$	2.9893	3.6742	1.8064	2.2816	5.4718	1.3542	2.2163
Maximum $\times 10^{-}16$	7.1507	9.9162	9.1738	8.2706	9.3396	3.8385	8.4899
Cost function evaluations see Fig. 6.40							
Minimum	1	1	1	1	1	1	1
Average	4	4	4	5	4	4	4
Maximum	17	16	15	20	18	15	14
Total for each algorithm	388	398	415	466	377	365	415

Table 6.9 T1S2, case study A - experiment summarization, part 1.

Algorithm	D1	D2	D3	D4	D5	D6	ES1	ES2
Pinning values see. Fig 6.37								
Minimum	1.8808	1.8809	1.8855	1.8846	1.8801	1.8795	1.8786	1.8786
Average	1.9960	1.9807	1.9855	2.0006	1.9940	2.0011	1.9840	1.9891
Maximum	2.1091	2.1126	2.1141	2.1128	2.1160	2.1117	2.1150	2.1120
Cost Values see. Fig 6.39								
Minimum	0.0004	0.0003	0.0033	0.0008	0.0003	0.0003	0.0017	0.0044
Average	0.0489	0.0496	0.0493	0.0479	0.0506	0.0525	0.0555	0.0508
Maximum	0.0981	0.0980	0.0980	0.0968	0.0997	0.0991	0.0999	0.0999
Cost function evaluations see Fig. 6.41								
Minimum	1	1	1	1	1	1	1	1
Average	21	19	24	20	25	16	17	27
Maximum	208	124	272	160	458	92	82	216
Total for each algorithm	2052	1858	2362	2044	2534	1631	1675	2732

Table 6.10 T1S2, case study A - experiment summarization, part 2.

Algorithm	G	SA1	SA2	S1	S2	S3	S4
Pinning values see. Fig 6.37							
Minimum	1.8787	1.8791	1.8839	1.8817	1.8786	1.8802	1.8825
Average	2.0033	1.9945	1.9917	2.0012	1.9903	1.9855	1.9949
Maximum	2.1157	2.1151	2.1144	2.1124	2.1127	2.1149	2.1138
Cost Values see. Fig 6.39							
Minimum	0.0005	0.0006	0.0008	0.0011	0.0003	0.0037	0.0027
Average	0.0483	0.0566	0.0534	0.0531	0.0465	0.0518	0.0507
Maximum	0.0998	0.0995	0.0982	0.0974	0.0999	0.0987	0.0977
Cost function evaluations see Fig. 6.41							
Minimum	1	1	1	1	1	1	1
Average	21	21	22	27	22	26	22
Maximum	88	198	90	302	100	96	97
Total for each algorithm	2086	2117	2162	2691	2188	2624	2175

Table 6.11 T1S1, case study B - experiment summarization, part 1.

Algorithm	D1	D2	D3	D4	D5	D6	ES1	ES2
Pinning sites see. Fig 6.42								
Minimum	1.02257	1.00343	1.00784	1.01996	1.00965	1.01471	1.00138	1.00225
Average	1.67993	1.73124	1.83038	1.80492	1.68419	1.75075	1.66654	1.72341
Maximum	2.4858	2.48778	2.4968	2.49605	2.47182	2.49547	2.4893	2.48723
Cost Values see. Fig 6.44								
Minimum	0	0	0	0	0	0	0	0
Average $\times 10^{-16}$	4.21885	3.86358	4.38538	5.25135	4.15223	4.75175	3.74145	5.01821
Maximum $\times 10^{-15}$	1.9984	2.44249	1.9984	2.22045	1.9984	2.44249	1.77636	2.22045
Cost function evaluations see Fig. 6.46								
Minimum	1	1	1	1	1	1	1	1
Average	6	8	7	6	7	6	6	5
Maximum	21	28	32	21	29	27	20	27
Total for each algorithm	559	762	664	584	564	578	620	516

Table 6.12 T1S1, case study B - experiment summarization, part 2.

Algorithm	G	SA1	SA2	S1	S2	S3	S4
Pinning sites see. Fig 6.42							
Minimum	1.0493	1.0064	1.0134	1.0096	1.0335	1.0047	1.0033
Average	1.8090	1.8221	1.7565	1.6742	1.7609	1.7357	1.7394
Maximum	2.4786	2.4886	2.4977	2.4713	2.4983	2.4979	2.4913
Cost Values see. Fig 6.44							
Minimum	0	0	0	0	0	0	0
Average $\times 10^{-16}$	3.8524	4.4408	4.2188	4.4631	4.8183	4.6407	4.9293
Maximum $\times 10^{-15}$	2.4424	2.6645	2.3314	2.1094	2.4424	2.4424	1.9984
Cost function evaluations see Fig. 6.46							
Minimum	1	1	1	1	1	1	1
Average	7	6	6	6	6	6	6
Maximum	27	36	31	26	24	29	22
Total for each algorithm	680	593	577	583	575	595	607

Table 6.13 T1S2, case study B - experiment summarization, part 1.

Algorithm	D1	D2	D3	D4	D5	D6	ES1	ES2
Pinning sites see. Fig 6.43								
Minimum	3.5103	3.5108	3.5100	3.5071	3.5065	3.5055	3.5003	3.5165
Average	4.0221	3.9555	3.9892	4.0533	3.9600	3.9621	3.9339	4.0353
Maximum	4.4924	4.4832	4.4997	4.4770	4.4969	4.4938	4.4986	4.4971
Cost Values see. Fig 6.45								
Minimum	0	0	0	0	0	0	0	0
Average	5.9×10^{-5}	2.1×10^{-4}	1.9×10^{-4}	8.9×10^{-12}	6.2×10^{-4}	1.3×10^{-6}	3.2×10^{-8}	4.1×10^{-5}
Maximum	3.4×10^{-3}	2.0×10^{-2}	1.9×10^{-2}	8.2×10^{-10}	6.1×10^{-2}	8.7×10^{-5}	3.2×10^{-6}	4.1×10^{-3}
Cost function evaluations see Fig. 6.47								
Minimum	1	1	1	1	1	1	1	1
Average	8	8	9	9	10	9	9	9
Maximum	51	41	41	41	54	44	42	41
Total for each algorithm	791	837	932	912	960	852	935	906

Table 6.14 T1S2, case study B - experiment summarization, part 2.

Algorithm	G	SA1	SA2	S1	S2	S3	S4
Pinning sites see. Fig 6.43							
Minimum	3.5039	3.5015	3.5208	3.5056	3.5012	3.5076	3.5037
Average	3.9498	3.9718	4.0012	4.0100	4.0362	3.9835	3.9643
Maximum	4.4454	4.4984	4.4843	4.4888	4.4940	4.4952	4.4998
Cost Values see. Fig 6.45							
Minimum	0	0	0	0	0	0	0
Average	2.5×10^{-8}	1.6×10^{-9}	1.3×10^{-5}	6.3×10^{-11}	1.2×10^{-5}	3.3×10^{-7}	1.2×10^{-8}
Maximum	2.5×10^{-6}	1.6×10^{-7}	9.6×10^{-4}	4.9×10^{-9}	1.2×10^{-3}	3.3×10^{-5}	1.2×10^{-6}
Cost function evaluations see Fig. 6.47							
Minimum	1	1	1	1	1	1	1
Average	8	9	8	9	10	9	10
Maximum	42	52	74	53	42	33	57
Total for each algorithm	768	911	793	886	963	935	968

Table 6.15 T1S1, case study C - experiment summarization, part 1.

Algorithm	D1	D2	D3	D4	D5	D6	ES1	ES2
Pinning sites see. Fig 6.48								
Minimum	1.503	1.5089	1.5242	1.512	1.5047	1.5217	1.5053	1.5024
Average	2.0244	1.9862	2.0589	1.9488	2.0465	1.9911	2.0068	2.0085
Maximum	2.491	2.4877	2.4871	2.4757	2.465	2.4823	2.4572	2.4954
Cost Values see. Fig 6.50								
Minimum +225	0	0	0	0	0	0	0	0
Average $\times 10^{-}11$ +225	1.5802	1.4836	1.4438	1.2931	1.5006	1.3102	1.3102	1.4267
Maximum $\times 10^{-}11$ +225	4.9993	4.6640	4.9993	4.9993	4.9993	4.9993	4.9993	5.329
Cost function evaluations see Fig. 6.52								
Minimum	1	1	1	1	1	1	1	1
Average	10	10	10	8	9	9	10	8
Maximum	61	60	57	26	36	52	45	42
Total for each algorithm	950	969	987	792	878	896	1011	751

Table 6.16 T1S1, case study C - experiment summarization, part 2.

Algorithm	G	SA1	SA2	S1	S2	S3	S4
Pinning sites see. Fig 6.48							
Minimum	1.5011	1.5012	1.5138	1.5218	1.5080	1.5349	1.5244
Average	1.9925	1.9750	1.9958	2.0538	1.9605	2.0220	2.0071
Maximum	2.4829	2.4976	2.4933	2.4974	2.4808	2.4972	2.4976
Cost Values see. Fig 6.50							
Minimum +225	0	0	0	0	0	0	0
Average $\times 10^{-}11$ +225	1.3756	1.5063	1.4267	1.5859	1.4637	1.3528	1.3983
Maximum $\times 10^{-}11$ +225	5.9941	3.9960	5.3290	5.6587	3.9960	4.9993	4.9993
Cost function evaluations see Fig. 6.52							
Minimum	1	1	1	1	1	1	1
Average	9	9	8	10	9	9	9
Maximum	41	30	33	47	29	76	37
Total for each algorithm	879	853	811	1049	857	926	859

Table 6.17 T1S2, case study C - experiment summarization, part 1.

Algorithm	D1	D2	D3	D4	D5	D6	ES1	ES2
Pinning sites see. Fig 6.49								
Minimum	3.5174	3.5067	3.5154	3.5015	3.5051	3.5122	3.5066	3.5001
Average	3.9792	3.9798	4.0314	4.0025	4.0154	4.0012	3.9725	3.9531
Maximum	4.4898	4.4968	4.4919	4.4821	4.4965	4.4929	4.4927	4.4896
Cost Values see. Fig 6.51								
Minimum	64.	64.	64.	64.	64.	64.	64.	64.
Average	65.9884	64.0479	64.	64.207	64.0153	64.0279	64.	64.2922
Maximum	180.838	68.7866	64.0007	81.3137	64.9063	66.7922	64.0013	93.1835
Cost function evaluations see Fig. 6.52								
Minimum	1	1	1	1	1	1	1	1
Average	7.86	9.52	9.16	9.48	7.97	9.43	8.04	7.65
Maximum	45	39	31	57	31	41	43	29
Total for each algorithm	786	952	916	948	797	943	804	765

Table 6.18 T1S2, case study C - experiment summarization, part 2.

Algorithm	G	SA1	SA2	S1	S2	S3	S4
Pinning sites see. Fig 6.49							
Minimum	3.5057	3.5167	3.5108	3.5017	3.5029	3.5040	3.5534
Average	3.9914	4.0025	4.0359	3.9847	4.0195	4.0074	4.0318
Maximum	4.4883	4.4901	4.4899	4.4992	4.4926	4.4884	4.4988
Cost Values see. Fig 6.51							
Minimum	64.	64.	64.	64.	64.	64.	64.
Average	64.4721	64.0003	64.	64.339	64.0002	64.0027	64.0418
Maximum	111.186	64.027	64.0001	97.7433	64.0218	64.2691	68.1787
Cost function evaluations see Fig. 6.52							
Minimum	1	1	1	1	1	1	1
Average	8.54	8.57	9.22	7.94	10.38	8.07	8.47
Maximum	32	54	90	34	55	55	42
Total for each algorithm	854	857	922	794	1038	807	847

Table 6.19 T1S1, case study D - experiment summarization, part 1.

Algorithm	D1	D2	D3	D4	D5	D6	ES1	ES2
Pinning sites see. Fig 6.54								
Minimum	1.5036	1.5218	1.5076	1.5044	1.5048	1.5032	1.5034	1.5143
Average	1.9670	2.0621	2.0152	1.9736	2.0192	1.9776	1.9433	2.0379
Maximum	2.4970	2.4987	2.4971	2.4953	2.4970	2.4938	2.4915	2.4933
Pinning value see. Fig 6.56								
Minimum	2.2686	2.3054	2.2658	2.2936	2.2713	2.2605	2.2659	2.2768
Average	2.8893	2.9488	2.8988	2.9181	2.8730	2.9060	2.8665	2.8506
Maximum	3.5222	3.5203	3.4891	3.5203	3.5122	3.5091	3.5195	3.5044
Cost Values see. Fig 6.58								
Minimum +225	0	0	0	0	0	0	0	0
Average +225	0.0089	0.0111	0.0069	0.0033	0.0056	0.0084	0.0094	0.0033
Maximum +225	0.2433	0.2994	0.2637	0.2807	0.2134	0.2604	0.2776	0.1012
Cost function evaluations see Fig. 6.60								
Minimum	1	1	1	1	1	1	1	1
Average	37	26	36	30	30	40	27	27
Maximum	632	152	294	192	234	408	256	231
Total for each algorithm	3715	2638	3585	2997	2984	3990	2733	2651

(according to eq. 6.12) on pinning value was calculated and is depicted in Fig. 6.36 - 6.41. Optimal pinning values were found in the interval [2.3, 3.5] for T_1S_1 (see Fig. 6.22) and in [1.87, 2.12] for T_1S_2 (see Fig. 6.26), (cost value is 0, i.e. minimal difference between CML behavior and desired behavior). Based on the results it can be stated that in all simulations suitable pinning values were estimated because according to [17] suitable pinning value (equal to 3) was used and here in each simulation the best values are around 2.9. Minimal and maximal values were observed in the interval with cost value 0, as depicted on Fig. 6.22. Important is the number of cost function evaluations. For T_1S_1 average number was only 4 while for T_1S_2 20, i.e. for T_1S_1 it seems to be useless to use EAs, while for T_1S_2 it is reasonable, because the number of cost function evaluations is bigger than needed for one generation, as in the case of standard EA (remember, population was 10). It is also important to remember, that individual length was only 1 to 2 and one of parameters was "frequency" of used pinning sites, while in [37] and [39] the individual length was equal to the number of pinning sites. In the case of [37] and [39] the number of cost function evaluations was much more higher.

Table 6.20 T1S1, case study D - experiment summarization, part 2.

Algorithm	G	SA1	SA2	S1	S2	S3	S4
Pinning sites see. Fig 6.54							
Minimum	1.5116	1.5053	1.5104	1.5029	1.5003	1.5012	1.5136
Average	1.9577	2.0183	2.0193	2.0193	2.0068	1.9595	2.0312
Maximum	2.4863	2.4984	2.4915	2.4987	2.4988	2.4618	2.4936
Pinning value see. Fig 6.56							
Minimum	2.2671	2.2740	2.2898	2.2716	2.2670	2.2684	2.2656
Average	2.9148	2.8685	2.8893	2.8538	2.8851	2.8318	2.8431
Maximum	3.5203	3.4947	3.5114	3.5233	3.5206	3.5124	3.4649
Cost Values see. Fig 6.58							
Minimum +225	0	0	0	0	0	0	0
Average +225	0.0074	0.0041	0.0047	0.0085	0.0111	0.0063	0.0034
Maximum +225	0.2491	0.1388	0.2028	0.2493	0.2846	0.1904	0.1877
Cost function evaluations see Fig. 6.60							
Minimum	1	1	2	1	1	1	1
Average	32	28	29	26	69	31	39
Maximum	176	153	117	442	2104	378	878
Total for each algorithm	3243	2753	2862	2607	6922	3148	3887

6.4.5.2 Case Study B – Pinning Sites Position Estimation

Based on results from the previous case study, this case study B was designed. EAs were used to estimate pinning sites for CML. Pinning values were a priori set to 3 (T_1S_1) and 2 (T_1S_2) for all estimated sites according to [17]. Calculation was again 100 times repeated and the best solution (pinning sites) from each simulation was used to create Fig. 6.42 - 6.47. From the figures is visible that for T_1S_1 each 1st or 2nd pinning site was estimated. For T_1S_2 only each 4th was estimated. Because pinning sites are "discrete" objects, and because evolution has been running on continuous parameter space, individual parameter representing pinning site has been rounded before it was used in CML control. Thus, when the rounded data in Fig. 6.42 and 6.43 is used, then the periodicity of the used pinning sites is 1 or 2 for T_1S_1 and 4 for T_1S_2. It was a good coincidence with values from Fig. 6.33 and Fig. 6.35. Comparison about the number of cost function evaluations is similar to that of case study A.

6.4.5.3 Case Study C – Minimal Pinning Sites Position Estimation

This case study was designed to improve the previous results from case study B. Cost function eq. 6.11 was modified to eq. 6.12. All other conditions were kept the

Table 6.21 T1S2, case study D - experiment summarization, part 1.

Algorithm	D1	D2	D3	D4	D5	D6	ES1	ES2
Pinning sites see. Fig 6.55								
Minimum	3.5001	3.5240	3.5096	3.5590	3.6315	3.5076	3.5271	3.5137
Average	4.0625	3.9971	4.0134	3.9488	3.9636	4.0138	4.0034	4.0318
Maximum	4.4628	4.4833	4.4996	4.4382	4.4946	4.4905	4.4939	4.4939
Pinning value see. Fig 6.57								
Minimum	1.9917	1.9917	1.9918	1.9926	1.9919	1.9917	1.9916	1.9918
Average	2.0000	1.9994	2.0004	2.0002	2.0015	1.9995	2.0006	1.9996
Maximum	2.0075	2.0082	2.0081	2.0068	2.0079	2.0081	2.0079	2.0080
Cost Values see. Fig 6.59								
Minimum	66.364	65.111	64.147	68.48	72.956	65.111	65.636	68.457
Average	147.18	141.75	131.28	139.16	166.8	131.15	139.	137.89
Maximum	223.06	222.96	221.72	201.11	219.24	223.35	225.03	221.
Cost function evaluations see Fig. 6.61								
Minimum	17	31	2	86	277	18	14	9
Average	2854	3182	3315	1252	2577	1764	23913	3008
Maximum	6322	7980	10242	2627	7905	9792	140590	7112
Total for each algorithm	59931	318214	314934	30053	25773	176418	2391314	300806

same. Again the same figures were created (Fig. 6.48 - Fig. 6.53. All algorithms had found in all 100 simulations a suitable number of pinning sites, for T_1S_1 each 2nd pinning site was estimated and for T_1S_2 each 4th (remember rounding). Cost function evaluations was similar to that of case study B.

6.4.5.4 Case Study D – Minimal Pinning Values and Sites Position Estimation

The last case study was dedicated to the estimation of minimal number of pinning sites and different pinning values. All simulations were repeated under the same conditions as in the case study C and the same kind of figures (Fig. 6.54 - 6.61) was created. Used pinning sites were estimated for T_1S_1 and each 2nd pinning site was estimated and for T_1S_2 each 4th. Pinning values were in [2.3, 3.5] for T_1S_1 and [1.99, 2.008] for T_1S_2. Concerning to the cost value, it is really important to note that it was not 0 because eq. 6.12 contain also heuristically set constants p_1 and p_2. If for example for T_1S_2 each 4th pinning site is recorded, then the total number of used pinning sites was 30 / 4 = 7, i.e. p_1=7 and p_2=100, see eq. 6.12. Lets say that the cost value was 150, then the real value of the cost value is $\frac{\sqrt{150-7}}{100}$ which is 0.119. Another point of discussion is the number of cost function evaluations. While in the

Table 6.22 T1S2, case study D - experiment summarization, part 2.

Algorithm	G	SA1	SA2	S1	S2	S3	S4
Pinning sites see. Fig 6.55							
Minimum	3.5082	3.5271	3.7477	3.7541	3.8251	3.6168	3.6415
Average	3.9625	3.9067	4.2846	3.8572	3.9636	4.0679	4.0846
Maximum	4.4880	4.4939	4.4276	4.4880	4.4624	4.4416	4.3741
Pinning value see. Fig 6.57							
Minimum	1.9917	1.9918	1.9918	1.9917	1.9926	1.9919	1.9917
Average	1.9999	2.0006	1.9992	2.0002	2.0008	2.0032	1.9999
Maximum	2.0078	2.0062	2.0085	2.0064	2.0032	2.0098	2.0073
Cost Values see. Fig 6.59							
Minimum	64.246	67.524	77.397	69.326	75.117	71.723	69.211
Average	134.4	147.16	159.8	141.15	149.	143.89	146.4
Maximum	224.18	218.66	215.85	221.18	219.04	223.92	222.8
Cost function evaluations see Fig. 6.61							
Minimum	29	153	96	180	91	54	75
Average	6976	2156	1986	3142	2723	3452	2541
Maximum	24864	8563	5743	4989	4981	7128	4782
Total for each algorithm	697604	238952	157863	95689	186597	248643	213721

previous three case studies, EAs are disputable (low number of CFE), in the case study D, it is visible, that especially in the case of T_1S_2, the use of EA is important with CFE being more than 5000 in average. As mentioned before, in [37] and [39], when different philosophy is taken into consideration for individual structure, a big number of CFE is needed for CML with only 10 sites.

6.4.5.5 Mutual Comparison

Complexity of all four case studies has increased. Based on data from all simulations two comparison can be done. The first one is from pinning sites point of view. As is depicted in all four case studies, all EAs are comparable in performance (with small deviations). It is also visible that changes in cost functions can significantly improve estimated solutions (case C/D) and [37] and [39]. To check that estimated pinning sites and pinning values really cause the CML to be stabilized, randomly selected figures (from 12000) were generated (4 cases $\times$ 2 regimes $\times$ (15×100 simulations)) based on data from all simulations. In all simulations CML was stabilized to desired behavior.

6.4.6 CML Real Time Control

Comparing to study reported in [37] and the extended modified study here, [39] has also done CML control in simulated realtime regime. The capability of EAs on such a blackbox processes control has been demonstrated. Comparing to non real-time CML as in [37], simulated realtime was in such a way that each configuration (individual) has been applied after n iterations without the possibility to start from initial conditions again. For demonstrative purposes, three figures are depicted here. Fig. 6.62 show typical controller output developed during evolution. Each change of its value is related to another, better individual, whose parameter (or one of them) was estimated controlled output. Fig. and show process of T_1S_2 stabilization. For more exact information, it is recommended to see [39].

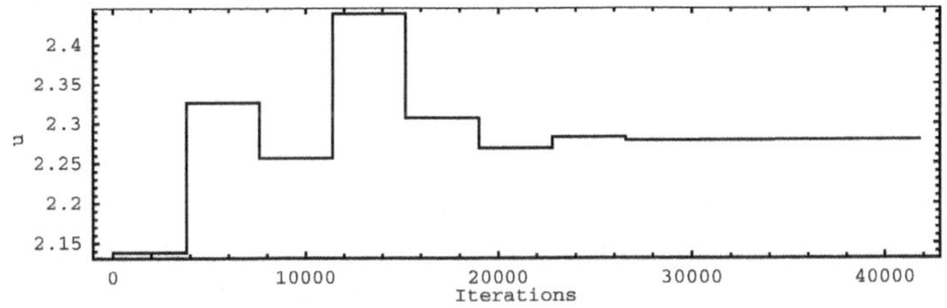

Fig. 6.62 An example of controller output

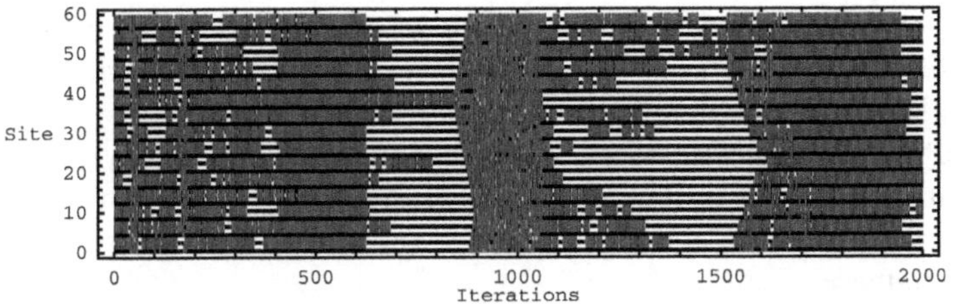

Fig. 6.63 Partial stabilization of realtime CML in T1S2 pattern - pattern is temporarily stabilized

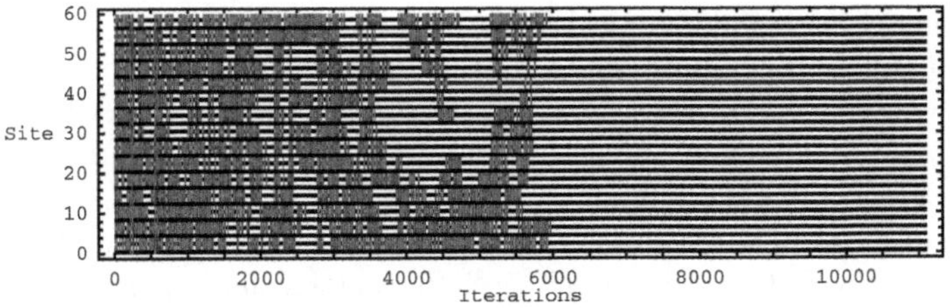

Fig. 6.64 Successful stabilization of realtime CML in T1S2 pattern - pattern is permanently stabilized

6.5 Conclusion

The method of evolutionary deterministic chaos control described here is relatively simple, easy to implement and easy to use. Based on its principles and its possible universality (it was tested with 5 evolutionary algorithms) it can be stated that evolutionary deterministic chaos control is capable to solve the class of CML deterministic chaos control problems. The main aim of this paper was to show how various CML control problems were solved by means of evolutionary algorithms. Evolutionary deterministic chaos control was used here in four basic comparative simulations. Each comparative simulation was 100 times repeated and all 12000 results (100 simulations for each of 15 algorithm and for each of 4 case studies and 2 CML regimes – stabilized states) were used to create graphs for evolutionary deterministic chaos control performance evaluation. For the comparative study, optimization algorithms such as Differential Evolution [28], SelfOrganizing Migrating Algorithm [35], Genetic Algorithms [16] and Simulated Annealing [21], [2] were used. They were chosen to show that evolutionary deterministic chaos control can be regarded as a "blackbox" method and that it can be implemented using arbitrary evolutionary algorithms. As a conclusion the following statements are presented:

- **Reached results:** based on the fact about cost function and its complexity (see 6.19 – 6.35), algorithm settings (Table 6.1 – 6.6) and results (Fig. 6.36 – 6.61, Table 6.22 - 6.7) it can be stated that all simulations give satisfactory results and thus evolutionary deterministic chaos control is capable of solving this class of problems. This statement is also "supported" by results reached in [37] and [39].
- **No. of successful experiments:** In all 12000 simulations, no failure had been observed (see Fig. 6.36 – 6.61), which means that all 15 algorithms had found a suitable solution before evolution has finished.
- **Mutual comparison:** when comparing all algorithms, it is visible that algorithms give good results. Parameter setting for all algorithms was based on heuristic approach and thus there is a possibility that better settings can be found there.
- **Penalization effect:** the penalization in the cost function, eq. 6.12, basically had little effect on the reached results (case D). Explanation is quite simple – evolution had found a suitable solution before effect of the penalization could be graphically visible, because individual structure has been simplified, comparing to simulations in [37].
- **Cost function evaluations:** were usually recorded as too small, as discussed in section 6.4.5.4, however for a little more complicated CML stabilized states, it dramatically increase (see Fig. 6.61), which confirm that EAs use on CML control is eligible.
- **No. of pinning sites:** the same principle applies as in the previous paragraph (case study B, C and D) In all cases the suitable site n period has been estimated (i.e. each n_{th} site was used). As written in the section 6.4.5.2 because pinning sites are "discrete" objects, and because evolution has been running on continuous parameter space, individual parameter representing pinning site

has been rounded before it was used in CML control. Thus, when rounded data in Fig. 6.42 and 6.43 is used, then periodicity of used pinning sites is 1 or 2 for T_1S_1 and 4 for T_1S_2. It was good coincidence with values from Fig. 6.33 and Fig. 6.35. Thus, the number of site period is depicted like a real number, however in reality only the rounded version has been used. Real number representation on the above mentioned figures is here only to show population and mainly results diversity.

- **Algorithms efficiency:** Based on the results obtained in this chapter, one can make conclusion that the problem itself is quite "simple" because such "simple and traditional" algorithms like SA or/and ES have demonstrated quite nice performance. To avoid such "misleading" conclusion, it is important remember that
  - problem itself is highly nonlinear, almost erratic (see Fig. 6.1, Fig. 6.2, Fig. 6.20 - 6.24 , Fig. 6.26 - 6.35) and thus classical optimization techniques would certainly fail,
  - CML is a source of very rich and complex behavior, see [31],
  - for example SA has also shown good performance on real time plasma reactor control [25], [26] which suggest that techniques like SA, ES,... are also capable to solve quite complex tasks.

According to the author's opinion, this is a promising area of research with many unanswered questions remaining. For example, it would be interesting to investigate the question as to why (and under what conditions) EAs are able to stabilize complex system such as CML. Answer can probably be based on the fact that EAs are from certain point of view like a filter which let pass to the new population (i.e. basically CML m_{th} iteration) only better individuals. Future research of evolutionary deterministic chaos control is still open. According to all results obtained, it is planned that during time the main activities would be focused on the expanding of this study for other chaotic systems.

Acknowledgements. This work was supported by grant No. MSM 7088352101 of the Ministry of Education of the Czech Republic and by grants of the Grant Agency of the Czech Republic GACR 102/09/1680.

References

1. Beyer, H.: Theory of Evolution Strategies. Springer, New York (2001)
2. Cerny, V.: Thermodynamical approach to the traveling salesman problem: An efficient simulation algorithm. J. Opt. Theory Appl. 45(1), 41–51 (1985)
3. Chen, G.: Controlling Chaos and Bifurcations in Engineering Systems. CRC Press, Boca Raton (2000)
4. Chen, G., Dong, X.: From Chaos to Order: Methodologies, Perspectives and Applications. World Scientific, Singapore (1998)
5. Clerc, M.: Particle Swarm Optimization. ISTE Publishing Company (2006)
6. Das, S., Konar, A.: A swarm intelligence approach to the synthesis of two-dimensional IIR filters. Eng. Appl. Artif. Intell. 20(8), 1086–1096 (2007)

7. Dashora, Y.: Improved and generalized learning strategies for dynamically fast and statistically robust evolutionary algorithms. Eng. Appl. Artif. Intel. (2007), doi:10.1016/j.engappai.2007.06.005
8. Davis, L.: Handbook of Genetic Algorithms. Van Nostrand Reinhold, Berlin (1996)
9. Deb, K.: Multi-Objective Optimization Using Evolutionary Algorithms. Wiley, Chichester (2001)
10. Deilami, M., Rahmani, C., Motlagh, M.: Control of spatio-temporal on-off intermittency in random driving diffusively coupled map lattices, Chaos, Solitons & Fractals, December 21 (2007)
11. Eberhart, R., Kennedy, J.: A new optimizer using particle swarm theory. In: Proceedings of the Sixth International Symposium on Micromachine and Human Science, Nagoya, Japan, pp. 39–43 (1995)
12. Gilmore, R., Lefranc, M.: The Topology of Chaos: Alice in Stretch and Squeezeland. Wiley-Interscience, New York (2002)
13. Grebogi, C., Lai, Y.C.: Controlling chaos. In: Schuster, H. (ed.) Handbook of Chaos Control. Wiley-VCH, New York (1999)
14. He, Q., Wang, L.: An effective co-evolutionary particle swarm optimization for constrained engineering design problems. Eng. Appl. Artif. Intell. 20(1), 89–99 (2007)
15. Hilborn, R.: Chaos and Nonlinear Dynamics. Oxford University Press, Oxford (1994)
16. Holland, J.: Adaptation in Natural and Artificial Systems. Univ. Michigan Press, Ann Arbor (1975)
17. Hu, G., Xie, F., Xiao, J., Yang, J., Qu, Z.: Control of patterns and spatiotemporal chaos and its application. In: Schuster, H. (ed.) Handbook of Chaos Control. Wiley-VCH, New York (1999)
18. Hwang, G.-H.-H., Dong-Wan, K., Jae-Hyun, L., Young-Joo, A.: Design of fuzzy power system stabilizer using adaptive evolutionary algorithm. Eng. Appl. Artif. Intell. 21(1), 86–96 (2007)
19. Just, W.: Principles of time delayed feedback control. In: Schuster, H. (ed.) Handbook of Chaos Control. Wiley-VCH, New York (1999)
20. Just, W., Benner, H., Reibold, E.: Theoretical and experimental aspects of chaos control by time-delayed feedback. Chaos 13, 259–266 (2003)
21. Kirkpatrick, S., Gelatt, C., Vecchi, M.: Optimization by simulated annealing. Science 220(4598), 671–680 (1983)
22. Liu, L., Liu, W., Cartes, D.: Particle swarm optimization-based parameter identification applied to permanent magnet synchronous motors. Eng. Appl. Artif. Intell. (2007), doi:10.1016/j.engappai.2007.10.002
23. Lorenz, E.: Deterministic nonperiodic flow. J. Atmos. Sci. 20(2), 130–141 (1963)
24. May, R.: Simple mathematical model with very complicated dynamics. Nature 261, 45–67 (1976)
25. Nolle, L., Goodyear, A., Hopgood, A.A., Picton, P.D., Braithwaite, N.StJ.: On Step Width Adaptation in Simulated Annealing for Continuous Parameter Optimisation. In: Reusch, B. (ed.) Fuzzy Days 2001. LNCS, vol. 2206, pp. 589–598. Springer, Heidelberg (2001)
26. Nolle, L., Zelinka, I., Hopgood, A., Goodyear, A.: Comparison of an self organizing migration algorithm with simulated annealing and differential evolution for automated waveform tuning. Adv. Eng. Software 36(10), 645–653 (2005)
27. Ott, E., Grebogi, C., Yorke, J.: Controlling chaos. Phys. Rev. Lett. 64, 1196–1199 (1990)
28. Price, K.: An Introduction to Differential Evolution. In: Corne, D., Dorigo, M., Glover, F. (eds.) New Ideas in Optimization, pp. 79–108. McGraw-Hill, London (1999)

29. Richter, H.: An evolutionary algorithm for controlling chaos: The use of multi-objective fitness functions. In: Guervós, J.J.M., Adamidis, P.A., Beyer, H.-G., Fernández-Villacañas, J.-L., Schwefel, H.-P. (eds.) PPSN 2002. LNCS, vol. 2439, pp. 308–317. Springer, Heidelberg (2002)
30. Richter, H., Reinschke, K.: Optimization of local control of chaos by an evolutionary algorithm. Physica D 144, 309–334 (2000)
31. Schuster, H.: Handbook of Chaos Control. Wiley-VCH, New York (1999)
32. Stewart, I.: The Lorenz attractor exists. Nature 406, 948–949 (2000)
33. Wang, X., Chen, G.: Chaotification via arbitrarily small feedback controls: Theory, method, and applications. Int. J. of Bifur. Chaos 10, 549–570 (2000)
34. Zahra, R., Motlagh, M.: Control of spatiotemporal chaos in coupled map lattice by discrete-time variable structure control. Phys. Lett. A 370(3-4), 302–305 (2007)
35. Zelinka, I.: SOMA - Self Organizing Migrating Algorithm. In: Babu, B., Onwubolu, G. (eds.) New Optimization Techniques in Engineering, pp. 167–218. Springer, New York (2004)
36. Zelinka, I.: Investigation on Evolutionary Deterministic Chaos Control. In: IFAC, Prague 2005 (2005a)
37. Zelinka, I.: Investigation on Evolutionary Deterministic Chaos Control - Extended Study. In: 19th International Conference on Simulation and Modeling, ECMS 2005, Riga, Latvia, June 1-4 (2005b)
38. Zelinka, I.: Investigation on realtime deterministic chaos control by means of evolutionary algorithms. In: Proc. First IFAC Conference on Analysis and Control of Chaotic Systems, Reims, France, pp. 211–217 (2006)
39. Zelinka, I.: Real-time deterministic chaos control by means of selected evolutionary algorithms. Eng. Appl. Artif. Intell (2008), doi:10.1016/j.engappai.2008.07.008
40. Zelinka, I., Nolle, L.: Plasma reactor optimizing using differential evolution. In: Price, K., Lampinen, J., Storn, R. (eds.) Differential Evolution: A Practical Approach to Global Optimization, pp. 499–512. Springer, New York (2006)
41. Zelinka, I., Senkerik, R., Navratil, E.: Investigation on Evolutionary Optimitazion of Chaos Control, Chaos, Solitons & Fractals (2007), doi:10.1016/j.chaos.2007.07.045
42. Zou, Y., Luo, X., Chen, G.: Pole placement method of controlling chaos in DC-DC buck converters. Chinese Phys. 15, 1719–1724 (2006)

Chapter 7
Chaotic Systems Reconstruction

Mohammed Chadli

Abstract. This chapter deals with the multiple model approach based chaotic systems reconstruction. The approach is based on the design of unknown inputs multiple observers using Linear Matrix Inequalities ($\mathcal{LMI}$) formulation. The objective is to estimate state variables of a multiple model subject to unknown inputs affecting both states and outputs of the system. Uncertainties affecting state matrices of the system are also considered for both continuous-time and discrete-time multiple models. In order to improve the performances of the observer, poles placement in an $\mathcal{LMI}$ region is also studied. Numerical examples are given to illustrate the effectiveness the given results. Application dealing with chaotic synchronization and message decoding are also given by considering chaotic multiple model subject to hidden message. The proposed approach can be also used in a chaotic cryptosystem procedure where the plaintext (message) is encrypted using chaotic signals at the drive system side. The resulting ciphertext is embedded to the output and/or state of the drive system and is sent via public channel to the response system. The plaintext is retrieved via the synthesis approach, i.e. the designed unknown input multiple observer.

7.1 Introduction

In last two decades, many studies concerning stability analysis and design of controllers and observers for a class of systems described by multiple model approach [20] are carried out. Such representation results from the interpolation of M local

Mohammed Chadli
University of Picardie Jules Verne, Laboratory of Modeling Information & Systems. 7, Rue du Moulin Neuf, 80000, Amiens, France
Tel.: +33(0)3 82227680
e-mail: mohammed.chadli@upicardie.fr

I. Zelinka et al. (Eds.): Evolutionary Algorithms and Chaotic Systems, SCI 267, pp. 237–264.
springerlink.com

LTI (linear time invariant) models throughout convex functions. These functions can be viewed as a weighted sum of local LTI models and quantify the relative contribution of each local model to the global model. The choice of the number of local models may be intuitively chosen by considering some operating regimes. Each LTI model can be obtained by using a direct linearization of an a priori nonlinear model around operating points, or alternatively by using an identification procedure [20, 5]. From a practical point of view, LTI model describes the system's local behavior around the ith, $i : 1...M$ regime. This approach includes Takagi-Sugeno fuzzy models [23] and PLDI representation [4]. Based on the Lyapunov method and Linear Matrix Inequalities ($\mathcal{LMI}$) formulation, sufficient conditions have been derived for stability analysis, controllers and observers design (see among others [7, 15, 16, 26, 24]). Recently, systems subject to unknown inputs are extensively considered in the literature. Unknown inputs can result either from model uncertainty, faults or due to the presence of unknown external excitation. This problem, usually referred as the unknown input observer design, has been considered actively for linear systems [13, 12, 27, 10, 22], for descriptor and nonlinear systems (see among other [19, 17, 14] and for multiple model approach (see for example [7, 6] and references therein). Based on unknown inputs observer design, many works have been carried out on secure communication and chaotic system reconstruction problem. The increasing need of secure communications leads to the development of many techniques which make difficult the detecting of transmitted message (see for example [18, 8, 3, 2, 1]. Indeed, the problem we are faced with consists of transmitting some coded message with a signal broadcasted by a communication channel. At the receiver side, the hidden signal is recovered by a decoding system. In this chapter, our goal is to show how to get chaotic multiple model from chaotic nonlinear system and how to design the proposed structure of observer for chaotic system reconstruction.

This chapter is organized as follows. In section 2, a considered unknown inputs multiple model in continuous-time case and his corresponding observer are given. Synthesis conditions for the proposed observer are given in $\mathcal{LMI}$ terms. Two cases are considered. The case of output signal not depending on the unknown inputs and the case when both state and output signal are affected by unknown inputs. To improve the performances of the proposed observer, the pole assignment in a $\mathcal{LMI}$ region is also studied. Unknown input estimation is given in section 3. Then these design conditions are extended to unknown inputs discrete-time multiple model in section 5. To illustrate the given synthesis $\mathcal{LMI}$ conditions, numerical examples and applications dealing with the chaotic system reconstruction for both continuous-time and discrete-time multiple model are proposed.

Throughout this chapter, $\mathbf{R}^n$ and $\mathbf{R}^{n\times m}$ denote, respectively, the n dimensional Euclidean space and the set of all $n \times m$ real matrices. Superscript "^{T}" denotes matrix transposition and the notation $X > Y$ where X and Y are symmetric matrices, means that $X - Y$ is positive definite. $\otimes$ is the Kronecker product, $\mathbf{I}$ is the identity matrix with compatible dimensions, the symbol $(*)$ denotes the transpose elements in the symmetric positions and $I_M = \{1, 2, \cdots, M\}$.

7.2 Unknown Inputs Multiple Observer Design

Consider a continuous-time multiple model with unknown inputs defined as follows

$$\begin{cases} \dot{x}(t) = \sum_{i=1}^{M} \mu_i(\xi(t))(\widehat{A}_i x(t) + B_i u(t) + D_i + R_i \bar{u}(t)) \\ y(t) = Cx(t) + F\bar{u}(t) \end{cases} \tag{7.1}$$

with

$$\mu_i(\xi(t)) \geq 0, \sum_{i=1}^{M} \mu_i(\xi(t)) = 1 \tag{7.2}$$

and

$$\widehat{A}_i = A_i + \Delta A_i(t) \tag{7.3}$$

M being the number of local LTI models, $x(t) \in \mathbf{R}^n$ the state vector, $u(t) \in \mathbf{R}^m$ the input vector, $\bar{u}(t) \in \mathbf{R}^q$, the unknown input and $y \in \mathbf{R}^p$ the measured outputs. $A_i \in \mathbf{R}^{n \times n}$, $B_i \in \mathbf{R}^{n \times m}$, $D_i \in \mathbf{R}^n$ and $C \in \mathbf{R}^{p \times n}$ define the ith local LTI model. Matrices $R_i \in \mathbf{R}^{n \times q}$ and $F \in \mathbf{R}^{p \times q}$ represent the influence of the unknown inputs. We assume that $q < p$ and, without loss of generality, that

Assumption 1: $rank(F) = q$ and $rank(R_i) = q$, i.e. F and R_i are full column ranks.

Assumption 2: $rank(C) = p$, i.e. C is full row rank.

$\Delta A_i(t)$ are time-varying matrices representing parametric uncertainties. These uncertainties are admissibly norm-bounded, structured and satisfy: $\Delta A_i = D_{A_i} F_{A_i} E_{A_i}$ with D_{A_i} and E_{A_i} are known real matrices with appropriate dimensions and F_{A_i} satisfies $F_{A_i}^{\top} F_{A_i} \leq \mathbf{I}$, $\forall\ i \in I_M$. The *activation functions* $\mu_i(.)$ depend on the so-called decision vector $\xi(t)$ assumed to depend on measurable variables.

In this section, we are concerned by the reconstruction of state variable $x(t)$ of multiple model (7.1) subject to unknown inputs, using only the available information, that is known input $u(t)$ and measured output $y(t)$.

The following lemma will be used in the rest of the paper.

Lemma 1. [28] : Let H and E be given matrices with appropriate dimensions and F satisfying $F^{\top} F \leq \mathbf{I}$. Then, we have for any $\varepsilon > 0$,

$$HFE + E^{\top} F^{\top} H^{\top} \leq \varepsilon H H^{\top} + \frac{1}{\varepsilon} E^{\top} E. \tag{7.4}$$

The considered unknown input multiple observer, for the unknown input multiple model (7.1), has the following structure

$$\begin{cases} \dot{z}(t) = \sum_{i=1}^{M} \mu_i(\xi(t)) \left(N_i z(t) + G_{i1} u(t) + G_{i2} + L_i y(t) \right) \\ \hat{x}(t) = z(t) - Ey(t) \end{cases} \tag{7.5}$$

The considered observer only uses known variables ($u(t)$ and $y(t)$) and the same functions $\mu_i(.)$ as used for the multiple model (7.1). The unknown inputs $\bar{u}(t)$ are considered non available.

In order to estimate the state of the unknown input multiple model (7.1), the variables $N_i \in \mathbf{R}^{n\times n}$, $G_{i1} \in \mathbf{R}^{n\times m}$, $G_{i2} \in \mathbf{R}^{n}$, $L_i \in \mathbf{R}^{n\times p}$ and $E \in \mathbf{R}^{n\times p}$ must be be determined such that the state estimation error:

$$\tilde{x}(t) = x(t) - \hat{x}(t) \tag{7.6}$$

satisfies $\tilde{x}(t) \to 0$ when $t \to \infty$. Multiple model subject to unknown inputs which affect state and outputs variables of the system are then studied. Poles placement in $\mathcal{LMI}$ region for the designed multiple observer is also considered.

7.2.1 Unknown Inputs Observer Design

This section addresses the case when only states are affected by unknown inputs $\bar{u}(t)$, i.e. the multiple model (7.1) with $F = 0$:

$$\begin{cases} \dot{x}(t) = \sum_{i=1}^{M} \mu_i(\xi(t))(\widehat{A}_i x(t) + B_i u(t) + D_i + R_i\bar{u}(t)), \\ y(t) = Cx(t) \end{cases} \tag{7.7}$$

The following result gives sufficient $\mathcal{LMI}$ conditions guaranteeing the global asymptotic convergence of the state estimation error (7.6).

Theorem 7.1. *The state estimation error between multiple observer (7.5) and unknown input multiple model (7.7) converges globally asymptotically towards zero, if there exists matrices $X > 0$, S and W_i and scalars ε_i such that the following conditions hold $\forall\, i \in I_M$:*

$$\begin{bmatrix} A_i^\top X + XA_i + A_i^\top C^\top S^\top + SCA_i - W_iC - C^\top W_i^\top + \varepsilon_i E_{A_i}^\top E_{A_i} & (X+SC)D_{A_i} \\ (*) & -\varepsilon_i \mathbf{I} \end{bmatrix} < 0 \tag{7.8a}$$

$$(X + SC)R_i = 0 \tag{7.8b}$$

Then multiple observer (7.5) is completely defined by:

$$E = X^{-1}S \tag{7.9a}$$

$$G_{i1} = (\mathbf{I} + X^{-1}SC)B_i \tag{7.9b}$$

$$G_{i2} = (\mathbf{I} + X^{-1}SC)D_i \tag{7.9c}$$

$$N_i = (\mathbf{I} + X^{-1}SC)A_i - X^{-1}W_iC \tag{7.9d}$$

$$L_i = X^{-1}W_i - N_iE \tag{7.9e}$$

Proof. From estimation error (7.6), the expression of $\hat{x}(t)$ given by multiple observer (7.5) and $x(t)$ given by (7.7), we get

$$\tilde{x}(t) = (\mathbf{I}+EC)x(t) - z(t) \tag{7.10}$$

The dynamic of state estimation error (7.10), taking account the expressions of $y(t)$ and $z(t)$ given in (7.7) and (7.5), is given by

$$\dot{\tilde{x}}(t) = \sum_{i=1}^{M} \mu_i(\xi) \Big(N_i\tilde{x}(t) + (T\widehat{A}_i - K_iC - N_i)x(t) + (TB_i - G_{i1})u(t) + (TD_i - G_{i2}) + TR_i\bar{u}(t) \Big) \tag{7.11}$$

with

$$K_i = N_iE + L_i, \quad T = \mathbf{I}+EC \tag{7.12}$$

The following change of variables

$$W_i = XK_i \tag{7.13a}$$
$$S = XE \tag{7.13b}$$

with $X > 0$ and (7.12) lead to the following expression

$$\dot{\tilde{x}}(t) = \sum_{i=1}^{M} \mu_i(\xi) \Big(N_i\tilde{x}(t) + (T\widehat{A}_i - K_iC - N_i)x(t) + ((\mathbf{I}+X^{-1}SC)B_i - G_{i1})u(t) + ((\mathbf{I}+X^{-1}SC)D_i - G_{i2}) + X^{-1}(X+SC)R_i\bar{u}(t) \Big) \tag{7.14}$$

Taking account (7.8b) and (7.9b-c), we get

$$\dot{\tilde{x}}(t) = \sum_{i=1}^{M} \mu_i(\xi) N_i\tilde{x}(t) \tag{7.15}$$

with

$$N_i = T\widehat{A}_i - K_iC \tag{7.16}$$

Then, the state estimation error (7.15) converges asymptotically to zero if there exist $X > 0$ such that $\forall\, i \in I_M$:

$$XN_i + N_i^{\top}X < 0 \tag{7.17}$$

With the same variable change (7.13), inequalities (7.17) are equivalent to

$$(X+SC)\widehat{A}_i - W_iC + ((X+SC)\widehat{A}_i - W_iC)^{\top} < 0 \tag{7.18}$$

By applying Lemma 1, with $\widehat{A}_i = A_i + \Delta A_i$ and $\Delta A_i = D_{A_i} F_{A_i} E_{A_i}$, the constraint (7.18) is equivalent to the existence of scalars $\varepsilon_i > 0$ such that

$$(X+SC)A_i + A_i^\top (X+SC)^\top - W_i C - C^\top W_i^\top + \varepsilon_i E_{A_i}^\top E_{A_i} + \varepsilon_i^{-1}(X+SC)D_{A_i}D_{A_i}^\top (X+SC)^\top < 0 \tag{7.19}$$

which is only the Schur complement of (7.8a). This completes the proof. □

For multiple model (7.7) without uncertainties, i.e. $\widehat{A}_i = A_i$, the following corollary gives sufficient $\mathcal{LMI}$ conditions for global asymptotic convergence of the state estimation error (7.6).

Corollary 1: The state estimation error between multiple observer (7.5) and unknown input multiple model (7.7) with $\Delta A_i = 0$ converges globally asymptotically towards zero, if there exists matrices $X > 0$, S and W_i such that the following conditions hold $\forall\, i \in I_M$:

$$(X+SC)A_i + A_i^\top (X+SC)^\top - W_i C - C^\top W_i^\top < 0 \tag{7.20a}$$

$$(X+SC)R_i = 0 \tag{7.20b}$$

Then multiple observer (7.5) is defined by (7.9).

Proof. It suffices to substitute $\widehat{A}_i$ by A_i in (7.18) to get (7.20a). The equalities constraints are not modified. □

It is important to note that equalities (7.8b)/(7.20b), with $X > 0$ and the change of variable (7.13), are equivalent to $(\mathbf{I} + EC)R_i = 0$, that is $ECR_i = -R_i \ \forall\, i \in I_M$. Lets notice that this condition contains the one for linear systems ($R_i = R$) where an solution E exits if and only if the rank constraint $rank(CR) = rank(R)$ holds (see for example [27, 10]). However, in contrast to linear systems, it is important to note that the condition on the rank is only a necessary condition for multiple model. Moreover, inequalities (7.20a) with $X > 0$ are equivalent to $X\big((\mathbf{I}+EC)A_i - K_iC\big) + \big((\mathbf{I}+EC)A_i - K_iC\big)^\top X < 0$. It is easy to note that these conditions contain the observability (detectability) conditions of $\big((\mathbf{I}+EC)A, C\big)$ for linear systems. Then, in order to assist the designer, the following procedure proposes to check two necessary conditions before solving conditions (7.8) or (7.20):

Procedure 1:

i) Check if $rank(CR_i) = rank(R_i) \ \forall\, i \in I_M$.

ii) Compute for each $i \in I_M$, a solution $E^{(i)} = -R_i(CR_i)^+$ and check the local observability of each pair $\big((\mathbf{I}+E^{(i)}C)A_i, C\big)$. Σ^+ denotes any generalized inverse of matrix Σ with $\Sigma\Sigma^+\Sigma = \Sigma$ [21].

If *(i)*-*(ii)* hold, then the designer can solve the sufficient $\mathcal{LMI}$ conditions (7.8)/(7.20) to design multiple observer (7.5).

7.2.2 $\mathcal{LMI}$ Design Conditions

This section considers the general structure of unknown inputs multiple model (7.1), that is when both the state and the output signal are affected by unknown inputs $\bar{u}(t)$. The following result gives sufficient $\mathcal{LMI}$ conditions guaranteeing the global asymptotic convergence of state estimation error (7.6).

Theorem 7.2. *The state estimation error between multiple observer (7.5) and unknown input multiple model (7.1) converges globally asymptotically towards zero, if there exists matrices $X > 0$, S and W_i and scalars ε_i such that the following conditions hold $\forall\, i \in I_M$:*

$$\begin{bmatrix} A_i^\top X + XA_i + A_i^\top C^\top S^\top + SCA_i - W_iC - C^\top W_i^\top + \varepsilon_i E_{A_i}^\top E_{A_i} & (X+SC)D_{A_i} \\ (*) & -\varepsilon_i \mathbf{I} \end{bmatrix} < 0 \tag{7.21a}$$

$$(X+SC)R_i = W_iF \tag{7.21b}$$

$$SF = 0 \tag{7.21c}$$

Then multiple observer (7.5) is completely defined by (7.9).

Proof. From estimation error (7.6) with the expression of $\hat{x}(t)$ given by observer (7.5) and multiple model (7.1), we obtain the following expression:

$$\tilde{x}(t) = (\mathbf{I}+EC)x(t) - z(t) + EF\bar{u}(t) \tag{7.22}$$

The dynamic of state estimation error is then given by

$$\dot{\tilde{x}}(t) = \sum_{i=1}^{M} \mu_i(\xi(t)) \Big(T\Big(\widehat{A}_i x(t) + B_i u(t) + R_i \bar{u}(t) + D_i\Big) - N_i z(t) - G_{i1}u(t) - G_{i2} - L_i y(t) \Big) + EF\dot{\bar{u}}(t) \tag{7.23}$$

where T is defined in (7.12). With the expressions of $y(t)$, $z(t)$ given in (7.1) and (7.5), we obtain

$$\dot{\tilde{x}}(t) = \sum_{i=1}^{M} \mu_i(\xi) \Big(N_i\tilde{x}(t) + \big(T\widehat{A}_i - K_iC - N_i\big)x(t) + \big(TB_i - G_{i1}\big)u(t) + (TD_i - G_{i2}) + \big(TR_i - K_iF\big)\bar{u}(t) \Big) + EF\dot{\bar{u}}(t) \tag{7.24}$$

with K_i defined in (7.12). Thus, using the same change of variable (7.13) with (7.9b-c) and (7.21b-c), we get

$$\dot{\tilde{x}}(t) = \sum_{i=1}^{M} \mu_i(\xi) N_i \tilde{x}(t) \tag{7.25}$$

where N_i is defined in (7.16). The rest of the proof is similar to the one of the theorem 1. This completes the proof. □

The following corollary obtained directly from theorem 2, gives sufficient linear conditions to design a multiple observer for multiple model (7.1) without uncertainties.

Corollary 2: The state estimation error between multiple observer (7.5) and unknown input multiple model (7.1) with $\Delta A_i = 0$ converges globally asymptotically towards zero, if there exists matrices $X > 0$, S and W_i such that the following conditions hold $\forall\, i \in I_M$:

$$(X+SC)A_i + A_i^\top (X+SC)^\top - W_i C - C^\top W_i^\top < 0 \tag{7.26a}$$

$$(X+SC)R_i = W_i F \tag{7.26b}$$

$$SF = 0 \tag{7.26c}$$

Then multiple observer (7.5) is defined by (7.9).

Remark 1. Classical numerical tools as the LMITOOL [25] may be used to solve the linear problem (7.8) on variables $X > 0$, S, W_i and scalars $\varepsilon_{i,\cdot}$. Examples are given in section 4 to illustrate the derived stability conditions.

Remark 2. Only uncertainties on matrices A_i are considered. Uncertainties on the other matrices lead to equalities constraints impossible to satisfy and not considered in this chapter.

Remark 3. The case of different multiple output matrices $C_i = C, \forall i \in I_M$ is not considered because it leads to non convex constraints not easy to resolve with existing numerical tools.

7.2.3 Pole Placement

In this part, we investigate how to improve the performances of the proposed observer (7.5) for multiple model (7.1). In order to achieve a desired transient performance, a pole placement should be considered. For many problems, exact pole assignment may not be necessary; it suffices to locate the pole in a sub-region of the complex left half plane [9]. This section discusses a pole assignment in $\mathcal{LMI}$ regions $S(\alpha,\beta)$.

Theorem 7.3. *A matrix $A \in \mathbf{R}^{n\times n}$ is D-stable if and only if there exists a symmetric positive definite matrix $X > 0$ such that*

$$M_D(A,D) = \alpha \otimes X + \beta \otimes (AX) + \beta^\top \otimes (AX)^\top < 0 \tag{7.27}$$

where $\alpha \in \mathbf{R}^{n\times n}$ and $\beta \in \mathbf{R}^{n\times n}$.

Since prescribed $\mathcal{LMI}$ region (7.27) will be added as supplementary constraint to these of theorem 1 or theorem 2, it should be noted that it only suffices to locate the poles of matrix $\sum_{i=1}^{M} \mu_i(\xi(t))N_i$ in prescribed $\mathcal{LMI}$ regions. Indeed, estimation error (7.25) is D-stable if there exists a matrix $X > 0$ such that

$$M_D(N_i,D) = \alpha \otimes X + \beta \otimes (N_i X) + \beta^\top \otimes (N_i X)^\top < 0 \tag{7.28}$$

With the same changes of variables (7.13) applied to inequalities (7.28), we obtain the following result.

Corollary 3: If there exit matrices $X > 0$, S and W_i such that the following conditions hold $\forall\, i \in I_M$:

$$\alpha \otimes X + \beta \otimes (X\widehat{A}_i + SC\widehat{A}_i - W_iC) + \beta^\top \otimes (X\widehat{A}_i + SC\widehat{A}_i - W_iC)^\top < 0 \tag{7.29a}$$

$$(X+SC)R_i = W_iF \tag{7.29b}$$

$$SF = 0 \tag{7.29c}$$

Then, multiple observer (7.5) is globally asymptotically convergent with the performance defined by complex region $S(\alpha,\beta)$. The multiple observer gains are as defined by (7.9).

For example, to ensure a given performance of the state estimation error, we define region $S_r(\alpha,\beta)$ as the intersection between a circle, of center $(0,0)$ and of radius β, and the left half plane limited by a vertical straight line of x-coordinate equal to $-\alpha < 0$. The corresponding $\mathcal{LMI}$ formulation of the corollary 3 is given by the following corollary.

Corollary 4: If there exit matrices $X > 0$, S, W_i and scalars ε_{i1} and ε_{i2} such that the following $\mathcal{LMI}$ conditions hold $\forall\, i \in I_M$:

$$\begin{bmatrix} -\beta X & XA_i + SCA_i - W_iC & (X+SC)D_{A_i} \\ (*) & -\beta X + \varepsilon_{i1} E_{A_i}^\top E_{A_i} & 0 \\ (*) & (*) & -\varepsilon_{i1}\mathbf{I} \end{bmatrix} < 0 \tag{7.30a}$$

$$\begin{bmatrix} A_i^\top X + XA_i + A_i^\top C^\top S^\top + SCA_i - W_iC - C^\top W_i^\top + 2\alpha X + \varepsilon_{i2} E_{A_i}^\top E_{A_i} & (X+SC)D_{A_i} \\ (*) & -\varepsilon_{i2}\mathbf{I} \end{bmatrix} < 0 \tag{7.30b}$$

$$(X+SC)R_i = W_iF \tag{7.30c}$$

$$SF = 0 \tag{7.30d}$$

Then, multiple observer (7.5) is globally asymptotically convergent with the performance defined by complex region $S_r(\alpha,\beta)$. The multiple observer gains are as defined by (7.9).

Proof. For the defined region $S_r(\alpha,\beta)$, constraints (7.29a) are equivalent to

$$\begin{bmatrix} -\beta X & X\widehat{A}_i + SC\widehat{A}_i - W_iC \\ (*) & -\beta X \end{bmatrix} < 0 \tag{7.31a}$$

$$X\widehat{A}_i + SC\widehat{A}_i - W_iC + (X\widehat{A}_i + SC\widehat{A}_i - W_iC)^\top + 2\alpha X < 0 \tag{7.31b}$$

Inequalities (7.31a) can be rewritten as follows

$$\begin{bmatrix} -\beta X & XA_i + SCA_i - W_iC \\ (*) & -\beta X \end{bmatrix} + \begin{bmatrix} (X+SC)D_{A_i} \\ 0 \end{bmatrix} F_{A_i} \begin{bmatrix} 0 & E_{A_i} \end{bmatrix} + \begin{bmatrix} 0 \\ E_{A_i}^\top \end{bmatrix} F_{A_i}^\top \begin{bmatrix} D_{A_i}^\top (X+SC)^\top & 0 \end{bmatrix} < 0 \tag{7.32}$$

Applying Lemma 1 to (7.32) and Schur complement to the result, we get $\mathcal{LMI}$ conditions (7.30a). $\mathcal{LMI}$ conditions (7.30b) are also obtained from (7.31b) using lemma 1. This completes the proof. □

Note that for the certain case, i.e. $\widehat{A}_i = A_i$, corollary 4 is rewritten as follows

$$\begin{bmatrix} -\beta X & XA_i + SCA_i - W_iC \\ (*) & -\beta X \end{bmatrix} < 0 \tag{7.33a}$$

$$XA_i + SCA_i - W_iC + (XA_i + SCA_i - W_iC)^\top + 2\alpha X < 0 \tag{7.33b}$$

$$(X+SC)R_i = W_iF \tag{7.33c}$$

$$SF = 0 \tag{7.33d}$$

7.3 Unknown Inputs Estimation

A lot of works have been considered for the unknown input estimation problem (see for example [7, 6, 11]). For example in [11], authors are proposed methods for detecting and reconstructing sensor faults using sliding mode observers whereas in [6] a method to simultaneously estimate unknown inputs and states for T-S fuzzy models is proposed.

The method proposed in this chapter is based on the hypothesis of the good estimation of the state variables [7]. Indeed, when the state estimation error is equal to zero; by replacing x by $\hat{x}$ in the equation (7.1) we obtain the following approximation:

$$\hat{y} = C\hat{x} + F\hat{u} \tag{7.34}$$

Since the assumption 1 holds, i.e. the matrix F is of full column rank, an estimation of unknown inputs can be carried out in a simpler way by

$$\hat{u} = (F^\top F)^{-1} F^\top (y - \hat{y}) \tag{7.35}$$

7.4 Simulation Examples

To illustrate the validness of the proposed results, two examples will be proposed. The first one illustrates the good estimation of both states and unknown input affecting multiple model in different case (without poles placement, with poles placement and uncertainties). The second one in section 4.2, deals with the chaotic system reconstruction. First by building chaotic multiple model. Then the design of a multiple observer for such chaotic multiple model is given. Simulation shows the good chaotic system reconstruction with the proposed design.

7.4.1 Academic Example

Now, consider the multiple model (7.1), where both the state and the output signal are affected by unknown inputs, with $M = 2$ and the following data:

$$A_1 = \begin{bmatrix} -2 & 1 & 1 \\ 1 & -3 & 0 \\ 2 & 1 & -6 \end{bmatrix} \quad A_2 = \begin{bmatrix} -3 & 2 & 2 \\ 5 & -8 & 0 \\ 0.5 & 0.5 & -4 \end{bmatrix}$$

$$B_1 = \begin{bmatrix} 1 \\ -0.5 \\ -0.5 \end{bmatrix} \quad B_2 = \begin{bmatrix} -0.5 \\ 1 \\ -0.25 \end{bmatrix} \quad F = \begin{bmatrix} 1 \\ 1 \end{bmatrix}$$

$$R_1 = \begin{bmatrix} 1 \\ -1 \\ 1 \end{bmatrix} \quad R_2 = \begin{bmatrix} 1 \\ 0.5 \\ -2 \end{bmatrix} \quad C = \begin{bmatrix} 1 & 1 & 1 \\ 1 & 0 & 1 \end{bmatrix}$$

and the functions

$$\begin{cases} \xi(t) = y_1(t) \\ \mu_1(\xi(t)) = \frac{1}{2}(1 - \tanh(\xi(t))) \\ \mu_2(\xi(t)) = 1 - \mu_1(\xi(t)) \end{cases}$$

The following subsections are dedicated to design a multiple observer of the form (7.5), firstly without pole placement in subsection 4.1.1 and then with pole placement in subsection 4.1.2. Uncertainties on state matrices are also studied in section 4.1.3.

7.4.1.1 Observer Design without Pole Placement

The resolution of conditions (7.26) lead to the following result:

$$\mathrm{X} = \begin{bmatrix} 0.1 & 0 & 0 \\ 0 & 0.1 & 0 \\ 0 & 0 & 0.1 \end{bmatrix} \quad W_1 = \begin{bmatrix} 0.22 & -0.10 \\ 0.33 & -0.37 \\ 0.30 & -0.15 \end{bmatrix} \quad W_2 = \begin{bmatrix} 0.49 & -0.40 \\ 0.17 & -0.15 \\ 0.49 & -0.72 \end{bmatrix}$$

From (7.9), we define completely multiple observer (7.5) as follows:

$$E=\begin{bmatrix}-0.18 & 0.18\\ -0.66 & 0.66\\ -0.57 & 0.57\end{bmatrix}\quad L_1=\begin{bmatrix}1.04 & 0.14\\ 0.76 & -1.09\\ -1.55 & 3.13\end{bmatrix}\quad L_2=\begin{bmatrix}3.70 & -2.8\\ -1.03 & 1.19\\ 4.05 & -6.34\end{bmatrix}$$

$$G_{11}=\begin{bmatrix}1.09\\ -0.16\\ -0.21\end{bmatrix}\quad G_{21}=\begin{bmatrix}-0.68\\ 0.33\\ -0.82\end{bmatrix}$$

With known input $u(t)$ and unknown input $\bar{u}(t)$ given in figures $1-2$ respectively and initial conditions $x_0=(1,0.5,0)^\top$ and $z_0=(-2,2,-3)^\top$, we obtain the simulation results given in figures $3-5$. As shown, the dynamic of the estimated state tends globally asymptotically to the model sate in spite of the presence of unknown input $\bar{u}(t)$. This allows to illustrate the effectiveness of the derived synthesis conditions.

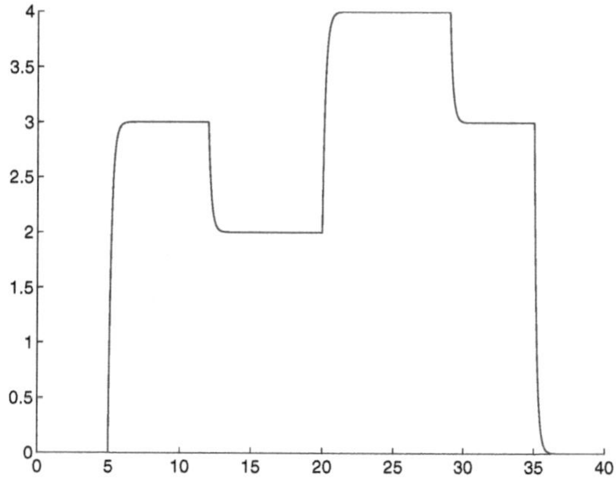

Fig. 7.1 Known input $u(t)$

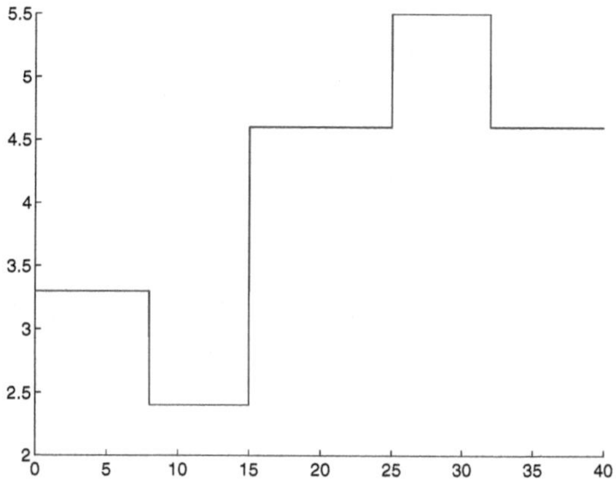

Fig. 7.2 Unknown input $\bar{u}(t)$

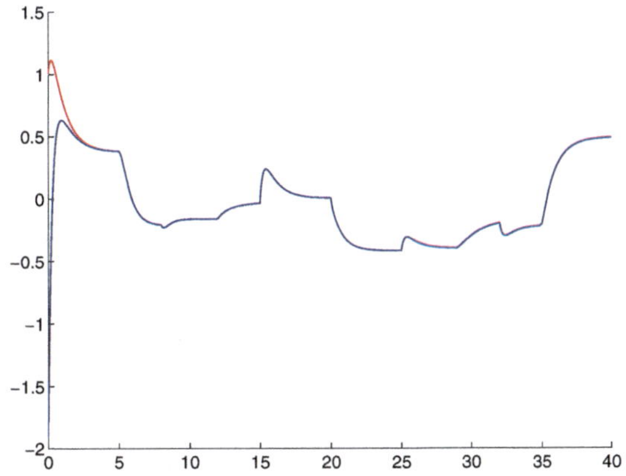

Fig. 7.3 State estimation without pole placement: $x_1(t)$ and $\hat{x}_1(t)$

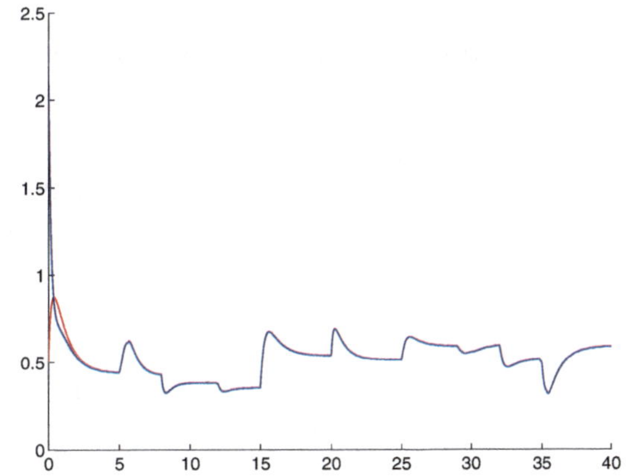

Fig. 7.4 State estimation without pole placement: $x_2(t)$ and $\hat{x}_2(t)$

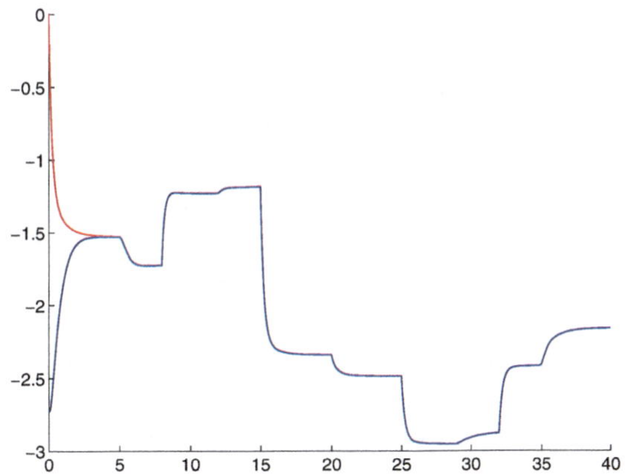

Fig. 7.5 State estimation without pole placement: $x_3(t)$ and $\hat{x}_3(t)$

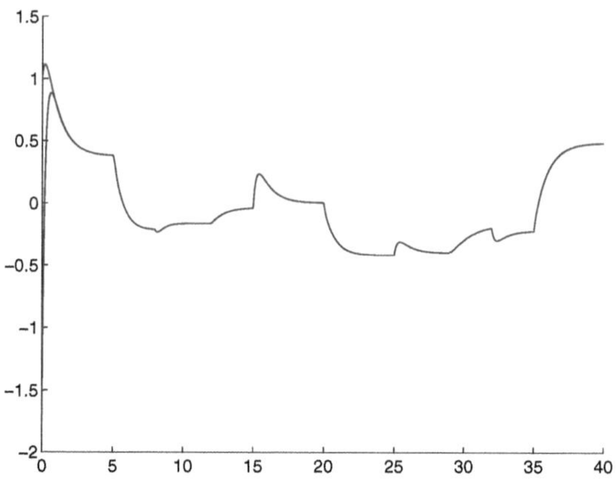

Fig. 7.6 State estimation with pole placement: $x_1(t)$ and $\hat{x}_1(t)$

7.4.1.2 Observer Design with Pole Placement

In order to improve performances of the above designed observer, the region $S_r(\alpha,\beta)$ defined in (7.33) is used. The considered region is an intersection between a circle, of center $(0,0)$ and of radius 10, and the left half plane limited by a vertical line at -2. The resolution of conditions (7.33) corresponding to the region $S_r(2,10)$ give:

$$X = \begin{bmatrix} 4.15 & 0 & -1.75 \\ 0 & 2.97 & 0 \\ -1.75 & 0 & 1.94 \end{bmatrix} \quad W_1 = \begin{bmatrix} 0.88 & 1.01 \\ 13.41 & -13.41 \\ 5.94 & -3.84 \end{bmatrix} \quad W_2 = \begin{bmatrix} 3.38 & 4.53 \\ 13.41 & -13.41 \\ 12.82 & -19.41 \end{bmatrix}$$

From (7.9), the corresponding multiple observer is given by

$$E = \begin{bmatrix} -0.47 & 0.47 \\ -1 & 1 \\ -1.42 & 1.42 \end{bmatrix} \quad L_1 = \begin{bmatrix} -0.13 & 1.6 \\ 0 & 0 \\ -7.5 & 10 \end{bmatrix} \quad L_2 = \begin{bmatrix} 4.64 & -3.88 \\ 0 & 0 \\ 8.17 & -10.88 \end{bmatrix}$$

$$G_{11} = \begin{bmatrix} 1.24 \\ 0 \\ 0.21 \end{bmatrix} \quad G_{21} = \begin{bmatrix} -0.97 \\ 0 \\ -1.67 \end{bmatrix}$$

With the same initial conditions, the known and unknown inputs given in figures $1-2$, we obtain the simulation result given in figures $6-8$. To show clearly the performance improvements of the designed multiple observer, the simulation of state estimation errors $\tilde{x}(t) = x(t) - \hat{x}(t)$ with and without pole assignment are presented in figures $9-11$.

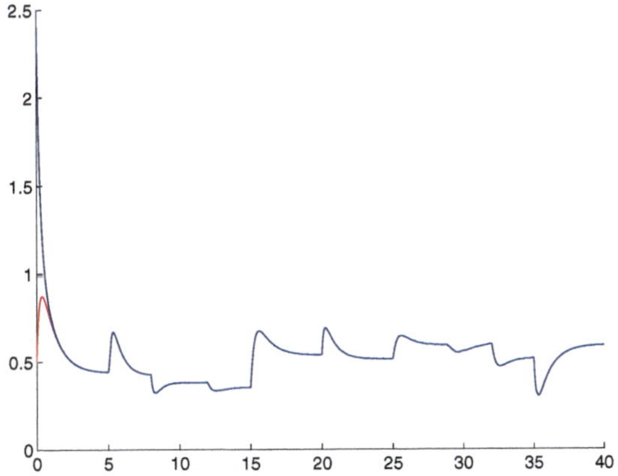

Fig. 7.7 State estimation with pole placement: $x_2(t)$ and $\hat{x}_2(t)$

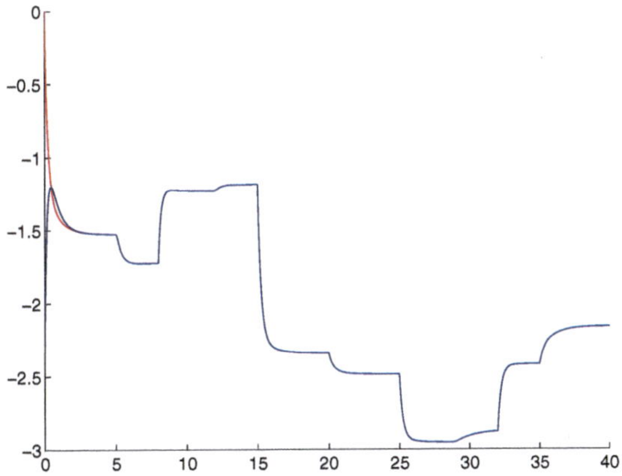

Fig. 7.8 State estimation with pole placement: $x_3(t)$ and $\hat{x}_3(t)$

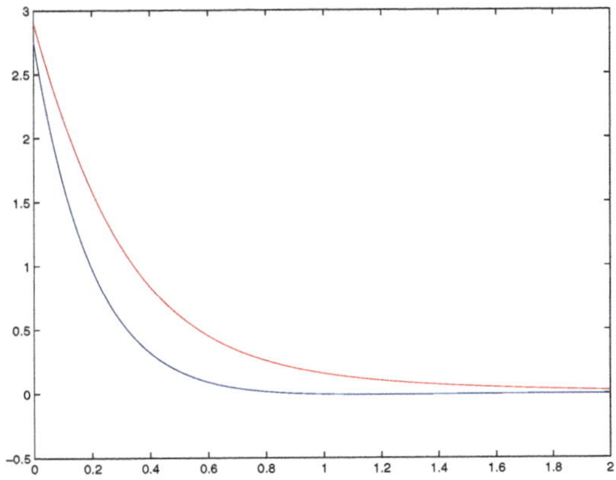

Fig. 7.9 Estimation errors with and without (red line) pole placement:$\tilde{x}_1(t) = x_1(t) - \hat{x}_1(t)$

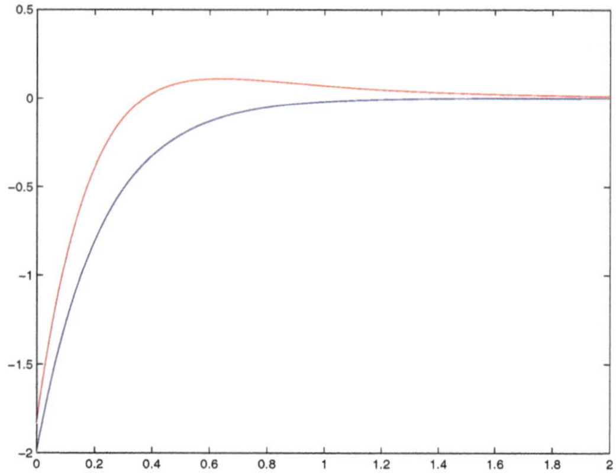

Fig. 7.10 Estimation errors with and without (red line) pole placement: $\tilde{x}_2(t) = x_2(t) - \hat{x}_2(t)$

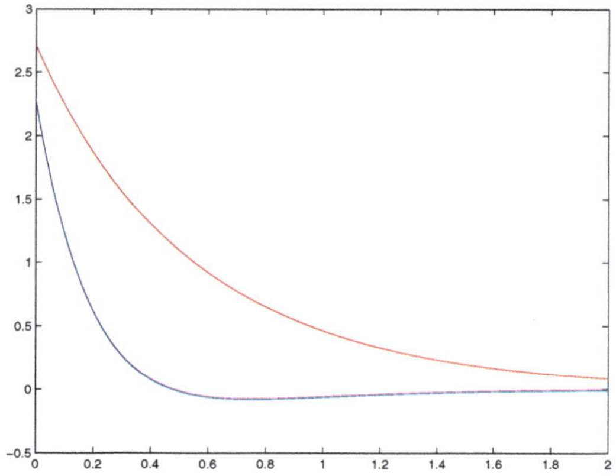

Fig. 7.11 Estimation errors with and without (red line) pole placement: $\tilde{x}_3(t) = x_3(t) - \hat{x}_3(t)$

7.4.1.3 Uncertainties on State Matrices

Now, to show the robustness of the proposed observer, consider the same example with uncertainties on state matrices as follows

$$D_{A_1} = D_{A_2} = \begin{bmatrix} 0.2 & 0 & 0 \\ 0 & 0.2 & 0 \\ 0 & 0 & 0.2 \end{bmatrix}, \quad E_{A_1} = \begin{bmatrix} 0 & 1 & 1 \\ 1 & 0 & 0 \\ 2 & 1 & -2 \end{bmatrix}, E_{A_2} = \begin{bmatrix} -2 & 2 & 2 \\ 5 & -4 & 0 \\ 0.5 & 0.5 & 0 \end{bmatrix}$$

The resolution of conditions (7.30) of corollary 4 corresponding to the same region $S_r(2,10)$ leads to feasible problem and gives:

$$X = \begin{bmatrix} 26.9004 & -0.4639 & -8.7213 \\ -0.4639 & 0.4138 & 0.0776 \\ -8.7213 & 0.0776 & 6.9058 \end{bmatrix}, S = \begin{bmatrix} -6.2098 & 6.2098 \\ -0.1912 & 0.1912 \\ -3.8747 & 3.8747 \end{bmatrix}$$

$$W_1 = \begin{bmatrix} 38.0992 & -13.2463 \\ 1.4271 & -2.0360 \\ 9.3190 & -7.3374 \end{bmatrix}, \; W_2 = \begin{bmatrix} 101.1735 & -60.1674 \\ 0.1294 & -0.6374 \\ 16.3630 & -40.7944 \end{bmatrix}$$

$$\varepsilon_{11} = 0.9609, \; \varepsilon_{21} = 0.1024, \; \varepsilon_{12} = 1.2122, \; \varepsilon_{22} = 0.1808$$

Then from (7.9), multiple observer (7.5) is defined with the following parameters

$$E = \begin{bmatrix} -0.7219 & 0.7219 \\ -0.9972 & 0.9972 \\ -1.4615 & 1.4615 \end{bmatrix}, \; G_{11} = \begin{bmatrix} 1.3609 \\ -0.0014 \\ 0.2308 \end{bmatrix}, \; G_{21} = \begin{bmatrix} -1.2219 \\ 0.0028 \\ -1.7115 \end{bmatrix}$$

$$N_1 = \begin{bmatrix} -4.4438 & -0.1141 & -0.7219 \\ 0.0057 & -6.1171 & 0.0028 \\ -1.9230 & -0.0381 & -8.4615 \end{bmatrix}, \; N_2 = \begin{bmatrix} -7.2484 & -0.0510 & 1.3609 \\ 0.0128 & -6.8255 & -0.0014 \\ -4.0768 & 0.0158 & -1.2692 \end{bmatrix}$$

$$L_1 = \begin{bmatrix} -1.0969 & 2.8188 \\ 0.0171 & -0.0200 \\ -8.3702 & 10.8317 \end{bmatrix}, \; L_2 = \begin{bmatrix} 4.5317 & -3.8926 \\ 0.0038 & -0.0024 \\ 7.3941 & -10.1249 \end{bmatrix}$$

It is important to note that the derived results can tolerate some level of uncertainties on the state matrices of the multiple model. The designed unknown input observer is proved to be robust against state matrices uncertainties.

7.4.2 Application to Chaotic System Reconstruction

Results developed in section 7.2 can be applied to reconstruct states of chaotic system in multiple model representation and also for a secure communication system. Indeed, the problem we are faced with consists of transmitting some coded message with a signal broadcasted by a communication channel. At the receiver side, the hidden signal is recovered by a decoding system. The increasing need of secure communications leads to the development of many techniques which make difficult the detecting of transmitted message (se for example [18, 8, 3, 2, 1]). In this section, our goal to show how the designed observer could be used in chaotic system reconstruction and in a secure communication scheme. For this purpose we use the nonlinear Lorenz model as chaotic systems represented by his equivalent chaotic multiple model. Consider the non linear Lorenz equation [1]:

$$\begin{cases} \dot{x}_1(t) = -ax_1(t) + ax_2(t) \\ \dot{x}_2(t) = cx_1(t) - x_2(t) - x_1(t)x_3(t) \\ \dot{x}_3(t) = x_1(t)x_2(t) - bx_3(t) \end{cases} \tag{7.36}$$

Which can be rewritten as follows

$$\dot{x}(t) = A(x(t))x(t) \tag{7.37}$$

with:

$$x = \begin{bmatrix} x_1 \\ x_2 \\ x_3 \end{bmatrix}, \quad A(x(t)) = \begin{bmatrix} -a & a & 0 \\ c & -1 & -x_1(t) \\ 0 & x_1(t) & -b \end{bmatrix}$$

and a, b, and c are constants. Assume that $x_1(t) \in [-d,d]$ with $d > 0$. Then, we can write $x_1(t) = -d.\mu_1(x_1(t)) + d.\mu_2(x_1(t))$ with $\mu_1(x_1(t)) + \mu_2(x_1(t)) = 1$ which leads to the following multiple model:

$$\dot{x}(t) = (\mu_1(x_1(t))A_1 + \mu_2(x_1(t))A_2)x(t) \tag{7.38}$$

where

$$A_1 = \begin{bmatrix} -a & a & 0 \\ c & -1 & -d \\ 0 & d & -b \end{bmatrix}, \quad A_2 = \begin{bmatrix} -a & a & 0 \\ c & -1 & d \\ 0 & -d & -b \end{bmatrix}$$

and

$$\mu_1(x_1(t)) = \frac{1}{2}\left(1 + \frac{x_1(t)}{d}\right) \quad \mu_2(x_1(t)) = \frac{1}{2}\left(1 - \frac{x_1(t)}{d}\right)$$

Note that the obtained multiple model exactly represents the nonlinear Lorenz model under $x_1(t) \in [-d,d]$.

In the following, we consider the chaotic multiple model (7.38) with $a = 10$, $b = 8/3$, $c = 28$ and $d = 30$ in his general form:

$$\begin{cases} \dot{x} = \sum_{i=1}^{2} \mu_i(y_1)\left(A_i x + R_i \bar{u}\right) \\ y = Cx + F\bar{u} \end{cases} \tag{7.39}$$

with:

$$A_1 = \begin{bmatrix} -10 & 10 & 0 \\ 28 & -1 & -30 \\ 0 & 30 & -8/3 \end{bmatrix} \quad A_2 = \begin{bmatrix} -10 & 10 & 0 \\ 28 & -1 & 30 \\ 0 & -30 & -8/3 \end{bmatrix}$$

$$B_1 = \begin{bmatrix} 0 \\ 0 \\ 0 \end{bmatrix} \quad B_2 = \begin{bmatrix} 0 \\ 0 \\ 0 \end{bmatrix} \quad C = \begin{bmatrix} 1 & 0 & 0 \\ 0 & 1 & 0 \end{bmatrix} \quad F = \begin{bmatrix} 1 \\ 1 \end{bmatrix}$$

The simulation of multiple model (7.39) without the unknown input $\bar{u}$ and with the initial value $x_0 = (1\ 1\ 1)^\top$ shows the chaotic behavior of the example plotted in the phase plan of the system (see figure 7.12).

The message to be encoded constitutes the so-called unknown input of the multiple model which plays the role of the encoder. The output of this model is transmitted using a public channel. On the receiving side, an unknown input multiple observer serves as a decoder in order to re-build the message. Clearly, the choice of LTI local models, their number, as well as the nature of the function $\mu_i(\xi(t)))$ are key elements for an external person to be able to decode the embodied crypted message from only the signal $y(t)$. The goal of the proposed example is only to show

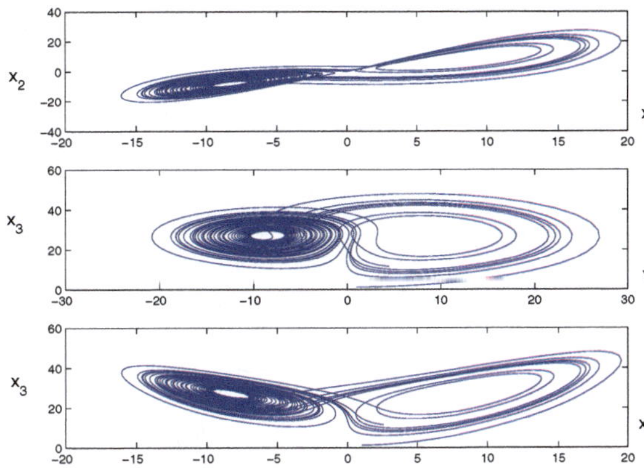

Fig. 7.12 Phase plan of the chaotic multiple model (7.38)

the feasibility of the proposed design in chaotic system reconstruction and in secure communication procedure.

Indeed, the unknown input can represent the hidden message to be transmitted. Thus the transmitted signal y is embedded with the hidden message $\bar{u}$ of the figure 7.14 .

The considered multiple observer for this application is

$$\begin{cases} \dot{z} = \sum_{i=1}^{2} \mu_i(y_1)\left(N_i z + L_i y\right) \\ \hat{x} = z - Ey \end{cases} \tag{7.40}$$

The resolution of conditions (7.26) with $B_1 = B_2 = (0,0,0)^\top$ lead to the following result:

$$X = \begin{bmatrix} 1.750 & 1.650 & -0.003 \\ 1.650 & 1.750 & -0.003 \\ -0.003 & -0.003 & 0.195 \end{bmatrix} \quad E = \begin{bmatrix} -3.05 & 3.05 \\ 3.99 & -3.99 \\ -0.004 & 0.004 \end{bmatrix}$$

$$N_1 = \begin{bmatrix} 33.66 & 47.89 & -91.72 \\ -36.18 & -45.78 & 89.96 \\ 61.12 & -32 & -2.79 \end{bmatrix} \quad L_1 = \begin{bmatrix} -16.44 & 17.44 \\ -14.94 & 15.94 \\ 253.9 & -252.9 \end{bmatrix}$$

$$N_2 = \begin{bmatrix} 35.06 & 49.54 & 91.72 \\ -37.82 & -47.14 & -89.96 \\ -62.08 & 31.20 & -2.53 \end{bmatrix} \quad L_2 = \begin{bmatrix} -19.38 & 17.32 \\ -13.67 & 17.67 \\ -252.38 & 253.37 \end{bmatrix}$$

Figure 7.13 represent the state estimation error with the initial conditions $x_0 = (1\,1\,1)^\top$ and $\hat{x}_0 = (0\,0\,0)^\top$. It shows the good reconstruction of chaotic system state. Figure 7.14 displays the hidden transmitted message and its estimate. Excepted around the time origin, the unknown input (transmitted message) is perfectly estimated.

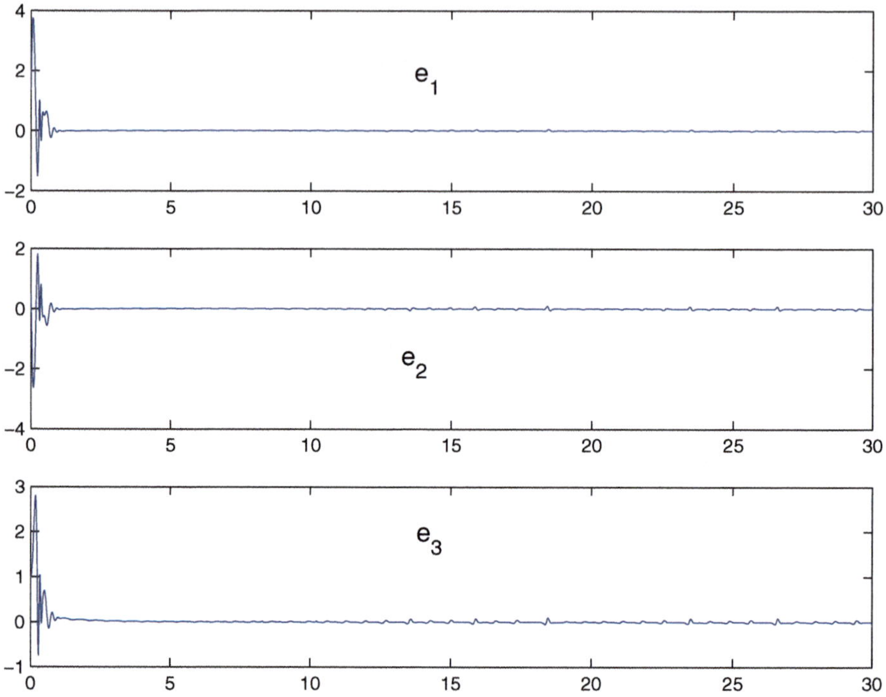

Fig. 7.13 Estimation errors $e_i = x_i - \hat{x}_i$, $i \in \{1,2,3\}$

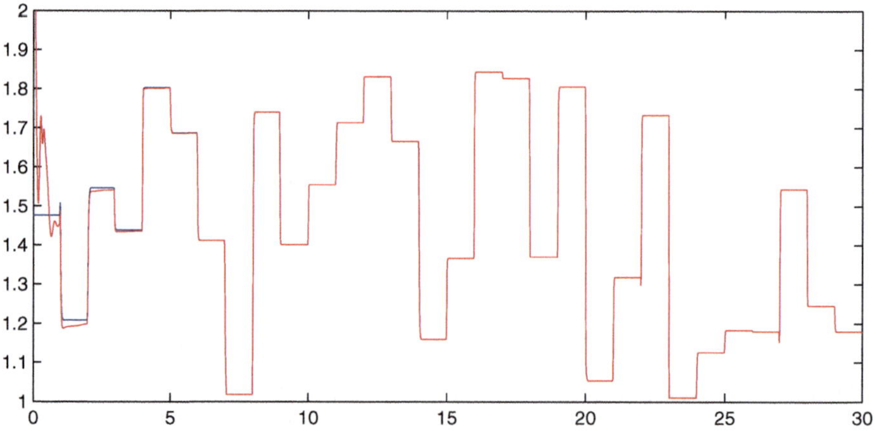

Fig. 7.14 Hidden message $\bar{u}$ and its estimate

7.5 Extension to Discret-Time Multiple Model

Consider the class of a nonlinear discrete-time system subject to unknown inputs represented by a discret-time multiple model:

$$\begin{cases} x(t+1) = \sum_{i=1}^{M} \mu_i(\xi(t))(A_i x(t) + B_i u(t) + R_i \bar{u}(t) + D_i) \\ y(t) = Cx(t) + F\bar{u}(t) \end{cases} \tag{7.41}$$

where $x(t) \in \mathbf{R}^n$ is the state vector, $u(t) \in \mathbf{R}^m$ the input vector, $\bar{u}(t) \in \mathbf{R}^q$, $q < n$, contains the unknown inputs and $y \in \mathbf{R}^p$ the measured outputs. Matrices $A_i \in \mathbf{R}^{n\times n}$ and $B_i \in \mathbf{R}^{n\times m}$. Matrices $R_i \in \mathbf{R}^{n\times q}$ and $F \in \mathbf{R}^{p\times q}$ are assumed to satisfy assumptions 1 and 2. The matrices $D_i \in \mathbf{R}^n$ are introduced to take into account the operating point of the system and $C \in \mathbf{R}^{p\times n}$ is the output matrix. Functions $\mu_i(\xi(t))$ are as defined in (7.2).

The considered structure of the multiple observer is

$$\begin{cases} z(t+1) = \sum_{i=1}^{M} \mu_i(\xi(t))\left(N_i z(t) + G_{i1} u(t) + G_{i2} + L_i y(t)\right. \\ \hat{x}(t) = z(t) - Ey(t) \end{cases} \tag{7.42}$$

where $N_i \in \mathbf{R}^{n\times n}$, $G_{i1} \in \mathbf{R}^{n\times m}$, $G_{i2} \in \mathbf{R}^n$, $E \in \mathbf{R}^{n\times p}$, $L_i \in \mathbf{R}^{n\times p}$ are the observer gains to be determined. The considered problem concerns both the reconstruction of state variable $x(t)$ and unknown input $\overline{u}(t)$, using only the available signal, that is known input $u(t)$ and measured output $y(t)$.

The following result gives sufficient conditions for the global asymptotic convergence of the state estimation error (7.6).

Theorem 7.4. *The state estimation error between multiple model (7.41) and unknown input multiple observer (7.42) converges globally asymptotically towards zero if there exists matrices $X > 0$, S and W_i such that the following conditions hold $\forall\, i \in I_M$:*

$$\begin{bmatrix} X & * \\ XA_i + SCA_i - W_i C & X \end{bmatrix} > 0 \tag{7.43a}$$

$$(X + SC)R_i = W_i F \tag{7.43b}$$

$$SF = 0 \tag{7.43c}$$

Multiple observer (7.42) is then completely defined by (7.9).

Proof. The prrof is obvious by using the same arguments used for proving theorem 1 and theorem 2. □

7.5.1 Pole Assignment

To improve performances of the multiple observer for better estimation of system state, dynamics of the multiple observer are constrained to be faster than that of the multiple model. As stated in section 2.3, it is possible to assign the poles to a specific sub-region is the complex plane [9]. For example if the prescribed region $S_r(\sigma, r)$ is a disk centered at $(\sigma, 0)$ and radius r, the $\mathcal{LMI}$ formulation of the previous problem is expressed by the following corollary.

Corollary 5: If there exist matrices X, S and W_i such that the following conditions hold $\forall\, i \in I_M$:

$$\begin{bmatrix} rX & * \\ X(A_i - \sigma\mathbf{I}) + SC(A_i - \sigma\mathbf{I}) - W_iC & rX \end{bmatrix} > 0 \tag{7.44a}$$

$$(X + SC)R_i = W_iF \tag{7.44b}$$

$$SF = 0 \tag{7.44c}$$

then the multiple observer (7.42) is globally asymptotically convergent with the performance defined by the complex region $S(\sigma, r)$. The observer parameters are as defined by (7.9).

Remark 4. Note that the $\mathcal{LMI}$ constraints (7.44) can be obtained from (7.43) by simply replacing the matrices A_i by $(A_i - \sigma\mathbf{I})/r$. Moreover if we are interested by the region $S_0(0, \alpha)$ it suffices to chose $\sigma = 0$ and $r = \alpha$.

Note that the $\mathcal{LMI}$ conditions (7.44) can be extended to incertain case, i.e. $\widehat{A}_i = A_i + \Delta A_i(t)$, as follows

Corollary 6: If there exit matrices $X > 0$, S, W_i and scalars ε_{i1} and ε_{i2} such that the following $\mathcal{LMI}$ conditions hold $\forall\, i \in I_M$:

$$\begin{bmatrix} rX & -XA_i - SCA_i + W_iC & -(X + SC)D_{A_i} \\ (*) & rX - \varepsilon_{i1}E_{A_i}^{\top}E_{A_i} & 0 \\ (*) & (*) & \varepsilon_{i1}\mathbf{I} \end{bmatrix} > 0 \tag{7.45a}$$

$$(X + SC)R_i = W_iF \tag{7.45b}$$

$$SF = 0 \tag{7.45c}$$

Then, multiple observer (7.42) is globally asymptotically convergent with the performance defined by complex region $S_r(\sigma, r)$. The multiple observer gains are as defined by (7.9).

Summarizing the estimation procedure, the design of multiple observer and the estimation of unknown inputs can be implemented as follows:

Procedure 2:

i) Solve the linear constraints (7.43) (or (7.45) for pole placement with uncertainties) with numerical tools such as the LMITOOL software [25],

ii) Deduce the observer parameters N_i, G_{i1}, G_{i2}, L_i and E of the multiple observer (7.42) using the equations (7.9).

iii) Under the assumption 1, estimate unknown input estimation using equation (7.35).

7.6 Application to Chaotic System Reconstruction

In this section, the proposed multiple observer is used to reconstruct states of chaotic systems and can be exploited in secure communication scheme. The message to be encoded is the unknown input of the chaotic multiple model.

Consider a chaotic discrete-time multiple model that results from the interpolation of two local models:

$$\begin{cases} x(t+1) = \sum_{i=1}^{2} \mu_i(\xi(t))\big(A_i x(t) + R_i \bar{u}(t)\big) \\ y(t) = Cx(t) + F\bar{u}(t) \end{cases} \tag{7.46}$$

The functions $\mu_i(.)$ depend on the multiple model output, $\xi(t) = y(t)$, and expressed by

$$\begin{cases} \xi(t) = y(t) \\ \mu_1(\xi(t)) = \frac{1}{2}(1 - \tanh(\xi(t))) \\ \mu_2(\xi(t)) = 1 - \mu_1(\xi(t)) \end{cases} \tag{7.47}$$

The numerical values of matrices are as follows:

$$A_1 = \begin{bmatrix} -1.1 & 0.5 \\ 0.3 & 0.7 \end{bmatrix}, \quad A_2 = \begin{bmatrix} 0.8 & -0.1 \\ 1 & 1.1 \end{bmatrix}, \quad C = \begin{bmatrix} 0.5 & 0.5 \end{bmatrix}, \quad F = 5$$

From the structure of multiple model (7.46), we can deduce the following values:

$$S = 0, \quad E = 0, \quad G_{i1} = 0, \quad G_{i2} = 0$$

For this example, since the encoding system (7.46) can be conceived at the same time as the decoding system (observer), the computation matrices R_i is then free. Thus, $\mathcal{LMI}$ (7.43a) can be solved without taking into account equalities (7.43b-c). The resolution of $\mathcal{LMI}$ (7.43a) gives

$$X = \begin{bmatrix} 1.6718 & -2.0563 \\ -2.0563 & 7.7169 \end{bmatrix} \quad W_1 = \begin{bmatrix} -3.9158 \\ 9.0362 \end{bmatrix} \quad W_2 = \begin{bmatrix} -2.3610 \\ 13.5810 \end{bmatrix}$$

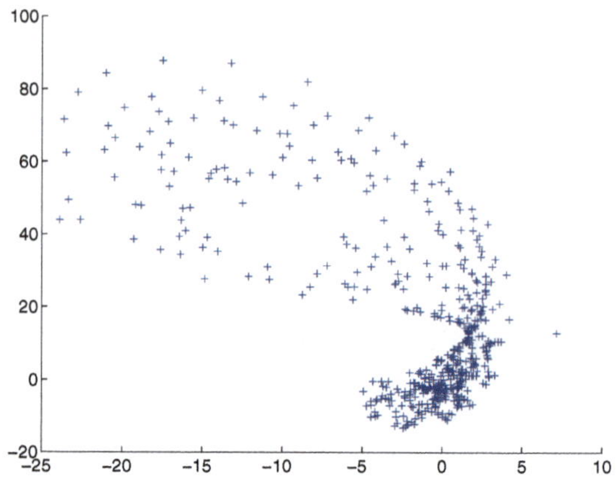

Fig. 7.15 Phase plan of the system

Using equations (7.9d-e), we obtain

$$L_1 = \begin{bmatrix} -1.3418 \\ 0.8134 \end{bmatrix} \qquad L_2 = \begin{bmatrix} 1.1192 \\ 2.0581 \end{bmatrix}$$

$$N_1 = \begin{bmatrix} -0.4291 & 1.1709 \\ -0.1067 & 0.2933 \end{bmatrix} \quad N_2 = \begin{bmatrix} 0.2404 & -0.6596 \\ -0.0291 & 0.0709 \end{bmatrix}$$

from (7.46b) with W_i and X, we compute the values of R_1 and R_2 as follows

$$R_1 = \begin{bmatrix} -6.7090 \\ 4.0671 \end{bmatrix}, \quad R_2 = \begin{bmatrix} 5.5959 \\ 10.2906 \end{bmatrix}$$

The designed unknown multiple observer can be applied in chaotic system reconstruction and also in a secure communication procedure. In this context, the problem consists of transmitting a resulting ciphertext embedded to the output by a communication channel. At the receiver side, the hidden signal (plaintext) is retrieved via the synthesis approach, i.e. the designed unknown input multiple. Concerning the transmission of a crypted message on a public channel of communication, one can wonder about the possibility of detecting and retrieving the message from the transmitted signal. In literature, some answers are given and one of them is satisfied here with a simple "visual appreciation" (see for example [18, 8, 2]). So, figure 7.15, plotted in the phase plan of the system, does not show any particular behavior of periodic type or with commutation. Obviously, these observations can not establish security on the inviolability of the transmitted signal. For simulation example, consider the message to be transmitted given by figure 7.16 and the resulting output (the encoded message) of the chaotic multiple model in figure 7.17. Figures 7.18 and 7.19 show the state variable of the chaotic system and its estimation which are perfectly superposed. Finally, figure 7.20 presents the estimated unknown input (message estimate) where the message is perfectly estimated except around the time origin.

Fig. 7.16 Message $\bar{u}(t)$

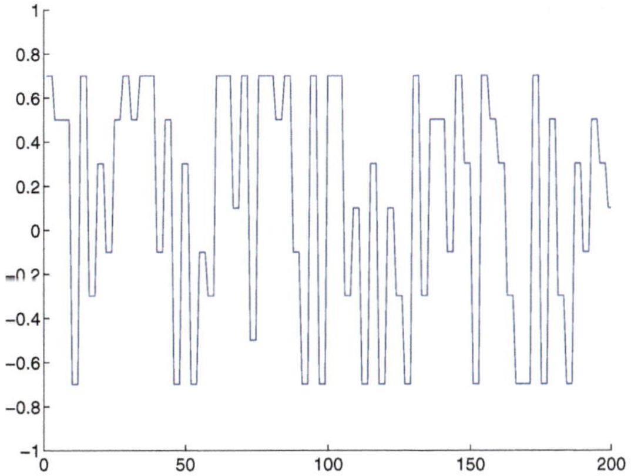

Fig. 7.17 Output $y(t)$

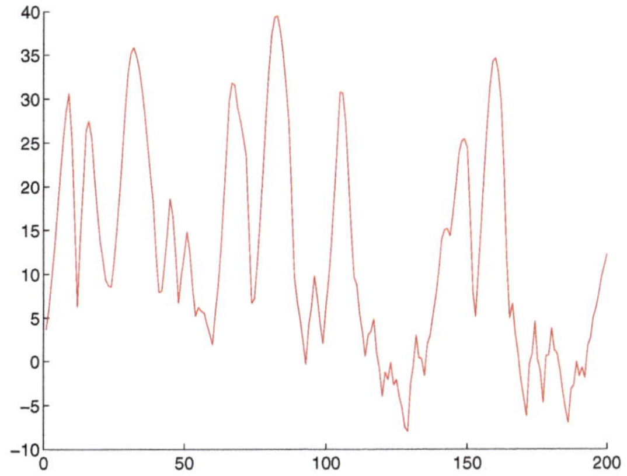

Fig. 7.18 $x_1(t)$ of chaotic multiple model and its estimate

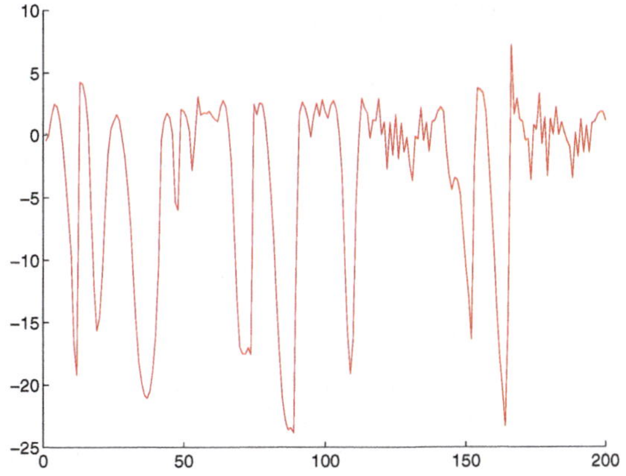

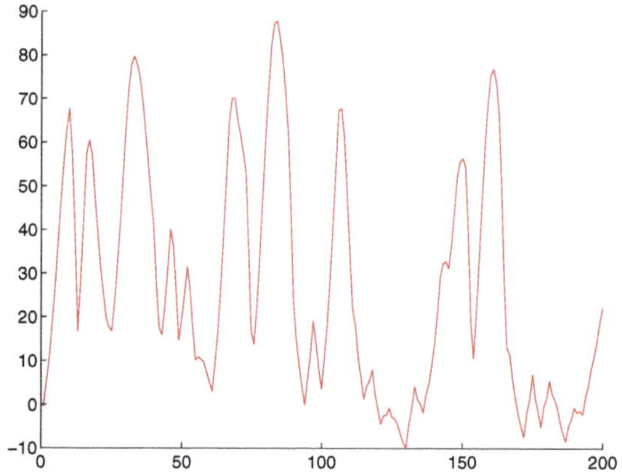

Fig. 7.19 $x_2(t)$ of chaotic multiple model and its estimate

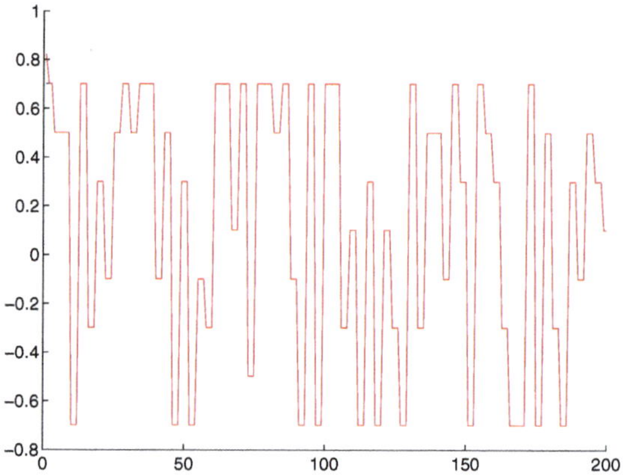

Fig. 7.20 Estimated message $\hat{u}(t)$

7.7 Conclusion

In this chapter we have shown how to use multiple model approach, multiple observer and $\mathcal{LMI}$ formulation for chaotic system reconstruction. Indeed, an approach to design an observer for multiple models with unknown inputs affecting both the state and the output of the system is proposed. Uncertainties on state matrices are also considered. Sufficient conditions to design the proposed structure of observer are given in $\mathcal{LMI}$ terms under linear equality constraints. To improve the performances of the proposed unknown inputs multiple observer, poles assignment in $\mathcal{LMI}$ regions is also addressed for continuous-time and discrete-time. It is shown that this approach can be used in chaotic communications in the sense of signal masking and encryption. Moreover, the hidden message may be embedded in the state/output of the drive (chaotic) system which enhances the design flexibility. The designed unknown inputs multiple observer is shown to be satisfactory for message (unknown input) estimation and for chaotic system reconstruction.

References

1. Akhenak, A., Chadli, M., Ragot, J., Maquin, D.: Unknown input multiple observer based approach: application to secure communication. In: 1st IFAC Conference on Analysis and Control of Chaotic Systems, Reims, France, June 28-30 (2006)
2. Alvares, G., Montoya, F., Romera, M., Pastor, G.: Breaking parameter modulated chaotic secure communication system. Chaos, Solitons & Fractals 21(4), 783–787 (2004)
3. Boutayeb, M., Darouach, M., Rafaralahy, H.: Generalized State-Space Observers for Chaotic Synchronization and Secure Communication. IEEE Trans Circ. Syst. Fund. Theor. Appl. 49(3), 345–349 (2002)
4. Boyd, S., Ghaoui, E., Feron, E., Balakrishnan, V.: Linear matrix inequalities in systems and control theory. SIAM, Philadelphia (1994)
5. Chadli, M., Maquin, D., Ragot, J.: An LMI formulation for output feedback stabilisation in multiple model approach. In: IEEE 41th Conference on Decision Control, USA, December 10-13 (2002)
6. Chadli, M., Akhenak, A., Ragot, J., Maquin, D.: On the Design of Observer for Unknown Iinputs Fuzzy Models. Int. J. Contr. Autom. Syst. 2(1), 113–125 (2008)
7. Chadli, M., Akhenak, A., Ragot, J., Maquin, D.: State and Unknown Input Estimation for Discrete Time Multiple Model. J. Franklin Inst. (2009), doi:10.1016/j.jfranklin.2009.02.011
8. Chen, M., Zhou, D., Shang, Y.: A new observer-based synchronization scheme for private communication. Chaos, Solitons & Fractals 24, 1025–1030 (2005)
9. Chilali, M., Gahinet, P.: H_∞ Design with pole placement constraints: an $\mathcal{LMI}$ approch. IEEE Transactions on Automatic Control 41(3), 358–367 (1996)
10. Darouach, M., Zasadzinski, M., Xu, S.: Full-order observers for linear systems with unknown inputs. IEEE Trans. Automatic Control 39, 606–609 (1994)
11. Edwards, C., Spurgeon, S., Patton, J.: Sliding mode observers for fault detection and isolation. Automatica 36(4), 541–553 (2000)
12. Floquet, T., Barbot, J.: A sliding mode approach of unknown input observers for linear systems. Decis. Contr., 1724–1729 (2004)
13. Guan, Y., Saif, M.: A novel approach to the design of unknown input observers. IEEE Trans. Automat. Contr. 36(5), 632–635 (1991)
14. Ha, Q., Trinh, H.: State and input simultaneous estimation for a class of nonlinear systems. Automatica 40(10), 1779–1785 (2004)
15. Johansson, M., Rantzer, A., Arzén, K.: Piecewise quadratic stability of fuzzy systems. IEEE Trans. Fuzzy Syst. 7(6), 713–722 (1999)
16. Kim, E., Lee, H.: New approaches to relaxed quadratic stability condition of fuzzy control systems. IEEE Trans. on Fuzzy Sets 8(5), 523–534 (2000)
17. Koenig, D.: Unknown input proportional multiple-integral observer design for descriptor systems: application to state and fault estimation. IEEE Transactions on Automatic Control 5(2), 213–217 (2005)
18. Li, C., Liao, X., Wong, K.: Lag synchronization of hyperchaos with application to secure communications. Chaos, Solitons & Fractals 23, 183–193 (2005)
19. Lin, S., Wang, P.: Unknown input observers for singular systems designed by eigenstructure assignment. J. Franklin Inst. 340(1), 43–61 (2003)
20. Murray-Smith, R., Johansen, T.: Multiple model approaches to modelling and control. Taylor & Francis, Abington (1997)
21. Rao, C., Mitra, S.: Generalized Inverse of Matrices and its Applications. Wiley, Chichester (1971)

22. Syrmos, V.: Computational observer design techniques for linear systems with unknown inputs using the concept of transmission zeros. IEEE Transactions on Automatic Control 38(5), 790–794 (1993)
23. Takagi, T., Sugeno, M.: Fuzzy identification of systems and its application to modelling and control. IEEE Trans. Syst. Man Cybern. 15(1), 116–132 (1985)
24. Tanaka, K., Wang, H.: Fuzzy Control Systems Design and Analysis: A linear Matrix Inequality Approach. John Wiley & Sons, Inc., Chichester (2001)
25. Vandenberghe, L., Boyd, S.: Semidefinite programming. SIAM Review 38(1), 49–95 (1996)
26. Xiaodiong, L., Qingling, Z.: New approach to H_∞ controller designs based on observers for T-S fuzzy systems via $\mathcal{LMI}$. Automatica 39, 1571–1582 (2003)
27. Yang, F., Wilde, R.: Observers for linear systems with unknown inputs. IEEE Trans. Automatic Control 33, 677–681 (1988)
28. Zhou, K., Doyle, J.: Essentials Of Robust Control. Prentice Hall, Englewood Cliffs (1998)

Chapter 8
Evolutionary Reconstruction of Chaotic Systems

Ivan Zelinka and Ales Raidl

Abstract. This chapter discusses the possibility of using evolutionary algorithms for the reconstruction of chaotic systems. The main aim is to show that evolutionary algorithms are capable of the reconstruction of chaotic systems without any partial knowledge of internal structure, i.e. based only on measured data. Five different evolutionary algorithms are presented and tested in a total of 13 and 12 versions in two different versions of experiments. System selected for numerical experiments here is the well-known logistic equation. For each algorithm and its version, 100 repeated simulations were conducted. According to obtained results it can be stated that evolutionary reconstruction is an alternative and a promising way as to how to identify chaotic systems.

8.1 Introduction

Identification of various dynamical systems is vitally important in theory and in practical applications. A rich set of various methods for dynamical system identification has been developed. In the case of chaotic dynamics, an example is the well-known reconstruction of chaotic attractor based on research of [35] who has shown that, after the transients have died out, one can reconstruct the trajectory on the attractor from the measurement of a single component. Because, the entire trajectory contains too much information, a series of papers by [12], [8] is introduced

Ivan Zelinka
Tomas Bata University in Zlin, Faculty of Applied Informatics, Nad Stranemi 4511, Zlin 76001, Czech Republic
and
VSB-TUO, Faculty of Electrical Engineering and Computer Science, 17. listopadu 15, 708 33 Ostrava-Poruba, Czech Republic
e-mail: zelinka@fai.utb.cz

Ales Raidl
Charles University, Faculty of Mathematics and Physics,V Holesovickach 2, 180 00 Prague 8, Czech Republic
e-mail: ales.raidl@mff.cuni.cz

I. Zelinka et al. (Eds.): Evolutionary Algorithms and Chaotic Systems, SCI 267, pp. 265–291.
springerlink.com

to show a set of averaged coordinate invariant numbers (generalized dimensions, entropies, and scaling indices) by which different strange attractors can be distinguished. The method presented here is based on evolutionary algorithms (EAs), see [1], which allows the reconstruction not only of chaotic attractors as a geometrical object, but also their mathematical description. All those techniques belong to the class of genetic programming techniques; see [17],[18]. Generally, when it is used on data fitting, these techniques are called symbolic regression (SR). The term symbolic regression (SR) represents a process, by which measured data is fitted by a suitable mathematical formula such as $x_2 + C$, $sin(x) + 1/e^x$, etc., Mathematically, this process is quite well known and can be used when data of an unknown process is obtained. Historically, SR has been in the preview of manual manipulation, however during the recent past, a large inroad has been made through the use of computers. Generally, there are two well-known methods, which can be used for SR by means of computers. The first one is called genetic programming (GP), [17], [18] and the other is grammatical evolution, [22], [31]. The idea as to how to solve various problems using SR by means of EA was introduced by John Koza, who used genetic algorithms (GA) for GP. Genetic programming is basically a symbolic regression, which is done by the use of evolutionary algorithms, instead of a human brain. The ability to solve very difficult problems is now well established, and hence, GP today performs so well that it can be applied, for example to synthesize highly sophisticated electronic circuits, [19]. In the last decade of the 20th century, C. Ryan developed a novel method for SR, called grammatical evolution (GE). Grammatical evolution can be regarded as an unfolding of GP due to some common principles, which are the same for both algorithms. One important characteristic of GE is that it can be implemented in any arbitrary computer language compared with GP, which is usually done (in its canonical form) in LISP. In contrast to other evolutionary algorithms, GE was used only with a few search strategies, for example with a binary representation of the populations in [23]. Another interesting investigation using symbolic regression was carried out by [14] working on Artificial Immune Systems or/and systems which are not using tree structures like linear genetic programming (full text is at https://eldorado.uni-dortmund.de/bitstream /2003/20098/2/Brameierunt.pdf) and another similar algorithm to AP, Multi Expression Programming (see http://www.mep.cs.ubbcluj.ro/). Simply put, evolutionary algorithm simulates Darwinian evolution of individuals (solutions of given problem) on a computer and are used to estimate-optimize numerical values of defined cost function. Methods of GP are able to synthesize in an evolutionary way complex structures like electronic circuits, mathematical formulas etc. from basic set of symbolic (nonnumeric) elements. In this chapter, analytic programming (AP) is applied, see [43], [36], [37], [41], [42] for the identification of selected chaotic system. Identification is not done on the "level" of strange attractor reconstruction, but it produces a symbolic-mathematical description of the identified system. Investigation reported here is a continuation of research done in [43] or extended study reported in Chapter 11.

Synthesis, identification and control of complex dynamical systems are usually extremely complicated. When classics methods are used, some simplification is required which tends to lead to idealized solutions that are far from reality. In contrast, the class of methods based on evolutionary principles is successfully used to solve this kind of problems with a high level of precision. In this chapter an alternative method of evolutionary algorithms, which has been successfully proven by many experiments like chaotic systems synthesis, neural network synthesis or electrical circuit synthesis. This chapter discusses the possibility of using evolutionary algorithms for the identification (reconstruction) of chaotic systems. The main aim of this work is to show that evolutionary algorithms are capable of reconstruction of chaotic systems without any partial knowledge of internal structure, i.e. based only on measured data. Five different evolutionary algorithms are presented and tested here in a total of 13 and 12 versions. The system selected for numerical experiments here is the well-known logistic equation for discrete systems and Lorenz attractor for continuous systems. For each algorithm and its version, 100 repeated simulations were conducted. According to obtained results it can be stated that evolutionary reconstruction is an alternative and promising way as to how to identify chaotic systems.

8.2 Motivation

Motivation of this investigation is quite simple. As mentioned in the introduction, evolutionary algorithms are capable of hard problem solving. Numerous examples on evolutionary algorithms can be easily found. Evolutionary algorithms use with chaotic systems is done for example in [30] where EAs has been used on local optimization of chaos, [27] for chaos control with use of the multi-objective cost function or in [28] and [29], where evolutionary algorithms have been studied on chaotic landscapes. A slightly different approach of evolutionary algorithms is presented in [43], where selected algorithms were used to synthesize artificial chaotic systems. In [39], [40], EAs has been successfully used for real-time chaos control and in [34] and [44] EAs was used for the optimization of Chaos Control. Other examples of evolutionary algorithms usage can be found in [6] which developed statistically robust evolutionary algorithms, and on the opposite side [11] used evolutionary algorithms for fuzzy power system stabilizer which has been applied on single-machine infinite bus system and multi-machine power system. Other research was done by [20]. Parameters of permanent magnet synchronous motors has been optimized by particle swarm algorithm and experimentally validated on the servomotor. In [5], swarm intelligence has been used for IIR filter synthesis and [26] applied co-evolutionary particle swarm optimization (CoPSO) approach for the design of constrained engineering problems, particularly for pressure vessel, compression spring and welded beam. The main question in the case of this chapter is if EAs are able to identify chaos in symbolic i.e. mathematical description. All experiments here were designed to check and either affirm or negate this idea.

8.3 Chaos System Reconstruction – Classical Methods

8.3.1 Reconstruction Based on Time Series Analysis

Control of chaos, calculation of quantifiers of chaos etc. requires the trajectories of the dynamic system in the phase space to be examined. This does not pose a serious problem if the control equations of the system are explicitly given. In actual experimental practice, however, one is often faced with the fact that the time sequence of one or, more favorably, more than one variable is measured while the equations of the system are unknown. For instance, only air temperature or air pressure is measured. Such a time series can be interpreted as a projection of the trajectory into one of the axes of the phase space. The task is then to reconstruct the trajectory in the phase space based on the measured time series of the scalar quantity. Under some assumptions, we are really able to reconstruct substantial properties of the system dynamics. The process of estimation of the chaos quantifier values thus separates into two main steps, i.e.:

- **reconstruction** of the trajectory in the phase space on a chaotic attractor, and
- **calculation** of the chaos descriptors itself.

Both steps can be subsequently used for additional modeling of the system, such as nonlinear prediction or reduction of noise in a signal, etc. For those purposes the so called reconstruction of the trajectory in the phase space is used. This will be, although a little bit imprecisely, abbreviated to "phase space reconstruction".

Time delay method is the approach to phase space reconstruction which is most frequently used. Let us start from a situation where a scalar quantity $x(t_i), i = 1,...,N$ is measured at uniformly distributed time moments $t_1, t_2 = t_1 + \Delta t, ..., t_i = t_1 + (i-1)\Delta t, ..., t_N$. In the early 1980s, Pakard, Crutchfield, Farmer and Shaw [24], Takens [35] and, according to [7], also Ruelle independently proposed constructing an m-dimensional signal as follows:

$$
\begin{aligned}
&X(t_1) = [x(t_1), x(t_1+\tau), x(t_1+2\tau), ..., x(t_1+(m-1)\tau)] \\
&X(t_2) = [x(t_2), x(t_2+\tau), x(t_2+2\tau), ..., x(t_2+(m-1)\tau)] \\
&\vdots \\
&X(t_i) = [x(t_i), x(t_i+\tau), x(t_i+2\tau), ..., x(t_i+(m-1)\tau)], \qquad i = 1,...,M
\end{aligned}
\tag{8.1}
$$

where m is the dimension of immersion, τ is a suitable time delay and $M = N-(m-1)\tau$. The quantities m and τ together are called immersion parameters. Under rather general assumptions, dynamics reconstructed through the system of eq. (8.1) is equivalent to the system dynamics on the attractor in the initial phase space. This equivalence is understood as the identity of characteristic invariants between the initial and reconstructed attractors. As regards to the full formulation of the immersion theorem, on which the reconstruction eq.(8.1) is based, the interested reader is referred to the original Takens' paper [35] or its extension presented by Sauer, York and Casdagli [33].

Unknown parameters in eq. (8.1) include time delay τ and immersion dimension m. For the latter it has been proven [35] that it is sufficient if $m \geq 2D+1$ and if the attractor is a smooth compact manifold of dimension D; in this case, D takes integer values. However, it is typical where an attractor which has no manifold and has a fractal structure is to be reconstructed. For such situations, Takens' theorem was generalized by the above authors - Sauer, York and Casdagli [33]. According to them, it is sufficient if $m > 2d_c$ where d_c is the capacity of the attractor. In some special cases, the condition for the immersion dimension can be made even less stringent. For the calculation of the correlation dimension d_2 it is even sufficient that $m > d_2$ [32]. As regards to the time delay τ, the immersion theorem does not put any special requirement on its choice, except for the necessity to exclude cases with periodic orbits with periods of τ , 2τ ... etc.

It should be noted that the immersion theorem works absolutely precisely with infinitely long time series. The above requirements for the time delay and for the immersion dimension apply to such cases. In reality, however, a researcher works with a finite volume of data involving some error - error of measurement and/or rounding error introduced by the computer. This should be borne in mind when determining immersion parameters based on experimental data. In fact, it appears that an inappropriate choice of the immersion parameters can affect the result of the reconstruction of the system dynamics substantially and can result in a wrong interpretation of the results of estimation of the attractor's characteristic invariants.

The time delay method represented by Scheme (8.1) is not the only way to construct a multidimensional signal from a one-dimensional time series. Immersion theorems [35], [32] enable us, within trajectory reconstruction in a phase space, to select from a wide choice of operations including a number of smooth transformations, both with the initial time series and with the reconstructed states. This allows a number of techniques to be used, such as principal component analysis, signal differentiation and integration, linear combination of time delayed coordinates or their filtration or utilization of variables simultaneously measured at different sites.

Let us describe the application of differentiated coordinates, which have a simple physical interpretation. For the Lorenz system [21] the system of 3 differential equations (8.2) for the 3 variables x, y, z could be replaced by a single differential equation for a single variable, x, which, however, is a 3rd order quantity. In this way it is possible to pass from the phase space formed by the coordinates (x, y, z) to new coordinates, viz. $(x, \frac{dx}{dt}, \frac{\partial^2 x}{\partial t^2})$. When working with a scalar time series $x(t_i)$, $i = 1, \ldots, N$ recorded in equidistant time intervals Δt, the 1st, 2nd, ... derivatives have to be estimated numerically, e.g.

$$\begin{aligned}
&\frac{dx}{dt}(t)_i \approx \frac{1}{2\Delta t}[x(t_i+\Delta t) - x(t_i-\Delta t)] \\
&\frac{d^2x}{dt^2}(t)_i \approx \frac{1}{\Delta t^2}[x(t_i+\Delta t) + x(t_i-\Delta t) - 2x(t_i)] \\
&\frac{d^3x}{dt^3}(t)_i \approx \frac{1}{2\Delta t^3}[x(t_i+2\Delta t) - 2x(t_i+\Delta t) + 2x(t_i-\Delta t) - x(t_i-2\Delta t)] \\
&\vdots
\end{aligned} \tag{8.2}$$

or, alternatively, higher order formulas have to be used. The coordinates of point $\mathbf{X}(t_i)$ in the phase space will then be

$$\mathbf{X}(t_i) = [x(t)_i, \frac{dx}{dt}(t)_i, \frac{d^2x}{dt^2}(t)_i, \frac{d^3x}{dt^3}(t)_i], \tag{8.3}$$

where m is, as usual, the dimension of the reconstructed phase space satisfying the condition $m > d_c$.

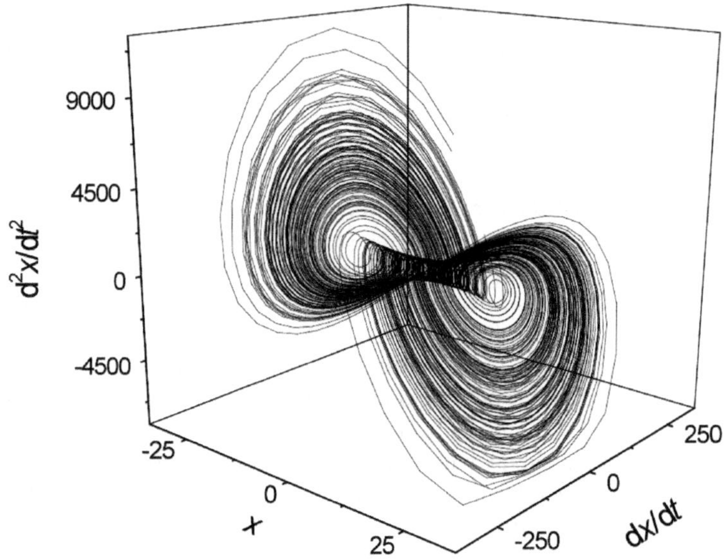

Fig. 8.1 Reconstruction of the Lorenz attractor by using differentiated coordinates.

Comparing the position of the point in the phase space so constructed with the reconstruction which was based on the time delay method (eq. (8.1)) and recalling that time delay τ is an integer multiple of the sampling step Δt, one can see that the differentiated coordinates are nothing more than a linear combination of coordinates obtained from the time delay method.

A different method of constructing the phase space, which obviates some problems encountered with differentiated coordinates and creates an orthogonal base of this space, is based on the principal component analysis approach. In the context of dynamic system analysis, this method was first used in the mid-1980s [3], [9], although its mathematical basis dates back to the early 20th century [10]. Currently, this method can be encountered under various names. Apart from the principal component analysis they include, for instance: decomposition into singular values, empirical orthogonal function, singular spectrum analysis and the Karhunen Loeve transformation [10], [15]. The different names given to the method actually reflect the different methods of covariant matrix estimation from relatively short

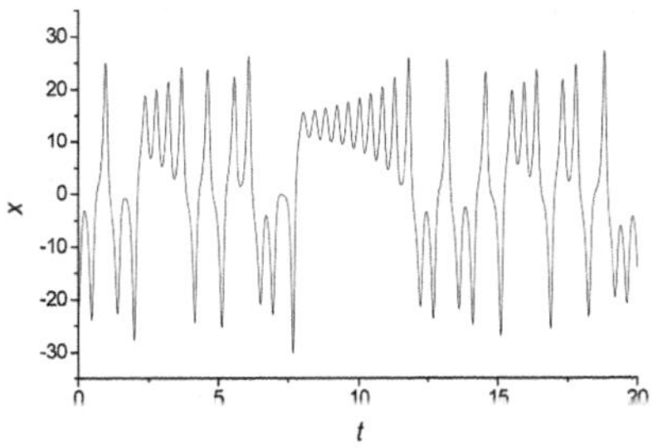

Fig. 8.2 Time development of the first four coordinates during reconstruction of the Lorenz system by using differentiated coordinates.

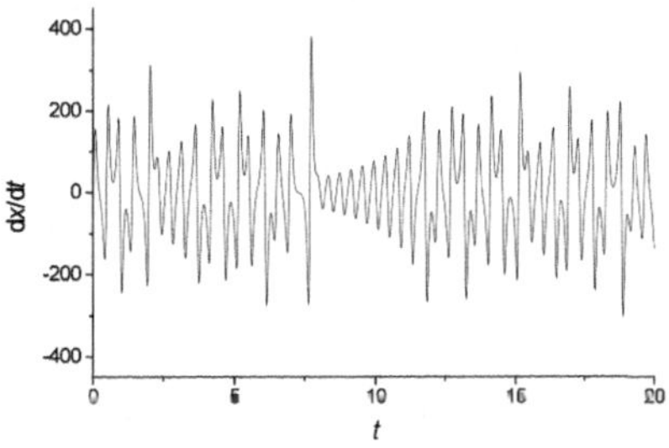

Fig. 8.3 Time development of the first four coordinates during reconstruction of the Lorenz system by using differentiated coordinates.

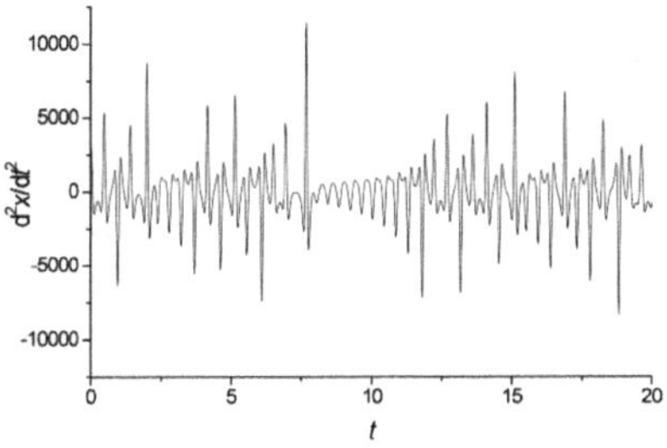

Fig. 8.4 Time development of the first four coordinates during reconstruction of the Lorenz system by using differentiated coordinates.

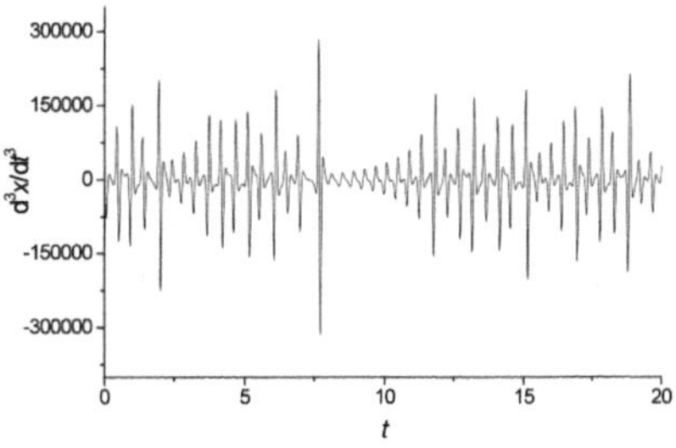

Fig. 8.5 Time development of the first four coordinates during reconstruction of the Lorenz system by using differentiated coordinates.

time series. The differences, however, are not appreciable provided that the immersion dimension m is substantially shorter than the length of the time series [15].

8.4 Evolutionary Reconstruction of Chaotic Systems

Another approach entirely different from classical methods (see previous section or Chapter 7), which is demonstrated in this chapter, is the use of evolutionary algorithms. They are applied on selected examples to demonstrate how evolutionary algorithms can be applied to the reconstruction of chaotic systems. The first example uses data from bifurcation diagram (discrete systems) to synthesize a suitable solution and the second one is using measured time series to partially reconstruct the mathematical description of the Lorenz attractor.

8.4.1 Problem Selection, Used Algorithms and Computer Technology

Based on previous successful experiments of [43], the well-known logistic equation (8.4) has been selected for experiments.

$$x_{n+1} = Ax_n(1 - x_n) \tag{8.4}$$

The selection has been made because its structure is simple, well studied and analyzed, however, this does not imply that other systems cannot be used. Main idea was to reconstruct mathematical description as described in detail in the Chapter 11, so that two bifurcation diagrams (original and synthesized solution) has been compared. The difference between them (see next section) is calculated like fitness and "says" of what quality the synthesized system is. For the experiments described here, stochastic optimization algorithms (see also Chapter 6), such as Differential Evolution (DE) [25], Self Organizing Migrating Algorithm (SOMA) [38], Genetic Algorithms (GA) [13], Simulated Annealing (SA) [16], [4] and Evolutionary Strategies (ES) [2] were selected. All experiments have been done on a special server consisting of 16 Apple XServer (2 x 2 GHz Intel Xeon, 1 GB RAM,), each with 4 CPU, so in total 64 CPUs were available for calculations. It is important to note here, that such technology was used to save time due to a large number of calculations (1300 simulations), however it must be stated that evolutionary reconstruction described here, is also solvable on a single PC. For all calculations and data processing, *Mathematica* version 7 was used.

8.4.2 The Cost Function

The cost function 8.5 has been designed so that its minimization should lead to the reconstruction of a system with the same behavior as the original system.

$$CV = \sum_{i=300}^{400} \sum_{j=200}^{300} \left| data_{i,j}^{L} - data_{i,j}^{ident} \right| \tag{8.5}$$

The cost function consists of two sums calculating the difference between two datasets. The first one, $data_{i,j}^{L}$, represents sorted data of the behavior of the logistic equation and the second one, $data_{i,j}^{ident}$, represent sorted data of the behavior of the identified system. The first sum ($i \in [300, 400]$) represents the fact that the synthesized systems has to be identified for the interval of the control parameter $A \in [3, 4]$ in which chaos is by eq. (8.4) generated. Parameter A has been changed by step 0.01, so 100 different time series was recorded. For each setting of A, 300 iterations has been done. Last 100 data-points (from 300 in total) were taken into calculation from each time series to calculate the final sum (or create bifurcation diagrams) - this is represented by the second sum ($j \in [200, 300]$). Based on previous facts, there were generated for each system, $100 \times 300 = 30\,000$ values and for cost value calculation, $100 \times 100 = 10\,000$ values were used. The minimal value that can be achieved by eq. (8.5) is 0, i.e. system with this cost value is probably an exact reconstruction of the original system. For all experiments a threshold has been set, which has been used for decision making, whether the identified system belongs to similar or exotic class of systems. System with cost value equal to 0 which were the exact reconstruction of the original system, with cost value $\in (0, 1500]$ are reported as similar reconstruction and with cost value > 1500 as exotic reconstruction.

8.4.3 Experiment Setup

Four versions of SOMA, six versions of DE, one version of GA, SA and ES have been applied in order with AP and were used for all simulations in this chapter. In Table 8.1 - Table 8.6 abbreviations of used algorithms and their setting is described. Parameters for the optimizing algorithm were set up in such a way as to reach approximately the same value of maximal cost function evaluations for all used versions. Each version of EAs has been applied 100× in order to synthesize

Table 8.1 Algorithms abbreviation

Algorithm	Version	Abbreviation
SOMA	AllToOne	S1
	AllToOneRandomly	S2
	AllToAll	S3
	AllToAllAdaptive	S4
Differential Evolution	DERand1Bin	D1
	DERand2Bin	D2
	DEBest2Bin	D3
	DELocalToBest	D4
	DEBest1JIter	D5
	DERand1DIter	D6
Genetic Algorithm		G
Evolutionary strategies (μ,λ)		ES
Simulated annealing		SA

Table 8.2 SOMA setting for 4 basic search strategies: S1, S2, S3 and S4

Algorithm	S1	S2	S3	S4
PathLength	3	3	3	3
Step	0.11	0.11	0.11	0.11
PRT	0.1	0.1	0.1	0.1
PopSize	200	200	40	40
Migrations	8	8	4	4
MinDiv	-0.1	-0.1	-0.1	-0.1
Individual Length	50	50	50	50
Max. CF Evaluations	42984	42984	42120	42120

Table 8.3 DE setting for 6 basic search strategies: D1, D2, D3, D4, D5 and D6

Algorithm	D1 - D6
NP	200
F	0.9
CR	0.3
Generations	200
Individual Length	50
Max. CF Evaluations	40000

Table 8.4 GA setting for canonical version of GA: G

Algorithm	G
PopSize	200
Mutation	0.4
Generations	100
Individual Length	50
Max. CF Evaluations	40000

Table 8.5 ES setting for search strategy: ES

Algorithm	ES
μ ,λ	200
σ	0.8
Iterations	200
Individual Length	50
Max. CF Evaluations	40000

Table 8.6 SA setting for search strategy: SA

Algorithm	SA
No. of particles	200
σ	0.5
k_{max}	66
T_{min}	0.0001
T_{max}	1000
α	0.93
Individual Length	50
Max. CF Evaluations	44600

an appropriate structure which can serve as models of the observed chaotic system. The primary aim here is not to show which version is better or worse, but to show that the EA can in reality be used for the reconstruction of chaotic systems without knowledge of internal structure or/and auxiliary information. The basic set of symbolic element (GFS) used for synthesis consist of : $A, x, +, -, *, /$.

Results from all experiments are reported in detail in the following sections. In totality, it can be stated that during all 1300 simulations (100%), original logistic equation has been identified on 73 occasions (5.6% from all simulations) and similar systems that less or more fit the behavior of the logistic equation on 186 occasions (14.3%). Therefore, in total 259 identified cases (19.92%), as given in Table 8.7.

Table 8.7 Experiment summarization

Note	Total value	%
Total number of simulations	1300	100
Exact reconstruction	73	5.6
Similar reconstruction	186	14.3
Total number of acceptable reconstruction	259	19.92

8.4.4 *Experimental Results*

8.4.4.1 Exact Reconstruction

During all simulations, the canonical version of the logistic equation has been synthesized 73× in total, (see Table 8.8). Logistic equation has been identified in 7 various versions which are clearly algebraic variation of its canonical version, i.e. after simple algebraic manipulations we get eq. 8.4, see eq. (8.6) - (8.10).

Table 8.8 Summarization of canonical version synthesis

Equation No.	Synthesized
(8.6)	14×
(8.7)	25×
(8.8)	9×
(8.9)	8×
(8.10)	11×
(8.11)	5×
(8.12)	2×
Total	73×

$$A(x - x^2) \tag{8.6}$$

$$x(A - Ax) \tag{8.7}$$

$$Ax - Ax^2 \tag{8.8}$$

$$A(1-x)x \tag{8.9}$$

$$-x(-A+Ax) \tag{8.10}$$

$$x(1-x)A \tag{8.11}$$

$$x^2(1/x-1)A \tag{8.12}$$

8.4.5 Reconstruction of Similar Systems

Beside the canonical version of the logistic equation, there has also been synthesized systems, which less or more fit the behavior of the original system. Selected examples of very good approximation of eq. (8.4) are for example systems eq. (8.13) and eq. (8.14), see for example Fig. 8.6. Significantly "worst" approximations are for example eq. 8.15 and 8.16, see Fig. 8.7. and Fig. 8.8. Corresponding cost values are given in Table 8.9 and 8.10. Minimal, maximal and average cost values of accepted similar systems (according to threshold) in this "category" are reported there. Behavior of other similar systems is reported in Fig. 8.9 - Fig. 8.16. From the given figures, it is visible that evolution has found really similar systems and their precise "evolutionary adjustment" to the logistic equation is probably only a question of better setting of evolutionary algorithm parameters.

Table 8.9 Similar systems – an overview

	Cost Value
Minimum	117.538
Average	1053.92
Maximum	1487.85

$$x\left(A-Ax+\frac{(1-A)x}{A^2(2A-x+Ax)}\right) \tag{8.13}$$

$$x\left(A-Ax+\frac{x^2}{A\left(\frac{1}{A}+A+\frac{A}{x}+Ax(A+x)\right)}\right) \tag{8.14}$$

$$\frac{A(1-x)x(x+(-A+x)(A+x))}{-A^2-x} \tag{8.15}$$

$$x\left(A-\frac{A^3x}{A+A^2+x}-\frac{A^2-2A(A-x)}{-\frac{A^2}{x}+2x}\right) \tag{8.16}$$

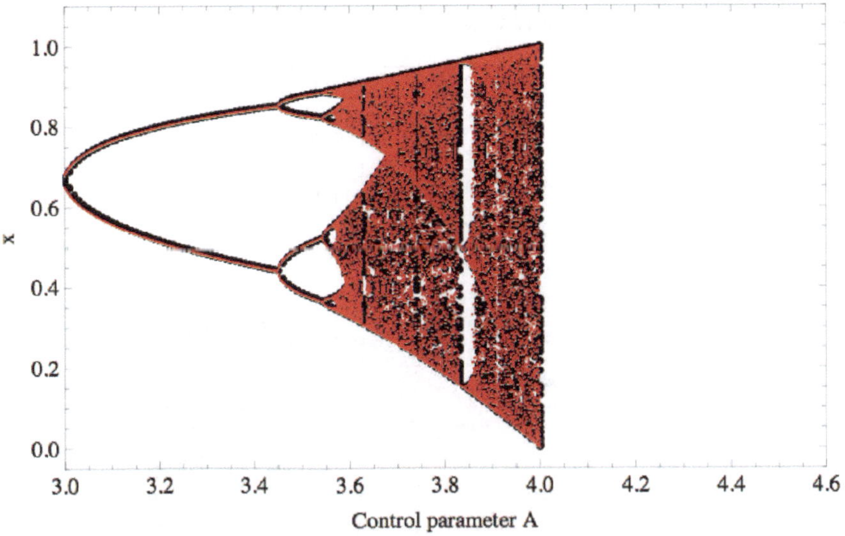

Fig. 8.6 The best synthesized solution, see eq. (8.13) and eq. (8.14). Red (thin) points represent the canonical logistic equation; black (thick) points represent the synthesized system.

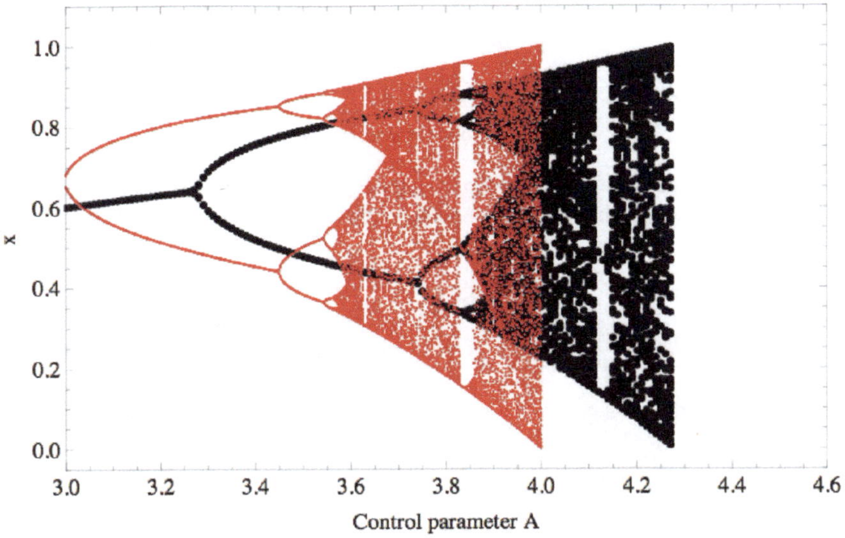

Fig. 8.7 Another solution, basically the same behavior of eq. (8.4), only shifted along axis x, see eq. (8.15). Red (thin) points represent the canonical logistic equation; black (thick) points represent the synthesized system.

Table 8.10 Cost values of similar systems

Equation No.	Cost Value
(8.13)	129.549
(8.14)	136.706
(8.15)	1048.72
(8.16)	993.346

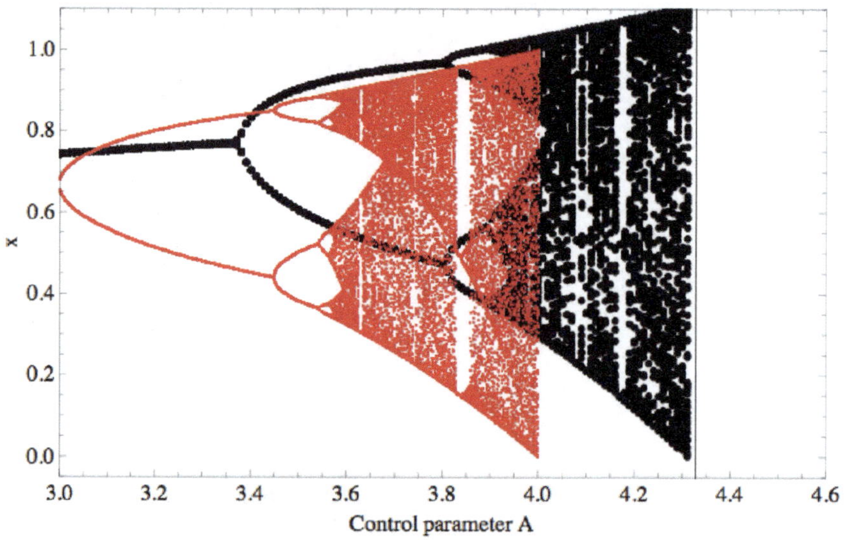

Fig. 8.8 Basically the same case as in Fig. 8.6, see eq. (8.16). Red (thin) points represent the canonical logistic equation; black (thick) points represent the synthesized system.

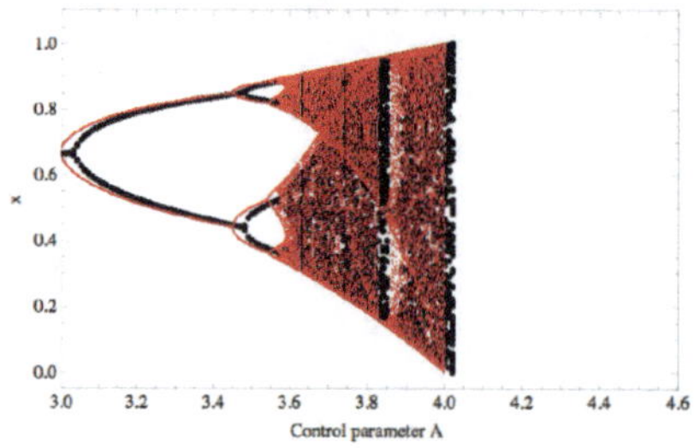

Fig. 8.9 Another similar solution.

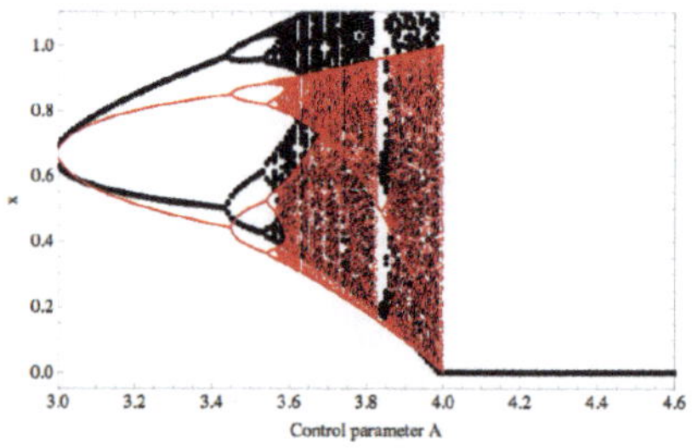

Fig. 8.10 Another similar solution.

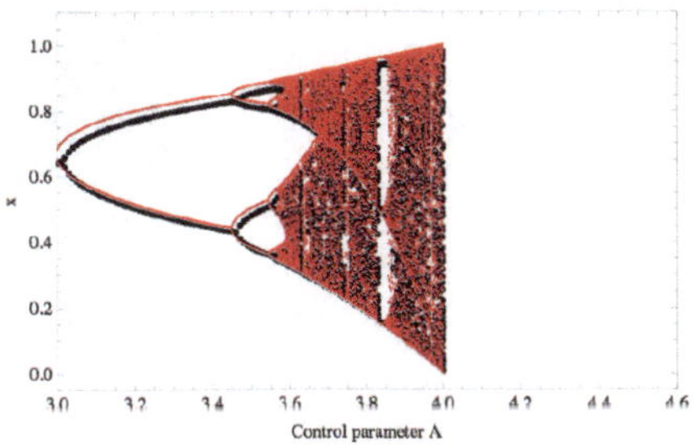

Fig. 8.11 Another similar solution.

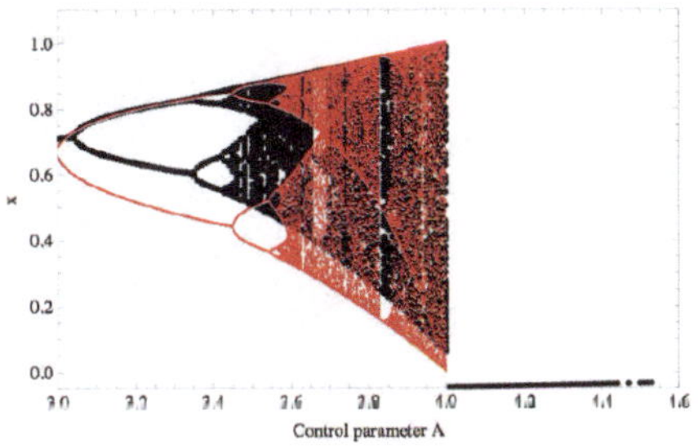

Fig. 8.12 Another similar solution.

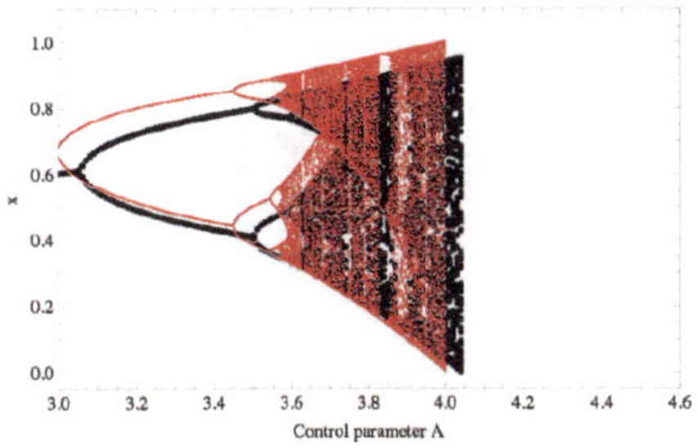

Fig. 8.13 Another similar solution.

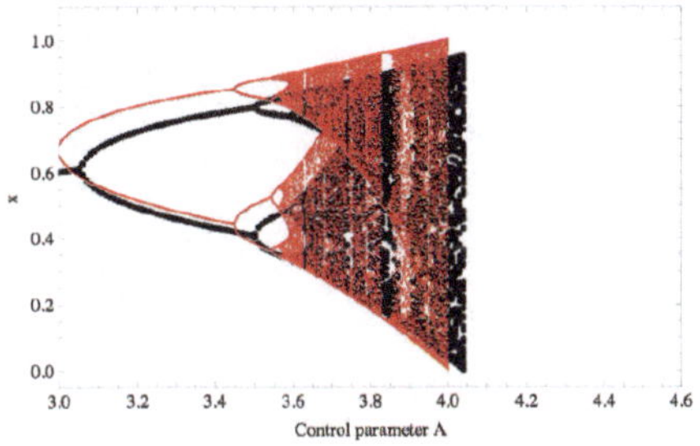

Fig. 8.14 Another similar solution.

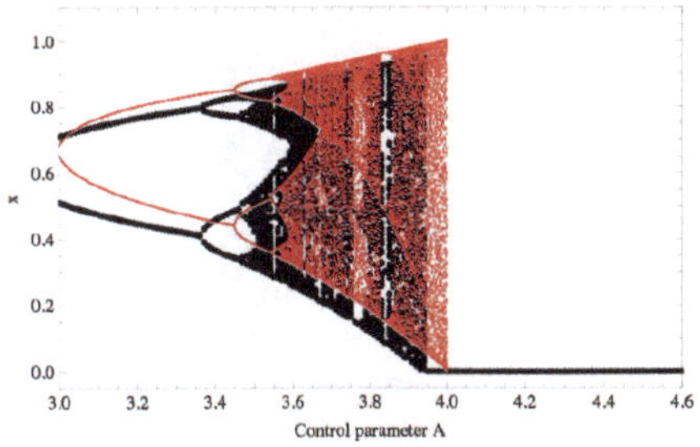

Fig. 8.15 Another similar solution.

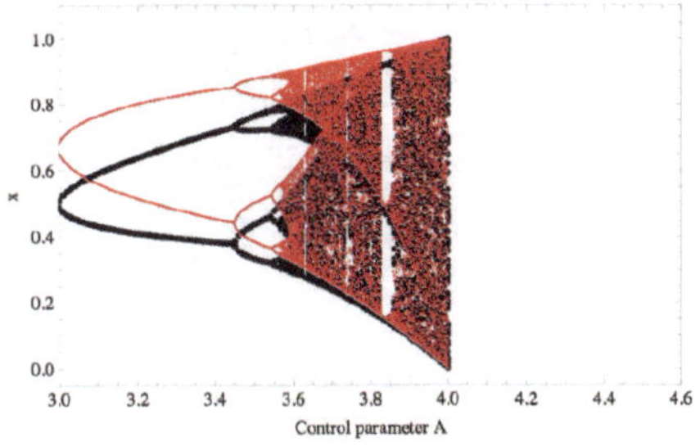

Fig. 8.16 Another similar solution.

8.4.6 *Unfinished Evolution*

During all simulations conducted, it been observed, that in many cases evolution would certainly need longer time to finish successfully the evolutionary reconstruction, like exact or similar reconstruction. Lets take a look on Fig. 8.17 - 8.24, or/and on eq. 8.17 - eq. 8.22. On the figures are depicted bifurcation diagrams, which are very similar to diagrams from logistic equation, they are only shifted along the x or/and y axes. It is clear that if evolution would run for a longer time, then the bifurcation diagrams (or better, the mathematical description in the background), like on

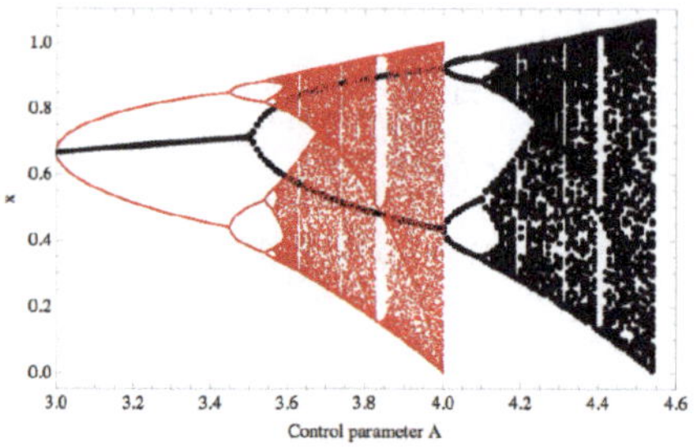

Fig. 8.17 Unfinished solution, eq. 8.17.

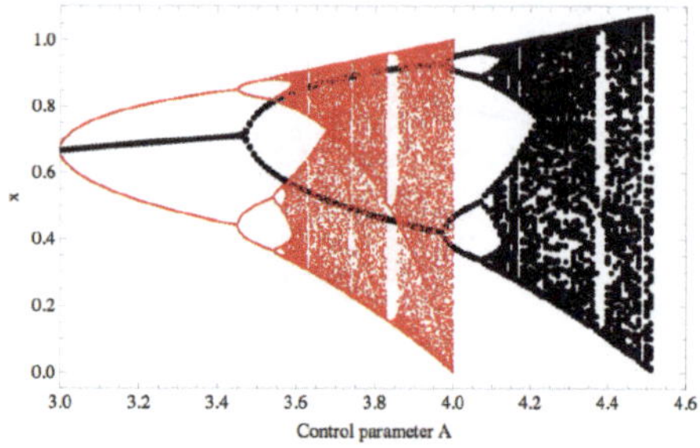

Fig. 8.18 Unfinished solution, eq. 8.18.

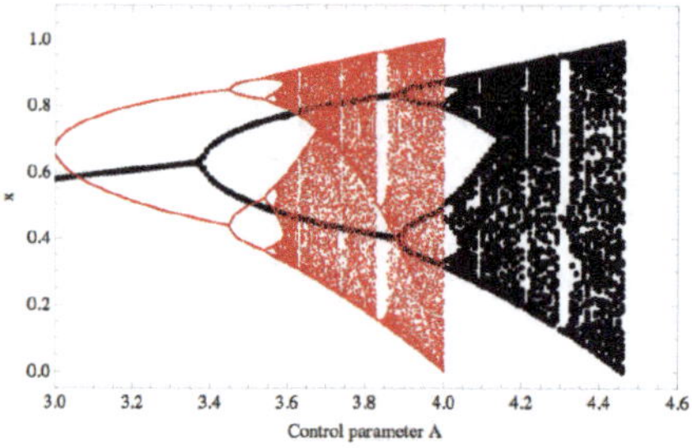

Fig. 8.19 Unfinished solution, eq. 8.19.

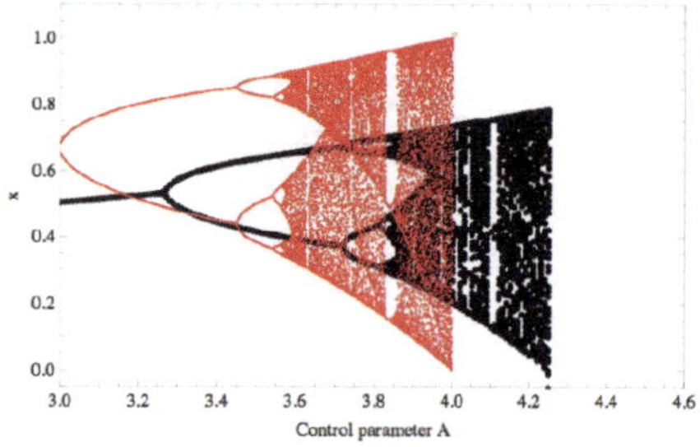

Fig. 8.20 Unfinished solution, eq. 8.20.

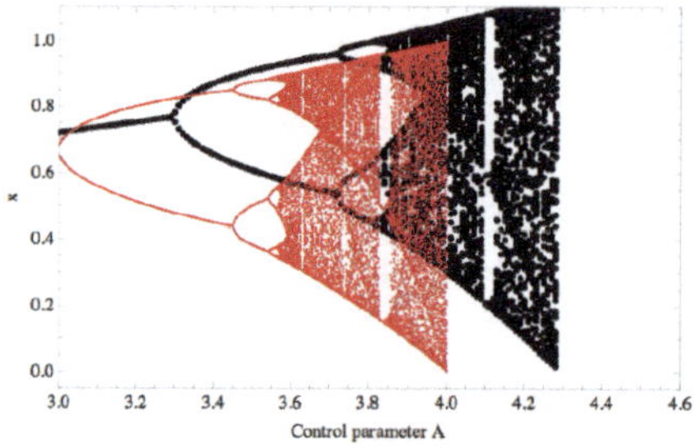

Fig. 8.21 Unfinished solution, eq. 8.21.

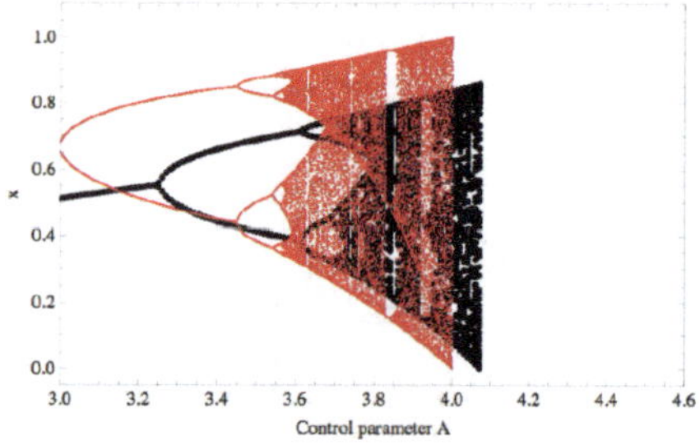

Fig. 8.22 Unfinished solution, eq. 8.22.

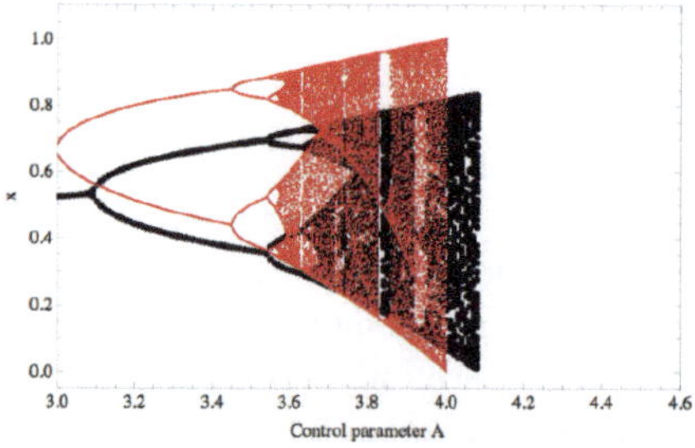

Fig. 8.23 Unfinished solution.

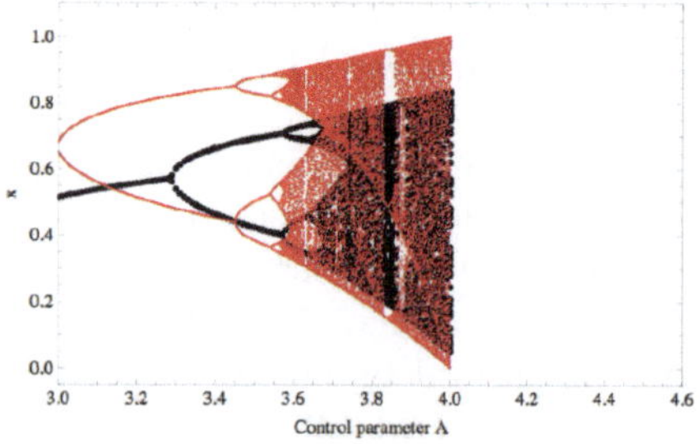

Fig. 8.24 Unfinished solution.

Fig. 8.17 - 8.24, would be better adapted to the identified one. Based on this, one can say that the above mentioned "similar" reconstruction are unfinished reconstruction with possibly very good quality (i.e. with low cost value).

$$x\left(-x(A-x)+\frac{x}{A}+A-x\right) \tag{8.17}$$

$$\frac{x\left(x-\left(\frac{A}{x}-A\right)(x-A)\right)\left(\frac{x}{(x-A)\left(-A+x^2-4x\right)}+x\right)}{A} \tag{8.18}$$

$$x\left(-x\left(-\frac{2x-Ax}{A\left(A^2-A+x\right)}+A-x\right)+A-x\right) \tag{8.19}$$

$$\frac{A\left(\frac{A}{x}+x\right)}{\frac{A\left(\frac{A}{A-x\left(\frac{x\left(\frac{x}{A}+A\right)}{2A+x}+A+x\right)}+x\right)}{x^2}+x} \tag{8.20}$$

$$-\frac{x(A+2x)\left(\frac{\frac{x}{A}-A-x}{\frac{x}{A}+A}+x\right)}{\frac{x}{A}+\frac{1}{A}+1} \tag{8.21}$$

$$x\left(\frac{\left(-Ax-\frac{x-A}{A}-A+x^2+x-1\right)\left(A(Ax+x)+A\right)}{2A^2}+A\right) \tag{8.22}$$

8.4.7 Exotic Solutions

Together with acceptable systems, other systems were also synthesized, which did not fit the threshold, mentioned in the section Cost function, i.e. its cost value was > 1500. This category is termed "exotic", i.e. systems that are very different from the logistic equation (by behavior and mathematical description), however, there is still visible a similar structure to the logistic equation. An example can be the systems given by eq. (8.23) and eq. (8.24), which had been synthesized during all 1300 experiments. For behavior of systems eq. (8.23) see Fig. 8.25, and for eq. (8.23) see Fig. 8.26. Another selected example is depicted in Fig. 8.27.

$$\frac{A-x+x^2-\frac{(A-x)x^3\left(-x+x^2\right)\left(x+A\left(2A+A^2+x\right)\right)}{A\left(1-\frac{A}{x}\right)}}{A} \tag{8.23}$$

$$\frac{x\left(2A-A\left(-x-\frac{x^2}{A}\right)\left(-2x+\frac{x}{A-\frac{1+2x}{x^2}}\right)\right)}{A} \tag{8.24}$$

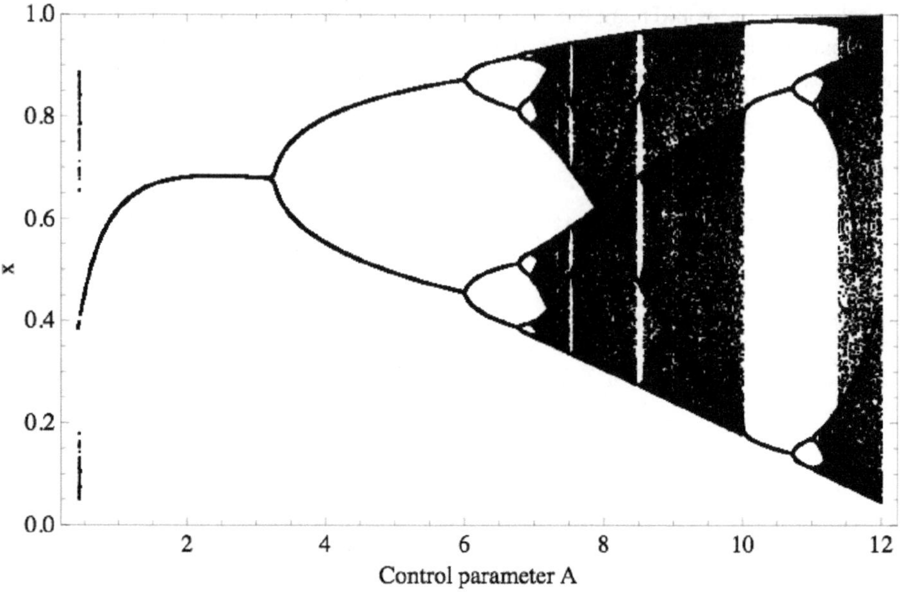

Fig. 8.25 Example of "exotic" solution, see eq. (8.23)

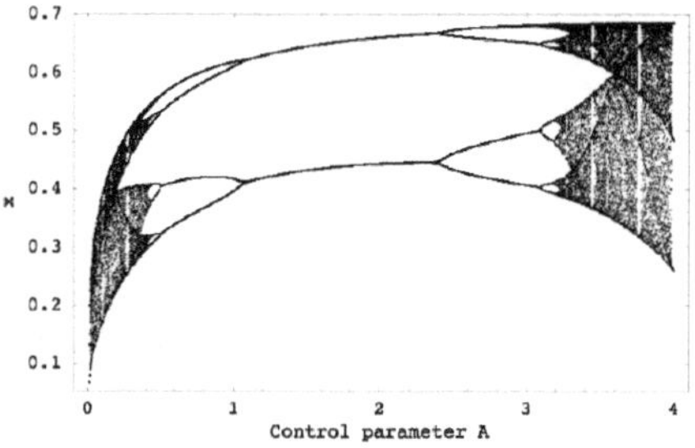

Fig. 8.26 Another "exotic" solution, see eq. (8.24).

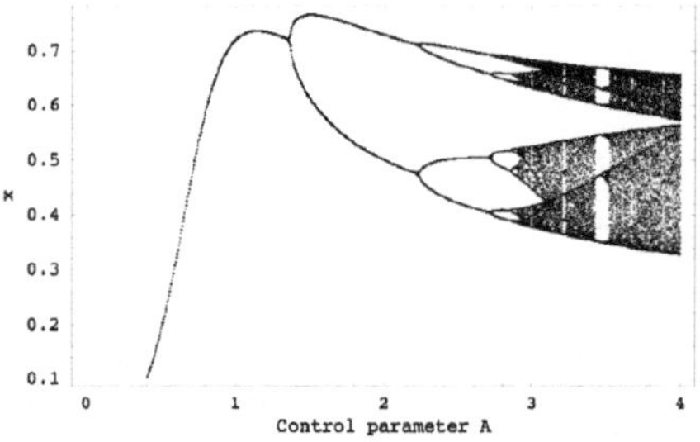

Fig. 8.27 Bifurcation diagram of Another "exotic" solution.

8.4.8 Continuous Systems: Preliminary Study

Evolutionary reconstruction of chaotic systems is certainly not restricted only to discrete systems. Methods of symbolic regression is general enough to be used also on reconstruction of continuous chaotic systems. To check this idea, the well known chaotic system has been selected - Lorenz equation, see eq. (8.25). To simplify this experiment for the first time, the third equation $\dot{z}$ has been selected to be synthesized, see eq. (8.25). Basic set of objects used in symbolic regression was $\{x(t), y(t), z(t), +, -, \times, /\}$. Total number of simulation has been set to 100 and 5 algorithms (DE, SOMA, GA, SA, ES) in all 12 versions were used in order to identify (reconstruct) by synthesis suitable solutions. In many cases the exact form of eq. $\dot{z}(t) = (x(t)y(t) - z(t))$ has been synthesized, see eq. 8.25. The remaining synthesized forms were of different form, see Table 8.18. Cost function was defined by eq. (8.26), as the difference between behavior of the original and identified system been calculated in the interval $t \in [0, 20]$ with randomly selected initial conditions. Cost value has been calculated in the interval $t \in [5, 20]$. Objective was to minimize this function to 0.

8.4.8.1 Experiment Setup

Four versions of SOMA, six versions of DE, one version of GA, and one of ES have been applied in order with AP and were used for all simulations. In Table 8.11 - Table 8.15 abbreviations of used algorithms and their setting is described. Parameters for the optimizing algorithm were set up in such a way as to reach approximately the same value of maximal cost function evaluations for all used versions. Each version of EAs has been applied 100× in order to synthesize an appropriate structure which can serve as models of the observed chaotic system.

Table 8.11 Algorithms abbreviation

Algorithm	Version	Abbreviation
SOMA	AllToAllAdaptive	S1
	AllToAll	S2
	AllToOne	S3
	AllToOneRandomly	S4
Differential Evolution	DERand1Bin	D1
	DERand2Bin	D2
	DEBest2Bin	D3
	DELocalToBest	D4
	DEBest1JIter	D5
	DERand1DIter	D6
Genetic Algorithm		G
Evolutionary strategies $(\mu + \lambda)$		ES2

Table 8.12 SOMA setting for 4 basic search strategies: S1, S2, S3 and S4

Algorithm	S1	S2	S3	S4
PathLength	3	3	3	3
Step	0.11	0.11	0.11	0.11
PRT	0.1	0.1	0.1	0.1
PopSize	100	100	100	100
Migrations	8	8	4	4
MinDiv	-0.1	-0.1	-0.1	-0.1
Individual Length	20	20	20	20

Table 8.13 DE setting for 6 basic search strategies: D1, D2, D3, D4, D5 and D6

Algorithm	D1 - D6
NP	100
F	0.9
CR	0.3
Generations	500
Individual Length	20

Table 8.14 GA setting for canonical version of GA: G

Algorithm	G
PopSize	100
Mutation	0.4
Generations	500
Individual Length	20

Table 8.15 ES setting for search strategies: ES

Algorithm	ES2
μ ,λ	100
σ	0.8
Iterations	500
Individual Length	20

8.4.8.2 Continuous Systems: Results

Results of this case study are depicted in Fig. 8.28 - 8.31 and Tables 8.16 - 8.18. Tables 8.16 - 8.17 refer to the number of cost function evaluations that has been used by EAs to obtain suitable solution. It is graphically reported in Fig. 8.28. Fig. 8.29 and 8.30 depicts two selected histograms of 15 in total to show typical performance of selected algorithms. The last figure 8.31 depict the success of used algorithms, i.e. how many times each algorithm fails or succeeds.

$$\begin{array}{l} \dot{x}(t) = -a(x(t) - y(t)) \\ \dot{y}(t) = bx(t) - x(t)z(t) - y(t) \\ \dot{z}(t) = identified\ part\ by\ EAs,\ see\ Table\ 8.18 \end{array} \tag{8.25}$$

$$CV = \sum_{t=0}^{t=20} \left|x_{t,Lorenz} - x_{t,Sythesized}\right| + \left|y_{t,Lorenz} - y_{t,Sythesized}\right| + \left|z_{t,Lorenz} - z_{t,Sythesized}\right| \tag{8.26}$$

It is clear that this approach is also usable, i.e. it can be used to synthesize continuous systems, however more extensive study is needed.

Table 8.16 Experiment summarization, continuous case, part 1.

Algorithm	D1	D2	D3	D4	D5	D6	ES2
Cost function evaluations see Fig. 8.28							
Minimum	25	16	14	74	35	11	32
Average	8601	8791	12010	13718	11787	11413	10516
Maximum	50870	39752	98324	57256	44956	50274	39602

Table 8.17 Experiment summarization, continuous case, part 2.

Algorithm	G	S1	S2	S3	S4
Cost function evaluations see Fig. 8.28					
Minimum	1	9	1048	61	59
Average	6516	11513	63907	12272	13246
Maximum	39043	79014	349478	51672	43202

Table 8.18 Exact / non-exact reconstruction.

Solution No.	$\dot{z}(t)$ reconstruction	Exact	Non-exact
1	$x(t)y(t)-z(t)$	1090	-
2	$y(t)\left(x(t)-\frac{z(t)}{y(t)}\right)$	-	58
3	$y(t)\left(x(t)-\frac{x(t)}{y(t)}\right)+x(t)-z(t)$	-	2
4	$y(t)(x(t)+z(t))-y(t)z(t)-z(t)$	-	4
5	$x(t)(x(t)+y(t))-x(t)^2-z(t)$	-	3
6	$x(t)(y(t)-x(t))+x(t)^2-z(t)$	-	2
7	$-y(t)\left(\frac{z(t)}{y(t)}-x(t)\right)$	-	5
8	$y(t)\left(x(t)-\frac{x(t)+z(t)}{y(t)}\right)+x(t)$	-	2
9	$-y(t)\left(\frac{z(t)-x(t)}{y(t)}-x(t)\right)-x(t)$	-	1
10	$y(t)\left(\frac{x(t)}{y(t)}+x(t)\right)-x(t)-z(t)$	-	2
11	$y(t)(x(t)-y(t))+y(t)^2-z(t)$	-	1
12	$(x(t)-1)y(t)+y(t)-z(t)$	-	3
13	$x(t)(y(t)-1)+x(t)-z(t)$	-	2
14	$-(1-x(t))y(t)+y(t)-z(t)$	-	1
15	$y(t)\left(x(t)-\frac{y(t)+z(t)}{y(t)}\right)+y(t)$	-	1
16	$(1-x(t))y(t)+2x(t)y(t)-y(t)-z(t)$	-	1
17	$(x(t)+1)y(t)-y(t)-z(t)$	-	3
18	$y(t)(x(t)-z(t))+y(t)z(t)-z(t)$	-	1
19	$x(t)(y(t)+z(t))-x(t)z(t)-z(t)$	-	1
20	$x(t)(y(t)+1)-x(t)-z(t)$	-	1
21	$y(t)(x(t)-y(t)-z(t))+y(t)(y(t)+z(t))-z(t)$	-	1
Total		**1080**	**95**

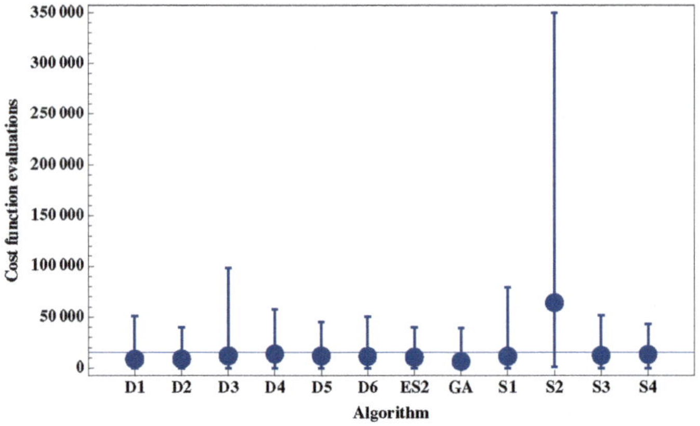

Fig. 8.28 Cost function evaluations. Thick dots are average values for each algorithm, horizontal line is average of all.

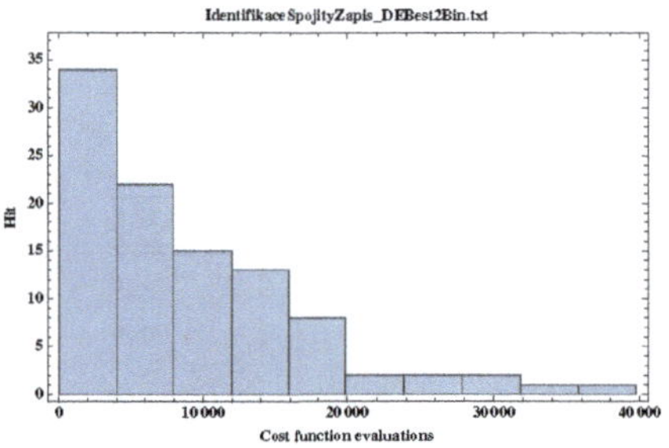

Fig. 8.29 Histogram for differential evolution algorithm, version DEBest2Bin.

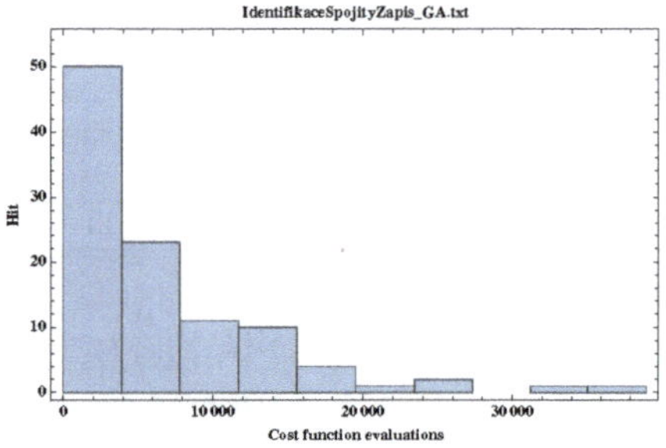

Fig. 8.30 Histogram for genetic algorithm.

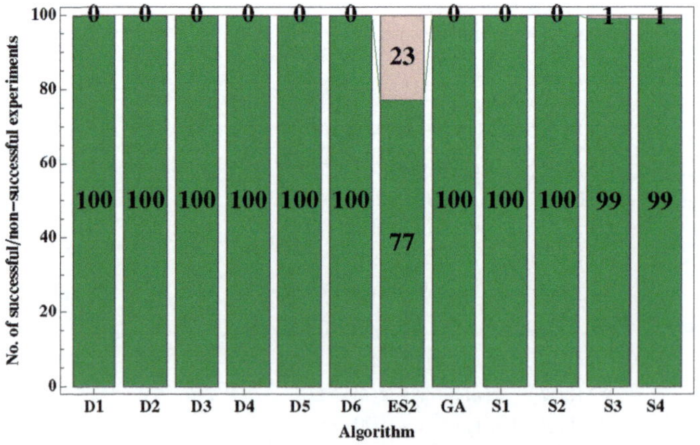

Fig. 8.31 No. of successful/non-successful reconstruction. Each bar is divided into two parts. The upper part represent number of non-successful reconstruction, the lower one successful reconstruction.

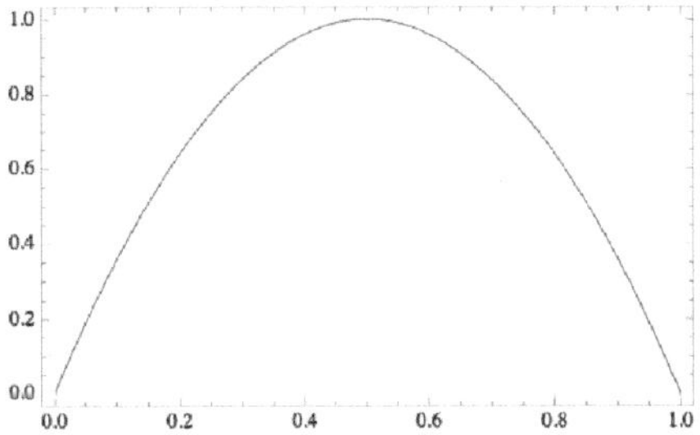

Fig. 8.32 Graph of the first part of eq. (8.27).

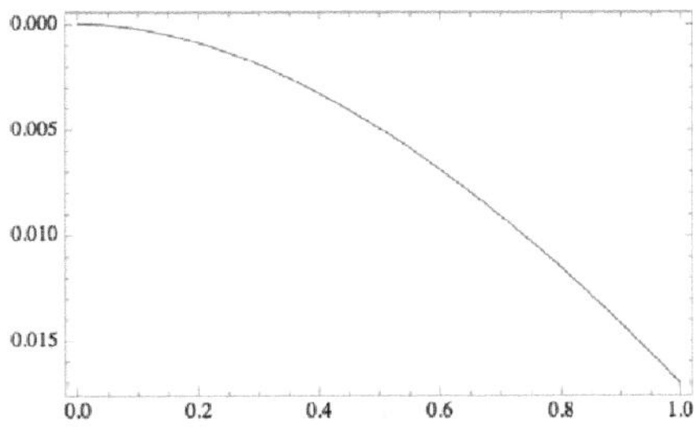

Fig. 8.33 Graph of the second part of eq. (8.27).

8.5 Conclusion

Based on recorded data and results, it may be stated that the simulations provided promising results, which shows that EAs are capable of model reconstruction of chaotic systems. In this chapter, five evolutionary algorithms in 13 (12 for continuous case) versions were used and tested. Exact descriptions of the identified systems (logistic equation, Lorenz system) as well as its variations have been identified from the results (see for example eq. (8.13), or Table 8.18). The question is why such complex equation like eq. (8.13) have similar behavior to eq. (8.4). The answer is simple. After expansion of eq. (8.13), equation (8.27) is obtained. The first part is basically the logistic equation (see Fig. 8.32). The remaining part participates on the final behavior without significant impact (see Fig. 8.33).

$$Ax - Ax^2 + \frac{x^2}{A^2\,(2A - x - Ax)} - \frac{x^2}{A\,(2A - x - Ax)} \tag{8.27}$$

Based on previously mentioned facts and all experimental results, conclusions and statements can be made for discrete system reconstruction as follows:

- **Experiment overview.** The cost function (8.5) consist of two sums where the total number of synthesized data-points was 10 000 from 30 000 (see section "Cost Function"). Based on the fact that 1300 experiments were conducted, 39 000 000 data-points and 13 000 000 of these points were used for the evaluation of all synthesized systems. The continuous case has similar behavior, see 8.26. This cost function calculates the difference between original behavior of Lorenz system and the just identified one in the time interval [0,20].
- **Number of successful reconstruction.** The results were divided into three categories for discrete system: exact, similar and exotic reconstructions. The representation is as follows: exact implies that logistic equation has been recovered in its canonical version (or its algorithmic variations), similar means that behavior of the synthesized systems was visually the same (or same and shifted along x axis) like that of the logistic equation, however with different mathematical description (see eq. (8.13), (8.14), (8.15) and eq. (8.16)). Exotic

reconstruction is partially similar to the original one. Based on all data analysis, it can be stated that a) exact form of logistic equation has been synthesized 73 times (see Table 8.8). Number of synthesized similar systems was 186. For general overview see Table 8.8.

In Table 8.18 results from continuous case are displayed. Exact description has been identified 1090× (see solution 1), another, say equivalent descriptions 58× (solution 2) etc. It is important to note that the behavior of Lorenz attractor was identical only in the interval $t \in [0,20]$. Outside this interval behavior of synthesized "Lorenz system" has been less or more divergent, which is of course obvious.

- **Used algorithms and experiment settings.** All algorithms has been initialized so that a) population size remained the same, b) cost function evaluations was similar amongst algorithms as much as possible. The first "condition" has not been followed for algorithms S3 and S4 compared to S1 and S2 (see Table 8.2). It is caused by the different internal algorithm structure for new individuals calculations. Due to this fact, condition b) has been kept with the highest priority for the chaotic discrete system identification. In the case of the Lorenz system we were interested mainly whether this idea will work or not.
- **Behavior preciseness.** It should be noted here that reconstruction has not been focused on exact behavior reconstruction for each time development of logistic equation, but on similarity of behavior via data used later for bifurcation diagrams, i.e. difference between bifurcation diagrams has been calculated for discrete systems. Despite the fact that some of them were precisely estimated, it is our duty to say that sometimes, very rarely and only for special setting of parameter A, trajectories of synthesized systems were running to infinity. To avoid this "side effect" the above-mentioned cost function should contain in future a penalization for such kinds of effects. From Fig. 8.9 - 8.16, it is also visible that a little bit longer time is needed for better estimation of system description. For some identified systems it has been observed that while $A \in [3,4]$ the behavior is identical or very similar with that of logistic equation was produced, whereas other values of A (for example negative) other chaotic behavior were generated, see for example $-A+(-(1/A)+A/x-2x)x+(A-x)x$ with $A \in [0,1]$. Concerning to the identified Lorenz system, as mentioned before, in the time interval $t \in [0,20]$ the difference between original and identified Lorenz was minimal.
- **Problem complexity and algorithm performance.** Lets take into consideration only discrete system. Based on the fact that individual can consist of 50 symbolic elements, there are 3.04×10^{64} possible combinations of synthesized structures - systems, including senseless combinations. This is of course only the theoretical number, because some combinations will be avoided due to the process of synthesis (only mathematically acceptable functions with appropriate number of arguments, ... structures are synthesized). However, in layman's terms, it can be stated that all 259 synthesized solutions (from 1300 in total) represents $8.51 \times 10^{-61}\%$ of such defined searched space. If we will follow maximal allowed number of cost function evaluations (see Table 8.2 -

Table 8.6, 534 808 cost function evaluations, i.e. tested solutions) then evolution searched maximally $1.74 \times 10^{-57}\%$ of the search space. Lets take a simplified time point of view. When for example MacBook, 2.33 MHz Intel Core Duo with 3 GB RAM is used, then one cost function evaluation needs (if we omit time needed for formula synthesis) approx. 0.3659 s. Then to evaluate all possible combinations by simple enumeration would take approx. 3.52×10^{56} years. This is 2.35×10^{46} longer than the expected lifespan of our universe. All those numbers clearly shows that EAs are powerful enough to handle such tasks and obtained results are not simply a matter of randomness.

- **Other evolutionary techniques.** In this chapter, the so-called analytic programming has been used, however we have to say that another and more well known techniques like genetic programming, see [17], [18] or grammatical evolution, see [22], should give similar results as reported here.

Conclusions about preliminary and simplified study of reconstruction of Lorenz system is:

- **Number of successful reconstruction.** The number of the same reconstructed form of the $\dot{z}(t)$ is reported in the Table 8.18 and is quite large. It shows that EAs were able to reconstruct $\dot{z}(t)$ in its exact form. On the other side, in Table 8.18 it is visible that EAs also has found another similar solutions, which, in the interval $t \in [0, 20]$ has fit the behavior of reconstructed system very well. Behind this interval, the trajectories of such systems usually runs out of attractor domain.
- **Simplifications.** For preliminary study on continuous system, such simplification that only $\dot{z}(t)$ has to be reconstructed has been used. Based on the performance of used algorithms in this and other chapters, it is logical to expect, that EAs should be able to identify all three part of Lorenz system. To confirm such statement, it is however necessary to do more extensive research.

In conclusion, it has to be stated that, a) EAs use on chaos identification is a promising direction of research; b) to increase the number of successful identifications (see Table 8.7, Table 8.8) the cost function or/and algorithm settings should be improved.

Acknowledgements. This work was supported by grant No. MSM 7088352101 of the Ministry of Education of the Czech Republic and by grants of the Grant Agency of the Czech Republic GACR 102/09/1680.

References

1. Back, T., Fogel, D., Michalewicz, Z.: Handbook of Evolutionary Computation. Institute of Physics, London (1997)
2. Beyer, H.: Theory of Evolution Strategies. Springer, New York (2001)
3. Broomhead, D., King, G.: Extracting qualitative dynamics from experimental data. Physica D 20, 217 (1986)

4. Cerny, V.: Thermodynamical approach to the traveling salesman problem: An efficient simulation algorithm. J. Opt. Theory Appl. 45(1), 41–51 (1985)
5. Das, S., Konar, A.: A swarm intelligence approach to the synthesis of two-dimensional IIR filters. Eng. Appl. Artif. Intell. 20(8), 1086–1096 (2007)
6. Dashora, Y., Sanjeev, K., Nagesh, S., Tiwarid, M.: Improved and generalized learning strategies for dynamically fast and statistically robust evolutionary algorithms. Eng. Appl. Artif. Intell (2007), doi:10.1016/j.engappai.2007.06.005
7. Drazin, P., Kind, G. (eds.): Interpretation of time series from nonlinear Systems. Special issue of Physica D 58 (1992)
8. Eckmann, J., Procaccia, I.: Fluctuation of Dynamical Scaling Indices in Non-Linear Systems. Phys. Rev. 34A, 659 (1986)
9. Fraedrich, K.: Estimating the dimension of weather and climate attractors. J. Atmom. Sci. 43, 419 (1986)
10. Galka, A.: Topics in nonlinear time series analysis with implications for EEG analysis. World Scientific, Singapore (2000)
11. Hwang, G.-H., Kim, D.-W., Lee, J.-H., An, Y.-J.: Design of fuzzy power system stabilizer using adaptive evolutionary algorithm. Eng. Appl. Artif. Intell. 21(1), 86–96 (2007)
12. Halsey, T., Jensen, M., Kadanoff, L., Procaccia, I., Schraiman, B.: Fractal Measures and Their Singularities: the Characterization of Strange Sets. Phys. Rev. 33A, 1141 (1986)
13. Holland, J.: Adaptation in Natural and Artificial Systems. Univ. Michigan Press, Ann Arbor (1975)
14. Johnson, C.: Artificial immune systems programming for symbolic regression. In: Ryan, C., Soule, T., Keijzer, M., Tsang, E., Poli, R., Costa, E. (eds.) EuroGP 2003. LNCS, vol. 2610, pp. 345–353. Springer, Heidelberg (2003)
15. Kantz, H., Schreiber, T.: Nonlinear time series analysis. Cambridge University Press, Cambridge (1997)
16. Kirkpatrick, S., Gelatt, C., Vecchi, M.: Optimization by simulated annealing. Science 220(4598), 671–680 (1983)
17. Koza, J.: Genetic Programming II. MIT Press, Cambridge (1998)
18. Koza, J., Bennet, F., Andre, D., Keane, M.: Genetic Programming III. Morgan Kaufmann, San Francisco (1999)
19. Koza, J., Keane, M., Streeter, M.: Evolving Inventions. Sci. Am., 40–47 (2003)
20. Liu, L., Wenxin, L., David, A.: Particle swarm optimization-based parameter identification applied to permanent magnet synchronous motors. Eng. Appl. Artif. Intell. (2007), doi:10.1016/j.engappai.2007.10.002
21. Lorenz, E.: Deterministic non-periodic flow. J. Atmos. Sci. 20, 20 (1963)
22. O'Neill, M., Ryan, C.: Grammatical Evolution. In: Evolutionary Automatic Programming in an Arbitrary Language. Kluwer Academic Publishers, Dordrecht (2002)
23. O'Sullivan, J., Conor, R.: An Investigation into the Use of Different Search Strategies with Grammatical Evolution. In: Foster, J.A., Lutton, E., Miller, J., Ryan, C., Tettamanzi, A.G.B. (eds.) EuroGP 2002. LNCS, vol. 2278, pp. 268–277. Springer, Heidelberg (2002)
24. Packard, N., Crutchfield, J., Farmer, D., Shaw, R.: Geometry from a time series. Phys. Rev. Lett. 45, 712 (1980)
25. Price, K.: An Introduction to Differential Evolution. In: Corne, D., Dorigo, M., Glover, F. (eds.) New Ideas in Optimization, pp. 79–108. McGraw-Hill, London (1999)
26. He, Q., Wang, L.: An effective co-evolutionary particle swarm optimization for constrained engineering design problems. Eng. Appl. Artif. Intell. 20(1), 89–99 (2007)
27. Richter, H.: An evolutionary algorithm for controlling chaos: The use of multi-objective fitness functions. In: Guervós, J.J.M., Adamidis, P.A., Beyer, H.-G., Fernández-Villacañas, J.-L., Schwefel, H.-P. (eds.) PPSN 2002. LNCS, vol. 2439, pp. 308–317. Springer, Heidelberg (2002)

28. Richter, H.: A study of dynamic severity in chaotic fitness landscapes, Evolutionary Computation, 2005. The IEEE Congress 3(2), 2824–2831 (2005)
29. Richter, H.: Evolutionary Optimization in Spatio-temporal Fitness Landscapes. In: Runarsson, T.P., Beyer, H.-G., Burke, E.K., Merelo-Guervós, J.J., Whitley, L.D., Yao, X. (eds.) PPSN 2006. LNCS, vol. 4193, pp. 1–10. Springer, Heidelberg (2006)
30. Richter, H., Reinschke, K.: Optimization of local control of chaos by an evolutionary algorithm. Physica D 144, 309–334 (2000)
31. Ryan, C., Collins, J.J., Neill, M.O.: Grammatical Evolution: Evolving Programs for an Arbitrary Language. In: Banzhaf, W., Poli, R., Schoenauer, M., Fogarty, T.C. (eds.) EuroGP 1998. LNCS, vol. 1391, p. 83. Springer, Heidelberg (1998)
32. Sauer, T., Yorke, J., Casdagli, M.: Embedology. J. Stat. Phys. 65, 579 (1991)
33. Sauer, T., Yorke, J.: How many delay coordinates do you need? Int. J. Bifurcat Chaos 3, 737 (1993)
34. Senkerik, R., Zelinka, I., Navratil, E.: Optimization of feedback control of chaos by evolutionary algorithms. In: 1st IFAC Conference on Analysis and Control of Chaotic Systems, Reims, France (2006)
35. Takens, F.: Detecting strange attractors in turbulence. Lect. Notes Math., pp. 366–381 (1981)
36. Zelinka, I.: Analytic programming by Means of Soma Algorithm. In: Proc. 8th International Conference on Soft Computing Mendel 2002, Brno, Czech Republic, pp. 93–101 (2002a)
37. Zelinka, I.: Analytic programming by Means of Soma Algorithm. In: ICICIS 2002, First International Conference on Intelligent Computing and Information Systems, Egypt, Cairo (2002b)
38. Zelinka, I.: SOMA - Self Organizing Migrating Algorithm. In: Babu, B., Onwubolu, G. (eds.) New Optimization Techniques in Engineering. Springer, Heidelberg (2004)
39. Zelinka, I.: Investigation on Realtime Deterministic Chaos Control by Means of Evolutionary Algorithms. In: 1st IFAC Conference on Analysis and Control of Chaotic Systems, Reims, France (2006)
40. Zelinka, I.: Real-time deterministic chaos control by means of selected evolutionary algorithms. Eng. Appl. Artif. Intell. (2008), doi:10.1016/j.engappai.2008.07.008
41. Zelinka, I., Oplatkova, Z.: Analytic programming - Comparative Study. In: The second International Conference on Computational Intelligence, Robotics, and Autonomous Systems, CIRAS 2003, Singapore (2003)
42. Zelinka, I., Oplatkova, Z.: Boolean Parity Function Synthesis by Means of Arbitrarry Evolutionary Algorithms - Comparative Study. In: 8th World Multiconference on Systemics, Cybernetics and Informatics, SCI 2004, Orlando, USA (2004)
43. Zelinka, I., Chen, G., Celikovsky, S.: Chaos Synthesis by Means of Evolutionary Algorithms. Int. J. Bifurcat Chaos Appl. Sci. Eng. 18(4), 911–942 (2008)
44. Zelinka, I., Senkerik, R., Navratil, E.: Investigation on Evolutionary Optimization of Chaos Control. Chaos, Solitons & Fractals (2007), doi:10.1016/j.chaos.2007.07.045

Chapter 9
Cryptography Based on Spatiotemporal Chaotic Systems

Ping Li, Zhong Li, Wolfgang A. Halang, and Guanrong Chen

Abstract. Chaos has been applied in cryptography in the past decades since there are tight relationships between chaos and cryptography. Especially, spatiotemporal chaotic systems can be used to design cryptosystems with satisfactory properties. The chapter focuses on applying a typical spatiotemporal chaotic system, i.e., a coupled map lattice (CML) in cryptography. Multiple-output pseudo-random bit generators (PRBGs) based on CMLs with various constructions and parameters values are designed. Their properties are investigated and compared to determine a certain CML with certain parameters from which the resulting PRBG have satisfactory properties. Additionally, a stream cipher based on the CML is designed and analyzed. It is shown that it has high security, high efficiency and low cost. Moreover, a multimedia cryptosystem based on the proposed stream cipher is constructed by using a field programmable gate array (FPGA). The effects of the encryptions of the text file, the audio file and the image file by using the cryptosystem is measured as effective.

9.1 Introduction

Over the past decades, there has been much interest in designing and analyzing chaos-based ciphers [16, 4]. The main reason for it is that chaotic systems are

Ping Li
Department of Electronic Engineering, Shunde Polytechnic, Kanton, P.R. China
e-mail: Kikiliping@hotmail.com

Zhong Li · Wolfgang A. Halang
Faculty of Electrical and Computer Engineering, FernUniversität in Hagen, 58084 Hagen, Germany
e-mail: zhong.li@fern-hagen.de

Guangrong Chen
Department of Electronic Engineering, City University of Hong Kong, Kowloon, Hong Kong SAR, P.R. China

I. Zelinka et al. (Eds.): Evolutionary Algorithms and Chaotic Systems, SCI 267, pp. 293–328.
springerlink.com

characterized by sensitive dependence on initial conditions and control parameters, random-like behavior and unstable periodic orbits with long periods, which are quite advantageous to ciphers [30, 13]. Till now, lots of chaos-based ciphers have been proposed, moreover, various techniques, such as using multiple chaotic systems [14], high-dimensional chaotic systems [5], multiple iterations of chaotic systems [33], have been proposed to improve chaos-based ciphers.

Among the proposed chaos-based ciphers, many ciphers are not applicable in practice, due to the following reasons. Firstly, for ciphers where the orbits of chaotic systems with simple constructions are directly used to encrypt plaintexts, useful information can be extracted from the chaotic orbits to break the ciphers. Secondly, there exists dynamical degradation of chaotic systems in their realization with digital computers, which threatens the security of the ciphers based on these chaotic systems [15]. Thirdly, some chaos-based ciphers have low implementation speeds [3], which makes the ciphers infeasible in practice. To overcome these drawbacks, multiple chaotic systems [14], high-dimensional chaotic systems [5], multiple iterations of chaotic systems [33], and perturbance-based algorithms [21] have been proposed to improve chaos-based ciphers.

More recently, a one-way coupled logistic-map lattice has been used to design ciphers [17]. Some modifications of these ciphers have been made to improve the security [26], performance [28] and robustness against channel noise [31]. These ciphers have high security and good performance since the following special inherent features of spatiotemporal chaos generated by the coupled logistic-map lattice.

1. The orbit of a spatiotemporal chaotic system has a long period even with dynamical degradation of digital chaos [27];
2. The randomness of the orbit of a spatiotemporal chaotic system is guaranteed by the complex dynamics with a large number of positive Lyapunov exponents;
3. There are multiple sites in a spatiotemporal chaotic system, which can generate independent keystreams simultaneously.

Therefore, using spatiotemporal chaotic systems in cryptography is a significant advance for improving chaos-based ciphers. It is meaningful to study spatiotemporal-chaos-based cryptography.

However, current ciphers based on spatiotemporal chaos adopt some conventional cryptographic techniques (such as S-box), and chaos synchronization. To study the benefit of using spatiotemporal chaos in cryptography, we concern about applying only spatiotemporal chaos into cryptography in the chapter.

In addition, the current research on spatiotemporal-chaos-based cryptography leads to the following problems to be addressed. Firstly, various coupled map lattices (CMLs) except for the one-way and diffusive coupled logistic-map lattice have never been applied in spatiotemporal-chaos-based ciphers. Thus, the issues of how to choose suitable spatiotemporal chaotic systems for cryptography are to be considered. Secondly, in the existing design of spatiotemporal-chaos-based ciphers, or even general chaos-based ciphers, the parameters of the nonlinear dynamical systems are fixed without explanation. Though the chaos-based ciphers with these parameters have acceptable properties from the cryptographic point of view, the question of how to choose the parameters remains to be answered.

This chapter is organized in the way of resolving the above listed problems, involving the following aspects of applying spatiotemporal chaotic systems in cryptography: design and analysis of CML-based pseudo-random bit generators (PRBGs), design and analysis of a CML-based stream cipher, design a multimedia cryptosystem based on the proposed stream cipher. Concretely, the chapter is sketched out in the following:

1. Design of a multiple-output PRBG using a diffusive coupled logistic-map lattice. The statistical properties, such as probability density function (PDF), linear complexity, auto-correlation and cross-correlation of the PRBGs based on various digitization methods are to be investigated. It will be shown that binary-representation method is the best one.
2. Analysis of the properties of the PRBGs based on various CMLs. To determine the CMLs, from which the resulting PRBGs have satisfactory properties, six PRBGs based on six different CMLs are investigated. The six CMLs consist of three simple chaotic systems, i.e., logistic map, skew-tent map and r-adic map, with two simplest coupling methods, i.e., one-way coupling and diffusive coupling, respectively. PDF, auto-correlation, cross-correlation, statistical test and cycle length of the six PRBGs with various parameters are to be investigated so as to determine the parameter intervals within which the PRBGs have satisfactory properties. It will be indicated that a one-way coupled logistic-map lattice with certain parameters has the best properties. This research results in criteria for designing PRBGs with proper performance.
3. Design of a stream cipher employing a one-way coupled logistic-map lattice with certain parameters. Only the last 32 bits are extracted from the chaotic orbit of each site in the CML to guarantee the pseudo-random bit sequences (PRBSs), which consist of the sequences of these 32bits, having perfect statistical properties. The PRBSs as keystreams are used to encrypt plaintexts by the "XOR" operation. The security of the stream cipher is to be tested by attacking it via typical attack methods and analyzing its cryptographic properties. Moreover, the efficiency of the stream cipher is to be analyzed. It will be shown that the cipher has higher security, higher efficiency and lower costs by comparing with Hu's stream cipher of a complicated configuration.
4. Design of a multimedia cryptosystem based on the proposed stream cipher. The cipher is to be implemented in a field programmable gate array (FPGA). The enhanced parallel port (EPP) is used to communicate data between a PC and the FPGA. A user-friendly interface is designed with Visual C++ 6.0, with which text, image and audio can be encrypted and decrypted successfully. The properties of the cryptosystem, such as the sensitivity to the key, speed and efficiency of the FPGA, are to be analyzed.

9.2 CML-Based Pseudo-Random-Bit Generators

A multiple-output PRBG is designed only based on a CML. The CML as a typical spatiotemporal chaotic system is introduced firstly. Digitization methods are used in

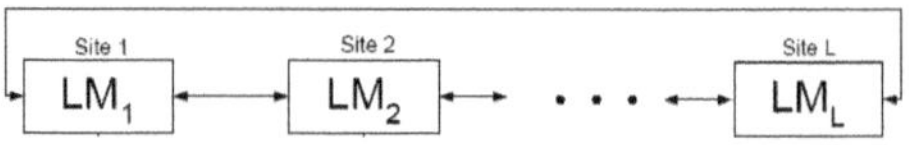

Fig. 9.1 A diffusive coupled map lattice

the PRBG and influence the properties of the PRBG. The statistical properties, such as probability density function (PDF), linear complexity, auto-correlation and cross-correlation of the PRBSs generated from the PRBGs based on various digitization methods are investigated and compared. It is indicated that the digitization method based on binary representation is good for a PRBG. Moreover, the configuration and parameters of the CML have affect on the properties of the PRBG. To determine suitable CMLs and corresponding parameter intervals for the PRBGs, six PRBGs are constructed by using the simplest coupling methods, i.e., one-way coupling and diffusive coupling, and three simple chaotic maps, i.e., logistic map, skew-tent map and r-adic map also named as sawtooth map. The statistical properties, periods and efficiency of these PRBGs with various parameters are investigated.

9.2.1 Coupled Map Lattice

CMLs are used as spatiotemporal chaotic systems in the chapter and introduced firstly. A spatiotemporal chaotic system is a spatially extended system, which can exhibit chaos in both space and time. It is often modeled by partial differential equations (PDE), coupled ordinary differential equations (CODE), or CML [23].

A CML is often adopted as the basic model of a spatiotemporal chaotic system. A CML is a dynamical system with discrete-time, discrete-space and continuous states. It consists of nonlinear maps located on the lattice sites, named as local maps. Each local map is coupled with other local maps in terms of certain coupling rules. Because of the intrinsic nonlinear dynamics of each local map and the diffusion due to the spatial coupling among local maps, the CML can exhibit spatiotemporal chaos. A CML has been extensively studied in the fields of bifurcation and chaos, pattern formation, physical biology and engineering since it was proposed by Kaneko in 1983 [7]. There are two main merits in using a CML as the model of a spatiotemporal chaotic system: one is that a CML captures the most essential features of spatiotemporal chaos; another is that a CML can be easily handled both analytically and numerically [23]. Further, by adopting various local maps and coupling methods [2], various CMLs can be constructed.

A diffusive coupling logistic-map lattice, one of the most popular CMLs, is described as

$$x_{n+1}^{j} = (1-\varepsilon)f(x_n^j) + \frac{\varepsilon}{2}[f(x_n^{j+1}) + f(x_n^{j-1})], \tag{9.1}$$

where $f(x) = rx(1-x)$ is the logistic map with $r \in (0,4]$, x_n^j represents the state variable for the jth site ($j = 1,2,...,L$, L is the number of the sites in the CML) at time n ($n = 0,1,2,...$), $\varepsilon \in (0,1)$ is the coupling strength. The periodic boundary condition, $x_n^0 = x_n^L$ for all n, is used in the CML. The CML can be illustrated in Fig. 9.1, where "LM" is the abbreviation of the local map.

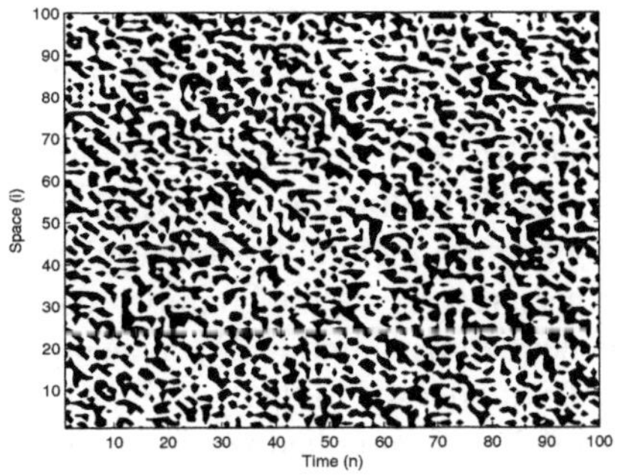

Fig. 9.2 The pattern of a CML

Consider the CML (9.1) with $\varepsilon = 0.9$, $r = 4$ and $L = 100$. By starting from random initial conditions and discarding 10^5 initial transients, 100 sequential output of each site of the CML are depicted in Fig. 9.2, where black points stands for ones which values are larger than 0.5 and white points stands for ones which values are smaller than 0.5. It is shown that the CML is chaotic in both time and space; that is, for a certain site, its output is random-like, for certain time, the output of all sites is also random-like.

Similarly, the CMLs in other configurations with certain parameters can exhibit spatiotemporal chaos.

9.2.2 Digitization Method

If a CML exhibits spatiotemporal chaos, then its output, x_n^i, can be regarded as a pseudo-random number, which means that $\{x_n^i\}(n = 1,2,...)$ can be used as a pseudo-random-number sequence (PRNS), denoted by $\mathrm{PRNS_i}$. Therefore, L PRNSs can be simultaneously generated from the CML with L sites.

By digitizing the PRNS of each site of the CML, PRBSs can be generated. There are three general methods to obtain PRBSs from PRNSs generated by chaotic maps as follows,

Method 1: By dividing the interval visited by a chaotic orbit, x_n, into m parts and labelling them with definite integers belonging to $[0, m-1]$, the nth pseudo-random number is assigned with an integer $r \in [0, m-1]$ when x_n enters the rth subinterval [25]. A special case is $m = 2$, that is, the interval $[a,b]$ is divided into two parts $[a,C]$ and $[C,b]$, where $x_n \in [a,b]$ and C is a threshold. This is the so-called threshold method proposed in [9] and applied in many PRBGs [32]. Then, a PRBS is defined as

$$s_n = \begin{cases} 1, & if \quad x_n \in [a,C] \\ 0, & if \quad x_n \in [C,b]. \end{cases}$$

Method 2: x_n can be represented as a binary sequence

$$x_n = 0.b_{n1}, b_{n2}, ..., b_{nP}, \tag{9.2}$$

where P stands for a certain precision. When a double-float precision is used in computer realization, P is equal to 52. Based on the binary representation, the digitization method is shown in the Fig. 9.3. It is seen that the mth bits in the binary representation comprise the mth PRBS. In maximum, P PRBSs can be generated from one PRNS by using this method. This method is proposed in [9] and widely used in PRBGs [32, 8].

Method 3: A modified version of Method 2 is proposed in [8], where a PRBS is generated as follow:

$$s_n = b_{n1} \oplus b_{n2} \oplus, ..., \oplus b_{nP}, \tag{9.3}$$

where $\oplus$ means XOR operation.

The digitization methods are applied here to get PRBSs from a PRNS. Thus, 3 PRBGs based on these three digitization methods are constructed, which are called PRBG1–3. In PRBG1 and PRBG3, only one PRBS is generated from one site. In PRBG2, the computation precision is assumed as 52, therefore, 52 PRBSs are generated from one site. Totally, $52L$ PRBSs can be generated at one time.

9.2.3 Statistical Properties

Such statistical properties as a uniform PDF, strong linear complexity, the δ-like auto-correlation and the close-to-zero cross-correlation are desirable for a PRBG to be applicable in cryptography [13].

- PDF
  Denote a PRBS generated from a PRBG as $S_N = \{b_1, b_2, ...b_i, ..., b_N\}$, where N is the iteration time. The uniform PDF of S_N means $P(b_i = 0) = P(b_i = 1)$. In other words, the ratio between the number of $\{s_i = 0\}$ and that of $\{s_i = 1\}$ is equal to 1.
- Strong linear complexity
  The linear complexity of S_N, denoted by L_n, is the length of the shortest LFSR that generates a sequence having S_N in its first n time. The sequence $\{L_n, n = 1, 2, ..., N\}$ is called the linear complexity profile of S_N, which can be computed using the Berlekamp-Massey algorithm [19]. By plotting the points (n, L_n) in the

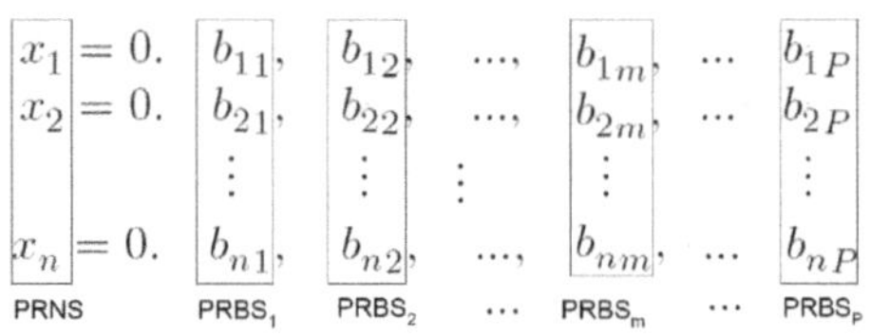

Fig. 9.3 Digitization method

$n \times L$ plane and then joining the successive points by a horizontal line followed by a vertical line, the linear complexity profile of S_N is graphed. The expected linear complexity of a PRBG should closely follow the line $L = N/2$.

- δ-like auto-correlation
  The auto-correlation of S_N measures the extent of similarity between the sequences S_N and a shift of S_N by t positions. The mean-removed auto-correlation, $C_{ii}(\tau)$, of a PRBS is given by

$$
\begin{aligned}
&C_{ii}(\tau) = \hat{C}_{ii}(\tau)/\hat{C}_{ii}(0),\\
&\hat{C}_{ii}(\tau) = \tfrac{1}{N}\textstyle\sum_{n=1}^{N}(b_n - \bar{b}_n)(b_{n+|\tau|} - \bar{b}_n),\\
&\bar{b}_n = \tfrac{1}{N}\textstyle\sum_{k=1}^{N} b_k,\\
&|\tau| = 0, 1, \ldots, N-1.
\end{aligned}
$$

- Close-to-zero cross-correlation
  It is known that PRBSs can be generated simultaneously from a PRBG. If being independent of each other with zero cross-correlation, they can be used to encrypt multiple plaintexts at one time.

The statistical properties of PRBG1, PRBG2 and PRBG3 are investigated and shown in Figs. 9.4, 9.5 and 9.6, respectively. Here, one of 52 PRBSs is randomly chosen for testing its distribution, linear complexity and auto-correlation. Additionally, two of 52 PRBSs are randomly chosen for testing their cross-correlation.

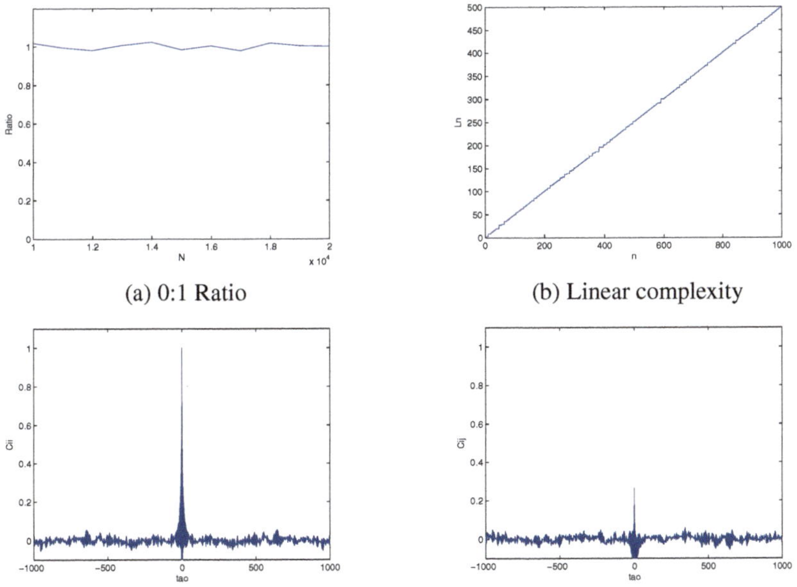

(a) 0:1 Ratio (b) Linear complexity

(c) Auto-correlation (d) Cross-Correlation

Fig. 9.4 Statistical properties of PRBG1

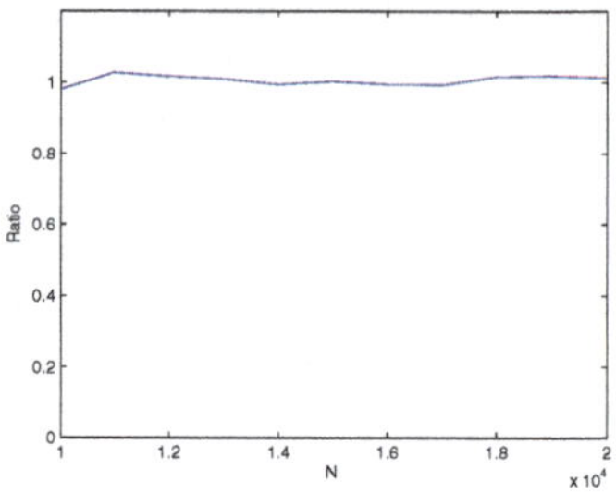

(a) 0:1 Ratio

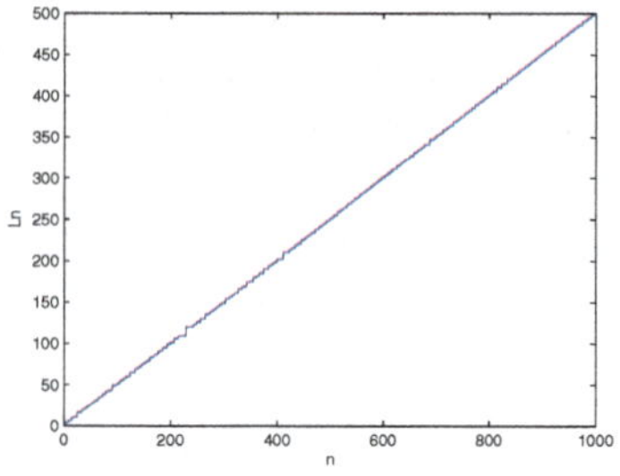

(b) Linear complexity

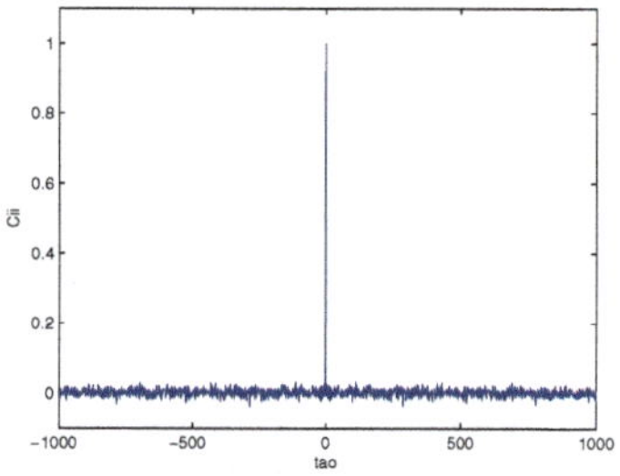

(c) Auto-correlation

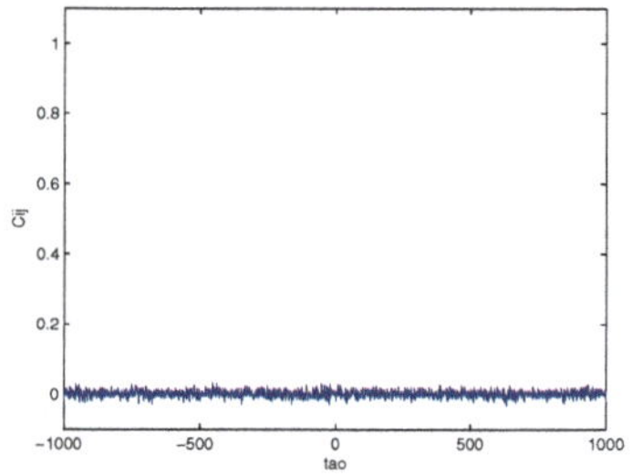

(d) Cross-Correlation

Fig. 9.5 Statistical properties of PRBG2

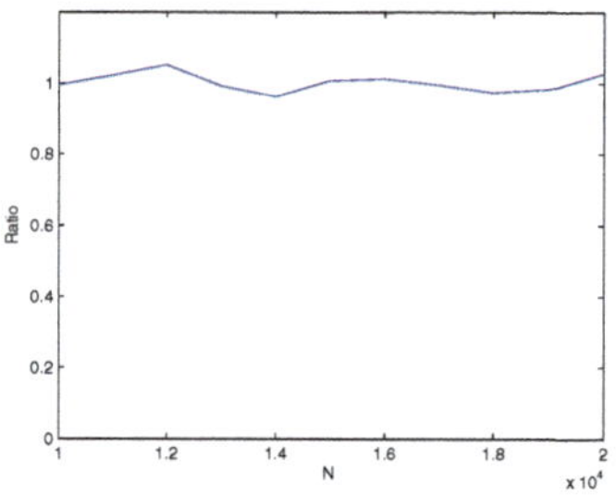

(a) 0:1 Ratio

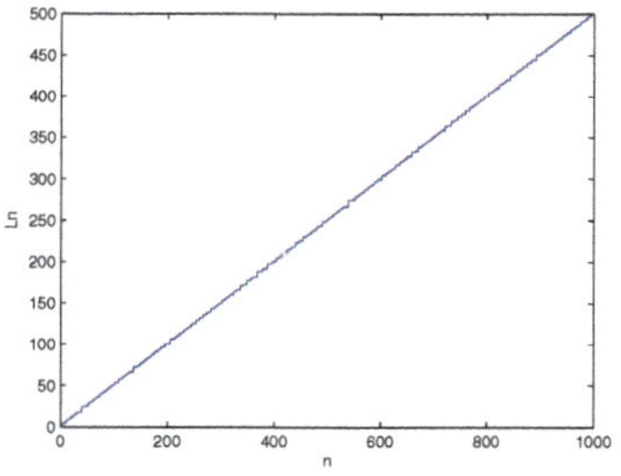

(b) Linear complexity

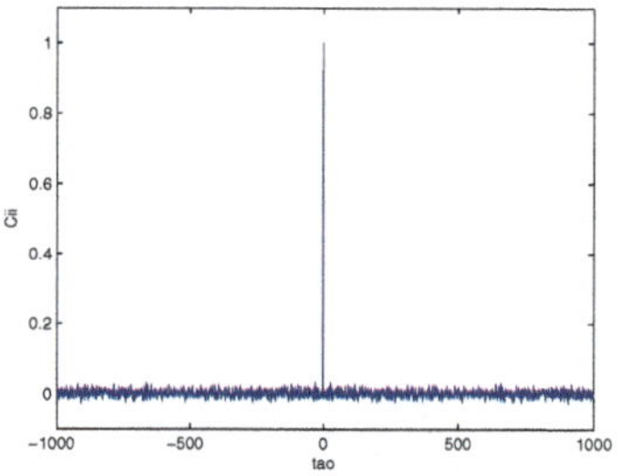

(c) Auto-correlation

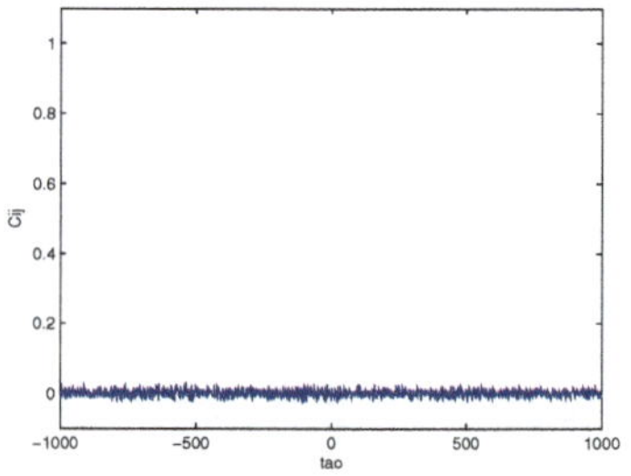

(d) Cross-Correlation

Fig. 9.6 Statistical properties of PRBG3

As we can see, different digitization methods result in different statistical properties of the PRBGs. A comparison of the statistical properties among the three PRBGs is concluded in Table. 9.1. The cross-correlation and the linear complexity of the three PRBGs meet the requirements of cryptography. It is remarked that the auto-correlation of PRBG1 is not δ-like, while the uniform distributions of PRBG1-2 are better than that of PRBG3; therefore, PRBG2 has the best overall statistical properties. The digitization method 2 is thus preferred in designing a PRBG.

Table 9.1 Comparison of statistical properties of Three PRBGs

PRBG	distribution	linear complexity	auto-correlation	cross-correlation
PRBG1	uniform	follows line $L = N/2$	not δ-like	close to zero
PRBG2	uniform	follows line $L = N/2$	δ-like	close to zero
PRBG3	almost uniform	follows line $L = N/2$	δ-like	close to zero

9.2.4 PRBGs Based on Various CMLs

A multiple-output PRBG based on a CML with certain parameters has good properties. This subsection concerns how to determine CMLs and their parameters for constructing PRBGs with satisfactory properties.

It is known that the PDF of the logistic map is equal to $\frac{1}{\pi\sqrt{x(1-x)}}$ [10], which is not uniform; whereas, any piecewise linear chaotic map $f : I \mapsto I$, $I = [a,b] \subset R$, have a uniform PDF, namely, $\frac{1}{b-a}$ [1]. The ununiformity of the PDF of the local map may have a negative effect on the PDF of the CML. Therefore, the CMLs based on a piecewise linear chaotic map, i.e., the skew tent map [18], is employed to construct a PRBG. The skew tent map is described as

$$f(x,p) = \begin{cases} x/p & , \quad x \in [0,p) \\ (x-p)/(1-p) & , \quad x \in (p,1] \end{cases} \; p \in (0.5,1).$$

In the following, the properties of the PRBG based on diffusive coupled skew-tent-map lattice (DCSTML) are analyzed.

9.2.4.1 Statistical Properties of the PRBGs

Since the parameters of the PRBG, p, ε and L, may have effects on the properties of the PRBG, the effect of each parameter on the properties of the PRBG is analyzed. PDF, auto-correlation and cross-correlation as the important statistical properties of the PRBGs with one varying parameter and others fixed are investigated. In the simulation, the lengths of the PRBSs are computed to be 10^4 and the parameters vary in the following way: firstly, increase p from 0.51 to 0.99 by 0.01 each time, while fix ε as 0.9 and L as 8; then increase ε from 0.01 to 0.99 by 0.02 each time, while fix p as 0.51 and L as 8; finally, increase L from 8 to 64 by 1 each time with fixing p as 0.51 and ε as 0.9. Due to the symmetric configuration of the CML, it is reasonable to analyze the statistical properties of the multiple PRBSs generated from an arbitrarily chosen PRNS.

PDF

To analyze the PDF of a PRBG, a scaled difference ΔP between $P\{(b_n)=0\}$ and $P\{(b_n)=1\}$ in each PRBS, i.e., $\Delta P=(N_1-N_0)/(N/2)$, (N_1, N_0, and N are the number of "1" and "0", and the length of the PRBS, respectively) is computed. Fig. 9.7(a) shows ΔP of the 52 PRBSs output from the PRBG with various p, where the x-axis is the index of the 52 PRBS, denoted by i, the y-axis denotes the various p and the z-axis stands for ΔP.

It is shown that ΔP of the first 4 PRBSs are much bigger than zero. Additionally, by setting a threshold of ΔP as 0.07, Figs. 9.7(a) and 9.7(d) are plotted in the following way: if ΔP of the PRBS is smaller than the threshold, the point corresponding to the index of the PRBS and p of the PRBG from which the PRBS output is drawn black, otherwise, the point is drawn white. In the same way, ΔP of the 52 PRBSs from the PRBG with various ε and various L are plotted in Figs. 9.7(b)(e) and in Figs. 9.7(c)(f), respectively. It is shown that the 5th–52nd PRBSs have uniform PDF whatever the parameters are.

Auto-correlation

Since δ-like auto-correlation means $C_{ii}(0)=1$ and $\{C_{ii}(\tau)\}(|\tau|=1,2,...,N-1)$ or the maximum of $\{C_{ii}(\tau)\}(|\tau|=1,2,...,N-1)$ is close to zero, the close-to-zero maximum auto-correlation of a PRBS is equivalent to the δ-like auto-correlation of a PRBS. The maximum auto-correlations of the 52 PRBSs of the PRBG with various parameters are computed and shown in Fig. 9.8. It is indicated that the maximum auto-correlations of the first 4 PRBSs are much far from zero and the rest except for the PRBSs from the PRBG with p close to 0.99 is close to zero.

Additionally, we set the threshold of the maximums as 0.05, and get Fig. 9.8(d)(e) and (f) in the same way as that of the previous section. It is shown that the maximum auto-correlations of all the 5th–52nd PRBSs output from the PRBG with any parameters values are smaller than 0.05. Therefore, the auto-correlation of the PRBG without the first 4 PRBSs satisfies the requirement of cryptography.

Cross-correlation

In order for the 5th-52nd PRBSs generated from the output of each site of the PRBG to be applicable in parallel, the cross-correlation between arbitrary two PRBSs should be close-to-zero. Maximum cross-correlations, denoted by $C_{i_1i_2}$, between $\mathrm{PRBS}^i_{m_1}$ and $\mathrm{PRBS}^i_{m_2}$ $(i,m_1,m_2\in N, i\in[1,L], m_1,m_2\in[5,52], m_1\neq m_2)$ output from the PRBG with various parameters are computed and shown in Fig. 9.9. Additionally, other maximum cross-correlations, denoted by $C_{i_1j_2}$, between $\mathrm{PRBS}^i_{m_1}$ and $\mathrm{PRBS}^j_{m_2}$ $(i,j,m_1,m_2\in N, i,j\in[1,L], i\neq j, m_1,m_2\in[5,52])$ of the PRBG with various parameters are shown in Fig. 9.10.

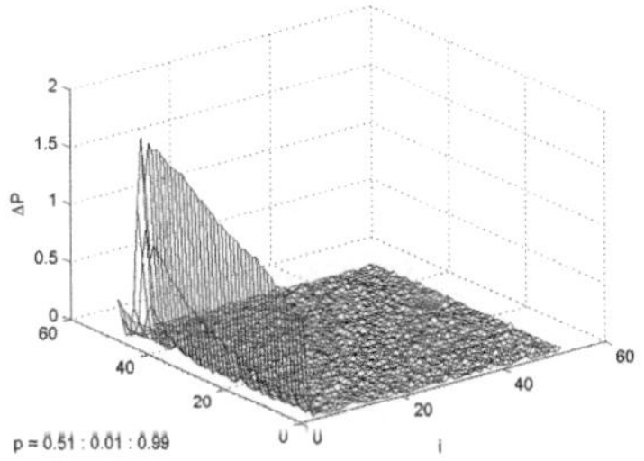

(a) ΔP of the PRBG with various p

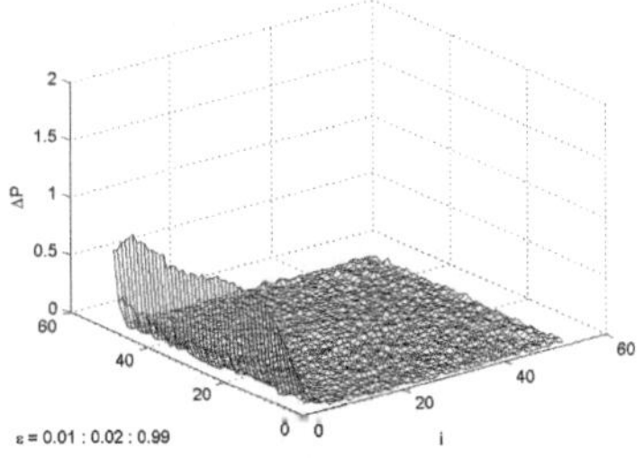

(b) ΔP of the PRBG with various ε

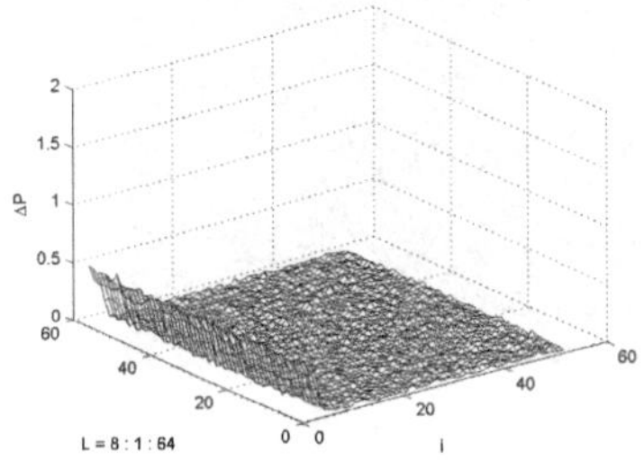

(c) ΔP of the PRBG with various L

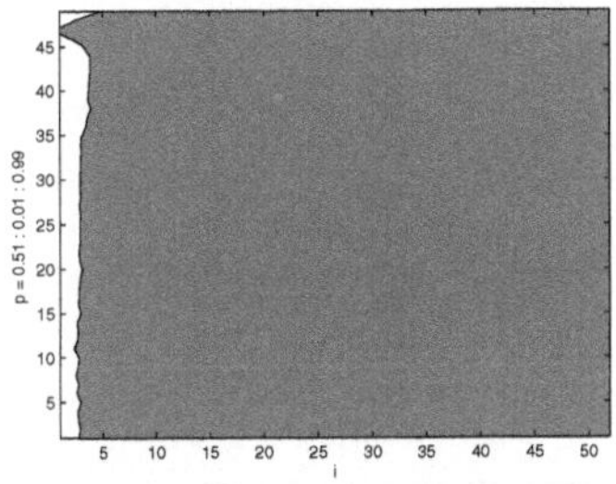

(d) The range of p within which the PRBG has uniform PDF

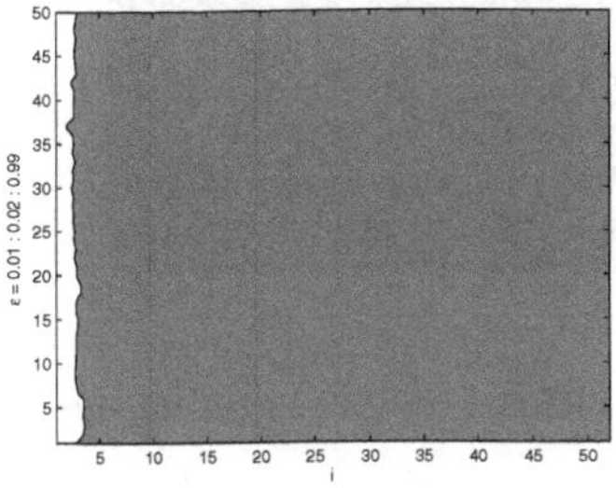

(e) The range of ε within which the PRBG has uniform PDF

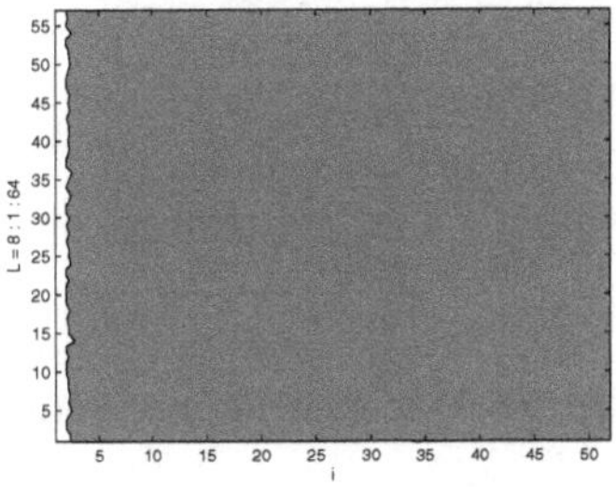

(f) The range of L within which the PRBG has uniform PDF

Fig. 9.7 ΔP of the PRBG with various parameters

We set the threshold of the maximum cross-correlation as 0.05 and find that the maximum cross-correlation between all pairs of PRBSs output from the PRBG with all parameters values are smaller than 0.05. Therefore, the cross-correlation of the PRBG is acceptable from the cryptographic point of view.

Statistical test

In practice, a statistical test is employed to investigate the randomness of PRBGs and thus to verify whether PRBGs are acceptable or not from the statistical point of view. There are many statistical test available, such as Diehard Battery of Tests, Knuth's

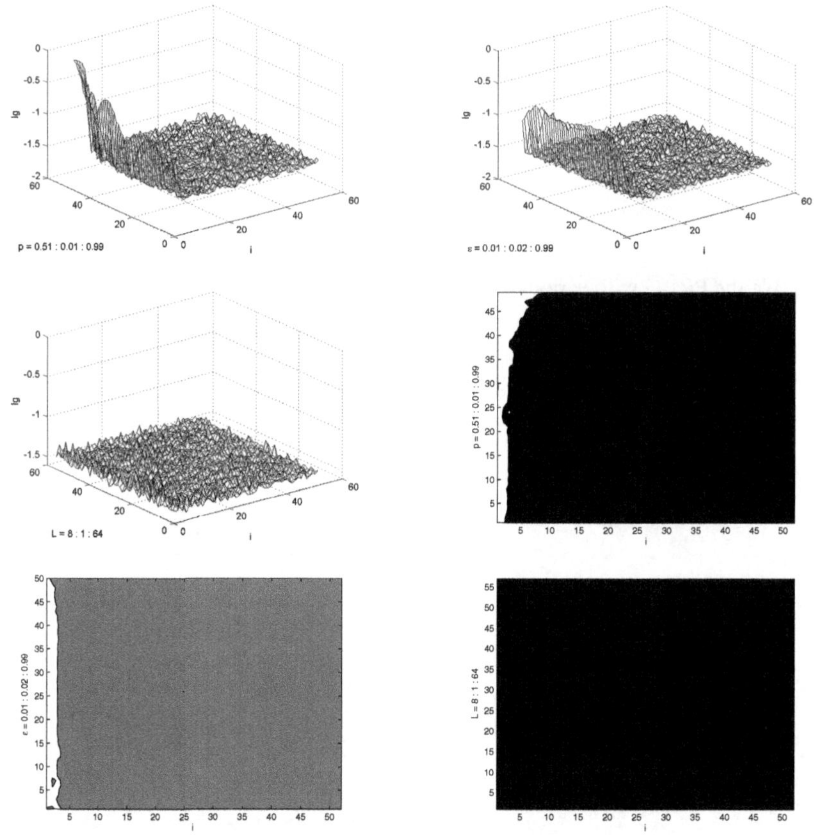

Fig. 9.8 The maximum auto-correlations of the PRBG with various parameters

Collection, FIPS 140-2 Statistical Test Suite, and NIST Statistical Test Suite [24]. Since the computations of the 52 PRBSs from the PRBG with various parameters are time-consuming, and among them, the FIPS 140-2 test takes the least computation time, therefore, in our work, the FIPS 140-2 [20] is used to evaluate the randomness of the proposed PRBG. The FIPS 140-2 specifies four statistical tests, i.e., monobit test, poker test, run test, and long run test, all of which should be passed if a PRBS passes the FIPS 140-2.

The 52 PRBSs generated from arbitrary one PRNS from the PRBG with various parameters are detected by using the FIPS 140-2. The results are shown in Fig. 9.11, where a black point corresponds to the index of the PRBS which passes the test and the parameter of the PRBG from which the PRBS is output. It is shown that the first 4 PRBSs have bad randomness with any parameter values. Therefore, these PRBSs should be discarded for a good PRBG. In addition, since the first 12 PRBSs

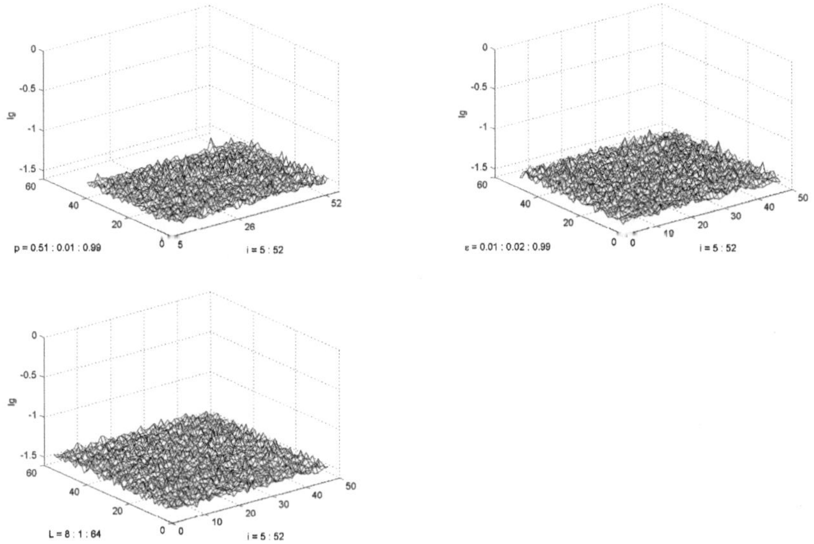

Fig. 9.9 The maximum $C_{i_1 i_2}$ of the PRBG with various parameters

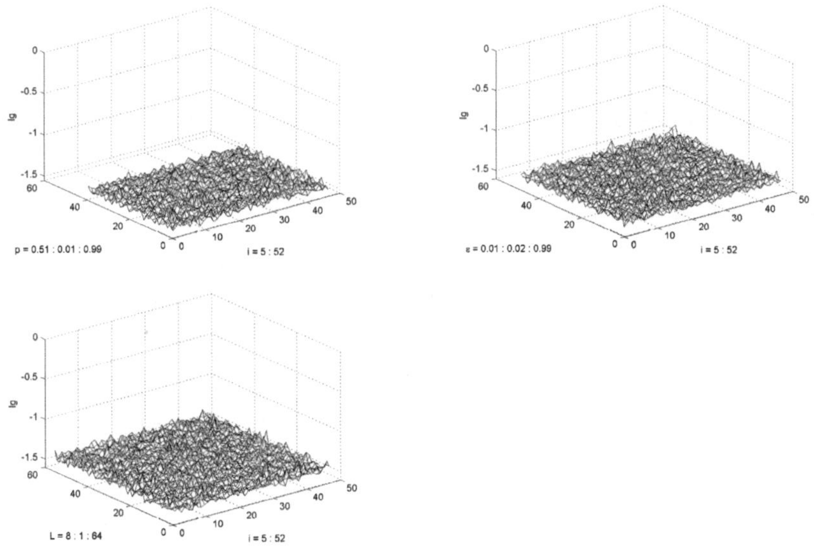

Fig. 9.10 The maximum $C_{i_1 j_2}$ of the PRBG with various parameters

from the PRBG with p close to 0.99 can not pass the FIPS 140-2 test, as shown in Fig. 9.11(a), p close to 0.99 should be avoided. According to Figs. 9.11(b) and 9.11(c), the 5th-52nd PRBSs can pass the FIPS 140-2 whatever ε and L are.

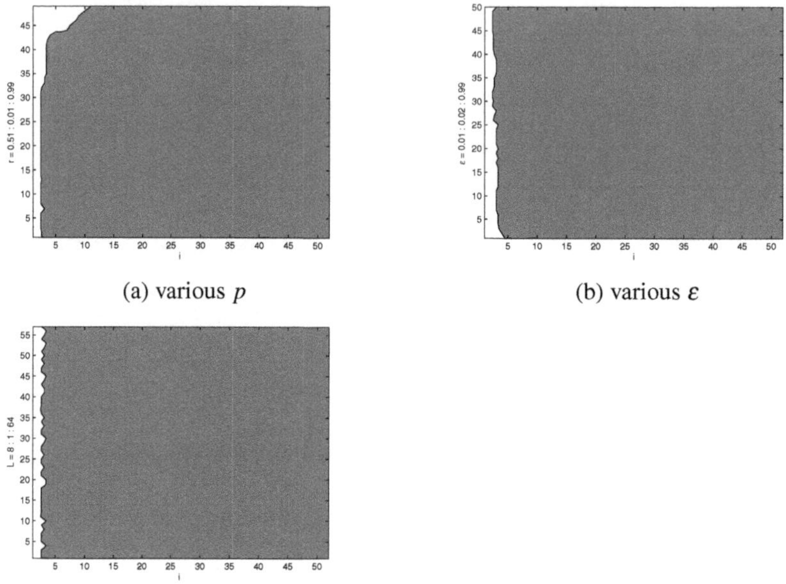

(a) various p

(b) various ε

(c) various L

Fig. 9.11 FIPS 140-2 of the PRBG with various parameters

Periodicity

Long period is an important cryptographic requirement to a PRBG. Similar to the method of estimating the periodicity of a PRNS employed in [27], the dependence of the transient time for the trajectory to enter the periodic circle and the period on the precision $\delta = 10^{-h}$ and L is investigated. $\tau = \sum_{i=1}^{N} \tau_i$ and $T = \sum_{i=1}^{N} T_i$ of any PRNS generated from the PRBG with $L = 2$ are derived by arbitrarily choosing N ($N = 10^h$ if h ≤ 6 or $N = 1$ if $h > 6$) different initial conditions of a PRBG and computing transient times τ_i and the periods T_i in the way as [27]. τ and T vs h, respectively, are plotted in Fig. 9.12, where the triangle and circle signs stand for τ and T, respectively.

It is shown that τ follows the solid curves well. Therefore, for a certain L, one has

$$\tau(h,L) \propto 10^{\alpha(L)+\beta(L)h}.$$

For different L, $\alpha(L)$, $\beta(L)$ have different values. According to Fig. 9.12(d), τ and T do not increase as L increases, and in addition, T is always smaller than τ.

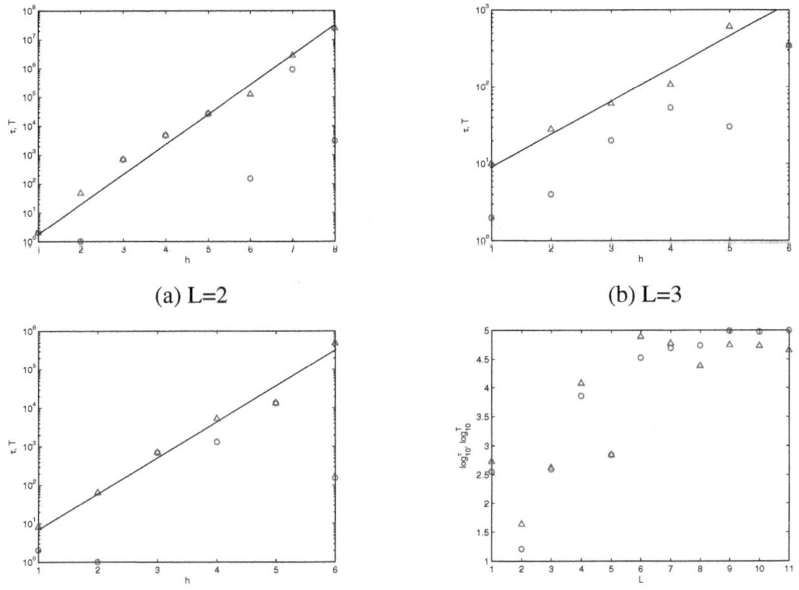

(a) L=2 (b) L=3

(c) L=4 (c) h=3

Fig. 9.12 τ and T of the PRBG with various L and h

Since the cryptographic properties of a PRBG are related to $T + \tau$ rather than τ, power law behavior of τ has an more important effect on the properties of the PRBG.

9.2.4.2 Comparison of PRBGs Based on Various CMLs

With different local maps and different coupling methods various CMLs can be constructed. In order to answer the question of how to determine the CMLs with certain parameters for constructing PRBGs with proper statistical properties, various CMLs with various local maps and coupling modes are used to construct PRBGs, and a comparison among the PRBGs is carried out in this section. Logistic map as the most well-known one-dimensional chaotic system, skew-tent map and r-adic map, described as $f(x,r) = rx \bmod 1$ $(r > 1)$ [2] as the simple piecewise linear chaotic maps are employed as the local maps of the CMLs; one-way coupling and diffusive coupling as the two simplest coupling modes are used as the coupling methods; thereby, six CMLs are obtained.

Similar to the analysis of the PRBG based on DCSTML, statistical properties of the PRBGs based on diffusive coupled logistic-map lattice (DCLML), diffusive coupled r-adic-map lattice (DCRML), one-way coupled logistic-map lattice (OWCLML), one-way coupled skew-tent-map lattice (OWCSTML), and one-way coupled r-adic-map lattice (OWCRML) with various parameters are investigated, respectively.

The results show that the PRBG based on one-way coupled logistic-map lattice with certain parameter has the most satisfactory statistical properties, largest period and highest efficiency among the six PRBGs.

9.3 CML-Based Stream Cipher

A one-way coupled logistic-map lattice (OWCLML) with certain parameters has the best cryptographic properties among the six simplest PRBGs, based on which a stream cipher is designed in the section.

9.3.1 Algorithm of the Cipher

Based on the OWCLML, a stream cipher can be constructed. The encryption is described as

$$\begin{aligned}
x_{n+1}^j &= (1-\varepsilon)f(x_n^j, a_j) + \varepsilon f(x_n^{j-1}, a_{j-1}),\\
f(x_n^j, a_j) &= (3.9 + 0.1a_j)x_n^j(1 - x_n^j),\\
K_n^j &= \text{int}[x_n^j \times 2^u] \mod 2^v,\\
C_n^j &= M_n^j \bigoplus K_n^j,
\end{aligned} \tag{9.4}$$

where $u, v \in N$, K_n^j, M_n^j, and C_n^j are keystream, plaintext and ciphertext, respectively, and $\bigoplus$ means bitwise XOR. Actually, the CML serves as a PRBG to produce L keystreams by imposing *int* and *mod* algebraic operations on the outputs of the CML. Plaintexts are bitwise XORed with keystreams to produce the ciphertext. Encryption keys are assumed as $a_j \in [0,1]$, denoted as a vector form $\mathbf{a} = \{a_1, a_2, ..., a_L\}$.

The configuration and parameters of the decryption are the same as those of the encryption, which is described as

$$\begin{aligned}
y_{n+1}^j &= (1-\varepsilon)f(y_n^j, a'_j) + \varepsilon f(y_n^{j-1}, a'_{j-1}),\\
f(y_n^j, a'_j) &= (3.9 + 0.1a'_j)y_n^j(1 - y_n^j),\\
K'^j_n &= \text{int}[y_n^j \times 2^u] \text{mod} 2^v,\\
M'^j_n &= C_n^j \bigoplus K'^j_n,
\end{aligned} \tag{9.5}$$

where $a'_j \in [0,1]$, denoted as $\mathbf{a}' = \{a'_1, a'_2, ..., a'_L\}$, are decryption keys. When $\mathbf{a}' = \mathbf{a}$ and $y_0^j = x_0^j$, these two CMLs are synchronized, i.e., $y_n^j = x_n^j$, thus producing identical keystreams, $K'^j_n = K_n^j$. As a result, the plaintext is decrypted, $M'^j_n = M_n^j$.

Remarks:

1. Self-synchronous chaotic ciphers have an advantage that they do not need an extra synchronization signal, but also a disadvantage that a ciphertext, which controls a keystream generator in a cipher, is accessible and thus can be used for cryptanalysis [6].

2. In terms of the statistical properties discussed in section 9.2, in order for the keystreams in the proposed cipher to have proper statistical properties [11], a_j $(j = 1,2,...,L)$ are set as keys to guarantee that the parameter of the logistic map falls in the range $[3.9, 4.0]$ and ε is fixed as 0.95.
3. The double floating-point arithmetic is used in the cipher. Since the number of the significant bits of the binary representation of double floating-point number in the computer is 52, u is set as 52.
4. v is assumed as 32 for the following reasons. First, the first 4 bits are discarded for their bad statistical properties. Second, the smaller v is, the harder it is to break the cipher with known-plaintext attack, which will be indicated in subsection 9.3.3. Finally, from the implementation point of view, the larger v is, the more efficient the cipher will be. Therefore, a tradeoff between efficiency and security leads to fix v as 32 by considering that common computers adopt 32 bits or 64 bits CPUs.
5. The determination of L lies in the following considerations. L has no evident influence on the cryptographic properties of the keystream [11] except for its period equal to about 10^{7L} [27]. Meanwhile, it does not influence the encryption speed, too, which will be indicated in subsection 9.3.4. Additionally, the cost of breaking the cipher is about 2^{40L}, which will be analyzed in detail in subsection 9.3.3. Therefore, in investigating a concrete cipher thereafter, L is assumed as 4 in order that the keystream has period of 10^{28} and the cost of breaking the cipher is up to 2^{160}, which are suitable from cryptographic point of view.

9.3.2 *Keyspace*

A keyspace is defined as a set of all possible keys [22], which should be studied in depth in designing a cipher. Error function [17] is used here to determine the keyspace of the cipher. When $\mathbf{a}' \neq \mathbf{a}$, the decrypted plaintext, M'^{j}_{n}, can be deviated from the original one, M^{j}_{n}. The error function is defined as

$$\begin{aligned} e(j,\Delta\mathbf{a}_t) &= \tfrac{1}{T}\textstyle\sum_{n=1}^{T} |m'^{j}_{n} - m^{j}_{n}|, j = 1,2,...L, \\ m'^{j}_{n} &= \tfrac{M'^{j}_{n}}{2^{32}}, m^{j}_{n} = \tfrac{M^{j}_{n}}{2^{32}}, \end{aligned} \tag{9.6}$$

where $\Delta\mathbf{a}_t = \{\Delta a_1, \Delta a_2, ..., \Delta a_t\}(\Delta a_i = a'_i - a_i, i = 1,2,...,t, t \leq L)$, and T is encryption times. The error function vs Δa_1 with $T = 10^5$ is plotted in Fig. 9.13. It is shown that the error function is not equal to zero but 0.25 even if Δa_1 takes an extremely small value 2^{-47}. In other words, the key a'_1 is sensitive to any differences equal to or larger than 2^{-47}. Similarly, the error function of $\Delta a_i (i = 2,3,...,L)$ are computed, and it is shown that the keys $a'_i (i = 2,3,...,L)$ are also sensitive to any differences equal to or larger than 2^{-47}. Therefore, the keyspace is 2^{47L}.

Fig. 9.13 Error function

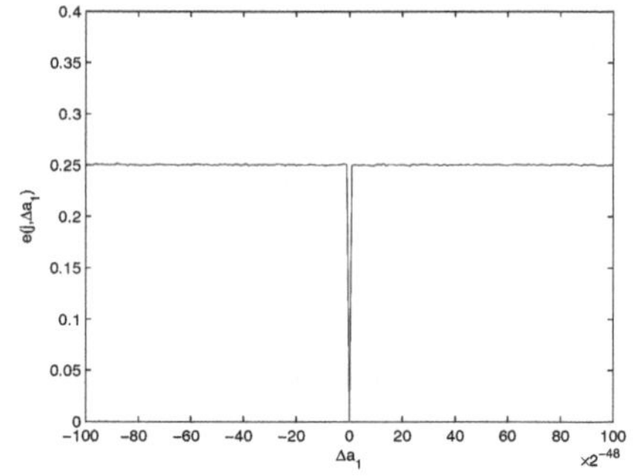

9.3.3 Cryptographic Properties of the Keystream

Since the ciphertext is generated by using directly bitwise XOR between the plaintext and the keystream in the cipher, the cryptographic properties of the keystream have significant effects on the security of the cipher. Due to the symmetric configuration of the CML, all keystreams have similar cryptographic properties. Some cryptographic properties of a keystream among the L ones, such as probability distribution, auto-correlation, and run probability, are numerically investigated in this section.

Probability distribution

The order-1 and order-2 probability distributions [17] of the keystreams in the cipher with random initial conditions, arbitrarily chosen plaintext, and $\mathbf{a} = 0.5\mathbf{I}$ ($\mathbf{I}$ is a L-vector with all elements equal to 1) are investigated.

The order-1 probability distribution of the keystream, $\rho(ks_n^j)(=\frac{\rho(K_n^j)}{2^{32}})$, is plotted in Fig. 9.14(a). The order-2 probability distribution of the keystream, $\rho(ks_n^j, ks_{n-1}^j)(=\frac{\rho(K_n^j, K_{n-1}^j)}{2^{32}})$, is plotted in Fig. 9.14(b). The length of the keystream is 10^6. It is shown that the probability distributions are uniform.

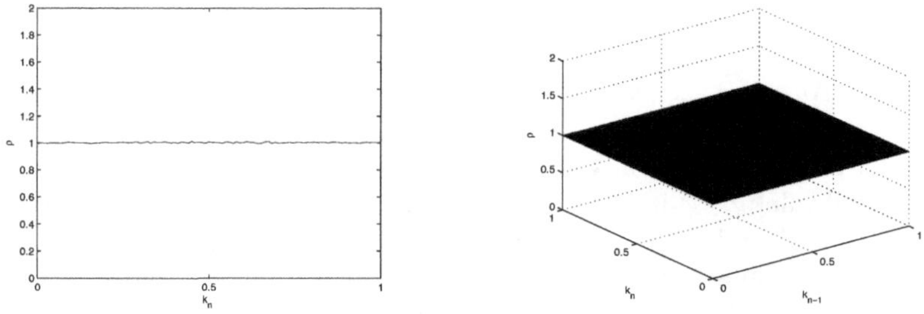

Fig. 9.14 Probability distribution of the keystream

Run

A run of a binary sequence s is another postulate of randomness, and defined as a subsequence of s consisting consecutive 0's or consecutive 1's that is neither preceded nor succeeded by the same symbol [19]. The probabilities of 0/1 runs of length $n (n = 1,2,...,N)$, denoted as $p_0(n)/p_1(n)$ or $p_{0/1}(n)$ of K_j, are investigated, where $p_{0/1}(n) = \frac{R_{0/1}(n)}{R_{0/1}}$ and $R_{0/1} = \sum_{n=1}^{N} R_{0/1}(n)$ with $R_{0/1}(n)$ being the number of 0/1 runs of length n. $p_{0/1}(n)$ vs n is plotted in Fig. 9.15. It is shown that $p_{0/1}(n)$ is directly proportional to n, which is the characteristic of a truly random binary sequence of an infinite length [17].

Auto-correlation

The δ-like auto-correlation is one of cryptographic requirements to keystream in a cipher. The mean-removed auto-correlation of the keystream with length $T = 10^6$ is plotted in Fig. 9.16. It is shown that the keystream has the δ-like auto-correlation.

In summary, according to the analysis above, L keystreams have satisfactory random-like statistic properties.

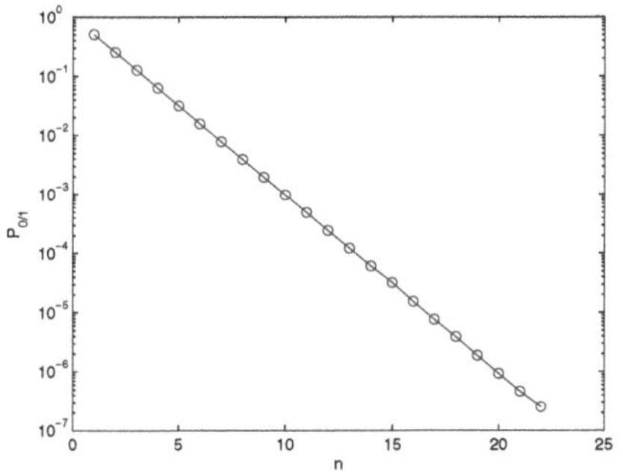

Fig. 9.15 Probability of the run of the keystream

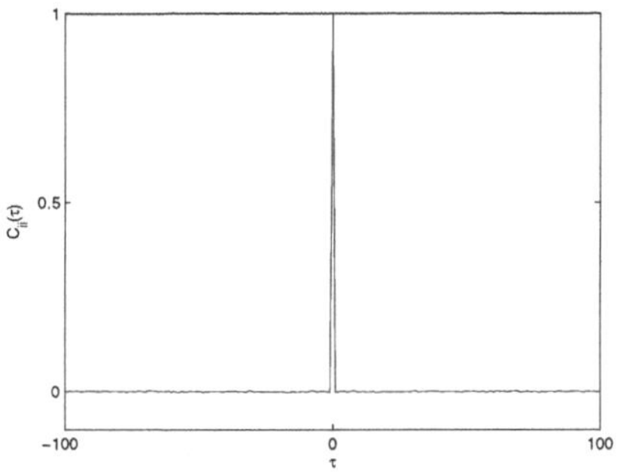

Fig. 9.16 Auto-correlation of the keystream

Security Analysis

In this section, the security of the cipher is evaluated by investigating its confusion and diffusion properties and using various typical attacks, such as the error function attack, the differential attack, the known-plaintext attack, the brute-force attack, and the chosen-plaintext/ciphertext attack.

9.3.3.1 Confusion and Diffusion

To resist common attacks, the cipher should have the following two basic cryptographic properties: confusion and diffusion. Confusion reflects the uniformity of all keys. To evaluate the confusion of the cipher, the independence of the probability distribution of the ciphertext on the exact value of a key is analyzed via $\rho(c|a)$ ($c = \frac{C}{2^{32}}$, $C = C_n(j;a)$, $\mathbf{a} = a\mathbf{I}$), which is shown in Fig. 9.17. The conditional probability distribution of the ciphertext is uniform for different keys. Therefore, the confusion of the cipher is guaranteed.

Diffusion reflects strong sensitivity of a key to tiny changes. In terms of the analysis of the error function described in subsection 9.3.2, the key of the cipher is even sensitive to a extremely small change 2^{-47}, which verifies the diffusion of the cipher.

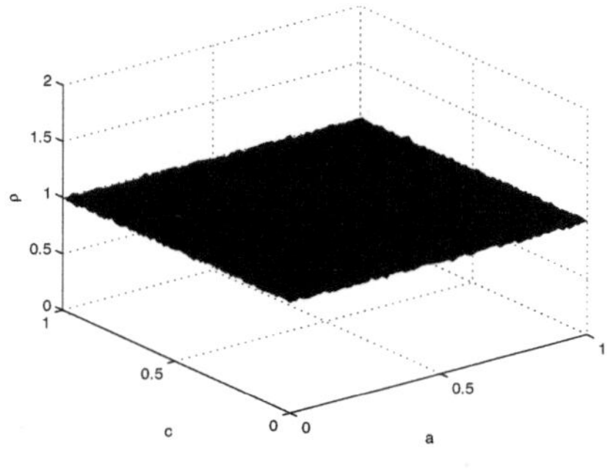

Fig. 9.17 Conditional probability distribution $\rho(c|a)$

9.3.3.2 Error Function Attack

The error function can also be used to break a cipher by an attacker, which is called Error Function Attack (EFA). For the proposed cipher, the cost of EFA is up to 2^{47L}. It is noted that the cost of EFA can be reduced by using some optimal adaptive searching methods if there exists certain tendency toward the key in the error function attack. To check if there is such a tendency, Fig. 9.13 is enlarged to Fig. 9.18. It is indicated that there is no tendency about the location of a_1 even if $a_1' - a_1$ is equal to 2^{-47}. Thus, any adaptive searching can hardly work without any tendencies. Therefore, the cost of EFA of the cipher is exactly equal to 2^{47L}.

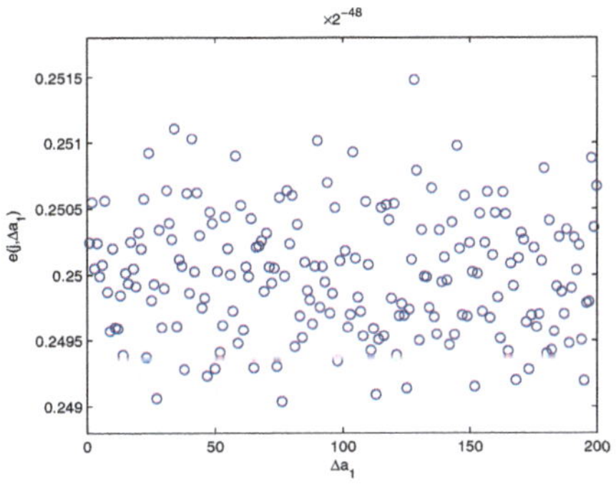

Fig. 9.18 Enlarged error function

9.3.3.3 Differential Attack

Some features of the differential relations between ciphertexts and plaintexts, such as some characteristic differential relations caused by any imperfect statistical properties of keystreams, can be used to break a cipher by a differential attack. To investigate whether there are such differential relations in the cipher, conditional probability of the ciphertext $\rho(\Delta c|\Delta m)(\Delta c = \frac{\Delta C}{2^{32}}, \Delta C = C_n(j;\hat{M}_n^j) - C_n(j;M_n^j))$ under the condition $\Delta m = \frac{\Delta M_n^j}{2^{32}}(\Delta M_n^j = \hat{M}_n^j - M_n^j)$ is shown in Fig. 9.19. The differential probability $\rho(\Delta c)$ is uniform whatever the differential probability $\rho(\Delta m)$ is. Therefore, the cipher is immune to the differential attack.

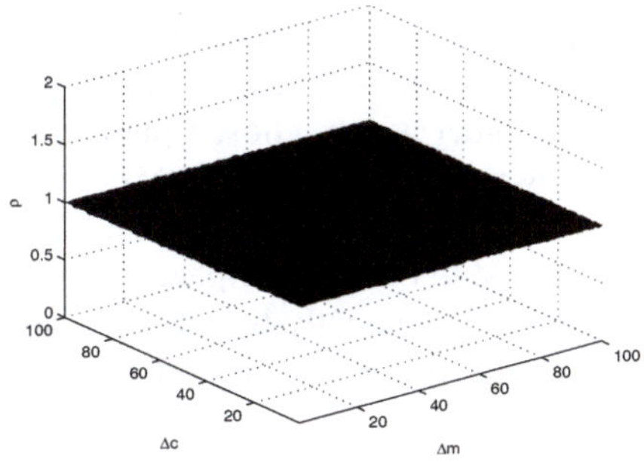

Fig. 9.19 Conditional differential probability of the ciphertext

9.3.3.4 Known-Plaintext Attack

With Kerchoffs' assumption, i.e., an attacker knows complete details of a cipher and implementation except keys, known-plaintext attack is to expose keys with public ciphertexts and known plaintexts. To break the proposed cipher, a known-plaintext attack is applied via an inverse analytical computation with known plaintexts and accessible ciphertexts, i.e., the keystreams K_n^j. The cost of the known-plaintext attack to the cipher can be estimated as follows. To simplify the conduction, $x_{n+1}^j = (1-\varepsilon)f(x_n^j, a_n^j) + \varepsilon f(x_n^{j-1}, a_n^{j-1})$ and $f(x_n^j, a_j) = (3.9 + 0.1a_j)x_n^j(1-x_n^j)$ in (9.4) can be recast as $x_{n+1}^j = G(x_n^j, x_n^{j-1}, a_j, a_{j-1})$. To obtain keys **a**, a set of L equations is given by

$$\begin{aligned} x_{n+1}^1 &= G(x_n^1, x_n^L, a_1, a_L), \\ x_{n+1}^2 &= G(x_n^2, x_n^1, a_2, a_1), \\ &\cdots \\ x_{n+1}^L &= G(x_n^L, x_n^{L-1}, a_L, a_{L-1}), \end{aligned} \tag{9.7}$$

where $2L$ variables, x_{n+1}^j and x_n^j $(j = 1, 2, ..., L)$, should be known to solve **a**. In addition, one x_n^j can be obtained from one K_n^j, however, one K_n^j corresponds to 2^{52-32} possible x_n^j. Consequently, the cost of the known-plaintext attack is no less than 2^{40L}.

A typical known-plaintext attack is the brute-force attack, where a cipher is attacked by trying every possible key one by one to decrypt plaintext with public ciphertext and checking whether the resulting plaintext is the original one. Since the keyspace is deduced as 2^{47L}, the cost of the brute-force attack of the cipher is 2^{47L}.

9.3.3.5 Chosen-Plaintext Attack and Chosen-Ciphertext Attack

In applying chosen-plaintext and chosen-ciphertext attacks to the cipher, an attacker chooses some special plaintexts and ciphertexts to capture certain keystreams. In the cases that certain keystreams correspond to certain keys, i.e., there exist some characteristic relations between keystreams and keys, those attacks are effective. The relation between the keystream $k(= \frac{K_n^j}{2^{32}})$ and the key $a(\mathbf{a} = a\mathbf{I})$ is investigated with conditional probability distribution of k under the condition a. $\rho(k|a)$ is plotted in Fig. 9.20. It is indicated that no characteristics of keys can be extracted from keystreams, consequently, no special plaintexts or ciphertexts can be chosen to break keys. In other words, chosen-plaintext/ciphertext attack has the same efficiency as the known-plaintext attack to this cipher.

In summary, the known-plaintext attack is the most effective attack to the cipher and its cost to break the cipher is 2^{40L}. Moreover, the security of the cipher can be increased conveniently by adding one lattice to the CML in the cipher with little more computation, which results in the cost of breaking the cipher rising 2^{40} times.

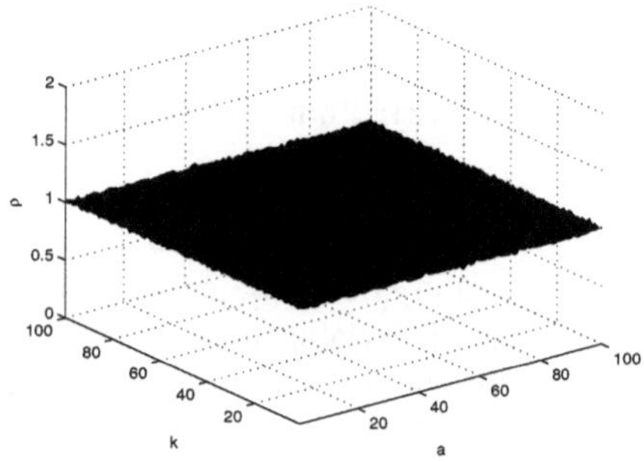

Fig. 9.20 Conditional probability distribution $\rho(k|a)$

9.3.4 High Efficiency

In addition to high security, the cipher is quite efficient. All the coupled maps of the CML in the cipher are used to encrypt plaintexts simultaneously. A close-to-zero cross-correlation among L keystreams guarantees the efficiency of the parallel L encryptions/decryptions in the cipher. The cross-covariance, i.e., mean-moved cross-correlation, is used here to analyze the cross-correlation of any two keystreams among L keystreams with length 10^6, K_n^i and K_n^j, which are described as

$$
\begin{aligned}
&C_{ij}(\tau) = \hat{C}_{ij}(\tau)/\sqrt{\hat{C}_{ii}(0)\hat{C}_{jj}(0)}, \tau = 0, 1, ..., T-1,\\
&\hat{C}_{ij}(\tau) = \tfrac{1}{T}\textstyle\sum_{n=1}^{T}(K_n^i - \overline{K}_n^i)(K_{n+\tau}^j - \overline{K}_n^j),\\
&\overline{K}_n^i = \tfrac{1}{T}\textstyle\sum_{n=1}^{T} K_n^i.
\end{aligned}
\tag{9.8}
$$

The result of the computation of the cross-covariance is plotted in Fig 9.21. It is shown that all keystreams are independent. Therefore, the parallel L keystreams can be used to effectively encrypt plaintexts at one time.

Due to the parallel operation, an around 700M bits plaintext can be encrypted per second in our computer with 1.8GHz CPU and 1.5GB RAM. In addition, the encryption speeds of the ciphers based on the CMLs of various sizes are similar, and this is indicated by the relation between the speeds and L as shown in Fig. 9.22. As a comparison, the encryption speed of the cipher proposed in [17] is computed

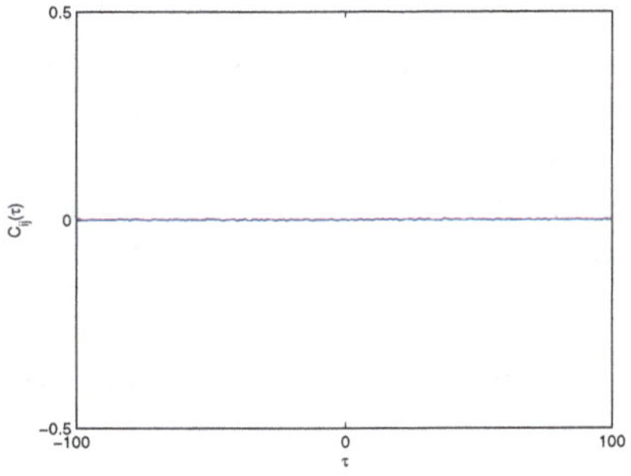

Fig. 9.21 Cross-correlation of any two keystreams

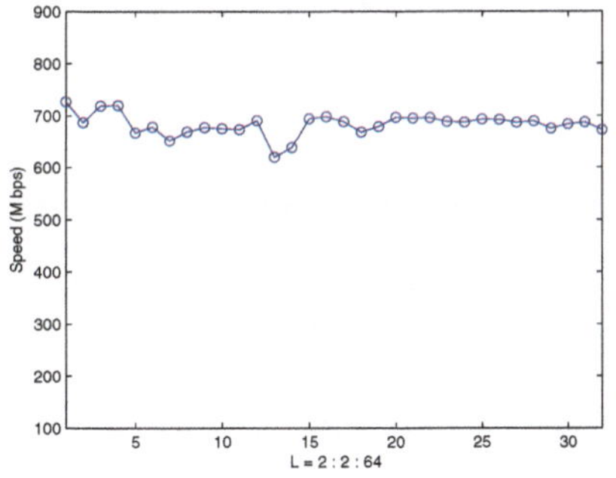

Fig. 9.22 The relation between the encryption speed and L

to be about 300M bits per second by our computer, which is much slower than the proposed cipher.

9.4 CML-Based Multimedia Cryptosystem

In this section, a multimedia cryptosystem based on a spatiotemporal chaotic system is proposed and implemented by a field programmable gate array (FPGA). The modification of the stream cipher proposed in section 9.3 is used in the cryptosystem. Since the generation processes of the multiple keystreams from the sites of the CML by a CPU of a computer are actually not in parallel, the cipher is implemented in an FPGA to generate the keysteams simultaneously. For implementation of the cipher in FPGA, the values of the cipher are digitized. The FPGA adopted in the cryptosystem is one Sparten-3 device produced by Xilinx company, i.e., XC3S400, because of its low cost and high efficiency of resource for implementing the digitized cipher. In designing the FPGA, a pipeline architecture is adopted to improve the usage efficiency of the FPGA. In the cryptosystem, data for encryption are input from a PC and transmitted to the FPGA. Data usually communicates via a serial port, a parallel port, a usual serial bus (USB), a peripheral component interconnect (PCI) bus, etc.. Within these communication modes, using a parallel port has some advantages: simple transfer protocol, high speed and easy implementation. Especially, an enhanced parallel port (EPP), which is a data transfer mode defined in IEEE 1284 standard, has simple transfer protocol and high transfer speed. Therefore, an EPP is applied in this system. Simulation shows that the data are communicated efficiently between a PC and an FPGA via the EPP. In addition, a user-friendly interface of the cryptosystem is designed with Visual C++. With the easy-to-handle interface, a user can encrypt/decryt text, image and audio files, and observe the results.

For testing the performance of the cryptosystem, the keyspace, the statistical properties, the security and the efficiency of the cryptosystem are investigated.

9.4.1 Design of CML-Based Multimedia Cryptosystem

The multimedia cryptosystem adopting a cipher based on a CML, which is implemented in a FPGA, is described in this section. The design of the cryptosystem consists of the FPGA implementation of the cipher, the EPP communication between a PC and the FPGA and a user-friendly interface, which are to be described in the following.

9.4.1.1 FPGA Implementation of the CML-Based Cipher

The multimedia cryptosystem adopts the cipher modified from that presented in section 9.3 [12]. Since the cipher (9.4) adopts double float precision, it should be digitized into integer domain for its implementation in an FPGA. In addition, to make the balance between efficiency and security of the cipher (9.4), the highest 32 bits in binary representations of the values are used as keystream. Similarly, only the highest 32-bit in binary representations of the values are used during iterations of the cipher. The modified cipher, named as digital CML-based cipher, is described as

$$\begin{aligned} x_{n+1}^{j} &= (((2^{32}-\varepsilon')f(x_n^j))_{h32}+(\varepsilon' f(x_n^{j-1}))_{h32})_{h32},\\ f(x_n^j) &= ((x_n^j << 2)_{h32}(2^{32}-x_n^j))_{h32},\\ K_n^j &= x_n^j\\ C_n^j &= M_n^j \oplus K_n^j, \end{aligned} \tag{9.9}$$

where "$(\cdot)_{h32}$" means to extract the highest 32 bits of the value in "()"; "$<< 2$" stands for right-shifting 2 bits since the parameter of the logistical map in each site is equal to 4 to achieve the strongest chaos of the CML [12]; $\varepsilon' = (\mathrm{int})\varepsilon \times 2^{32}$; ε is assumed as 0.95 to make the keystream possessing good statistical properties [11].

The FPGA implementation of each site in the cipher is similar. Taking the 1st site as an example, its implementation at one round is realized in the following four steps:

1. compute $1-x_n^1$ and $1-x_n^4$,
2. compute $f(x_n^1)$ and $f(x_n^4)$,
3. compute $(1-\varepsilon)f(x_n^1)$ and $\varepsilon f(x_n^4)$,
4. compute x_{n+1}^1,

To achieve the highest speed of the cipher with the limited resource of the FPGA, a pipeline architecture is adopted. Meantime, to use effectively the resource of the FPGA with the pipeline architecture, the CML in the cipher adopts 8 sites. Thus, 16 18×18 multipliers of Spaten3 series are needed simultaneously for four multiple arithmetic of two 32-bit integers. Therefore, the FPGA in the cryptosystem adopts the type XC3S400 of Spaten3 series. The simulation done in Modelsim 5.8 environment shows that all multipliers of the FPGA which are the key resource of a FPGA work simultaneously and one datum can be output from the CML at any clock cycle.

Moreover, to evaluate the efficiency of using the FPGA resources, the conceptual VHDL design definition is synthesized to generate the logical or physical representation for the targeted silicon device. Xilinx Synthesis Technology (XST) is used to synthesize the CML-based cipher by choosing the device xc3s400pq208. The result is shown in Table. 9.2, which indicates that all multipliers are used, therefore, the device xc3s400 can execute the algorithm well.

Table 9.2 Performance of the FPGA implementation of the CML-based cipher

CLK	Slices	Flip Flops	LUT-4	GCLK	MULT18X18s
150.060MHz	504 (14%)	545 (7%)	677(9%)	1(12%)	16 (100%)

9.4.1.2 EPP Communication between PC and FPGA

In the cryptosystem, data for encryption are input from a PC and transmitted to the FPGA. Data communication can be usually via a serial port, a parallel port, USB, a PCI bus, etc.. Within these communication modes, since the EPP has relatively simple transfer protocol and high transfer speed, the data between a PC and an FPGA are transferred via EPP in the presented cryptosystem. The implementation of EPP communication includes designing a state machine and optimizing.

Design of state machine: A state machine is used to realize the communication protocols of EPP in the FPGA. The adopted state machine is the so-called mealy state machine consisting of 5 states.

Optimization of communication: Since the transistor-transistor logic (TTL) of an EPP adopts $0 \sim 5v$ voltage standard, which can be easily perturbed, the output signal from a PC, such as a read/write control signal, a data selection signal and 8-bit signal in the data bus, should be filtered before entering into the FPGA. A low-pass filter using two-stage D flip-flops is employed here as filtering function.

The cryptosystem is simulated in Modelsim 5.8. The simulation result indicates that the data are successfully communicated between a PC and the FPGA, and encrypted/decrypted correctly.

9.4.1.3 Implementation of the Cryptosystem

An interface for a user to manipulate the cryptosystem is designed with Visual C++ 6.0, as shown in Fig. 9.23. The interface includes the buttons of inputting encryption/decryption keys and sending the keys. For text encryption/decryption, it supports the windows for inputting plaintexts and displaying ciphertexts and decrypted texts. For image encryption/decryption, plain-image, cipher-image and decrypted-image can be visible by choosing the corresponding check boxes. The cipher-image and decrypted image can also be saved by clicking the **Save** buttons. For audio encryption/decryption, by clicking **Read In Audio** button, one audio file can be chosen or a recorder can be open for recording audio files to be encrypted. Plain-audio, cipher-audio and decrypted-audio can be played by clicking corresponding buttons.

With the user-friendly interface, the sender can input plain-media and encryption key, encrypt the plain-media and display cipher-media in his PC; the receiver can input decryption keys, and display cipher-media sent by the sender and decrypted-media in his PC.

9.4.2 Performance Analysis

For measuring the performance of the cryptosystem, its cryptographic properties, security, speed and its effects of encrypting multimedia is investigated quantitively in this section.

9.4.2.1 Properties of the Cryptosystem

Since the cipher of the cryptosystem is modified from that in [12], the properties of the cryptosystem may be different from those of the CML-based cipher in [12]. Similar to the way in [12], the keyspace, the statistical properties of the keystreams,

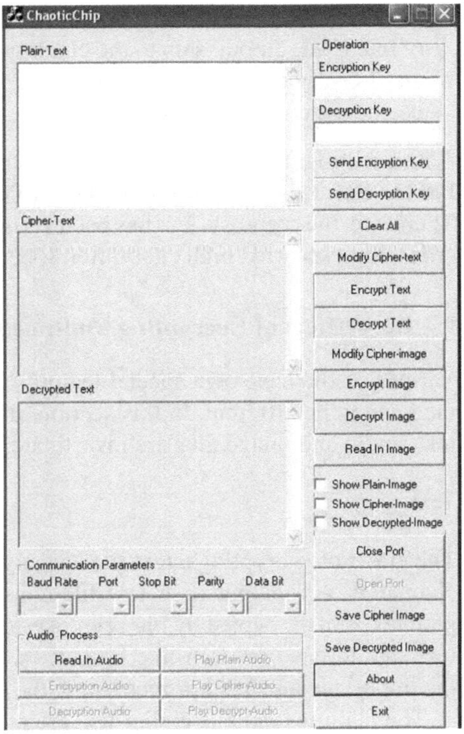

Fig. 9.23 The system interface

the security and the efficiency of the cryptosystem are analyzed, which results are described as follows.

Keyspace: Error function is used to determine the keyspace of the cryptosystem, which is equal to $2^{8\times 32} = 2^{256}$.

Statistical properties of the keystream: The important statistical properties, such as probability distribution, auto-correlation and run probability, of the keystream generated from the cryptosystem are numerically investigated. It is shown that the cryptosystem has uniform 1-order and 2-order probability distribution, δ-like auto-correlation and random-like run-probability. Additionally, NIST test suite is used to measure the randomness of the keystream. 20 keystream sequences of length 10^6 are tested here by the NIST statistical test suite. All passing rate of tests except for AET being 95% are equal to 100%, which indicates that the keystreams pass the NIST test suite.

Security: The security of the cryptosystem is evaluated by investigating its confusion and diffusion properties and using various typical attacks, such as the error function attack, the differential attack, the known-plaintext attack, the brute-force attack and the chosen-plaintext/ciphertext attack. It is shown that the brute force attack is the most efficient attack, which costs 2^{256}.

Efficiency: Since the FPGA adopts the pipeline architecture, the generation speed of one CML output is up to the clock speed in the FPGA. In the cases of adopting an FPGA with the clock frequency of 150.060MHz, a plaintext of 4.8G bits can be encrypted per second.

In addition, the cost of the cryptosystem is nearly equal to that of the adopted FPGA. Since the cost of the adopted FPGA is low, the cryptosystem is low-cost. Therefore, the cryptosystem has been verified to possess satisfactory statistical properties, high security, high encryption speed and low-cost.

9.4.2.2 Effect of Encrypting Multimedia

Various media have own special features, so the effect of encrypting various multimedia may be different. In this section, the effect of the cryptosystem's encrypting text, audio and image files are investigated, respectively.

Text

The effect of encrypting a text file is analyzed by taking a special file with identical characters as a plaintext. A text file consisting of 10000 "1" in byte is used as a plaintext and encrypted by our cryptosystem. To resist statistical attacks, the ciphertext should have random-like properties. The distribution and auto-correlation, as the important statistical properties, of the ciphertext are investigated.

The plain-text and its cipher-text are plotted in Figs. 9.24(a) and 9.24(b), respectively. Their histograms with 256 bins are obtained in Figs. 9.24(c) and 9.24(d), respectively. Since the text file consists of identical characters, its distribution is the worst. However, the ciphertext has nearly uniform distribution. To measure the uniformity of the generated ciphertext, the χ^2 test of the ciphertext's histogram is applied [29].

The statistic of the χ^2 test is described as

$$\chi^2 = \sum_{i=1}^{K} \frac{(o_i - e_i)^2}{e_i}, \tag{9.10}$$

where o_i, e_i and K are an observed frequency, an expected (theoretical) frequency and the number of distinct events, respectively. Here, K is equal to 255 due to the histogram having 256 bins, then $e_i = (int)\frac{10000}{256}$ and χ^2 of the histogram is computed as 223.88. With a significance level of 0.05, it is found that $\chi^2_{255;0:05} = 293.2478 > 223.88$, which implies that the distribution of the ciphertext is uniform.

In addition, the auto-correlations of the plaintext and the ciphertext are obtained and plotted in Figs. 9.24(e) and 9.24(f), respectively. It is seen that the ciphertext has δ-like auto-correlation, which is similar to random texts.

To check the diffusion with respect to the plaintext, the cross-correlation of the two ciphertexts from two plain-texts with the difference of only 1 bit is computed and plotted in Fig. 9.25.

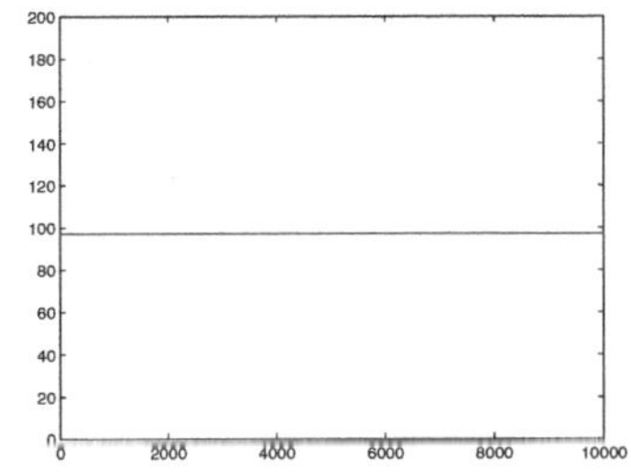

(a) Plain-text

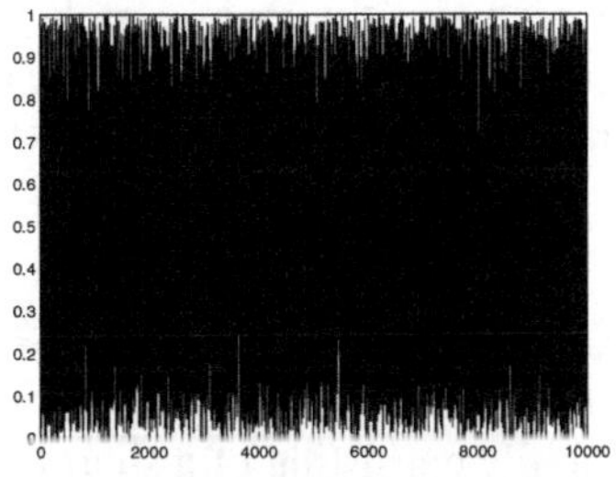

(b) Cipher-text

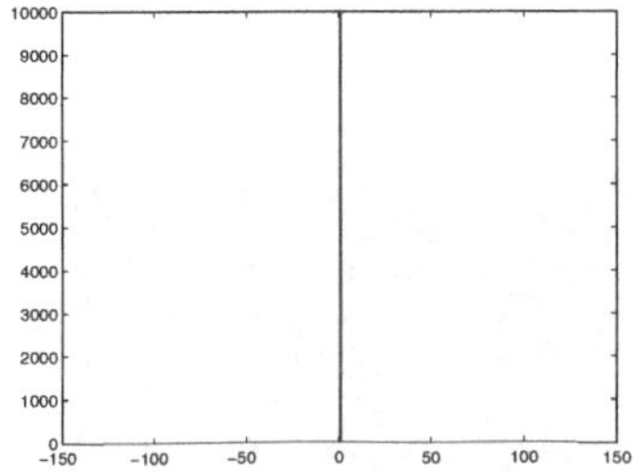

(c) Distribution of Plain-text

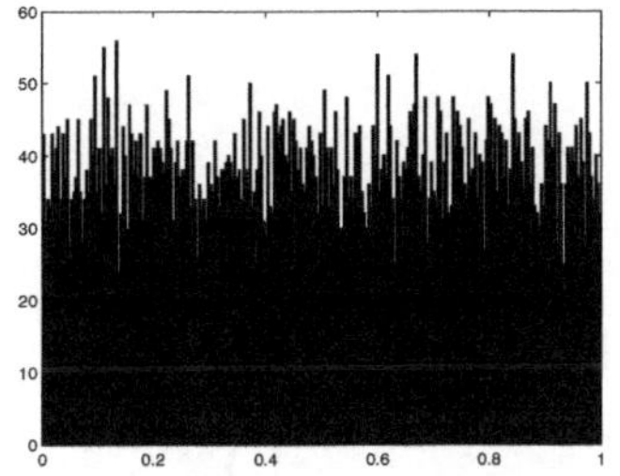

(d) Distribution of Cipher-text

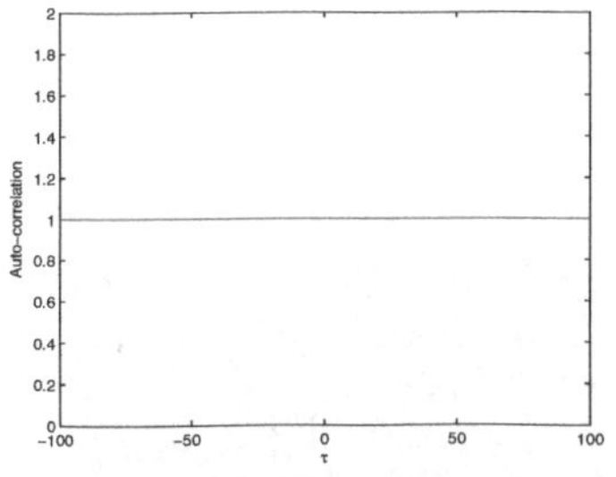

(e) Auto-correlation of Plain-text

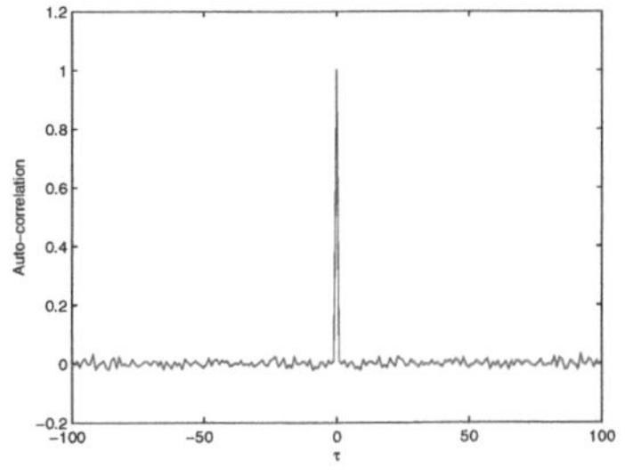

(f) Auto-correlation of Cipher-text

Fig. 9.24 Plain-text and cipher-text, and their distributions and Auto-correlation

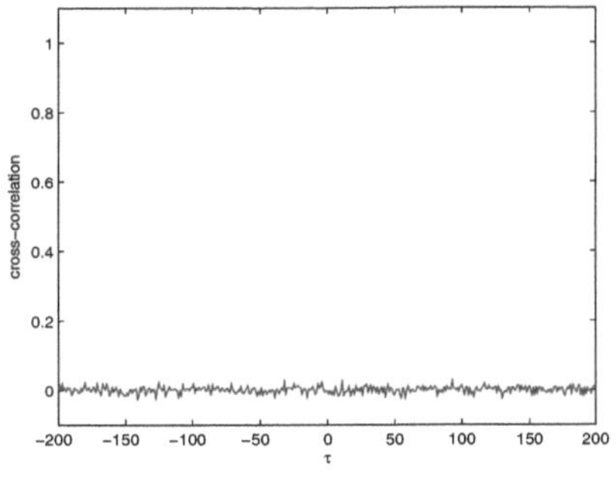

Fig. 9.25 The correlation of two cipher-texts from two plain-texts with tiny difference

It is seen that the cross-correlation is close-to-zero, which indicates that small change in a plain-text can influence over whole the cipher-text. Therefore, the cryptosystem can encrypt texts effectively.

Audio

Generally, the correlation of adjacent data in an audio file is stronger than that of a text file. To check the effect of encrypting an audio file with this cryptosystem, an audio file consisting of a short time of silence and several same words is tested. The plain-audio and its cipher-audio are plotted in Figs. 9.26(a) and 9.26(b), respectively. Their histograms with 1024 bins are obtained in Figs. 9.26(c) and 9.26(d),

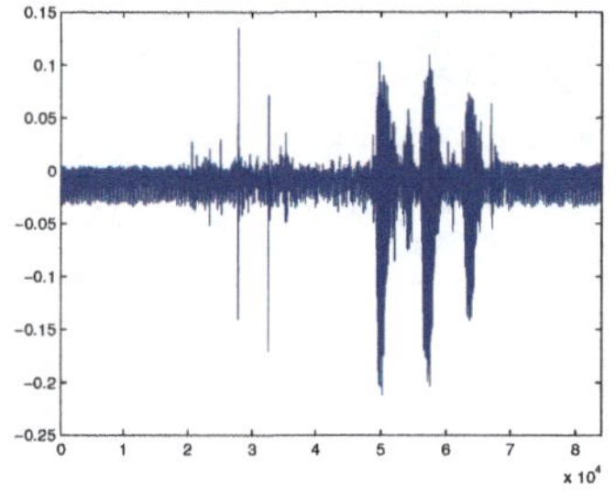

(a) Plain-audio

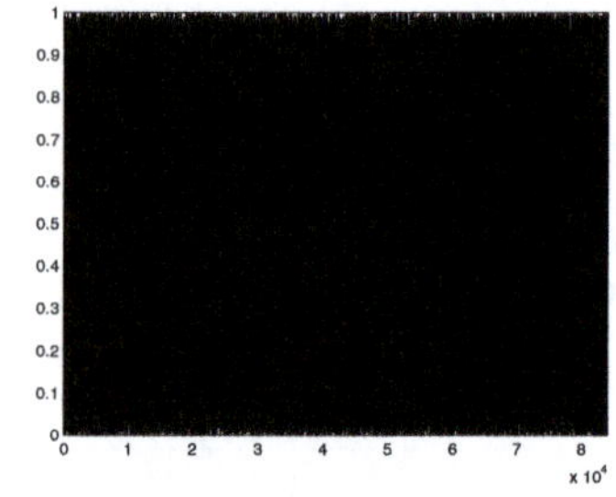

(b) Cipher-audio

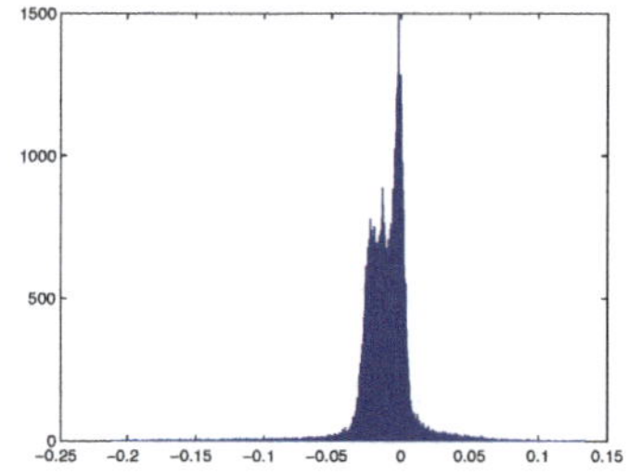

(c) Distribution of Plain-audio

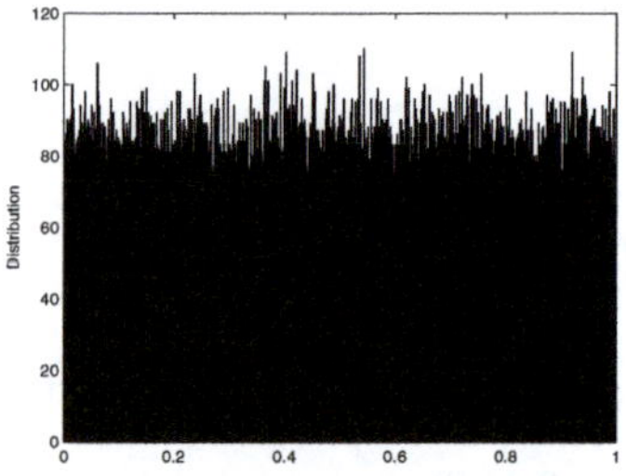

(d) Distribution of Cipher-audio

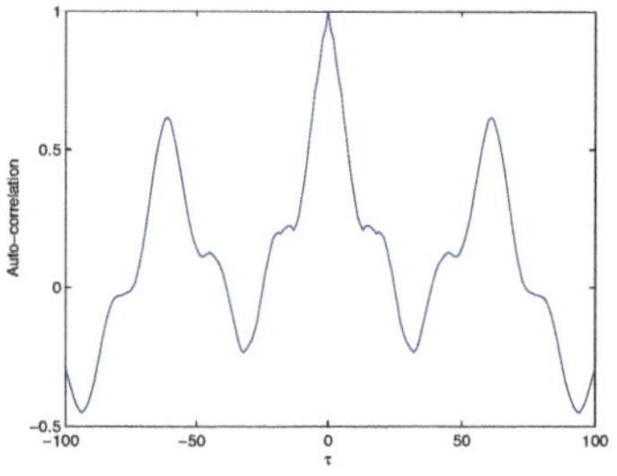

(e) Auto-correlation of Plain-audio

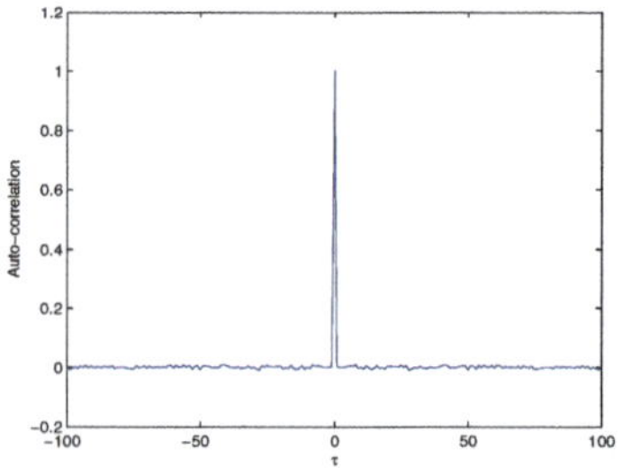

(f) Auto-correlation of Cipher-audio

Fig. 9.26 Plain-audio and cipher-audio, and their distributions and Auto-correlation

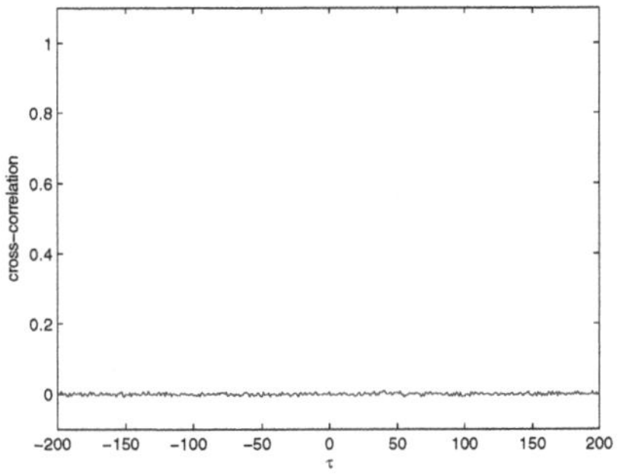

Fig. 9.27 The correlation of two cipher-Audios from two plain-Audios with tiny difference

respectively. It is shown that the cipher-audio has much more uniform distribution than the plain-audio. χ^2 test is also used here to check whether the distribution of the cipher-audio is uniform. χ^2 of the distribution of the cipher-image is computed as 926.54 and smaller than $\chi^2_{1023;0:05} = 1098.5208$ with a significance level of 0.05. Therefore, the distribution of the cipher-audio is regarded as uniform. In addition, the auto-correlations of the plain-audio and cipher-audio are obtained and plotted in Figs. 9.26(e) and 9.26(f), respectively. It is indicated that the cipher-audio has δ-like auto-correlation.

Moreover, the cross-correlation of two cipher-audios from two plain-audios with tiny difference is computed and plotted in Fig. 9.27.

The close-to-zero cross-correlation indicates the complete diffusion with respect to a plain-audio. As a result, the cryptosystem can encrypt an audio file effectively.

Image

It is known that adjacent pixels in an image have high correlation. In order to resist the statistical attacks, a cipher-image should possess uniform distribution and close-to-zero correlation of adjacent pixels. Moreover, to avoid the known-plaintext attack and the chosen-plaintext attack, the changes in the cipher-image should be significant even with a small change in the original plain-image; that is, the influence of one-pixel change in the plain-image should be on whole the cipher-image. Two measures, i.e., the number of pixel change rate (NPCR) and unified average changing intensity (UACI) [4], can be adopted to test the influence. The NPCR is used to measure the number of different pixels between two images. UACI is to measure the average intensity difference between two images. In the following, by taking the "Lena" gray image with 256×256 pixels, which is a popular image for general image analysis, as a plain-image, the distribution and correlation of adjacent pixels of the plain-image and its cipher-image, and NPCR and UACI of the cipher-images are analyzed.

Distribution: The Lena image and its histogram are plotted in Figs. 9.28(a) and 9.28(b), respectively.

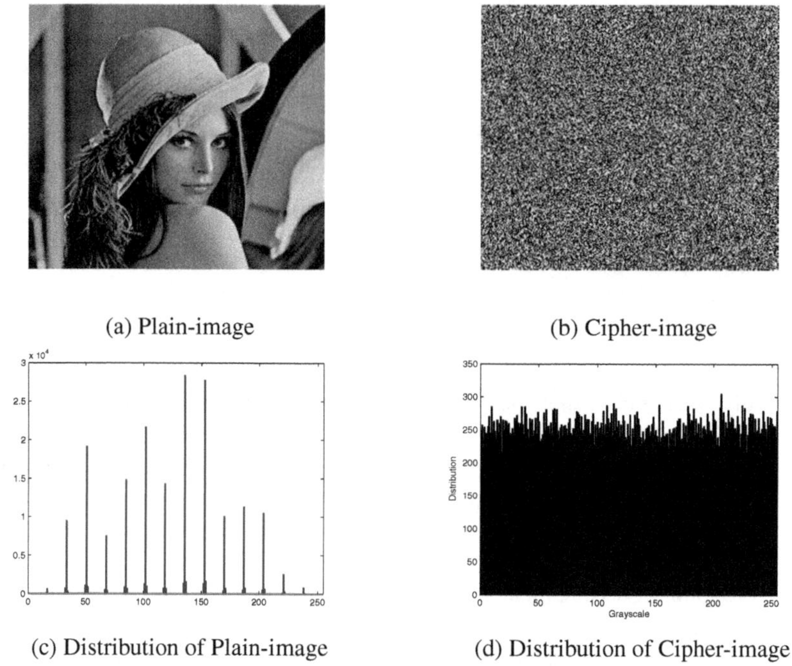

Fig. 9.28 Plain-image, cipher-image and their distributions

It is seen that the distribution of the plain-image is not uniform. In addition, the cipher-image and its gray histograms are plotted in Figs. 9.28(c) and 9.28(d), respectively. It is shown that the distribution of the cipher-image becomes uniform. This uniformity is further justified by the χ^2 test. According to Eq. (9.10), o_i is the observed occurrence frequencies of each gray level, which belongs to the range $[0,255]$; e_i is expected occurrence frequencies of each gray level and equal to 256 for the image with 256×256 pixels; k is 255 due to the number of gray levels as 256. χ^2 of the distribution of the cipher-image is computed as 256.09 and smaller than $\chi^2_{255;0:05} = 293.25$ with a significance level of 0.05, which verifies that the distribution of the cipher-image is uniform.

Correlation of adjacent pixels: For an ordinary image, each pixel is usually highly correlated with its adjacent pixels either in horizontal, vertical or diagonal directions. The correlation property in horizontal, vertical and diagonal direction can be quantified by the auto-correlation of the sequence consisting of pixels queuing row by row, column by column and diagonal row by diagonal row, respectively. The three correlation coefficients of the Lena image and those of its cipher-image are plotted in Fig. 9.29. It can be observed that the cipher-image obtained from the cryptosystem retains small correlation coefficients in all directions, which are similar to those of random image.

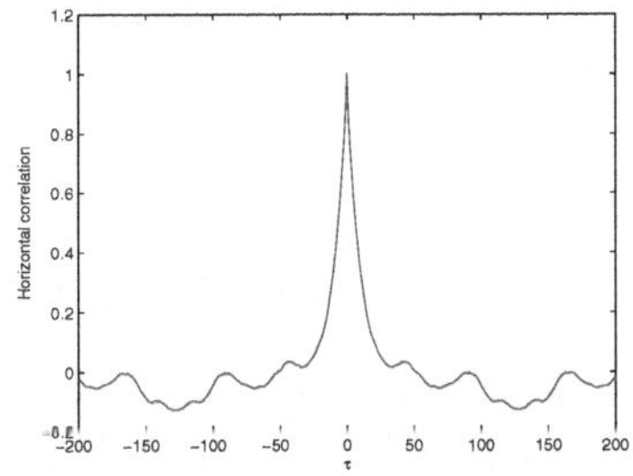

(a) Horizontal correlation of plain-image

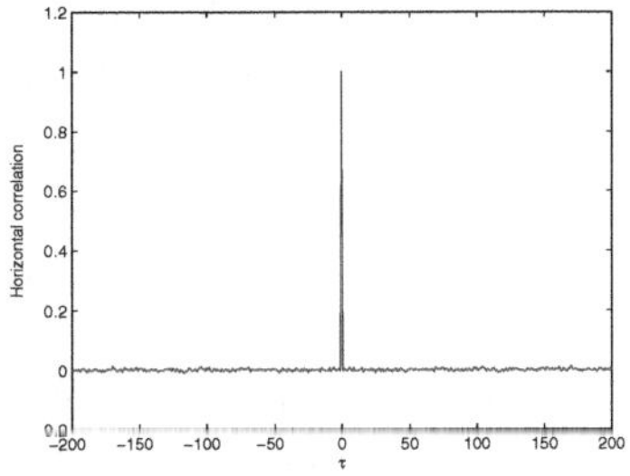

(b) Horizontal correlation of cipher-image

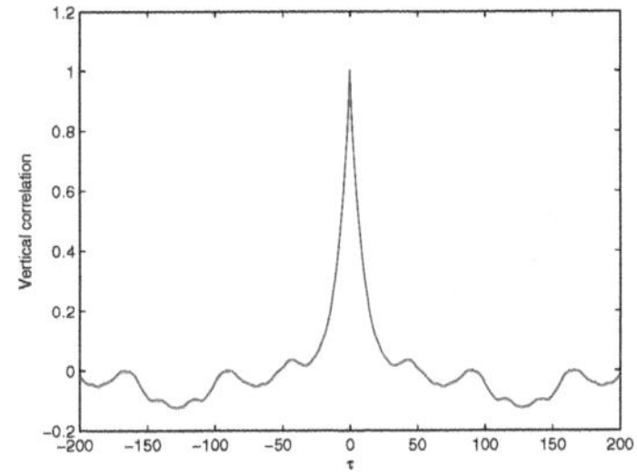

(c) Vertical correlation of plain-image

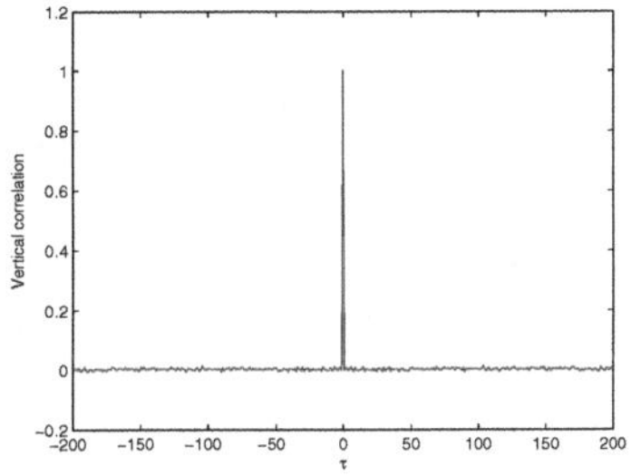

(d) Vertical correlation of cipher-image

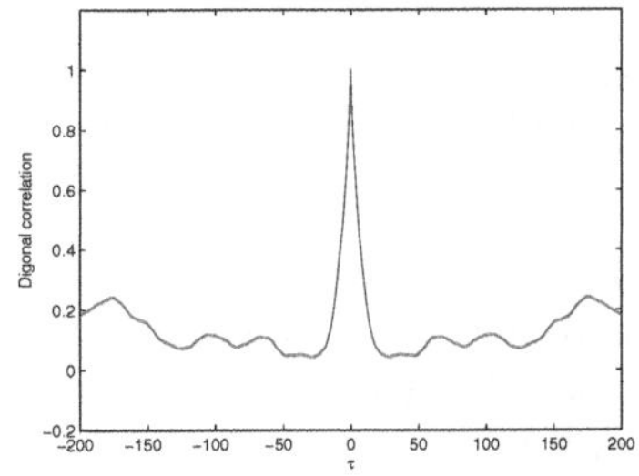

(e) Diagonal correlation of plain-image

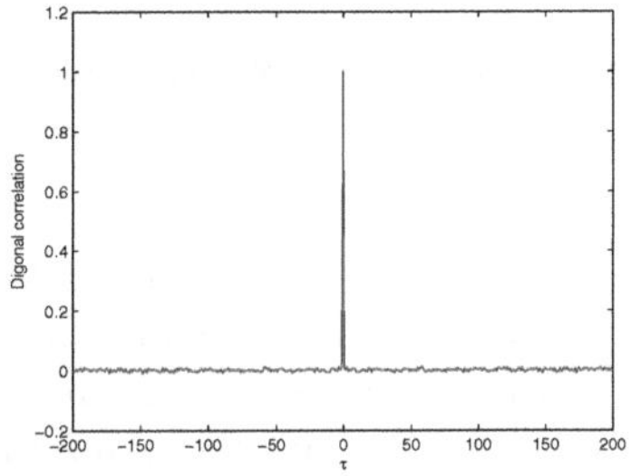

(f) Diagonal correlation of cipher-image

Fig. 9.29 Horizontal, vertical and diagonal correlation of plain-image and cipher-image

NPCR: Let $C(i,j)$ and $C'(i,j)$ be the pixel in the ith row and jth column of two images C and C', respectively, the NPCR can be defined as

$$\mathrm{NPCR} = \frac{\sum_{i,j}^{N} D(i,j)}{N} \times 100\%,$$

where N is the total number of pixels in the image and $D(i,j)$ is defined as

$$D(i,j) = \begin{cases} 0, & C(i,j) = C'(i,j) \\ 1, & C(i,j) \neq C'(i,j) \end{cases}$$

To check whether the NPCR of a cryptosystem is similar to that of random image, the NPCR of a random image is computed, which is given by

$$NPCR_R = 1 - 2^S,$$

where s is the number of bits representing 256 gray scales, therefore, equal to 8; that is, $NPCR_R = 0.9961$.

The NPCR of the cipher-images encrypted from two images with only one bit difference via the cryptosystem is computed as 0.996048, which is similar to that of random images.

UACI: UACI is defined as

$$\text{UACI} = \frac{1}{N}(\sum_{i,j}^{N} \frac{|C(i,j) - C'(i,j)|}{255}) \times 100\%$$

Similarly, to check whether the UACI of the cryptosystem is similar to that of random image, the UACI of a random image is computed, which is given by

$$UACI_R = \frac{\frac{1}{2^{2S-1}} \sum_{i=1}^{2^S-1} i(i+1)}{2^S - 1} = 0.3346354,$$

The UACI of the cipher-images encrypted from two images with only one bit difference via the cryptosystem is computed as 0.335383, which is similar to that of random images.

Therefore, the cryptosystem can also encrypt an image well.

In summary, since the plain-media with worse or the worst statistical properties can be encrypted to the cipher-media with random-like statistical properties, the cryptosystem is able to encrypt multimedia effectively.

9.5 Conclusion

Spatiotemporal chaos has advantages to ciphers because of tis inherent characteristics. This chapter has applied a typical spatiotemporal chaotic system, i.e., a CML, in cryptography. Firstly, a multiple-output PRBG using a CML is designed. The statistical properties, such as probability density function (PDF), linear complexity, auto-correlation and cross-correlation of the PRBGs based on various digitization methods have been investigated. It has shown that binary-representation method is the best one. To determine the CMLs, from which the resulting PRBGs have satisfactory properties, six PRBGs based on six different CMLs have been investigated. The six CMLs consist of three simple chaotic systems, i.e., logistic map, skew-tent map and r-adic map, with two simplest coupling methods, i.e., one-way coupling and diffusive coupling, respectively. PDF, auto-correlation, cross-correlation, statistical test and cycle length of the six PRBGs with various parameters have been investigated so as to determine the parameter intervals within which the PRBGs have satisfactory properties. It has been indicated that a one-way coupled logistic-map lattice with certain parameters has the best properties. This research results in criteria for designing PRBGs with proper performance. Secondly, a stream cipher employing a one-way coupled logistic-map lattice with certain parameters has been designed. The security of the stream cipher has been tested by attacking it via typical attack methods and analyzing its cryptographic properties. Moreover, the efficiency of the stream cipher has been analyzed. It has shown that the cipher has higher security, higher efficiency and lower costs by comparing with Hu's stream cipher of a complicated configuration. Finally, a multimedia cryptosystem based on the proposed stream cipher has been designed and implemented in a FPGA. The EPP is used to communicate data between a PC and the FPGA. A user-friendly interface is designed with Visual C++ 6.0, with which text, image and audio can be encrypted and decrypted successfully. The properties of the cryptosystem, such as the sensitivity to the key, speed and efficiency of the FPGA, have been analyzed to be satisfactory.

References

1. Baranovsky, A., Daems, D.: Design of one-dimensional chaotic maps with prescribed statistical properties. Int. J. Bifurcat Chaos Appl. Sci. Eng. 5, 1585–1598 (1995)
2. Batista, A.M., Pinto, S.E., Viana, R.L., Lopes, S.R.: Lyapunov spectrum and synchronization of piecewise linear map lattiecs with power-law coupling. Phys. Rev. E 65 (2002)
3. Baptista, M.S.: Cryptography with chaos. Phys. Lett. A 240, 50–54 (1999)
4. Chen, G., Mao, Y., Chui, C.: A symmetric image encryption scheme based on 3rd chaotic cat maps. Chaos, Solitons & Fractals 21, 749–761 (2003)
5. Garcia, P., Parravano, A., Cosenza, M., Jimenez, J., Marcano, A.: Coupled map networks as communication schemes. Phys. Rev. E 65, 195–201 (2002)
6. Gotz, M., Kelber, K., Schwarz, W.: Discrete-time chaotic encryption systems-part i: Statistical design approach. IEEE Trans. Circ. Syst. Fund. Theor. Appl. 44(10), 963–970 (1997)
7. Kaneko, K.: Theory and Application of Coupled Map Lattices. John Wiley and Sons, New York (1993)
8. Kocarev, L., Jakimoski, G.: Pseudorandom bits generated by chaotic maps. IEEE Trans. Circ. Syst. Fund. Theor. Appl. 50, 123–126 (2003)
9. Kohda, T., Tsuneda, A.: Pseudonoise sequence by chaotic nonlinear maps and their correlation properties. IEICE Trans. Commun. E76-B, 855–862 (1993)
10. Lasota, A., Mackey, M.C.: Chaos, Fractals, and Noise: stochastic aspects of dynamics. Springer, New York (1997)
11. Li, P., Li, Z., Halang, W.A., Chen, G.R.: Analysis of a multiple output pseudo-random-bit generator based on a spatiotemporal chaotic system. Int. J. Bifurcat Chaos Appl. Sci. Eng. 16(10), 2949–2963 (2006)
12. Li, P., Li, Z., Halang, W.A., Chen, G.R.: A stream cipher based on a spatiotemporal chaotic system. Chaos, Solitons & Fractals 32(5), 1867–1876 (2007)
13. Li, S.J.: Analyses and New Designs of Digital Chaotic Ciphers. Ph.D thesis, School of Electronics and Information Engineering, Xi'an Jiaotong University, Xi'an, China (2003)

14. Shujun, L., Xuanqin, M., Yuanlong, C.: Pseudo-random bit generator based on couple chaotic systems and its applications in stream-cipher cryptography. In: Pandu Rangan, C., Ding, C. (eds.) INDOCRYPT 2001. LNCS, vol. 2247, pp. 316–329. Springer, Heidelberg (2001)
15. Li, S.J., Chen, G.R., Qin, M.: On the dynamical degradation of digital piecewise linear chaotic maps. Int. J. Bifurcat Chaos Appl. Sci. Eng. 15(10), 3119–3151 (2005)
16. Li, S., Álvarez, G., Chen, G.R.: Breaking a chaos-based secure communication scheme designed by an improved modulation method. Chaos, Soliton & Fractals 25, 109–120 (2005)
17. Lu, H., Wang, S., Li, X., Tang, G., Kuang, J., Ye, W., Hu, G.: A new spatiotemporally chaotic cryptosystem and its security and performance analyses. Chaos 14(3), 617–629 (2004)
18. Masuda, N., Aihara, K.: Cryptosystems with discretized chaotic maps. IEEE Trans. Circ. Syst. Fund. Theor. Appl. 49(1), 28–40 (2002)
19. Menezes, A., Oorschot, P.V., Vanstone, S.: Handbook of Applied Cryptography. CRC Press, Boca Raton (1997)
20. NIST, Security requirements for cryptographic modules (FIPS pub 140-2) (2001), `http://csrc.nist.gov/publications/fips/fips140-2`
21. Sang, T., Wang, R., Yan, Y.: Clock-controlled chaotic keystream generators. Electronics Letters 34(20), 1932–1934 (1998)
22. Schneier, B.: Applied Cryptography: Protocols, algorithms, and source code in C. John Wiley and Sons, New York (1996)
23. Schuster, H.G.: Handbook of Chaos Control. WILEY-VCH, Weinheim (1999)
24. Soto, J.: Statistical testing of random number generators (1999), `http://csrc.nist.gov/rng/rng5.html`
25. Stojanovski, T., Kocarev, L.: Chaos-based random number generators-part i: Analysis. IEEE Trans. Circ. Syst. Fund. Theor. Appl. 48(3), 281–288 (2001)
26. Tang, G., Wang, S., Lu, H., Hu, G.: Chaos-based cryptograph incorporated with S-box algebraic operation. Phys. Lett. A 318, 388–398 (2003)
27. Wang, S., Liu, W., Lu, H., Kuang, J., Hu, G.: Periodicity of chaotic trajectories in realizations of finite computer precisions and its implication in chaos communications. Int. J. Mod. Phys. B 18(17-19), 2617–2622 (2004)
28. Wang, S., Ye, W., Lu, H., Kuang, J., Li, J., Luo, Y., Hu, G.: A spatiotemporal-chaos-based encryption having overall properties considerably better than advanced encryption standard. Comm. Theor. Phys. 40, 57–60 (2003)
29. Wikipedia (2006) Chi Test, `http://en.wikipedia.org/wiki/Pearson%7s-chi-square-test`
30. Yang, T.: A survey of chaotic secure communication systems. International Journal of Computational Cognition 2(2), 81–130 (2004)
31. Ye, W., Dai, Q., Wang, S., Lu, H., Kuang, J., Zhao, Z., Zhu, X., Tang, G., Huang, R., Hu, G.: Experimental realization of a highly secure chaos communication under strong channel noise. Phys. Lett. A 330, 75–84 (2004)
32. Zhang, H., Wang, H., Chen, W.: Oversampled chaotic binary sequences with good security. J. Circ. Syst. Comput. 11, 173–185 (2002)
33. Zhou, H., Ling, X.: Problems with the chaotic inverse system encryption approach. IEEE Trans. Circ. Syst. Fund. Theor. Appl. 44(3), 268–271 (1997)

Chapter 10
Evolutionary Decryption of Chaotically Encrypted Information

Ivan Zelinka and Roman Jasek

Abstract. This chapter introduces the concept of decryption of chaotically encrypted information. Five evolutionary algorithms have been used for chaos synchronization here: differential evolution, self-organizing migrating algorithm, genetic algorithm, simulated annealing and evolutionary strategies in a total of 15 versions. The main aim was to ascertain if evolutionary algorithms are able to identify the "key" (control parameter) of the chaotic system, which was used to encrypt information. The proposed scheme is based on the extended map of Clifford strange attractor, where each dimension has a specific role in the encryption process. Investigation consists of one case study. All the algorithms was 100 times repeated in order to show and check robustness of the proposed methods and experiment configurations. All data were processed in order to get summarized results and graphs.

10.1 Introduction

Chaotic systems are extremely sensitive to initial conditions and this feature can be very helpful in the field of cryptography. Various encryption schemes use chaotic systems for encryption key generation and this key is then used for pixel permutation and pixel diffusion. But chaotic systems and their maps can be used directly

Ivan Zelinka
Tomas Bata University in Zlin, Faculty of Applied Informatics, Nad Stranemi 4511, Zlin 76001, Czech Republic
and
VSB-TUO, Faculty of Electrical Engineering and Computer Science, 17. listopadu 15, 708 33 Ostrava-Poruba, Czech Republic
e-mail: zelinka@fai.utb.cz

Roman Jasek
Tomas Bata University in Zlin, Faculty of Applied Informatics, Nad Stranemi 4511, Zlin 76001, Czech Republic
e-mail: jasek@fai.utb.cz

I. Zelinka et al. (Eds.): Evolutionary Algorithms and Chaotic Systems, SCI 267, pp. 329–343.
springerlink.com

for encryption purpose. The proposed scheme is based on the extended map of the Clifford strange attractor, where each dimension has a specific role in the encryption process. Two dimensions are used for pixel permutation and the third dimension is used for pixel diffusion. The theoretical and simulation results prove many properties of this scheme such as large key space and high security.

Image is a multimedia signal providing the most information to a person. For this reason the question appears as to which way can be signal be secured against unauthorized reading e.g. in medicine or military fields. Position permutation and diffusion of the pixels belongs to basic methods of image encryption. Their combination leads to better security against known attacks and is very often has found practical usage. However, these methods remains open for various encryption algorithms and that is why the knowledge of chaotic systems can be useful. These systems are extremely sensitive to initial conditions and thus they are suitable candidates in the field of cryptography. Many papers have been written on this theme for that very reason.

Chaos-based image encryption is discussed in detail in the previous chapter as well as in [14]. Most of the papers have described the process of the generation of the time series based on a chaotic map. These series are used for the creation of the binary sequence as an encryption key and pixels of plain image are then rearranged and XOR operated with this key. For example in [6] three logistic maps are used in key stream generator and this improved the linear complexity of key stream. Each paper proposed various type of key generator or improvements of chaotic encryptions in terms of security and speed [17], [2], [10] but only a few of them show a different way of encryption, such as using hyper-chaotic system for confusing the relationship between the plain-image and the cipher-image [7], Lorentz system for key-stream generation [5] or S-box algebraic operations [11], [1]. When other methods such as image encryption in wavelet domain proposed in [16] are used with chaos-based encryption scheme, we should expect interesting results. The results for the audio signals are presented in [8], where the wavelet coefficients were modified and the audio signal becomes inaudible.

The aim of this chapter is to show that evolutionary algorithms are capable, at least under strong simplifications, of decrypting information, which has been encrypted by chaotic dynamics. We have used results from [9]. The proposed scheme in this chapter uses the formula of a strange attractor for encryption purposes. It does not create any encryption key but uses attractor map for pixel permutation and diffusion directly. Parameters of attractor map play the role of encryption keys here and key-space is very large due to their non-integer character. Chaos based encryption has been done by means of so called Clifford system (attractor), which is depicted in Fig. 10.1 - 10.3.

$$\begin{aligned} x_n &= \sin(ay_n) + c\cos(ax_n) \\ y_n &= \sin(bx_n) + d\cos(by_n) \end{aligned} \tag{10.1}$$

$$\begin{aligned} x_n &= \sin(ay_n) + c\cos(ax_n) \\ y_n &= \sin(bx_n) + d\cos(by_n) \\ z_n &= \sin(ey_n) + f\cos(ez_n) \end{aligned} \tag{10.2}$$

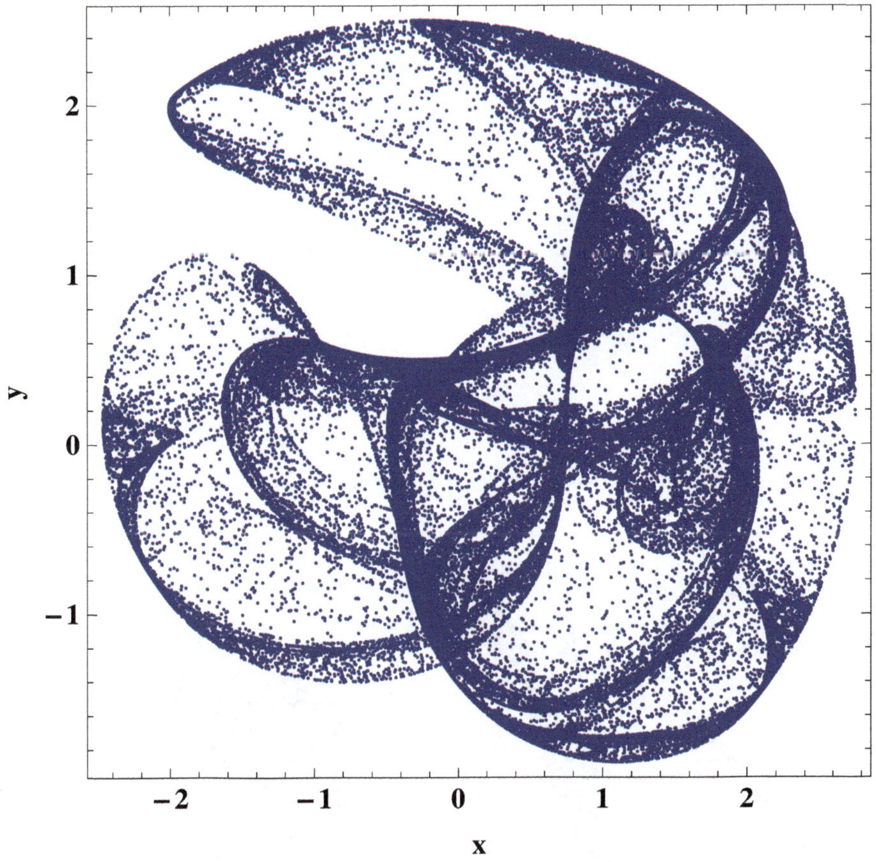

Fig. 10.1 Clifford attractor according to eq. (10.1).

In the research work of [9], encryption and its robustness with Clifford chaotic system use has been tested. The tested "message" for encryption were two pictures, see Fig. 10.4 and Fig. 10.5. In the encryption scheme, the Clifford attractor was used. It belongs to the trigonometric strange attractors and is described by eq. (10.1) and eq. (10.2). Fig. 10.6 shows the flowchart of this encryption scheme. The main parameters (key) are used for the iterative process of the Clifford system. Pixel of image is used as the initial value of Clifford system. New positions and modification value is gained after iterations and quantization. These positions are then used for pixel permutation and the modification value is XOR operated with original pixel value and the value of the previous pixel. Encrypted pixel is gained this way.

When the proposed schema of encryption (for more see [9]) is used, then one can obtain a picture as in Fig. 10.7. Histograms related to the original Lena and its encrypted version are depicted in Fig. 10.8 and Fig. 10.9. Both kind of pictures shows (of course there is also rigorous mathematical background) that the picture is really well encrypted.

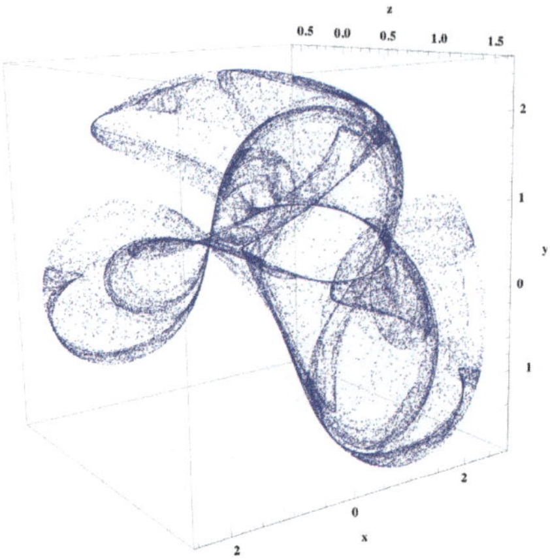

Fig. 10.2 Clifford attractor according to eq. (10.2)...

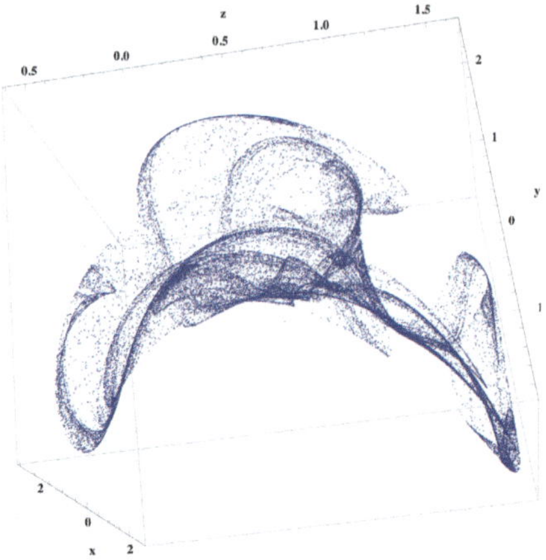

Fig. 10.3 ... and another 3D view.

Fig. 10.4 Lena

Fig. 10.5 Man with camera.

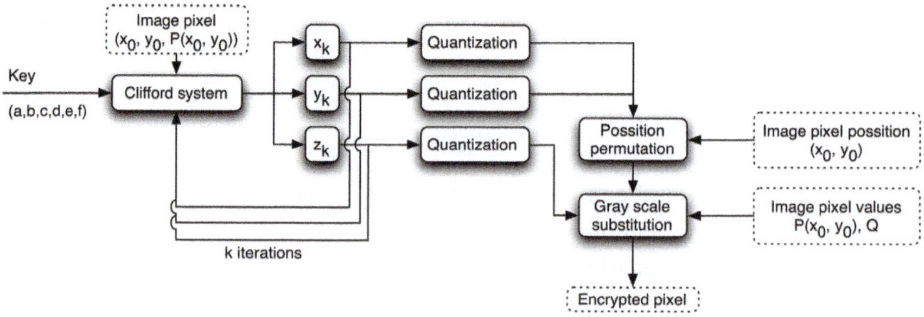

Fig. 10.6 Scheme with encoding.

Fig. 10.7 Encrypted Lena.

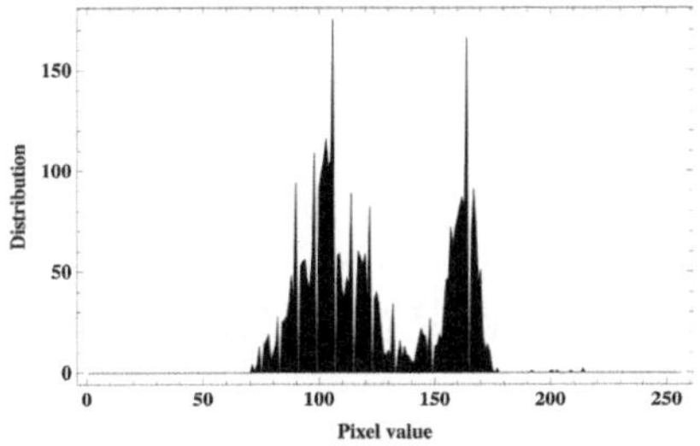

Fig. 10.8 Histogram of original Lena figure...

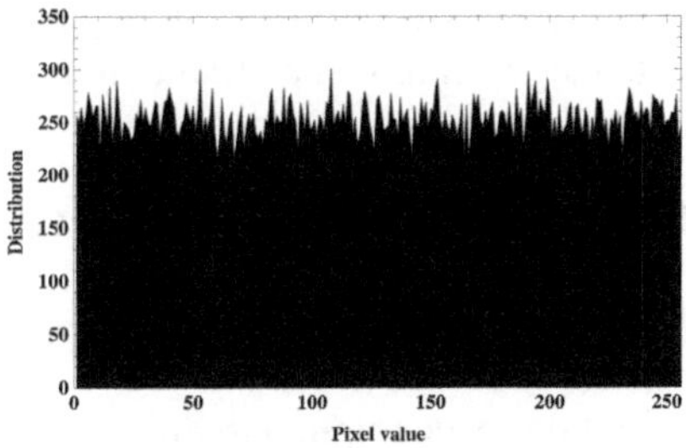

Fig. 10.9 ...and after encryption.

10.2 Motivation

Motivation of this research is very simple. Chaos based encryption is under intensive research attention today and encryption itself is vitally important for various communities, from industrial to government. We would like to ascertain if it is possible to identify key used for encryption by means of evolutionary algorithms.

Good encryption scheme must be resistant against any brute-force attacks, so the key space must be too large. The total precision of a common PC processor is 16 decimal digits, therefore the number of different combinations of one parameter is 10^{16} and it corresponds approximately to 253 size key space. Six attractor parameters are used in the proposed scheme; hence the key space is enlarged to 2^{318}. Also, the number of iterations k of Clifford system eq. (10.2) and the number of encryption rounds m can be considered as keys. Thus, the key space of the proposed scheme is large enough to make the classical brute-force attack infeasible. Table 10.1 shows comparison of key spaces of various encryption schemes. Our proposed encryption scheme has the largest key space.

Table 10.1 Key space comparison

Encryption scheme	Key space
Proposed in [9]	2^{318}
[14]	2^{128}
[6]	2^{158}
[7]	2^{232}
[8]	2^{256}

10.3 Selected Evolutionary Algorithm – A Brief Introduction

For the numerical and symbolic experiments described here, stochastic optimization algorithms such as Differential Evolution (DE) [15], Self Organizing Migrating Algorithm (SOMA) [18], Genetic Algorithms (GA) [12], Simulated Annealing (SA) [13], [4] and Evolutionary Strategies (ES) [3] were selected. Description of all selected algorithms can be found in the mentioned references or in Chapter 6.

10.4 Evolutionary Decryption

10.4.1 Used Hardware, Problem Selection and Case Studies

Evolutionary decryption in this case study has been done on a specialized grid computer. This grid computer consist of two special Apple servers (for pictures, see Chapter 6). A total of 78 CPUs were available for computation. This grid has been used for calculations so that each CPU has been used like a single processor and thus a rich set of statistically repeated experiments were possible which are not time dependent. Typical parallel computing has been avoided in experiments described here.

10.4.2 Cost Function

The Lena picture has been encrypted by eq. (10.2) with parameters defined as: a = -1.85, b = 1.48, c = -1.55, d = -1.87, e = -4.32, f = 0.63. In [9] the encrypted picture was successfully tested for key sensitivity. The set of keys in [9] are very similar, only one parameter is different with minimal divergence ($b = 1.4800001$). This small difference is enough to get after the decryption of a noisy picture as in Fig. 10.7. To test key sensitivity, which is based on encryption of the "Lena"

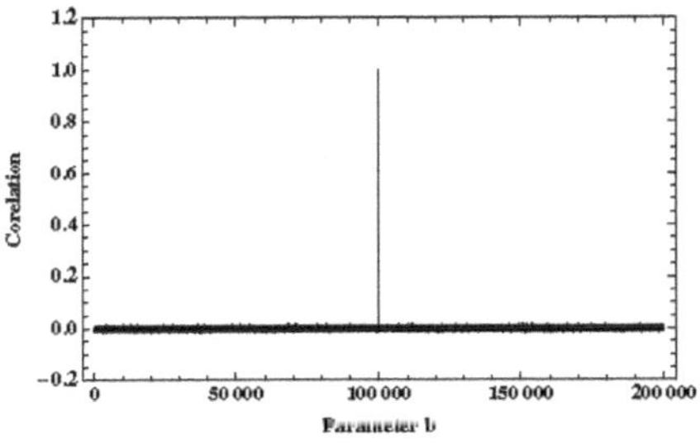

Fig. 10.10 Correlation of Lena picture, sharp peak at position 100000 represent solution - right estimation of the parameter b.

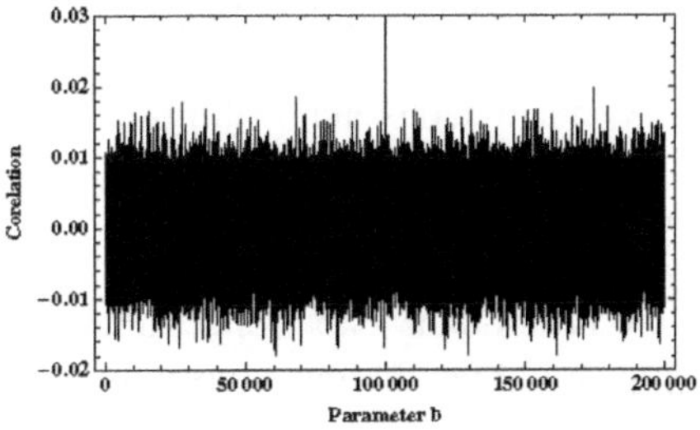

Fig. 10.11 Detail view.

image by mentioned setting, cross-correlation of their encrypted forms was then computed. Fig. 10.10 and Fig. 10.11 shows cross-correlation of images encrypted by these two different set of keys. Correlation value does not exceed 0.02. This implies very low correlation and very low similarity of images and their pixels. In general, adjacent pixels of the most plain-images are highly correlated. One of the requirements of an effective image encryption process is to generate encrypted image with low correlation of adjacent pixels. Correlation between two horizontally, vertically and diagonally adjacent pixels of original and encrypted image was also analyzed in [9].

The fitness (cost function) has been calculated very simply. In fact it was a simple search for such a value of parameter b so that cross-correlation value was equal to 1, i.e. right parameter of b was found. In total, 6 parameters were used like the key. To simplify the situation and make calculation time shorter, only parameter b has been selected for evolutionary estimation. Also, another important note should be mentioned here: numerical accuracy. Parameter b has been estimated with different level on numerical precision. The largest was for $\Delta b = 1 \times 10^{-15}$, which was used to generate in total 200 000 data points around the right value of b. For this type of precision, a tiny region of parameter b has been explored.

Comparing to another case studies, reported in this book, similarity between two kind of behavior and other parameters was not measured. Only similarity, via cross correlation, has been measured. Due to the chaotic nature of cost function landscape (Fig. 10.11), it was near to random search, thanks to the sophisticated search process. The cost function can be simply described by eq. (10.3). From that viewpoint, it behaves as a blind search, because on cross-correlation graphs, the general trend is completely flat.

$$\begin{array}{l} if \\ \quad b\ is\ such\ that\ cross-correlation = 1\ then\ stop \\ else \\ \quad continue\ in\ evolution \\ end \end{array} \tag{10.3}$$

10.4.3 Parameter Setting

The control parameter settings have been found empirically and are given in Tables 10.2 - 10.7. Number of cost function evaluations was not an objective in this study. Only one objective was there - to successfully estimate part of the encrypting key. We would like to note that settings of all used algorithms here, has been based on our preliminary experiences and certainly can be improved. However, this topic is quite numerically time consuming, so we let this topic open for future research.

Table 10.2 Algorithms abbreviation

Algorithm	Version	Abbreviation
Differential Evolution	DEBest1JIter	D1
	DEBest2Bin	D2
	DELocalToBest	D3
	DERand1Bin	D4
	DERand1DIter	D5
	DERand2Bin	D6
Evolutionary strategies	(μ,λ)	ES1
Evolutionary strategies	$(\mu+\lambda)$	ES2
Genetic Algorithm		G
Simulated annealing with elitism		SA1
Simulated annealing without elitism		SA2
SOMA	AllToAllAdaptive	S1
	AllToAll	S2
	AllToOne	S3
	AllToOneRandomly	S4

Table 10.3 DE setting.

Parameter	Value
NP	500
F	0.9
CR	0.3
Generations	500
Individual Length	1

Table 10.4 ES setting.

Parameter	Value
μ ,λ	500
σ	1
Iterations	100
Individual Length	1

Table 10.5 GA setting.

Parameter	Value
Population size	500
Mutation	0.4
Generations	561
Individual Length	

Table 10.6 SA setting.

Parameter	Value
No. of particles	500
σ	0.5
$k_{\max}$	66
T_{min}	0.0001
T_{max}	1000
α	0.9
Individual Length	1

Table 10.7 SOMA setting.

Parameter	Value
PathLength	3
Step	.11
PRT	1
PopSize	500
Migrations	10
MinDiv	-0.1
Individual Length	1

All algorithms (SOMA, DE, SA, GA, ES) have been evaluated 100 times in order to find the optimum of both case studies. The primary aim of this comparative study is not to show which algorithm is better and worst, but to show whether evolutionary synchronization can be used for decryption of chaotically encrypted information. Comparing to the other case studies reported in this book, population size is in this application is set to quite a high number (500). This number has not been selected randomly, but was obtained after a very simple set of simulations. Before the population size has been determined, a simple investigation on how dependent successful decryption is on population size was conducted. Fig. 10.12 captures this dependance. Straightforward dependance on population size is clearly visible there. When population size is more than 300, then evolutionary algorithms are capable of finding a larger number of successful decryptions, compared to unsuccessful ones. This is the reason why 500 individuals has been set for each algorithm.

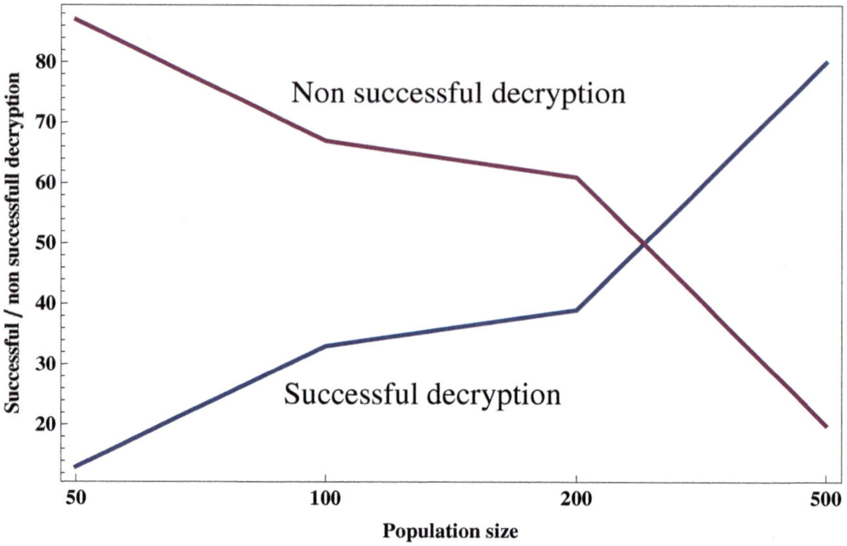

Fig. 10.12 Dependance of the number of successful decryptions on population size.

10.4.4 Experimental Results

Outputs of all simulations is depicted in Fig. 10.13 and Fig. 10.14, which shows results of all 100 simulations. In Fig. 10.13 one can see minimal, average as well as maximal number of cost function evaluations to get successful decryption. We have

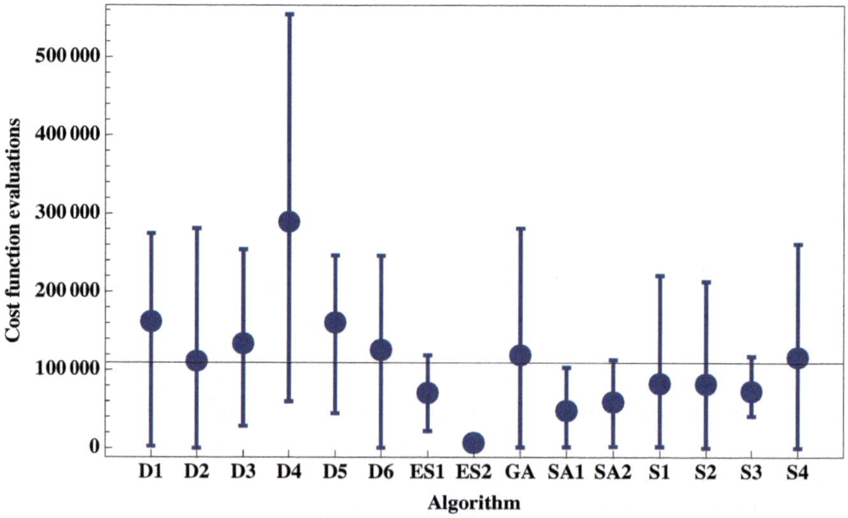

Fig. 10.13 Number of cost function evaluations needed to successfully decrypt figure of Lena. Horizontal line is an average of all.

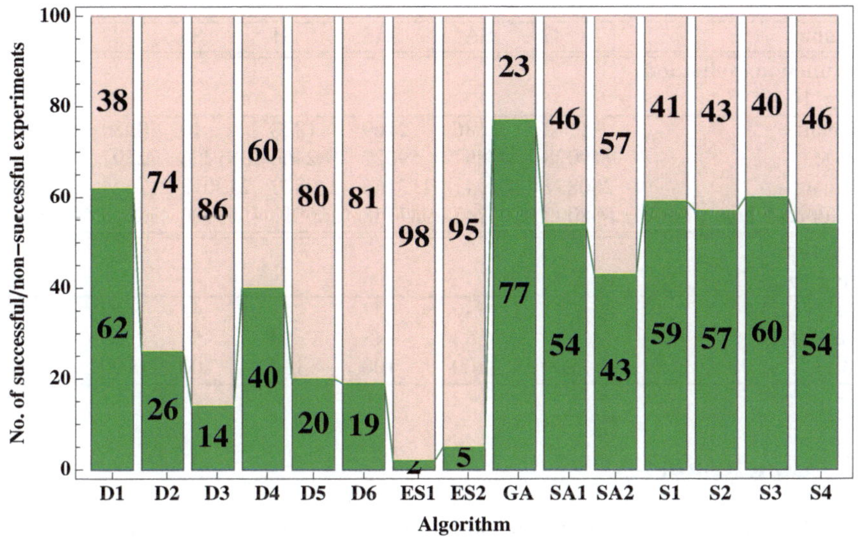

Fig. 10.14 No. of successful/non-successful decryptions. Each bar is divided into two parts. The upper part represent number of non-successful decryption, the lower one successful decryption.

to note that in this figure are reported **only** positive results. All results (positive / negative; successful / non successful;...) are reported in Fig. 10.14. All informations together are summarized in Tables 10.8 - 10.9.

Table 10.8 Experiment summarization, part 1.

Algorithm	D1	D2	D3	D4	D5	D6	ES1	ES2
Cost function evaluations								
see Fig. 10.13								
Minimum	2552	51	28282	59560	44722	460	22001	192
Average	161918	111398	134070	289378	160836	125863	70601	6904
Maximum	274100	280484	253892	554632	246312	245866	119201	12009
Total for each algorithm	10038904	2896337	670348	10706974	1769198	1636222	141202	55231
Decryption								
see. Fig 10.14								
Non-successful	38	74	86	60	80	81	98	95
Successful	62	26	14	40	20	19	2	5
Total	100	100	100	100	100	100	100	100

Table 10.9 Experiment summarization, part 2.

Algorithm	G	SA1	SA2	S1	S2	S3	S4
Cost function evaluations see Fig. 10.13							
Minimum	1112	1586	2169	1828	10	40980	135
Average	119028	47897	59022	82466	81756	72597	116153
Maximum	280837	102941	113166	220822	213012	117624	260978
Total for each algorithm	8808093	2203247	1947736	4865512	4660108	435580	5226899
Decryption see. Fig 10.14							
Non-successful	23	46	57	41	43	40	46
Successful	77	54	43	59	57	60	54
Total	100	100	100	100	100	100	100

10.5 Conclusion

In this chapter, we have studied the possibility of evolutionary decryption of encrypted information, based on chaotic systems. Compared to other case studies, only one "case study" is reported here as given in figures above (Lena decryption). All details about it are discussed below. As a conclusion, summarizing all previous informations, it can be stated that:

- **Usability of evolutionary algorithms.** In experiments reported here, two **very strong** simplifications has been taken into consideration. The first one was that only one parameter b of 6 (a - f from eq. (10.2)) has been estimated. The second one is partially done by restriction based on computer and used software accuracy. Part of decryption (cross-correlations "landscape", where EAs were searching for optimal value of b) has been made in C++ programming language, thus parameter b has been estimated with level of numerical precision of $\Delta b = 1\times10^{-15}$. Further, evolutionary search has been restricted to a **tiny** region which was used to generate in total - only 200 000 data points around the right value of b. For this kind of precision, a tiny region of parameter b has been explored. From figures and tables above, it seems that evolutionary search was quite successful, however, it is very logical to expect that if more than one parameter in a wider intervals would be estimated, then evolutionary algorithms would certainly fail.
- **Effectiveness** of used algorithms and proposed methods can be evaluated from two viewpoints. The first one is, that we can evaluate each algorithm separately, according to Fig. 10.14 and Tables 10.8 and 10.9. If we take into consideration the fact that there was 200 000 possible points to search through, it seems that evolutionary algorithms give good performance, because according to Fig. 10.13 all average values (excluding one - DE4) are below 200 000, which is better than a "brute force" method. On the other side, it is important to note that this conclusion is valid only when when brute force (i.e. all

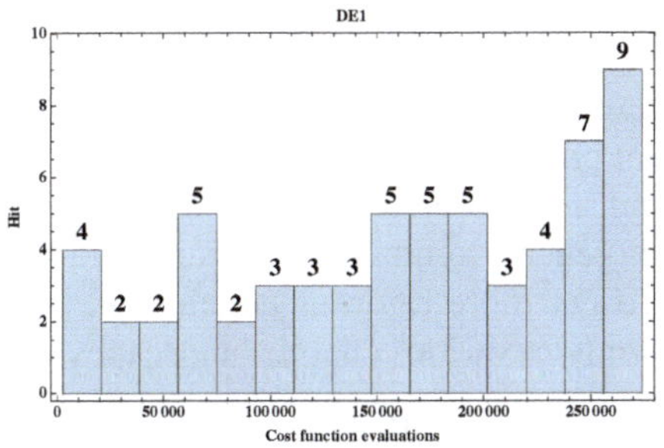

Fig. 10.15 Histogram of successful decryptions for DE1.

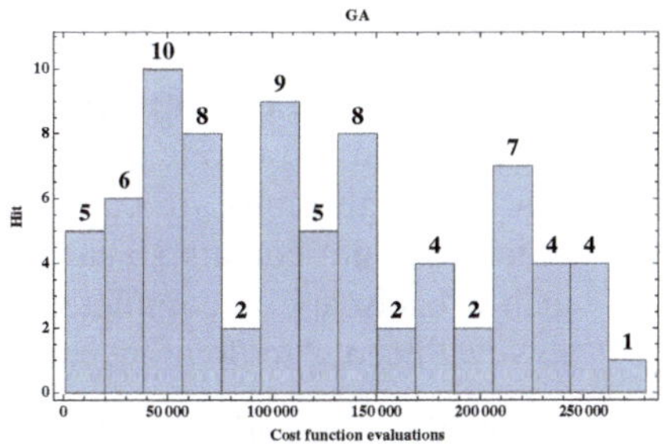

Fig. 10.16 Histogram of successful decryptions for GA.

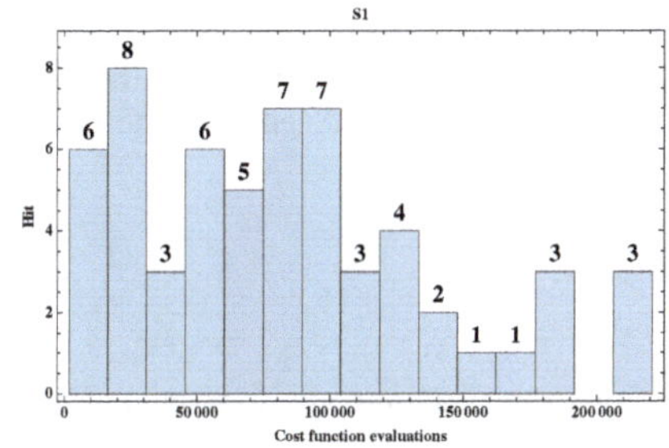

Fig. 10.17 Histogram of successful decryptions for S1.

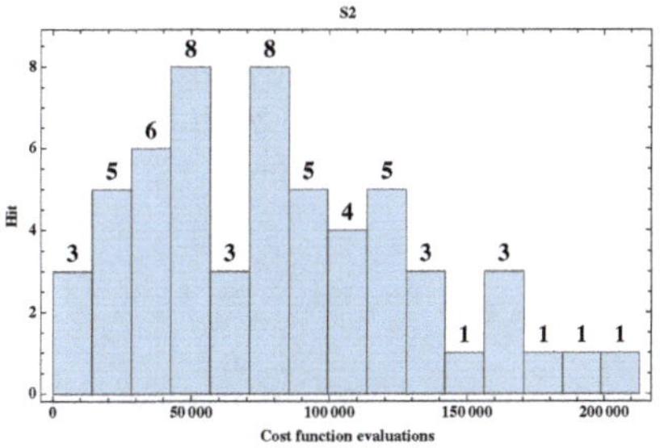

Fig. 10.18 Histogram of successful decryptions for S2.

possible solutions are investigated) is used so that each solution is randomly selected. If each solution would be selected consequently in order (i.e. the first, the second, ...), then the performance of evolution would be overwhelmed on 100 000 (remember, that is the position of the right value of b). If random search is compared, then result would be similar. Algorithms with values below 100 000 are ES1, ES2, SA1-S3.

The second point of view is that when we evaluate all results of all algorithms together, as reported in Fig. 10.14. In that case, unfortunately it appears that average effectiveness is almost random.

- **Performance-** misleading conclusion can also be made when we forget that **only** positive results are repoted in Fig. 10.13. For example, algorithm ES2 seems to be absolutely excellent, however when one takes a closer look in Fig. 10.14, then it is easily visible that values reported in Fig.10.13. are based on **5 successful** results. Cost function evaluations in all 5 cases are low and probably are a matter of “randomness”, i.e. from only 5 cases we can hardly deduced any statistics. Another point of view can be obtained when separate histograms are reported for each algorithm, for example in Figs. 10.15 - 10.18. It is clearly visible that some algorithms has found more positive results below 100 000 and 200 000, and for some of them it is just simply a uniform distribution.

- **Ability to locate extreme on chaotic landscape.** All results plotted and discussed above shows one quite important preliminary fact. More or less, evolutionary algorithms are capable to find an extreme on chaotic landscapes, which does not contain so called trend (general trend, average trend, ...), i.e. such a landscape is completely flat. To get more rigorous conclusion, it is however needed to do more extensive study in various chaotic landscapes.

Based on all results and their analysis, we can conclude, that chaos based encryption is still very safe and is probably not solvable by such techniques as evolutionary algorithms. On the other side, results reported here seems to be an inspiration (at least for us) for more extensive study on how effective evolutionary decryption is, when more individuals and decrypted parameters are taken into account.

Acknowledgements. This work was supported by grant No. MSM 7088352101 of the Ministry of Education of the Czech Republic and by grants of the Grant Agency of the Czech Republic GACR 102/09/1680.

References

1. Asim, M., Jeoti, V.: On Improving an Image Encryption Scheme based on Chaotic Logistic Map. In: ICIAS (2007)
2. Asim, M., Jeoti, V.: Hybrid Chaotic Image Encryption Scheme based on S-box and Ciphertext Feedback. In: ICIAS (2007)
3. Beyer, H.-G.: Theory of Evolution Strategies. Springer, New York (2001)
4. Cerny, V.: Thermodynamical approach to the traveling salesman problem: An efficient simulation algorithm. J. Opt. Theory Appl. 45(1), 41–51 (1985)
5. Fu, C., Zhang, Z., Cao, Y.: An Improved Image Encryption Algorithm Based on Chaotic Maps. In: ICNC (2007)
6. Fu, C., Zhang, Z., Chen, Z., Wang, X.: An Improved Chaos-Based Image Encryption Scheme. In: ICCS 2007. Springer, Berlin (2007)
7. Gao, T., Chen, Z.: A new image encryption algorithm based on hyper-chaos. ScienceDirect (2007)
8. Giesl, J., Vlcek, K.: Audio signal encryption in wavelet domain based on chaotic maps. In: Mendel 2008, Brno (2008)
9. Giesl, J., Vlcek, K.: Image Encryption Based on Strange Attractor. ICGST-GVIP Journal 9(2), 19–26 (2009)
10. Gu, G., Han, G.: An Enhanced Chaos Based Image Encryption Algorithm. In: ICICIC (2006)
11. He, X., Zhu, Q., Gu, P.: A New Chaos-Based Encryption Method for Color Image. Springer, Berlin (2006)
12. Holland, J.: Adaptation in Natural and Artificial Systems. Univ. Michigan Press, Ann Arbor (1975)
13. Kirkpatrick, S., Gelatt, C., Vecchi, M.: Optimization by simulated annealing. Science 220(4598), 671–680 (1983)
14. Mao, Y., Chen, G.: Chaos-Based Image Encryption. Springer, Berlin (2003)

15. Price, K.: An Introduction to Differential Evolution. In: Corne, D., Dorigo, M., Glover, F. (eds.) New Ideas in Optimization, pp. 79–108. McGraw-Hill, London (1999)
16. Seo, Y.-H., Kim, D.-W., Yoo, J.-S., Dey, S., Agrawal, A.: Wavelet Domain Image Encryption by Subband Selection and Data Bit Selection. Springer, Berlin (2003)
17. Wong, K., Kwok, B., Law, W.-S.: A Fast Image Encryption Scheme based on Chaotic Standard Map. Springer, Berlin (2006)
18. Zelinka, I.: SOMA – Self Organizing Migrating Algorithm. In: Babu, B., Onwubolu, G. (eds.) New Optimization Techniques in Engineering, pp. 167–218. Springer, New York (2004)

Chapter 11
Chaos Synthesis by Evolutionary Algorithms

Ivan Zelinka, Guanrong Chen, and Sergej Celikovsky

Abstract. This chapter introduces the notion of chaos synthesis by means of evolutionary algorithms and develops a new method for chaotic systems synthesis. This method is similar to genetic programming and grammatical evolution and is applied alongside evolutionary algorithms: differential evolution, self-organizing migrating, genetic algorithm, simulated annealing and evolutionary strategies. The aim of this investigation is to synthesize new and "simple" chaotic systems based on some elements contained in a pre-chosen existing chaotic system and a properly defined cost function. The investigation consists of two case studies based on the aforementioned evolutionary algorithms in various versions. For all algorithms, 100 simulations of chaos synthesis were repeated and then averaged to guarantee the reliability and robustness of the proposed method. The most significant results are carefully selected, visualized and commented in this chapter.

Ivan Zelinka
Tomas Bata University in Zlin, Faculty of Applied Informatics, Nad Stranemi 4511, Zlin 76001, Czech Republic
and
VSB-TUO, Faculty of Electrical Engineering and Computer Science, 17. listopadu 15, 708 33 Ostrava-Poruba, Czech Republic
e-mail: `zelinka@fai.utb.cz`

Guanrong Chen
Department of Electronic Engineering, City University of Hong Kong, Kowloon, Hong Kong SAR, P.R. China
e-mail: `eegchen@cityu.edu.hk`

Sergej Celikovsky
Institute of Information Theory and Automation,
Academy of Sciences of the Czech Republic, Faculty of Electrical Engineering,
Czech Technical University in Prague
e-mail: `celikovs@utia.cas.cz`

I. Zelinka et al. (Eds.): Evolutionary Algorithms and Chaotic Systems, SCI 267, pp. 345–382.
springerlink.com

11.1 Introduction

Deterministic chaos, discovered by E. Lorenz [26] is a fairly active area of research in the last few decades. The Lorenz system produces a well-known chaotic attractor in a simple three-dimensional autonomous system of ordinary differential equations [26], [41]. For discrete chaos, there is another famous chaotic system, called logistic equation [27], which was found based on a predator-prey model showing complex dynamical behaviors. These simple models are widely used in the study of chaos today, while other similar models exist (e.g., canonical logistic equation [14] and 1D or 2D coupled map lattices [40]). To date, a large set of nonlinear systems that can produce chaotic behaviors have been observed and analyzed. Chaotic systems thus have become a vitally important part of science and engineering at the theoretical as well as practical levels of research. The most interesting and applicable notions are, for example, chaos control and chaos synchronization related to secure communications, among others.

Recently, the study of chaos is focused not only along the traditional trends but also on the understanding and analyzing principles, with the new intention of controlling and utilizing chaos toward real-world applications as demonstrated in [6], [45] and many references therein. The term chaos control was first used by Ott, Grebogi and Yorke in 1990. It represents a process in which a control law is derived and used such that the original chaotic behavior can be stabilized on a constant level of output value or a periodic cycle. Since the first experimental report on chaos control, many control methods have been developed and some are based on the first approach [33], including pole placement [15], [56] and delay feedback [36], [20], [21], to name just a couple. Many methods were adapted to spatiotemporal chaos represented by Coupled Map Lattices (CML). Control laws derived for CML are usually based on existing system structures [40], or using an external observer [5], etc. Evolutionary approach for control was also successfully developed in, for example, [38], [37], [50].

Simultaneously with these research activities, some new chaotic systems were found, like the chaotic Chen system [7], [42], which according to [44], [43] is a dual system of the Lorenz system. Other well-known chaotic systems were discovered, including dissipative ones like Lozi, Tinkerbell, Ikeda, Sinai, Burger, Duffing systems, and conservative ones like Chirikov, Arnold, Baker, Nose-Hoover and Henon-Heiles systems. These chaotic systems were mostly derived from some known physical systems. A typical example is the logistic equation obtained based on the predator-prey system or the Lorenz system derived from an atmospheric model. On the contrary, another direction of research was evolving about chaotic systems synthesis. As a few representative examples, [55] investigated an algorithm for computing heteroclinic orbits with possible use in chaos synthesis. This investigation partly used ideas on chaos synthesis and synchronization from [1]. In [12] there was another investigation, which is based mostly on hardware to generate multiple-scroll strange attractors. Basically very similar research was also carried out in [11], [10].

Methods used in generating new chaotic systems from physical systems or from "manipulations" (e.g., control and parameter estimation [40], [18]) are based on deterministic mathematical analysis. Along with these classical methods, there are also numerical methods based partly on deterministic and partly on stochastic methods, called evolutionary algorithms (EAs) [2]. Evolutionary algorithms were used in searching solutions in many computationally hard problems including classes of P and NP problems [13]. In chaos studies, they have been used for chaos control [49], [50], [37], among others.

The aim of this chapter is to show that EA-based symbolic regression (i.e., handling with symbolic objects to create more complex structures) is capable of synthesizing chaotic behavior in the sense that the mathematical descriptions of chaotic systems are synthesized symbolically by means of evolutionary algorithms. The ability of EAs to successfully solve this kind of black-box problems has been proven (see, for example, [51], [28]), and is reinforced once again here in this chapter.

The paper is organized as follows. The first part outlines the motivation of the research. This is followed by a brief survey of evolutionary algorithms, along with a brief description of symbolic regression methods used with evolutionary algorithms. Next, the method of symbolic regression, called analytic programming, is described in more detail, which will be used in the rest experiments. Evolutionary synthesis of chaos is then studied, and finally experimental results are reported, followed by the conclusion.

11.2 Motivation

In recent years, interests in soft computing methods are increasing, including in particular evolutionary algorithms. These algorithms are based on similar principles of biological evolution in the real world. The aim of EAs is to solve computationally hard problems which are too complex to be solved by conventional methods. In its canonical form, EAs can be used only for numerical estimation of parameters (usually, arguments of a given cost function). Together with EAs in the canonical form, another modification allows to use EAs as a symbolic "constructors", i.e., a processor, for synthesizing complex structures in a symbolic way, based on some predefined simple elements (mathematical operators or electronic elements like diode, transistor, etc.). The term "symbolic way" stipulates that mathematical structures and equations, electronic systems, etc., are generated from those simple elements just mentioned.

Given the above background, the main motivation of this research was the question *"Is it possible to synthesize the mathematical description of a new chaotic system, based on simple and elementary mathematical objects, by means of evolutionary computation?"* This question was also based partially on the fact that in engineering applications, it is very often vitally important to know not only when chaos can be generated but also how to generate it [6], [34]. This is extremely important in cryptography, for example, where chaotic systems are often used in the design. From a mathematical point of view, it is quite clear that there are some classes of chaotic systems which can be represented by one canonical form (one class – one canonical form) [14]. However, generally speaking, it is not so easy to exactly synthesize a chaotic system with specified features by means of classical mathematical methods. A positive answer to the question mentioned above would open possibilities to synthesize not only a set of not-yet-described chaotic systems, but also some chaotic systems with predefined features. It is believed that such possibilities would have an important impact on engineering design of various complex nonlinear systems, especially chaotic systems.

11.3 Brief Review of the Selected Evolutionary Algorithm

For both numerical and symbolic experiments described below, stochastic optimization algorithms such as Differential Evolution (DE) [35], Self Organizing Migrating Algorithm (SOMA) [48], Genetic Algorithms [17], [4] for Simulated Annealing (SA) and [9] or [3] for Evolutionary Strategies (ES) are selected to use. For detail description of selected evolutionary algorithms see Chapter 6.

11.4 Symbolic Regression – An Introduction

The term "symbolic regression" represents a process during which measured data sets are fitted thereby a corresponding mathematical formula is obtained in an analytical way. An output of the symbolic expression could be, for example, $\sqrt[N]{x^2 + \frac{y^3}{k}}$, and the like. For a long time, symbolic regression was a domain of human calculations but in the last few decades computers as generally used for symbolic computation.

The initial idea of symbolic regression by means of a computer program was proposed in Genetic Programming (GP) [22], [23]. The other approaches are Grammatical Evolution (GE) developed in [39] and Analytic Programming (AP) in [53]. Other interesting investigations using symbolic regression were carried out in [19] on Artificial Immune Systems and Probabilistic Incremental Program Evolution (PIPE), which generates functional programs from an adaptive probability distribution over all possible programs. As an extension of GE to the another algorithms is also [30], where DE was used with GE. Symbolic regression is schematically depicted in Fig. 11.1. Generally speaking, it is a process which combines, evaluates and creates more complex structures based on some elementary and noncomplex objects, in an evolutionary way. Such elementary objects are usually simple mathematical operators $(+, -, \times, ...)$, simple functions (*sin*, *cos*, *And*, *Not*, ...), user-defined functions (simple commands for robots – MoveLeft, TurnRight, ...), etc. An output of symbolic regression is a more complex "object" (formula, function, command,...), solving a given problem like data fitting of the so-called Sextic and Quintic problem described by eq. (11.1) [25], [52], randomly synthesized function by eq. (11.2) [52], Boolean problems of parity and symmetry solution (basically logical circuits

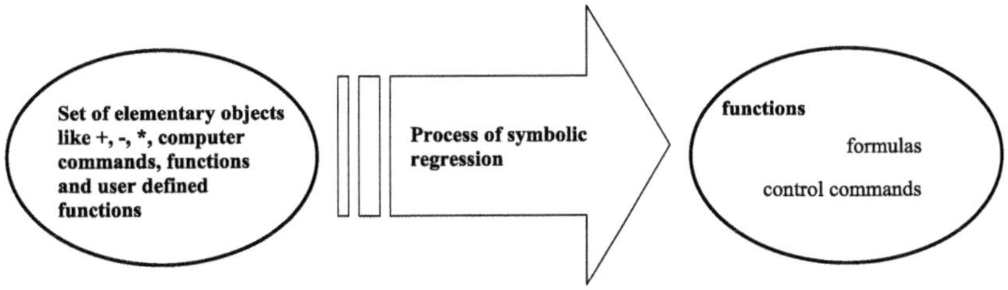

Fig. 11.1 Symbolic regression - schematicall view

synthesis) by eq. (11.3) [24], [53], or synthesis of quite complex robot control command by eq. (11.4) [23], [32]. Equations (11.1)–(11.4) mentioned here are just a few samples from numerous repeated experiments done by AP, which are used to demonstrate how complex structures can be produced by symbolic regression in general for different problems.

$$x\left(K_1+\frac{(x^2K_3)}{K_4\,(K_5+K_6)}\right)*(-1+K_2+2x(-x-K_7)) \tag{11.1}$$

$$\sqrt{t}\left(\frac{1}{\log(t)}\right)^{\sec^{-1}(1.28)}\log^{\sec^{-1}(1.28)}(\sinh(\sec(\cos(1)))) \tag{11.2}$$

$$\begin{aligned}&Nor[(Nand[Nand[B||B,B\&\&A],B])\&\&C\&\&A\&\&B,\\&Nor[(!C\&\&B\&\&A||!A\&\&C\&\&B||!C\&\&!B\&\&!A)\&\&\\&(!C\&\&B\&\&A||!A\&\&C\&\&B||!C\&\&!B\&\&!A)||\\&A\&\&(!C\&\&B\&\&A||!A\&\&C\&\&B||!C\&\&!B\&\&!A),\\&(C||!C\&\&B\&\&A||!A\&\&C\&\&B||!C\&\&!B\&\&!A)\&\&A]]\end{aligned} \tag{11.3}$$

$$\begin{aligned}&\text{Prog2[Prog3[Move, Right, IfFoodAhead[Left, Right]],}\\&\text{IfFoodAhead[IfFoodAhead[Left, Right], Prog2[IfFoodAhead[}\\&\text{IfFoodAhead[IfFoodAhead[Left, Right], Right], Right],}\\&\text{IfFoodAhead[Prog2[Move, Move], Right]]]]}\end{aligned} \tag{11.4}$$

11.4.1 Genetic Programming

Genetic programming was the first tool for symbolic regression carried out by means of computers instead of humans. The main idea comes from genetic algorithms (GA), which was used in GP [22], [23]. Its ability to solve very difficult problems is well proven; for example, GP performs so well that it can be applied to synthesize highly sophisticated electronic circuits [24].

The main principle of GP is based on GA, which is working with populations of individuals represented in LISP programming language. Individuals in a canonical

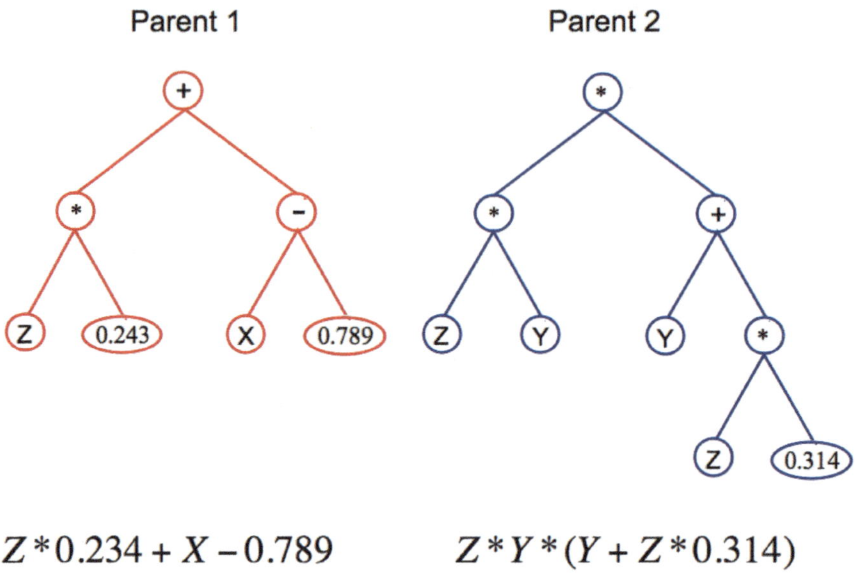

Fig. 11.2 Parental trees

form of GP are not binary strings, different from GA, but consist of LISP symbolic objects (commands, functions, ...), etc. These objects come from LISP, or they are simply user-defined functions. Symbolic objects are usually divided into two classes: functions and terminals. Functions were previously explained and terminals represent a set of independent variables like x, y, and constants like π , 3.56, etc.

The main principle of GP is usually demonstrated by means of the so-called trees (basically graphs with nodes and edges, as shown in Fig. 11.2 and Fig. 11.3, representing individuals in LISP symbolic syntax). Individuals in the shape of a tree, or formula like $0.234Z+X-0.789$, are called programs. Because GP is based on GA, evolutionary steps (mutation, crossover, ...) in GP are in principle the same as GA. As an example, GP can serve two artificial parents – trees on Fig. 11.2 and Fig. 11.3, representing programs $0.234Z+X-0.789$ and $ZY(Y+0.314Z)$. When crossover is applied, for example, subsets of trees are exchanged. Resulting offsprings of this example are shown in Fig. 11.3.

Thereafter, offspring fitness is calculated such that the behavior of the just-synthesized and evaluated individual-tree should be as much as possible similar to the desired behavior. Here, desired behavior can be understood to be like a measured data set from some process (a program that should fit them as well as possible) or like an optimal robot trajectory, i.e., when the program is realizing a sequence of robot commands (TurnLeft, Stop, MoveForward,...) leading as close as possible to the final position. This is basically the same for GE.

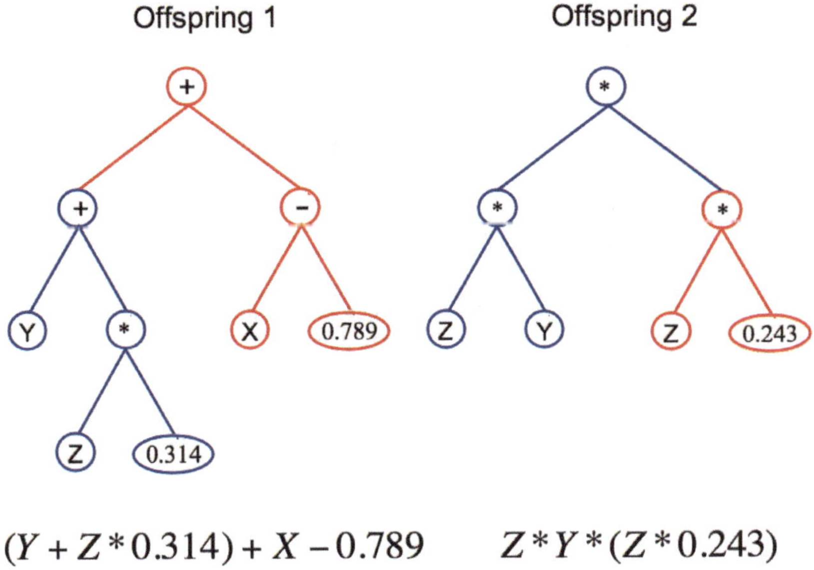

Fig. 11.3 Offsprings

For detailed description of GP, see [23], [25] and for on-line working example, visit [http://evonet.lri.fr/CIRCUS2/node.php?node=56].

11.4.2 Grammatical Evolution

Another method for the same task in view of the resulting program alike GP was developed in [29] called grammatical evolution (GE). GE has one advantage compared to GP and this is the ability to use arbitrary programming languages, not only LISP as in the case of the canonical version of GP. In contrast to other evolutionary algorithms, GE was used only with a few search strategies, and with a binary representation of the populations [29]. Last successful experiment with DE applied on GE was also done in [30]. Grammatical evolution is in its canonical form based on GA, thanks to a few important changes it has in comparison with GP. The main difference is in individual coding.

While GP manipulates in LISP symbolic expressions, GE uses individuals based on a binary strings. These are transformed into integer sequences and then mapped into a final program in the Backus-Naur form (BNF) [29], as explained by the following artificial example. Let $T = \{+, -, \times, /, x, y\}$ be a set of operators and terminals and let F = {epr, op, var} be the so-called nonterminals. In this case, the grammar used for final program synthesis is given in Table 11.1. The rule used for individuals transforming into a program is based on eq. (11.5) below. Grammatical evolution is based on binary chromosome with a variable length, divided into the so-called codons (range of integer values, 0-255), which is then transformed into an integer domain according to Table 11.2.

$$\begin{aligned} &\text{unfolding} = \text{codon mod rules} \\ &\text{where rules is number of rules for given nonterminal} \end{aligned} \tag{11.5}$$

Table 11.1 Grammatical evolution - rules

Nonterminals		Unfolding	Index
expr	::=	op expr expr	0
		var	1
op	::=	+	0’
		-	1’
		*	2’
		/	3’
var	::	X	0”
		Y	(1”)

Table 11.2 Grammatical evolution - codon

Chromosome	Binary	Integer	BNF index
Codon 1	101000	40	0
Codon 2	11000011	162	2’
Codon 3	1100	67	1
Codon 4	10100010	12	0”
Codon 5	1111101	125	1
Codon 6	11100111	231	1”
Codon 7	10010010	146	Unused
Codon 8	10001011	139	Unused

Synthesis of an actual program is described as the following: start with a nonterminal object expr. Because the integer value of Codon 1 (see Table 11.2) is 40, according to eq. (11.5) one has an unfolding of *expr* = *op expr expr* (40 mod 2, 2 rules for *expr*, i.e., 0 and 1). Consequently, Codon 2 is used for the unfolding of *op* by * (162 mod 4), which is terminal and thus the unfolding for this part of program is closed. Then, it continues in unfolding of the remaining nonterminals (*expr expr*) till the final program is fully closed by terminals. If the program is closed before the end of the chromosome is reached, then the remaining codons are ignored; otherwise, it continues again from the beginning of the chromosome. The final program based on the just-described example is in this case $x \cdot y$ (see Fig. 11.4). For a fully detailed description of GE principles, see [29] or consult [http://www.grammaticalevolution.org/].

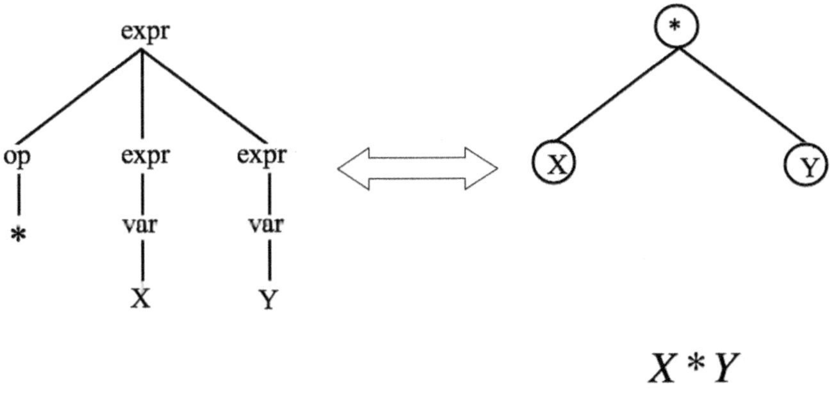

Fig. 11.4 Final program by GE

11.4.3 Analytic Programming

The final method described here and used for experiments in this chapter, is called Analytic Programming (AP), which has been compared to GP with very good results (see, for example, [52], [31], [53], [32]), [54] or visit the online university website *www.ivanzelinka.eu*, subpage *sites*.

The basic principles of AP were developed in 2001 and first published in [46],[47]. AP is also based on the set of functions, operators and terminals, which are usually constants or independent variables alike, for example:

- **functions**: *sin, tan, tanh, And, Or,...*
- **operators**: +, -, ×, /, dt,...
- **terminals**: 2.73, 3.14, t,...

All these objects create a set, from which AP tries to synthesize an appropriate solution. Because of the variability of the content of this set, it is called a general functional set (GFS). The structure of GFS is nested, i.e., it is created by subsets of functions according to the number of their arguments (Fig. 11.5). The content of GFS is dependent only on the user. Various functions and terminals can be mixed together. For example, GFS_{all} is a set of all functions, operators and terminals, GFS_{3arg} is a subset containing functions with maximally three arguments, GFS_{0arg} represents only terminals, etc. (Fig. 11.5).

AP, as further described later, is a mapping from a set of individuals into a set of possible programs. Individuals in population and used by AP consist of non-numerical expressions (operators, functions,...), as described above, which are in the evolutionary process represented by their integer position indexes (Fig. 11.6, Fig. 11.7, see also Chapter 2). This index then serves as a pointer into the set of expressions and AP uses it to synthesize the resulting function-program for cost function evaluation.

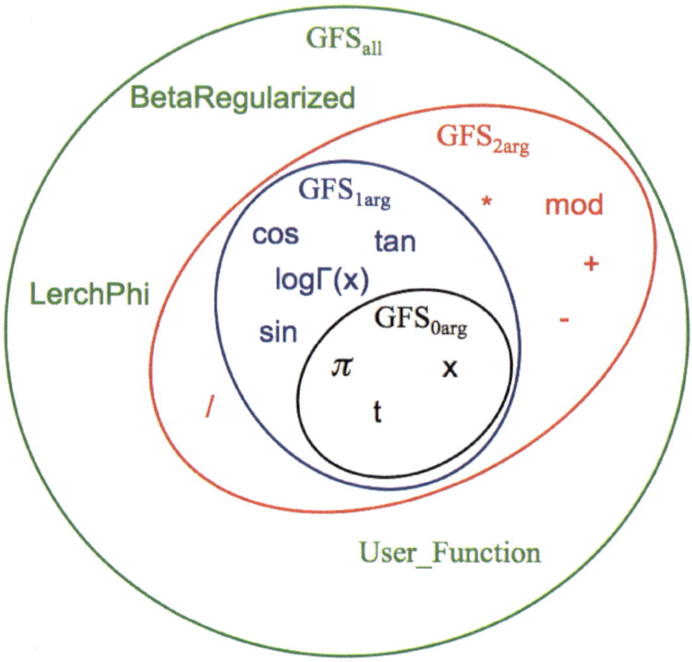

Fig. 11.5 Hierarchy in GFS

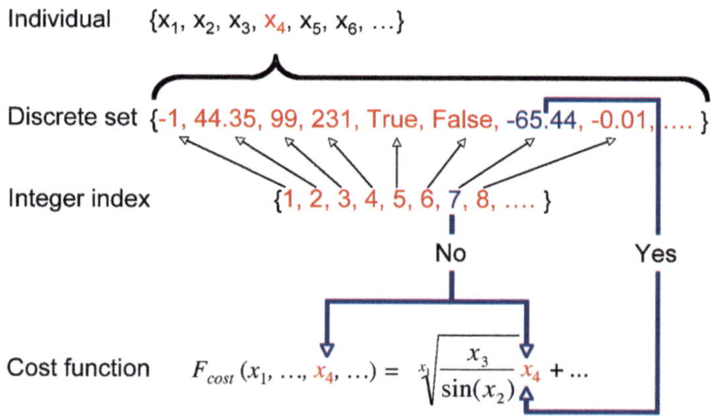

Fig. 11.6 DSH-Integer index, see Chapter 2

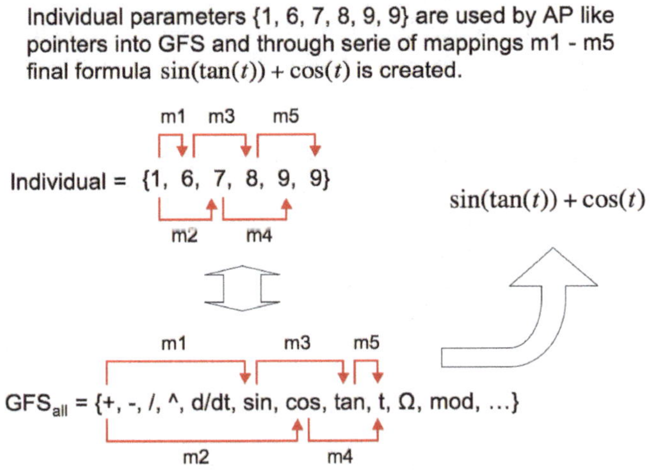

Fig. 11.7 Principle of mapping from GFS to programs

Fig. 11.7 demonstrates an artificial example as how a final function is created from an integer individual via DSH. Number 1 in the position of the first parameter means that the operator "+" from GFS_{all} is used (the end of the individual is far enough). Because the operator "+" must have at least two arguments, the next two index pointers 6 (*sin* from GFS) and 7 (*cos* from GFS) are dedicated to this operator as its arguments. The two functions, *sin* and *cos*, are one-argument functions, so the next unused pointers 8 (*tan* from GFS) and 9 (*t* from GFS) are dedicated to the *sin* and *cos* functions. As an argument of *cos*, the variable *t* is used, so this part of the resulting function is closed (*t* is zero-argument) in its AP development. The one-argument function *tan* remains, and because there is one unused pointer 9, *tan* is mapped on *t* which is on the 9th position in GFS.

To avoid synthesis of pathological functions, a few security "tricks" are used in AP. The first one is that GFS consists of subsets containing functions with the same or a smaller number of arguments. The nested structure (see also Fig. 11.5) is used in the special security subroutine, which measures how far the end of an individual is and, according to this, mathematical elements from different subsets are selected to avoid pathological functions synthesis. More precisely, if more arguments are desired then a possible function (the end of the individual is near) will be replaced by another function with the same index pointer from the subset with a smaller number of arguments. For example, it may happen that the last argument for one function will not be a terminal (zero-argument function). If the pointer is longer than the length of subset, e.g., a pointer is 5 and is used GFS_0, then the element is selected according to the "formula" element = pointer_value mod number_of_elements_in_GFS_0. In this example, the selected element would be the variable *t* (see GFS_0 in Fig. 11.5).

GFS doesn't need to be constructed only from clear mathematical functions as demonstrated above, but may also be constructed from other user-defined functions, e.g., logical functions, functions which represent elements of electrical circuits or robot movement commands, linguistic terms, etc.

During the evolution of a population, a different set of operators are used, such as crossover and mutation. In comparison with GP or GE, evolutionary operators like mutation, crossover, tournament, and selection are fully used in the competence of the established evolutionary algorithm. AP does not contain them in any scale of its internal structure. AP is created to be like a superstructure of EA for symbolic regression independent of their algorithmic structures. Operations used in EA are not influenced by AP and vice versa. For example, if DE is used for symbolic regression in AP, then all evolutionary operations are completed according to the DE rules and only by DE. AP just transforms individuals into formulas.

During the evolution, more or less appropriate individuals are synthesized. Some of these individuals are used to reinforce the evolution towards a better solution synthesis. The main idea of reinforcement is based on the addition of the just-synthesized and partly successful program into an initial set of terminals. Reinforcement is based on a user-defined criterion used in decision as to which individual will be used as an addition into the initial set of terminals. A criterion for the decision is in fact a threshold, i.e., by a user-defined cost value, under which conditions are synthesized solutions added into GFS. For example, if the threshold is set to 5, and if the fitness of all individuals (programs in the population) is bigger than 5, then evolution is running on the basic, i.e., initially defined, GFS. When the best individual in the actual population is less than 5, then it is entirely added into the initial GFS and is marked as a terminal. Since this moment, evolution is running on the enriched GFS containing a partially successful program. Thanks to this advantage, evolution is able to synthesize the final solutions much faster than the AP without reinforcement. This fact has been repeatedly verified by simulations on different problems. When the program is added into GFS, the threshold is also set to its fitness. If furthermore an individual with lower fitness than the just-reset threshold is synthesized, then the old one is rewritten by the better one, and the threshold is rewritten by a new fitness value again.

Adding a partially successful program as a terminal, just for simplicity, can avoid programming difficulties if it would be added like a new function. It is quite similar to automatically defined functions (ADF) for GP; however, the set of functions and terminals in GP can contain more than one ADF, which of course at least theoretically increases the complexity of the search space to the order of $n!$), including properly defined arguments of these ADF and critical situation checking (selfcalling,...). This is not a problem of AP reinforcement, because adding a program into the initial GFS is regarded as a terminal (or a terminal structure), i.e., no function, no arguments, no selfcalling, etc., and the cardinality of the initial GFS set increases only by one.

For more exact description with more details, we recommend to read [54].

11.5 Experiment Design

11.5.1 Parameter Setting

The control parameter settings of used EAs (with abbreviation in Table 11.3) have been found empirically and are given in Table 11.4 (SOMA), Table 11.5 (DE), Table 11.6 (GA), Table 11.7 (ES) and Table 11.8 (SA) respectively. The main criterion for this setting is to keep the setting of parameters as much the same as possible for all simulations and, of course, the same number of cost function evaluations as well as the same population sizes (parameter PopSize for SOMA and GA, and NP for DE, etc). Individual length represents the number of optimized parameters, i.e., in the case of this research a maximal number of integer indexes will be used in evolution in order to synthesize a new chaotic system.

AP, being applied on the evolutionary algorithms in 13 versions, was used for the experimentation. Symbolic objects (e.g., variables, constants,...) for manipulation and complex structure synthesis were selected from the well-known logistic equation:

$$x_{n+1} = Ax(1-x) \tag{11.6}$$

Table 11.3 Algorithms abbreviation

Algorithm	Version	Abbreviation
SOMA	AllToOne	A
	AllToOneRandomly	B
	AllToAll	C
	AllToAllAdaptive	D
Differential Evolution	DERand1Bin	E
	DERand2Bin	F
	DEBest2Bin	G
	DELocalToBest	H
	DEBest1JIter	I
	DERand1DIter	J
Genetic Algorithm		K
Evolutionary strategies (μ,λ)		L
Simulated annealing		M

Table 11.4 SOMA setting for 4 basic search strategies: A, B, C, D

Algorithm	A	B	C	D
PathLength	3	3	3	3
Step	0.11	0.11	0.11	0.11
PRT	0.1	0.1	0.1	0.1
PopSize	200	200	200	200
Migrations	10	10	10	10
MinDiv	-0.1	-0.1	-0.1	-0.1
Individual Length	50	50	50	50

Table 11.5 DE setting for 6 basic search strategies: E, F, G, H, I, J

Algorithm	E	F	G	H	I	J
NP	200	200	200	200	200	200
F	0.9	0.9	0.9	0.9	0.9	0.9
CR	0.3	0.3	0.3	0.3	0.3	0.3
Generations	200	200	200	200	200	200
Individual Length	50	50	50	50	50	50

Table 11.6 GA setting for canonical version of GA: K

Algorithm	K
PopSize	200
Mutation	0.4
Generations	100
Individual Length	50

Table 11.7 ES setting for search strategy: L

Algorithm	L
μ ,λ	200
σ	0.8
Iterations	100
Individual Length	50

Table 11.8 SA setting for search strategy: M

Algorithm	M
No. of particles	200
σ	0.5
$k_{\max}$	66
T_{min}	0.0001
T_{max}	1000
α	0.95
Individual Length	50

This selection was based on the fact that the logistic equation is a well-known simple system that can produce chaotic behavior. This equation is also well analyzed. It was expected that evolutionary search would be able to synthesize the logistic equation, which was in fact a source of elements for GFS. Evolutionary synthesis of logistic equation was actually observed, as further discussed later. Another reason behind the selection of the logistic equation is that results from designed experiments can be easily compared, verified and analyzed.

The basic set of objects used in symbolic regression are $\{x, A, +, -, \times, /\}$. It is also important to note that experiments provided here, i.e., evolutionary synthesis of chaotic systems, are not restricted to one-dimensional chaotic maps but can be applied in principle to synthesis of higher-dimensional and more complex chaotic systems. This declaration is based on many other successful complex examples accomplished by GP, GE and AP in the past.

In this research, a total of 1300 independent simulations were completed, 100 trials by each of the 13 algorithms. Each simulation was started at randomly selected initial conditions (i.e., each initial population was randomly generated).

11.5.2 Cost Function

The cost function was in fact a little bit complex decision function with multiple "If" conditions.

The cost function used for chaos synthesis, comparing with other problems like chaos control [49], [50] or black-box optimization [28], is quite a complex structure which cannot be easily described by a few simple mathematical equations. Instead, it is described by the following procedure:

1. Take a synthesized function and evaluate it for 500 iterations with a sampling step of Δ A = 0.1.
2. Check if each value of A for all 500 iterations is unique or if some data are repeated in the series (the first check for chaos, indirectly). If the data are not unique, then go to step 5, else go to step 3.
3. Take the last 200 values, and for each value of A, calculate its Lyapunov exponent.
4. Check the Lyapunov exponent: If the Lyapunov exponent is positive, write all important data (synthesized functions, number of cost function evaluations, etc.) into a file. Then, repeat the simulation for another synthesized system by going to step 1.
5. If the data are not unique, i.e., if the Lyapunov exponent is not positive, return an individual fitness, and sum all values whose occurrences in the dataset from step 1 are more than 1 (simply, it returns the occurrences of periodicity, quasi-periodicity – higher penalization of an individual in the evolution).

More brief and simple description of above algorithmically defined cost function can be also done as in eq.11.7.

$$\begin{gathered} Data[f_{synt,\,1},\dots,f_{synt,\,500}] := f_{synt,\,k+1} = f_{synt,\,k}(x_{start}),\ k \in [1,\ 500] \\ \begin{cases} if\ Data[f_{synt,\,1}] \neq Data[f_{synt,\,2}] \neq \dots\ \neq Data[f_{synt,\,500}] \\ then\ \{\text{calculate } \lambda \text{ for } Data[f_{synt,\,300},\cdots,f_{synt,\,500}],\ if\ \lambda > 0 \text{ write all to file} \\ else \text{ penalize individual} \end{cases} \end{gathered} \tag{11.7}$$

The input to this cost function is a synthesized function and the output is the fitness (quality) of the synthesized function (i.e., the individual in the population).

In the cost function, it was tested twice to ensure that the behavior of the just-synthesized formula is really chaotic. The first test was done in step 2 (unique appearance in the data series) and the second one, in step 4, where the Lyapunov exponent was tested numerically [16].

The reason as to why in step 5, the sum of the non-unique data appearances was returned is based on the fact that the evolution is searching for minimal values. In this case, the value 2 means that some data element appears in the 500-data series twice, and 1 would means that there is no periodicity and thus synthesized system is a possible candidate for chaos.

To ensure that the results obtained are correct, all written synthesized functions were used for automatic generation of bifurcation diagrams and Lyapunov exponents, as further discussed below.

11.5.3 Case Studies

Two case studies are presented in this chapter. The first and main one (discrete systems) is the continuation of research done in [54]. Simulations has been enlarged (compare with [54]) for other evolutionary algorithms (GA, SA, ES). The second one is focused on how to synthesize simple discrete systems based on user demand.

11.5.3.1 Discrete Systems: Simulations and Results

All algorithms (SOMA, DE, GA, ES and SA) in 13 versions have been applied for 100 times in order to find artificially synthesized functions that can produce chaos. All of these experiments were done using the *Mathematica* software. The primary aim of this comparative study is not to show which algorithm is better or worst, but to show that symbolic regression is able to synthesize some new (at least in the sense of mathematical description and behavior) chaotic systems.

Based on the results from experiments, two different sets of figures were created. The first set (Fig. 11.8 - Fig. 11.10) shows the performances of different algorithms from different points of views, the second set (Fig. 11.13 - Fig. 11.88) shows behaviors of the selected synthesized programs, i.e., bifurcation diagrams. The synthesized programs are also appended to each figure in the form of the mathematical formula. Fig. 11.11 shows an example of the so-called program length histogram, generated from 100 simulations. Program length (in *Mathematica* command: LeafCount, denoted as LC) means a number of elements that create a mathematical formula.

As an example, the logistic equation (11.6), for which LC = 8, seems at the first glance to be false (equation contains $A, x, 1$ and $\times$). However, with a closer look at this equation via the *Mathematica* command *TreeForm*, one gets formula 11.8 (see also Fig. 11.12), and LC = 8 is thus clear: ($\times$, A, +, 1, $\times$, -1, x, x).

$$Times[A, Plus[1, Times[-1, x]], x] \tag{11.8}$$

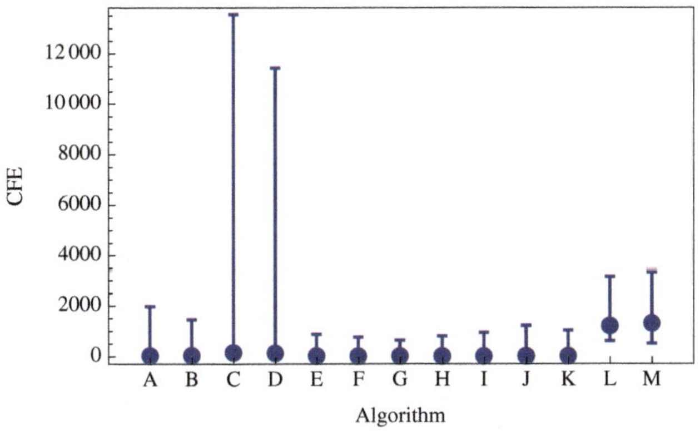

Fig. 11.8 Mutual comparison of all algorithms

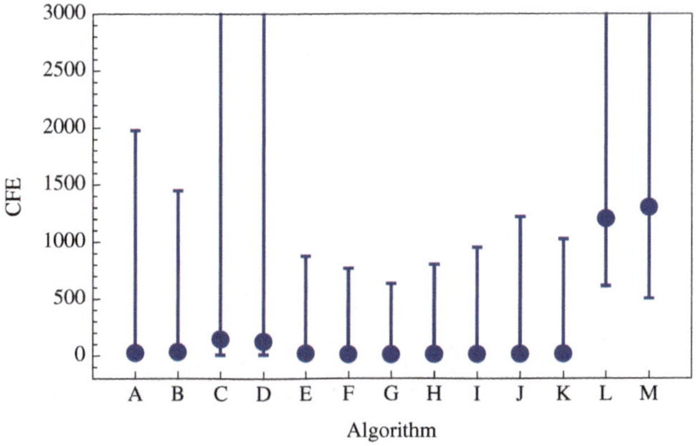

Fig. 11.9 Mutual comparison of all algorithms - detail

There is an explanation for the contradiction between the fact that the length of an individual was set to 50 (Table 11.4 - Table 11.8) and Fig. 11.10, where one can observe programs of lengths larger than 50. The explanation is that when a symbolic string like "$Ax(1-x)$" is transformed into an expression, it becomes a formula 11.8, i.e., it is "artificially" enhanced due to some *Mathematica* internal programming reasons.

For a better overview of the performances of all such algorithms and the lengths of the synthesized programs, Fig. 11.9 was generated and displayed, where for all 13 algorithms, the corresponding minimal, average and maximal values are depicted. Almost the same quality in LC is observed for all of them.

When evolutionary techniques are used, usually their performances are evaluated via cost function evaluations [2], [48], i.e., how many times the cost function has to be re-calculated in order to reach a suitable solution. Fig. 11.8 and Fig. 11.9 are

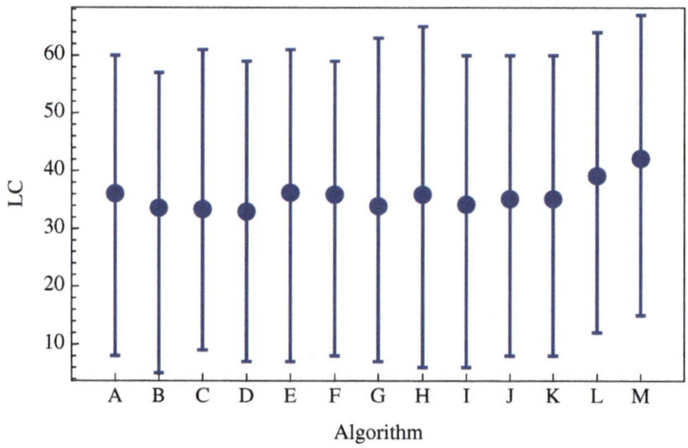

Fig. 11.10 Mutual comparison of LC of all algorithms

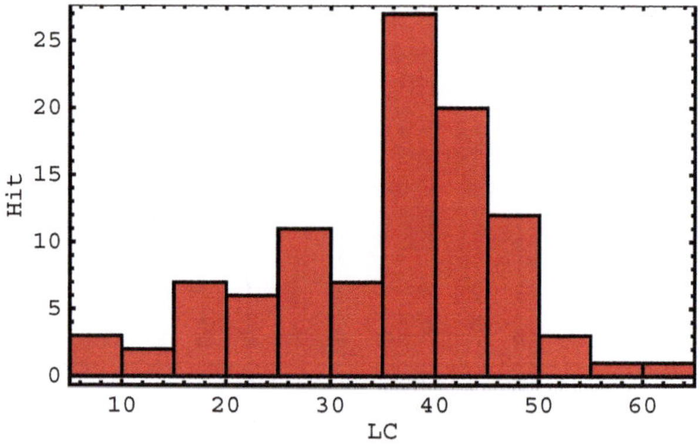

Fig. 11.11 Histogram of LC for DERand1Bin

displayed for this purpose. From both pictures, it appears that only algorithms C and D have less performance. But this is not true, because the SOMA versions C and D have larger numbers of cost function evaluations (see [48]). Because EA was stopped only according to a defined number of cost function evaluations (see [48] or Section 11.5.1 Parameter Setting), these two versions of SOMA logically differ from each other, as shown in Fig. 11.9 .

For mutual comparisons of algorithm performances in successfully generating chaotic systems, Fig. 11.13 - Fig. 11.88 and corresponding equations show selected (visually the best) results of all 1300 simulations in all case studies. With each figure is joined equations of the synthesized systems. To be "sure" that those bifurcation diagrams are true, Lyapunov exponents were also generated for each bifurcation diagram (not completely reported here).

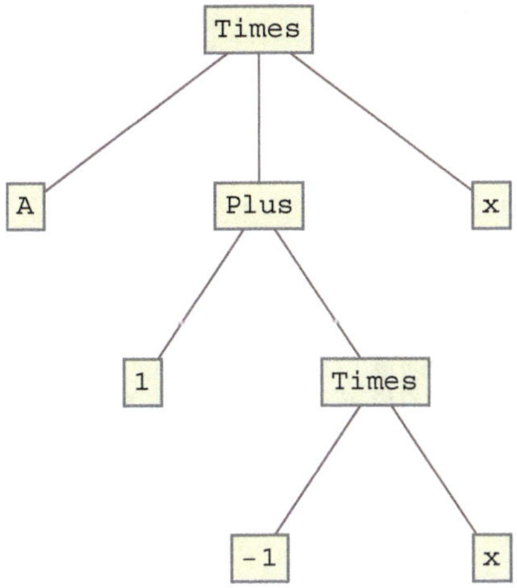

Fig. 11.12 Tree representation of eq. (11.8)

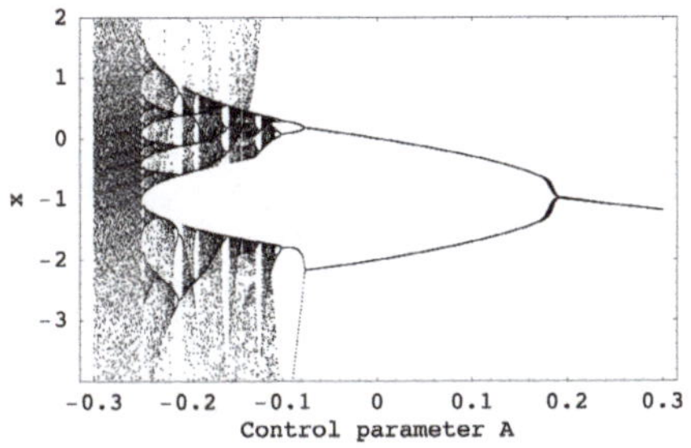

Fig. 11.13 Bifurcation diagram of $\frac{2A(2x-1)}{A+x^2}$ ...

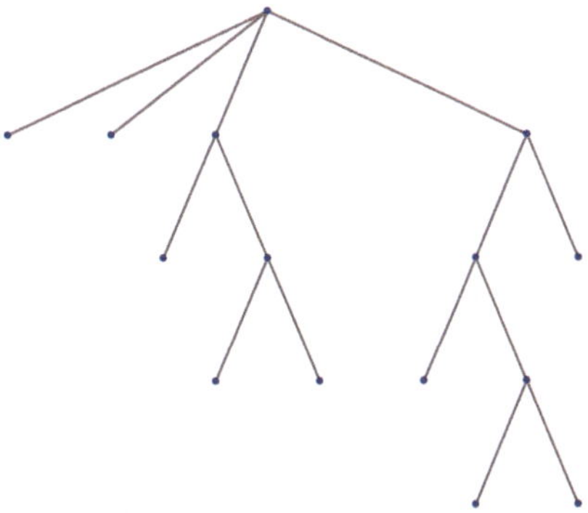

Fig. 11.14 ... and its tree representation.

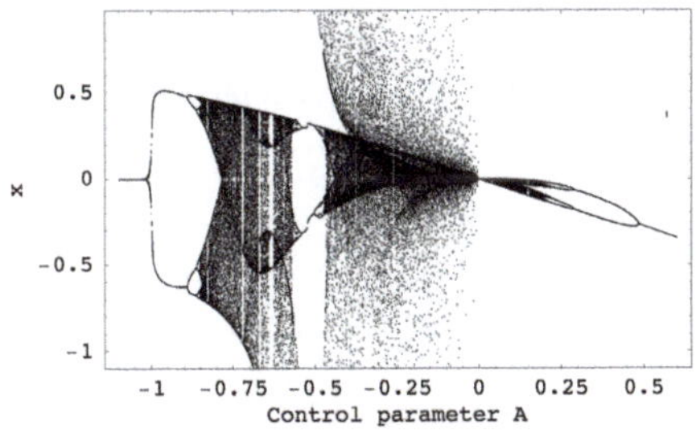

Fig. 11.15 Bifurcation diagram of $\frac{x}{\frac{x^3}{2A^3(x-A)(A+x)}+A}$ ...

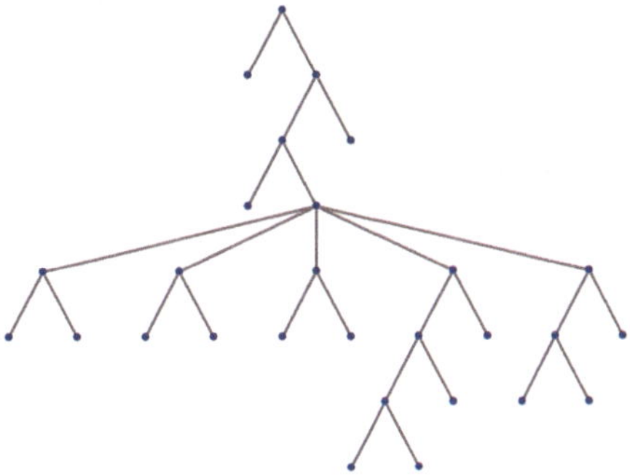

Fig. 11.16 ... and its tree representation.

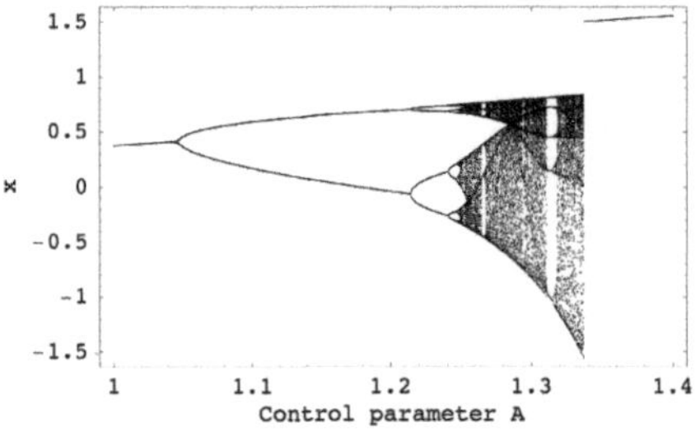

Fig. 11.17 Bifurcation diagram of $A-\frac{A}{x\left(A(-x)+\frac{A}{x}-\frac{x(Ax+A+x)+1}{A}+A\right)+A}\ldots$

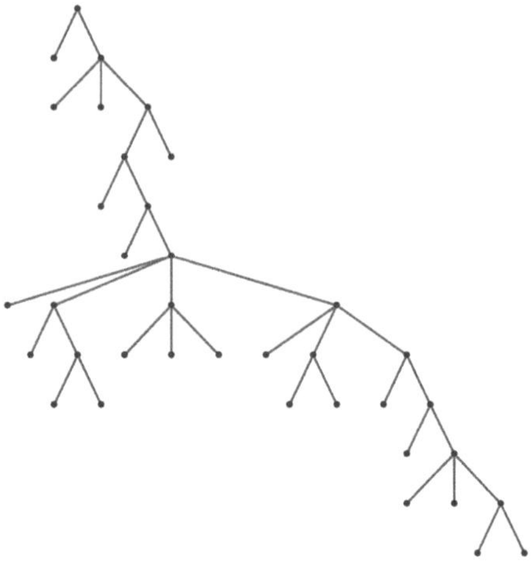

Fig. 11.18 ... and its tree representation.

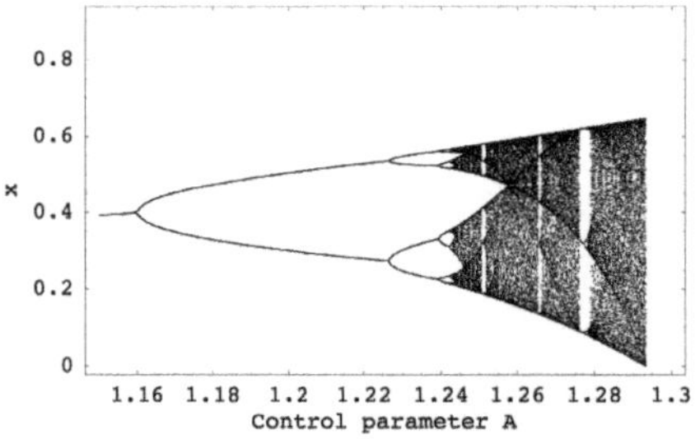

Fig. 11.19 Bifurcation diagram of $\frac{A\left(\frac{A+x}{2A+1}+2A\right)}{\frac{1}{x(A-2x)}+A}\ldots$

Fig. 11.20 ... and its tree representation.

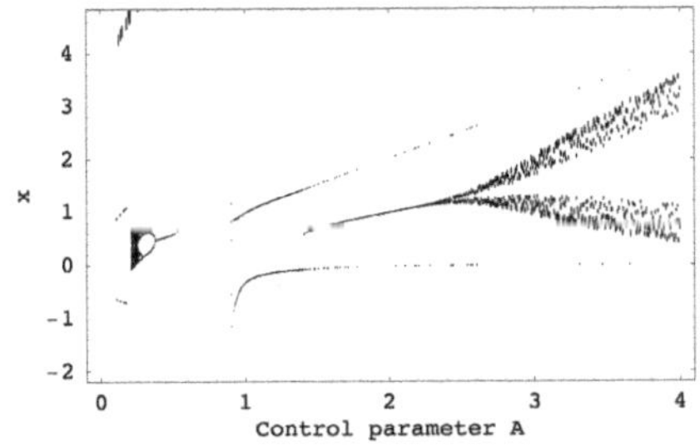

Fig. 11.21 Bifurcation diagram of $-\frac{x}{(A+x)\left(A^2+\frac{x}{A}\right)(2(A-1)x+x^2)-A}+A-x$ ...

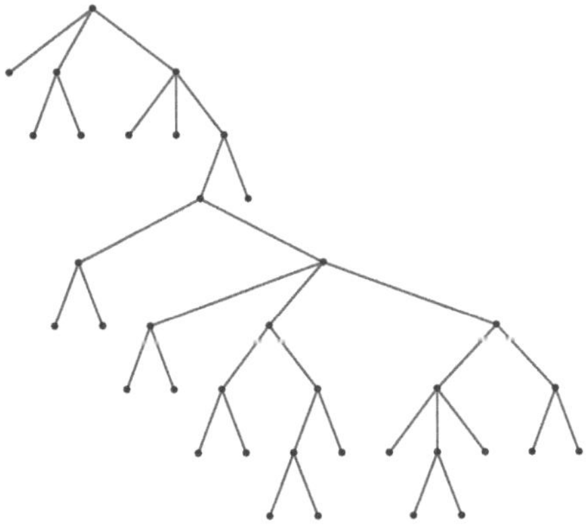

Fig. 11.22 ... and its tree representation.

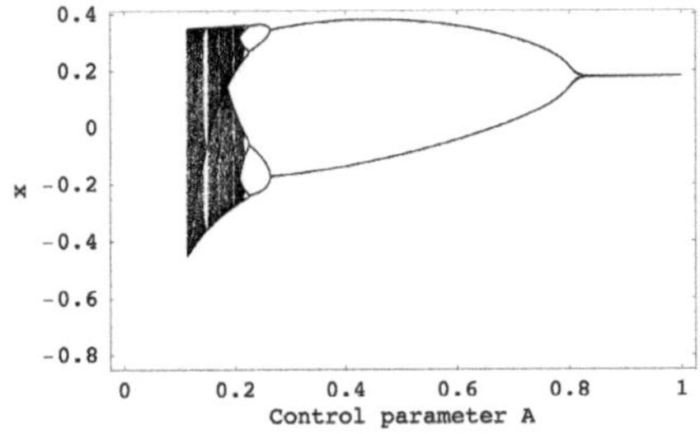

Fig. 11.23 Bifurcation diagram of $\frac{x\left(3Ax^2-Ax+\frac{A}{x}-2A-2x\right)}{3A+x^2}$ ...

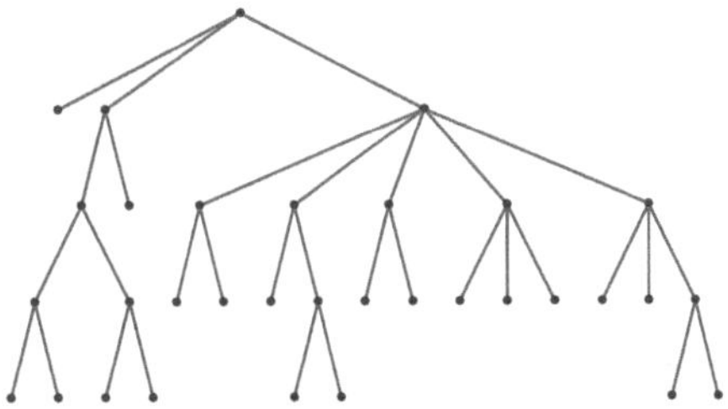

Fig. 11.24 ... and its tree representation.

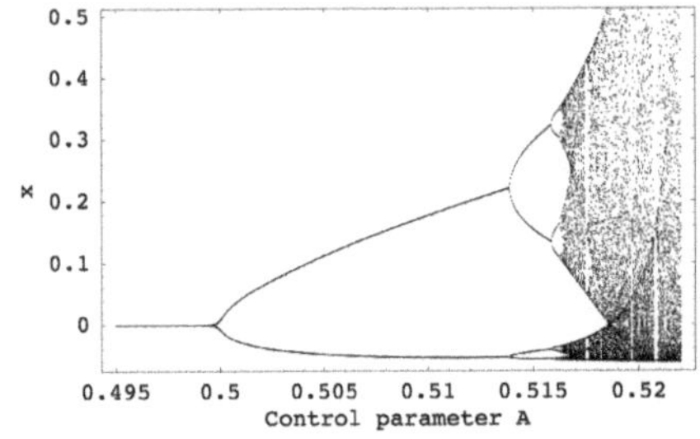

Fig. 11.25 Bifurcation diagram of $\frac{x}{A(-x)+\frac{x(-A-x+1)+A}{A+x}-\frac{1}{A(A-x)}+4A-x}$ ...

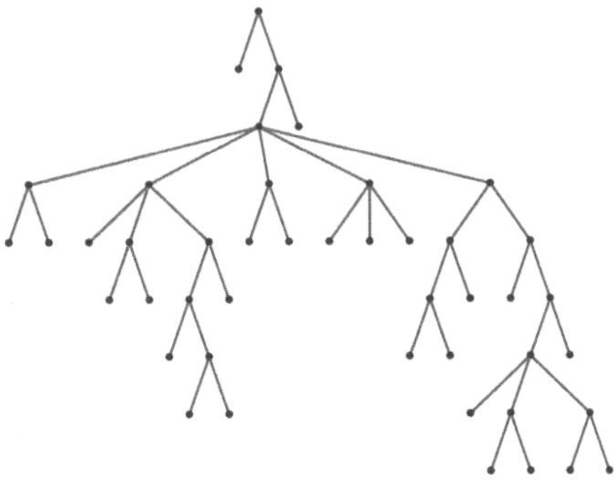

Fig. 11.26 ... and its tree representation.

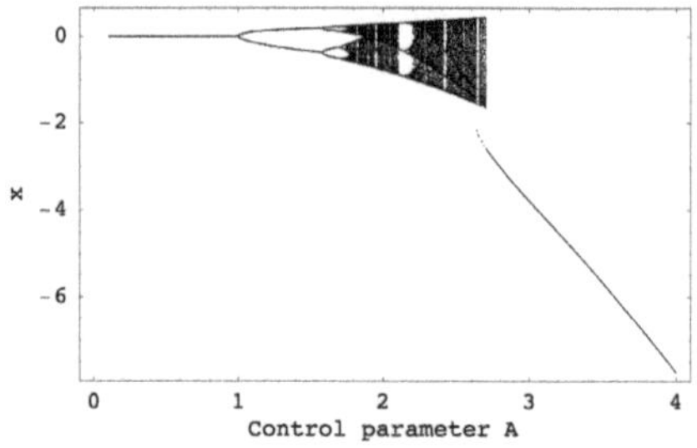

Fig. 11.27 Bifurcation diagram of $\frac{A(A(-x)-A-x)}{\frac{A}{x}+\frac{x}{A}+2x}$ ...

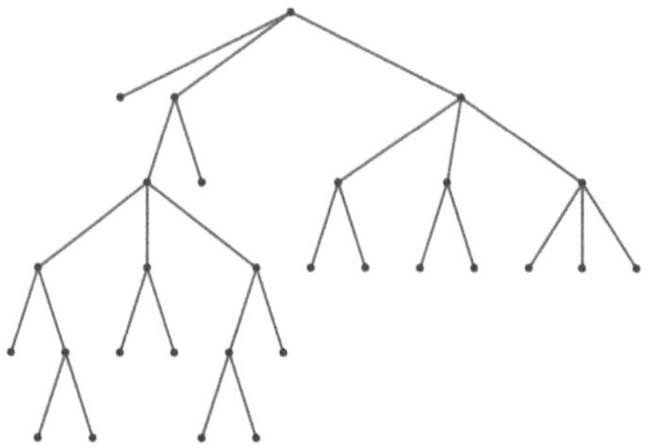

Fig. 11.28 ... and its tree representation.

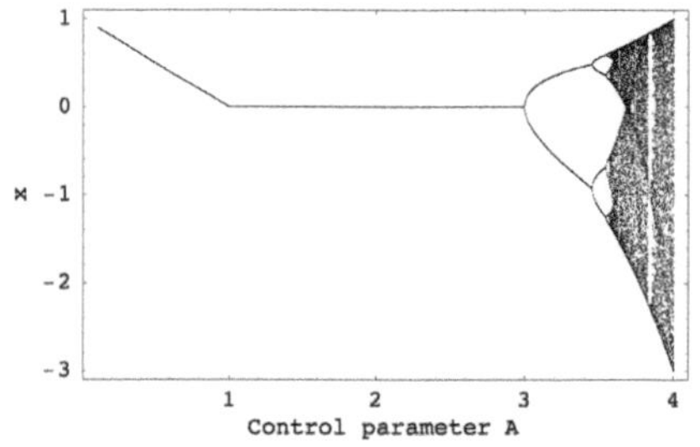

Fig. 11.29 Bifurcation diagram of $2x - x(A+x)$ ...

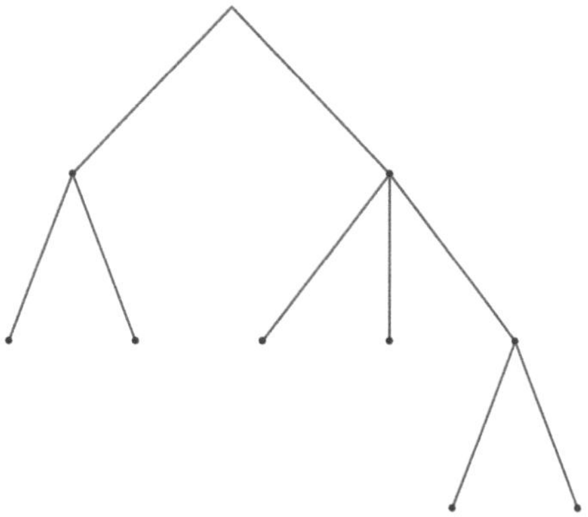

Fig. 11.30 ... and its tree representation.

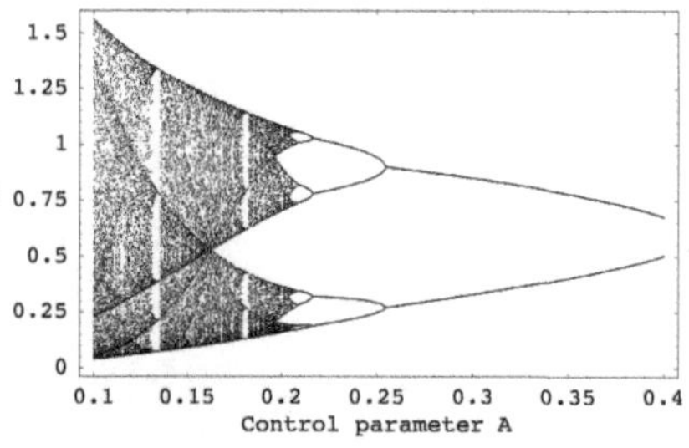

Fig. 11.31 Bifurcation diagram of $\frac{A}{\frac{x\left(\frac{A^2}{x^2}+Ax^2\right)}{-x(A+x)+\frac{x}{A}+2A-x}+x^2}$ ...

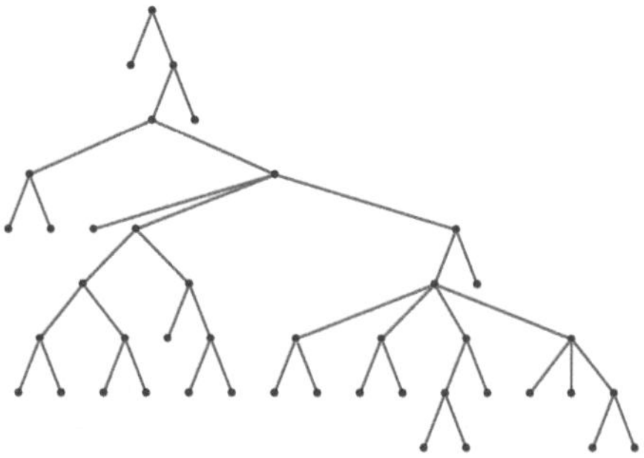

Fig. 11.32 ... and its tree representation.

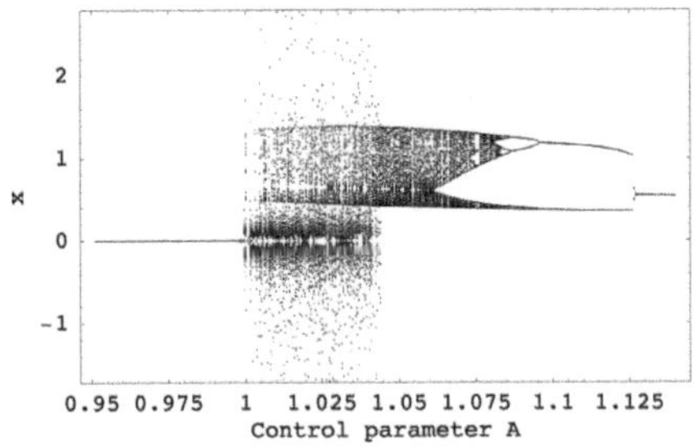

Fig. 11.33 Bifurcation diagram of $\frac{A}{\frac{-\frac{-A^2+\frac{2A+x}{A}-x}{x}-\frac{x}{3A-2x}+A}{A}+Ax+A+x}$ ...

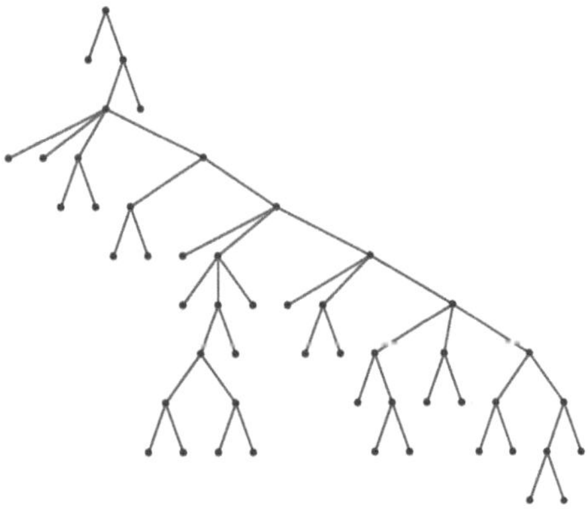

Fig. 11.34 ... and its tree representation.

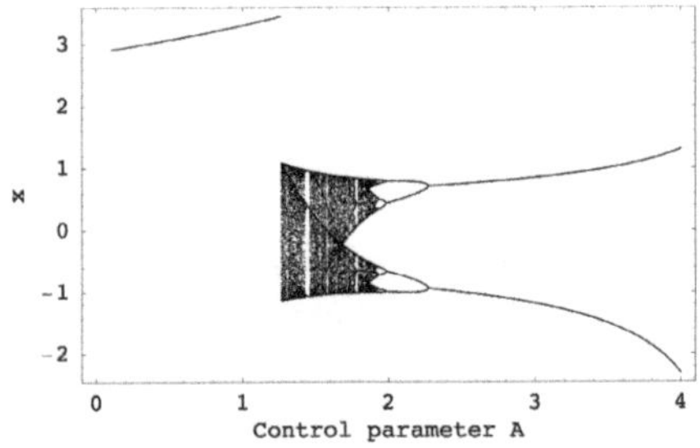

Fig. 11.35 Bifurcation diagram of $\frac{-A+3x-\frac{1}{x}}{\frac{A}{x}-A+x}$ ...

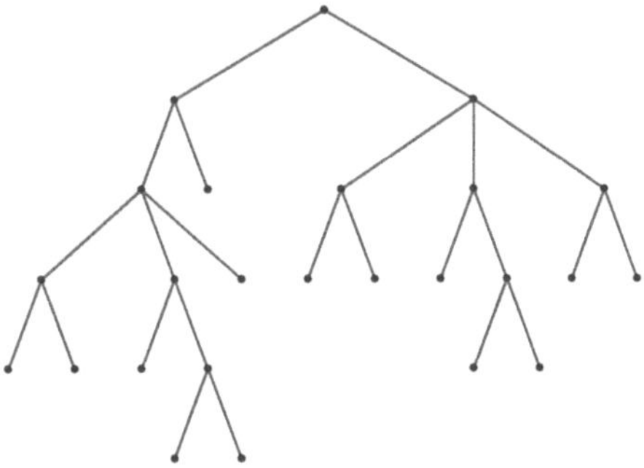

Fig. 11.36 ... and its tree representation.

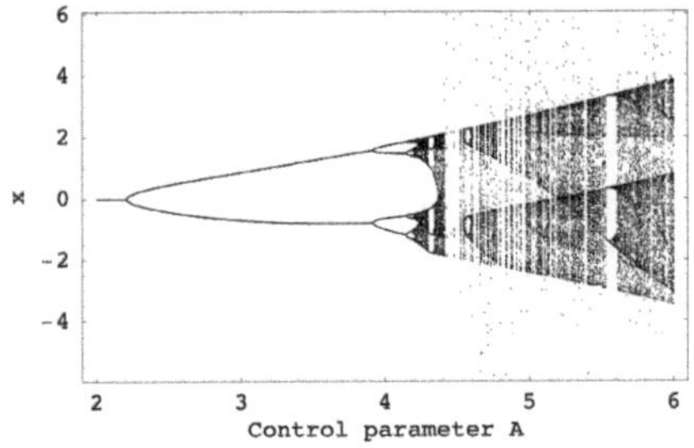

Fig. 11.37 Bifurcation diagram of $\frac{A^2}{-\frac{x^2}{A}+\frac{1}{2x(x^2-A)}-\frac{A}{x}-2x}+x$ ...

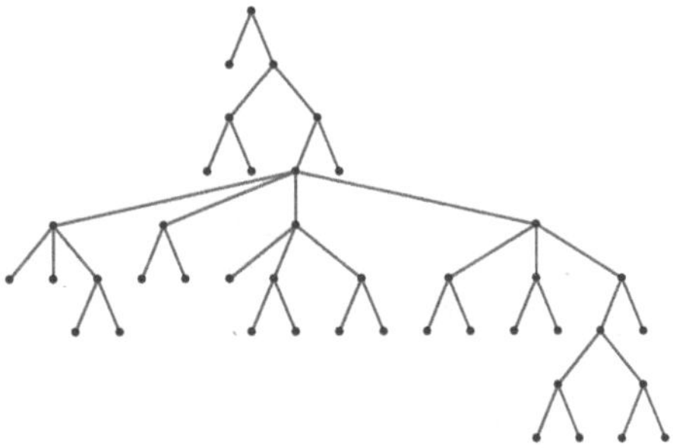

Fig. 11.38 ... and its tree representation.

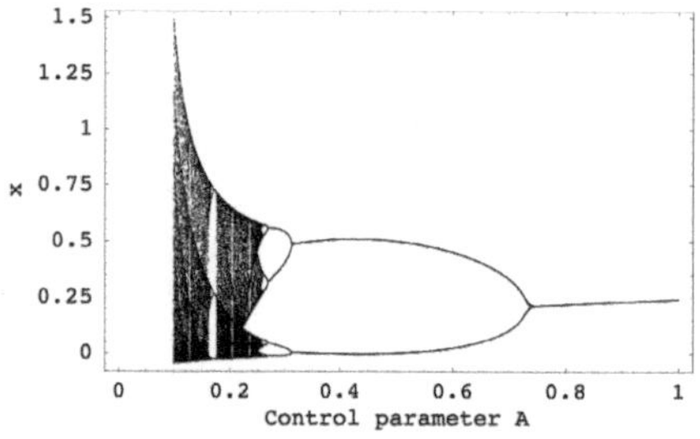

Fig. 11.39 Bifurcation diagram of $\frac{(A-x)\left(-\frac{Ax^2}{A+x}+A-x^2\right)}{(A+x)(2A+x)}$ ...

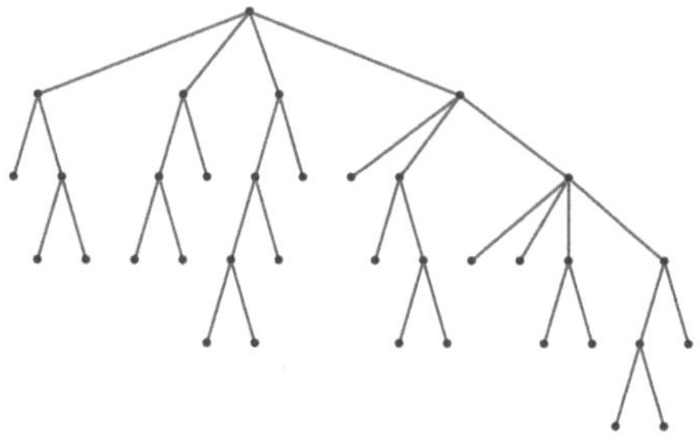

Fig. 11.40 ... and its tree representation.

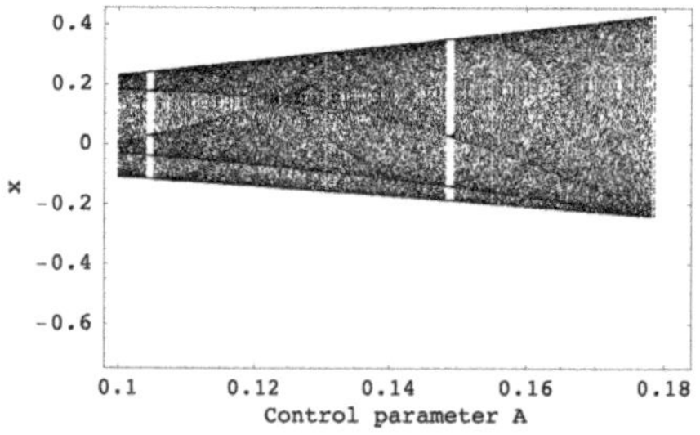

Fig. 11.41 Bifurcation diagram of $\frac{(-Ax-x)(Ax-2A+x)}{\frac{Ax}{A+x}+x^2}$ ...

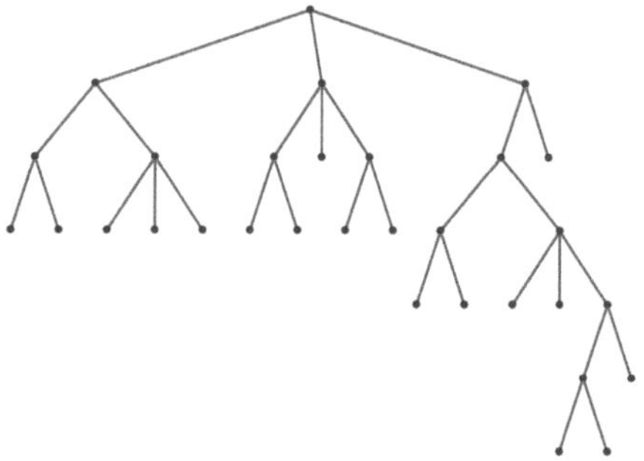

Fig. 11.42 ... and its tree representation.

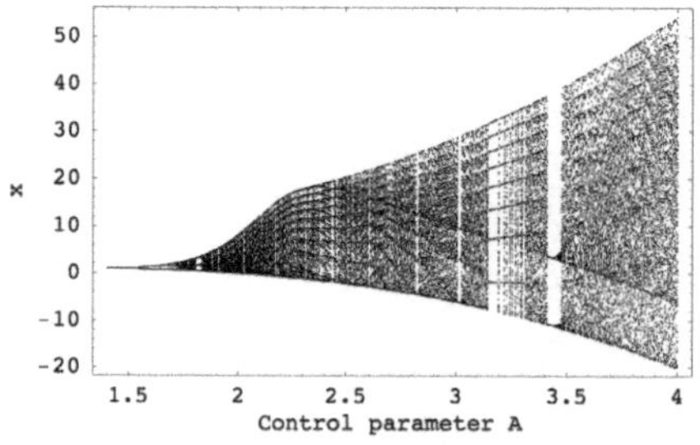

Fig. 11.43 Bifurcation diagram of $\frac{A^2(A-Ax)}{\frac{x\left(-\frac{A}{x}+x+1\right)}{A}+A}+x$ ...

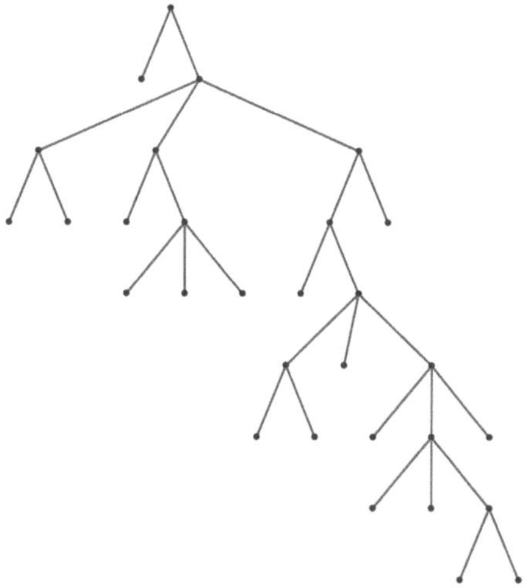

Fig. 11.44 ... and its tree representation.

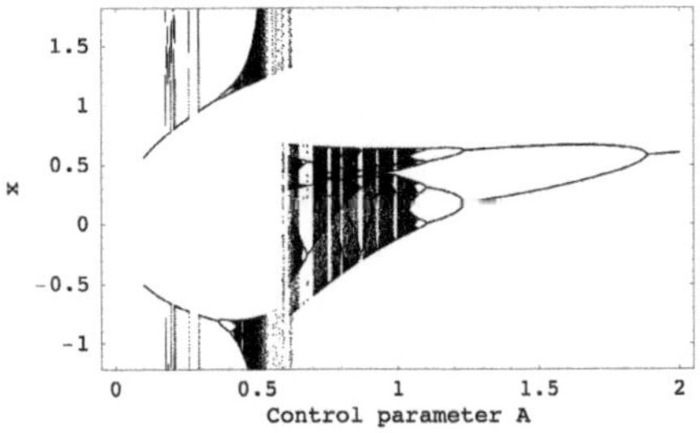

Fig. 11.45 Bifurcation diagram of $\frac{A}{\frac{x^2}{A}-x\left(\frac{x}{x-A}+A-x\right)-\frac{3x-A}{A+2x}+A}-x\ldots$

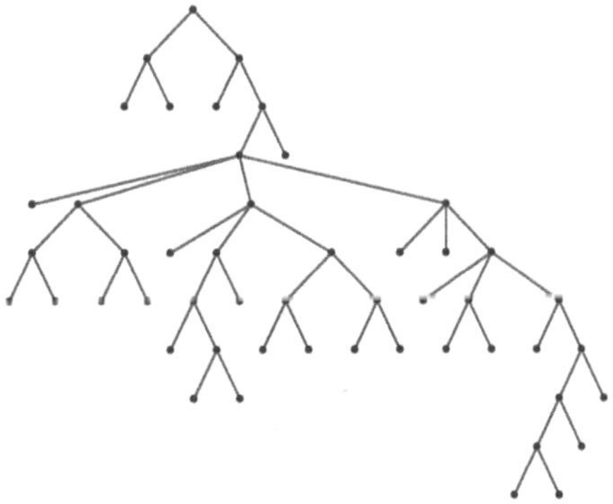

Fig. 11.46 ... and its tree representation.

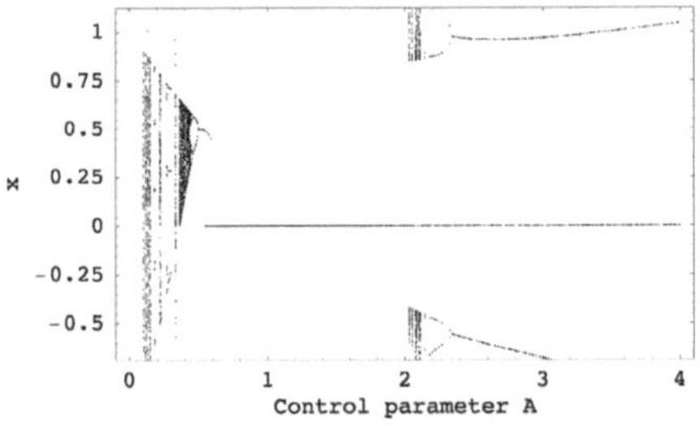

Fig. 11.47 Bifurcation diagram of $\frac{x}{\frac{x\left((A-x)^2-2x\right)}{-\frac{2A}{\frac{x}{A}-1}-A}+A}-x\ldots$

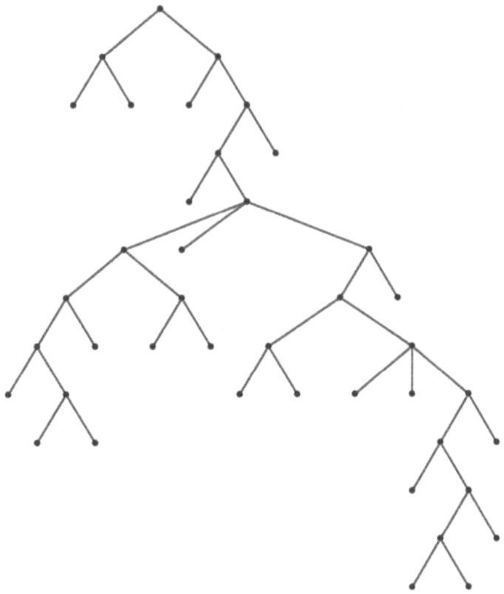

Fig. 11.48 ... and its tree representation.

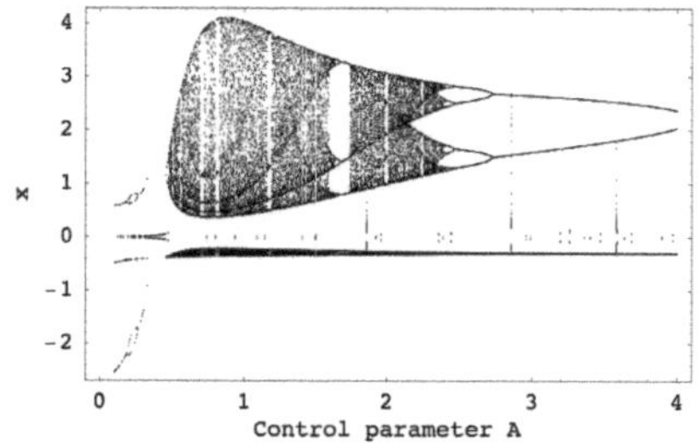

Fig. 11.49 Bifurcation diagram of $\frac{A+x}{-\frac{A}{x}-x\left(\frac{3A+x}{x}+x\right)}$ ...

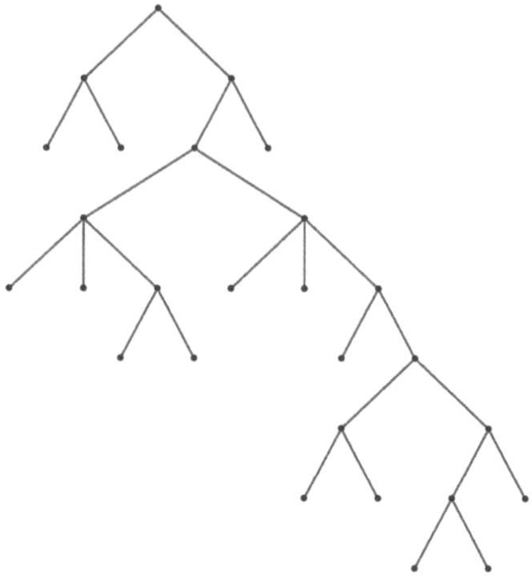

Fig. 11.50 ... and its tree representation.

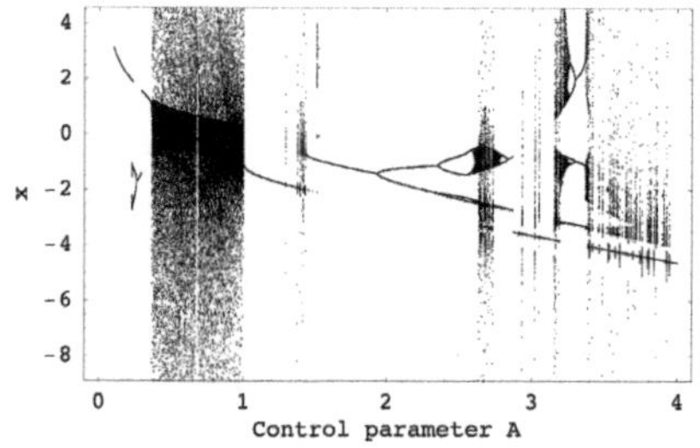

Fig. 11.51 Bifurcation diagram of $\frac{\left(Ax+\frac{x}{A}+A+x^2\right)\left(-\frac{2A}{A+x}+A+x\right)}{x(Ax-x(A-x)+2A+x)}-A$ ...

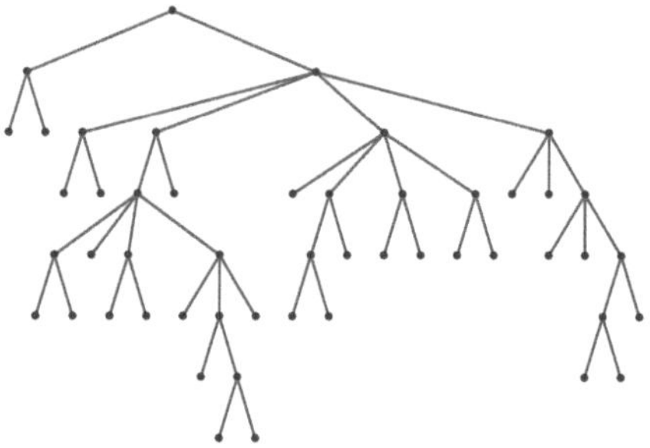

Fig. 11.52 ... and its tree representation.

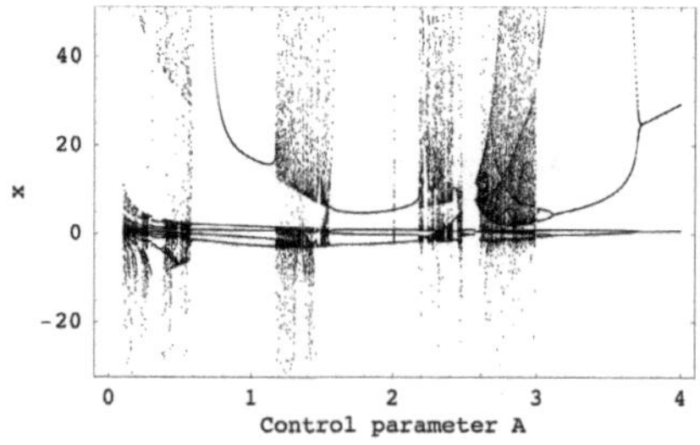

Fig. 11.53 Bifurcation diagram of $\frac{\frac{A^2(A-x)}{x(A(-x)+A+2x)}-2A+x-1}{x}$ ...

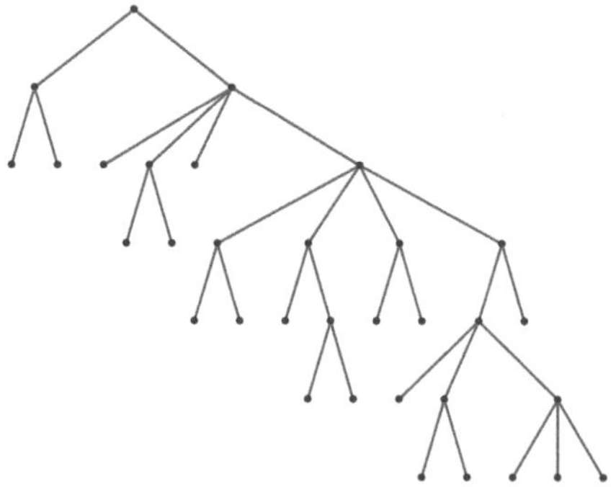

Fig. 11.54 ... and its tree representation.

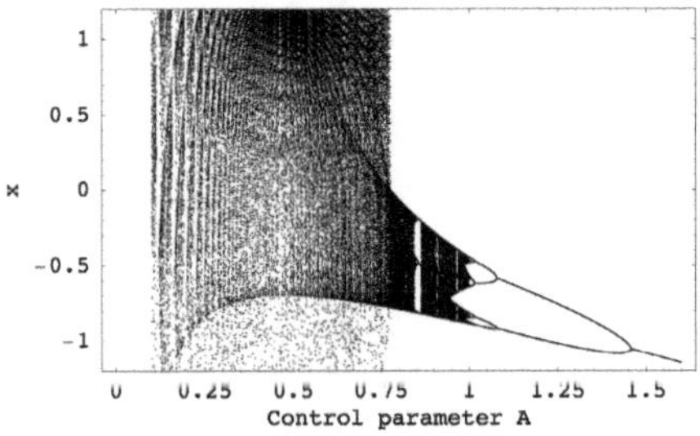

Fig. 11.55 Bifurcation diagram of $\frac{A+\frac{1}{2A}-3x}{\frac{A}{2x}-\frac{A}{A+x}-A-x}$ ...

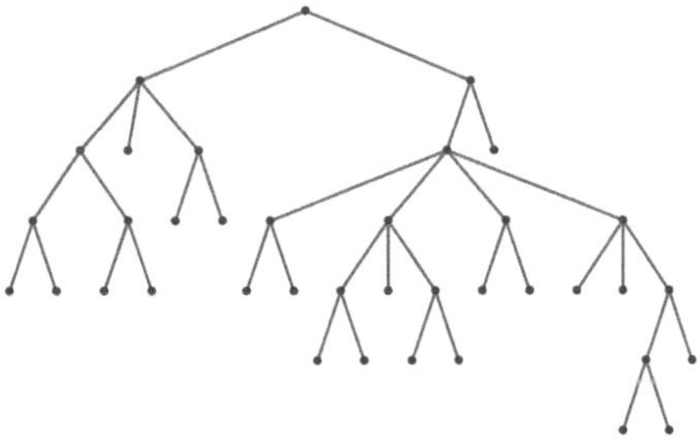

Fig. 11.56 ... and its tree representation.

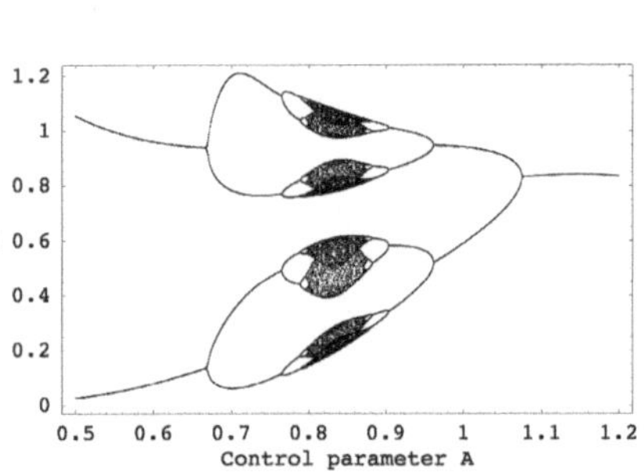

Fig. 11.57 Bifurcation diagram of $-\frac{A}{2x\left(-\frac{x^3}{A^3}-\frac{A^2}{x}+A\right)}$ ...

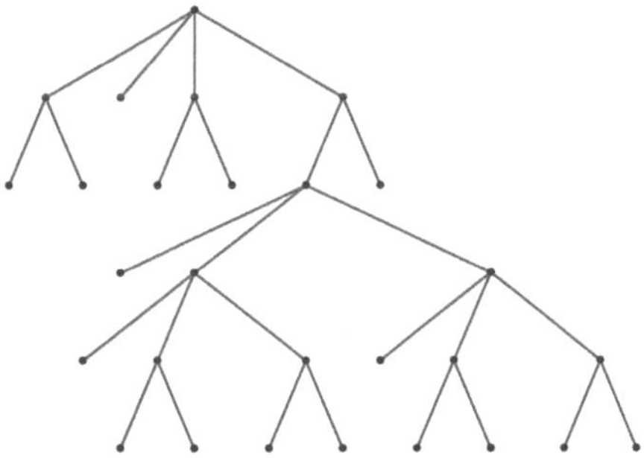

Fig. 11.58 ... and its tree representation.

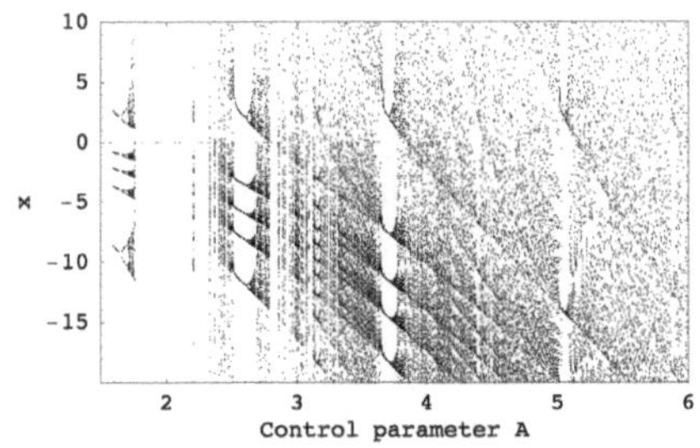

Fig. 11.59 Bifurcation diagram of $\frac{A^2x\left(A^2+A+2x-1\right)}{(-A-x+1)\left(\frac{A}{x}-A+x+\frac{1}{x}\right)}$ ...

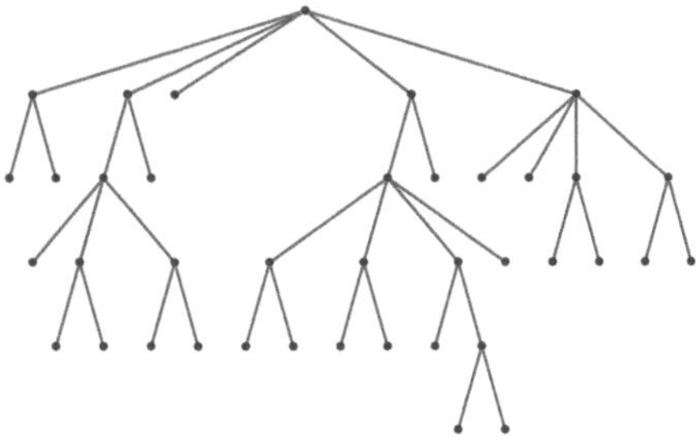

Fig. 11.60 ... and its tree representation.

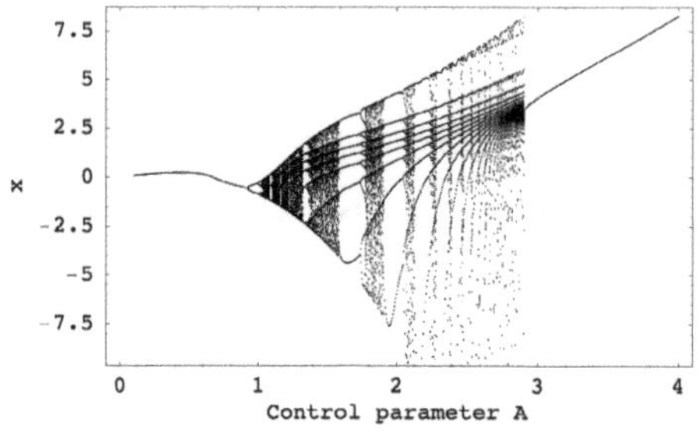

Fig. 11.61 Bifurcation diagram of $A - \frac{A\left(3A-x^2\right)}{\frac{A-2x(-A-x)}{A}+1}$ ...

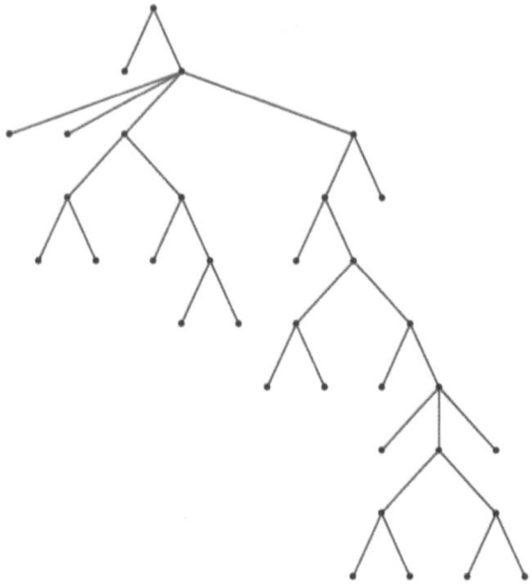

Fig. 11.62 ... and its tree representation.

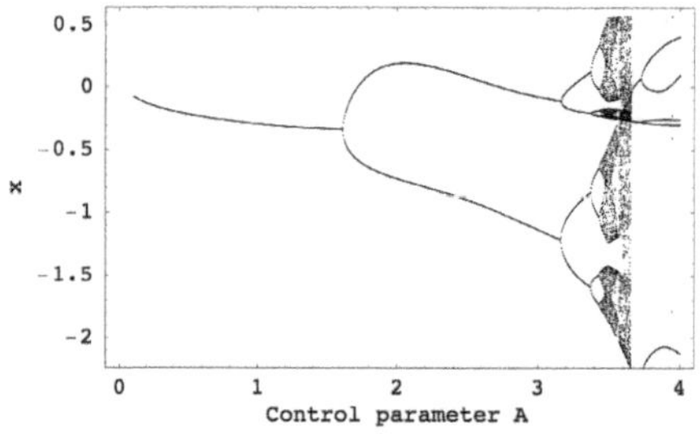

Fig. 11.63 Bifurcation diagram of $\frac{A}{\frac{\frac{A(A+x)}{2x}+\frac{A}{Ax+1}}{\frac{1}{x}+1}-2A+x-1}$ ...

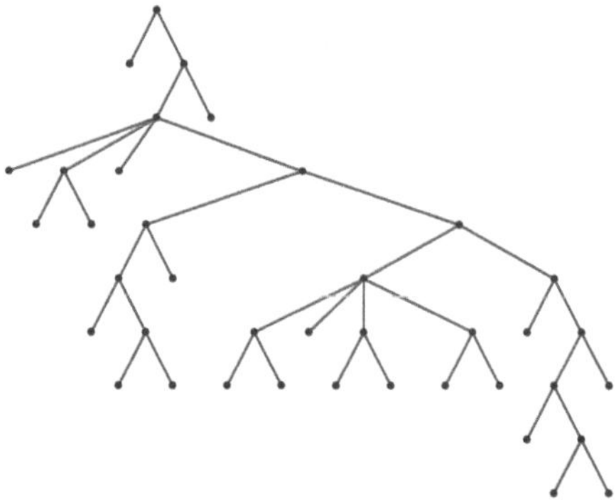

Fig. 11.64 ... and its tree representation.

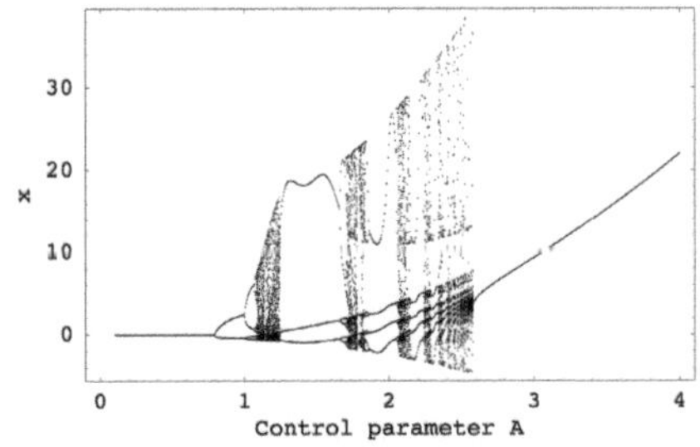

Fig. 11.65 Bifurcation diagram of $-\frac{2A^2x(A(-x)+A+x)}{Ax^2+2x+1}$ ...

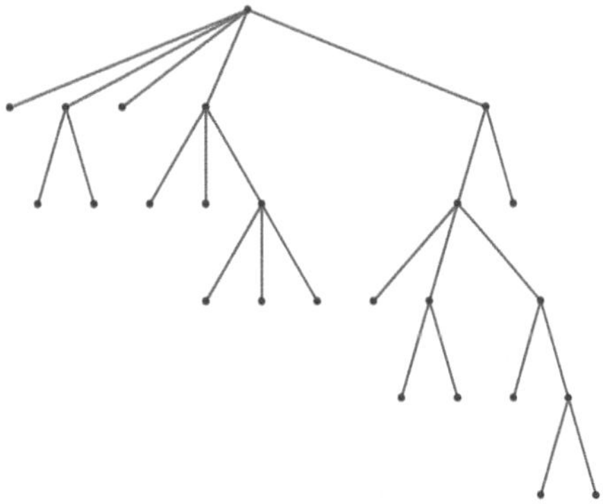

Fig. 11.66 ... and its tree representation.

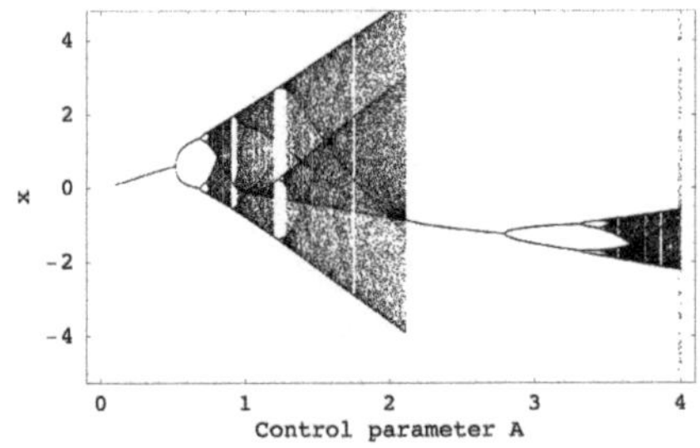

Fig. 11.67 Bifurcation diagram of $\frac{\frac{A}{x}+2A-x^2}{\frac{A+x}{2Ax}+x}$ ...

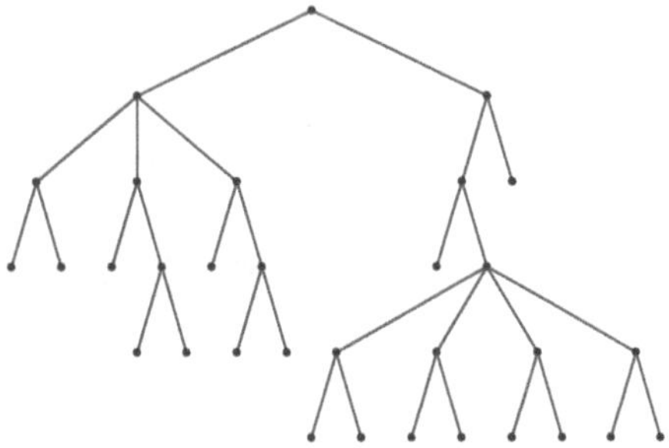

Fig. 11.68 ... and its tree representation.

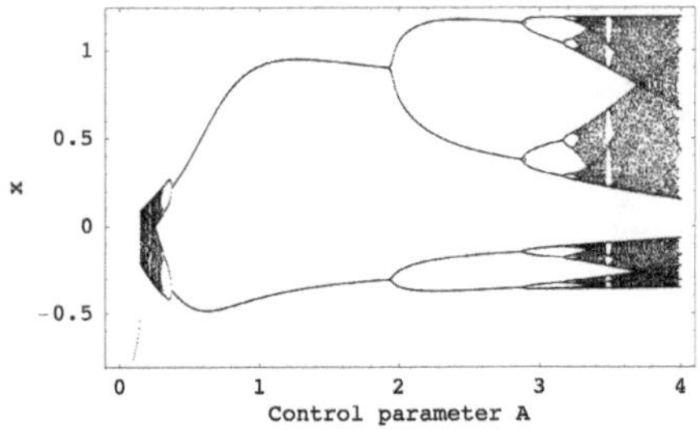

Fig. 11.69 Bifurcation diagram of $\frac{Ax-A-x}{\frac{A}{2x}-\frac{A(A-x)}{Ax+2A-x+1}+A+x}$ ...

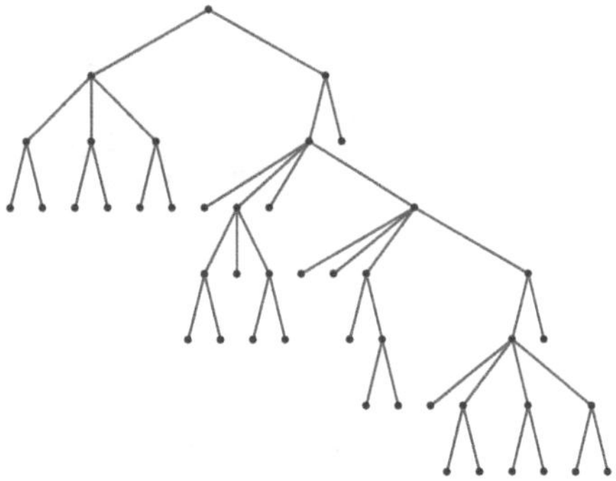

Fig. 11.70 ... and its tree representation.

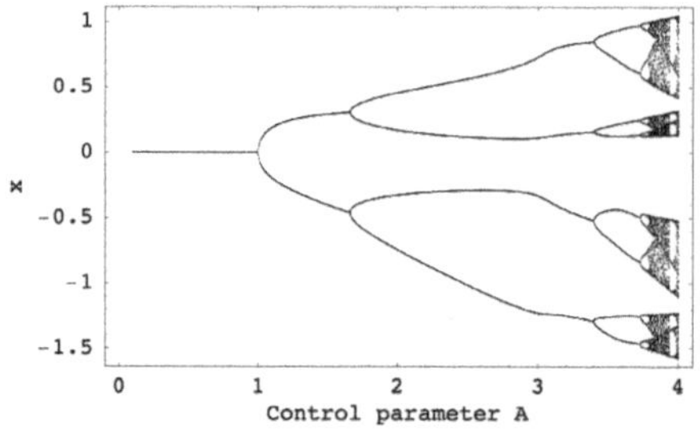

Fig. 11.71 Bifurcation diagram of $\frac{A}{\frac{x^2\left(Ax+\frac{x}{A}\right)}{x-A}-\frac{4Ax}{A+2x}+2x-\frac{1}{x}}$ ...

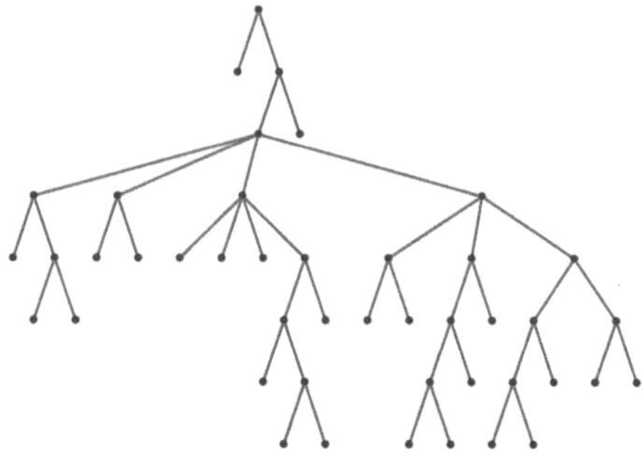

Fig. 11.72 ... and its tree representation.

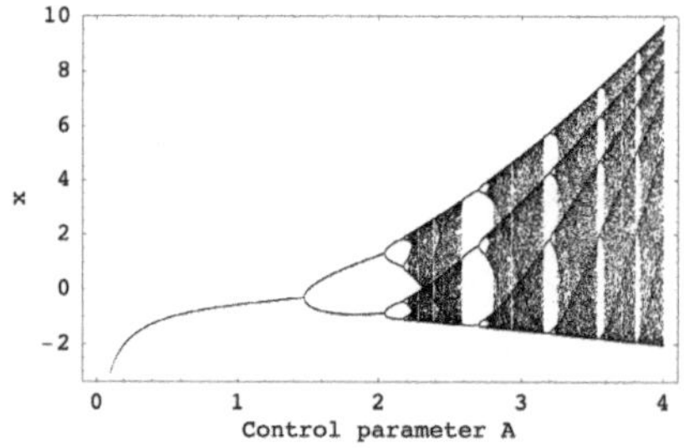

Fig. 11.73 Bifurcation diagram of $-\frac{A(A-x)(2x-A)}{2A+\frac{1}{A}+x^2}-A+x$ ...

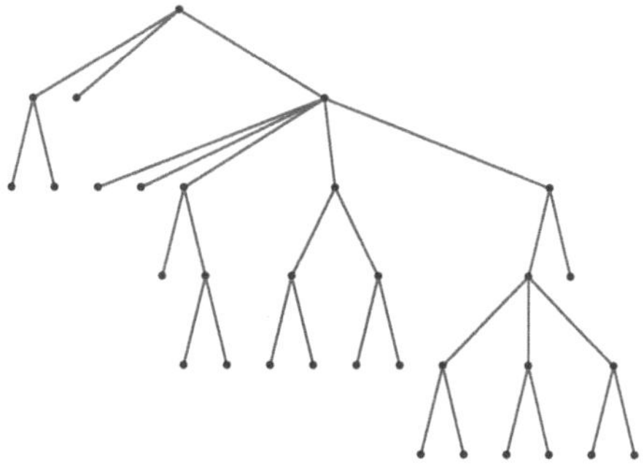

Fig. 11.74 ... and its tree representation.

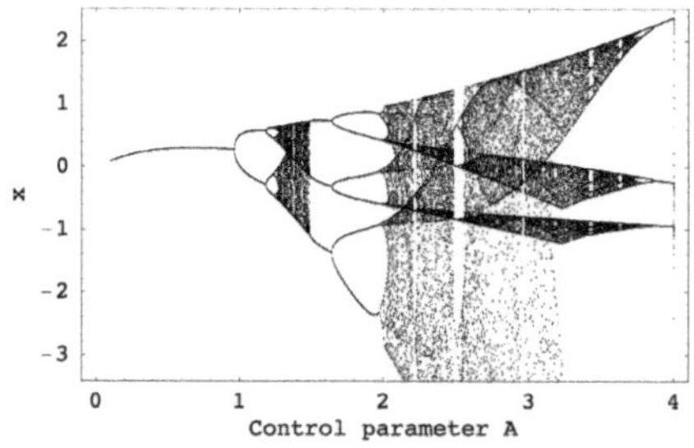

Fig. 11.75 Bifurcation diagram of $\frac{x-Ax^2(Ax+A+x)}{x^2\left(A+\frac{1}{x^2}-x\right)\left(\frac{x}{A}+x\right)}$ ...

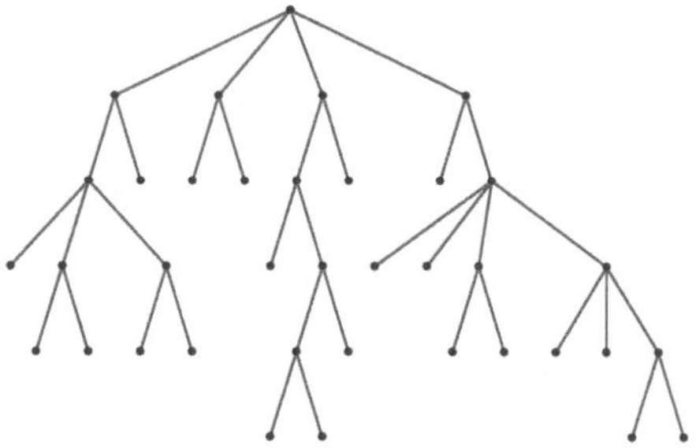

Fig. 11.76 ... and its tree representation.

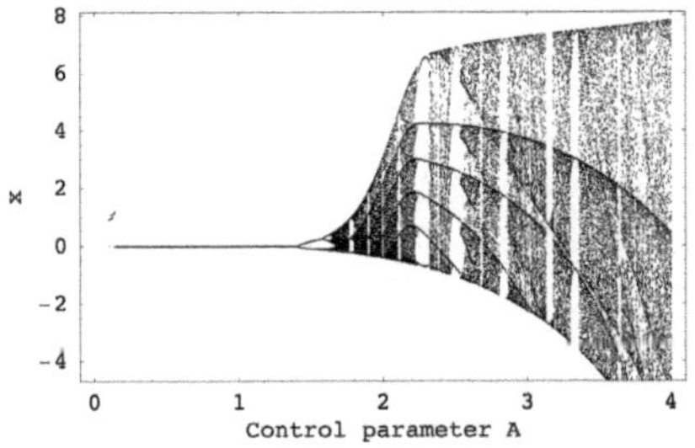

Fig. 11.77 Bifurcation diagram of $\frac{A^3}{A^2-\frac{A}{x}-\frac{(-A-x)(x-3A)}{A}-A-x}+x$ ...

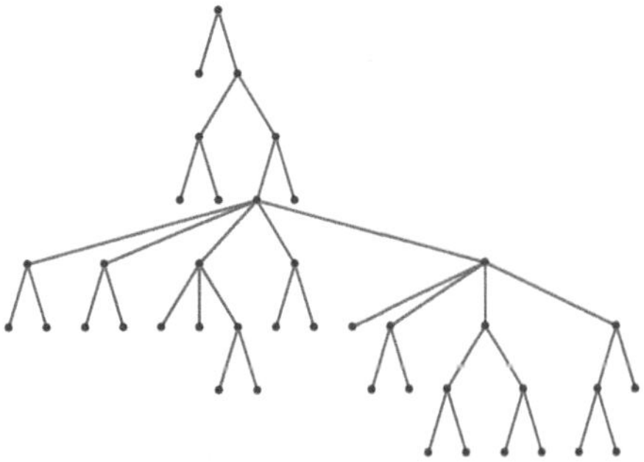

Fig. 11.78 ... and its tree representation.

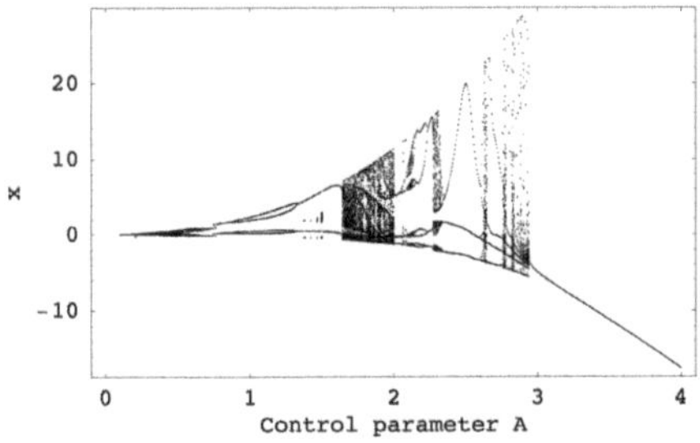

Fig. 11.79 Bifurcation diagram of $\frac{(x-A)\left(A(-x)-\frac{A}{x}-A+2x\right)}{x\left(\frac{A}{x^2\left(Ax+\frac{2x}{A}+2A\right)}+\frac{x}{2A}\right)}$ ...

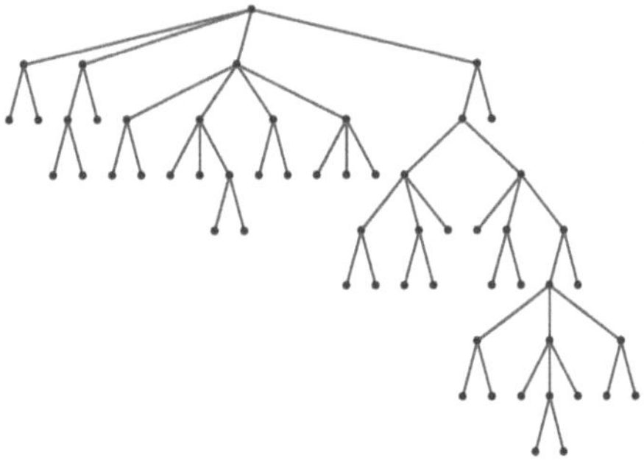

Fig. 11.80 ... and its tree representation.

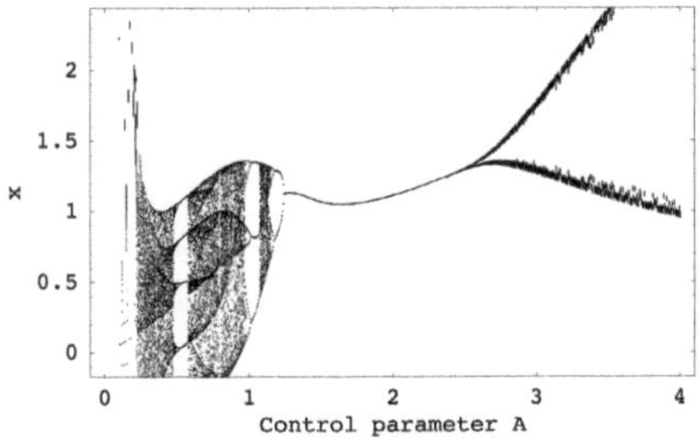

Fig. 11.81 Bifurcation diagram of $\frac{A}{A^3x(A-x)+\frac{x^3}{A^2}+\frac{A(A-x)}{2x^3}}+A-x$ ...

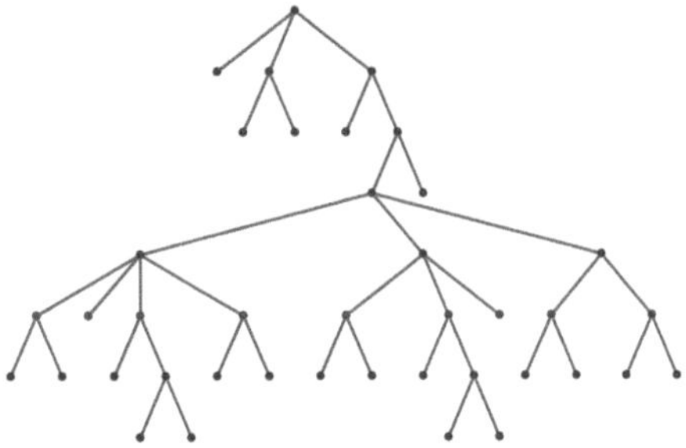

Fig. 11.82 ... and its tree representation.

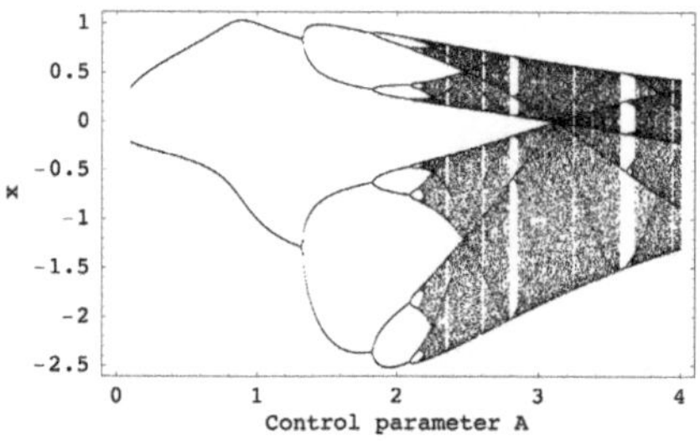

Fig. 11.83 Bifurcation diagram of $\frac{Ax-2x+2}{x^2-\frac{\frac{A}{x}+x}{A}}$ ...

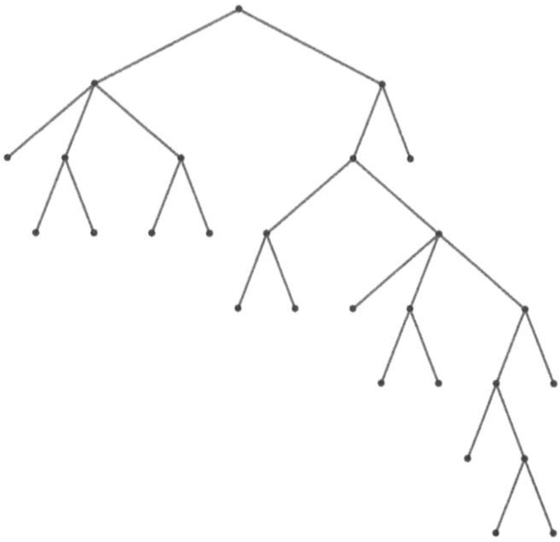

Fig. 11.84 ... and its tree representation.

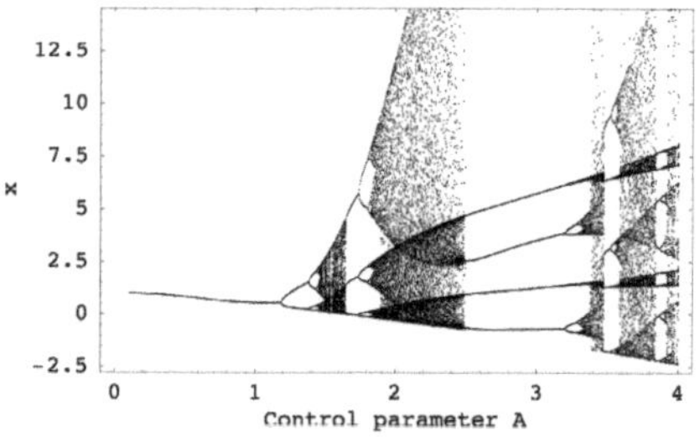

Fig. 11.85 Bifurcation diagram of $\frac{A-2A\left((A-x)^2-x\right)}{x(A-x)-\frac{2(Ax+A+x)}{A}}+1$ ...

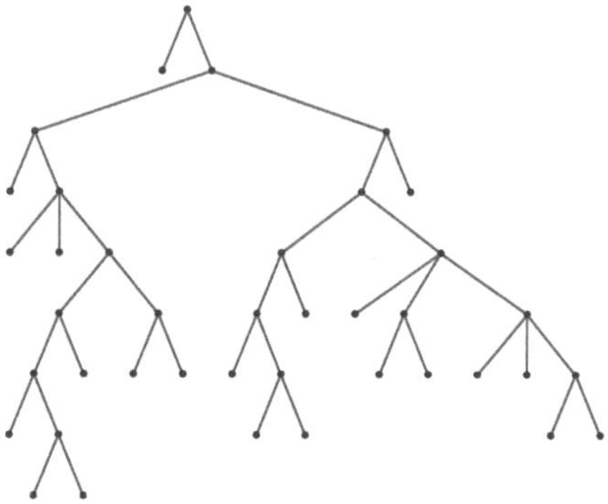

Fig. 11.86 ... and its tree representation.

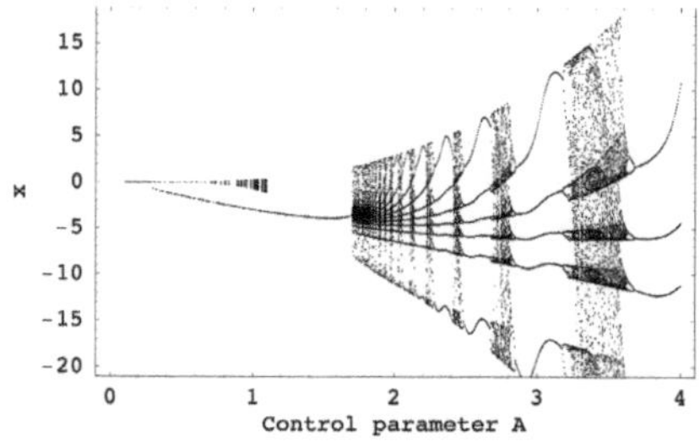

Fig. 11.87 Bifurcation diagram of $\frac{2Ax\left(-A^2-2x\right)}{A+x^2}$ ...

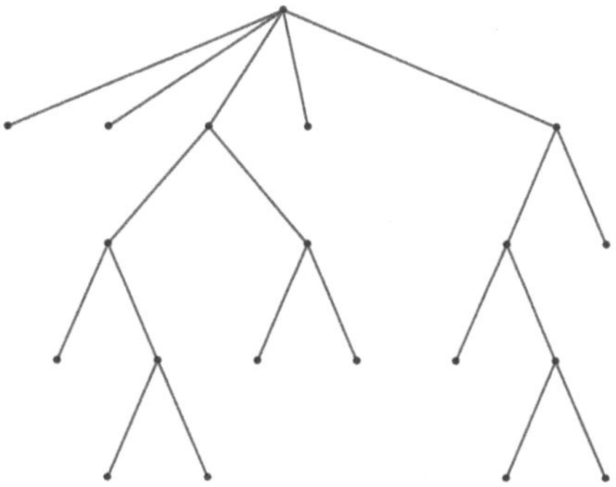

Fig. 11.88 ... and its tree representation.

11.5.3.2 Engineering Design: Preliminary Study

The same principle has been used to synthesize discrete chaotic systems with user defined conditions, i.e. simple engineering design of chaotic systems has been done in this part. In this case study, systems with chaotic behavior in the interval [2, 3], has been accepted while search has been running in interval [0, 4] (overlapping during search was also allowed). In this preliminary study only DE and SOMA algorithms have been used in 50 repeated simulations. Selected demonstrative results are depicted on Fig. 11.89 - 11.96. Basic set of objects used in symbolic regression was again $\{x, A, +, -, \times, /\}$. Bifurcation diagrams of selected systems (Fig. 11.89 - 11.96) are in a few cases overlapping interval [2, 3] and are different in amplitude. It is reasonable to expect that if restriction would be applied on amplitude, then such defined systems would also be synthesized.

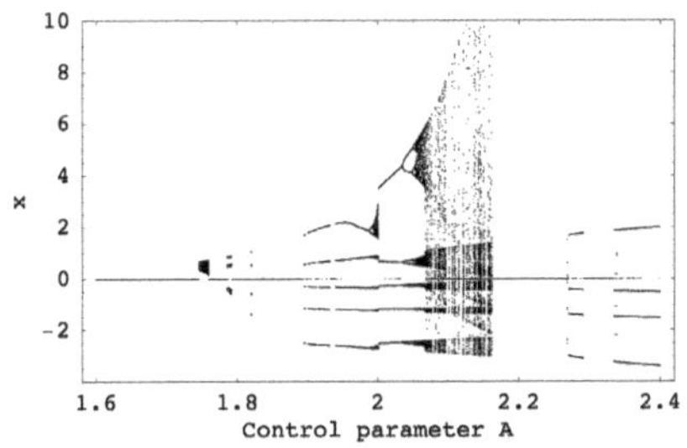

Fig. 11.89 Engineering design: bifurcation diagram of
$$-\frac{A(A+2x)}{2\left(\frac{A\left(Ax+\frac{A}{x}\right)}{\left(A^2-A+1\right)(Ax+2x)(A(x-A)+A)}+x\right)}.$$

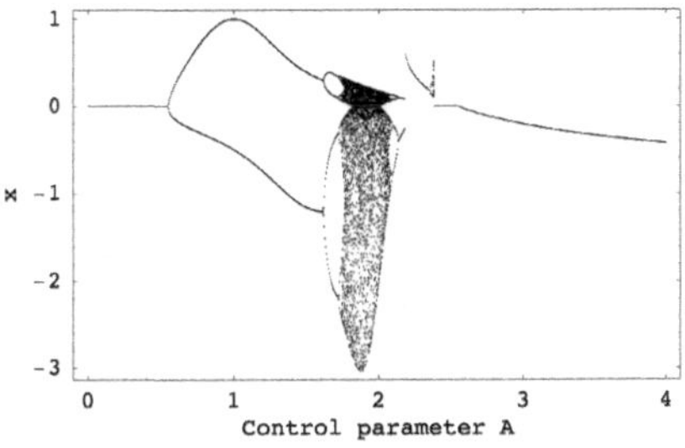

Fig. 11.90 Engineering design: bifurcation diagram of $\frac{A\left(-A^2+A+x\right)+A^2+A-x}{A^2(-x)+A(A^2+x)-\frac{A+x}{Ax}-A-2x}$.

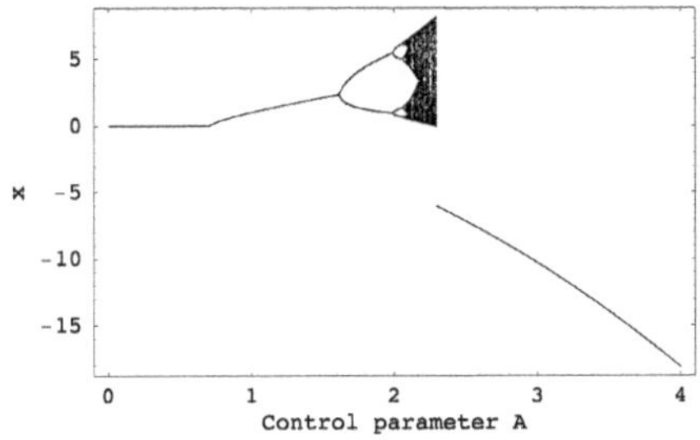

Fig. 11.91 Engineering design: bifurcation diagram of $-\frac{Ax\left(Ax-A^2(x-2A)\right)}{A(-A-x^2+x)-x}$.

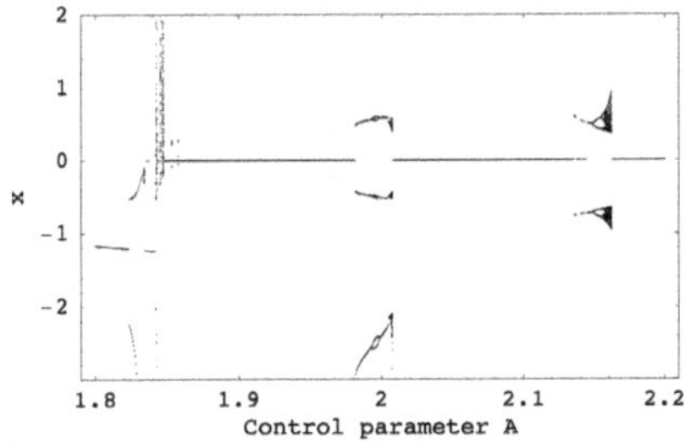

Fig. 11.92 Engineering design: bifurcation diagram of
$$\frac{Ax}{\left((A+x)\left(-A^2+A+x\right)+A+1\right)\left(A-\frac{A\left(\frac{x}{2\left(-\frac{A}{x}+A-x\right)(A+x)}+\frac{x+1}{A}\right)}{2x}\right)}.$$

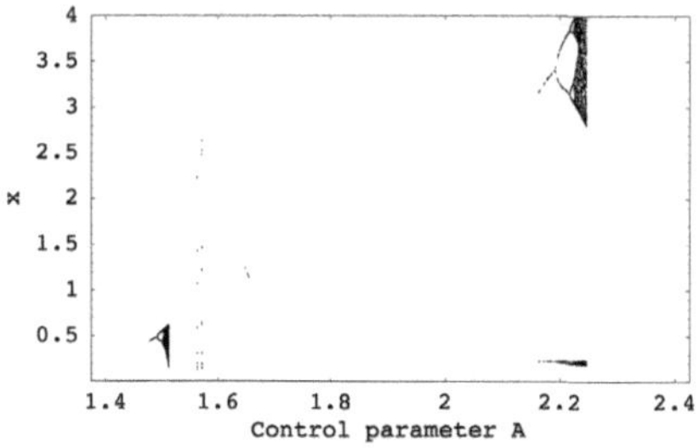

Fig. 11.93 Engineering design: bifurcation diagram of $\frac{A\left(2A(x-A)+\frac{1}{A-x}+A\right)}{-\frac{1}{A^2x^2}+\frac{4x^3}{A}-x(x-A)-x}$.

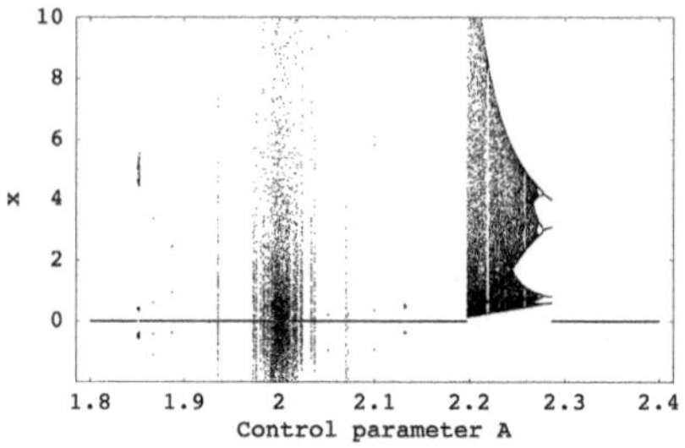

Fig. 11.94 Engineering design: bifurcation diagram of $-\frac{Ax^2}{-A^2+2A-x^3+x}$.

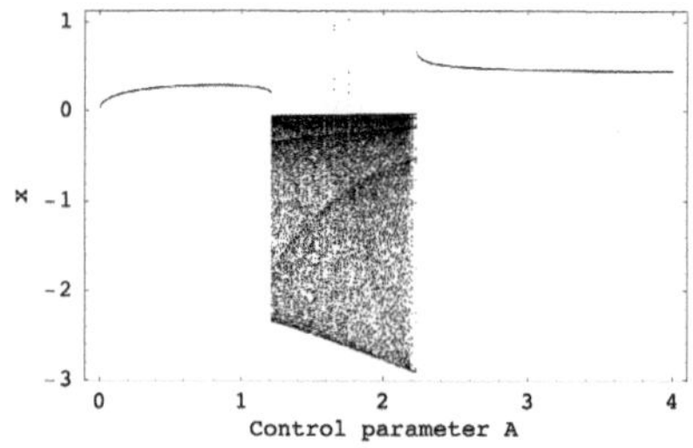

Fig. 11.95 Engineering design: bifurcation diagram of $\frac{A}{x\left(\frac{A}{-A^2+Ax+A-x^2+x}+Ax^2-x\right)+\frac{A}{2x}+A+x}$.

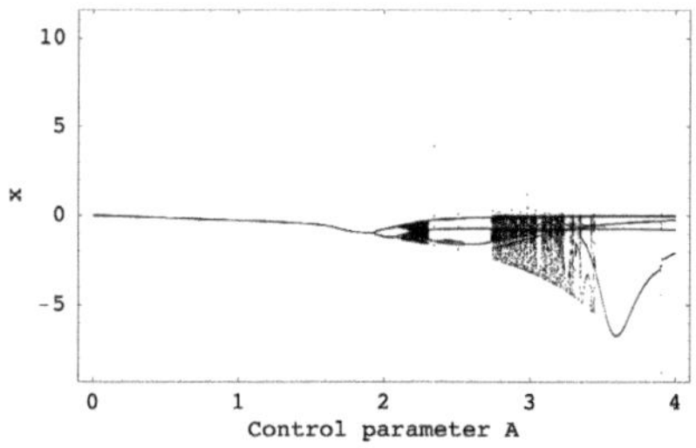

Fig. 11.96 Engineering design: bifurcation diagram of $\frac{A^2}{\left(Ax(A-x)+\frac{x}{A}+A-3x\right)\left(\left(\frac{x}{A}+A-x^2\right)\left(Ax^2+x\right)+\frac{Ax\left(\frac{x}{2A+2x}-Ax\right)}{\frac{A}{x(A+x)}-\frac{x}{A-x}}-A\right)-\frac{A}{A+x}-A}$.

11.6 Conclusion

The aim of this paper is to show how various chaotic systems can be synthesized by means of evolutionary algorithms. Evolutionary synthesis of chaotic systems has been applied to 13 basic comparative simulations in this chapter. Each comparative simulation was repeated 100 times and all 1300 results (100 simulations for each algorithm) were used to create Fig. 11.14 - Fig. 11.88 for overall performance evaluation of evolutionary chaos synthesis. The results look quite promising and convincing.

For comparative studies, five algorithms was used - Differential Evolution (DE) [35], Self Organizing Migrating Algorithm (SOMA) [48], Genetic Algorithms [17], [4], Simulated Annealing (SA) and [9] Evolutionary Strategies (ES) [3]. They were chosen to show that evolutionary synthesis of chaos by AP can be implemented via any evolutionary algorithm and that they all give reasonable results.

The method of symbolic regression described in this paper is relatively simple, but feasible to implement and easy to use. Based on its principles and its possible universality (as just mentioned, it was tested with 5 evolutionary algorithms – SOMA, DE, GA, ES and SA in 13 versions), symbolic regression seems quite capable of synthesizing new dynamical systems for generating chaos.

As a summary, the following statements are presented:

- **Result verification.** To be sure that the results as presented in this chapter are correct, all written synthesized functions were used for automatic generation of bifurcation diagrams and Lyapunov exponents.
- **Simulation results.** Based on the results (Fig. 11.14 - Fig. 11.88) and the selected bifurcation diagrams, it can be stated that all simulations give satisfactory results and that evolutionary synthesis of chaos is capable of solving this class of problems.
- **Range of chaos and interval of observation.** During evolutions, chaos was searched by focusing on interval [0, 4], based on the a priori known behavior of the logistic equation, whose elements were used in the evolution. Despite the a priori known information, a few chaotic systems were located outside of this interval. That was due to the fact that a part of the chaotic behavior was inside the interval [0, 4] and thus EA was able to identify it. From these facts, it is clear that EA are able to locate chaos in a wider range than those expected from some textbook exemplary systems.
- **Exemplary system synthesis.** Based on the fact that the logistic equation (its elements and range) is used for chaos synthesis, it is logical to expect that during evolution (if repeated for many times) the original system should also be synthesized. That event was also observed for a few times, exactly in the mathematical form eq. (11.6).
- **Mutual comparison.** When comparing all algorithms, it is obvious that these algorithms produced good results. Parameter setting for the algorithms was based on a heuristic approach and thus there is a possibility that better settings can be found there. Based on these results, it is clear that for symbolic synthesis via analytic programming any evolutionary algorithm can be used.
- **Engineering design.** It is quite clear that evolutionary synthesis of chaos can be applied to engineering design of devices based on chaos (signal transmission via chaos, chaos-based encryption, and so on). Based on principles and results reported in this paper, it should be possible to synthesize systems with some precisely defined chaotic features and attributes.

Future research is being carried on under the framework of evolutionary synthesis of chaos. It is expected that all 13 EAs will also be used for synthesis of chaotic systems that are not restricted to the simple logistic equation. Based on the results reported in this chapter and our experience with EAs, it is believed that symbolic regression based on EAs is also able to synthesize various chaotic systems according to some predefined characteristics and conditions. It can be foreseeable that the possibility of synthesizing such artificial systems would have an impact on engineering applications dealing with chaos (signal transmission, cryptography, etc.), which is worth further investigation.

Acknowledgements. This work was supported by grant No. MSM 7088352101 of the Ministry of Education of the Czech Republic and by grants of the Grant Agency of the Czech Republic GACR 102/09/1680.

References

1. Alvarez, J., Puebla, H., Cervantes, I.: Stability of observer-based chaotic communitcation for a class of Lur'e systems. Int. J. Bifurcat Chaos Appl. Sci. Eng. 7, 1605–1618 (2002)
2. Back, T., Fogel, D., Michalewicz, Z.: Handbook of Evolutionary Computation, Institute of Physics, London (1997)
3. Beyer, H.: Theory of Evolution Strategies. Springer, New York (2001)
4. Cerny, V.: Thermodynamical approach to the traveling salesman problem: An efficient simulation algorithm. J. Opt. Theory Appl. 45(1), 41–51 (1985)
5. Chen, G.: Controlling Chaos and Bifurcations in Engineering Systems. CRC Press, Boca Raton (2000)
6. Chen, G., Dong, X.: From Chaos to Order: Methodologies, Perspectives and Applications. World Scientific, Singapore (1998)
7. Chen, G., Ueta, T.: Yet another chaotic attractor. Int. J. Bifurcat Chaos Appl. Sci. Eng. 9, 1465–1667 (1999)
8. Davis, L.: Handbook of Genetic Algorithms. Van Nostrand Reinhold, Berlin (1996)
9. Deb, K.: Multi-Objective Optimization Using Evolutionary Algorithms. Wiley, Chichester (2001)
10. Dmitriev, A., Panas, A., Starkov, S.: Ring oscillating systems and their application to the synthesis of chaos generators. Int. J. Bifurcat Chaos Appl. Sci. Eng. 6(5), 851–865 (1996)
11. Dmitriev, A., Efremova, E., Kuzmin, L., Anagnostopoulos, A.: High dimensional RC – oscillators of chaos. In: International Symposium on Nonlinear Theory and its Applications, Miyagi, Japan (2001)
12. Eguchi, K., Inoue, T., Tsuneda, A.: Synthesis and analysis of a digital chaos circuit generating multiple-scroll strange attractors. IEICE Trans. Fundamentals E82-A(6), 965–972 (1999)
13. Garey, M., Johnson, D.: Computers and Intractability: A Guide to the Theory of NP-Completeness. Freeman, New York (1979)
14. Gilmore, R., Lefranc, M.: The Topology of Chaos: Alice in Stretch and Squeezeland. Wiley Interscience, New York (2002)
15. Grebogi, C., Lai, Y.: Controlling chaos. In: Schuster, H. (ed.) Handbook of Chaos Control. Wiley-VCH, New York (1999)
16. Hilborn, R.: Chaos and Nonlinear Dynamics. Oxford University Press, UK (1994)
17. Holland, J.: Adaptation in Natural and Artificial Systems. Univ. Michigan Press, Ann Arbor (1975)
18. Hu, G., Xie, F., Xiao, J., Yang, J., Qu, Z.: Control of patterns and spatiotemporal chaos and its application. In: Schuster, H. (ed.) Handbook of Chaos Control. Wiley-VCH, New York (1999)
19. Johnson, C.: Artificial immune systems programming for symbolic regression. In: Ryan, C., Soule, T., Keijzer, M., Tsang, E.P.K., Poli, R., Costa, E. (eds.) EuroGP 2003. LNCS, vol. 2610, pp. 345–353. Springer, Heidelberg (2003)
20. Just, W.: Principles of time delayed feedback control. In: Schuster, H. (ed.) Handbook of Chaos Control. Wiley-VCH, New York (1999)

21. Just, W., Benner, H., Reibold, E.: Theoretical and experimental aspects of chaos control by time-delayed feedback. Chaos 13, 259–266 (2003)
22. Koza, J.: Genetic Programming: A paradigm for genetically breeding populations of computer programs to solve problems. Stanford University, Computer Science Department, Technical Report STAN-CS-90-1314 (1990)
23. Koza, J.: Genetic Programming. MIT Press, Boston (1998)
24. Koza, J., Keane, M., Streeter, M.: Evolving inventions. Scientific American, 40–47 (2003)
25. Koza, J., Bennet, F., Andre, D., Keane, M.: Genetic Programming III. Morgan Kaufnamm, New York (1999)
26. Lorenz, E.: Deterministic nonperiodic flow. Journal of the Atmospheric Sciences 20(2), 130–141 (1963)
27. May, R.: Simple mathematical model with very complicated dynamics. Nature 261, 45–67 (1976)
28. Nolle, L., Zelinka, I., Hopgood, A., Goodyear, A.: Comparison of an self organizing migration algorithm with simulated annealing and differential evolution for automated waveform tuning. Adv. Eng. Software 36(10), 645–653 (2005)
29. O'Neill, M., Ryan, C.: Grammatical Evolution. In: Evolutionary Automatic Programming in an Arbitrary Language. Springer, New York (2003)
30. O'Neill, M., Brabazon, A.: Grammatical Differential Evolution. In: Proc. International Conference on Artificial Intelligence (ICAI 2006), pp. 231–236. CSEA Press (2006)
31. Oplatkova, Z.: Optimal trajectory of robots using symbolic regression. In: Proc. 56th International Astronautics Congress 2005, Fukuoka, Japan, paper nr. IAC-05-C1.4.07 (2005)
32. Oplatkova, Z., Zelinka, I.: Investigation on artificial ant using analytic programming. In: Proc. Genetic and Evolutionary Computation Conference 2006, Seattle, WA, pp. 949–950 (2006)
33. Ott, E., Grebogi, C., Yorke, J.: Controlling chaos. Phys. Rev. Lett. 64, 1196–1199 (1990)
34. Perruquetti, W., Barbot, J.: Chaos in Automatic Control. CRC, Bota Raton (2005)
35. Price, K.: An Introduction to Differential Evolution. In: Corne, D., Dorigo, M., Glover, F. (eds.) New Ideas in Optimization, pp. 79–108. McGraw-Hill, London (1999)
36. Pyragas, K.: Continuous control of chaos by self-controlling feedback. Phys. Lett. A 170, 421–428 (1992)
37. Richter, H.: An evolutionary algorithm for controlling chaos: The use of multi-objective fitness functions. In: Guervós, J.J.M., Adamidis, P.A., Beyer, H.-G., Fernández-Villacañas, J.-L., Schwefel, H.-P. (eds.) PPSN 2002. LNCS, vol. 2439, pp. 308–317. Springer, Heidelberg (2002)
38. Richter, H., Reinschke, K.: Optimization of local control of chaos by an evolutionary algorithm. Physica D 144, 309–334 (2000)
39. Ryan, C., Collins, J., O'Neill, M.: Grammatical evolution: Evolving programs for an arbitrary language. In: Banzhaf, W., Poli, R., Schoenauer, M., Fogarty, T.C. (eds.) EuroGP 1998. LNCS, vol. 1391, p. 83. Springer, Heidelberg (1998)
40. Schuster, H.: Handbook of Chaos Control. Wiley-VCH, New York (1999)
41. Stewart, I.: The Lorenz attractor exists. Nature 406, 948–949 (2000)
42. Ueta, T., Chen, G.: Bifurcation analysis of Chen's attractor. Int. J. Bifurcat Chaos Appl. Sci. Eng. 10, 1917–1931 (2000)
43. Vanecek, A., Celikovsky, S.: Chaos synthesis via root locus. IEEE Trans. on Circ. and Systems 41, 59–60 (1994)
44. Vanecek, A., Celikovsky, S.: Control Systems: From Linear Analysis to Synthesis of Chaos. Prentice-Hall, London (1996)

45. Wang, X., Chen, G.: Chaotification via arbitrarily small feedback controls: Theory, method, and applications. Int. J. Bifurcat Chaos Appl. Sci. Eng. 10, 549–570 (2000)
46. Zelinka, I.: Analytic programming by Means of new evolutionary algorithms. In: Proc. 1st International Conference on New Trends in Physics 2001, Brno, Czech Republic, pp. 210–214 (2001)
47. Zelinka, I.: Analytic programming by means of soma algorithm. In: ICICIS 2002, First International Conference on Intelligent Computing and Information Systems, Cairo, Egypt, pp. 148–154 (2002)
48. Zelinka, I.: SOMA – Self Organizing Migrating Algorithm. In: Babu, B.V., Onwubolu, G. (eds.) New Optimization Techniques in Engineering, pp. 167–218. Springer, New York (2004)
49. Zelinka, I.: Investigation on evolutionary deterministic chaos control. In: Proc. IFAC, Prague, Czech Republic, paper No. 03187 (2005)
50. Zelinka, I.: Investigation on realtime deterministic chaos control by means of evolutionary algorithms. In: Proc. First IFAC Conference on Analysis and Control of Chaotic Systems, Reims, France, pp. 211–217 (2006)
51. Zelinka, I., Nolle, L.: Plasma reactor optimizing using differential evolution. In: Price, K., Lampinen, J., Storn, R. (eds.) Differential Evolution: A Practical Approach to Global Optimization, pp. 499–512. Springer, New York (2005)
52. Zelinka, I., Oplatkova, Z.: Analytic programming – Comparative study. In: Proc. the Second International Conference on Computational Intelligence, Robotics, and Autonomous Systems, Singapore, paper No. PS04-2-04 (2003)
53. Zelinka, I., Oplatkova, Z., Nolle, L.: Analytic programming – Symbolic regression by means of arbitrary evolutionary algorithms. Int. J. of Simulation, Systems, Science and Technology 6(9), 44–56 (2005)
54. Zelinka, I., Guanrong, C., Celikovsky, S.: Chaos Synthesis by Means of Evolutionary algorithms. Int. J. Bifurcat Chaos Appl. Sci. Eng. 18(4), 911–942 (2008)
55. Zhou, T., Chen, G., Celikovský, S.: An algorithm for computing heteroclinic orbits and its application to chaos synthesis in the generalized Lorenz system. In: Proc. 16th World Congress of the International Federation of Automatic Control [CD-ROM], Praha, Czech Republic (2005)
56. Zou, Y., Luo, X., Chen, G.: Pole placement method of controlling chaos in DC–DC buck converters. Chinese Phys. 15, 1719–1724 (2006)

Chapter 12
Evolutionary Synchronization of Chaotic Systems

Ivan Zelinka and Ales Raidl

Abstract. This chapter introduces a simple investigation on deterministic chaos synchronization by means of selected evolutionary techniques. Five evolutionary algorithms has been used for chaos synchronization here: differential evolution, self-organizing migrating algorithm, genetic algorithm, simulated annealing and evolutionary strategies in a total of 15 versions. Experiments in this chapter has been done with two different coupled systems (master - slave) - Rössler-Lorenz and Lorenz-Lorenz. The main aim of this chapter was to show that evolutionary algorithms, under certain conditions, are capable of synchronization of, at least, simple chaotic systems, when the cost function is properly defined as well as the parameters of selected evolutionary algorithm.This chapter consists of two different case studies. For all algorithms each simulation was 100 times repeated to show and check the robustness of proposed methods and experiment configurations. All data were processed to obtain summarized results and graphs.

12.1 Introduction

Synchronization is a dynamical process during which one system (synchronized, slaved) is remoted by another (synchronizing, master) so that the synchronized system is in a certain manner following the behavior of the master system. The

Ivan Zelinka
Tomas Bata University in Zlin, Faculty of Applied Informatics, Nad Stranemi 4511, Zlin 76001, Czech Republic
and
VSB-TUO, Faculty of Electrical Engineering and Computer Science, 17. listopadu 15, 708 33 Ostrava-Poruba, Czech Republic
e-mail: zelinka@fai.utb.cz

Ales Raidl
Charles University, Faculty of Mathematics and Physics,V Holesovickach 2, 180 00 Prague 8, Czech Republic
e-mail: ales.raidl@mff.cuni.cz

I. Zelinka et al. (Eds.): Evolutionary Algorithms and Chaotic Systems, SCI 267, pp. 383–407.
springerlink.com

word "synchronization" come from the Greek word "synchronos" ($\sigma\upsilon\nu\chi\rho o\nu o\varsigma$) in which $\sigma\upsilon\nu$ (syn) means the same (common,...) and $\chi\rho o\nu o\varsigma$ (chronos) means the "time". Synchronization can be divided into the following classes [9], [4], [12]:

- **Identical synchronization.** This synchronization may occur when two identical chaotic oscillators are mutually coupled (unidirectional or bidirectional coupling), or when one of them drives the other, which is the case of numerical study A (Lorenz-Lorenz), reported in this chapter. Basically, if $\{x_1, x_2, ...x_n\}$ is a set of state (dynamical) variables of the master system as well as $\{x_1', x_2', ...x_n'\}$ of the slave system, then both systems are synchronized if under certain initial conditions and $t \to \infty$ is true that $|x_1 - x_1'| \to 0$. This states that nothing more than large time is enough for dynamics of both systems in a good approximation. This kind of synchronization is usually called identical synchronization.
- **Generalized synchronization** differ from the previous case by the fact that coupled chaotic oscillators are different and that the dynamical state of one of the oscillators is completely determined by the state of the other. This is a case of numerical study B (Rössler-Lorenz), reported in this chapter.
- **Phase synchronization** is another case of synchronization which occurs when the oscillators coupled are not totally identical and the amplitudes of the oscillator remain unsynchronized, while only oscillator phases evolve in synchrony. There is a geometrical interpretation of this case of synchronization. It is possible to find a so called plane in phase space in which the projection of the trajectories of the oscillator follows a rotation around a well-defined center. The phase is defined by the angle $\varphi(t)$, described by the segment which is joining the center of rotation and the projection of the trajectory point onto the plane.
- **Anticipated and lag synchronization.** Lets say that we have a synchronizing system with state variables $\{x_1, x_2, ...x_n\}$ and a synchronized system with state variables $\{x_1', x_2', ...x_n'\}$. Anticipated and lag synchronization occurs when $x_1'(t) = x_1(t + \tau)$ holds true. This relation, in fact, states that the dynamics of one of the systems follows, or anticipates, the dynamics of the other and whose dynamics is described by delay differential equations.
- **Amplitude envelope synchronization** is a kind of synchronization which may appear between two weakly coupled chaotic oscillators. Comparing with another cases of synchronization, there is no correlation between phases or amplitudes. One can observe a periodic envelope that has the same frequency in the two systems. Magnitude of that envelope has the same order of the difference between the average frequencies of oscillation of both systems. It is important to note that phase synchronization can develops from amplitude one, when the strength of the coupling force between two amplitude envelope synchronized oscillators increases in time.

A rich amount of literature of working with synchronization exist. We can recommend as a representative literature [9], [4] and [12]; all three books are well written and highly readable. Other research works are [13], [2] (synchronization based on time series analysis), [11] (robustness of synchronized systems), and many others. Avery good starting reference can be found in the above mentioned books [9], [4] and [12].

The main aim of this research is to show that evolutionary algorithms (EA) are capable of synchronizing simple chaotic systems, without the knowledge of internal system structure. The ability of EAs to successfully work with black box type of problems have been proven; see for example real-time control of plasma reactor [7], [8], [20] or CML non real-time control by evolutionary algorithms [17], [18], [21] and Chapter 6 in this book. This chapter is organized as follows. The first part outlines the motivation of EAs use on synchronization. This is followed by a very brief note about used evolutionary algorithms whose detailed description is presented in Chapter 6. Evolutionary synchronization is then studied, and finally experimental results are reported, followed by conclusion.

12.2 Motivation

Motivation of this research is quite simple. As mentioned in the introduction and also in the previous chapters, evolutionary algorithms are capable of hard problem solving. A lot of examples about evolutionary algorithms can be easily found like their use in control, artificial intelligence, electronic devices design and setting etc. For more, see for example mentioned references in Chapter 6, section Motivation. The main question in the case of this chapter was if EAs are able to synchronize simple chaotic systems. Main attention has been paid to continuous chaotic systems, i.e. to Rössler and Lorenz systems. All experiments here were designed to confirm or reject this idea and were designed to be as simple as possible, to show the methodology of evolutionary algorithms use.

12.3 Selected Evolutionary Algorithm – A Brief Introduction

For the numerical and symbolic experiments described here, stochastic optimization algorithms such as Differential Evolution (DE) [10], Self Organizing Migrating Algorithm (SOMA) [16], Genetic Algorithms (GA) [5], Simulated Annealing (SA) [6], [3] and Evolutionary Strategies (ES) [1] were selected. Description of all these algorithms can be found in mentioned references or in Chapter 6.

12.4 Evolutionary Synchronization

12.4.1 Used Hardware, Problem Selection and Case Studies

Synchronization in this case study has been done on a special grid computer, comparing to simple PC as in [19]. This grid computer consist of two special Apple servers (for pictures, see Chapter 6). In total 78 CPUs are available. Emanuel has been used for calculations such that each CPU has been used like a single processor and thus a rich set of statistically repeated experiments was possible which was not time consuming. Typical parallel computing has been avoided in experiments described here.

Chaotic systems used in this chapter, has been selected from continuous domain, especially the Rössler and Lorenz systems. This selection has been done because in the remaining part of this book are used mostly discrete chaotic systems (1D as well as CML systems) prior to continuous systems. We would like to show that EAs are not restricted only to discrete domain, so this was the main reason of Rössler and Lorenz system use.

Two case studies (A and B) were defined and used. In the first a coupled system Lorenz-Lorenz (case A, eq. (12.1)) was used. Synchronization has been done via parameter d and coupling "part" $d\,(x_1(t)-x_2(t))$ in eq. (12.1). The cost function in this case has been calculated according to eq. (12.3), i.e. difference in master-slave system in all three system variables.

In the second case study (B), Rössler-Lorenz system was used as described by eq. (12.2). Synchronization has been done via parameter c and coupling "part" $c\,(y_1(t)-y_2(t))$ in eq. (12.2). Cost function in this case has been calculated according to eq. (12.4), i.e. only, comparing with eq. (12.3), for one system variable difference in master-slave system.

12.4.2 Cost Function

The fitness (cost function) has been calculated, as mentioned in the last paragraph, according to the distance between desired synchronizing system state and actual synchronized system state. The minimal value of this cost function, guarantees the best solution. The aim of all simulations was to find the best solution, i.e. a solution that returns the cost value as small as possible. The difference between eq. (12.3) and eq. (12.4) is in the number of used state variables. In the case of eq. (12.3) it is logical to expect that all three state variables will be synchronized perfectly, while in the case of the eq. (12.4) we expected that only synchronized variable, in this case $y_2(t)$, will be synchronized in an acceptable manner. Results depicted later in various figures has confirmed all these presumptions. The cost value was in fact the absolute value of summarization of grey areas between synchronizing and synchronized system output (time series) as demonstrated in Fig. 12.1.

$$
\begin{aligned}
&\textbf{Lorenz}-\textbf{Lorenz synchronization}\\
&\textbf{Lorenz system (master)}:\\
&x_1'(t) = -a\,(x_1(t)-y_1(t))\\
&y_1'(t) = -x_1(t)z_1(t)+bx_1(t)-y_1(t)\\
&z_1'(t) = x_1(t)y_1(t)-cz_1(t)\\
&\\
&\textbf{Lorenz system (slave)}:\\
&x_2'(t) = -\mathbf{a_2}\,(x_2(t)-y_2(t))+\mathbf{d}\,(x_1(t)-x_2(t))\\
&y_2'(t) = -x_2(t)z_2(t)+bx_2(t)-y_1(t)\\
&z_2'(t) = x_2(t)y_2(t)-z_2(t)
\end{aligned}
\tag{12.1}
$$

$$
\begin{aligned}
&\textbf{Rössler – Lorenz synchronization}\\
&\textbf{Rössler system (master)}:\\
&x_1'(t) = -y_1(t) - z_1(t)\\
&y_1'(t) = -x_1(t) - \tfrac{y_1(t)}{5}\\
&z_1'(t) = (x_1(t) - 5.7)\, z_1(t) + 0.2\\
&\textbf{Lorenz system (slave)}:\\
&x_2'(t) = -\mathbf{a}\,(x_2(t) - y_2(t))\\
&y_2'(t) = -x_2(t) z_2(t) + \mathbf{b} x_2(t) + \mathbf{c}\,(y_1(t) - y_2(t))\\
&z_2'(t) = x_2(t) y_2(t) - z_2(t)
\end{aligned}
\tag{12.2}
$$

$$
CF_{LL}(a_2, d) = \int_0^{100} |x_1(t) - x_2(t)| + |y_1(t) - y_2(t)| + |z_1(t) - z_2(t)|\, dt \tag{12.3}
$$

$$
CF_{RL}(a, b, c) = \int_0^{200} |(y_1(t) - y_2(t)|\, dt \tag{12.4}
$$

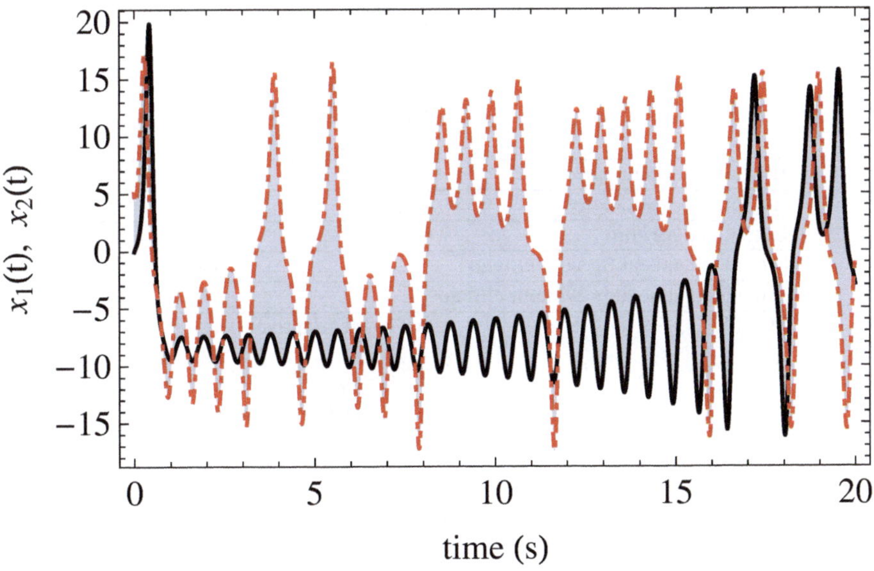

Fig. 12.1 Principle of cost value calculation. For all three variables $x(t)$, $y(t)$ and $z(t)$ has been calculated difference between behavior of synchronizing and synchronized system (light grey area).

12.4.3 Parameter Setting

The control parameter settings have been found empirically and are given in Tables 12.1 - 12.6. The main criterion for this setting was to keep the same setting of parameters as much as possible alongside the same number of cost function evaluations as well as the population size. Individual length represents the number of optimized parameters (in this case coupling parameter d (L-L system) or c (R-L system)).

We would like to note that settings of all used algorithms has been based on our preliminary experiences and certainly can be improved. However this topic is quite numerically time consuming, so we let this topic open for future research.

All algorithms (SOMA, DE, SA, GA, ES) have been applied 100 times in order to find the optimum of both case studies. The primary aim of this comparative study is not to show which algorithm is better and worst, but to show that evolutionary synchronization can be really used for chaotic systems. Outputs of all simulations are depicted in Fig. 12.6 - 12.12, and Fig. 12.34 - 12.36, which shows results of all 100 simulations for each case study.

Table 12.1 Algorithms abbreviation

Algorithm	Version	Abbreviation
Differential Evolution	DEBest1JIter	D1
	DEBest2Bin	D2
	DELocalToBest	D3
	DERand1Bin	D4
	DERand1DIter	D5
	DERand2Bin	D6
Evolutionary strategies	(μ,λ)	ES1
Evolutionary strategies	$(\mu+\lambda)$	ES2
Genetic Algorithm		G
Simulated annealing with elitism		SA1
Simulated annealing without elitism		SA2
SOMA	AllToAllAdaptive	S1
	AllToAll	S2
	AllToOne	S3
	AllToOneRandomly	S4

Table 12.2 DE setting for case studies A and B

Case Study	A	B
NP	100	100
F	0.9	0.9
CR	0.3	0.3
Generations	500	500
Individual Length	2	3

Table 12.3 ES setting for case studies A and B

Case Study	A	B
μ ,λ	100	100
σ	1	1
Iterations	100	100
Individual Length	2	3

Table 12.4 GA setting for case studies A and A

Case Study	A	B
Population size	100	100
Mutation	0.4	0.4
Generations	500	500
Individual Length	2	3

Table 12.5 SA setting for case studies A and B

Case Study	A	B
No. of particles	100	100
σ	0.5	0.5
$k_{\max}$	66	66
T_{min}	0.0001	0.0001
T_{max}	1000	1000
α	0.95	0.95
Individual Length	2	3

Table 12.6 SOMA setting for case studies A and B

Case Study	A	B
PathLength	3	3
Step	.11	.11
PRT	0.1	0.1
PopSize	20	20
Migrations	10	10
MinDiv	-0.1	-0.1
Individual Length	2	3

12.4.4 Experimental Results

Two main case studies has been done in this chapter. Case study A (synchronization of Lorenz-Lorenz system), and B (synchronization of Rössler-Lorenz system). In both cases attention was paid to parameter estimation, obtained cost value as well as to cost function evaluations needed to reach acceptable setting of given synchronization. All data has been processed in order to get viable statistics about the evolutionary dynamics behind these experiments. These statistics has been processed into figures, which shows the performance of evolutionary techniques from different point of views. Average values are depicted as horizontal line in each figure.

12.4.4.1 Case Study A: Lorenz - Lorenz Synchronization

In the first case study, we have used for synchronization two identical systems Lorenz - Lorenz systems (eq. (12.1), Fig. 12.2). Synchronization has been done by coupling of variables $x_{1,2}(t)$ via parameter d. The difference between them has

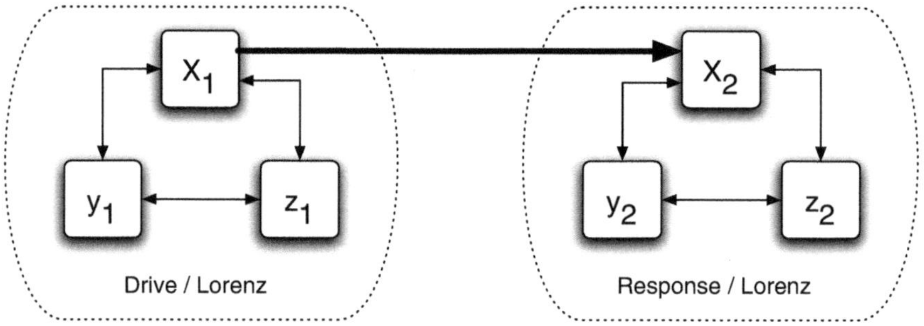

Fig. 12.2 Schematic of Lorenz-Lorenz synchronization. Variable x_2 has been directly synchronized.

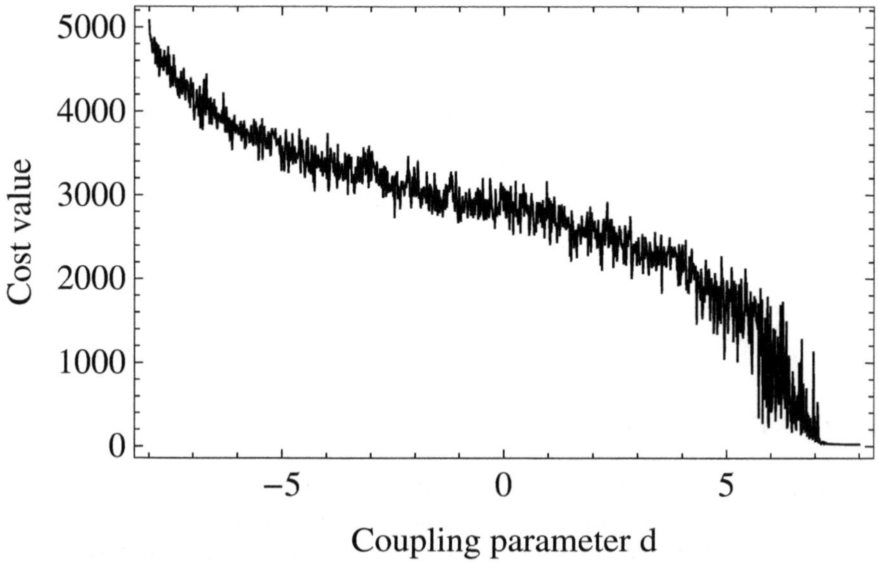

Fig. 12.3 Dependance of cost function value on coupling parameter d.

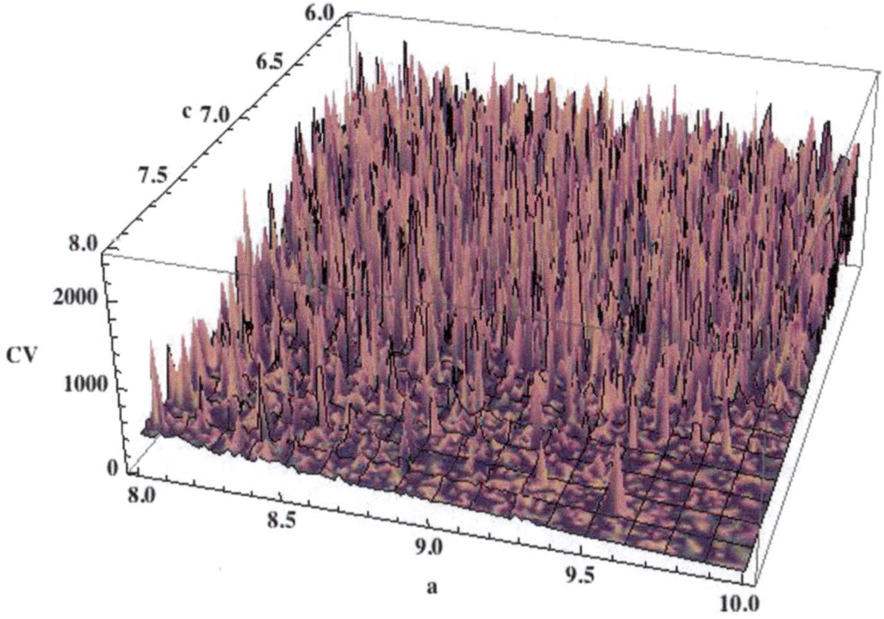

Fig. 12.4 Dependance of cost function value on coupling parameter d and parameter a_2.

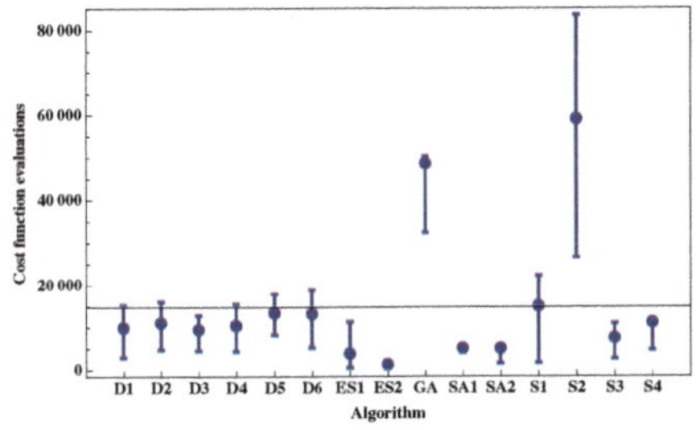

Fig. 12.5 Cost function evaluations...

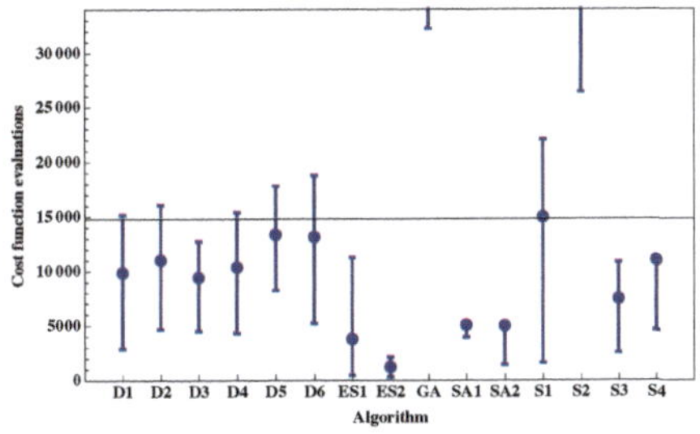

Fig. 12.6 ... and detail view; horizontal line is an average value of all.

been calculated and multiplied by parameter d, see $(d\,(x_1(t) - x_2(t)))$ in eq. (12.1). In fact, the search for optimal parameter setting has been done with dependance on two parameters: on parameter d and a_2 in eq. (12.1).

Complexity of cost function landscape, is depicted on Fig. 12.3 for dependance only on coupling parameter d and Fig. 12.4 for dependance at both parameters. All parameters were varied around nominal values, as referred in the literature. Complexity is very high, as anyone can easily see. In the chaotic landscape a linear-like trend is visible. Thanks to this trend, it is visible that the minimum can be expected at position $\{d, a_2\} = \{8, 10\}$.

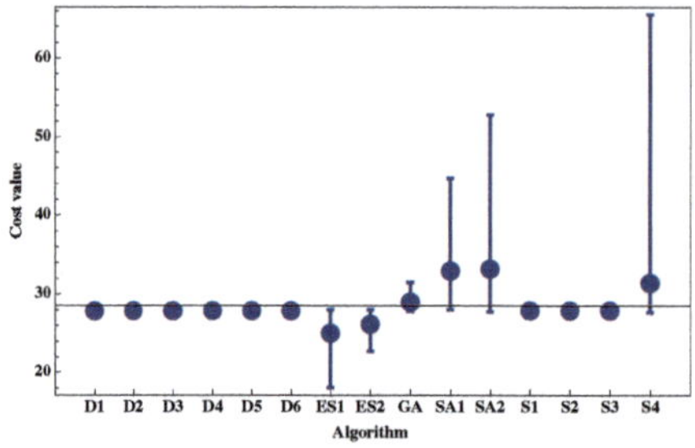

Fig. 12.7 Cost value...

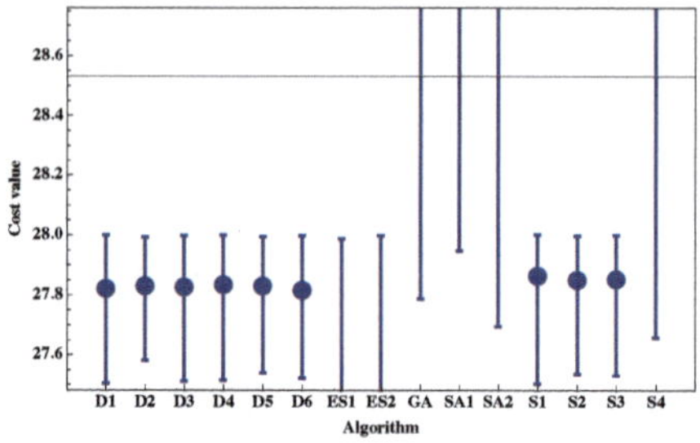

Fig. 12.8 ... and detail view; horizontal line is an average value of all.

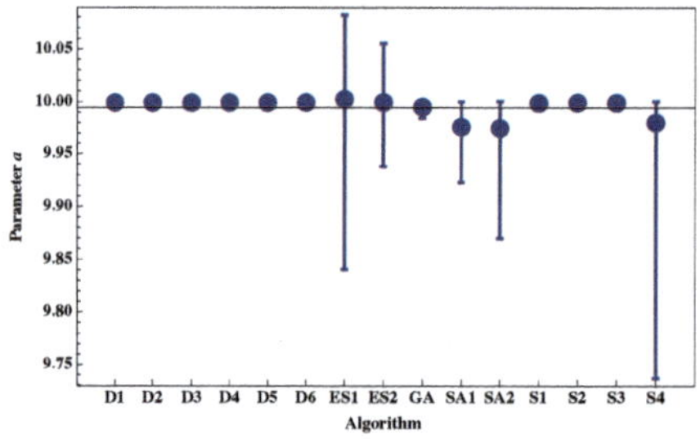

Fig. 12.9 Estimation of parameter *a* ...

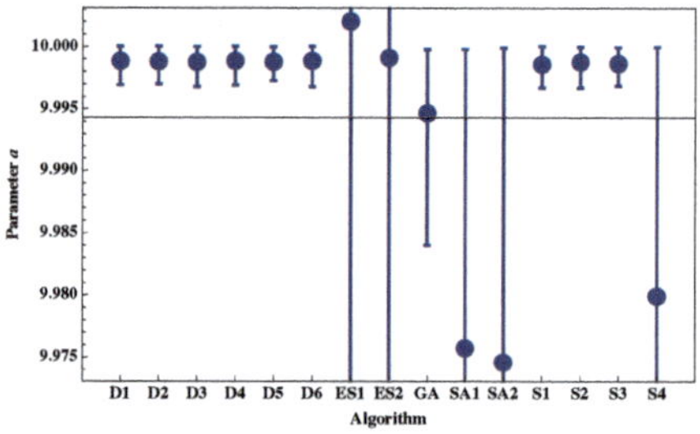

Fig. 12.10 ... and detail view; horizontal line is an average value of all.

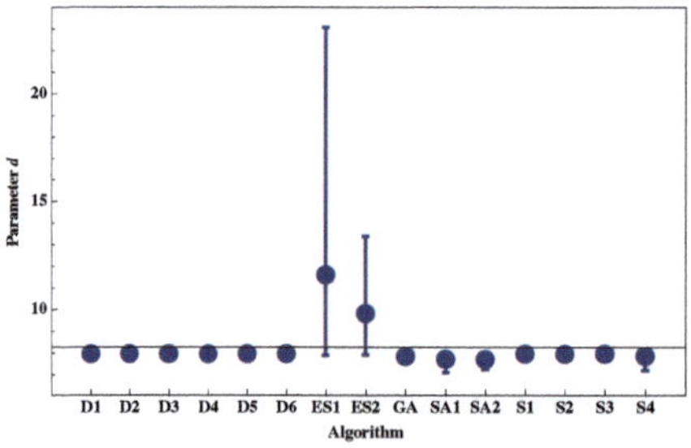

Fig. 12.11 Estimation of parameter *d* ...

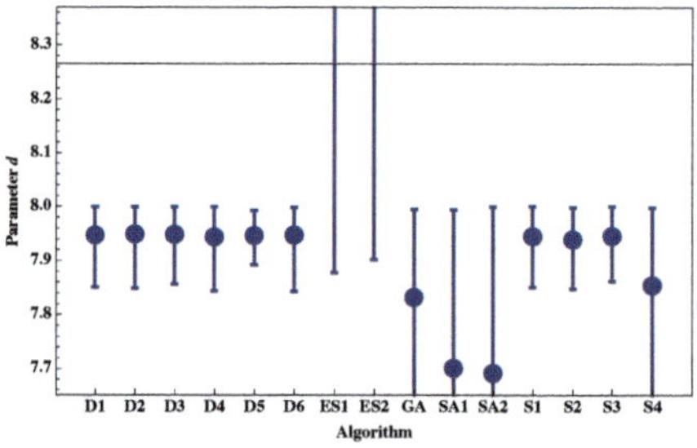

Fig. 12.12 ... and detail view; horizontal line is an average value of all.

During this case study 5 algorithms has been used in 15 versions. Attention was focused on quality from the cost function evaluations point of view, further on cost value, parameter a_2 setting as well as setting of the coupling parameter d. Parameters a_2 and d has been estimated simultaneously, i.e. individual dimension was 2. All these results are reported in Figs. 12.5 - 12.12. Together with minimal, maximal and average values, average vaue of all algorithms (horizontal line) is also depicted.

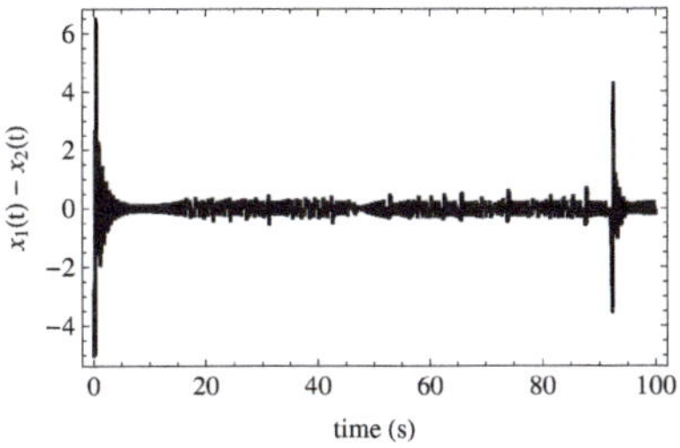

Fig. 12.13 Difference between $x_1(t)$ and $x_2(t)$.

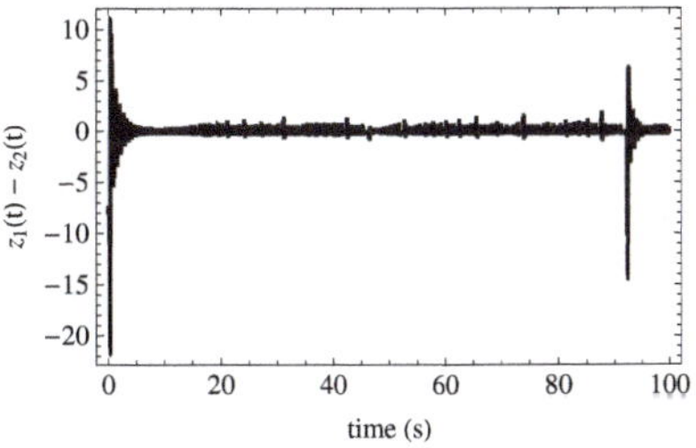

Fig. 12.14 Difference between $z_1(t)$ and $z_2(t)$.

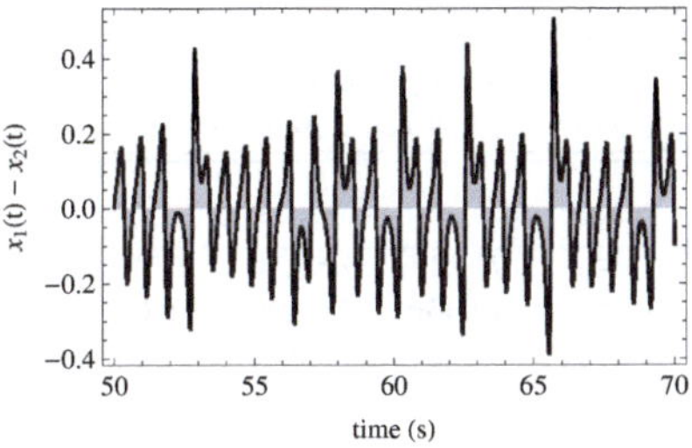

Fig. 12.15 Difference between $x_1(t)$ and $x_2(t)$ - detail...

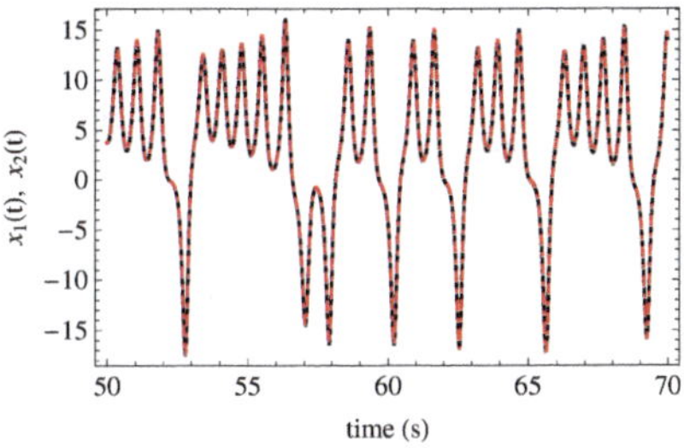

Fig. 12.16 ...and total view on both variables $x_1(t)$ and $x_2(t)$. Both tme series are almost identical.

On Fig. 12.5 and Fig. 12.6 the performance (how many cost function evaluations was needed to find acceptable setting) of all algorithms is reported. Fig. 12.7 and Fig. 12.8 show the same for cost value related to used estimated setting (i.e. how much differs the behavior of both systems), Fig. 12.9 and Fig. 12.10 show how well the parameter a_2 has been estimated and Fig. 12.11 and Fig. 12.12 the give the same for parameter d. Difference of estimated parameters is still in the range of acceptable values (i.e. slaved system has been synchronized very well) ism shown in Fig. 12.14, as it is visible in Fig. 12.13 and 12.14. Difference between the worst and the best behavior of $x_{1,2}$ and $z_{1,2}$ is depicted there. The biggest impact on cost value come from interval [0, 7s], before systems are well synchronized. Another anomaly is at position 92s, which is just a sharp peak with little impact on cost value. Middle part is depicted in Fig. 12.15 and related time series of both variables $x_{1,2}$ is shown in Fig. 12.16.

From all figures it is visible that all EAs has demonstrated almost the same performance. Only a few differences has been recorded thanks to "outliers" (values "far" away from average), probably caused by non-optimal setting of selected algorithm. The problem of finding real optimal setting of used algorithms is quite complex and time consuming process and it was not an objective of this study.

Table 12.7 Experiment summarization, Lorenz - Lorenz, part 1.

Algorithm	D1	D2	D3	D4	D5	D6	ES1	ES2
Cost function evaluations see Fig. 12.5								
Minimum	2820	4588	4428	4240	8180	5150	401	237
Average	9828.24	10977.46	9386.76	10330.74	13300.32	13088.38	3702.85	1145.04
Maximum	15120	16020	12704	15356	17748	18714	11201	2055
Total for each algorithm	982824	1097746	938676	1033074	252706	1308838	370285	114504
Cost Values see. Fig 12.7								
Minimum	27.50	27.58	27.50	27.51	27.53	27.51	18.03	22.69
Average	27.82	27.82	27.82	27.83	27.82	27.81	24.98	26.13
Maximum	27.99	27.99	27.99	27.99	27.99	27.99	27.98	27.99
Parameter a setting see. Fig 12.9								
Minimum	9.9969	9.9969	9.9967	9.9968	9.9972	9.9967	9.8399	9.9376
Average	9.9988	9.9987	9.9987	9.9988	9.9987	9.9988	10.002	9.999
Maximum	9.9999	9.9999	9.9999	9.9999	9.9999	9.9999	10.081	10.055
Parameter d setting see. Fig 12.11								
Minimum	7.8500	7.8476	7.8553	7.8434	7.8911	7.8424	7.8772	7.9012
Average	7.9464	7.9482	7.9476	7.9430	7.9456	7.9464	11.589	9.8122
Maximum	7.9987	7.9992	7.9990	7.9991	7.9917	7.9979	23.048	13.365

Table 12.8 Experiment summarization, Lorenz - Lorenz, part 2.

Algorithm	G	SA1	SA2	S1	S2	S3	S4
Cost function evaluations see Fig. 12.5							
Minimum	32180	3844	1346	1508	26338	2458	4534
Average	48490.67	4972.66	4909.23	14908.32	58855.69	7371.88	10993.90
Maximum	50100	4984	4984	22006	83430	10794	11220
Total for each algorithm	872832	497260	412375	1490832	3766764	737188	1099390
Cost Values see. Fig 12.7							
Minimum	27.78	27.94	27.69	27.50	27.53	27.53	27.65
Average	28.93	32.89	33.15	27.86	27.84	27.85	31.36
Maximum	31.48	44.65	52.80	27.99	27.99	27.99	65.48
Parameter a setting see. Fig 12.9							
Minimum	9.9839	9.9225	9.8695	9.9966	9.9966	9.9968	9.7364
Average	9.9946	9.9757	9.9745	9.9985	9.9987	9.9986	9.9799
Maximum	9.9997	9.9997	9.9998	9.9999	9.9999	9.9999	9.9999
Parameter d setting see. Fig 12.11							
Minimum	7.5454	7.0813	7.2120	7.8497	7.8472	7.8618	7.1813
Average	7.8311	7.7000	7.6908	7.9442	7.9383	7.9450	7.8539
Maximum	7.9942	7.9929	7.9989	7.9995	7.9979	7.9997	7.9979

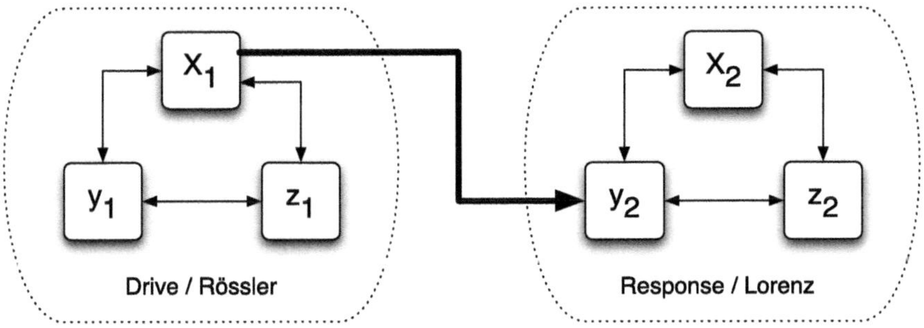

Fig. 12.17 Schematic of Rössler-Lorenz synchronization. Variable y_2 has been directly synchronized.

12.4.4.2 Case study B: Rössler - Lorenz Synchronization

In this case study we have used the selected evolutionary algorithms on synchronization of two different systems (Fig. 12.17), i.e. on synchronization of the Rössler - Lorenz systems (see Fig. 12.18 and Fig. 12.19). Synchronization has been done by the coupling of variables $y_{1,2}(t)$ so that the difference between them has been calculated and multiplied by parameter c, see $(c\,(y_1(t) - y_2(t)))$ in eq. (12.2). Typical difference between non-synchronized and synchronized behavior is depicted in Fig. 12.20 - Fig. 12.25. The effect of synchronization on synchronized system is depicted in Fig.12.26 and Fig. 12.27 (compare with Fig. 12.19).

To estimate the complexity of cost function landscape, a series of figures showing dependance on three parameters a, b and c in eq. (12.2), has been generated. All three parameters were varied around nominal values referred in the literature. For each parameter change, the behavior of master-slave system has been generated and the cost value calculated. In Fig. 12.28 and 12.29 dependance on various values of parameter a is depicted, in Fig. 12.30 and 12.31 on parameter b and in Fig. 12.32 and 12.33 on parameter c. It is clear that the cost function landscape (three dimensional surface (three variables a, b and c) is in four dimensional space - fourth dimension is cost value) is very complex, nonlinear and almost erratic. Thus the use of evolutionary computation is again acceptable in this case.

Similarly, like in the previous case study, 5 algorithms in 15 versions have been used. Attention, like in the previous case, was focused also on cost value, parameter a, b setting as well as the setting of the coupling parameter c. All three parameters has been estimated simultaneously, i.e. individual dimension was 3. All these results are reported in Figs. 12.34 - 12.39. Together with minimal, maximal and average values, the average vaue of all algorithms (horizontal line) is also depicted. Fig. 12.37 show the same for the cost value related to used estimated setting (i.e. how much differs behavior of both systems), Fig. 12.34 show how well parameter a has been estimated and the same is done in Fig. 12.35 for parameter b and Fig. 12.36 for parameter c. The number of cost function evaluations needed for each algorithm is reported in Fig. 12.38 and 12.39. From all figures, it is visible that all EAs have

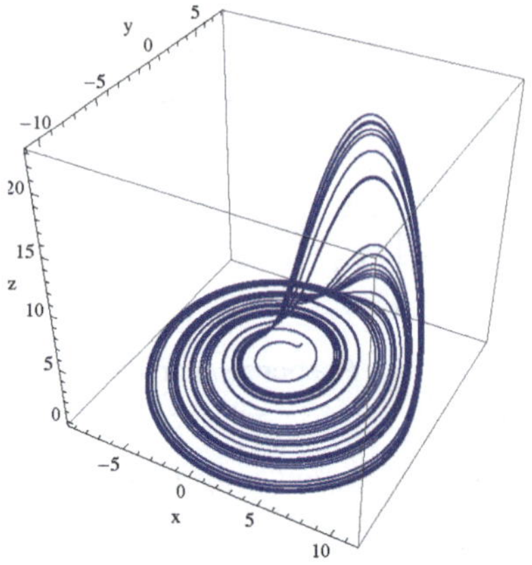

Fig. 12.18 Rössler attractor (master), see. eq. (12.2)

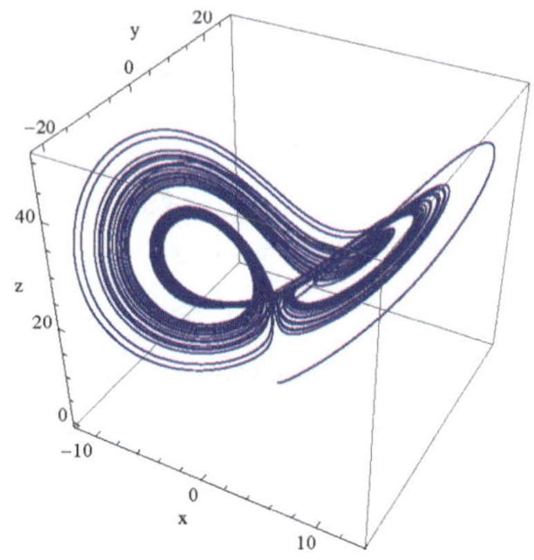

Fig. 12.19 Lorenz attractor (slave), see. eq. (12.2)

demonstrated almost the same performance and generally works as well as in the previous case.

After evolution, it was discovered that EAs had found different setting than is reported in literature, i.e. $\{a,b,c\} = \{0.13089, 3.35025, 69.9999\}$ (cost value was 7.69753) instead of $\{a,b,c\} = \{3, 26.5, 69\}$ as used in Fig. 12.20 - Fig. 12.25. The behavior of evolutionary synchronized R-L system is depicted in Fig. 12.41 - Fig. 12.44. From figures is clearly visible that variable $y_2(t)$ has been synchronized very well while another state variables were almost supressed. Fig. 12.44 shows the difference between $y_1(t)$ and $y_2(t)$. When comparing evolutionary setting of parameter a with Fig. 12.28, then it is clear that EAs have found different settings (extreme on function) than is visible in this figure (around a = 1). Value $a = 0.13089$ signalize, that there is more deeper extreme, which is depicted in Fig. 12.40. It is located almost on the left side. Despite the fact that its location is on the border of the searched space and there are another extremes (including many of chaotic) EAs has successfully found this setting repeatedly.

12.5 Conclusion

In this chapter we have studied the possibility on synchronization with evolutionary estimation of coupling parameters as well as internal parameters of selected chaotic systems. Two kind of synchronized systems were used: Lorenz - Lorenz (L-L) and Rössler - Lorenz (R-L) systems. Attention has been paid on synchronization of variable $x_2(t)$ for L-L synchronization and state variable $y_2(t)$ in the case of R-L synchronization. For the comparative study optimization algorithms such as Differential Evolution (DE) [10], Self Organizing Migrating Algorithm (SOMA) [16], Genetic Algorithms (GA) [5], Simulated Annealing (SA) [6], [3] and

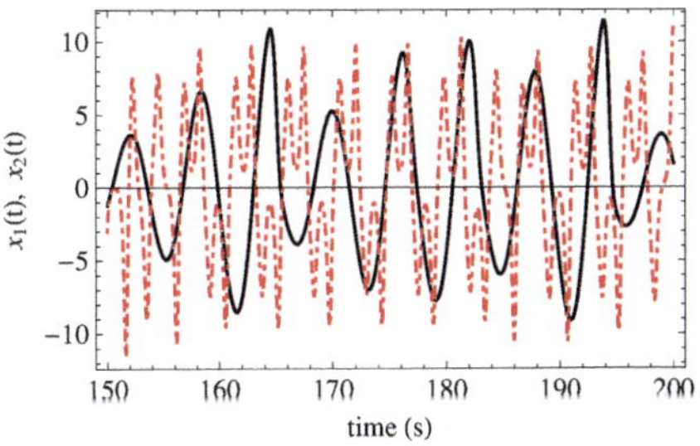

Fig. 12.20 Rössler-Lorenz system for variables $x_1(t)$ and $x_2(t)$ (eq. (12.2)) with c = 0 (not synchronized)...

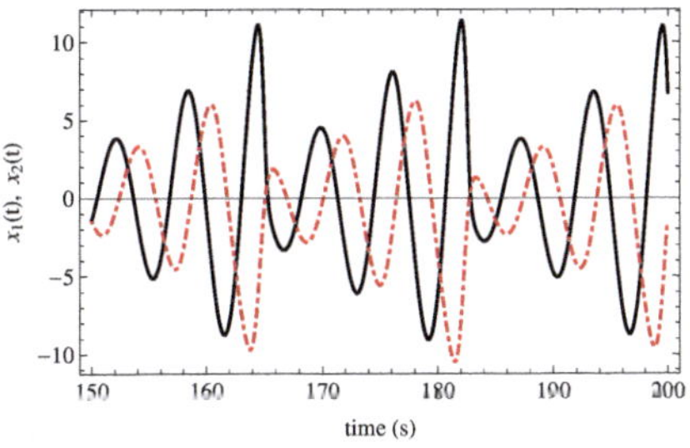

Fig. 12.21 ... and under synchronization (c = 69.4458). Lorenz system is dotted light red curve.

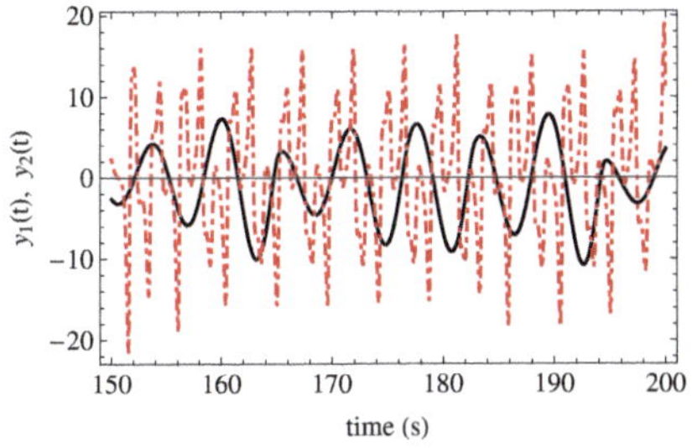

Fig. 12.22 Rössler-Lorenz system for variables $y_1(t)$ and $y_2(t)$ (eq. (12.2)) with c = 0 (not synchronized)...

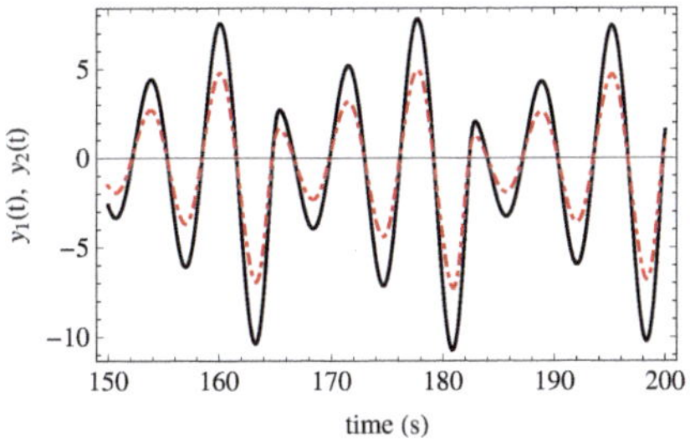

Fig. 12.23 ... and under synchronization (c = 69.4458). Lorenz system is dotted light red curve.

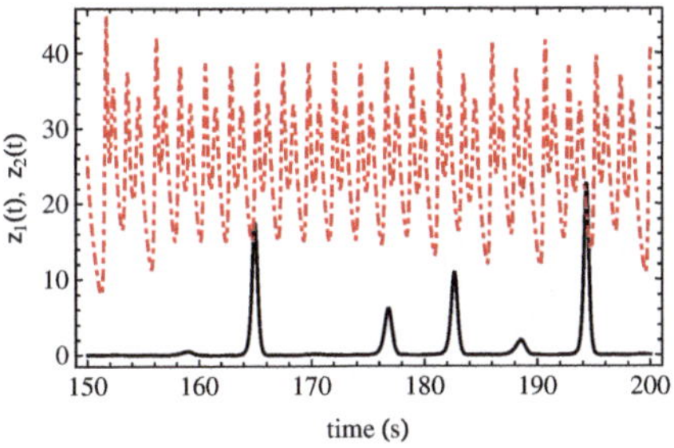

Fig. 12.24 Rössler-Lorenz system for variables $z_1(t)$ and $z_2(t)$ (eq. (12.2)) with c = 0 (not synchronized)...

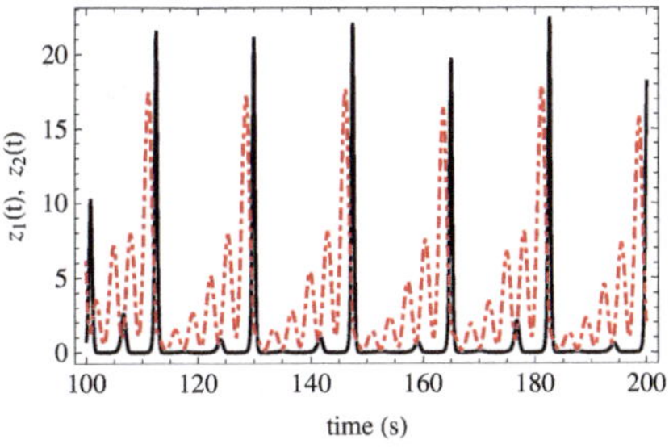

Fig. 12.25 ... and under synchronization (c = 69.4458). Lorenz system is dotted light red curve.

Evolutionary Strategies (ES) [1] were selected. As a conclusion the following statements are presented:

- **Algorithm performance.** Based on all informations and figures reported in this chapter, it can be stated, that all EAs showed good performance. In both

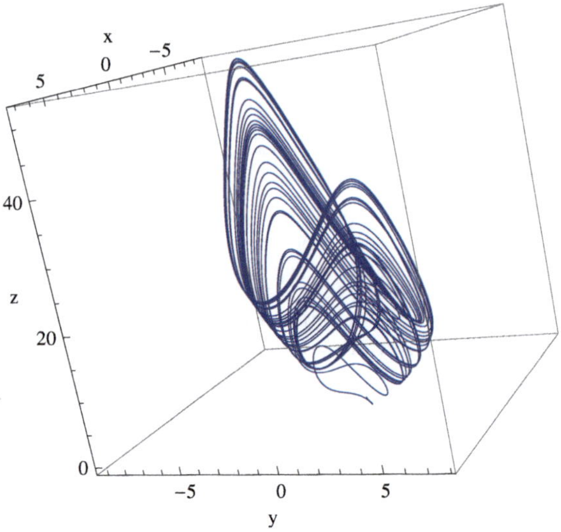

Fig. 12.26 Synchronized Lorenz attractor

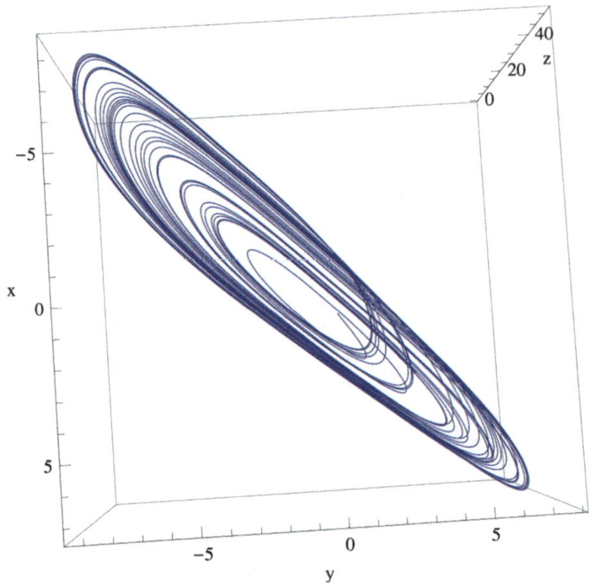

Fig. 12.27 Synchronized Lorenz attractor, another view.

case studies the averages of each algorithm were "almost" on the same value (average of all) which is depicted by a horizontal line. For L-L synchronization see Fig. 12.5 - 12.16; for R-L synchronization see Fig. 12.34 - 12.39. Values far from average can be explained by the fact that **a)** algorithms are

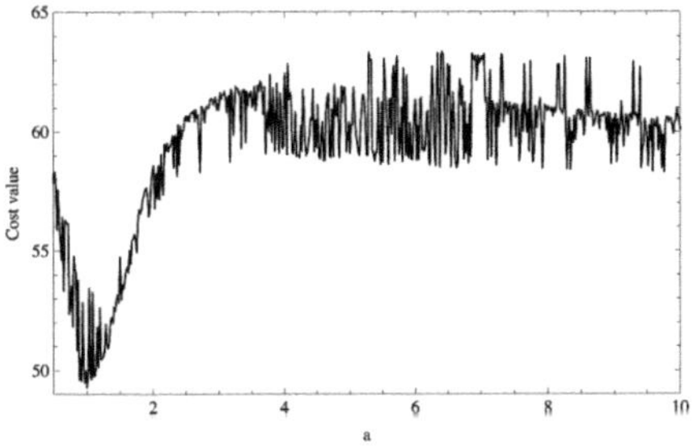

Fig. 12.28 Dependance of parameter *a* in Rössler-Lorenz system (eq. (12.2))...

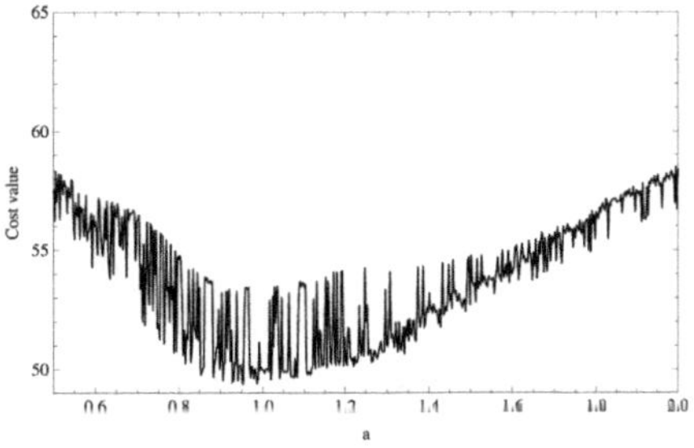

Fig. 12.29 ... and its detail.

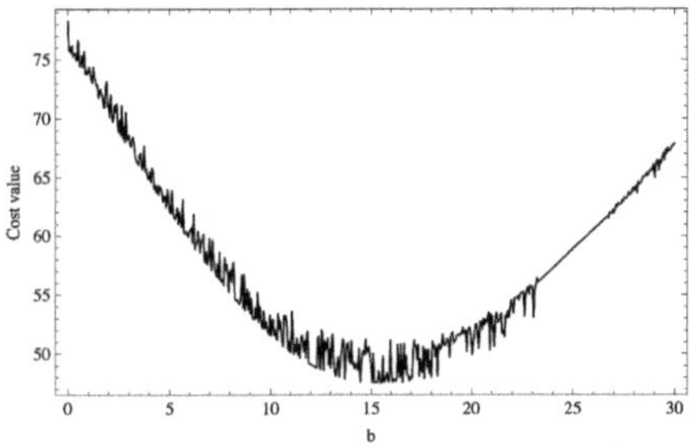

Fig. 12.30 Dependance of parameter *b* in Rössler-Lorenz system (eq. (12.2))...

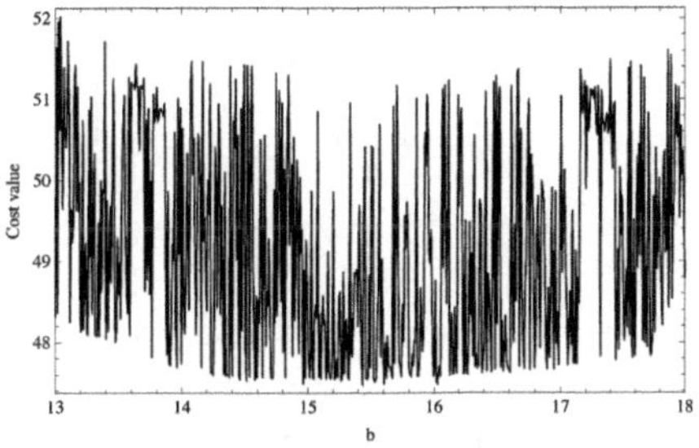

Fig. 12.31 ... and its detail.

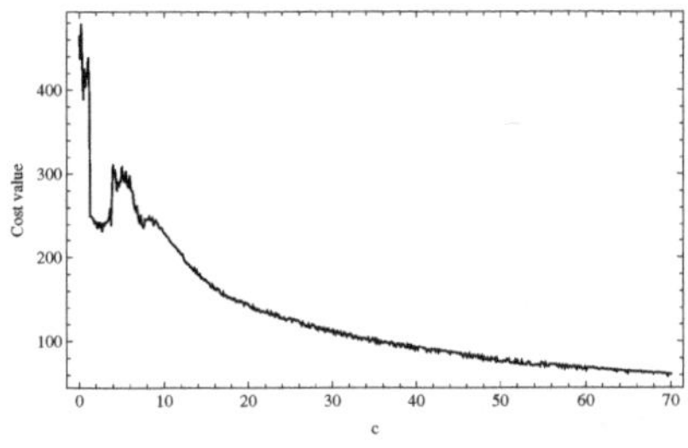

Fig. 12.32 Dependance of parameter *c* in Rössler-Lorenz system (eq. (12.2))...

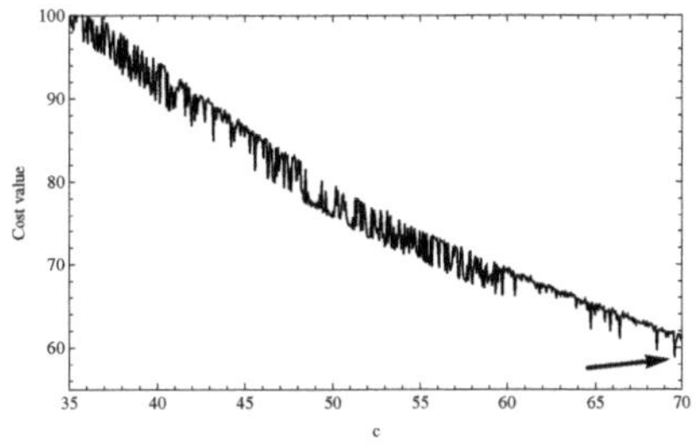

Fig. 12.33 ... and its detail.

of evolutionary (pseudorandom, etc...) nature; **b)** settings of algorithm control parameters is not optimal; **c)** algorithms need longer time to get better values. However the most simplest explanation is that all those extreme values are just "outliers", i.e. only a few values estimated not so optimally, due to randomness of evolutionary algorithms.Two randomly selected histograms showing outliers are depicted in Fig. 12.45 and 12.46. From Fig. 12.45 it is clearly visible that cost function evaluation (approx. 30000, see Fig. 12.38, algorithm S4)

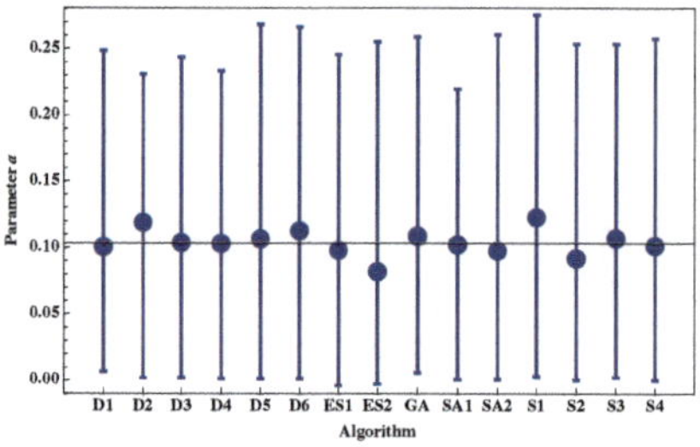

Fig. 12.34 Estimation of parameter *a* ...

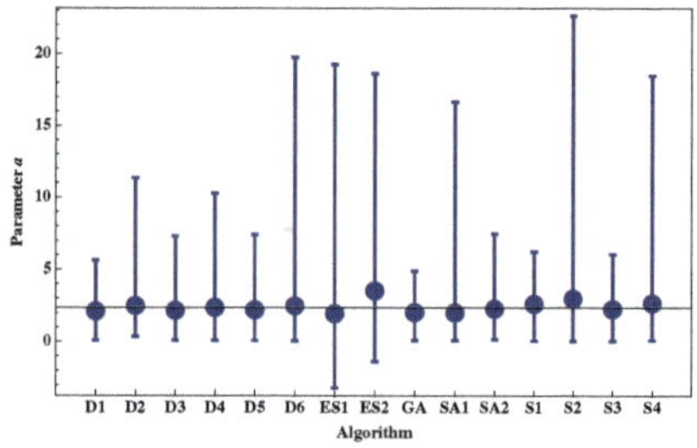

Fig. 12.35 ... and of parameter *b* .

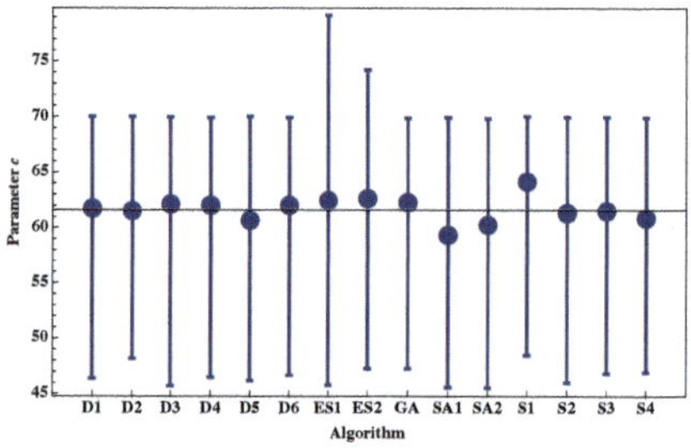

Fig. 12.36 Estimation of parameter *c* ...

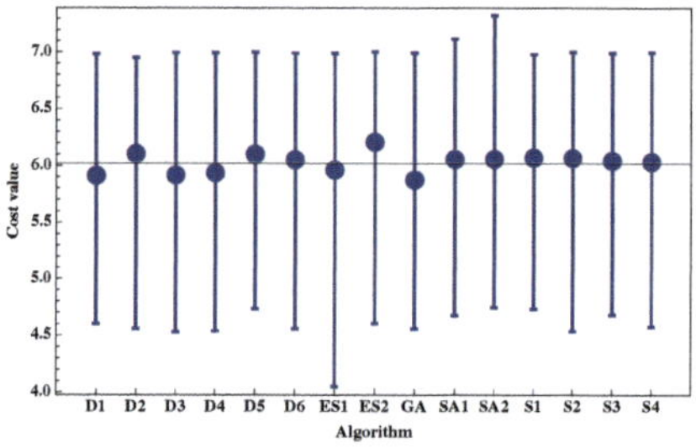

Fig. 12.37 ... and summarization of cost values.

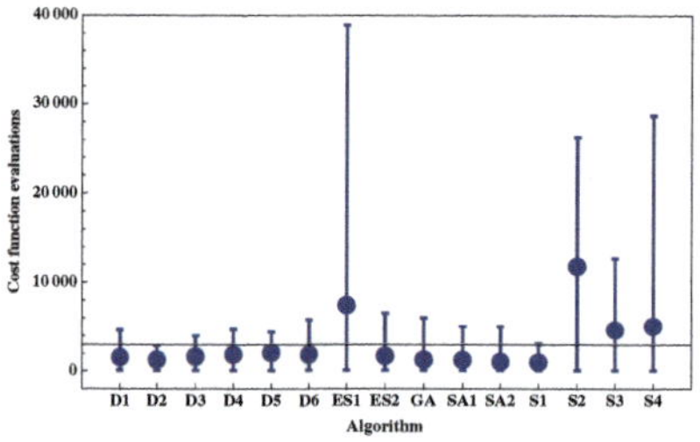

Fig. 12.38 Cost function evaluations...

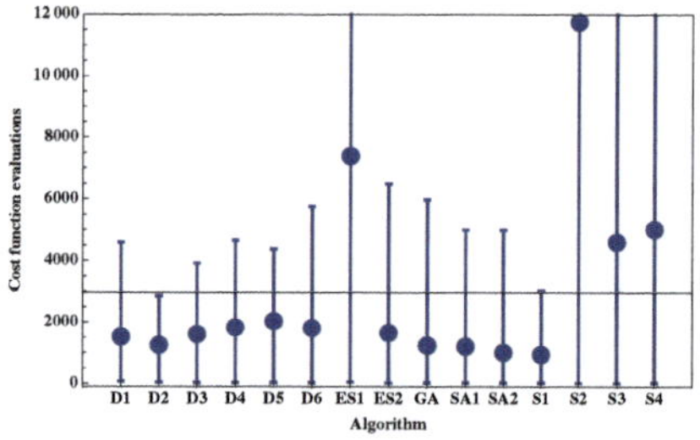

Fig. 12.39 ... and detail view.

is really not a rule, but an outlier. The same is visible in Fig. 12.46, compared with Fig. 12.11, algorithm ES1.

- **Statistical robustness.** In the frame of this case study 30 (2×15) simulations has been done, and each has been 100 $\times$ repeated. Thus the total number of simulations was 3000 simulations equal to 14975294 / 4387854 (L-L / R-L) cost function evaluations, see for more detailed description see Table 12.7 and 12.8 (L-L system) and Tables 12.9 - 12.10 (R-L system). All those calculations led to positive results, i.e. systems have been synchronised succesfully. All calculations has been done on a grid computer which consist of two special Apple servers: 16 XServers, each 2x2 GHz Intel Xeon, 1 GB RAM, 80 GB

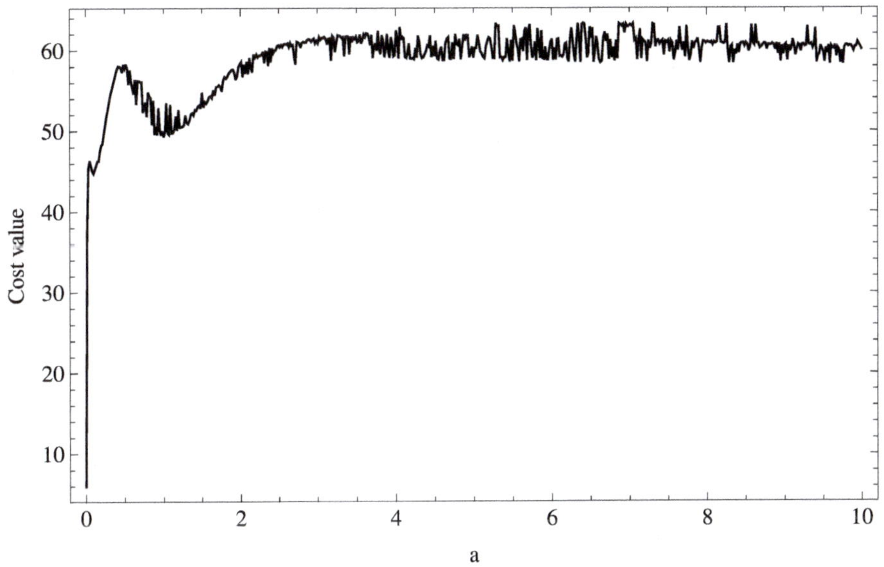

Fig. 12.40 Total view of cost value dependance on parameter a

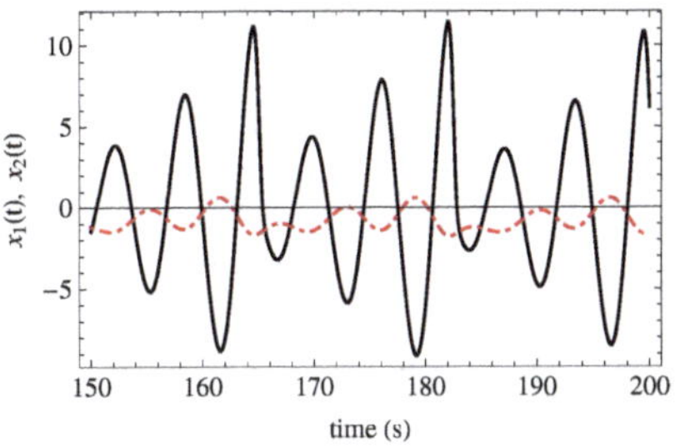

Fig. 12.41 Synchronization of $x_{1,2}(t)$ according to the best estimated setting.

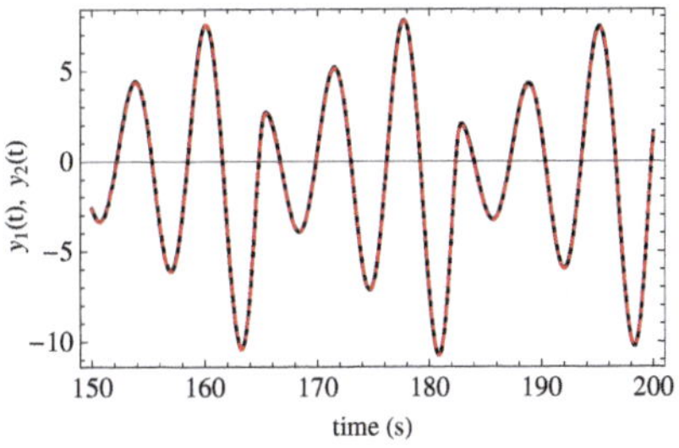

Fig. 12.42 Synchronization of $y_{1,2}(t)$ according to the best estimated setting.

HD i.e. 64 CPUs) and 7 Apple Minimacs CoreDuo i.e. number of accessible CPUs is 14. In total 78 CPUs there was available for computation.

- **Results divergence.** As mentioned before, results in both case studies are slightly different. Obtained averages are mostly on the same level, however their divergence is for algorithms like ES and SA is different. As mentioned before, probably better setting should be applied. On the other side, there is so called "No Free Lunch" theorem, see [14], according to which universal algorithm does not exist, i.e. some of selected evolutionary algorithm is not much suitable for this task. But this is probably not a case of the algorithms used here.

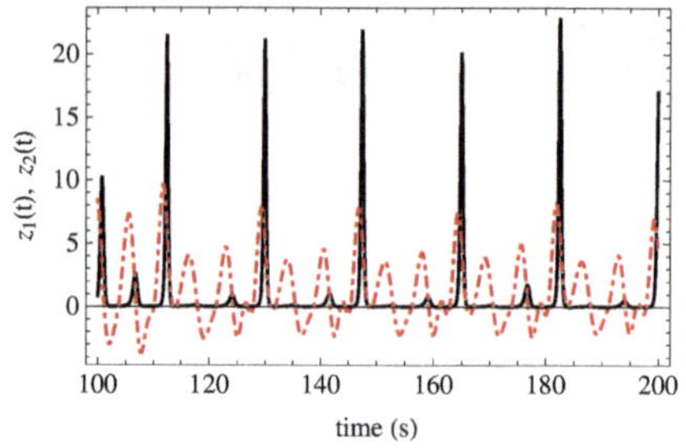

Fig. 12.43 Synchronization of $z_{1,2}(t)$ according to the best estimated setting.

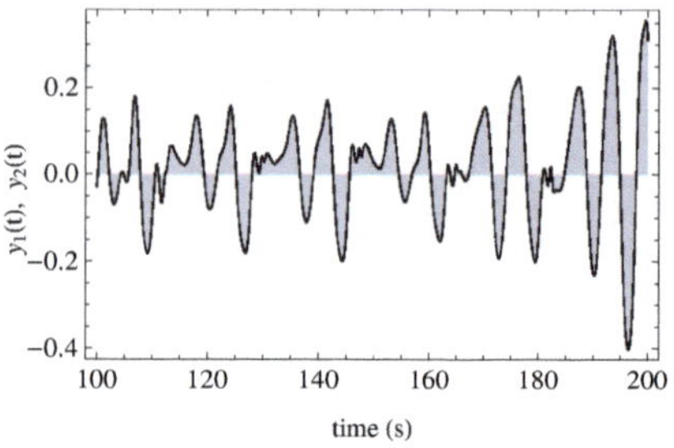

Fig. 12.44 Synchronization: difference between $y_1(t) - y_2(t)$.

Table 12.9 Experiment summarization, Rössler - Lorenz, part 1.

Algorithm	D1	D2	D3	D4	D5	D6	ES1	ES2
Cost function evaluations see Fig. 12.5								
Minimum	54	19	11	7	6	8	36	1
Average	1500	1234	1589	1806	2013	1791	7376	1637
Maximum	4582	2832	3900	4644	4364	5744	38861	6477
Total for each algorithm	149963	56767	158851	180646	201257	179055	737582	158797
Cost Values see. Fig 12.37								
Minimum	4.5966	4.5531	4.5254	4.5314	4.7290	4.5533	4.0446	4.5991
Average	5.9067	6.0994	5.9115	5.9307	6.0956	6.0467	5.9568	6.2024
Maximum	6.9800	6.9453	6.9906	6.9888	6.9933	6.9855	6.9829	6.9992
Parameter *a* setting see. Fig 12.34								
Minimum	0.0063	0.0015	0.0018	0.0008	0.001	0.0008	-0.004	-0.003
Average	0.1001	0.1184	0.1033	0.1027	0.1062	0.1122	0.0975	0.0815
Maximum	0.2482	0.2304	0.2433	0.2328	0.2679	0.2658	0.2451	0.2548
Parameter *b* setting see. Fig 12.35								
Minimum	0.0534	0.3000	0.0416	0.0391	0.0313	0.0104	-3.257	-1.413
Average	2.0676	2.4464	2.1227	2.3213	2.1520	2.4390	1.9063	3.4911
Maximum	5.6232	11.309	7.2874	10.248	7.3880	19.717	19.221	18.584
Parameter *c* setting see. Fig 12.36								
Minimum	46.333	48.138	45.681	46.458	46.144	46.607	45.726	47.220
Average	61.686	61.474	62.077	61.985	60.619	61.969	62.397	62.576
Maximum	69.984	69.990	69.941	69.873	69.995	69.890	79.150	74.171

Table 12.10 Experiment summarization, Rössler - Lorenz, part 2.

Algorithm	G	SA1	SA2	S1	S2	S3	S4
Cost function evaluations see Fig. 12.5							
Minimum	5	8	12	9	3	3	7
Average	1222	1188	989	923	11738	4588	5002
Maximum	5957	4984	4984	3008	26220	12656	28628
Total for each algorithm	122169	118765	98879	92258	1173843	458846	500176
Cost Values see. Fig 12.37							
Minimum	4.5533	4.6710	4.7431	4.7287	4.5311	4.6754	4.5709
Average	5.8689	6.0519	6.0542	6.0657	6.0624	6.0390	6.0297
Maximum	6.9866	7.1153	7.3234	6.9764	6.9965	6.9898	6.9938
Parameter *a* setting see. Fig 12.34							
Minimum	0.0053	0.0004	0.0004	0.0026	0.	0.0019	0.0001
Average	0.1083	0.1017	0.0971	0.1223	0.0914	0.1065	0.1009
Maximum	0.2586	0.2191	0.2602	0.2749	0.253	0.2533	0.2572
Parameter *b* setting see. Fig 12.35							
Minimum	0.0403	0.0617	0.1234	0.0261	0.0157	0.0167	0.0674
Average	1.9918	1.9770	2.2398	2.5879	2.9517	2.2215	2.637
Maximum	4.8671	16.629	7.4493	6.2005	22.604	6.0407	18.449
Parameter *c* setting see. Fig 12.36							
Minimum	47.188	45.505	45.489	48.398	45.961	46.766	46.868
Average	62.236	59.287	60.202	64.109	61.287	61.465	60.811
Maximum	69.834	69.892	69.797	69.996	69.920	69.911	69.864

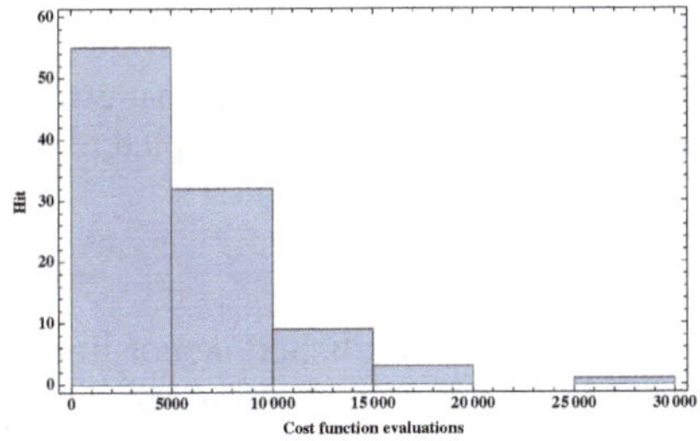

Fig. 12.45 Histogram of cost function evaluations; R-L system.

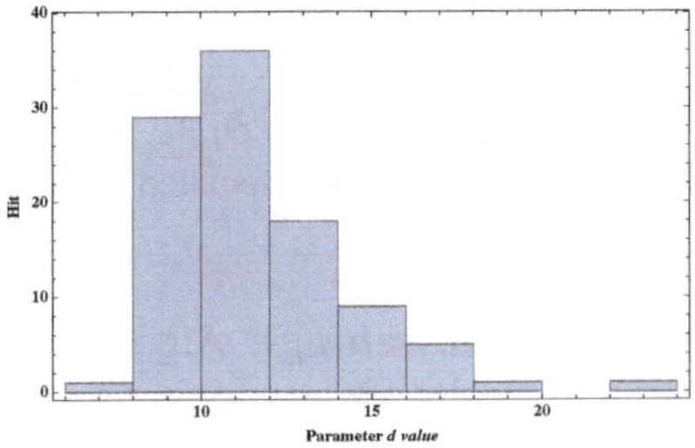

Fig. 12.46 Histogram for parameter d; L-L system.

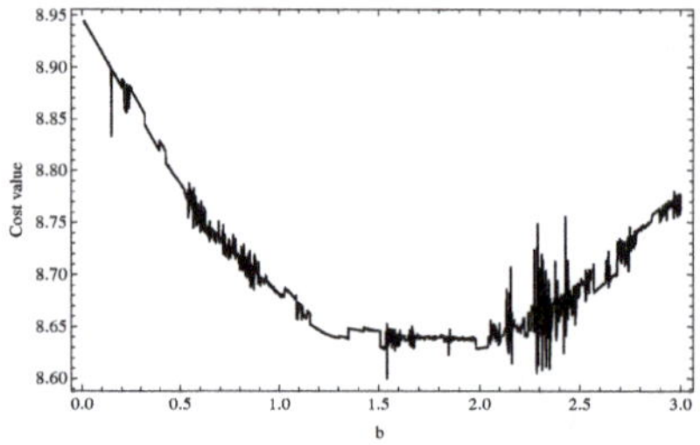

Fig. 12.47 Dependance of cost value on the parameter *b* for $a = 0.1$ and $c = 69.4458$ ...

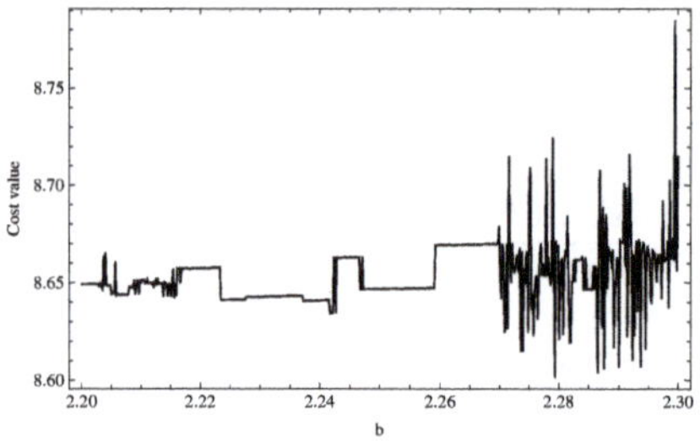

Fig. 12.48 ... and detail view.

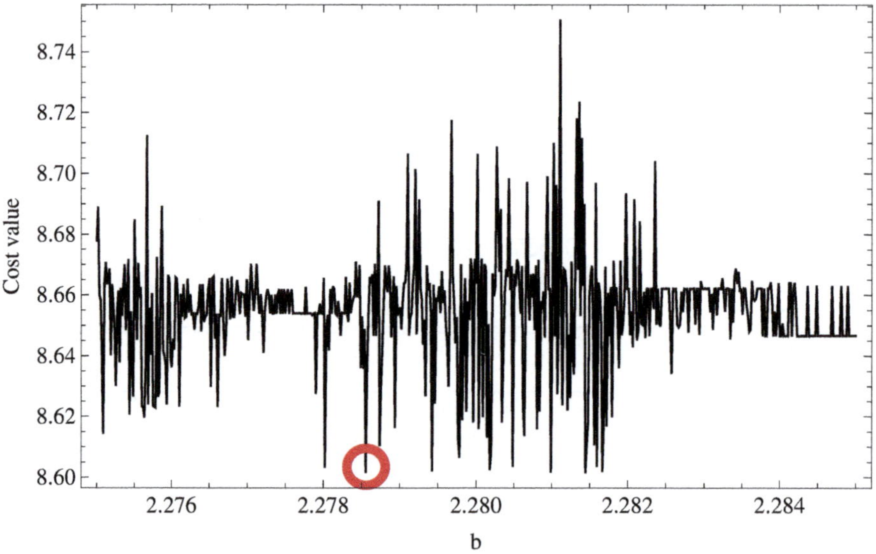

Fig. 12.49 Dependance of cost value on the parameter *b* in very tiny region. Global extreme is marked by red (light grey) circle. Its estimation is approximate, because this picture (and dataset used for) has been calculated with certain, limiting accuracy.

- **Algorithm settings.** Algorithm setting has been established according to heuristically known setting for each algorithm as well as on our own experiences. We would like to remind that it does not mean that there is no better settings for any of used algorithms. Main aim was not focused on speed of used algorithms but on successful synchronization.
- **Synchronization settings.** During all simulations different setting for estimated parameters has been found, comparing with literature and our heuristically obtained setting ($a = 3$, for $b = 26.5$ and $c = 70$ for R-L system). Compare Fig. 12.21 with Fig. 12.41, Fig. 12.23 with Fig. 12.42 and Fig. 12.25 with Fig. 12.43. Thus EAs have found better setting. Differences of

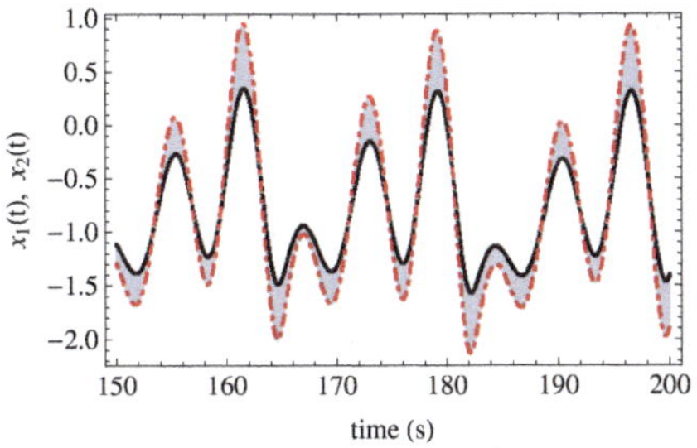

Fig. 12.50 Difference between $x_{2-best}(t)$ (solid line) and $x_{2-worst}(t)$ (dotted line).

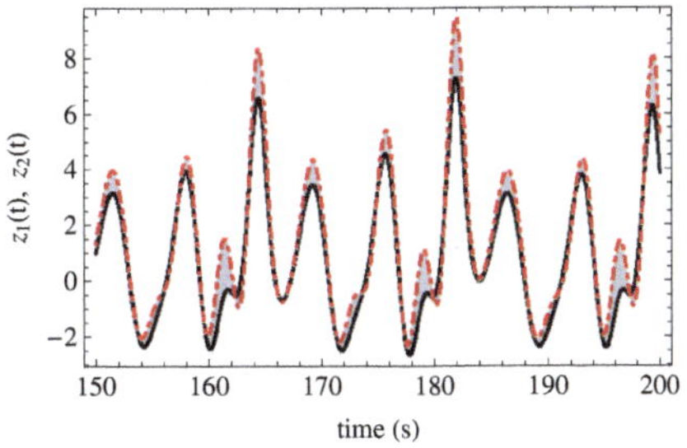

Fig. 12.51 Difference between $z_{2-best}(t)$ (solid line) and $z_{2-worst}(t)$ (dotted line).

behavior between synchronizing and synchronized variables are also depicted in Fig. 12.13, 12.14, 12.15 and Fig. 12.16.

- **Problem complexity.** Problem complexity, represented by the cost function landscape, is depicted in Fig. 12.3, Fig. 12.4, Fig. 12.28 - 12.33 and Fig. 12.40. It is clearly visible that cost function landscape is very erratic, nonlinear and multimodal. Another view on its complexity is given in Fig. 12.47 - 12.49. Compare Fig. 12.47 with Fig. 12.30. The extreme of parameter b dependance has moved for different a and c from positions (approx.) 15 to 2.2785. In other words, optimal setting cannot be found simply by visual checking of each parameter dependance, due to their mutual influence. Thus naturally, problem of synchronization, is suitable for evolutionary algorithms.
- **Different results.** When comparing R-L and L-L systems, then one can see, that there is visible difference in the range of the parameter value estimation, as well as in cost function evaluations and cost values. It is obvious because **a)** both systems are different, **b)** in both systems is estimated by different number of parameters, **c)** cost values are differently calculated, see eq. (12.3) and (12.4). For L-L system the cost value (eq. (12.3)) is calculated in interval $t \in [0, 100]$ like difference between all three variables, while in R-L system (eq. (12.4)) for only $y_{1,2}(t)$ variable and $t \in [0, 200]$. Another important point is, that because in R-L system $y_1(t) - y_2(t) -$ is minimized only, then the variables $x_2(t)$ and $z_2(t)$ were not pressed by evolution into exact values. Variables $x_2(t)$ and $z_2(t)$ has thus a "freedom" to reach different values. It is visible in Fig. 12.50 and 12.51. A difference between the best and the worst estimated synchronization is depicted there for R-L system and variables $x_2(t)$ and $z_2(t)$. Gray area represent "space" for all possible values of both variables.
- **Estimated parameters.** In the synchronization experiments coupling parameters are usually estimated. In this numerical study, the internal parameters of chaotic systems has been selected for optimization are also estimated. It can reflect, for certain systems, situation that some of physical parameters (pressure, current, ...) can be remoted by an external observer. Because the parameters has some certain physical meaning (they are not abstract

numbers), one has to carefully work with them and sometimes this is simply not allowed. We did not follow this idea in numerical studies here.

- **Negative values.** In the reported results it can easily be found that for some values, negative values are returned, especially for evolutionary strategies. It was caused by the fact that in evolutionary strategies procedure "watching" has not been applied whether evolutionary process overstepped the allowable search space or not. Thus, evolutionary strategies, has also searched a little bit behind of searchable space borders.

According to the author's opinion, this is a promising area of evolutionary algorithms use. Experiments designed, numerically simulated and reported here were one of the most simplest. Based on results obtained here and also in other chapters, it is possible to say that EAs are viable and should also work on more complicated cases of synchronization, for example the CML systems.

Acknowledgements. This work was supported by grant No. MSM 7088352101 of the Ministry of Education of the Czech Republic and by grants of the Grant Agency of the Czech Republic GACR 102/09/1680.

References

1. Beyer, H.: Theory of Evolution Strategies. Springer, New York (2001)
2. Brown, R., Rulkov, N., Tracy, E.: Modeling and synchronization chaotic system from time-series data. Phys. Rev. E 49, 3784 (1994)
3. Cerny, V.: Thermodynamical approach to the traveling salesman problem: An efficient simulation algorithm. J. Opt. Theory Appl. 45(1), 41–51 (1985)
4. Gonzalez-Miranda, J.: Synchronization and Control of Chaos. An introduction for scientists and engineers. Imperial College Press (2004)
5. Holland, J.: Adaptation in Natural and Artificial Systems. Univ. Michigan Press, Ann Arbor (1975)
6. Kirkpatrick, S., Gelatt, C., Vecchi, M.: Optimization by simulated annealing. Science 220(4598), 671–680 (1983)
7. Nolle, L., Goodyear, A., Hopgood, A., Picton, P., Braithwaite, N.StJ.: On Step Width Adaptation in Simulated Annealing for Continuous Parameter Optimisation. In: Reusch, B. (ed.) Fuzzy Days 2001. LNCS, vol. 2206, pp. 589–598. Springer, Heidelberg (2001)
8. Nolle, L., Zelinka, I., Hopgood, A., Goodyear, A.: Comparison of an self organizing migration algorithm with simulated annealing and differential evolution for automated waveform tuning. Adv. Eng. Software 36(10), 645–653 (2005)
9. Pikovsky, A., Rosemblum, M., Kurths, J.: Synchronization: A Universal Concept in Nonlinear Sciences. Cambridge University Press, Cambridge (2001)
10. Price, K.: An Introduction to Differential Evolution. In: Corne, D., Dorigo, M., Glover, F. (eds.) New Ideas in Optimization, pp. 79–108. McGraw-Hill, New York (1999)
11. Rulkov, N., Sushchik, M.: Robustness of synchronized chaotic oscillations. Int. J. Bifurcat Chaos Appl. Sci. Eng. 7, 625 (1997)
12. Schuster, H. (ed.): Handbook of Chaos Control. Wiley-VCH, New York (2007)
13. Sushchik, M., Rulkov, N., Tsimring, L., Abarbanel, H.: Generalized synchronization of chaos in directionally coupled chaotic systems. In: Proceedings of Intl. Symp. on Nonlinear Theory and Appl., vol. 2, pp. 949–952. IEEE, Los Alamitos (1995)

14. Wolpert, D., Macready, W.: No Free Lunch Theorems for Search, Technical Report SFI-TR-95-02-010, Santa Fe Institute (1995)
15. Wolpert, D., Macready, W.: No Free Lunch Theorems for Optimization. IEEE Trans. Evol. Comput. 1(67) (1997)
16. Zelinka, I.: SOMA – Self Organizing Migrating Algorithm. In: Babu, B., Onwubolu, G. (eds.) New Optimization Techniques in Engineering, pp. 167–218. Springer, New York (2004)
17. Zelinka, I.: Investigation on Evolutionary Deterministic Chaos Control. In: IFAC, Prague (2005a)
18. Zelinka, I.: Investigation on Evolutionary Deterministic Chaos Control – Extended Study. In: 19th International Conference on Simulation and Modeling (ECMS 2005), Riga, Latvia, June 1-4 (2005b)
19. Zelinka, I.: Real-time deterministic chaos control by means of selected evolutionary algorithms. Eng. Appl. Artif. Intell (2008), doi:10.1016/j.engappai.2008.07.008
20. Zelinka, I., Nolle, L.: Plasma reactor optimizing using differential evolution. In: Price, K., Lampinen, J., Storn, R. (eds.) Differential Evolution: A Practical Approach to Global Optimization, pp. 499–512. Springer, New York (2006)
21. Zelinka, I., Senkerik, R., Navratil, E.: Investigation on Evolutionary Optimitazion of Chaos Control. Chaos, Solitons & Fractals (2007), doi:10.1016/j.chaos.2007.07.045

Chapter 13
Evolutionary Optimization and Dynamic Fitness Landscapes
From Reaction–Diffusion Systems to Chaotic CML

Hendrik Richter

Abstract. Evolutionary algorithms are a promising option for solving dynamic optimization problems. These problems have fitness landscapes whose topological features change dynamically with the run–time of the evolutionary algorithm. In this chapter, we study these landscapes by analyzing and quantifying their properties using topological and dynamical landscape measures such as modality, ruggedness, information content, dynamic severity and two types of dynamic complexity measures, Lyapunov exponents and bred vector dimension. Here, our main focus is on dynamic fitness landscapes that exhibit spatio–temporal chaotic behavior. We further discuss evolutionary algorithms and modifications needed to make them fit to perform in dynamic landscapes and present numerical experiments showing the algorithms' performances. These results allow us to link the landscape measures to the behavior of the evolutionary algorithms.

13.1 Introduction

An evolutionary algorithm is a stochastically driven but systematic search method for solving optimization problems. All of its three main operators, selection, recombination and mutation, depend on random elements. In other words, an evolutionary algorithm, just as its biological inspiration and namegiver natural evolution, is a phenomenon of chance, albeit the effect of chance is directed, mainly as a result of the selection process. However, due to the heavy influence of chance in the working of the algorithm, it is a challenge to establish some sound theory of evolutionary computation. A corner–stone in such a theory is the conceptional framework of fitness landscapes. The concept of fitness landscapes was introduced in the context of

Hendrik Richter
HTWK Leipzig, Fakultät Elektrotechnik und Informationstechnik,
Institut Mess–, Steuerungs– und Regelungstechnik, D–04251 Leipzig, Germany
e-mail: `richter@fbeit.htwk-leipzig.de`

I. Zelinka et al. (Eds.): Evolutionary Algorithms and Chaotic Systems, SCI 267, pp. 409–446.
springerlink.com

theoretical biology by Wright in the early 1930s [65] and later became an important tool in theoretical studies in evolutionary optimization [19, 27, 32, 55, 57]. A fitness landscape combines a search space with a notation of fitness for every point in it, which for instance can be obtained by a genotype–to–fitness mapping or more generally by encoding the set of all possible solutions of an optimization problem and assorting a fitness value each. So, the fitness landscape appears as a potential function on which the individuals of the population may move. This permits to pose the question of how properties of the fitness landscape reflect, explain and allow to predict the behavior of the evolutionary algorithm, and vice versa [27]. It can also be studied how the population dynamics of the search algorithm (that is, the flow of the individuals of the evolutionary algorithm in the landscape) interrelate with topological and dynamical features of the fitness landscape.

A traditional field of application and theoretical study for evolutionary algorithms is to consider static optimization problems. These problems have a fitness landscape that does not change its topological features while the evolutionary algorithm is running. In recent years, we saw an increasing interest in solving dynamic optimization problems [9, 24, 37, 44, 69]. Here, the fitness landscape has topological features that change dynamically with the run–time of the evolutionary algorithm. Hence, such dynamic fitness landscapes can be viewed as spatially extended dynamical systems. So, our main topics here are how dynamic fitness landscapes can be formulated mathematically, how they relate to spatio–temporal dynamical systems considered in nonlinear dynamics, what their properties are and how these properties correspond to the behavior of evolutionary algorithms used to do optimization in these landscapes.

In this chapter, we consider dynamic fitness landscapes and study their topological and dynamical properties. We show in Sec. 13.2 how these landscapes can be constructed from reaction–diffusion systems modelled by partial differential equations (PDE) and also from coupled map lattices (CML). With this we intend to establish relationships between these different kinds of description. In this context, our main emphasis is on dynamic fitness landscapes that exhibit spatio–temporal chaotic behavior. In Sec. 13.3 the study of topological and dynamical properties of fitness landscapes is formalized and we present different types of landscape measures. We consider the topological landscape measures modality, ruggedness and information content and the dynamical landscape measure severity and two types of dynamic complexity measures, Lyapunov exponents and bred vector dimensions. The evolutionary algorithm and modifications needed to make them fit to perform in dynamic landscapes are discussed in Sec. 13.4. Four types of implementation are considered, hyper–mutation, self–adaption, and two types of memory schemes, direct and abstract memory. We present numerical experiments with the evolutionary algorithm implementations and the fitness landscapes in Sec. 13.5. We use these experiments to evaluate the performance of the algorithms and to link these results to the landscape measures studied before. The chapter ends with concluding remarks and a pointer at further problems.

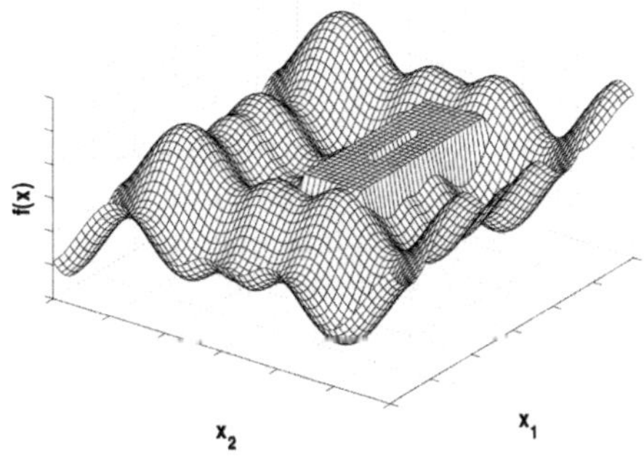

Fig. 13.1 Static fitness landscape in $\mathbb{R}^2$ as mountainous region with peaks, valleys, ridges and plateaus.

13.2 Constructing Dynamic Fitness Landscapes from Reaction–Diffusion Systems and CML

13.2.1 Static and Dynamic Fitness Landscapes

In this chapter we will define dynamic fitness landscapes and provide a framework for posing dynamic optimization problems. According to [27, 57] a static fitness landscape Λ_S is given by

$$\Lambda_S = (S, n, f), \tag{13.1}$$

where S is the search space that can be constructed from a genotype–to–fitness mapping[1] or more generally from encoding the set of all possible solutions of an optimization problem. The neighborhood structure $n(x)$ is a function that assigns to every $x \in S$ a set of neighbors.[2] The fitness function $f(x) : S \to \mathbb{R}$ gives the fitness value for every point in the search space. In Fig. 13.1 a typical fitness landscape is shown over a two–dimensional search space. In this special case with $S = \mathbb{R}^2$ the metaphorical meaning of a fitness landscape as a mountainous region with peaks, valleys, ridges and plateaus becomes particularly apparent. As each point of the search space is characterized by a unique fitness value, solving an optimization problem translates into finding the highest peak (or lowest valley). Hence, a static optimization problem is

---

[1] In theoretical biology a finer distinction is drawn between genotype and phenotype, e.g. [57], which leads to a genotype–to–phenotype–to–fitness mapping. Genotype here stands for the genetic make–up of a generic individual, i.e. its total genetic information, the sum of all (genetically) possible individuals of a species. The phenotype characterizes a particular individual, i.e. a specific instance of the generic, genotypical individual. In biology this distinction is necessary because genetic fluctuations by mutation can only happen on the level of genotypes, while fitness can only be assigned to phenotypical individuals. In evolutionary computation, genotype can be thought of as standing for the search space, phenotype for the individuals of an evolutionary algorithm, and fitness remains the same.

[2] If the search space is a metric space (for instance a Hilbert (or Banach) space which is frequently taken to define spatially extended systems properly), this neighborhood structure is inherent and there is no need to define it additionally.

$$f_S = \max_{x \in S} f(x), \tag{13.2}$$

which is finding the maximal fitness value f_S and its location $x_S = arg\, f(x_S)$.[3]

If an evolutionary algorithm is employed to solve the problem, the fitness landscape concept becomes once more useful as a population intended to find the optimum can be viewed as though living on the landscape's surface. Moreover, generational change in the population means movement on the surface with the aim to ascent a peak. The picture furthermore illustrates a dynamic optimization problem in that the landscape is to change dynamically beneath the individuals of the population. To describe such a kind of problem, we need the concept of a dynamic fitness landscape[4] which we consider next.

A dynamic fitness landscape Λ_D can be defined by

$$\Lambda_D = (S, n, \Gamma, F, \phi), \tag{13.3}$$

where S is the search space and again represents all possible solutions $x \in S$ of the optimization problem and $n(x)$ is an equivalent neighborhood structure; footnote 2 applies likewise. Γ is a time set (transition semi–group) that defines a measuring and ordering scale for the changes; F is the set of fitness functions and every $f \in F$ with $f : S \times \Gamma \to \mathbb{R}$ depends on time and provides a fitness value to every point in the search space and any element of the time set Γ. The transition map $\phi : F \times S \times \Gamma \to F$ describes how the fitness function changes over time. Further, the map must satisfy the temporal identity and composition conditions, that is $\phi(f,x,0) = f(x,0)$ and $\phi(f,x,t_1+t_2) = \phi(\phi(f,x,t_1),x,t_2)$, $\forall f \in F$, $\forall x \in S$, $\forall t_1, t_2 \in \Gamma$ and the spatial boundary conditions $\phi(f, x_{bound}, t) = f(x_{bound}, t)$, $\forall f \in F$, $\forall t \in \Gamma$ and x_{bound} being the boundary set of search space S. The transition map can depend on continuous and/or discrete values conditional to whether time and/or space possess that property. So, we can put the continuous and discrete number sets, $\mathbb{R}$ and $\mathbb{Z}$, to the time sets and search spaces. For a discrete search space there is $S \subseteq \mathbb{Z}^n$ and for a continuous one $S \subseteq \mathbb{R}^n$, where n is its dimensionality. We use $x \in S \subseteq \mathbb{R}^n$ and $i \in S \subseteq \mathbb{Z}^n$ to specify a point in continuous or discrete search space. The time variables become $\Gamma = \mathbb{Z}$ for discrete and $\Gamma = \mathbb{R}$ for continuous, where we

[3] Optimization problems can be either maximization or minimization problems. As shown with (13.2), we only consider maximization problems here. Between maximization or minimization problems there is the relationship $max\, f(x) = -min\, f(x)$, so this is without loss of generality.

[4] Instead of the term *dynamic fitness landscape* we also find *dynamic environment* or even *non–stationary environment* in the literature. *Environment* and *fitness landscape* are rather synonymous, but we prefer fitness landscape as there is a substantial mathematical theory on fitness landscapes available, which appears to be useful in the context of dynamic optimization. Statistically speaking, the term *non–stationary* implies more than dynamics, namely that the dynamics is generated by a stochastic process and the expected value of the process changes over time. Hence, it should only be used if this is indeed the focus of the dynamics considered.

use $k \in \mathbb{Z}$ and $t \in \mathbb{R}$ to label specific points in time. With these preliminaries, we can formulate the dynamic optimization problem

$$f_S(t) = \max_{x \in S} f(x,t), \qquad \forall t \geq 0, \tag{13.4}$$

which yields the temporarily highest fitness $f_S(t)$ and its solution trajectory[5]

$$x_S(t) = arg\, f_S(t), \qquad \forall t \geq 0. \tag{13.5}$$

For calculating the time evolution of all fitness values in the search landscape, it can be convenient to have an iterative generation law describing how a fitness value at $f(x,t)$ evolves into $f(x,t+\delta t)$ with δt a small time increment. In the dynamic fitness landscape (13.3), this time evolution of a point x not only depends on time and the fitness values of the point itself, but also on the fitness values of surrounding points, that is $f(x+\delta x,t)$ with $\delta x = (\delta x_1, \delta x_2, \ldots, \delta x_n)$. So, a general evolution law becomes

$$f(x,t+\delta t) = \Psi\left(f(x,t), f(x_1+\delta x_1,t), f(x_2+\delta x_2,t), \ldots, f(x_n+\delta x_n,t)\right), \tag{13.6}$$

with Ψ being the generator mapping.

It is noteworthy that such a definition is closely related to the standard definition for dynamical systems, see e.g. [1, 30], Ch. 1. In addition to the elements there, the notation of a time–depended fitness function replaces the state space variables in order to tackle the proposed dynamic optimization problem.

13.2.2 Hierarchy of Fitness Landscapes

For the class of spatially extended systems, a hierarchy of spatio–temporal dynamics has been suggested [15, 26] which stems from the decision of discretization of space and time. We adapt this hierarchy for discussing different kinds of static and dynamic fitness landscapes, see Tab. 13.1.[6] The given classes indicate an increasing degree of complexity which relates to the amount of information required to specify a unique fitness value and hence to one type of scale for the expected difficulty in solving the posed optimization problem. The classes 1 and 2 are static combinatorial and continuous optimization problems which are the topic of a widely ramified and extensive literature in the context of evolutionary computation, e.g. [3, 16, 35]. The

[5] For the dynamic optimization problem in discrete time, we replace formally k for t in (13.4) and (13.5).

[6] In addition to the discretization of space and time, for spatio–temporal dynamics a discretization of the local state variable has been suggested [15, 26], particularly to capture dynamics where states can only have a finite number of different values as for instance described by cellular automata. In our field of application, such a discretization would mean to have discrete fitness values. Such discrete fitness values sometimes occur, for instance in using surrogate models for the fitness function evaluation, but generally, fitness landscapes have rarely this property and so we do not consider such a distinction here.

Table 13.1 Hierarchy of fitness landscapes; S static, D discrete, C continuous

Class	Space	Time	Model
1	D	S	Discrete fitness function
2	C	S	Continuous fitness function
3	D	D	Coupled map lattices (CML)
4	C	D	Continuous fitness function with external discrete dynamics
5	D	C	Lattice of coupled ordinary differential equation (ODE)
6	C	C	Partial differential equation (PDE)

classes 3 and 6 will be closer looked at below. The models in class 3 and 4 include dynamic optimization problems that received much attention in form of continuous or discrete fitness functions whose selected topological features change with a discrete time regime, usually generated by some external source of dynamics.

A well–known example of a class 4 problem is the so–called moving peak benchmark [8, 38] which uses as fitness function $f(x) : S \to \mathbb{R}$ an n–dimensional "field of cones on a zero plane", where the cones have randomly chosen heights and slopes and are distributed across the landscape. So, we write

$$f(x) = \max\left\{0, \max_{1 \le i \le N} [h_i - s_i \|x - c_i\|]\right\}, \tag{13.7}$$

where N is the number of cones in the landscape, c_i are the coordinates of the i–th cone, and h_i, s_i specify its height and slope, see Fig. 13.2 for a typical landscape in $\mathbb{R}^2$. The given specification of dynamics requires to move N cones in terms of coordinates, heights and slopes. By defining dynamic sequences for coordinates $c(k)$, heights $h(k)$ and slopes $s(k)$, a dynamic fitness landscape

$$f(x,k) = \max\left\{0, \max_{1 \le i \le N} [h_i(k) - s_i(k)\|x - c_i(k)\|]\right\} \tag{13.8}$$

can be obtained. In studies of the dynamic fitness landscape (13.8) three main types of dynamics regarding the coordinates $c_i(k)$, heights $h(k)$ and slopes $s(k)$ of the cones have been considered: (i.) regular dynamics usually generated by analytic coordinate transformations, for instance cyclic dynamics where each $c_i(k)$, $h(k)$, $s(k)$ repeats itself after a certain period of time or translatory dynamics where the quantities ascribe a pre–defined track or tour, (ii.) chaotic dynamics generated by a chaotic discrete–time system, for instance the generalized Hénon map, see [44, 45] for details of the generation process, and (iii.) random dynamics with each $c_i(k)$, $h(k)$, $s(k)$ for each k being an independent realization of, for example, a normally or uniformly distributed random variable.

A similar and also popular dynamic fitness landscape is the XOR-generator by Yang [67, 69], which is a class 3 problem. This generator can be constructed from

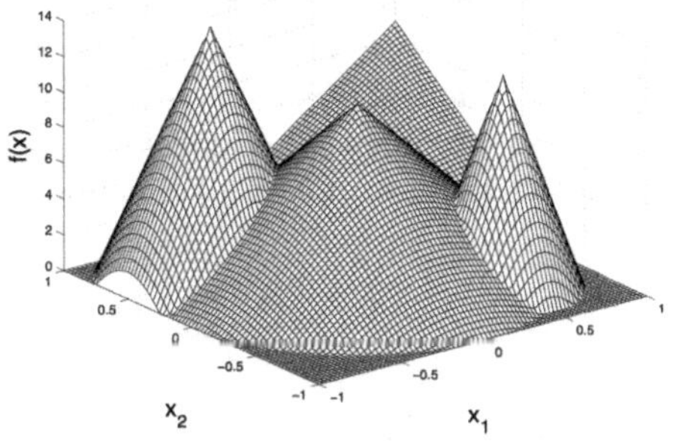

Fig. 13.2 Typical fitness landscape (13.7) for $n = 2$ and $N = 4$.

any binary–encoded stationary function $f(x)$ as follows. For each environment k, an XORing mask $M(k)$ is incrementally generated by

$$M(k) = M(k-1) \oplus T(k), \tag{13.9}$$

where "$\oplus$" is a bitwise exclusive-or (XOR) operator (i.e., $1 \oplus 1 = 0$, $1 \oplus 0 = 1$, and $0 \oplus 0 = 0$) and $T(k)$ is an intermediate binary template generated for environment k. $T(k)$ is generated with $\rho \times l$ ($\rho \in (0.0, 1.0]$) random loci set to 1 while the remaining loci are set to 0. For the initial environment $k = 1$, $M(1)$ is set to a zero vector, i.e., $M(1) = 0$.

To summarize, the majority of the literature on evolutionary computation in dynamic fitness landscapes focusses on class 3 and 4 landscapes. In contrast, the class 5 dynamic fitness landscape does not play a major role in studies. It corresponds to a combinatorial optimization problem, where the fitness function changes with continuous time. Even if a practical optimization problem would have such features, we would most likely model discrete time behavior in the dynamic landscape for reasons discussed right afterwards.

A class 3 problem, a CML–based dynamic fitness landscape and its relationship to both a class 4 and class 6 problem, the latter is PDE–based, is the main topic of this chapter. In modelling physical systems, we usually consider continuous changes in both space and time. So, a general dynamic fitness landscape may describe the evolution of fitness values in a search space where the landscape may undergo changes continuously in both space and time. Such a dynamic evolution has to be modelled by a PDE. On the other hand, to facilitate efficient computing, an appropriate discretization is needed, the more so as numerical effort in solving the dynamic optimization problem by an evolutionary algorithm scales with the number of fitness function evaluations. Such a discretization of space and time can be obtained by the CML formalism, in particular for reaction–diffusion systems and surface growth. It is important to note that by doing so essential features of the dynamics are preserved, e.g. [25, 31, 42, 59]. Moreover, as we focus on dynamic fitness functions in which an evolutionary algorithm is used for solving an optimization problem, and as in evolutionary algorithms time is counted by generations and is hence discrete, it appears to be sensible to have dynamic fitness landscapes that change at discrete points in time, too. As mentioned before we put for the CML–based landscape $S \subseteq \mathbb{Z}^n$ and

$\Gamma = \mathbb{Z}$ and for the PDE–based $S \subseteq \mathbb{R}^n$ and $\Gamma = \mathbb{R}$, where n is dimensionality of the search space. Note that this implies $i = (i_1, i_2, \ldots, i_n)$ and $x = (x_1, x_2, \ldots, x_n)$ for the discrete and continuous spatial variables and k and t for the discrete and continuous temporal variables. So, from the generator mapping (13.6) we obtain for discrete time and space the CML–like mapping

$$f(i,k+1) = \Psi\left(f(i,k), \sum_{j_1=1}^{J_1} f(j_1, i_2, \ldots, i_n, k), \sum_{j_2=1}^{J_2} f(i_1, j_2, \ldots, i_n, k), \ldots\right) \quad (13.10)$$

and for continuous time and space the PDE[7]

$$\frac{\partial f(x,t)}{\partial t} = \Psi\left(f(x,t), \frac{\partial f}{\partial x_1}, \frac{\partial f}{\partial x_2}, \ldots, \frac{\partial f}{\partial x_n}, \frac{\partial f^2}{\partial x_1 x_2}, \ldots, \frac{\partial f^2}{\partial x_1 x_n}, \ldots, \frac{\partial f^2}{\partial x_1^2}, \ldots\right). \quad (13.11)$$

With these mathematical descriptions, dynamic fitness landscapes of class 3 and 6 are specified in a very general way. In order to analyze both types and their relationship, we next consider the 2D cases, that is $n = 2$.

13.2.3 Relationships between Coupled Map Lattices and Reaction–Diffusion Systems

For a two–dimensional search space S a dynamic fitness landscape can be viewed as the time evolution of the surface over a 2D plane at point x and time t. Such a general dynamic 2D fitness landscape[8] that describes the dynamics of the fitness value $f(x_1, x_2, t)$ with continuous spatial variables (x_1, x_2) and continuous time t can be modelled by the parabolic PDE

$$\frac{\partial f}{\partial t} = a_1 \left(\frac{\partial^2 f}{\partial x_1^2} + \frac{\partial^2 f}{\partial x_2^2}\right) - a_2 g_1 \left(\frac{\partial f}{\partial x_1}, \frac{\partial f}{\partial x_2}\right) + g_2(f), \quad (13.12)$$

where a_1, a_2 are coefficients and g_1, g_2 are mappings. With (13.12), we describe a class 6 dynamic fitness landscape. It can be interpreted as a reaction–diffusion system with an additional nonlinear term and is a special case of the general description (13.11). This type of PDE has close resemblance to the Kardar–Parisi–Zhang (KPZ) equation [29], which has been proposed to model surface growth. The main difference is that the KPZ equation includes an explicit stochastic (Gaussian noise) term. Recently, the KPZ equation has been intensively studied [31, 33, 36] while particularly the relation to Coupled Map Lattices (CML) has been a central question. Clearly, both are models of extended dynamical systems. Also and as mentioned before, a numerical solution of a PDE always requires to have some kind of

[7] In doing so, we assume that the space S and the mapping ϕ enjoy properties that guarantee existence and uniqueness of such a spatio–temporal evolution.

[8] To emphasize that the dynamics of the fitness landscape is that of an extended dynamical system, also the (synonymous) term *spatio–temporal fitness landscape* is used.

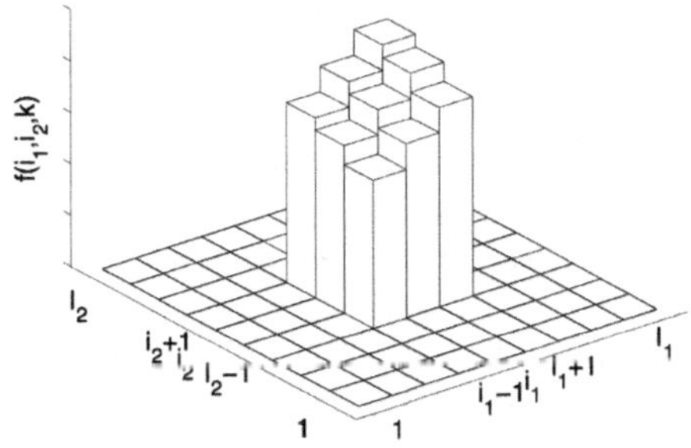

Fig. 13.3 The coupled map lattice (CML) (13.14) as building block for class 3 and 4 fitness landscapes.

discretization of space and time. So, an alternative strategy to a study by any of the methods for numerically solving the PDE, appears to consist of a study of the corresponding CML and their mutual dynamical properties.

Recently, a fitness landscape based on a CML has been studied [46, 47], which is of class 3. We will relate this fitness landscape to the PDE–based landscape (13.12). For the CML, we lay out a lattice grid with $I_1 \times I_2$ equally sized cells, which builds a $2D$–structure. For every discrete time step k, $k = 0, 1, 2, \ldots$, each cell is characterized by its height

$$f(i_1, i_2, k), \quad i_1 = 1, 2, \ldots, I_1, \quad i_2 = 1, 2, \ldots, I_2, \tag{13.13}$$

where (i_1, i_2) denote the spatial indices in vertical and horizontal directions, respectively, see Fig. 13.3. We interpret this height $f(i_1, i_2, k)$ as fitness according to the geometrical metaphor of a fitness landscape. It is subject to changes over time, which are described by the two–dimensional CML with nearest–neighbor coupled interaction [12, 25]

$$\begin{aligned} f(i_1, i_2, k+1) = (1-\varepsilon) g(f(i_1, i_2, k)) + \frac{\varepsilon}{4} \Big[& g(f(i_1 - 1, i_2, k)) + g(f(i_1 + 1, i_2, k)) \\ & + g(f(i_1, i_2 - 1, k)) + g(f(i_1, i_2 + 1, k)) \Big], \end{aligned} \tag{13.14}$$

where $g(f(i_1, i_2, k))$ is a local mapping function and ε is the diffusion coupling strength. As local mapping function we use the logistic map

$$g(f(i_1, i_2, k)) = \alpha f(i_1, i_2, k)(1 - f(i_1, i_2, k)). \tag{13.15}$$

It is a nonlinear map with the parameter $0 < \alpha < 4$ which is defined for the unit interval $f \in [0, 1]$. For some parameter α, the map (13.15) exhibits chaotic behavior, for instance in the parameter interval $\alpha \in [3.57, 4]$. This local chaotic behavior is distributed to other areas of the lattice by coupling. So, it is the source of spatio–temporal chaos in the extended dynamical system.

Finally, we need to set the period boundary conditions

$$\begin{aligned} f(I_1 + 1, i_2, k) &= f(1, i_2, k), \\ f(i_1, I_2 + 1, k) &= f(i_1, 1, k). \end{aligned} \tag{13.16}$$

Initialization of the CML is done by initial heights $f(i_1,i_2,0)$ being realizations of a random variable uniformly distributed on $[0,1]$. The spatio–temporal behavior of the CML depends on the lattice size $I_1 \times I_2$ and two parameters, the coupling strength ε and the nonlinear parameter α. The CML (13.14) can be seen as a special case in two dimensions of the general CML–like spatio–temporal mapping (13.10).

The CML are known to exhibit a rich spatio–temporal behavior, including different types of spatio–temporal periodicity and chaos, quasi–periodicity and pattern formation. So, the CML are an instructive example for the principle of generating high–dimensional complex spatio–temporal dynamics by using local chaos created by a low–dimensional mechanism that is transmitted to a spatial extension by coupling.

We now link the discrete space and time fitness landscape $f(i_1,i_2,k)$ to the landscape with continuous space and time $f(x_1,x_2,t)$ according to eq. (13.12). We take the continuum limit of the CML and employ the following discretizations: a forward difference of the time derivative

$$\frac{\partial f(x_1,x_2,t)}{\partial t} \leftrightarrow \frac{f(i_1,i_2,k+1)-f(i_1,i_2,k)}{\delta t} \tag{13.17}$$

and the central differences of the space derivatives

$$\frac{\partial f(x_1,x_2,t)}{\partial x_1} \leftrightarrow \frac{1}{2}\frac{f(i_1+1,i_2,k)-f(i_1-1,i_2,k)}{\delta x_1}, \tag{13.18}$$

$$\frac{\partial f(x_1,x_2,t)}{\partial x_2} \leftrightarrow \frac{1}{2}\frac{f(i_1,i_2+1,k)-f(i_1,i_2-1,k)}{\delta x_2} \tag{13.19}$$

and the second derivatives

$$\frac{\partial^2 f(x_1,x_2,t)}{\partial x_1^2} \leftrightarrow \frac{f(i_1+1,i_2,k)-2f(i_1,i_2,k)+f(i_1-1,i_2,k)}{(\delta x_1)^2}, \tag{13.20}$$

$$\frac{\partial^2 f(x_1,x_2,t)}{\partial x_2^2} \leftrightarrow \frac{f(i_1,i_2+1,k)-2f(i_1,i_2,k)+f(i_1,i_2-1,k)}{(\delta x_2)^2}, \tag{13.21}$$

with the time step δt and the spatial steps $\delta x_1, \delta x_2$ being equal to one in the used system of units. So, we obtain the PDE

$$\frac{\partial f}{\partial t} = \frac{\alpha\varepsilon}{4}\left(\frac{\partial^2 f}{\partial x_1^2}+\frac{\partial^2 f}{\partial x_2^2}\right)-\frac{\alpha\varepsilon}{2}\left(\left(\frac{\partial f}{\partial x_1}\right)^2+\left(\frac{\partial f}{\partial x_2}\right)^2\right)+(\alpha-1)f-\alpha f^2, \tag{13.22}$$

where $k\delta t \to t$, $i_1\delta x_1 \to x_1$, $i_2\delta x_2 \to x_2$ and $f(i_1,i_2,k) \to f(x_1,x_2,t)$. With eq. (13.22), we have a parabolic PDE of the reaction–diffusion type (13.12).

As shown in [47] from the CML (13.14), a fitness landscape of class 4 with continuous space and the search space variable x can been defined by setting scaling factors $s_1,s_2 \in \mathbb{R}_+$ and by imposing a rounding condition, so that

$$\begin{pmatrix} \lceil s_1 x_1 \rceil \\ \lceil s_2 x_2 \rceil \end{pmatrix} = \begin{pmatrix} i \\ j \end{pmatrix}. \tag{13.23}$$

So, we obtain the dynamic fitness function for the two–dimensional CML (13.14) as

$$f(x,k) = \left\{ \begin{matrix} f(\lceil s_1 x_1 \rceil, \lceil s_2 x_2 \rceil, k) \text{ for } & \begin{matrix} 1 \le \lceil s_1 x_1 \rceil \le I_1 \\ 1 \le \lceil s_2 x_2 \rceil \le I_2 \end{matrix} \\ 0 & \text{otherwise} \end{matrix} \right\}, \; k \ge 0. \tag{13.24}$$

This dynamic fitness landscape, see Fig. 13.4, will be the test bed for the numerical experiments reported in Sec. 13.5. For (13.24), we can pose a dynamic optimization problem

$$f_S(k) = \max_{x \in \mathbb{R}^2} f(x,k) = \left\{ \max_{\substack{1 \le \lceil s_1 x_1 \rceil \le I_1 \\ 1 \le \lceil s_2 x_2 \rceil \le I_2}} f(\lceil s_1 x_1 \rceil, \lceil s_2 x_2 \rceil, k) \right\}, \; k \ge 0, \tag{13.25}$$

which yields a sequence $f_S(k)$ of the highest fitness. Solving the dynamic optimization problem defines a solution trajectory

$$x_S(k) = \arg\max_{x \in \mathbb{R}^2} f(x,k) = \arg \left\{ \max_{\substack{1 \le \lceil s_1 x_1 \rceil \le I_1 \\ 1 \le \lceil s_2 x_2 \rceil \le I_2}} f(\lceil s_1 x_1 \rceil, \lceil s_2 x_2 \rceil, k) \right\}, \; k \ge 0, \tag{13.26}$$

which we intend to find by using an evolutionary algorithm.

To summarize, we have shown the mathematical relationship between a CML–based dynamic fitness landscape and a class of parabolic PDEs. In doing so, we have created a link between a computational efficient and hence numerically experimentable model and a description of real physical phenomena such as surface growth and reaction–diffusion dynamics. So, the dynamic fitness landscape (13.24) can be considered to be a more realistic description of real–world dynamic optimization problems than the benchmarks such as (13.8) or (13.9).

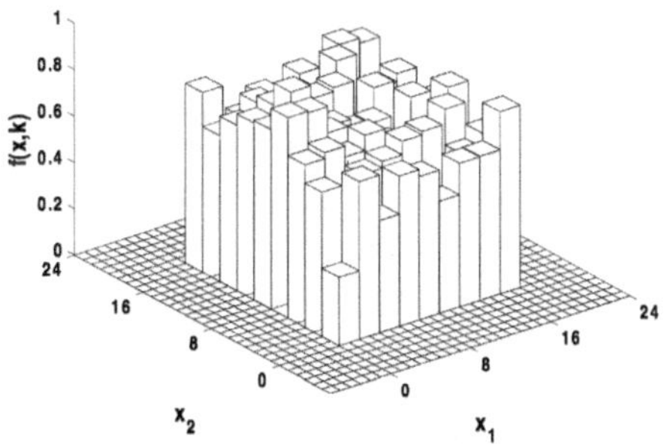

Fig. 13.4 CML–based fitness landscape (13.24) for $I_1 = I_2 = 9$ and $s_1 = s_2 = 0.5$.

13.3 Properties of Dynamic Fitness Landscapes

A main concern in evolutionary computation is developing, testing and applying algorithms that solve optimization problems. From a theoretical perspective, it is therefore desirable to classify the tackled optimization problems in order to compare the approaches and also to suggest possibilities for improving the algorithms. A first and rather structural classification is provided by the hierarchy of fitness landscapes given in Sec. 13.2.2. An important aspect of the concept of fitness landscapes is that it provides a theoretical framework for describing, evaluating, comparing and quantifying the difficulty of a given optimization problem. Based on this notation, it is intended to obtain an estimation for the behavior an evolutionary algorithm is expected to have. Within the conceptional framework this question is addressed by concepts and quantifiers for measuring fitness landscapes [21, 27, 57, 62]. These landscape measures can also be seen as an attempt to define some types of metric for fitness landscapes. Some of these measures have been suggested for static fitness landscapes and hence account for *topological properties* of the fitness landscape. Besides, for dynamic fitness landscape also *dynamical properties* have to be considered.

13.3.1 Topological Properties and Topological Problem Difficulty

Unfortunately, even for a static fitness landscape the question of how difficult a certain optimization problem is for an evolutionary algorithm is not easy to answer. If we view a fitness landscape as in the Figs. 13.1 or 13.2, we see a collection of hills and valleys that can be accompanied by ridges, plateaus, etc. The optimization task is constituted by finding the highest hill (or lowest valley). The evolutionary algorithm puts individuals into these landscapes, while in the generational circle they should trawl the search space and finally find the optimum. In this process, the only feedback from the landscape comes from the fitness values of all member of the population. Moreover, a movement towards an optimum can only be expected if either an individual with high fitness pulls other individuals to itself in the recombination step or random fluctuations working on the individuals during the mutation step put them nearer to the optimum. However, as both movements are censored by the selection step, if either one or the other movement leads to a decreasing fitness, it becomes futile. From these thoughts it is clear why a single sphere of ever increasing fitness with a single highest value, see Fig. 13.5, is a particularly easy problem to solve. There is no distraction for the evolutionary process. The individuals just move up the single hill. However, in more complex landscapes with more than one hill also their number, size, form and distribution constitute difficulty in the search process of stochastically driven search procedures as evolutionary algorithms. In terms of an optimization task these features correspond to the number of optima, how they are distributed and what the space in–between the optima looks like.

It is easily understood and intuitively reasonable that the difficulty of finding the global optimum among several local optima depends on their number; the larger the

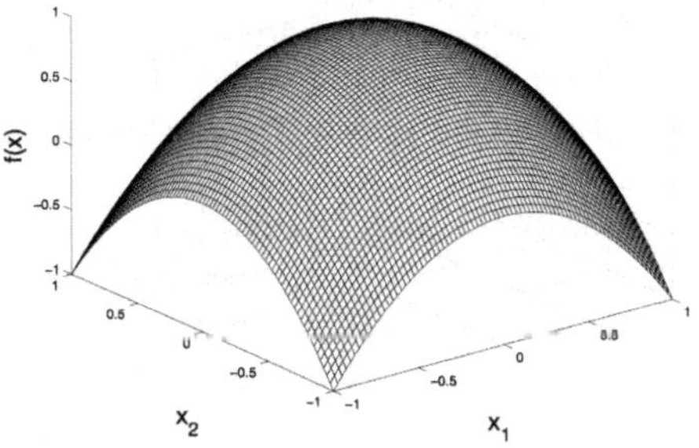

Fig. 13.5 Sphere–like fitness landscape of continuously increasing fitness towards the single optimum.

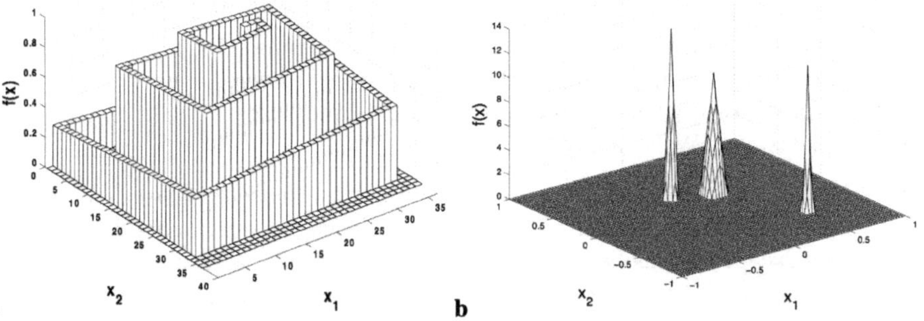

Fig. 13.6 Different types of fitness landscapes: a) long–path problem, b) neutrality with spike–like peaks.

number of local optima, the more difficult the problem is [27, 57]. Second, distribution of the optima relates to problem difficulty. The problem becomes different if the optima are either grouped in one subset of the search space or if they are scattered widely. In the former case, the population only has to find the area of the optima and can then jump from lower optima to higher optima using the stochastic drive in the algorithm. If the optima are distributed widely, the population might split and some areas might not be searched at all. Furthermore, as pointed out above an evolutionary algorithm is using differences in the fitness of individuals populating the fitness landscape in its search process. That means not only the wider surrounding of the optima counts, but also the nearer neighborhood is of interest. For instance, the problem belongs to a different category, if the optima consist of peaks that gradually slope down into all directions, see Fig. 13.5, or the peak can only be approached by a narrow single path of monotonically increasing fitness (so–called long path problems [23]), see Fig. 13.6a, or there are slim and distant peaks on an otherwise plain surface of equal or nearly equal fitness (so–called neutrality [55]), see Fig. 13.6b. These geometrically motivated features are addressed by the notion of accessability (or basin of attraction) of optima and they are a third major factor of landscape topology that contributes to problem hardness.

However, from these main ingredients of problem difficulty, it cannot be easily deduced how in a certain landscape the number of local optima, their distribution and their accessability balance each other in terms of problem difficulty [55, 62]. Clearly, modality, which expresses the number of local optima, is a primary factor and might in addition have the advantage that it can be assigned rather straightforwardly by enumeration. However, apart from the sheer number of optima, it is the interplay of all three aspects that defines hardness of optimizing in a fitness landscape for an evolutionary algorithm. This is the reason why landscape measures have been suggested that are aimed to catch problem hardness more generally rather than just by accounting for a single aspect of the landscape topology. In addition, these measures allow to weight the three factors.

For evaluating topological problem difficulty the topological landscape measures

- modality = number and distribution of local maxima,
- ruggedness = analysis of the static correlation structure,
- information content = an entropic landscape measure,
- epistasis = a Walsh analysis

have been suggested and for static landscapes, these measures have been studied intensively [19, 57, 62]. In [22, 46, 47], these measures were applied to dynamic fitness landscapes. It has been shown that dynamic fitness landscapes inherit topological aspects of problem difficulty from their static counterparts. On the other hand, the features of the dynamics in the landscape contribute in their own way. The definitions of the topological measures and the results obtained for the CML–based dynamic fitness landscape (13.24) are briefly recalled in Sec. 13.3.3 to have a reference and comparison to the dynamical measures considered next. Only the Walsh epistasis measure is omitted as it had been shown in [47] that it poorly reflects problem hardness for the CML–based dynamic landscape considered here.

13.3.2 Dynamical Properties and Dynamical Problem Difficulty

In a dynamic fitness landscape not only topological features constitute problem difficulty, there is also a contribution of features of the involved dynamics [9, 24, 37]. Similar to the situation with topological properties, there is no simple classification. If we look at a dynamic fitness landscape (imagine a landscape as in the Figs. 13.1 or 13.2, where now the hills, valleys and plateaus are changing their position and shape and move around the plane that forms the search space) then again some intuitively comprehensible factors that make finding the moving optima easy or hard can be seen. An evolutionary algorithm carries out a parallelized population–based search in which detecting the optimum depends on improvements over a certain number of generations; with the more generations available, the better for problem solving. Therefore, finding the optimum in just one generation is highly improbable and generally speaking controverts the fundamental idea of evolutionary search. Dynamical problems that can be solved robustly by an evolutionary algorithm should involve a change pattern that allows the algorithm at least a certain number of generations.

So, the speed at which the landscape changes must have some influence. Generally it must hold that the faster the speed (more changes per time interval) is, the more complicated is the dynamical problem. In defining the speed of the landscape changes, the time scale of the dynamic fitness landscape needs to be related to the computation time of the evolutionary algorithm, which counts time by generations. In general, a generation of the evolutionary algorithm results from the computation time for the fitness evaluation, which needs to be carried out for all the individuals of the population and usually is the main contribution, and the time needed for executing the evolutionary operators such as selection, recombination and mutation, which is a minor part. This gives an estimate for the time required to calculate one generation. Note that for a given implementation and hardware, this time can be converted into real CPU time. So, for a population size that is constant over the evolutionary run, for every generation the (approximately) same time interval should go by. For a dynamic fitness landscape that also has a continual change pattern, this means that both time scales relate linearly.

We can describe the dynamics of an evolutionary algorithm by the generation transition function ψ, see e.g. [3], p. 64–65, which can be interpreted as a nonlinear probabilistic dynamical system that maps and transforms a population $P(\tau)$ at generation $\tau \in \mathbb{N}_0$ into a population $P(\tau+1)$ at generation $\tau+1$,

$$P(\tau+1) = \psi(P(\tau)), \tau \geq 0 \tag{13.27}$$

by using the evolutionary operators selection, recombination and selection (and possibly some additional operators such as memory, hyper–mutation and so on). Starting from an initial population $P(0)$, eq. (13.27) describes the population dynamics in the search space. With the proposed linear scale between both the time scales of the evolutionary algorithm τ and the time scale of the dynamic fitness landscape k, we obtain a relation by the change frequency $\gamma \in \mathbb{N}$.[9] There is

$$\tau = \gamma k \tag{13.28}$$

with γ being constant.[10] The quantity γ can be interpreted as the computation time that the algorithm needs to solve the problem and hence is an estimate of the required

[9] Instead of the term *change frequency*, we can also find *change period* in the literature. Change period is motivated by interpreting γ as time interval, change frequency because γ indicates after how many generations the landscape changes. In the following, we prefer the latter interpretation.

[10] The relation (13.28) links the time scales for dynamic fitness landscapes of class 3 and 4 with discrete time as for instance given by (13.8) or (13.10). For dynamic fitness landscapes of class 6 as modelled by (13.11) and (13.12) the changes happen continuously. This means changes in the dynamic fitness landscape occur in–between generations or several (in theory an infinite number of) times within one generation. But as fitness evaluation in an evolutionary algorithm usually takes place just once in a generation, these changes would probably not come into effect before the next generation, that is the next synchronization point between $t \in \mathbb{R}$ and $\gamma^{-1}\tau$. Therefore, the discussion above applies to continuous dynamic fitness landscapes in the same way.

time between changes of the fitness landscape.[11] However, numerical experiments in evolutionary computation of dynamic fitness landscapes usually view the change frequency γ slightly differently. They consider γ an adjustable parameter that can be used to evaluate and compare different types and implementations of evolutionary algorithms. This view is justified by the fact that γ can indeed be adjusted by modifications done on the parameters and implementation of the evolutionary algorithm and the hardware on which the algorithm runs. Note that this view also means that the change frequency is no longer a property of the dynamic fitness landscape but is seen to be independent of it. In the numerical experiments, we will adopt this view and consider change frequency as something that can be adjusted. Note further that by doing so, change frequency has a unique role among all other topological and dynamical properties defining problem hardness. While all the other properties belong to the dynamic fitness landscape considered, change frequency is defined by the evolutionary algorithm and hence assumed to be freely settable (at least within certain limits).

Aside from the (relative) speed of the landscape changes, a second major dynamical influence on problem hardness addresses the spatial distance that the optimum moves if the landscape changes, that is the (relative) strength of the landscape changes. As optimum finding for an evolutionary algorithm implies to trawl the search space for a certain time, time restrictions as those coming from a changing landscape mean that the average distance between subsequent optima is a good measure for problem hardness. This dynamical property is called dynamic severity for which there are several notations [9, 45, 63]. They all have in common that they measure the (relative) magnitude of the changes by comparing the landscape at subsequent points in time, for instance between k to $k+1$ or t to $t+\delta t$. In terms of the dynamic fitness landscapes of class 3 and 4, dynamic severity means to evaluate the (average) distance from the highest peak's coordinates $x_S(k) = arg f_S(k)$ before and after a change, as given by (13.5).[12] With eq. (13.44), this is applied to the CML–based dynamic fitness landscape considered here.

In dynamic optimization, we sometimes find a discrimination between gradual and abrupt changes. What distinguishes gradual from abrupt changes is basically understood as different degrees of dynamic severity, but somehow change frequency also contributes and is intertwined with it. Our view is this. For landscapes with discrete time the situation is rather straightforward. As in discrete time the changes happen one after the other to distinct points in time, an abrupt change is one with a large severity, a gradual one has a low severity, no matter what the change frequency is. For discrete time dynamic fitness landscape of class 3 and 4, the landscape has

[11] Usually, γ is considered to be constant for all generations τ, but it might also be a function of k and even be different (for instance a positive integer realization of a random process) for every k.

[12] Using a similar argumentation as for change frequency, see footnote 10, dynamic severity can be defined for class 6 dynamic fitness landscapes as for instance (13.11) in the likewise fashion. Therefore, the highest peak's coordinates as given from the solution trajectory of the dynamic optimization problem (13.4) must be compared for a time lapse between t and $t+\delta t$.

no speed by itself, the only base for comparison is the generational dynamics of the evolutionary algorithm linearly linked via the change frequency (13.28). For changes in the fitness landscape in continuous time, a more elaborated discussion is necessary; for those dynamic fitness landscapes we need a notation of speed on its own. The (direction–less) speed $v_S(t)$ of the optimum $x_S(t)$ at time t can be defined by

$$v_S(t) = \lim_{\delta t \longrightarrow 0} \frac{\|x(t+\delta t) - x(t)\|}{\delta t}. \tag{13.29}$$

For δt being small and constant, the average speed of the optimum $\langle v_S(t) \rangle$ can be calculated by

$$\langle v_S(t) \rangle = \frac{1}{K} \sum_{k=0}^{K-1} \frac{\|x(t+(k+1)\delta t) - x(t+k\delta t)\|}{\delta t} \tag{13.30}$$

with K sufficiently large. The nominator term is dynamic severity (cf. eq. (13.44) so that for continuous time dynamic fitness landscape an abrupt change is indicated by a $\langle v_S(t) \rangle \delta t$ above a certain limit, while a gradual change is characterized by a small value of this quantity.

Dynamic severity is an intrinsic property of the fitness landscape. This also applies to a third dynamical property of fitness landscapes that tries to capture the complexity of the dynamics. In this context, complexity refers to limits in the long–term predictability of the spatio–temporal evolution, even if explicit stochastic elements in the describing equations are absent. So, studying dynamic complexity is highly linked to and conceptionally as well as methodically overlapping with the study of deterministic chaos.[13] A first and again rather structural classification of dynamic complexity is to categorize dynamics as either *regular*, *chaotic* or *random*. Here, regular dynamics is completely predictable and usually generated by analytic coordinate transformations. It might, for instance, be cyclic, where we have a periodic recurrence of all topological features of the landscape after a certain time interval or translatory, where the topological features follow a pre–defined track or tour. Chaotic dynamics is generated by deterministic chaotic systems (that might be locally interacting with the spatially distributed landscape) and is predictable only for a short term. Random dynamics stems from a stochastic process, that is from realizations of a random variable, and is unpredictable even for short terms.

Apart from this verbal assignment and in order to have a quantification, we can resort to quantifiers of dynamics used and established in the field of nonlinear dynamics, such as Lyapunov exponents and vectors, different types of entropies and

[13] An alternative (and complimentary to the degree of predictability) approach to define complexity is by using concepts from algorithmic information theory, e.g. [5, 14]. Accordingly, algorithmic complexity of a spatio–temporal evolution is defined by the length of the smallest algorithm capable of specifying the evolution. As chaotic evolutions are nonperiodic and oscillatory, their algorithmic complexity is large. However, algorithmic complexity only superficially allows to separate chaos and random, as all chaotic behavior is algorithmically complex, but not all evolutions that are algorithmically complex, are chaotic, too.

information flows, correlations and related quantities as bred vector dimensions, or fractal dimensions. All these quantities are widely used as an analyzing tool in nonlinear dynamical systems theory. However, they were developed to deal with low–dimensional nonlinear (possibly chaotic) dynamical systems described by ordinary differential equations (ODEs). Only for such systems, these quantities are unambiguously meaningful and the relationships between the quantities are largely understood.[14] In recent years, several attempts have been made to extend the theory of low–dimensional dynamical systems to (infinite–dimensional) spatio–temporal systems and by doing so to establish quantities similar to the conventional Lyapunov exponents, dimensions and entropies. Still, this work is in its infancy and the power of the theory is confined to certain limits. Also, while some quantities as Lyapunov exponents or bred vector dimensions have shown to be meaningful in quantifying patterns of space–time dynamics, others are rather ambiguous and allusive. Moreover, the relationships between the quantities are still far from being clear.

In the following, we consider the dynamic landscape measures

- change frequency = speed of fitness landscape changes relative to EA,
- dynamic severity = distance between subsequent optima,
- dynamic complexity = predicability of spatio–temporal evolution:
  - Lyapunov exponents = divergence rate between nearby evolutions,
  - bred vector dimensions = analysis of the dynamic correlation structure.

In the next section, we will study dynamical landscape measures for the CML–based fitness landscape. As discussed above, change frequency is here considered as a parameter to be set. On the other hand, dynamic severity and the quantities to measure dynamic complexity, that are Lyapunov exponents and bred vector dimensions, will be looked at as depending on the dynamic landscape.

13.3.3 Topological and Dynamical Landscape Measures for the CML–Based Landscape

13.3.3.1 Topological Measures

Topological landscape measures have been intensively studied for the CML–based dynamic fitness landscape (13.24) in [46, 47] and we therefore only briefly recall the definitions and some of the results.

Modality. For the fitness landscape (13.24), the topological landscape measure modality accounts for the average number of local maxima and can be assigned by enumeration. We consider as neighborhood structure the surrounding heights. That means the neighborhood structure $N_{\pm 1}(i_1,i_2)$ of the (i_1,i_2)–cell is

$$N_{\pm 1}(i_1,i_2) = (i_1+\delta_1, i_2+\delta_2), \tag{13.31}$$

[14] For instance, there are the relationships between Lyapunov exponents and fractal dimensions via the Kaplan–Yorke dimension [13, 18, 49] or between Lyapunov exponents and entropies via the Kolmogovov–Sinai entropy and the Pesin entropy formula, see e.g. [4].

where (δ_1, δ_2) are taken as disjunction of the permutations over the set $\mathbb{S} = \{-1, 0, 1\}$ that is

$$(\delta_1, \delta_2) = (-1,-1) \wedge (-1,0) \wedge (-1,1) \wedge (0,-1) \\ \wedge (0,1) \wedge (1,-1) \wedge (1,0) \wedge (1,1). \quad (13.32)$$

Additionally, the cell specified by $(\delta_1, \delta_2) = (0,0)$ is excluded from the neighborhood structure $N_{\pm 1}(i_1, i_2)$. Here, $(i_1, i_2)^T = (\lceil s_1 x_1 \rceil, \lceil s_2 x_2 \rceil)^T$. Hence, the fitness function possesses a local maximum at point (i_1, i_2) and time k if

$$f(i_1, i_2, k) \geq f(N(i_1, i_2), k). \quad (13.33)$$

We denote $\#_{LM}(k)$ the number of local maxima at time k. As this quantity may change over time, we consider its time average $\langle \#_{LM}(k) \rangle$, which is

$$\langle \#_{LM}(k) \rangle = \lim_{K \to \infty} \frac{1}{K} \sum_{k=0}^{K-1} \#_{LM}(k). \quad (13.34)$$

For computing an approximate value of the time average number of local maxima, the $\langle \#_{LM}(k) \rangle$ is replaced by $\#_{LM} = \frac{1}{K} \sum_{k=0}^{K-1} \#_{LM}(k)$ with K sufficiently large.

Ruggedness. The topological landscape measure ruggedness can be analyzed by the static correlation structure. This method works by performing a random walk on the landscape and calculating its random walk correlation function. For the dynamic fitness landscape (13.24), this begins with generating a time series

$$f(\tau_s, k) = f(i_1(\tau_s), i_2(\tau_s), k), \quad \tau_s = 1, 2, \ldots, T \quad (13.35)$$

of the heights $f(\lceil s_1 x_1 \rceil, \lceil s_2 x_2 \rceil, k)$ with $(i_1, i_2)^T = (\lceil s_1 x_1 \rceil, \lceil s_2 x_2 \rceil)^T$. For doing the random walk, we create $2 \times T$ independent realizations (t_{i_1}, t_{i_2}) of an integer random variable uniformly distributed on the set $\mathbb{S} = \{-1, 0, 1\}$. Starting from an initial cell $(i_1(1), i_2(1))^T$, the next cell indices $(i_1(\tau_s + 1), i_2(\tau_s + 1))^T$ on the walk are obtained by adding the two independent realizations of the random variable to the current cell indices:

$$(i_1(\tau_s + 1), i_2(\tau_s + 1))^T = (i_1(\tau_s) + t_{i_1}, i_2(\tau_s) + t_{i_2})^T. \quad (13.36)$$

In addition, the boundary condition (13.16) is observed. From the random walk in the two spatial dimensions that is specified by $(i_1(\tau_s), i_2(\tau_s))^T$, we obtain the needed time series on the dynamic fitness landscape by recording the heights $f(\tau_s, k) = f(i_1(\tau_s), i_2(\tau_s), k)$ at time k. For this time series, the spatial correlation can be calculated. The spatial correlation is widely used in determining ruggedness of static landscape [19, 56, 64]. It is an estimate $r(t_L, k)$ of the autocorrelation function of the time series with time lag t_L, also called random walk correlation function:

$$r(t_L,k) = \frac{\sum\limits_{\tau=1}^{T-t_L} \left(f(\tau_s,k) - \bar{f}(k)\right)\left(f(\tau_s+t_L,k) - \bar{f}(k)\right)}{\sum\limits_{\tau_s=1}^{T} \left(f(\tau_s,k) - \bar{f}(k)\right)^2}, \tag{13.37}$$

where $\bar{f}(k) = \frac{1}{T}\sum\limits_{\tau_s=1}^{T} f(\tau_s,k)$ and $T \gg t_L > 0$. The spatial random walk correlation function measures the correlation between different regions of the fitness landscape for a fixed k. As $r(t_L,k)$ changes over time, we consider its time average $\langle r(t_L,k)\rangle$, for which we calculate numerically an approximated value $r(t_L)$ similarly as for the average number of maxima. From this quantity, the correlation of the lag t_L

$$\lambda_R(t_L) = -\frac{1}{\ln\left(|r(t_L)|\right)} \tag{13.38}$$

can be obtained. Among the correlations of the lag t_L, it has been shown that ruggedness is best expressed by the correlation length [56]

$$\lambda_R = -\frac{1}{\ln\left(|r(1)|\right)}, \tag{13.39}$$

which is the correlation of the lag $t_L = 1$. The lower the value of λ_R, the more rugged is the landscape. This kind of evaluating fitness landscapes relies upon the assumption that the landscape is statistically isotropic [19, 20]. This means that the value of $r(t_L,k)$ obtained from the random walk does not depend on the specific random walk used and particularly not on the chosen initial cell. Our numerical results have shown that this holds for the landscapes considered here.

Information content. The topological landscape measure information content can be accounted for by entropic measures [58, 61, 62]. Starting point for this method to evaluate landscapes is again, as for the correlation structure considered above, a time series (13.35), $f(\tau_s,k)$, which is generated by a random walk on the dynamic landscape for a fixed time k. From this time series, we code the difference in fitness between two consecutive walking steps by the symbols $s_{\tau_s} \in \mathbb{S}$, $\tau_s = 1,2,\ldots,T-1$, taken from the set $\mathbb{S} = \{-1,0,1\}$. These symbols are calculated by

$$s_\tau(e,k) = \begin{cases} -1, & \text{if} \quad f(\tau_s+1,k) - f(\tau_s,k) \ < e \\ 0, & \text{if} \quad |f(\tau_s+1,k) - f(\tau_s,k)| \le e \\ 1, & \text{if} \quad f(\tau_s+1,k) - h(\tau_s,k) \ > e \end{cases} \tag{13.40}$$

for a fixed $e \in [0,L]$, where L is the maximum difference between two fitness values. The obtained symbols are concatenated to a string

$$S(e,k) = s_1 s_2 \ldots s_{T-1}. \tag{13.41}$$

The parameter e defines the sensitivity by which the string $S(e,k)$ accounts for differences in the fitness values. For $e = 0$, the string $S(e,k)$ contains the symbol zero only if the random walk has reached a strictly flat area. It hence discriminates very sensitively between increasing and decreasing fitness values. On the other hand, for $e = L$, the string only contains the symbol zero, which makes evaluating the structure of the landscape pointless. In this way, a fixed value of e with $0 < e < L$ defines a level of detail of information about the landscape's structure. The string (13.41) represents this information depending on e and codes it by subblocks over the set $\mathbb{S}$. In other words, varying e allows to zoom in on or to zoom out of the information structure of the landscape.

For defining entropic measures of the landscape, we look at the distribution of subblocks of length two, $s_{\tau_s} s_{\tau_s+1}$, $\tau_s = 1, 2, \ldots T-2$, within the string (13.41). These subblocks stand for local patterns in the landscape. We denote the probability of the occurrence of the pattern $\delta_1 \delta_2$ with $\delta_1, \delta_2 \in \mathbb{S}$ and $\delta_1 \neq \delta_2$ by $p_{\delta_1 \delta_2}$. For numerical calculation, we approximate this probability by the relative frequency of the patterns within the string $S(e,k)$. As the set $\mathbb{S}$ consists of three elements, we find six different kinds of subblock $s_{\tau_s} s_{\tau_s+1} = \delta_1 \delta_2$ with $\delta_1 \neq \delta_2$ within the string. From their probabilities at a fixed time k and a given sensitivity level e we calculate the entropic measure [62]

$$H_{IC}(e,k) = - \sum_{\substack{\delta_1, \delta_2 \in \mathbb{S} \\ \delta_1 \neq \delta_2}} p_{\delta_1 \delta_2}(e,k) \log_6 \left(p_{\delta_1 \delta_2}(e,k) \right), \tag{13.42}$$

which is called information content of the fitness landscape. Note that by taking the logarithm in Eq. (13.42) with the base 6, the information content is scaled to the interval $[0,1]$. As for the other landscape measures, for evaluating dynamic fitness landscapes, we consider the time average $\langle H_{IC}(e,k) \rangle$ for which we numerically calculate an approximated value $H_{IC}(e)$. In the numerical calculation, we set the value $e = 0$, that is we consider the information content of highest sensitivity. As was shown in [47], epistasis measured by Walsh analysis is not a particularly meaningful quantity for the CML–based fitness landscape with chaotic behavior. Therefore, we do not consider it here.

Fig. 13.7 shows the topological landscape measures modality $\#_{LM}$, ruggedness λ_R and information content H_{IC} for varying ε, $\alpha = 3.999$, $s_1 = s_2 = 1$ and constant lattice sizes. We see that in large areas of the parameter space of ε there are similar characteristics, which in the case of λ_R scales in an inverse manner. Further, there are rarely intervals in ε where the measures remain constant. Also, only for the modality measure $\#_{LM}$ we find real differences for varying lattice sizes. The other two measures are very similar no matter if the quadratic lattice size is $I_1 = I_2 = 12$ or $I_1 = I_2 = 20$. Moreover, for $0.1 \leq \varepsilon \leq 0.3$, all topological measures behave differently from the other parameter values; we find a rather erratic characteristic. These results can be understood by considering the spatio–temporal behavior of the CML that defines the fitness landscape. In this parameter range, the CML are known to possess spatio–temporal periodic patterns, while elsewhere the system exhibits spatio–temporal chaos [66].

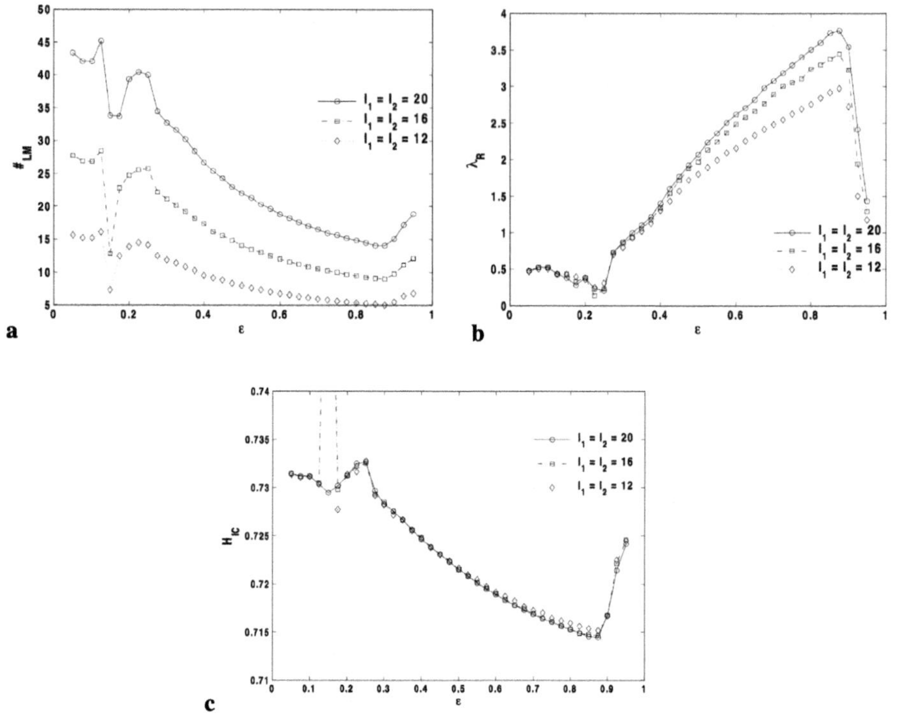

Fig. 13.7 The topological landscape measures for varying ε, $\alpha = 3.999$ and constant lattice sizes: a) modality $\#_{LM}$, b) ruggedness λ_R and c) information content H_{IC}.

13.3.3.2 Dynamical Measures

We next consider dynamic landscape measures and start with dynamic severity as change frequency is treated as a property of the evolutionary algorithms, but not of the dynamic fitness landscape.

Dynamic severity. As discussed above, dynamic severity accounts for the average distance from the highest peak's coordinates $x_S(k) = arg f_S(k)$ before and after a change.

Hence, dynamic severity σ can be calculated for the CML–based fitness landscape (13.24) by

$$\sigma(k+1) = \|x_S(k+1) - x_S(k)\| \tag{13.43}$$

with

$$x_S(k) = \arg\left\{ \max_{\substack{1 \leq \lceil s_1 x_1 \rceil \leq l_1 \\ 1 \leq \lceil s_2 x_2 \rceil \leq l_2}} f(\lceil s_1 x_1 \rceil, \lceil s_2 x_2 \rceil, k) \right\}$$

being the solution of the dynamic optimization problem (13.26). As this quantity may vary with time k, we consider the time average severity

$$\langle \sigma(k) \rangle = \lim_{K \to \infty} \frac{1}{K} \sum_{k=0}^{K-1} \sigma(k) \tag{13.44}$$

and calculate an approximative value σ similarly as done for the other measures.

Dynamic complexity measure: Lyapunov exponents. A first method to measure dynamic complexity is by using the concept of Lyapunov exponents. The Lyapunov exponents give the divergence or convergence rates between a time evolution of the system and its nearby evolution that results from being displaced from the original one by an infinitesimal perturbation [28, 40]. For calculating the (largest) Lyapunov exponent for the CML–based fitness landscape (13.24), we therefore consider the time evolution itself, $f(\lceil s_1 x_1 \rceil, \lceil s_2 x_2 \rceil, k)$ and its neighboring evolution $\Delta f(\lceil s_1 x_1 \rceil, \lceil s_2 x_2 \rceil, k)$, which is obtained by linearizing the system along the evolution $f(\lceil s_1 x_1 \rceil, \lceil s_2 x_2 \rceil, k)$. The linearized system determined along $f(\lceil s_1 x_1 \rceil, \lceil s_2 x_2 \rceil, k)$ describes exactly the result of an infinitesimal displacement and allows to observe if both time evolutions diverge (positive Lyapunov exponent) or converge (negative Lyapunov exponent). A positive Lyapunov exponent generally indicates chaos, and the magnitude of the positive Lyapunov exponent can be seen as a dynamic complexity measure.

We write the linearization of the CML–based fitness landscape (13.24) as follows:

$$\begin{aligned} \Delta \tilde{f}(i_1, i_2, k) &= \frac{d}{df} g(f(i_1, i_2, k)) \\ \Delta f(i_1, i_2, k+1) &= (1-\varepsilon)\tilde{f}(i_1, i_2, k) + \frac{\varepsilon}{4} \Big[\tilde{f}(i_1 - 1, i_2, k) + \tilde{f}(i_1 + 1, i_2, k) \\ &\quad + \tilde{f}(i_1, i_2 - 1, k) + \tilde{f}(i_1, i_2 + 1, k) \Big], \end{aligned} \tag{13.45}$$

where $(i_1, i_2)^T = (\lceil s_1 x_1 \rceil, \lceil s_2 x_2 \rceil)^T$. This linearization is the tangential system to the 2D CML and has dimension $I_1 \times I_2$. For its calculation, the boundary conditions (13.16) have to be taken into account. From the linearization (13.45), we can define the (largest) Lyapunov exponent λ_L as

$$\lambda_L = \lim_{k \to \infty} \frac{1}{k} \ln \frac{\|\Delta f(i_1, i_2, k)\|}{\|\Delta f(i_1, i_2, 0)\|}, \quad \forall \quad 1 \leq i_1 \leq I_1, 1 \leq i_2 \leq I_2. \tag{13.46}$$

Calculation of the Lyapunov exponent can be done using standard methods, for instance using QR–factorization, which is known to be computationally efficient, reliable and robust, see e.g. [11].

Dynamic complexity measure: bred vector dimension. As a second measure for dynamic complexity, we consider bred vector dimensions. Bred vector dimensions are a concept for evaluating the dynamic correlation structure of spatially extended systems [17, 41, 43]. It has been used to identify local regions where this correlation is high and which can therefore be used for prediction and short term forecast. Here, we will use bred vector dimensions as a general measure for the dynamic correlation structure of fitness landscapes and so as another dynamic landscape measure.

For calculating the bred vector dimension, we again consider a neighborhood structure as in (13.31), but with a difference: $N_{\pm\ell}(i_1,i_2)$ is now the disjunction of the permutations over the set $\mathbb{S} = \{-\ell, -\ell+1, \ldots, -1, 0, 1, \ldots, \ell-1, \ell\}$ and $(\delta_1, \delta_2) = (0,0)$ remains included. So, we specify $(2\ell+1)^2$ cells around and including the cell (i_1,i_2) by this neighborhood structure. For doing so around every (i_1,i_2), again the boundary conditions (13.16) need to be satisfied. As there are maximally $I_1 \times I_2$ cells with (possibly) different fitness values in the CML–based fitness landscape and because taking into account the same spatial subsection twice would bias the results, we need to limit $\ell \leq \frac{1}{2}(\sqrt{I_1 I_2} - 1)$. We now calculate the dynamic fitness landscape for a point in time k, which should be large enough so that the transients starting from $f(i_1,i_2,0)$ are removed. At this point in time, we disturb κ times the fitness landscape $f(i_1,i_2,k)$ within the neighborhood structure $N_{\pm\ell}(i_1,i_2)$ by a small Gaussian noise with mean zero and standard deviation std, that is

$$f_j(N_{\pm\ell}(i_1,i_2),k) := f_j(N_{\pm\ell}(i_1,i_2),k) + \mathcal{N}(0, std), \quad j = 1, 2, \ldots, \kappa. \tag{13.47}$$

Then, we define a time lag k_L and calculate the time evolution of all κ disturbed landscapes for further k_L iterations. The obtained $f_j(N_{\pm\ell}(i_1,i_2),k+k_L)$ are formed into a vector $\hat{f}_j$ of length $(2\ell+1)^2$, which is called the bred vector. It is normalized to unity and hence contains the normalized fitness values for the entire neighborhood structure of the disturbed landscape. From the κ bred vectors $\hat{f}_j$ as columns, we build a matrix B, which is of dimension $(2\ell+1)^2 \times \kappa$. The matrix $B^T B$ can be regarded as its corresponding covariance matrix. It expresses the local linear independence of the κ local bred vectors. A measure for this independence can be obtained from the singular values σ_i of the matrix B, which are the roots of the eigenvalues of the covariance matrix $B^T B$, $\sigma_i = \sqrt{eig_i(B^T B)}$. These singular values are a measure for the amount of variance in the set of bred vectors. In other words, they account for the degree of difference imposed by the disturbances applied to the fitness landscape k_L time steps before. From these singular values we finally calculate the quantity

$$\psi_B = \frac{(\sum_{i=1}^{\kappa} \sigma_i)^2}{\sum_{i=1}^{\kappa} \sigma_i^2}, \tag{13.48}$$

which is called the bred vector dimension [41]. As each bred vector, which forms one of the columns of the matrix B, is normalized to unit length, ψ_B can have values $1 \leq \psi_B \leq \kappa$. The value $\psi_B = 1$ indicates that all bred vectors are equal, meaning that the correlation between them is maximal. Any $1 < \psi_B \leq \kappa$ expresses differences in the bred vectors with the magnitude of ψ_B being a measure of the amount of difference. An integer ψ_B can even be interpreted as to relate to the dimensionality of the subspace spanned by the bred vectors. So, by fixing a time lag k_L, we obtain a quantity for the degree of temporal divergence and correlation that disturbances in the fitness landscape cause and hence a dynamic correlation structure related to

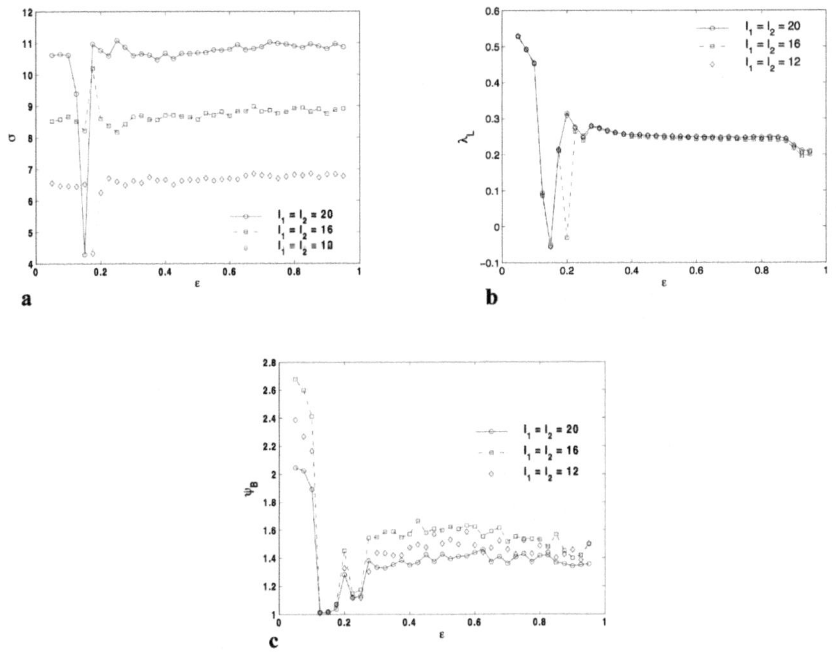

Fig. 13.8 The dynamical landscape measures for varying ε, $\alpha = 3.999$ and constant lattice sizes: a) dynamic severity σ, b) Lyapunov exponent λ_L and c) bred vector dimension ψ_B.

dynamic complexity. Note also that the bred vector dimension is a measure that is conceptionally similar to the Lyapunov exponents and Lyapunov vectors considered above. The main difference is that Lyapunov vectors account for the effect of infinitesimal perturbations to the time evolution, while bred vectors evaluate the result of finite perturbations. In the numerical calculation, the values $k_L = 25$, $\kappa = 5$ and $std = 0.0001$ have been taken. As reference cell in the center of the neighborhood structure, the cell with the maximum fitness value at time k was used.

Fig. 13.8 shows the dynamical landscape measures dynamic severity σ and the dynamic complexity measures Lyapunov exponent λ_L and bred vector dimension ψ_B for varying ε, $\alpha = 3.999$, $s_1 = s_2 = 1$ and constant lattice sizes. The most striking feature is that all three dynamical landscape measures show large parameter intervals in ε, particularly for $\varepsilon > 0.3$ where the values vary only slightly. This is a characteristic not found in the topological measures, see Fig. 13.7. Also, only dynamic severity shows a strong dependency on the lattice size. Again, as for the topological measures, the parameter interval with spatio–temporal periodic patterns is clearly visible. For the Lyapunov exponents, we obtain negative values, indicating an absence of chaos.

The dynamical measure severity not only depends on the lattice size but also on the scaling factors s_1 and s_2. As shown in Fig. 13.8a, the quantity σ obtained for the quadratic lattice and $(i_1, i_2)^T = (\lceil x_1 \rceil, \lceil x_2 \rceil)^T$ (that is $s_1 = s_2 = 1$) can be regarded as constant for a given lattice size and a large majority of values of ε. So, dynamic severity of the CML–based fitness landscape with $s_1 = s_2 = 1$ is a dynamic property depending on the choice of α, ε and $I_1 \times I_2$. This property can be linearly scaled and adjusted by s_1 and s_2. In general, for quadratic lattices with $I_1 = I_2$, dynamic severity scales as $\sigma \sim \sqrt{(s_1^2 + s_2^2)} \cdot I_1$, while for rectangular lattices, we have $\sigma \sim \sqrt{s_1^2 I_1^2 + s_2^2 I_2^2}$.

13.4 Evolutionary Optimization

As we have discussed in the previous section, problem hardness in a dynamic optimization problem not only depends on topological features of the associated fitness landscape, but also on its dynamical properties. In this context, it is interesting to note that one of the driving forces behind the development of evolutionary algorithms was exactly to have search methods for topological difficult problems. So, it has been shown that evolutionary algorithms are remarkably successful in solving static optimization problems with a high degree of problem difficulty. In recent years it has further been shown that these problem solving abilities can also be used to tackle dynamic optimization problems. However, certain modifications in the algorithmic structure of the evolutionary algorithm are necessary to make it work in dynamic fitness landscapes. The working principle of evolutionary algorithms is to maintain and evolve a population of candidate solutions through selection, recombination and mutation. The next generation's population is generated by first selecting relatively fitter individuals from the current population and then applying changes to the selected individuals. This is sequentially done by either more deterministic means (recombination) and or more stochastic (mutation). With these steps the new off–spring of the next generation are created. In the normal working mode the individuals in the population will eventually converge to the optimal solution due to the selection pressure. This convergence property, when happening at a proper pace, is intended and expected from the evolutionary algorithm in order to locate the optimal solution of the static problems.

For solving dynamic optimization problems, however, this convergence property causes big problems for the evolutionary algorithm because it deprives the population of genetic diversity. Dynamic optimization means not longer to find one optimal solution, but to track the movement of the optimal solution with time. Consequently, when the fitness landscape changes, it is hard for the population to escape from the old optimal solution in order to search for the new one, if its diversity is low. The algorithm's convergence is simultaneously corroding its genetic diversity that after a change is exactly needed to explore the search space and to find the optimum again. This situation means that the algorithm must be equipped with some additional schemes which can control, maintain and occasionally enhance the population's diversity.[15] For achieving this goal, several approaches have been suggested.

[15] This often requires to detect the point in time where a change in the landscape occurs, see [48] for a discussion of the involved problems.

One way to classify them is by looking at what element of the evolutionary algorithm is modified and if therefore rather stochastic or deterministic means are used. With this classification, there are four groups of schemes for diversity management to make evolutionary algorithms fit to perform in dynamic fitness landscapes:

- on the level of the algorithm's individuals by mainly stochastic means, as for instance hyper–mutation [39] or random immigrants [60],
- on the level of the algorithm's population by mainly deterministic means, as for instance with different types of memory [8, 50, 53, 68] and multi–population approaches [10],
- on the level of the algorithm's parameters, as for instance by stochastic self–adaption of the mutation [2, 7],
- on the level of the algorithm's operators with additional and completely different operators, as for instance for anticipating and predicting the dynamics [6, 51, 54].

In a specific implementation, there can be hybrid types of the above mentioned schemes. One of those is an abstract memory scheme [50, 51], which combines a memory with a prediction of dynamics. Apart from the abstract memory scheme, we will consider a hyper–mutation scheme, a standard direct memory and a self–adaptive mutation scheme in the numerical experiments with the CML–based dynamic fitness landscape.

The considered schemes work as follows, see Tab. 13.2 for details of the parameter settings. With the hyper–mutation scheme, we have a standard evolutionary algorithm with selection, recombination and mutation. In the implementation here, we use a population size μ, a tournament selection of tournament size 2, a fitness–related intermediate sexual recombination (which is operated μ times and for each recombination two individuals are chosen randomly to produce offspring that is the fitness–weighted arithmetic mean of both parents), and a standard mutation with mutation probability p_m and base mutation rate bm (that means a mutated individual x' differs from an un–mutated individual x by $(x'-x) \sim bm \cdot \mathcal{N}(0,1)$). For the initial population, individuals are generated whose elements are realizations of a random variable normally distributed on $[0, \omega^2]$. The hyper–mutation now increases the mutation strength if a change in the fitness landscape has occurred for a limited number of generations (usually one or two). Therefore, the base mutation rate is multiplied by the hyper–mutation rate hm, so that the hyper–mutated individuals are $(x'-x) \sim bm \cdot hm \cdot \mathcal{N}(0,1)$. In this way, for a certain generational lag, the need for an abrupt increase in genetic diversity is satisfied.

In self–adaption, we use the standard operators of an evolutionary algorithm and have used the same implementation as for hyper–mutation, that is population size μ, tournament selection and fitness–related intermediate sexual recombination. In contrast, the mutation rate itself is a parameter that undergoes a permanent optimization and adaption process. Therefore, the mutation rate is considered as an additional subject to optimize. For every individual in the population, its mutation rate becomes an extra component. In other words, the mutation rate becomes an additional dimension in the fitness landscape. So, the mutation rates mr are subject to the selection process and are continuously changed by an own mutation (and hence adaption) process, see e.g. [2, 3, 34]. They are adapted every generation with $mr(\tau+1) = mr(\tau) \cdot \exp(\tau_A \mathcal{N}(0,1))$, where τ_A is an adaption rate and for generation $\tau = 0$, $mr(0) = mr_0$. Based on this self–adapting rate, the individuals are mutated according to the magnitude that the quantity has in a particular generation, that is $(x' - x) \sim mr(\tau+1) \cdot \mathcal{N}(0,1)$. With such a design of the mutation process, the mutation rate is no longer a parameter to be set before the evolutionary run and controlled by the experimenter. So, the mutation rates might converge to some (optimal) values or might be oscillating between certain values. As the rates are subject to the internal optimization and adaption process, they often have well–fitting values. On the other hand, this may also mean that the rates take values that are poorly suited. To counter this effect, in the mutation step, which follows the adaption of the mutation rates, the number λ of individuals generated by mutation, called offspring candidates, exceeds the population size of the parents. In a second selection process only the best offspring candidates are picked, eliminating ill–fitting mutation rates in the process.

The principle of memory schemes is to store useful information from the old fitness landscape and reuse it later in a new one. Therefore, a memory with the same representation as the population is set up that splits an extra storage space to explicitly store information from a current generation. This information is employed later, for instance by merging the stored individuals with the population at that time. This is known as direct memory [8, 53, 68]. The memory has the size μ_{mem} and is feeded by individuals selected for their high fitness. If a change in the landscape occurs, the stored individuals are inserted in the population, replacing individuals with low fitness. Since memory space is usually limited, we need to update the information stored in the memory. A general strategy is to select one memory space to be replaced by the best individual of the population.

A second example for a memory that additionally realizes some elements of predicting the dynamics of the landscape is the abstract memory scheme [50, 51]. The basic idea of abstract memory is that the solutions are not stored directly, that is as individuals representing points in search space, but as their abstraction. We understand as an abstraction of a good solution its approximate location in the search space, which is therefore partitioned with a grid of size ε_G. In an abstract storage process, μ_{stor} individuals with high fitness are taken and sorted according to the partition in the search space which they represent. Each individual sorted increases a counter belonging to that partition and indicates how often a good solution has occurred in exactly this subsection of the search space. In the abstract retrieval process, we fix a number of individuals to be inserted in the population by μ_{retr} and create these individuals randomly such that their statistical distribution regarding the partition matches that stored in the memory. Hence, abstract memory combines ideas from memory such as saving individuals for future re–insertion with attempts to predict the time evolution of the dynamic fitness landscape: storing the abstraction of good solutions, that is to use their approximate location in the search space, allows to deduce a probabilistic model for the spatial distribution of future solutions of the problem.

Table 13.2 Parameter of the tested evolutionary algorithms

Considered Scheme	Design parameter	Symbol	Value
All schemes	Population size	μ	50
	Initial population width	ω^2	5
Hyper–mutation	Base–mutation rate	bm	0.1
	Hyper–mutation rate	hm	30
	Mutation probability	p_m	0.25
Self–adaption	Offspring size	λ	50
	Initial mutation rates	mr_0	1.5
	Adaption rate	τ_A	1.25
Direct memory	Memory size	μ_{mem}	10
Abstract memory	Grid size	ε_G	1.0
	Individuals to memory	μ_{stor}	3
	Individuals from memory	μ_{retr}	20

The parameters used in the implementations that are the subjects of the numerical experiments reported in the next section, are summarized in Tab. 13.2.

13.5 Numerical Experiments

The performance of the algorithms is measured by the Mean Fitness Error (MFE), defined as below:

$$MFE = \frac{1}{R}\sum_{r=1}^{R}\left[\frac{1}{T}\sum_{\tau=1}^{T}\left(f(x_s(k),k) - \max_{x_j(\tau)\in P(\tau)} f(x_j(\tau),k)\right)\right]_{k=\lfloor\gamma^{-1}\tau\rfloor}, \tag{13.49}$$

where $\max_{x_j(\tau)\in P(\tau)} f\left(x_j(\tau),\lfloor\gamma^{-1}\tau\rfloor\right)$ is the fitness value of the best–in–generation individual $x_j(\tau)\in P(\tau)$ at generation τ, $f\left(x_S(\lfloor\gamma^{-1}\tau\rfloor),\lfloor\gamma^{-1}\tau\rfloor\right)$ is the maximum fitness value at generation τ, T is the number of generations used in the run, and R is the number of consecutive runs. We set $R = 150$ and $T = 1500$ in all experiments. Note that $f\left(x_s(\lfloor\gamma^{-1}\tau\rfloor),\lfloor\gamma^{-1}\tau\rfloor\right)$ and $\max_{x_j(\tau)\in P(\tau)} f\left(x_j(\tau),\lfloor\gamma^{-1}\tau\rfloor\right)$ change every γ generations according to Eq. (13.28). The MFE serves as a performance criterion and as behavior data for the evolutionary optimization in the dynamic fitness landscape (13.24). Fig. 13.9 show the MFE for the four considered evolutionary algorithm implementations hyper–mutation, self–adaption, direct and abstract memory and the landscape parameters $\alpha = 3.999$, $I_1 = I_2 = 16$, $s_1 = s_2 = 1$ and different values of ε. For these values the dynamic fitness landscape shows mostly spatio–temporal chaotic behavior, but also spatio–temporal periodic patterns. The figures give the MFE and its 95% confidence interval for the change frequencies $\gamma = 10$, $\gamma = 20$ and $\gamma = 30$. We observe that for most values of ε, the curves for each γ are distinct with the smallest γ leading to the largest MFE, and vice versa.

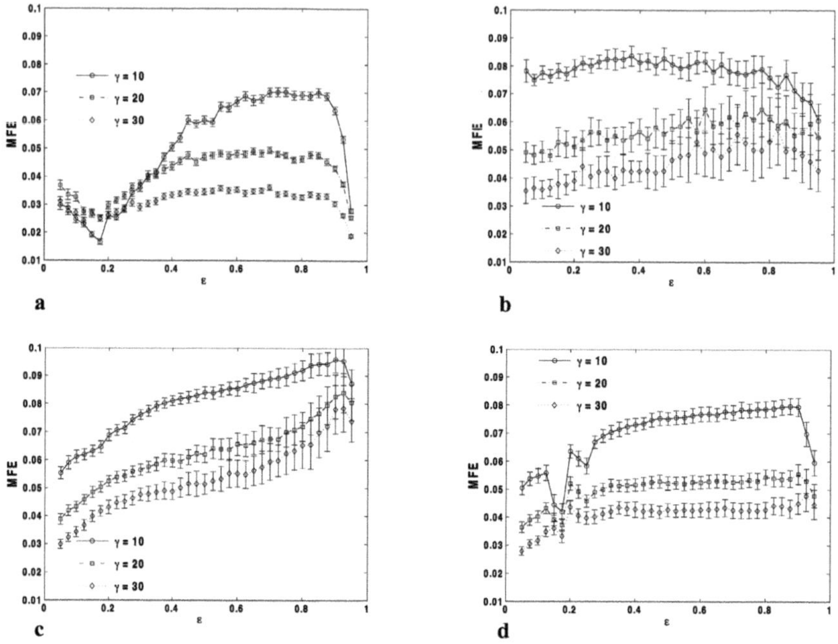

Fig. 13.9 Behavior of different implementations of the evolutionary algorithm expressed as the mean fitness error MFE (13.49) for the CML–based fitness landscape (13.24) with $I_1 = I_2 = 16$ and $\alpha = 3.999$: a) hyper-mutation, b) self–adaption, c) direct memory, d) abstract memory.

Looking the results, we see some differences between the four implementations. Although all results are for 150 runs, the confidence intervals that reflect to what degree the given mean value can be expected to be indeed the mean value if an infinite number of runs would have been carried out, are much higher for self–adaption and direct memory than for hyper–mutation and abstract memory. This means that for both implementations, for self–adaption even to a larger degree than for direct memory, some performance results are much better than the mean, but others are much worse. This result may possibly be explained for self–adaption by the fact that the mutation rates evolve towards optimal values after a change, but sometimes exactly these optimal values become unfavorable after the next change. As there is only the feedback via the individuals' fitness and the selection process (including potentially ill–posed mutation rates), it might take a certain time until optimal mutation rates are obtained again. This point of view seems to be confirmed by the observation that the confidence intervals are particularly large for small γ, that is for a landscape that changes fast. For direct memory, the effect might be that the stored individuals inserted after a change do not help in that particular new environment. The retrieved individuals might be favorable after another change, but not after the given one. Also, it is interesting to note that some implementations, particularly hyper–mutation, but to some extent also abstract memory, react on the presence of spatio–temporal periodic patterns at $0.1 \leq \varepsilon \leq 0.3$ with a drop in the MFE.

However, our interest here is neither to discuss possible performance enhancing alterations in the algorithms' parameters or implementation details or more generally the optimal design of the algorithms, nor to argue about which implementation is superior over another. To do so, the extent of the presented results is much to small, and also we find it more illuminating to study the underlying working principles of the algorithms. In other words, in theoretical studies the behavior of the evolutionary algorithm is sometimes much more interesting than the actual performance record. Therefore, we next approach the question of which implementation fulfils the expectations with respect to the landscape measures consider above. To get a metric of the strength of the relationship between landscape measures and the algorithms' behavior, we apply a parametric and a nonparametric correlation analysis, e.g. [52]. In particular, we study the Pearson product–moment correlation and the Kendall rank–order correlation. So, we run tests for relationships between these quantities, these are linear relation (Pearson) and piecewise linear relation (Kendall). With this, we intend to establish how reliable a linear or a piecewise linear relation between the landscape measures and the algorithms' behavior is. We would like to stress that this correlation analysis cannot imply any simple causation between landscape properties and the algorithm's behavior. As frequently in correlation analysis, to claim causation from observed correlation can be questionable or even misleading. Clearly, the landscape measures reflect different aspects of the landscape's problem difficulty, and this problem difficulty, in turn, must affect the flow of individuals in the landscape and hence the behavior of the evolutionary algorithm. However, each topological and dynamical measure emphasizes a specific aspect in problem hardness and there is no guarantee that the considered measures do not ignore properties that are important for details of the algorithms' behavior in exactly that landscape.

In the Figs. 13.10 and 13.11, the Pearson correlation coefficient ρ_P^2 (also known as "Pearson's r") and the Kendall correlation coefficient ρ_K^2 (also known as "Kendall's τ") are given. We write $\rho_P^2(MFE,\#_{LM})$ for the squared Pearson correlation between MFE and $\#_{LM}$, $\rho_K^2(MFE,\#_{LM})$ for the squared Kendall correlation and so on. The square of the correlation coefficients ρ_P^2 is also known as the coefficient of determination and can be interpreted as follows. The squared correlation coefficient represents the fraction of variance that is expressed by the fit between the MFE and the landscape measures. If the data from the landscape measures and the behavior data from the evolutionary algorithm were used in a statistical model, the quantity ρ_P^2 would be a metric of how well this model is able to predict further data. Hence, we view ρ_P^2 as a measure of reliability, strength and predictive power of the relationship between the landscape measures and the evolutionary algorithms' behavior. For the correlation coefficient ρ_K^2 a likewise interpretation is possible.

The results in the Figs. 13.10 and 13.11 show some clear trends. A first is that the topological landscape measures modality, ruggedness and information content have a stronger correlation with the behavior of the evolutionary algorithm expressed by

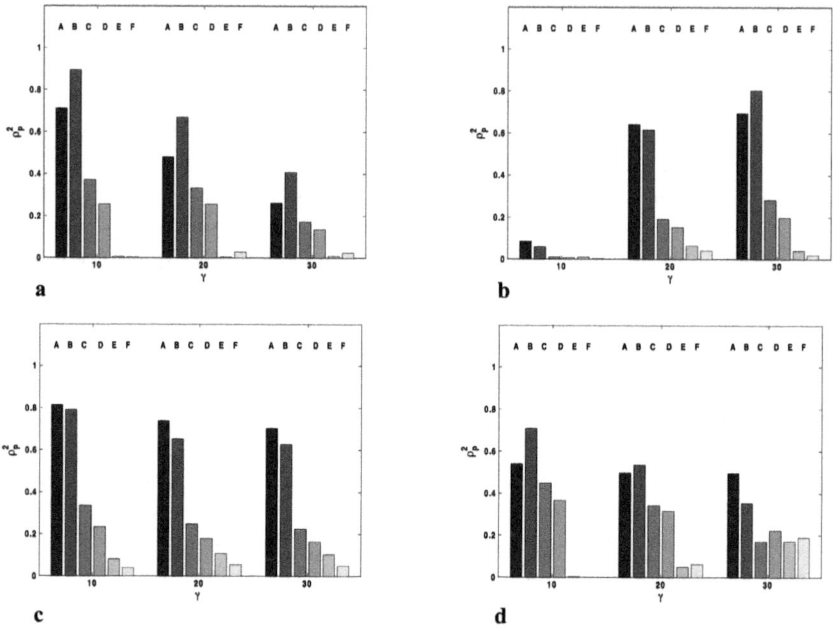

Fig. 13.10 Correlation analysis using Pearson product–moment correlation between topological and dynamical landscape measures and the behavior of the evolutionary algorithm expressed by the mean fitness error MFE for different implementations: a) hyper-mutation, b) self–adaption, c) direct memory, d) abstract memory. A - $\rho_P^2(MFE,\#_{LM})$, B - $\rho_P^2(MFE,\lambda_R)$, C - $\rho_P^2(MFE,H_{IC})$, D - $\rho_P^2(MFE,\sigma)$, E - $\rho_P^2(MFE,\lambda_L)$, F - $\rho_P^2(MFE,\psi_B)$.

the performance measure MFE, as compared to the dynamical landscape measures. A second trend is that the correlation of the topological measures decrease with increasing change frequency γ. A possible explanation is that change frequency seems to be the fundamental factor determining the MFE, see Fig. 13.9, where γ seems to set the level of the curves but not their form, which is done by ε and so by the other landscape measures. If γ gets larger, that is, the evolutionary algorithm has more generations to find the optimum, then the algorithms' performance is no longer so heavily influenced by topological problem hardness. The performance of an evolutionary algorithm having enough time to search for the maxima is weaker affected by the difficulties that are accounted for by the topological landscape measures. For the dynamical landscape measures severity, Lyapunov exponents and bred vector dimensions the correlations are generally much weaker but such a ceasing relationship for increasing γ is also not obtained. On the contrary, in some cases these measures seem to predict the algorithms' behavior even stronger for larger γ than for smaller ones. A comparison between the two types of correlation coefficients considered here yields rather equivocal results, although Pearson correlation ρ_P^2 seems to give slightly stronger indications, particularly for the topological landscape measures.

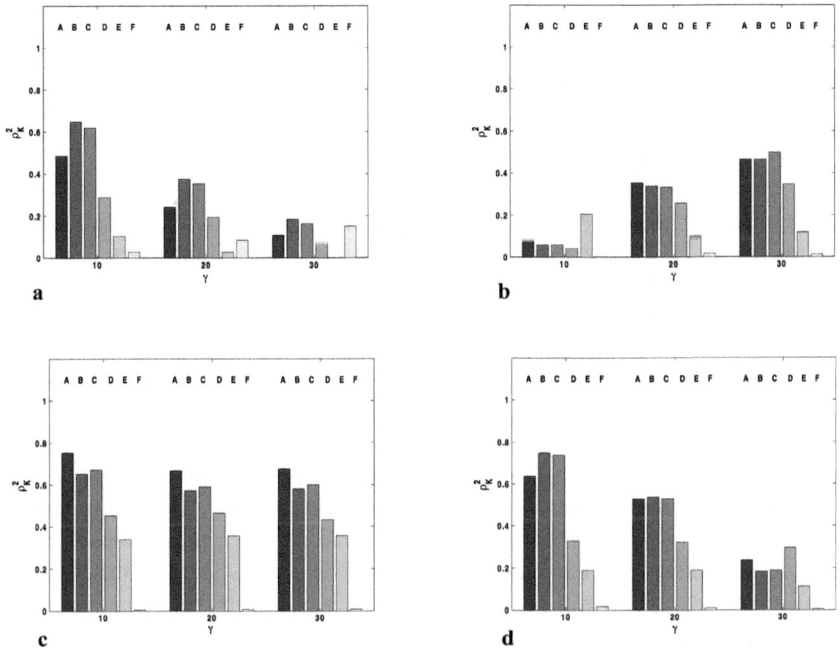

Fig. 13.11 Correlation analysis using Kendall rank–order correlation between topological and dynamical landscape measures and the behavior of the evolutionary algorithm expressed by the mean fitness error MFE for different implementations: a) hyper-mutation, b) self–adaption, c) direct memory, d) abstract memory. A - $\rho_K^2(MFE, \#_{LM})$, B - $\rho_K^2(MFE, \lambda_R)$, C - $\rho_K^2(MFE, H_{IC})$, D - $\rho_K^2(MFE, \sigma)$, E - $\rho_K^2(MFE, \lambda_L)$, F - $\rho_K^2(MFE, \psi_B)$.

So, the question if linear relationships (Pearson) are to prefer over piecewise linear (Kendall) is not definitely answerable.

A general exception to these rules are the results for self–adaption. Here, an implementation detail might offer some clues. Self–adaption is the only of the four implementations that does not directly and externally–triggered react on a landscape change. The other three implementations detect a change and immediately response with diversity enhancing actions such as hyper–mutation or inserting individuals from the memory. Self–adaption, on the other hand, relies upon the mutation rates to adjust themselves over a certain number of generations as the result of the self–adaption process. This might be a reason why the problem solving abilities of the self–adaption scheme, particularly for small γ, are less determined by the problem hardness accounted for by the landscape measures, while the actual performance results are comparable to the other implementations. For self–adaption this seems to mean that the population's diversity is high enough all the time. So, while the other three schemes experience jump–like increases in their diversity as a result of the actions carried out after a detected change, but also a rapidly deterioration of diversity afterwards, self–adaption includes a continuing and not dwindling diversity management that makes it more independent from the landscape properties.

13.6 Concluding Remarks

In this chapter of the book, we have considered the behavior of evolutionary algorithms in dynamic fitness landscapes that exhibit spatio–temporal chaos. These landscapes can be constructed from reaction–diffusion systems or from coupled map lattices (CML) and both kinds of description are related to each other. We have analyzed and quantified their properties using topological and dynamical landscape measures such as modality, ruggedness, information content, dynamic severity and two types of dynamic complexity measures, Lyapunov exponents and bred vector dimension. Four types of evolutionary algorithm implementations, hyper–mutation, self–adaption, and two kinds of memory schemes, direct and abstract memory, were studied and their performance in the spatio–temporal chaotic fitness landscapes was recorded. We used these performance data to relate the algorithms' behavior to the landscape measures using a correlation analysis. So, it was shown that the topological landscape measures correlate stronger with the algorithms' behavior, particularly for landscapes that change frequently. This correlation tends to cease for landscapes with a slower change pattern. Albeit dynamical landscape measures do show weaker correlations, they tend to remain preserved for varying change frequency.

As initially stated, a main point in a theoretical approach to evolutionary computation is to study how properties of the fitness landscape reflect, explain and allow to predict the behavior of the evolutionary algorithm, and vice versa. This question was posed here specifically for dynamical fitness landscapes and our hope is that the given approach might be useful as a starting point for a more general theory of dynamic fitness landscapes, which still is in its infancy. For further developing such a theory it might be helpful to go on taking inspiration from both the theory of static fitness landscapes and of spatially extended dynamical systems. We believe that only by bringing these fields together (which is a variation of the overall topic of the present book) substantial progress can be achieved.

References

1. Aoki, N., Hiraide, K.: Topological Theory of Dynamical Systems. North-Holland, Amsterdam (1994)
2. Arnold, D., Beyer, H.: Optimum tracking with evolution strategies. Evol. Comput. 14, 291–308 (2006)
3. Bäck, T.: Evolutionary Algorithms in Theory and Practice: Evolution Strategies. In: Evolutionary Programming, Genetic Algorithms. Oxford University Press, Oxford (1996)
4. Barreira, L., Pesin, Y.: Nonuniform Hyperbolicity: Dynamics of Systems with Nonzero Lyapunov Exponents. Cambridge University Press, Cambridge (2007)

5. Batterman, R., White, H.: Chaos and algorithmic complexity. Found. Phys. 26, 307–336 (1996)
6. Bosman, P.: Learning and anticipation in online dynamic optimization. In: Yang, S., Ong, Y., Jin, Y. (eds.) Evolutionary Computation in Dynamic and Uncertain Environments, pp. 129–152. Springer, Heidelberg (2007)
7. Boumaza, A.: Learning environment dynamics from self-adaptation. In: Yang, S., Branke, J. (eds.) GECCO Workshops 2005, pp. 48–54 (2005)
8. Branke, J.: Memory enhanced evolutionary algorithms for changing optimization problems. In: Angeline, P., Michalewicz, Z., Schoenauer, M., Yao, X., Zalzala, A. (eds.) Proc. Congress on Evolutionary Computation, IEEE CEC 1999, pp. 1875–1882. IEEE Press, Piscataway (1999)
9. Branke, J.: Evolutionary Optimization in Dynamic Environments. Kluwer Academic Publishers, Dordrecht (2001)
10. Branke, J., Kau, T., Schmidt, C., Schmeck, H.: A multi–population approach to dynamic optimization problems. In: Parmee, I. (ed.) Proc. of the 4th Int. Conf. on Adaptive Computing in Design and Manufacturing, pp. 299–308 (2000)
11. Bremen, H., Udwadia, F., Proskurowski, W.: An efficient method for the computation of Lyapunov numbers in dynamical systems. Physica D 101, 1–16 (1997)
12. Chazottes, J., Fernandez, B.: Dynamics of Coupled Map Lattices and of Related Spatially Extended Systems. Springer, Heidelberg (2005)
13. Chlouverakis, K., Sprott, J.: A comparison of correlation and Lyapunov dimensions. Physica D 200, 156–164 (2005)
14. Chrianti, A., Falconi, M., Mantic, G., Vulpiani, A.: Applying algorithmic complexity to define chaos in the motion of complex systems. Phys. Rev. E 50, 1959–1967 (1994)
15. Crutchfield, J., Kaneko, K.: Phenomenology of spatiotemporal chaos. In: Hao, B. (ed.) Directions in Chaos, vol. 1, pp. 272–353. World Scientific, Singapore (1987)
16. Eiben, A., Smith, J.: Introduction to Evolutionary Computing. Springer, Heidelberg (2003)
17. Francisco, G., Muruganandam, P.: Local dimension and finite time prediction in spatiotemporal chaotic systems. Phys. Rev. E 67, 066204–1–5 (2003)
18. Frederickson, P., Kaplan, P., Yorke, E., Yorke, J.: The Lyapunov dimension of strange attractors. J. Diff. Equations 49, 185–203 (1983)
19. Hordijk, W.: A measure of landscapes. Evol. Comput. 4, 335–360 (1996)
20. Hordijk, W.: Correlation analysis of the synchronizing–CA landscape. Physica D 107, 255–264 (1997)
21. Hordijk, W., Stadler, P.: Amplitude spectra of fitness landscapes. Adv. Complex Systems 1, 39–66 (1998)
22. Hordijk, W., Kauffman, S.: Correlation analysis of coupled fitness landscapes. Complexity 10, 42–49 (2005)
23. Horn, J., Goldberg, D., Deb, K.: Long path problems. In: Davidor, Y., Männer, R., Schwefel, H.-P. (eds.) PPSN 1994. LNCS, vol. 866, pp. 149–158. Springer, Heidelberg (1994)
24. Jin, Y., Branke, J.: Evolutionary optimization in uncertain environments – A survey. IEEE Trans. Evol. Comput. 9, 303–317 (2005)
25. Kaneko, K.: The coupled map lattice. In: Kaneko, K. (ed.) Theory and Application of Coupled Map Lattices, pp. 1–49. John Wiley, Chichester (1993)
26. Kaneko, K., Tsuda, I.: Complex Systems: Chaos and Beyond. Springer, Heidelberg (2001)

27. Kallel, L., Naudts, B., Reeves, C.: Properties of fitness functions and search landscapes. In: Kallel, L., Naudts, B., Rogers, A. (eds.) Theoretical Aspects of Evolutionary Computing, pp. 177–208. Springer, Heidelberg (2001)
28. Kantz, H., Schreiber, T.: Nonlinear Time Series Analysis. Cambridge University Press, Cambridge (1999)
29. Kardar, M., Parisi, G., Zhang, Y.: Dynamic scaling of growing interfaces. Phys. Rev. Lett. 56, 889–892 (1986)
30. Katok, A., Hasselblatt, B.: Introduction to the Modern Theory of Dynamical Systems. Cambridge University Press, Cambridge (1995)
31. Katzav, E., Cugliandolo, L.: From coupled map lattices to the stochastic Kardar–Parisi–Zhang equation. Physica A 371, 96–99 (2006)
32. Kauffman, S.: The Origin of Order: Self–Organization and Selection in Evolution. Oxford University Press, Oxford (1993)
33. Ma, K., Jianga, J., Yanga, C.: Scaling behavior of roughness in the two-dimensional Kardar–Parisi–Zhang growth. Physica A 378, 194–200 (2007)
34. Meyer, S., Nieberg, B.H.: Self–adaptation in evolutionary algorithms. In: Lobo, F., Lima, C., Michalewicz, Z. (eds.) Parameter Setting in Evolutionary Algorithm, pp. 47–75. Springer, Heidelberg (2007)
35. Michalewicz, Z.: Genetic Algorithms + Data Structures = Evolution Programs. Springer, Heidelberg (1996)
36. Miranda, V., Aarão Reis, F.: Numerical study of the Kardar–Parisi–Zhang equation. Phys. Rev. E 77, 031134–1–6 (2008)
37. Morrison, R.: Designing Evolutionary Algorithms for Dynamic Environments. Springer, Heidelberg (2004)
38. Morrison, R., De Jong, K.: A test problem generator for non–stationary environments. In: Angeline, P., Michalewicz, Z., Schoenauer, M., Yao, X., Zalzala, A. (eds.) Proc. Congress on Evolutionary Computation, IEEE CEC 1999, pp. 2047–2053. IEEE Press, Piscataway (1999)
39. Morrison, R., De Jong, K.: Triggered hypermutation revisited. In: Zalzala, A., Fonseca, C., Kim, J., Smith, A., Yao, X. (eds.) Proc. Congress on Evolutionary Computation, IEEE CEC 2000, pp. 1025–1032. IEEE Press, Piscataway (2000)
40. Ott, E.: Chaos in Dynamical Systems. Cambridge University Press, Cambridge (1993)
41. Patil, D., Hunt, B., Kalnay, E., Yorke, J., Ott, E.: Local low dimensionality of atmospheric dynamics. Phys. Rev. Lett. 86, 5878–5881 (2001)
42. Pesin, Y., Yurchenko, A.: Some physical models of the reaction–diffusion equation and coupled map lattices. Russ. Math. Surv. 59, 481–513 (2004)
43. Primo, C., Szendro, I., Rodríguez, M., López, J.: Dynamic scaling of bred vectors in spatially extended chaotic systems. Europhys. Lett. 765, 767–773 (2006)
44. Richter, H.: Behavior of evolutionary algorithms in chaotically changing fitness landscapes. In: Yao, X., Burke, E.K., Lozano, J.A., Smith, J., Merelo-Guervós, J.J., Bullinaria, J.A., Rowe, J.E., Tiňo, P., Kabán, A., Schwefel, H.-P. (eds.) PPSN 2004. LNCS, vol. 3242, pp. 111–120. Springer, Heidelberg (2004)
45. Richter, H.: A study of dynamic severity in chaotic fitness landscapes. In: Corne, D. (ed.) Proc. Congress on Evolutionary Computation, IEEE CEC 2005, pp. 2824–2831. IEEE Press, Piscataway (2005)
46. Richter, H.: Evolutionary optimization in spatio–temporal fitness landscapes. In: Runarsson, T.P., Beyer, H.-G., Burke, E.K., Merelo-Guervós, J.J., Whitley, L.D., Yao, X. (eds.) PPSN 2006. LNCS, vol. 4193, pp. 1–10. Springer, Heidelberg (2006)

47. Richter, H.: Coupled map lattices as spatio–temporal fitness functions: Landscape measures and evolutionary optimization. Physica D 237, 167–186 (2008)
48. Richter, H.: Detecting change in dynamic fitness landscapes. In: Tyrrell, A. (ed.) Proc. Congress on Evolutionary Computation, IEEE CEC 2009, pp. 1613–1620. IEEE Press, Piscataway (2009)
49. Richter, H.: Can a polynomial interpolation improve on the Kaplan-Yorke dimension? Phys. Lett. A 372, 4689–4693 (2008)
50. Richter, H., Yang, S.: Memory based on abstraction for dynamic fitness functions. In: Giacobini, M., Brabazon, A., Cagnoni, S., Di Caro, G.A., Drechsler, R., Ekárt, A., Esparcia-Alcázar, A.I., Farooq, M., Fink, A., McCormack, J., O'Neill, M., Romero, J., Rothlauf, F., Squillero, G., Uyar, A.Ş., Yang, S., et al. (eds.) EvoWorkshops 2008. LNCS, vol. 4974, pp. 597–606. Springer, Heidelberg (2008)
51. Richter, H., Yang, S.: Learning behavior in abstract memory schemes for dynamic optimization problems. Soft Computing 13, 1163–1173 (2009)
52. Sheskin, D.: Handbook of Parametric and Nonparametric Statistical Procedures. CRC Press, Boca Raton (1997)
53. Simões, A., Costa, E.: Variable-size memory evolutionary algorithm to deal with dynamic environments. In: Giacobini, M., et al. (eds.) EvoWorkshops 2007. LNCS, vol. 4448, pp. 617–626. Springer, Heidelberg (2007)
54. Simões, A., Costa, E.: Evolutionary algorithms for dynamic environments: Prediction using linear regression and Markov chains. In: Rudolph, G., Jansen, T., Lucas, S., Poloni, C., Beume, N. (eds.) PPSN 2008. LNCS, vol. 5199, pp. 306–315. Springer, Heidelberg (2008)
55. Smith, T., Husbands, P., Layzell, P., O'Shea, M.: Fitness landscapes and evolvability. Evol. Comput. 10, 1–34 (2002)
56. Stadler, P.: Landscapes and their correlation functions. J. Math. Chem. 20, 1–45 (1996)
57. Stadler, P., Stephens, C.: Landscapes and effective fitness. Comm. Theor. Biol. 8, 389–431 (2003)
58. Teo, J., Abbass, H.: An information–theoretic landscape analysis of neuro–controlled embodied organisms. Neural Comput. Appl. 13, 80–89 (2004)
59. Tereshko, V.: Selection and coexistence by reaction–diffusion dynamics in fitness landscapes. Phys. Lett. A 260, 522–527 (1999)
60. Tinós, R., Yang, S.: A self–organizing random immigrants genetic algorithm for dynamic optimization problems. Genet. Program Evol. Mach. 8, 255–286 (2007)
61. Vassilev, V.: Information analysis of fitness landscapes. In: Husbands, P., Harvey, I. (eds.) Proc. Fourth European Conference on Artificial Life, pp. 116–124. MIT Press, Cambridge (1997)
62. Vassilev, V., Fogarty, T., Miller, J.: Information characteristics and the structure of landscapes. Evol. Comput. 8, 31–60 (2000)
63. Weicker, K.: An analysis of dynamic severity and population size. In: Schoenauer, M., Deb, K., Rudolph, G., Yao, X., Lutton, E., Merelo Guervós, J., Schwefel, H. (eds.) Parallel Problem Solving from Nature–PPSN VI, pp. 159–168. Springer, Heidelberg (2000)
64. Weinberger, E.: Correlated and uncorrelated fitness landscapes and how to tell the difference. Biol. Cybern. 63, 325–336 (1990)
65. Wright, S.: The roles of mutation, inbreeding, crossbreeding and selection in evolution. In: Jones, D. (ed.) Proc. of the Sixth International Congress on Genetics, pp. 356–366 (1932)

66. Xie, F., Hu, G.: Spatio–temporal periodic and chaotic pattern in a two–dimensional coupled map lattice system. Phys. Rev. E 55, 79–86 (1997)
67. Yang, S.: Non–stationary problem optimization using the primal-dual genetic algorithm. In: Sarker, R., Reynolds, R., Abbass, H., Tan, K., Essam, D., McKay, R., Gedeon, T. (eds.) Proc. Congress on Evolutionary Computation, IEEE CEC 2003, pp. 2246–2253. IEEE Press, Piscataway (2003)
68. Yang, S.: Associative memory scheme for genetic algorithms in dynamic environments. In: Rothlauf, F., Branke, J., Cagnoni, S., Costa, E., Cotta, C., Drechsler, R., Lutton, E., Machado, P., Moore, J.H., Romero, J., Smith, G.D., Squillero, G., Takagi, H., et al. (eds.) EvoWorkshops 2006. LNCS, vol. 3907, pp. 788–799. Springer, Heidelberg (2006)
69. Yang, S., Yao, X.: Experimental study on population-based incremental learning algorithms for dynamic optimization problems. Soft Computing 9, 815–834 (2005)

Chapter 14
Controller Parameters Optimization on a Representative Set of Systems Using Deterministic-Chaotic-Mutation Evolutionary Algorithms

Donald Davendra and Ivan Zelinka

Abstract. The application of Differential Evolution (DE) to the task of PID controller optimization is explored in this chapter. DE is applied in two variants; canonical and chaos mutated. DE canonical version uses a random number generator where as the chaos mutated version uses chaotic maps as the mutation generator. These two variants are applied to three different systems in order to gauge their effectiveness. The results present two main points. The first is the effectiveness of DE over tuning algorithms and other metaheuristics. The second is the competitiveness and effectiveness of embedding chaotic systems in DE.

14.1 Introduction

The PID controller is the most common form of feedback. It was an essential element of early governors and it became the standard tool when process control emerged in the 1940s. In process control today, more than 95% of the control loops are of PID type, most loops are actually PI control. PID controllers are today found in all areas where control is used. The controllers come in many different forms.

Donald Davendra
Tomas Bata University in Zlin, Faculty of Applied Informatics, Nad Stranemi 4511, Zlin 76001, Czech Republic
e-mail: davendra@fai.utb.cz

Ivan Zelinka
Tomas Bata University in Zlin, Faculty of Applied Informatics, Nad Stranemi 4511, Zlin 76001, Czech Republic
and
VSB-TUO, Faculty of Electrical Engineering and Computer Science, 17. listopadu 15, 708 33 Ostrava-Poruba, Czech Republic
e-mail: zelinka@fai.utb.cz

I. Zelinka et al. (Eds.): Evolutionary Algorithms and Chaotic Systems, SCI 267, pp. 447–480.
springerlink.com

There are stand-alone systems in boxes for one or a few loops, which are manufactured by the hundred thousands yearly. PID control is an important ingredient of a distributed control system. The controllers are also embedded in many special-purpose control systems. PID control is often combined with logic, sequential functions, selectors, and simple function blocks to build the complicated automation systems used for energy production, transportation, and manufacturing. Many sophisticated control strategies, such as model predictive control, are also organized hierarchically. PID control is used at the lowest level; the multivariable controller gives the set-points to the controllers at the lower level. The PID controller can thus be said to be the "bread and butter" of control engineering. It is an important component in every control engineers tool box [1].

PID controllers have survived many changes in technology, from mechanics and pneumatics to microprocessors via electronic tubes, transistors, integrated circuits. The microprocessor has had a dramatic influence on the PID controller. Practically all PID controllers made today are based on microprocessors. This has given opportunities to provide additional features like automatic tuning, gain scheduling, continuous adaptation and more recently evolutionary tuning [6].

One of the recent approaches of the tuning of these controllers is in the usage of evolutionary algorithms (EA's). This chapter discusses two such unique techniques.

The first is the use of Differential Evolution (DE) for PID tuning. DE is a global optimizer which uses a stochastic random search strategy.

The second technique is the embedding of chaotic maps inside DE to act as mutation value generators in order to have a deterministic search pattern in the solution space. This strategy is implemented in order to compare the effectiveness of random and chaotic search systems.

This chapter is divided into the following sections. Section 14.2 introduces the PID controller. Section 14.3 discusses the various controller tuning parameters whereas section 14.4 gives a background on the system specifications of PID controller. DE is introduced in section 14.5 and the Chaotic Systems in section 14.6. The experimentations conducted to gauge the effectiveness of these approaches is given in section 14.7. Section 14.8 concludes the chapter.

14.2 PID Controller

The PID controller contains three unique parts; proportional, integral and derivative controller [1]. The following sections describe the different components.

14.2.1 Proportional Algorithm

The mathematical representation is given as,

$$\frac{mv(s)}{e(s)} = k_c \tag{14.1}$$

in the Laplace domain or as,

$$mv(t) = mv_{ss} + k_c e(t) \tag{14.2}$$

in the time domain.

The proportional mode adjusts the output signal in direct proportion to the controller input (which is the error signal, **e**). The adjustable parameter to be specified is the controller gain, k_c. The larger the k_c, the more the controller output will change for a given error [1].

The time domain expression also indicates that the controller requires calibration around the steady-state operating point. This is indicated by the constant term *mvss*. This represents the 'steady-state' signal for the *mv* and is used to ensure that at zero error, the *cv* is at **setpoint**. In the Laplace domain this term disappears, because of the deviation 'variable' representation [1].

A proportional controller reduces error but does not eliminate it (unless the process has naturally integrating properties), i.e. an offset between the actual and desired value will normally exist [1].

14.2.2 Proportional Integral Algorithm

The mathematical representation is given as,

$$\frac{mv(s)}{e(s)} = k_c \left[1 + \frac{1}{T_i s}\right] \tag{14.3}$$

in the Laplace domain or as,

$$mv(t) = mv_{ss} + k_c \left[e(t) + \frac{1}{T_i}\int e(t)\,dt\right] \tag{14.4}$$

in the time domain.

The additional integral mode corrects for any offset (error) that may occur between the desired value (setpoint) and the process output automatically over time. The adjustable parameter to be specified is the integral time (T_i) of the controller [1].

14.2.3 Proportional Integral Derivative Algorithm

The mathematical representation is given as,

$$\frac{mv(s)}{e(s)} = k_c \left[1 + \frac{1}{T_i s} + T_D s\right] \tag{14.5}$$

in the Laplace domain or as,

$$mv(t) = mv_{ss} + k_c \left[e(t) + \frac{1}{T_i}\int e(t)\,dt + T_D \frac{de(t)}{dt}\right] \tag{14.6}$$

Derivative action anticipates where the process is heading, by looking at the time rate of change of the controlled variable (its derivative). T_D is the 'rate time' and this characterizes the derivative action (with units of minutes). In theory, derivative action should always improve dynamic response and it does in many loops. In others, however, the problem of noisy signals makes the use of derivative action undesirable (differentiating noisy signals can translate into excessive *mv* movement).

Derivative action depends on the slope of the error, unlike P and I. If the error is constant, derivative action has no effect [1].

The parallel PID controller is given as in Equation 14.7.

$$mv(s) = k_c e(s) + \frac{1}{T_i s} e(s) + T_D s e(s) \tag{14.7}$$

A further alternative simplified form is given in Equation 14.8.

$$G(s) = K\left(1 + \frac{1}{sT_i} + sT_d\right) \tag{14.8}$$

The PID form most suitable for analytical calculations is given in Equation 14.9.

$$G(s) = k + \frac{k_i}{s} + sk_d \tag{14.9}$$

The parameters are related to the standard form through: $k = K$, $k_i = \frac{K}{T_i}$ and $k_d = KT_d$.

The advantage is that the parameters appear linearly and it is possible to obtain pure proportional, integral, or derivative action by finite values of the parameters.

The overall PID controller is given in Fig.14.1.

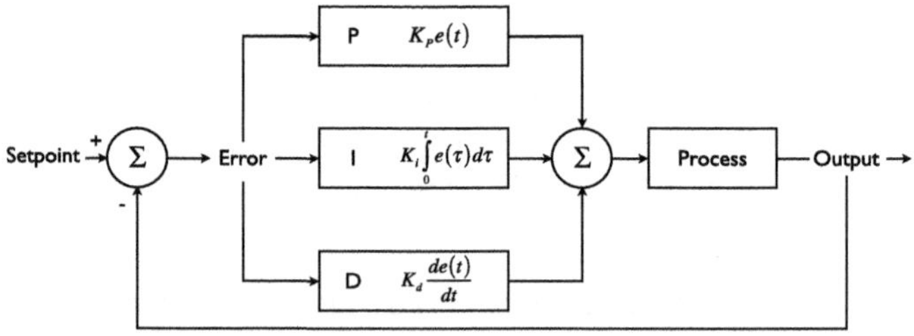

Fig. 14.1 PID Controller

14.3 Controller Tuning

Controller tuning involves the selection of the best values of k_c, T_i and T_D. This is often a subjective procedure and is certainly process dependent. When tuning a PID algorithm, generally the aim is to match some preconceived "ideal" response profile for the closed loop system. The following response profiles are typical [6].

Overshoot: this is the magnitude by which the controlled "variable swings" past the setpoint. 5/10% overshoot is normally acceptable for most loops.
Rise time: the time it takes for the process output to achieve the new desired value. One-third the dominant process time constant would be typical.
Decay ratio: this is the ratio of the maximum amplitude of successive oscillations.
Settling time: the time it takes for the process output to die to between, say +/- 5% of setpoint.

14.3.1 Ziegler Nichols Closed Loop Method

The Ziegler Nichols method is straightforward. First, set the controller to P mode only. Next, set the gain of the controller (k_c) to a small value. Make a small setpoint (or load) change and observe the response of the controlled variable. If k_c is low, the response should be sluggish. Increase k_c by a factor of two and make another small change in the setpoint or the load. Keep increasing k_c (by a factor of two) until the response becomes oscillatory. Finally, adjust k_c until a response is obtained that produces continuous oscillations. This is known as the ultimate gain (k_u). Note, the period of the oscillations is P_u. The control law settings are then obtained as given in Table 14.1.

Table 14.1 Ziegler Nichols Tuning

	k_c	T_i	T_D
P	$\frac{k_u}{2}$		
PI	$\frac{k_u}{2.2}$	$\frac{P_u}{2.2}$	
PID	$\frac{k_u}{1.7}$	$\frac{P_u}{2}$	$\frac{P_u}{8}$

It is unwise to force the system into a situation where there are continuous oscillations as this represents the limit at which the feedback system is stable.

Generally, it is a good idea to stop at the point where some oscillation has been obtained. It is then possible to approximate the period P_u and if the gain at this point is taken as the ultimate gain k_u, then this will provide a more conservative tuning regime.

14.4 System Specifications

14.4.1 Sensitivity Specifications

In order to measure the system, it becomes imperative to first discuss the specifications of the system [1]. Consider a system given in Fig 14.2.

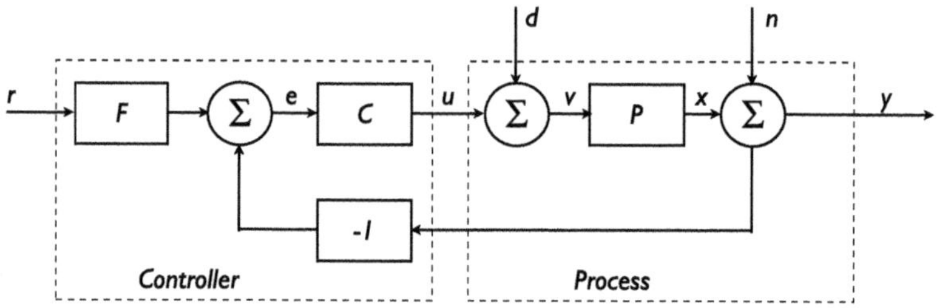

Fig. 14.2 Block diagram of basic feedback loop.

A block diagram of a basic feedback loop is shown in Fig 14.2. The system loop is composed of two components, the process P and the controller. The controller has two blocks, the feedback block C and the feedforward block F. There are two disturbances acting on the process, the load disturbance d and the measurement noise n. The load disturbance represents disturbances that drive the process away from its desired behavior. The process variable x is the real physical variable that we want to control. Control is based on the measured signal y, where the measurements are corrupted by measurement noise n. Information about the process variable x is thus distorted by the measurement noise. The process is influenced by the controller via the control variable u. The process is thus a system with three inputs and one output. The inputs are: the control variable u, the load disturbance d and the measurement noise n. The output is the measured signal. The controller is a system with two inputs and one output. The inputs are the measured signal y and the reference signal r and the output is the control signal u. Note that the control signal u is an input to the process and the output of the controller and that the measured signal is the output of the process and an input to the controller.

Taking Laplace transforms and simplifying the systems in terms of input and output, the following gang of four characteristics emerge when the system has pure error of F=1 [1].

$$\frac{PC}{1+PC} \tag{14.10}$$

$$\frac{P}{1+PC} \tag{14.11}$$

$$\frac{C}{1+PC} \tag{14.12}$$

$$\frac{1}{1+PC} \tag{14.13}$$

Equation 14.10 is the *complementary sensitivity function*, Equation 14.11 is the *load disturbance sensitivity function*, Equation 14.12 is the *noise sensitivity function* and Equation 14.13 is the *sensitivity function*. In this chapter, Equation 14.10 and 14.13 are used as system measure.

14.4.2 Optimization Specifications

The properties of the transfer functions can also be based on integral criteria [3]. Let $e(t)$ be the error caused by reference values or disturbances and let $u(t)$ be the corresponding control signal. The performance index is calculated over a time interval; T, normally in the region of $0 \le T \le t_s$ where t_s is the settling time of the system. The following performance indices are of note:

14.4.2.1 Integral of Time Multiplied by Absolute Error (ITAE)

$$I_{ITAE} = \int_0^T t\left|e\left(t\right)\right|dt \tag{14.14}$$

The ITAE weights the error with time and hence emphasizes the error values later on in the response rather than the initial large errors.

14.4.2.2 Integral of Absolute Magnitude of the Error (IAE)

$$I_{IAE} = \int_0^T \left|e\left(t\right)\right|dt \tag{14.15}$$

IAE gets the absolute value of the error to remove negative error components. IAE is good for simulation studies.

14.4.2.3 Integral of the Square of the Error (ISE)

$$I_{ISE} = \int_0^T e^2\left(t\right)dt \tag{14.16}$$

The ISE squares the error to remove negative error components. ISE discriminates between over-damped and under damped systems, i.e. a compromise minimizes the ISE.

14.4.2.4 Mean of the Square of the Error (MSE)

$$I_{MSE} = \frac{1}{n}\sum_{i=1}^{n}(e(t))^2 \tag{14.17}$$

MSE reflects all variation and deviation from the target value.

14.5 Differential Evolution Algorithm

Differential evolution (DE) is one of the evolutionary optimization methods proposed by Price [8] to solve the Chebychev polynomial fitting problem. DE is a population-based and stochastic global optimizer, and has proven to be a robust technique for global optimization.

In order to describe DE, a schematic is given in Fig 14.3.

$$\begin{array}{l}
1.\text{Input}: D, G_{\max}, NP \geq 4, F \in (0, 1+), CR \in [0,1], \text{and initial bounds}: x^{(lo)}, x^{(hi)}. \\
2.\text{Initialize}: \left\{ \begin{array}{l} \forall i \leq NP \wedge \forall j \leq D : x_{i,j,G=0} = x_j^{(lo)} + rand_j[0,1] \bullet \left(x_j^{(hi)} - x_j^{(lo)}\right) \\ i = \{1,2,...,NP\}, j = \{1,2,...,D\}, G = 0, rand_j[0,1] \in [0,1] \end{array} \right. \\
\left\{ \begin{array}{l}
3.\text{While} \quad G < G_{\max} \\
\forall i \leq NP \left\{ \begin{array}{l}
4. \quad \text{Mutate and recombine}: \\
4.1 \quad r_1, r_2, r_3 \in \{1,2,....,NP\}, \\
\quad \text{randomly selected, except}: r_1 \neq r_2 \neq r_3 \neq i \\
4.2 \quad j_{rand} \in \{1,2,...,D\}, \text{randomly selected once each } i \\
4.3 \quad \forall j \leq D, u_{j,i,G+1} = \left\{ \begin{array}{l} x_{j,r_3,G} + F \cdot (x_{j,r_1,G} - x_{j,r_2,G}) \\ \text{if} \quad (rand_j[0,1] < CR \vee j = j_{rand}) \\ x_{j,i,G} \quad \text{otherwise} \end{array} \right. \\
5. \quad \text{Select} \\
x_{i,G+1} = \left\{ \begin{array}{l} u_{i,G+1} \text{ if } \quad f(u_{i,G+1}) \leq f(x_{i,G}) \\ x_{i,G} \quad \text{otherwise} \end{array} \right.
\end{array} \right. \\
G = G + 1
\end{array} \right.
\end{array}$$

Fig. 14.3 Canonical Differential Evolution Algorithm

There are essentially five sections to the code. Section 1 describes the input to the heuristic. D is the size of the problem, $G_{\max}$ is the maximum number of generations, NP is the total number of solutions, F is the scaling factor of the solution and CR is the factor for crossover. F and CR together make the internal tuning parameters for the heuristic.

Section 2 outlines the initialization of the heuristic. Each solution $x_{i,j,G=0}$ is created randomly between the two bounds $x^{(lo)}$ and $x^{(hi)}$. The parameter j represents the index to the values within the solution and i indexes the solutions within the population. So, to illustrate, $x_{4,2,0}$ represents the second value of the fourth solution at the initial generation.

After initialization, the population is subjected to repeated iterations in section 3.

Section 4 describes the conversion routines of DE. Initially, three random numbers r_1, r_2, r_3 are selected, unique to each other and to the current indexed solution i in the population in 4.1. Henceforth, a new index j_{rand} is selected in the solution. j_{rand} points to the value being modified in the solution as given in 4.2. In 4.3, two solutions, $x_{j,r_1,G}$ and $x_{j,r_2,G}$ are selected through the index r_1 and r_2 and their values subtracted. This value is then multiplied by F, the predefined scaling factor. This is added to the value indexed by r_3.

However, this solution is not arbitrarily accepted in the solution. A new random number is generated, and if this random number is less than the value of CR, then the new value replaces the old value in the current solution. Once all the values in the solution are obtained, the new solution is vetted for its fitness or value and if this improves on the value of the previous solution, the new solution replaces the previous solution in the population. Hence the competition is only between the new *child* solution and its *parent* solution.

Price [8] has suggested ten different working strategies. It mainly depends on the problem on hand for which strategy to choose. The strategies vary on the solutions to be perturbed, number of differing solutions considered for perturbation, and finally the type of crossover used. The following are the different strategies being applied.

Strategy 1: DE/best/1/exp: $u_{i,G+1} = x_{best,G} + F \bullet (x_{r_1,G} - x_{r_2,G})$
Strategy 2: DE/rand/1/exp: $u_{i,G+1} = x_{r_1,G} + F \bullet (x_{r_2,G} - x_{r_3,G})$
Strategy 3: DE/rand−best/1/exp: $u_{i,G+1} = x_{i,G} + \lambda \bullet (x_{best,G} - x_{r_1,G}) + F \bullet (x_{r_1,G} - x_{r_2,G})$
Strategy 4: DE/best/2/exp: $u_{i,G+1} = x_{best,G} + F \bullet (x_{r_1,G} - x_{r_2,G} - x_{r_3,G} - x_{r_4,G})$
Strategy 5: DE/rand/2/exp: $u_{i,G+1} = x_{5,G} + F \bullet (x_{r_1,G} - x_{r_2,G} - x_{r_3,G} - x_{r_4,G})$
Strategy 6: DE/best/1/bin: $u_{i,G+1} = x_{best,G} + F \bullet (x_{r_1,G} - x_{r_2,G})$
Strategy 7: DE/rand/1/bin: $u_{i,G+1} = x_{r_1,G} + F \bullet (x_{r_2,G} - x_{r_3,G})$
Strategy 8: DE/rand−best/1/bin: $u_{i,G+1} = x_{i,G} + \lambda \bullet (x_{best,G} - x_{r_1,G}) + F \bullet (x_{r_1,G} - x_{r_2,G})$
Strategy 9: DE/best/2/bin: $u_{i,G+1} = x_{best,G} + F \bullet (x_{r_1,G} - x_{r_2,G} - x_{r_3,G} - x_{r_4,G})$
Strategy 10: DE/rand/2/bin: $u_{i,G+1} = x_{5,G} + F \bullet (x_{r_1,G} - x_{r_2,G} - x_{r_3,G} - x_{r_4,G})$

The convention shown is DE/x/y/z. DE stands for Differential Evolution, x represents a string denoting the solution to be perturbed, y is the number of difference solutions considered for perturbation of x, and z is the type of crossover being used (exp: exponential; bin: binomial).

DE has two main phases of crossover: binomial and exponential. Generally, a child solution $u_{i,G+1}$ is either taken from the parent solution $x_{i,G}$ or from a mutated donor solution $v_{i,G+1}$ as shown : $u_{j,i,G+1} = v_{j,i,G+1} = x_{j,r_3,G} + F \bullet (x_{j,r_1,G} - x_{j,r_2,G})$.

The frequency with which the donor solution $v_{i,G+1}$ is chosen over the parent solution $x_{i,G}$ as the source of the child solution is controlled by both phases of crossover. This is achieved through a user defined constant, crossover CR which is held constant throughout the execution of the heuristic.

The *binomial* scheme takes parameters from the donor solution every time that the generated random number is less than the CR as given by $rand_j[0,1] < CR$, else all parameters come from the parent solution $x_{i,G}$.

The *exponential* scheme takes the child solutions from $x_{i,G}$ until the first time that the random number is greater than CR, as given by $rand_j[0,1] < CR$, otherwise the parameters comes from the parent solution $x_{i,G}$.

To ensure that each child solution differs from the parent solution, both the exponential and binomial schemes take at least one value from the mutated donor solution $v_{i,G+1}$.

14.5.1 Tuning Parameters

Outlining an absolute value for CR is difficult. It is largely problem dependent. However a few guidelines have been laid down [8]. When using binomial scheme, intermediate values of CR produce good results. If the objective function is known to be separable, then $CR = 0$ in conjunction with binomial scheme is recommended. The recommended value of CR should be close to or equal to 1, since the possibility or crossover occurring is high. The higher the value of CR, the greater the possibility of the random number generated being less than the value of CR, and thus initiating the crossover.

The general description of F is that it should be at least above 0.5, in order to provide sufficient scaling of the produced value.

The tuning parameters and their guidelines are given in Table 14.2.

Table 14.2 Guide to choosing best initial control variables

Control Variables	Lo	Hi	Best?	Comments
F: Scaling Factor	0	1.0+	0.3 – 0.9	$F \geq 0.5$
CR: Crossover probability	0	1	0.8 – 1.0	CR = 0, separable CR = 1, epistatic

14.6 Chaotic Systems

Chaos theory has its manifestation in the study of dynamical systems that exhibit certain behavior due to the perturbation of the initial conditions of the systems. A number of such systems have been discovered and this branch of mathematics has been vigorously researched for the last few decades.

The area of interest for this chapter is the embedding of chaotic systems in the form of *chaos number generator* for an evolutionary algorithm.

The systems of interest are *discrete dissipative* systems. The two common systems of Lozi map and Delayed Logistic (DL) were selected as mutation generators for the DE heuristic.

14.6.1 Lozi Map

The Lozi map is a two-dimensional piecewise linear map whose dynamics are similar to those of the better known Henon map [4] and it admits strange attractors.

The advantage of the Lozi map is that one can compute every relevant parameter exactly, due to the linearity of the map, and the successful control can be demonstrated rigorously.

The Lozi map equations are given in Equations 14.18 and 14.19.

$$y_1(t+1) = 1 - a|y_1(t)| + y_2(t) \tag{14.18}$$

$$y_2(t+1) = by_1(t) \tag{14.19}$$

The parameters used in this work are $a = 1.7$ and $b = 0.5$ as suggested in [2]. The Lozi map is given in Figure 14.4.

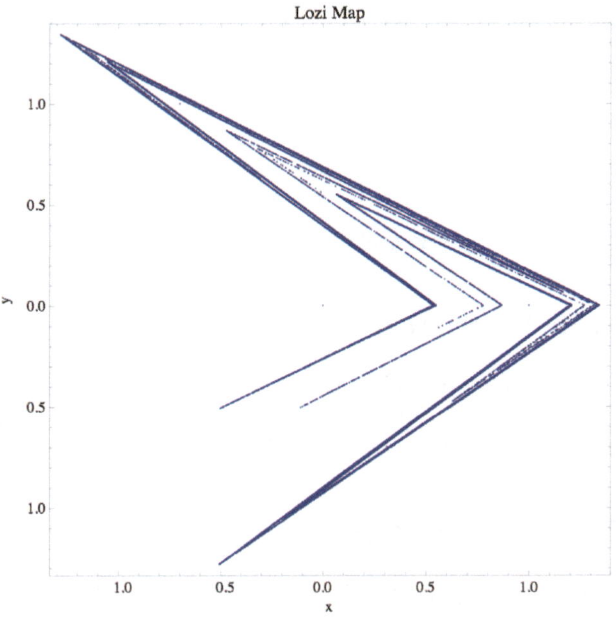

Fig. 14.4 Lozi map

14.6.2 Delayed Logistic Map

The Delayed Logistic (DL) map equations are given in Equations 14.20 and 14.21.

$$y_1(t+1) = y_2 \tag{14.20}$$

$$y_2(t+1) = ay_2(1 - y_1) \tag{14.21}$$

The parameters used in this work is $a = 2.27$. The DL map is given in Figure 14.5.

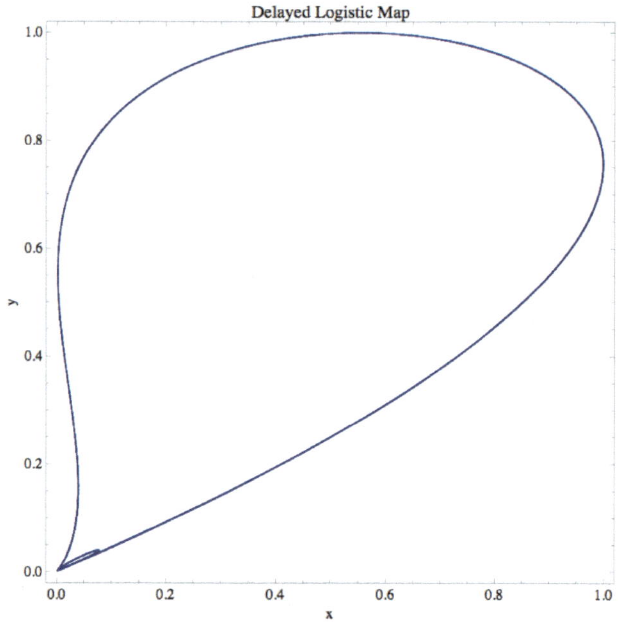

Fig. 14.5 Delayed Logistic

14.7 Problem Description

Three unique control systems problems are evaluated using DE and its chaos variant DE_{chaos}. These are a fourth order ball and hoop system [9], a generic third order system and an electric DC motor system of [7].

14.7.1 Fourth Order System

The first problem selected is a fourth order system comparable with the ball and hoop system [9] and presented in [3].

The Open Loop Transfer Function (OLTS) of this system [3] is given in Equation 14.22.

$$G(s) = \frac{1}{s^4 + 6s^3 + 11s^2 + 6s} \tag{14.22}$$

14.7.1.1 Ziegler Nichols Designed Controller

The Ziegler-Nichols tuning method using root-locus was the "conventional" method used to evaluate the PID gains for the system. The root locus diagram is given in Figure 14.6.

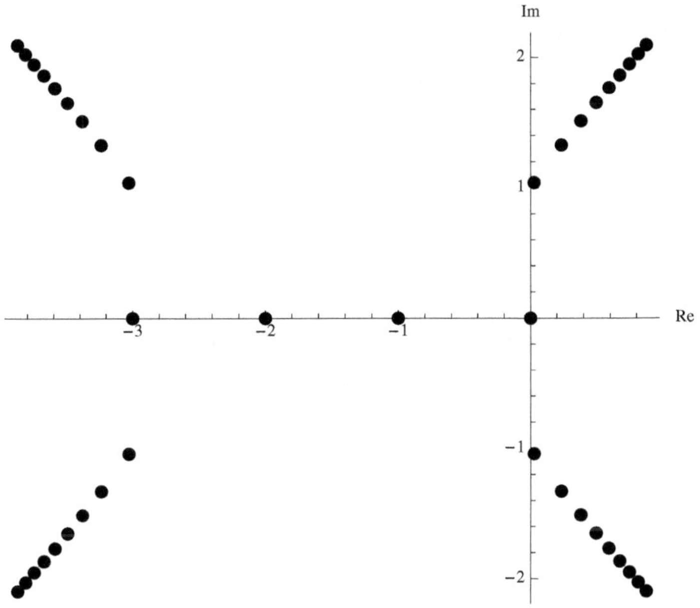

Fig. 14.6 Root Locus plot for G(s)

The crossover and gain for the system is **j1** and **10** respectively. With a frequency w_c of 1rad/s, the period T_c is calculated as:

$$T_c = \frac{2\pi}{w_c} = 6.28\,\text{sec}$$

Placing the values of k_c and T_c into Table 14.3

Table 14.3 Ziegler Nichols PID tuning parameters for 4th order system

Controller	k_p	T_i	T_D
PID	$0.6k_c$	$T_c/2$	$T_c/8$

which gives the tuning values in Table 14.4.
Using the relationships:

$$k_i = \frac{k_p}{T_i} \tag{14.23}$$

$$k_D = k_p T_D \tag{14.24}$$

the PID gain can be calculated as in Table 14.5.

Table 14.4 Ziegler Nichols PID tuning parameters for 4th order system

Controller	k_p	T_i	T_D
PID	6	3.14	0.79

Table 14.5 Ziegler Nichols PID Gain values for 4th order system

Controller	k_p	k_i	k_D
PID	6	1.91	4.74

Following Equation 14.24, the transfer function for the PID controller is now given as:

$$PID = 6 + \frac{1.97}{s} + 4.74s$$

Connecting the controller in series with the plant, the transfer function can be visualized as the controller-plant:

$$CP = \frac{(6 + 1.97/\mathrm{s} + 4.74\mathrm{s})}{(6\mathrm{s} + 11\mathrm{s}^2 + 6\mathrm{s}^3 + \mathrm{s}^4)}$$

Using the *sensitivity function* in Equation 14.13, the step response of the system can be seen in Figure 14.7.

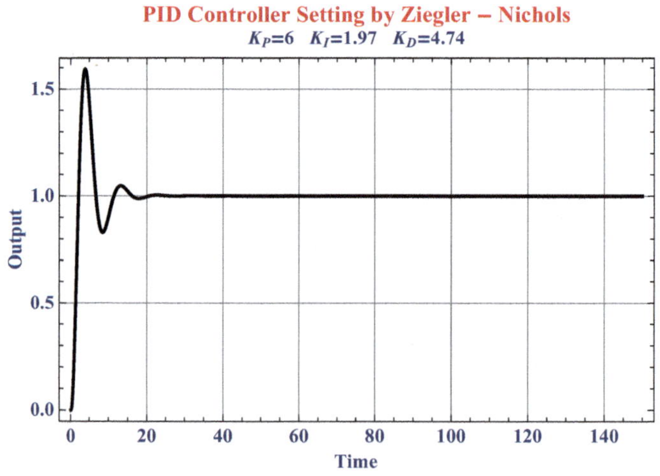

Fig. 14.7 System response for PID controller

The steady state characteristics of the controlled system is given in Table 14.6.

Table 14.6 Steady state responses

Title	Rise Time	Overshoot	Settling Time
Root Locus	2.935	59.244	15.05

14.7.1.2 Evolutionary Synthesis

The most important aspect of evolutionary synthesis is the formulation of the objective function. EA works of the premise of a black-box system, whether to maximize or minimize a system.

In order to effectively optimize a relevant set of PID controllers, the ideal approach is to minimize the error of the system. These specifications are given in Section 14.4.2.

Four distinct objective functions are created, each to minimize one of the stated errors. DE operates on trying to minimize these four unique errors.

The following Table 14.7 gives the results of the four different error analysis.

Table 14.7 Objective function values

Parameters	Values	k_p	k_i	k_D
IAE	7.818	5.856	0.0043	11.835
ISE	6.4102	5.204	0.1568	20.804
ITAE	15.19	4.8436	0.00025	7.0235
MSE	0.0312	5.204	0.1568	20.804

Figures 14.8 - 14.11 give the system response of the four different errors.

The combined system responses for DE are given in Fig 14.12.

The system specifications are given in Table 14.8. From the values, it is obvious that the best controller was designed by ITAE. In term of overshoot, ITAE has the optimal value of 6.715, ISE and MSE have the lowest rise time of 1.31sec and IAE has the lowest settling time of 5.19sec.

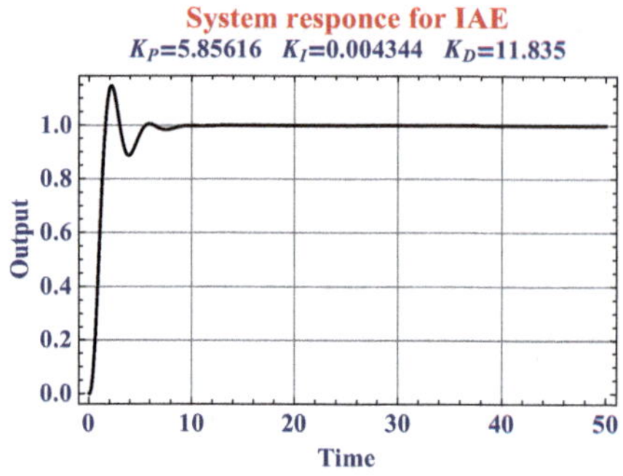

Fig. 14.8 System response for IAE

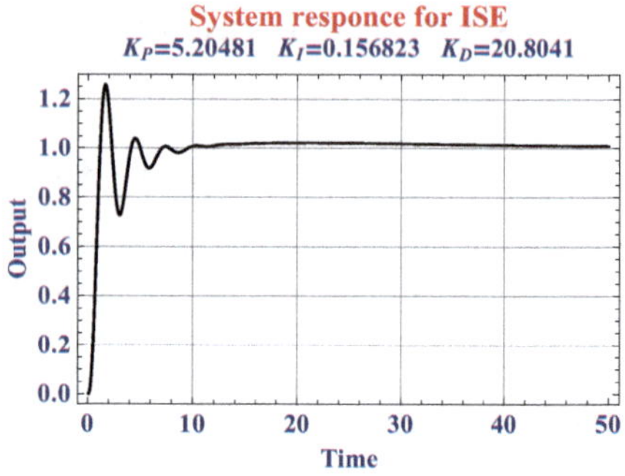

Fig. 14.9 System response for ISE

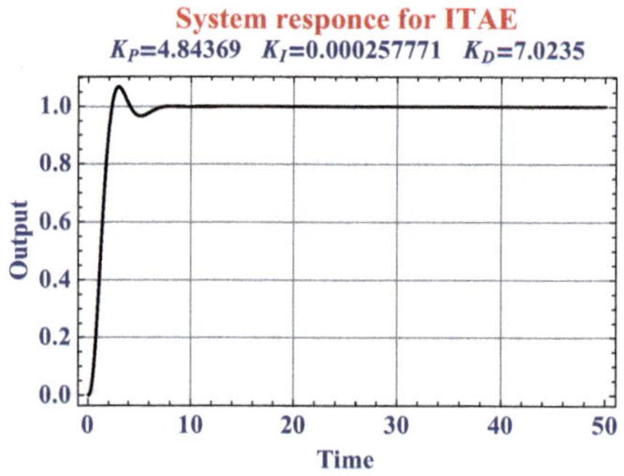

Fig. 14.10 System response for ITAE

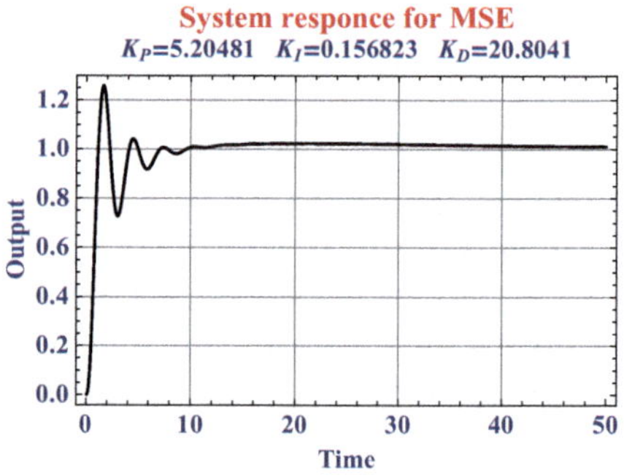

Fig. 14.11 System response for MSE

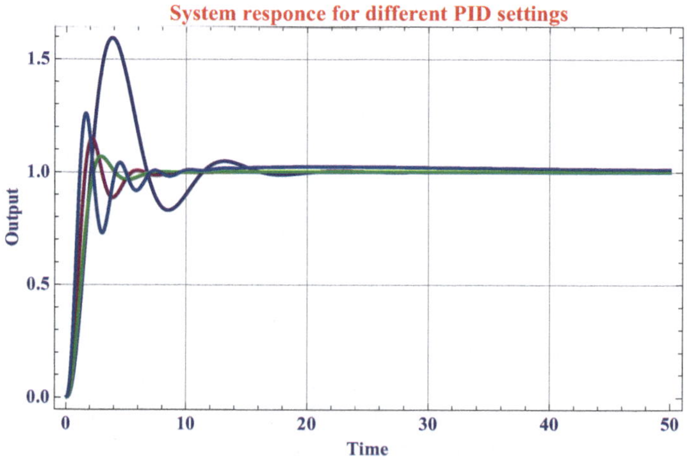

Fig. 14.12 Combined system response for DE

Table 14.8 Steady state responses of DE for fourth order system

Title	Ziegler Nichols	IAE	ISE	ITAE	MSE
Overshoot	59.244	14.5	24.52	**6.715**	24.52
Rise Time	2.935	1.665	**1.31**	2.155	**1.31**
Settling Time	15.05	**5.19**	9.22	6.095	9.22

14.7.1.3 Evolutionary Synthesis with Chaos

The two uniques chaos maps of Lozi and DL are swapped with the random generator in DE. The results are given in Tables 14.9 and 14.10 for the Lozi and DL systems respectively.

Table 14.9 DE_{chaos} **Lozi** objective function values

Parameters	Values	k_p	k_i	k_D
IAE	0.0811	5.856	0.0043	11.835
ISE	0.156	5.204	0.1568	20.804
ITAE	0.065	4.8436	0.00025	7.0235
MSE	31.224	5.204	0.1568	20.804

Table 14.10 DE_{chaos} **DL** objective function values

Parameters	Values	k_p	k_i	k_D
IAE	0.0811	5.856	0.0043	11.835
ISE	0.156	5.204	0.1568	20.804
ITAE	0.065	4.8436	0.00025	7.0235
MSE	31.224	5.204	0.1568	20.804

The interesting feature is that the values obtained for both the systems is identical. Since this has now effectively become a deterministic search pattern, the trajectory of the two systems encompass the same local optimal value. The steady state responses of the two systems is given in Table 14.11.

Table 14.11 Steady state responses of DE_{chaos} **Lozi/DL** for fourth order system

Specification	Ziegler Nichols	IAE	ISE	ITAE	MSE
Overshoot	59.244	14.5	24.52	**6.715**	24.52
Rise Time	2.935	1.665	**1.31**	2.155	**1.31**
Settling Time	15.05	**5.19**	9.22	6.095	9.22

As the responses of both the chaos map systems is the same, the plots of the different system measures are given in Figures 14.13 - 14.16. The combined plot is given in Figure 14.17.

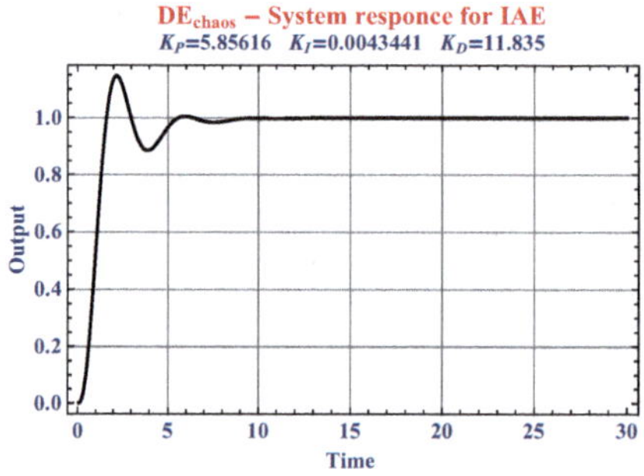

Fig. 14.13 System responce for IAE

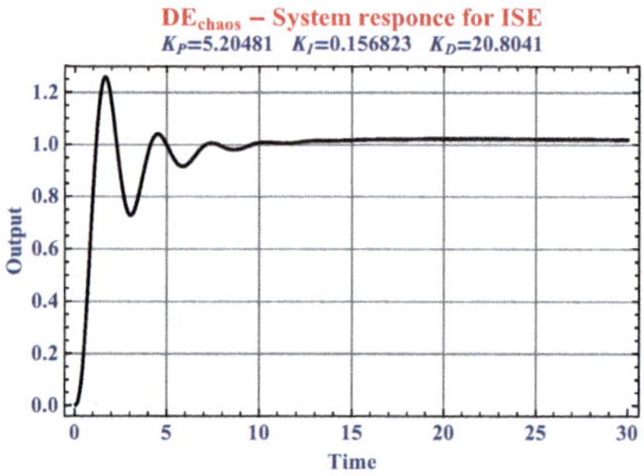

Fig. 14.14 System responce for ISE

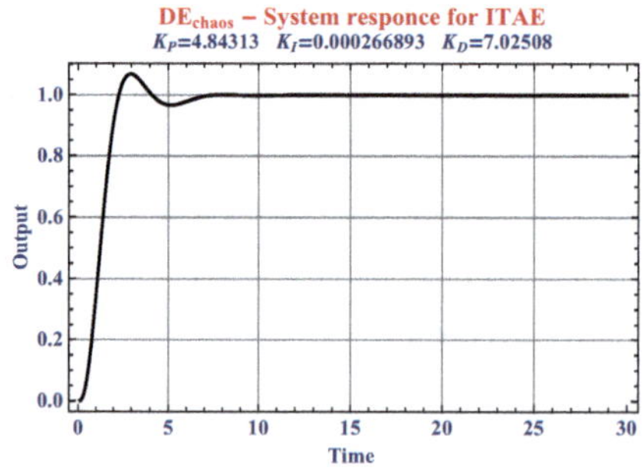

Fig. 14.15 System responce for ITAE

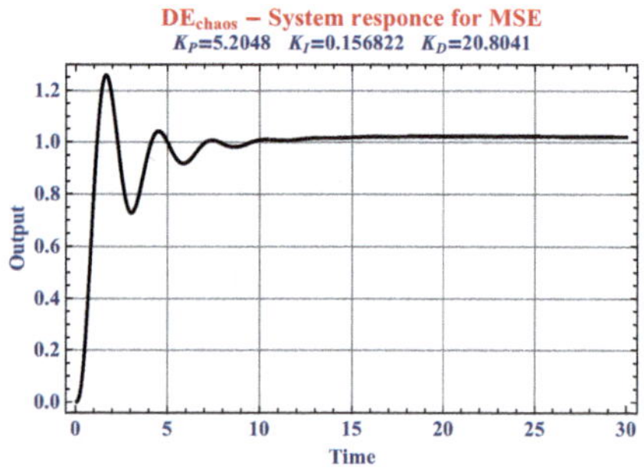

Fig. 14.16 System responce for MSE

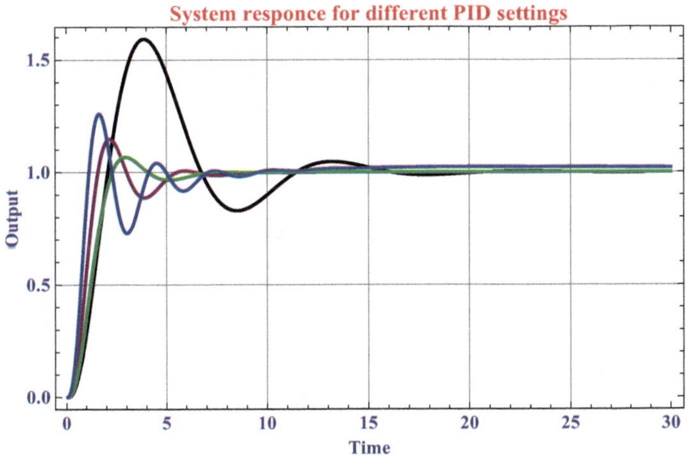

Fig. 14.17 Combined system responce for DE_{chaos}

14.7.1.4 Analysis

The results obtained for DE and DE_{chaos} are compared with Genetic Algorithm (GA) of [3] for the same problem. The results are presented in Tables 14.12 and 14.13. The results obtained for both DE and DE_{chaos} are identical, and apart from the rise time, the DE variants dominates all other steady state responses over GA.

Table 14.12 Comparison of DE_{chaos}, DE and GA for ISE and IAE

Specification	Ziegler	IAE			ISE		
Specification	Ziegler Nichols	GA	DE	DE_{chaos}	GA	DE	DE_{chaos}
Overshoot	59.244	44.97	**14.5**	**14.5**	28.804	**24.52**	**24.52**
Rise Time	2.935	**1.2**	1.665	1.665	**1.2**	1.31	1.31
Settling Time	15.05	9.3	**5.19**	**5.19**	20.4	**9.22**	**9.22**

Table 14.13 Comparison of DE_{chaos}, DE and GA for ITAE and MSE

Specification	Ziegler	ITAE			MSE		
	Nichols	GA	DE	DE_{chaos}	GA	DE	DE_{chaos}
Overshoot	59.244	57.19	**6.715**	**6.715**	28.59	**24.52**	**24.52**
Rise Time	2.935	**1.3**	2.155	2.155	**1.2**	1.31	1.31
Settling Time	15.05	8.2	**6.095**	**6.095**	20.4	**9.22**	**9.22**

14.7.2 Third Order System

The second problem is a third order problem. The plant transfer function is given as:

$$G(s) = \frac{0.1}{s(3s+1)(0.8s+1)} \tag{14.25}$$

14.7.2.1 Ziegler Nichols Designed Controller

The root locus plot of the system $G(s)$ is given in Figure 14.18.

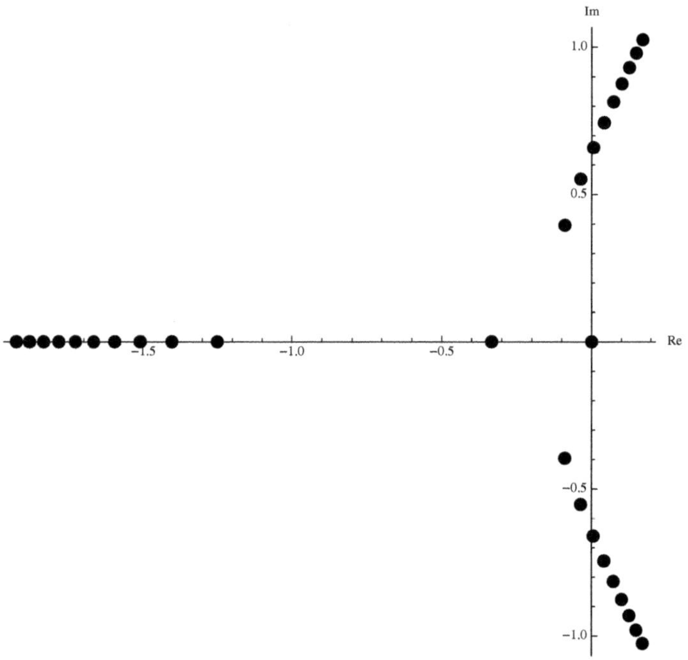

Fig. 14.18 Root locus for third order system

The crossover and gain for the system is **0.6449i** and **1.58** respectively. The operating values for the Zeigler Nichlos is given in Table 14.14.

Table 14.14 Ziegler Nichols PID tuning parameters for 3rd order system

Controller	k_p	T_i	T_D
PID	6	3.14	0.79

Using the *sensitivity function* in Equation 14.13, the step response of the system can be seen in Figure 14.19.

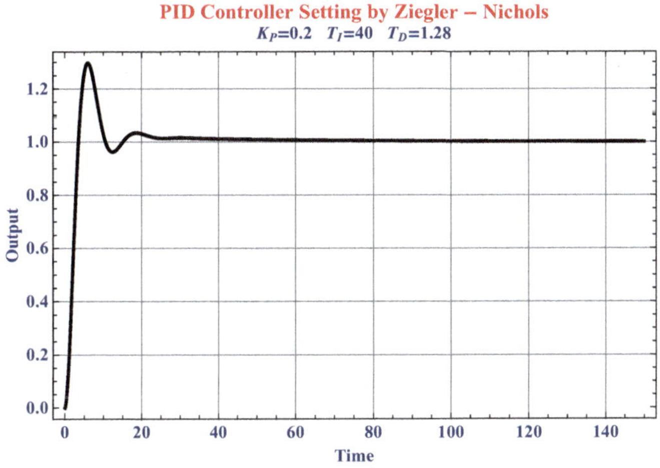

Fig. 14.19 System response for PID controller with Ziegler Nichols

14.7.2.2 Evolutionary Synthesis

The operating parameters of DE is given in Table 14.15.

Table 14.15 Operating Parameters of DE

Parameter	Value
Population Size	500
Generations	500
F	0.82
CR	0.8
Specimen	200 - 500

Two separate simulations was conducted. For the first simulation, the specimen size was kept at 200, whereas for the second simulations, the size was set at 500.

The results obtained from the evolutionary synthesis of the problem is given in Table 14.16.

Table 14.16 Objective function values

Experiment	Parameters	Values	k_P	k_i	k_D
Specimen=200	IAE	4.866	27.457	0.4939	199.999
	ISE	1.013	0.3196	1.7648	199.999
	ITAE	2.731	41.6181	0.1386	199.999
	MSE	0.005	0.3196	1.7648	199.999
Specimen=500	IAE	8.351	33.584	2.1051	475.966
	ISE	2.949	0.3089	4.1444	499.999
	ITAE	13.106	74.7689	0.656	482.150
	MSE	0.01474	0.3089	4.1444	499.999

The assignment of the operating parameters for DE is very important. As shown in Table 14.15, specimen has two different values; 200 and 500. The different error responses are given in Figures 14.20 to 14.27.

Figures 14.20 and 14.21 give the responses for IAE for the different specimen parameter setting of 200 and 500.

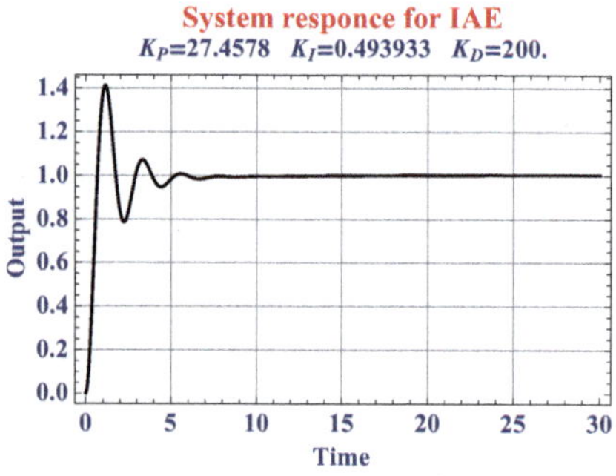

Fig. 14.20 System response for Specimen = 200

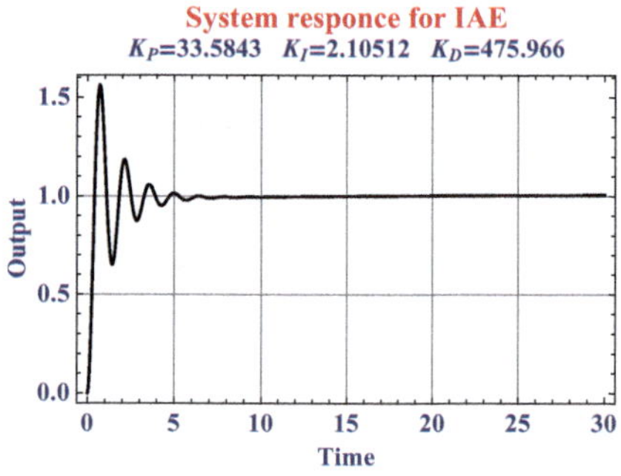

Fig. 14.21 System response for Specimen = 500

Figures 14.22 and 14.23 give the responses for ISE for the different specimen parameter setting of 200 and 500.

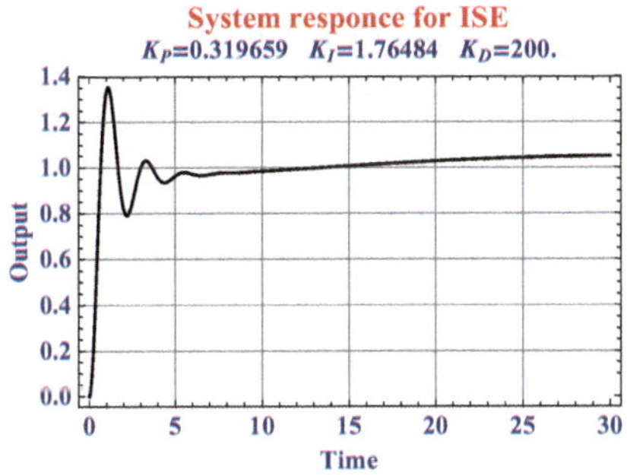

Fig. 14.22 System response for Specimen = 200

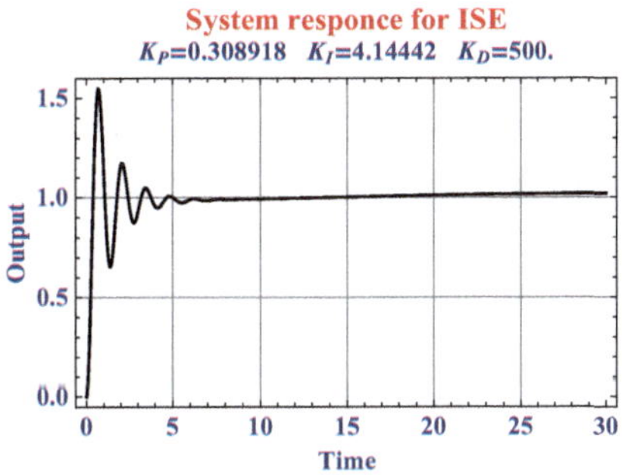

Fig. 14.23 System response for Specimen = 500

Figures 14.24 and 14.25 give the responses for ITAE for the different specimen parameter setting of 200 and 500.

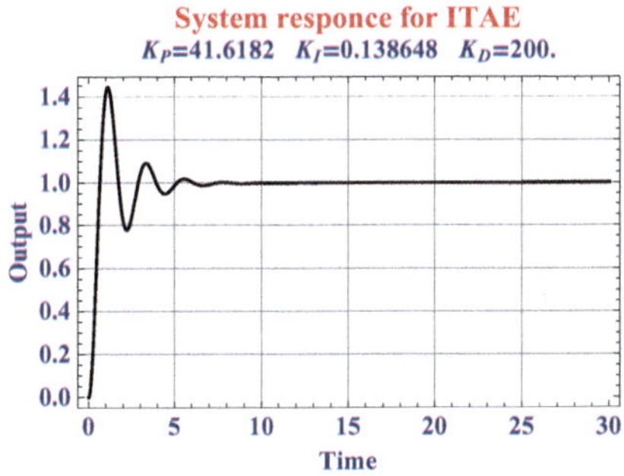

Fig. 14.24 System response for Specimen = 200

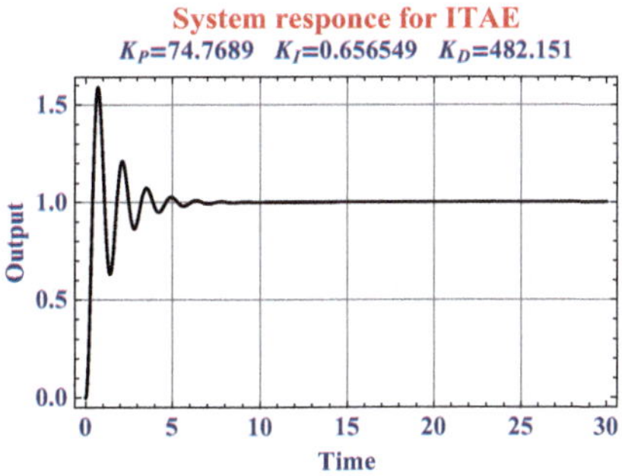

Fig. 14.25 System response for Specimen = 500

Figures 14.26 and 14.27 give the responses for MSE for the different specimen parameter setting of 200 and 500.

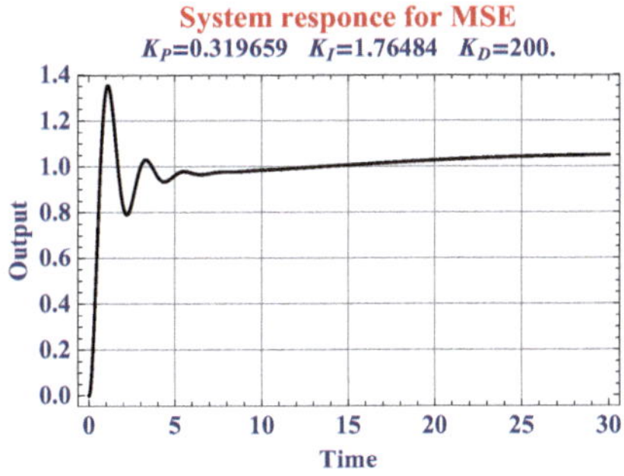

Fig. 14.26 System response for Specimen = 200

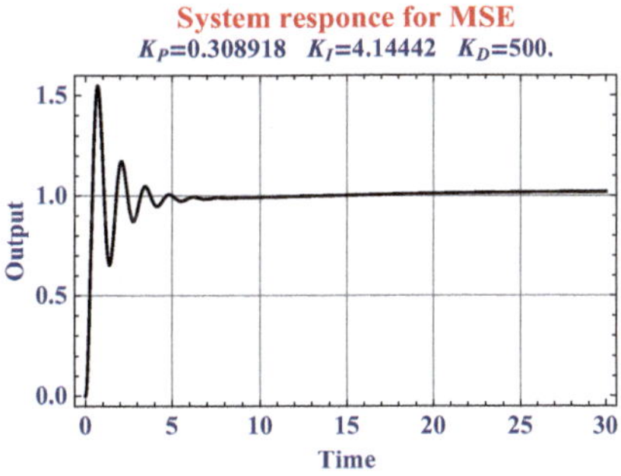

Fig. 14.27 System response for Specimen = 500

Comparing the results between the two settings, it is obvious that both specimen settings offer different parameter performances. The rise time of the 500 specimen is less, however, the peak overshoot and number of oscillations are less in specimen setting of 200. The settling time is almost identical for both systems.

The system plots of both the experiments is given in Figures 14.28 and 14.29.

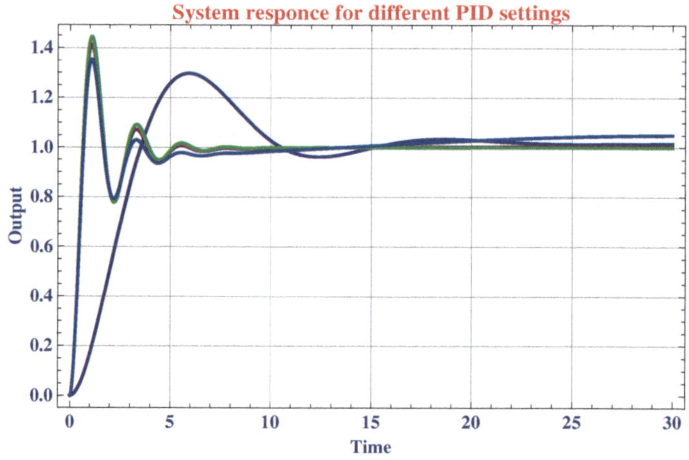

Fig. 14.28 Comparison results with Ziegler Nichols and DE with Specimen=200

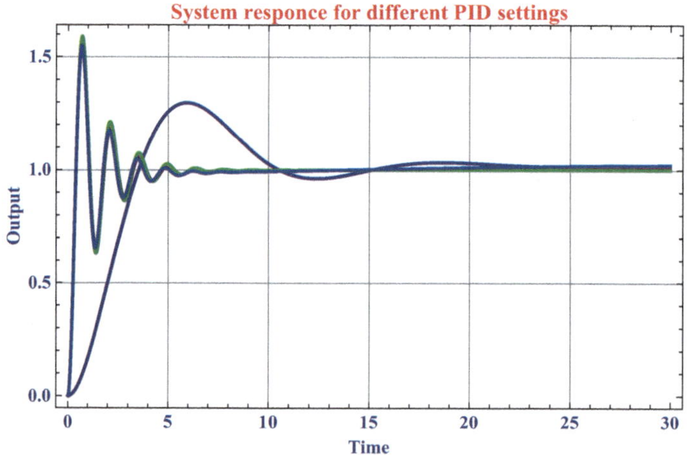

Fig. 14.29 Comparison results with Ziegler Nichols and DE with Specimen=500

The steady state responses for the two different systems is given in Table 14.17 and 14.18. The major drawback for DE is the overshoot, which is significantly higher than that of Ziegler Nichols. However, in all other aspects, DE proves a better tuning heuristic. The rise time is marginally lower and the settling time is significantly lower.

Table 14.17 Steady state responses for third order system with Specimen = 200

Title	Ziegler Nichols	IAE	ISE	ITAE	MSE
Overshoot	**27.7602**	40.7344	28.905	44.3401	28.905
Rise Time	4.315	**0.85**	**0.84**	**0.85**	**0.84**
Settling Time	20.315	6.75	20.26	**4.94**	20.26

Table 14.18 Steady state responses for third order system with Specimen = 500

Specification	Ziegler Nichols	IAE	ISE	ITAE	MSE
Overshoot	**27.7602**	55.502	55.211	58.751	55.211
Rise Time	4.315	0.55	**0.535**	0.55	**0.535**
Settling Time	20.315	5.885	37.22	**5.76**	37.22

14.7.2.3 Evolutionary Synthesis with Chaos

The results of the two chaotic mutation systems is given in Tables 14.19 and 14.20. As with the previous results of the fourth order system, the two results are identical, stipulating that the solution lies on the trajectory of the attractors.

Table 14.19 DE_{chaos} **Lozi** objective function values for third order system

Parameters	Values	k_p	k_i	k_D
IAE	0.0811	5.856	0.0043	11.853
ISE	0.1561	5.204	0.1568	20.804
ITAE	0.0658	4.843	0.00025	7.0234
MSE	31.224	5.2048	0.1568	20.8041

Table 14.20 DE_{chaos} **DL** objective function values for third order system

Parameters	Values	k_p	k_i	k_D
IAE	0.0811	5.856	0.0043	11.853
ISE	0.1561	5.204	0.1568	20.804
ITAE	0.0658	4.843	0.00025	7.0234
MSE	31.224	5.2048	0.1568	20.8041

The steady state responses for the two systems in given in Table 14.21. When compared to the Ziegler Nichols method, DE_{chaos} obtains better performance indices for all the specifications. IAE obtains a better settling time whereas ISE and MSE attain better overshoot and rise time.

Table 14.21 DE_{chaos} - **Lozi/DL** steady state responses for third order system

Specification	Ziegler Nichols	IAE	ISE	ITAE	MSE
Overshoot	27.7602	10.441	**2.561**	14.36	**2.561**
Rise Time	4.315	4.33	**3.335**	5.58	**3.335**
Settling Time	20.315	**10.735**	26.825	13.085	26.825

The system responses for DE_{chaos} for the four different errors is given in Figures 14.30 – 14.33.

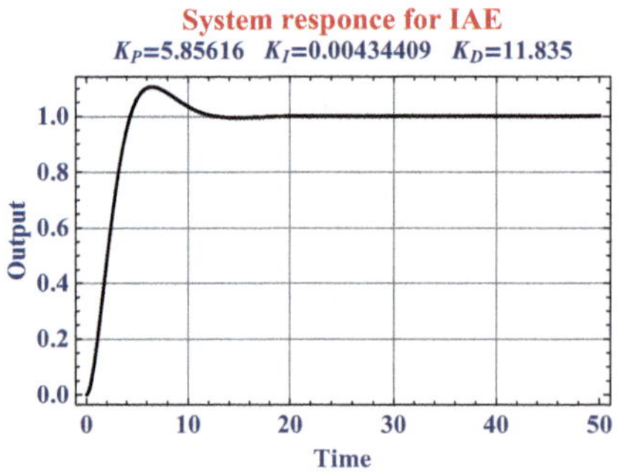

Fig. 14.30 System response for IAE

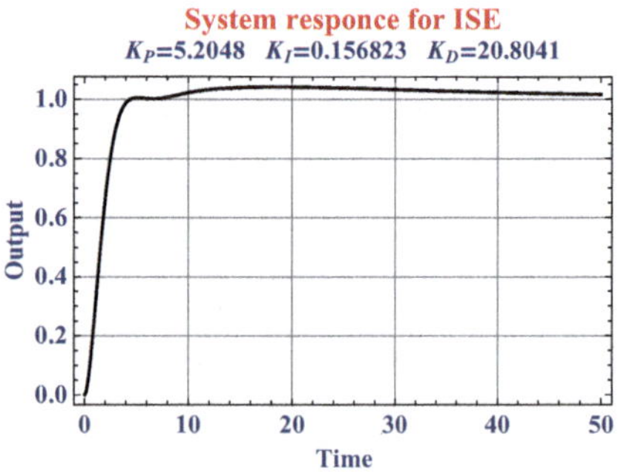

Fig. 14.31 System response for ISE

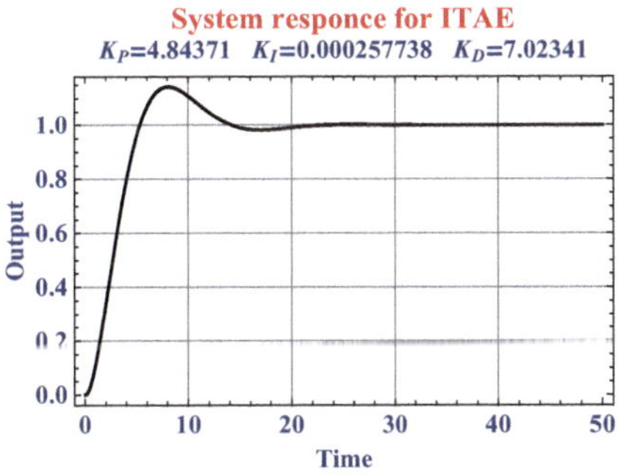

Fig. 14.32 System response for ITAE

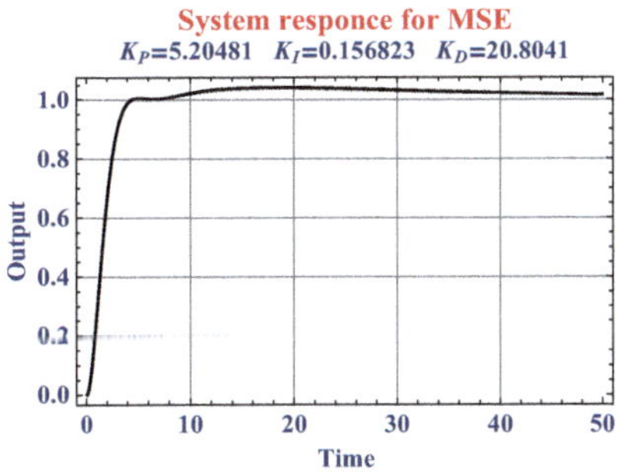

Fig. 14.33 System response for MSE

The combined results of the four different error specifications is given in Figure 14.34.

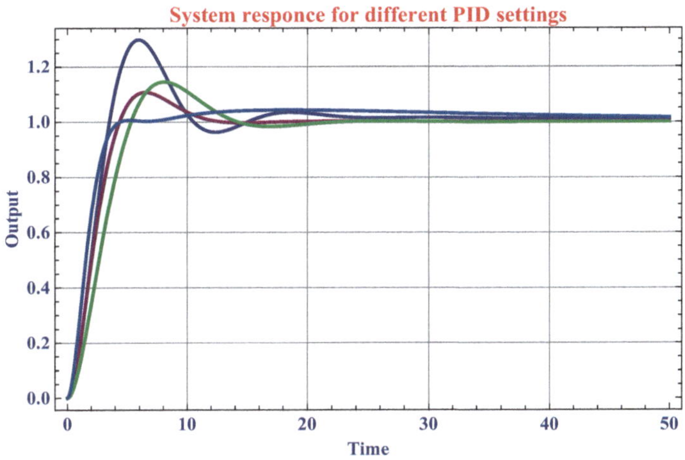

Fig. 14.34 Combined system response for DE_{chaos}

14.7.2.4 Analysis

Analysis is done with the results obtained for the different DE variants. The optimal results from the experimentation of DE is compared with that of DE_{chaos} as given in Table 14.22. The most striking feature is that DE_{chaos} is able to produce significantly lower overshoot than all the other systems. This is seen across all the different error specifications. DE is however better performing when rise time and settling time is concerned.

Table 14.22 Comparison of DE_{chaos} and DE for third order system.

Specification	Ziegler Nichols	IAE		ISE		ITAE		MSE	
		DE	DE_{chaos}	**DE**	DE_{chaos}	**DE**	DE_{chaos}	**DE**	DE_{chaos}
Overshoot	57.19	40.734	**10.441**	28.905	**2.561**	44.34	**14.36**	28.905	**2.561**
Rise Time	2.935	**0.85**	4.33	**0.84**	3.335	**0.85**	5.58	**0.84**	3.335
Settling Time	15.05	**6.75**	10.735	**20.26**	26.825	**4.94**	13.085	**20.26**	26.825

14.7.3 Electric DC Motor

The third problem is the Electric DC Motor [7]. The transfer function is given in Equation 14.26.

$$G(s) = \frac{K}{L_a J s^3 + (R_a J + B L_a) s^2 + (K^2 + R_a B) s} \tag{14.26}$$

where

$$\begin{aligned} L_a &= \text{armature inductance} \\ R_a &= \text{armature resistance} \\ K &= \text{motor constant} \\ J &= \text{motor of inertia} \\ B &= \text{mechanical friction} \end{aligned}$$

The values of the parameters are given as:

$$\begin{aligned} L_a &= 0.025 \\ R_a &= 5 \\ K &= 0.9 \\ J &= 0.042 \\ B &= 0.01625 \end{aligned}$$

The transfer function is now given as in Equation 14.27:

$$G(s) = \frac{0.9}{0.0005s^3 + 0.2104s^2 + 0.8913s} \tag{14.27}$$

The root locus plot of the plant is given in Figure 14.35.

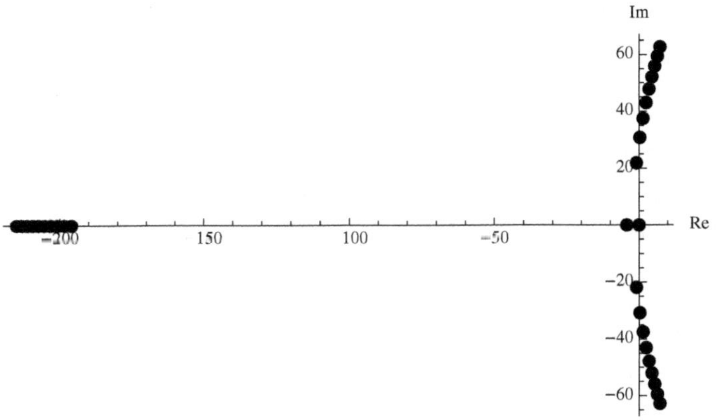

Fig. 14.35 Root Locus Plot for Electric DC Motor transfer function

14.7.3.1 Evolutionary Synthesis

DE produced two distinct results when applied to this problem. The obtained values are given in Table 14.23.

Table 14.23 Objective function values

Parameters	Values	k_p	k_i	k_D
IAE	$1.31783 * 10^{-13}$	418.184	$5.648 * 10^{-11}$	96.58
ISE	$3.91643 * 10^{-18}$	418.185	$1.805 * 10^{-6}$	96.58
ITAE	$6.1959 * 10^{-10}$	65.254	$9.925 * 10^{-12}$	15.0712
MSE	$1.95822 * 10^{-20}$	418.185	$1.802 * 10^{-6}$	96.58

Table 14.24 gives the steady state responses of the different DE heuristics. A notable feature is the very low overshoot of DE_1. This is quite unique since its it almost a hundred times lower than Ziegler Nichols. In terms of settling time and rise time, DE_1 and DE_2 produce better results.

The optimal result in terms of settling time and peak overshoot is shown in Figure 14.36.

The second result, which has a better rise time is shown in Figure 14.37 together with the first result.

Table 14.24 Comparison of steady state responses of DE and Ziegler Nichols method

Specification	Ziegler Nichols	DE_1	DE_2
Overshoot	41.4	**0.4727**	28.255
Settling time	2.56	**0.04**	**0.04**
Rise Time	0.242	0.03	**0.01**

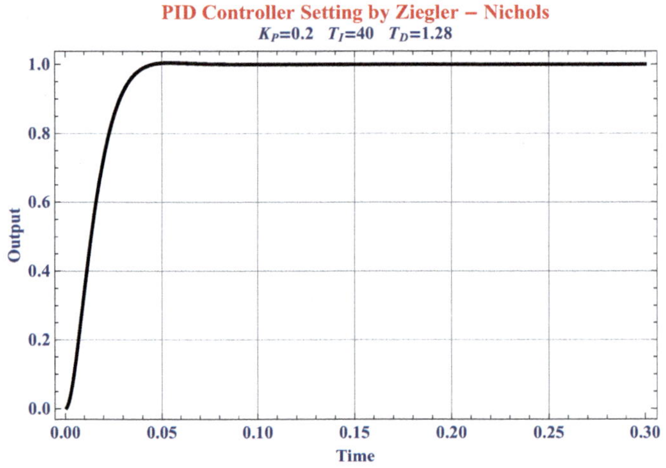

Fig. 14.36 System response for evolved PID settings

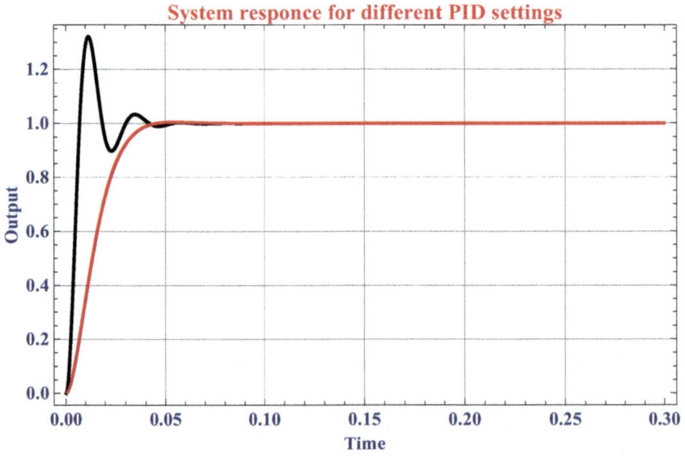

Fig. 14.37 System response for different evolved PID settings

14.7.3.2 Evolutionary Synthesis with Chaos

The results obtained with the chaotic maps is given in Tables 14.25 and 14.26. As has been the case with the other experiments, the results obtained for both the systems are identical.

Table 14.25 DE_{chaos} **Lozi** objective function values

Parameters	Values	k_p	k_i	k_D
IAE	$3.20351 * 10^{8}$	158.059	$3.333 * 10^{-10}$	36.505
ISE	$2.5533 * 10^{17}$	158.059	$2.57536 * 10^{-7}$	36.505
ITAE	$1.3302 * 10^{9}$	158.059	$8.2886 * 10^{-10}$	36.505
MSE	$5.1066 * 10^{19}$	158.059	$2.5711 * 10^{-7}$	36.505

Table 14.26 DE_{chaos} **DL** objective function values

Parameters	Values	k_p	k_i	k_D
IAE	$7.8537 * 10^{7}$	158.060	$7.3295 * 10^{-7}$	36.505
ISE	$2.3876 * 10^{17}$	158.059	$3.4944 * 10^{-7}$	36.505
ITAE	$1.2072 * 10^{8}$	158.051	$2.9060 * 10^{-8}$	36.503
MSE	$4.7151 * 10^{19}$	158.059	$1.07143 * 10^{-6}$	36.505

The steady state responses of the two systems is given in Table 14.27. In all the specifications, DE_{chaos} obtains better performance indices.

Table 14.27 DE_{chaos} - **Lozi/DL** steady state responses for DC Motor system

Specification	Ziegler Nichols	IAE	ISE	ITAE	MSE
Overshoot	41.4	**11.97**	**11.97**	**11.97**	**11.97**
Rise Time	2.56	**0.015**	**0.015**	**0.015**	**0.015**
Settling Time	0.242	**0.035**	**0.035**	**0.035**	**0.035**

The system responses plots of the four error specifications is given in Figures 14.38 - 14.41 and the combined system response for DE_{chaos} is given in Figure 14.42.

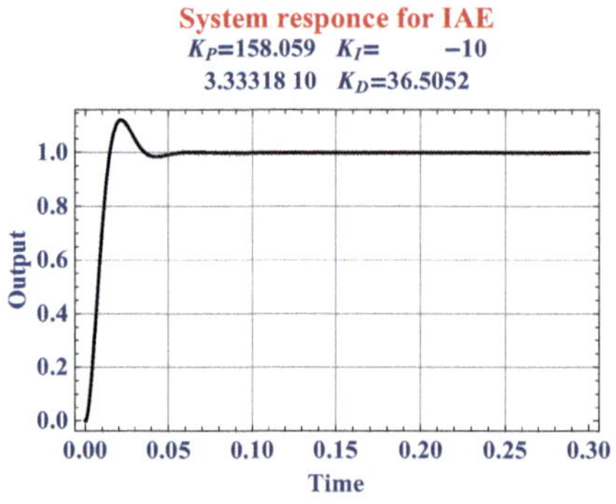

Fig. 14.38 System response for IAE

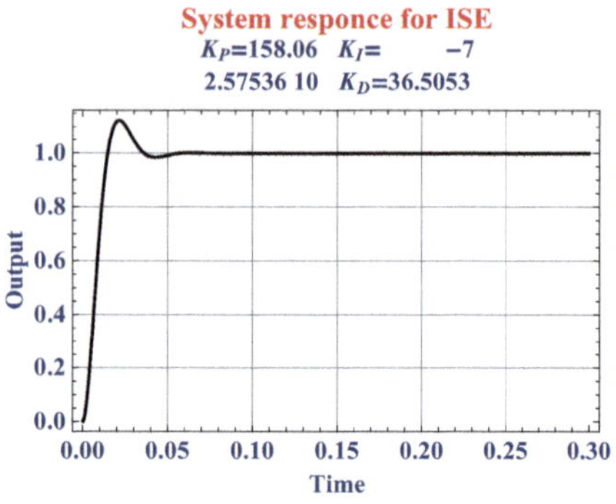

Fig. 14.39 System response for ISE

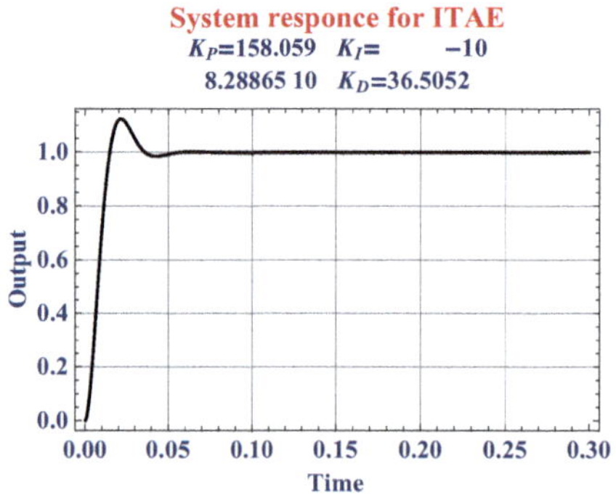

Fig. 14.40 System response for ITAE

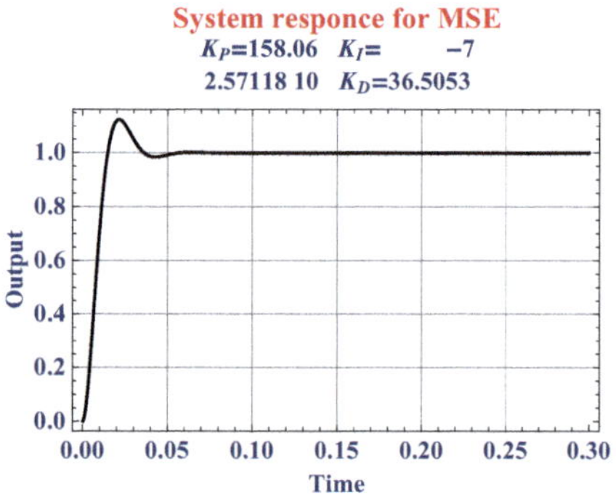

Fig. 14.41 System response for MSE

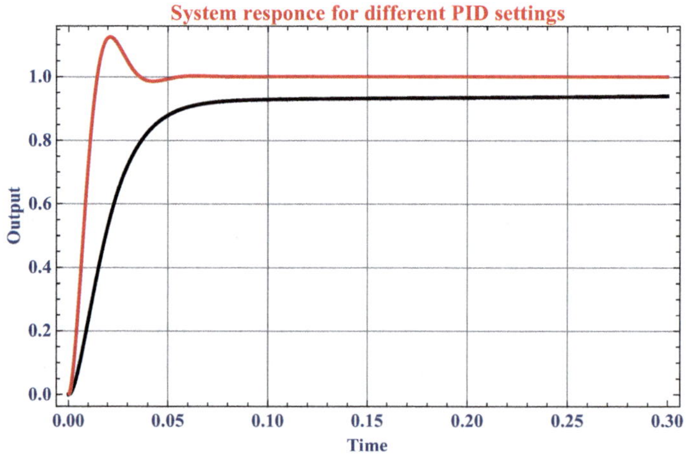

Fig. 14.42 Combined system response for DE_{chaos}

14.7.3.3 Analysis

DE and DE_{chaos} is compared with the tuning algorithms of *Ziegler-Nichlos* and *Continuous Cycling (CC)* and metaheuristics of *Genetic Algorithm (GA)*, *Evolutionary Programming (EP)* and *Particle Swarm Optimization (PSO)* of [7].

Table 14.28 gives the steady state responses of all the different heuristics. DE and DE_{chaos} are able to produce better results for all the compared specifications. DE obtains the best overshoot value and rise time, whereas DE_{chaos} has the optimal settling time.

Table 14.28 Comparison of steady state responses of different heuristics for the DC motor system.

Specification	**Ziegler Nichols**	**CC**	**EP**	**GA**	**PSO**	DE_1	DE_2	DE_{chaos}
Overshoot	41.4	87.6	8.81	13	12.9	**0.472**	28.25	11.97
Settling time	2.56	4.31	0.205	0.324	1.15	0.04	0.04	**0.015**
Rise Time	0.242	0.0474	0.014	0.0317	0.0317	0.03	**0.01**	0.035

14.8 Conclusion

DE is shown as a robust tuning algorithm for PID controller. In the three problems attempted, it has generally proven more successful than that of Ziegler Nichols tuning method and the other heuristics of GA and PSO with which it is compared against.

The two variants of DE; the stochastic canonical (*DE*) and deterministic chaotic (DE_{chaos}) are equally successful. The most striking feature is the consistency of the results produced by DE_{chaos}. For all the experimentations with the two unique chaos maps, the result obtained were identical. This led to the claim that the result lay in the path of the attractors which were mapping the solution space.

Through the obtained results and its analysis, this research reinforces the claim that chaos can be utilized alongside traditional metaheuristics in the task of global optimization.

Acknowledgements. The following two grants are acknowledged for the financial support provided for this research.

1. Grant Agency of the Czech Republic **GARC 102/09/1680**
2. Grant of the Czech Ministry of Education **MSM 7088352102**

References

1. Astrom, K.: Control System Design. University of California, Santa Barbra (2002)
2. Caponetto, R., Fortuna, L., Fazzino, S., Xibilia, M.: Chaotic sequences to improve the performance of evolutionary algorithms. IEEE Trans. Evol. Comput. 7(3), 289–304 (2003)
3. Griffin, I.: On-line pid controller tuning using genetic algorithm. Master's thesis, Dublin City University, Ireland (2003)
4. Hennon, M.: A two-dimensional mapping with a strange attractor. Commun. Math. Phys. 50, 69–77 (1979)
5. Ibrahim, S.: The pid controller design using genetic algorithms. Master's thesis, University of Southern Queensland, Australia (2005)
6. Landau, Y.: Digital Control Systems. Springer, London (2006)
7. Nagraj, B., Subha, S., Rampriya, B.: Tuning algorithms for pid controller using soft computing techniques. International Journal of Computer Science and Network Security 8, 278–281 (2008)
8. Price, K.: New ideas in Optimisation. McGraw Hill, New York (1999)
9. Wellstead, P.: Ball and hoop (2008), `http://www.control-systems-principles.co.uk`

Chapter 15
Chaotic Attributes and Permutative Optimization

Donald Davendra, Ivan Zelinka, and Godfrey Onwubolu

Abstract. Population dynamics and its relation to chaotic systems is analyzed in this Chapter. Using basic chaotic principles of attractors and edges, a dynamic population is developed, which is used to induce and retain diversity in a metaheuristic population. Simulation is done with Genetic Algorithm, Differential Evolution and Self-Organizing Migrating Algorithm on the combinatorial problem of Quadratic Assignment with promising results.

15.1 Introduction

One of the most challenging optimization problems is *permutative based* combinatorial optimization. This class of problem harbors some of the most famous optimization problems like traveling salesman and vehicle routing problem.

Another very important branch is that of scheduling, to which a number of manufacturing problems are associated. The most realized and of interest are the shop scheduling problems of flow shop and job shop.

Donald Davendra
Tomas Bata University in Zlin, Faculty of Applied Informatics, Nad Stranemi 4511, Zlin 76001, Czech Republic.
e-mail: davendra@fai.utb.cz

Ivan Zelinka
Tomas Bata University in Zlin, Faculty of Applied Informatics, Nad Stranemi 4511, Zlin 76001, Czech Republic
and
VSB-TUO, Faculty of Electrical Engineering and Computer Science, 17. listopadu 15, 708 33 Ostrava-Poruba, Czech Republic
e-mail: zelinka@fai.utb.cz

Godfrey Onwubolu
Knowledge Management & Mining, Inc., Richmond Hill, Ontario, Canada
e-mail: onwubolu_g@dsgm.ca

I. Zelinka et al. (Eds.): Evolutionary Algorithms and Chaotic Systems, SCI 267, pp. 481–517.
springerlink.com

What makes a permutative problem complex is that the solution representation is very concise, since it must have a discrete number of values, and each occupied variable in the solution is unique. Given a problem of size n, a representation can be described as $x = \{x_1, x_2, x_3, ..., x_n\}$, where each value x_i in the solution is unique and the entire set of solutions is an integer representation from $\{1, n\}$.

From an optimization point of view, this represents a number of problems. Firstly, the search space is discrete and a number of validations inevitably have to be conducted in order to have a viable solution. Secondly, the search space is very large, to the scale of $n!$. Consequently, these problems are generally termed *NP* or *NP Hard* [16].

The usual approach is to explore the search space in the neighborhood of good solutions in order to find better solutions. This unfortunately has the effect of converging the population, which then leads to stagnation. The usual term is *local optima convergence/stagnation.* Local minima regions acts as attractor basins, where solutions converge. Diversity in the population decreases and possibility of future evolution diminishes.

This chapter presents the research that looks at the *diversity* of the population in order to aid the application of metaheuristics. A permutative solution and its representation present some advantages to this effect. The usual measure of a solution is its fitness, in respect to the problem being solved. In a permutative solution, the distinct ordering of values gives the opportunity to have other measures of diversity.

The following sections describe the approach developed within this scope of research. The first section analyzes chaos and its implications in metaheuristics. The second section outlines the need for having distinct initial population, and chaotic attributes of attractors and edges. The third section describes some fields of measure of diversity followed by the approach of clustering of the solutions. Three unique crossover paradigms are then presented to which the clustered population has been applied. The subsequent sections describe the extensive experimentation conducted on the Quadratic Assignment Problem (QAP) and an analysis of the dynamic nature of clustered populations.

15.2 Chaotic Signature in Population Dynamics

Population and its application to chaotic systems is well documented. Populations viewed as dynamical systems was first discussed by [24]. Subsequent work by [15], further chronicled the work of viewing populations as number systems. The logistic map, the simplest chaotic system is also used for the modeling of population dynamics [24]. Another system is the Voltterra-Lotka equations of biological models.

Chaos in optimization has been largely explored through Neural Networks [18]. The core approach has been to avoid regions of "local optima" or "stagnation" in order to find better solutions. The basic concept has been that chaotic dynamics have been able to search for solutions along the formation of a strange attractor which has fractal structures. These structures are then used to search for solutions in state space along such fractal attractors who's Legesgue measure is zero.

Nozawa [26] modified the Hopfield-Neural network by the Eulers method to create an equivalent to the chaotic neural network of [2]. A 10 city problem is solved with better results than stochastic models.

Yamada and Aihara [38] solved the Traveling Salesman Problem with chaotic neural networks by computing the largest Lyapunov exponent. They showed that the solving abilities are very high when the largest Lyapunov exponent is near zero, which implies that "an edge of chaos " could have high performance to solve combinatorial problems.

Maintenance scheduling problems were solved by a chaotic simulated annealing approach by [5]. It was also proven of the existence of chaotic dynamics in solving combinatorial problems using chaotic neural networks.

A further exploration of chaos in optimization was done by [19], who proposed a new network model of chaotic potts spin. Using this method the constraint term is always satisfied and feasible solutions are always obtained.

This chapter takes a similar approach to the ones described, as the main aspect is the avoidance of "local optima" regions in the search space. However, we look upon the population as the driving system behind the convergence of the population.

The usual approach is to visualize the population as a fitness landscape, where solutions transverse towards global optimal solutions. This approach takes a different view of the population. A population is looked upon as an information base, a "genetic code" base where each solution occupies a distinct place in the information space. An example is given in Figure 15.1.

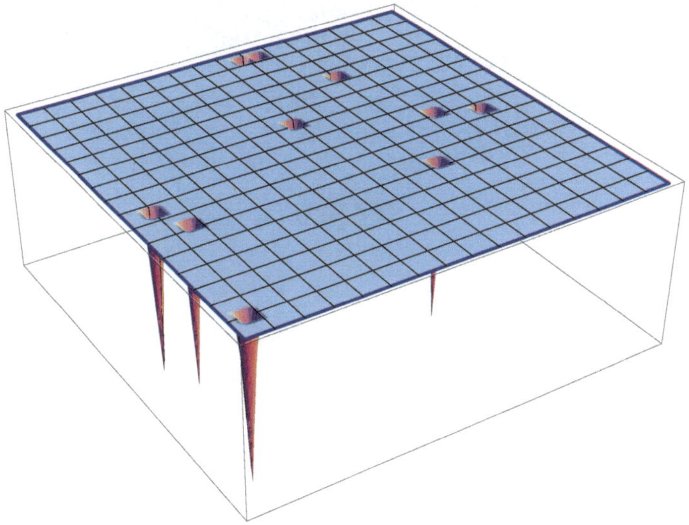

Fig. 15.1 Population representation

During successive generations, solutions are mated together, and an exchange of information takes place. Based on selection criteria of different algorithms, new and better performing solutions are accepted in the population after each generation.

However, during evolution, solutions tend to converge towards each other. What this in effect does, is reduce the amount of information available in the information plane of the collective information gene pool. Even if the solution converges towards the global minima, the information left in the population is usually marginal. This is what is termed as "local optima stagnation".

The main input in this research is the creation of a dynamic population which is kept on the threshold of information viability and which can be used by any number of metaheuristics as a population paradigm.

15.3 Population Dynamics

Each solution in a population contains certain information, its own "genetic code" which is used for replication. A way to visualize it is to see a solution as occupying a certain point in the information space as given in Figure 15.2.

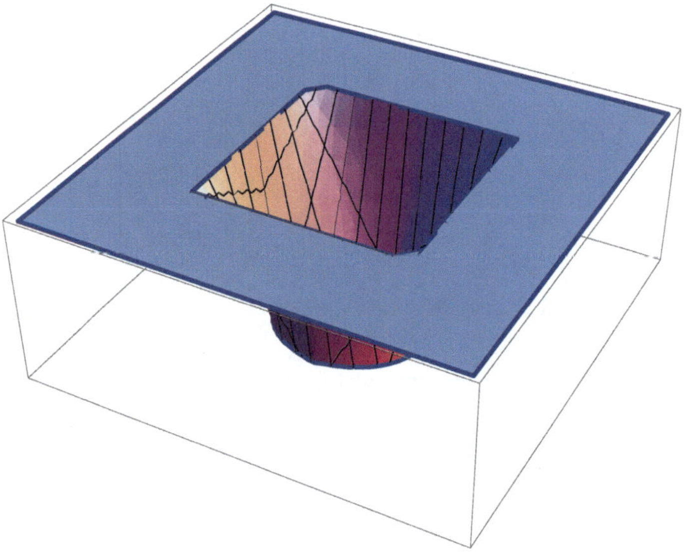

Fig. 15.2 Solution in information space

The *basin* or trough that the solution occupies is dependent on the number of solutions which occupy the same basin. The basin boundaries are not exactly linear, but rather a contour. This presents the possibility/probability for entry and escape from this specific point as given in Figure 15.3.

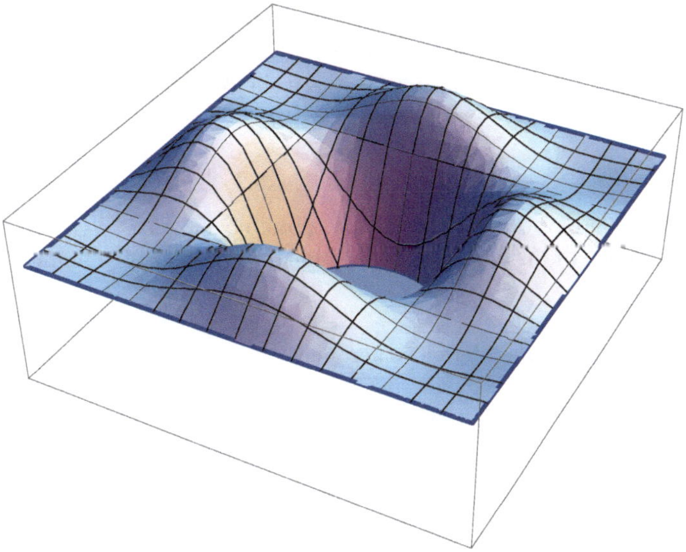

Fig. 15.3 Solution boundary in information space

As population evolves, the information is shared within the evolving solutions. Within a number of generations, a number of solutions can occupy the same information space. The size of the "basin" increasing and its attraction energy also increases. As more and more solutions are replicated, the number of "evolutionary channels" which exists between the solutions decreases. This gives rise to stagnation of the population, where no new solutions with new/better information is produced.

15.3.1 Initial Population

The main reason for random population is to provide an initial loose mapping of the solution space. For permutative problems, where solution ordering is stringent, it is often the case that adjacent values are required. A typical approach of using local search heuristics to search in the neighborhood of the solutions usually yields closely aligned solutions.

The initial population P, for this heuristic is partially stochastic and partly deterministic. The population is divided into two sub-populations, SPs, one randomly generated (SP_{rand}) and the other structurally generated (SP_{struct}).

The formulation for SP_{rand} is fairly simple. A random permutative string is generated for each solution till a specified number given as P_{size}.

The structured population SP_{struct} is somewhat more complex. It is made of two parts. In the first part, an initial solution is generated with ascending values given as $x_{ascending} = \{1,2,3,..,n\}$, where n is the size of the problem. In order to obtain a structured solution, the first solution is segmented and recombined in different orders to produce different combinations. The first segmentation occurs at ${}^{n}/_{2}$, and the two halfś are swapped to produce the second solution. The second fragmentation occurs by the factor 3; $^{n}/_{3}$ Three regions of solutions now exist. The number of possible recombination's that can exist is $3! = 6$. At this point there are nine solutions in the SP_{struct}. The general representation is given as:

$$k \geq 1 + 2! + 3! + ... + z! \tag{15.1}$$

where z is the total number of permutations possible and k is $^{P_{size}}/_{2}$.

The pseudocode of the clustered population generation is given in Figure 15.4.

Algorithm for Clustered Population Generation

Assume a population given as P which is divided equally into two sub-populations; one random SP_r and one structured SP_s. The schedule size is n and population size is NP. The maximum catenation of the schedule is given as c and the permutation rate is given as $p_r = c!$. Generate random population.

1. For $i = 1, 2, ..., NP/2$ do the following:
   a. Create a *random* solution schedule $\exists! x_i : SP_r := \{x_1, .., x_i.., x_n\}; i \in Z^+$
2. Create structured population.
   a. Calculate the truncation point and number as $t_p = \lfloor ^{n}/_{c} \rfloor$.
   b. Generate two schedules, one *forward biased* $X_f = \{1, 2, ..., n\}$ and the other *reverse biased* $X_r = \{n, n-1, ..., 1\}$.
   c. Generate permutation list for forward bias given as: $\{X_f\} = \{\{1, .., x_{t_p}\}, \{x_{t_p} + 1, ..., 2 \bullet x_{t_p}\}, ..., \{c \bullet x_{t_p}, ..., n\}\}$ and reverse bias as $\{X_r\} = \{\{n, .., c \bullet x_{t_p}\}, \{2 \bullet x_{t_p}, ..., x_{t_p} + 1\}, ..., \{x_{t_p}, ..., 1\}\}$.
   d. $i = 1, 2, ..., p_r$ do the following:
      i. Generate a permutative list based on the truncation points in the solution.
3. Output $P = SP_r \cup SP_s$ as the final population.

Fig. 15.4 Algorithm for Clustered Population Generation

15.3.2 Solution Dynamics

A solution represented as $x = \{x_1, x_2, ..., x_n\}$, where n is the number of variables, within a population has a number of attributes. Usually the most visible is its fitness value, by which it is measured within the population. This approach is not so viable in order to measure the diversity of the solution in the population. In retrospect, a single solution is assigned a number of attributes for measure, as given in Table 15.1.

Table 15.1 Solution Parameters

Parameter	Description	Activity
Deviation	Measure of the deviation of the solution	Control
Spread	Alignment of the solution	Control
Life	Number of generation cycles	Selection
Offspring	Number of successful offspring′s produced	Selection

The most important attribute is the *deviation* (the difference between successive values in a solution). Since we are using only permutative solutions, deviation or *ordering* of the solution is important. This is due to the fact that each value in the solution is unique. Each value in the solution has a unique footprint in the search space. The formulation for deviation is given as:

$$\delta = \left(\frac{\sum_{i=1}^{n-1} |x_i - x_{i+1}|}{n} \right) x_i \in \{x_1, x_2, ..., x_n\} \tag{15.2}$$

Spread of a solution gives the alignment of the solution. Each permutative solution has a specific ordering, whether it is *forward* aligned or *reverse* aligned. Whereas deviation measures the distance between adjacent solutions, spread is the measure of the hierarchy of subsequent solutions given as:

$$\partial = \begin{cases} +1 & if\,(x_{i+1} - x_i) \geq 1 \\ -1 & if\,(x_{i+1} - x_i) \leq 1 \end{cases} \quad i \in \{1, 2, ..., n\} \tag{15.3}$$

The generalization of *spread* is given in Table 15.2.

Table 15.2 Spread generalization

Spread	Generalization
> 0	Forward spread
0	Even spread
< 0	Reverse spread

Life is the number of generations the solution has survived in the population and *Offspring* is the number of viable solutions that have been created from that particular solution. These two variables are used for evaluating the competitiveness of different solutions.

The pseudocode is given in Figure 15.5.

Algorithm for Solution Dynamics

Assume a problem of size n, and a schedule given as $X = \{x_1,..,x_n\}$. There are NP schedules in the population. Initialize $X_{sprd} = 0$.

1. For $i = 1,2,....,NP$ do the following:

   a. Calculate **deviation**: $X_{i,dev} = \sum_{1}^{n-1} \frac{|x_i - x_{i+1}|}{n}$

   b. Calculate **spread**: $X_{i,sprd} = X_{sprd} + 1 \Leftrightarrow \sum_{1}^{n-1} (x_i - x_{i+1}) > 1$ and $X_{i,sprd} = X_{sprd} - 1 \Leftrightarrow \sum_{1}^{n-1} (x_i - x_{i+1}) < 1$

Fig. 15.5 Algorithm for Solution Dynamics

15.3.3 Chaotic Features

Within the population, certain solutions are seen to exhibit attracting features. These points are usually local optima regions, which draw the solutions together. The approach utilized is to subdivide the population in clusters, each cluster a distinct distance from another.

Figure 15.6 shows a “deviation” space with three clusters. Each cluster contains “n” solutions. At any one time “n” clusters will be in the population, and these clusters share information to create new solutions.

Two controlling parameters are now defined which control the clusters.

Chaos Attractor C_A: The distance that each segment of solution has to differ from each other. The C_A is given in (15.4).

$$C_A \in [0.1, 1+) \tag{15.4}$$

Within the population indexed by the *deviation*, solutions with similar deviation are clustered together, and each cluster is separated by at least a single C_A as seen in (15.5).

$$\begin{aligned} &(\delta_1, \delta_2, ..., \delta_i) \overset{C_A}{\leftrightarrow} (\delta_{i+1}, \delta_{i+2}, ..., \delta_{2i}) \overset{C_A}{\leftrightarrow} \\ &... \overset{C_A}{\leftrightarrow} (\delta_{3i+1}, \delta_{3i2}, ..., \delta_{4i}) \end{aligned} \tag{15.5}$$

The second controlling factor is the ***Chaos Edge*** C_E . Whereas C_A is the mapping of individual solutions, C_E is the measure of the entire population. Figure 15.7 shows

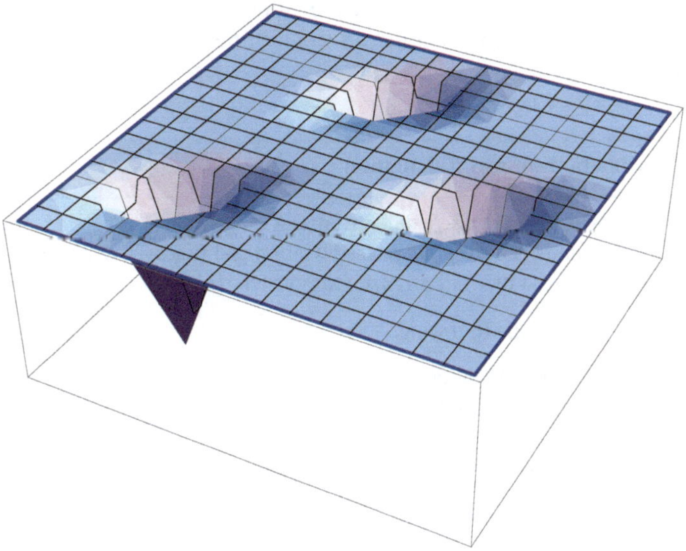

Fig. 15.6 Clusters in deviation space

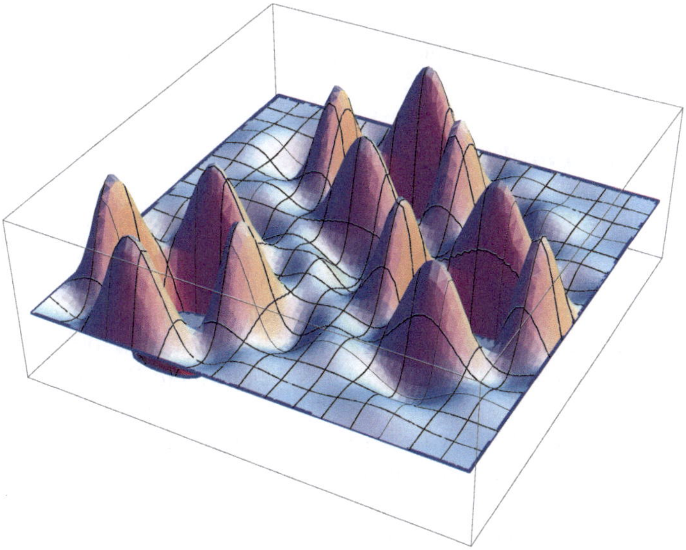

Fig. 15.7 Boundary of the clusters

the deviation space with the boundary outline. The entire "active" solution space is within the region of the outer contours. This is the "chaotic edge" of the current information space.

C_E is the measure of the deviation of the fitness of the population and is used to prevent the population from stagnating to any fitness minima. The algorithm is given in Figure 15.8.

Algorithm for Chaotic Features Calculation

Assume a problem of size n, and a schedule given as $X = \{x_1, .., x_n\}$. There are NP schedules in the population $\{P\}$ and each schedule has a deviation and fitness given by X_{devi} and X_{sprd}. The cluster distance is given by C_A. Initialize four clusters $\{C_1\}$, $\{C_2\}$, $\{C_3\}$ and $\{C_4\}$.

1. For $i = 1, 2, ..., NP$ do the following:
   a. Sort the $\{P\}$ in ascending order of X_{devi}.
   b. Divide the population into the four clusters based on X_{devi}.
   c. For $j = 1, .., 4$ do the following:
      i. Calculate the difference between boundary solutions of each cluster $\{C\}$. $C_{A,j} = X_{\max[X_{devi}],C_j} - X_{\min[X_{devi}],C_j}$
      ii. **IF** $C_{A,j} < C_A$
         A. Dynamic clustering of the boundary solutions of each cluster.
2. Output $\{P_C\}$ as the clustered population.

Fig. 15.8 Algorithm for Chaotic Features Calculation

15.3.4 Selection and Deletion

Selection of the next generation is based on a tier-based system. If the new solution improves on the global minima, it is then accepted in the solution. Otherwise, competing clusters jokey for the new solution. Initially the solution is mapped for its deviation. This deviation is then mapped to the corresponding cluster.

Within the cluster, the placement of the solution is evaluated. If the new solution corresponds to an existing solution, or reduces the threshold C_A value of the cluster, then it is discarded.

The solution is accepted if it improves on the C_A value of the cluster (hence improving diversity) and also to some extent keeps the balance of the C_E . If the cluster has less than average solutions, then the new solution is admitted.

Table 15.3 gives the selection criteria.

Once the solution is added to the cluster, another solution can be discarded. This solution is usually elected from the middle placed solutions in the cluster, whose fitness is not in the top 5% of the population. If no such solutions exist, then the average rated solution is removed. Solution with high *Life* and low *Offspring* are discarded, since they are considered dormant within the cluster. The algorithm for selection is given in Figure 15.9 and the algorithm for deletion is given in Figure 15.10.

Table 15.3 Selection criteria

Variables	Criteria
Fitness	Improves clusters best solution
C_A	Increases the value of C_A
C_E	Problem dependent

Algorithm for Selection

Assume a problem of size n, and a new schedule given as $X_{new} = \{x_1, .., x_n\}$. There are NP schedules in the population $\{P\}$ and each schedule has a deviation and fitness given by X_{devi} and X_{sprd}. The cluster distance is given by C_A.

1. Calculate the deviation and spread of the solution X_{new} as $X_{new,devi}$ and $X_{new,devi}$.
2. Find the associated cluster $P_{C,X}$ of the new solution X_{new} based on $X_{new,devi}$: $X_{new,devi} \in C$.
3. Calculate the fitness of the new solution $f(X_{new})$.
4. **IF** $X_{new} \to \{P_{C,X}\} \, \| X_{new,devi} \cup \{P_{C,X}\} > C_{A,X}$

   a. Insert the new solution in the associated cluster $X_{new} \to \{P_{C,X}\}$.
   b. Update the life X_{life} and offspring $X_{offspring}$ value of the parent solution.
   c. Calculate the $C_{E,X}$ of the new cluster.

Fig. 15.9 Algorithm for Selection

Algorithm for Deletion

Assume a problem of size n, and a new schedule given as $X_{new} = \{x_1, .., x_n\}$. There are NP schedules in the population $\{P\}$ and each schedule has a deviation and fitness given by X_{devi} and X_{sprd} and life and offspring given as X_{life} and $X_{offspring}$. The cluster distance is given by C_A and the Edge is given as C_E. The active cluster is given as $P_{C,A}$.

1. Randomly select a boundary solution as in the active cluster X_A. If the solution has poor offspring and long life in comparison to the average values of the cluster, it is deleted from the population.
2. **IF** $X_{A,offspring} < avg\left[P_{C,offspring}\right] \, \| X_{A,life} > avg\left[P_{C,life}\right]$

   a. Delete X_A.

   If the selected solution increases the C_A value between the clusters, it is selected for deletion.
3. **ELSE IF** $(X_A \not\subset \{P_{C,X}\}) > C_A$

   a. Delete X_A.
4. Calculate the $C_{E,X}$ of the new cluster.

Fig. 15.10 Algorithm for Deletion

Table 15.4 gives the deletion criteria.

Table 15.4 Deletion criteria

Variables	Criteria
Life	High
Offspring	Low
C_A	Decreases

15.3.5 Dynamic Clustering

The selection and crossover criteria have now been outlined. After each generation / migration, the clusters are reconfigured. Since, in all heuristics, there is a tendency to converge, it is imperative to keep the solutions unique.

The procedure is to calculate the deviation of the new solutions. Since a mesh of solutions may exist, it is feasible to reconfigure certain boundary solutions. Figure 15.11 can be a representation of a sub-population (SP).

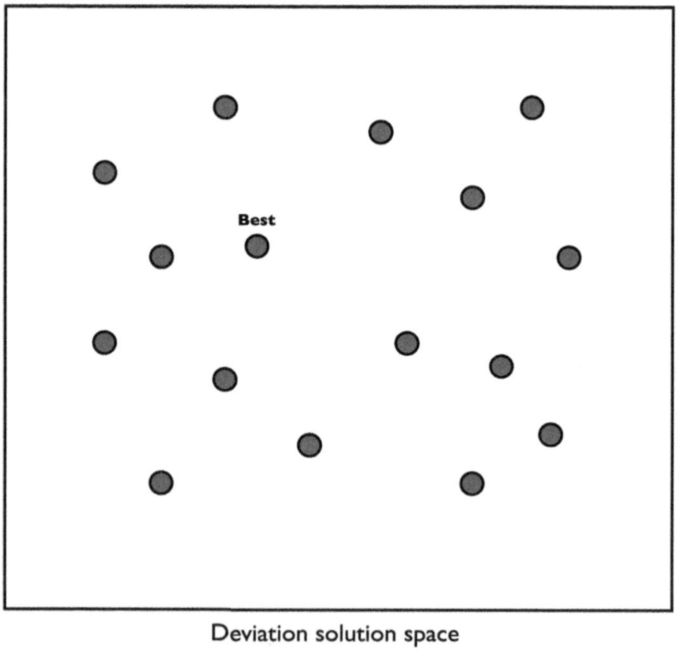

Fig. 15.11 Solution space after migration

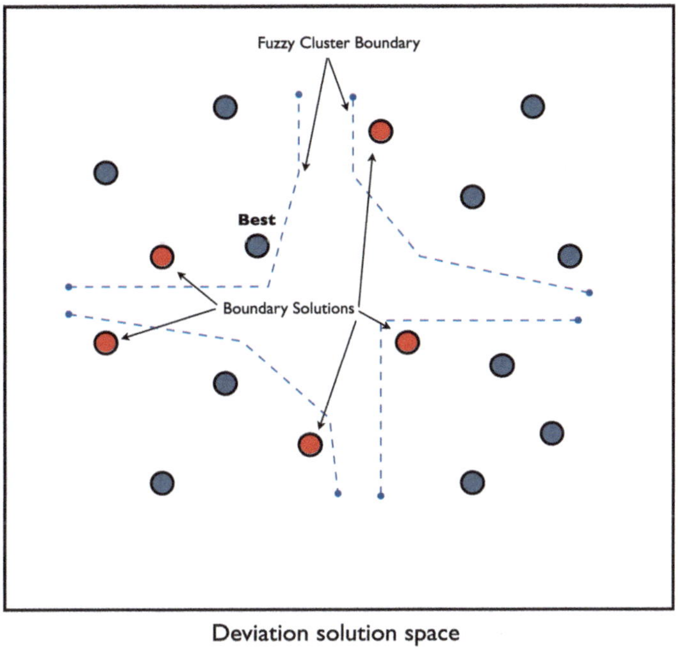

Fig. 15.12 Fuzzy clustering and boundary solution isolation

A mutation routine is used to reconfigure a solution. By altering certain positions within the solution it is possible to realign the deviation and spread of the solution. Boundary values within the solutions (usually represented by the upper and lower bound of the solution) are swapped. Another approach is to have two random positions generated and the values in these positions swapped. An illustration is given to describe this process in Table 15.5, Figure 15.12 and Figure 15.13.

Table 15.5 Swap of boundary values

Solution	Deviation	Spread
10 9 6 5 2 **1** 8 7 4 3	2.1	-7
1 9 6 5 2 **10** 8 7 4 3	3.0	-5

Once the boundary values are re-aligned, the second migration/generation loop occurs. The entire process pseudocode is given in Figure 15.14.

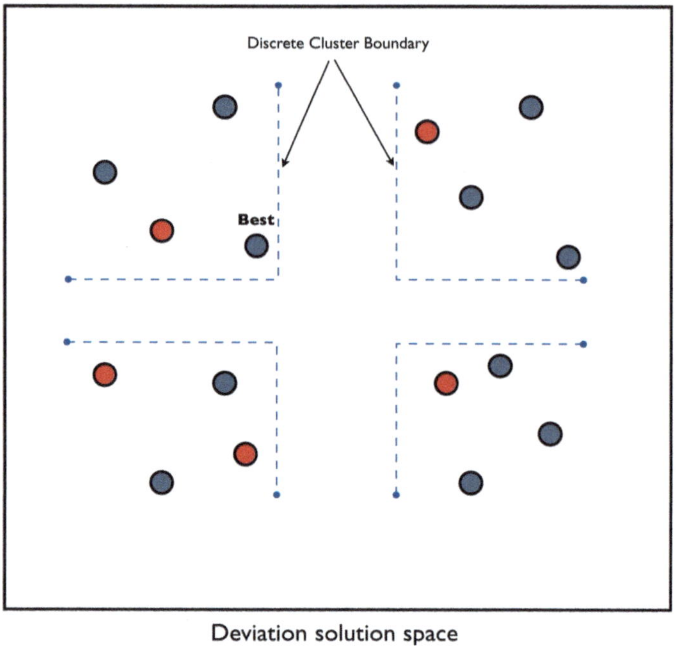

Fig. 15.13 Realigned solutions into discrete clusters

Algorithm for Dynamic Clustering

Assume four clusters C_1 - C_4, each with separation distance $C_{A,i}$, where i refers to the corresponding cluster. Each schedule has n variables.

1. Isolate each schedule in a cluster which has a separation value less than that of C_A: $X_{devi} < C_{A,X}$.
2. **DO**
   a. Randomly select two unique random indices on the schedule $Rnd[r_1, r_2] \in n$.
   b. Using these indices exchange the values in the solution: $x_{r_1} \Leftrightarrow x_{r_2}$.
   c. Calculate new deviation of the solution $X_{new,devi}$.
   d. **IF** $X_{devi} > C_{A,X}$
      i. Accept new schedule in the solution $X_{new} \rightarrow \{P_{C,X}\}$
3. **WHILE** new schedule NOT accepted in cluster

Fig. 15.14 Algorithm for Dynamic Clustering

15.4 Metaheuristics

The clustered population is designed to be used by any metaheuristic. This is the advantage of this approach, since it is not tied down to a specific method. This section discusses three different heuristics of Genetic Algorithm (GA), Differential Evolution (DE) algorithm and finally Self-Organising Migrating Algorithm (SOMA). Each of these heuristics has been applied to the QAP problem. The DE approach is taken from [11], and the SOMA approach is of [12]. The GA approach was created to bring completeness to the strategies used.

In each of the heuristics used, the canonical population was removed and replaced with the clustered population and its integrated features.

15.4.1 Genetic Algorithms

Genetic Algorithm (GA) is an adaptive heuristic search algorithm premised on the evolutionary ideas of natural selection and genetics. GA is designed to simulate processes in natural system necessary for evolution, specifically those that follow the principles first laid down by Charles Darwin of survival of the fittest. As such, they represent an intelligent exploitation of a random search within a defined search space to solve a problem [17].

A number of variants of GA exist. For this research, a two-point crossover approach was used as the crossover methodology for the propagation of the population.

A two-point crossover approach is simple to execute. Two solutions from *different clusters* are randomly selected. These solutions are checked to ensure that their *spread* is not equal. This is done to map more diversified solutions. Two crossover positions are randomly selected in the solutions given as $\{CP_1, CP_2\} = Random[n]$, and the two solutions are mated with a possibility of six unique offspring's being created. An illustration of the selection and crossover is given in Figure 15.15.

An example of this process can be shown by having the two values of crossover given as $CP_1 = 2$ and $CP_2 = 4$. The two solutions selected for crossover can be represented as $x_1 = \left\{2,5,\underset{2}{|}4,3,\underset{4}{|}1,6\right\}$ and $x_2 = \left\{3,4,\underset{2}{|}1,2,\underset{4}{|}6,5\right\}$. Three regions exist within each solution. By swapping alternate regions, a total number of possible solutions is now given as in Table 15.6.

With this crossover process, infeasible solutions are usually created. An effective repairment routine is described in the following section that was used to repair the solutions.

Once all the solutions are repaired, their fitness is evaluated and the solution with the best fitness is selected for possible adaptation into the population.

15.4.1.1 Repairment

The repairment process is given in a number of routines. The first routine is to check the entire solution for repeated values. These repeated values and their positions are

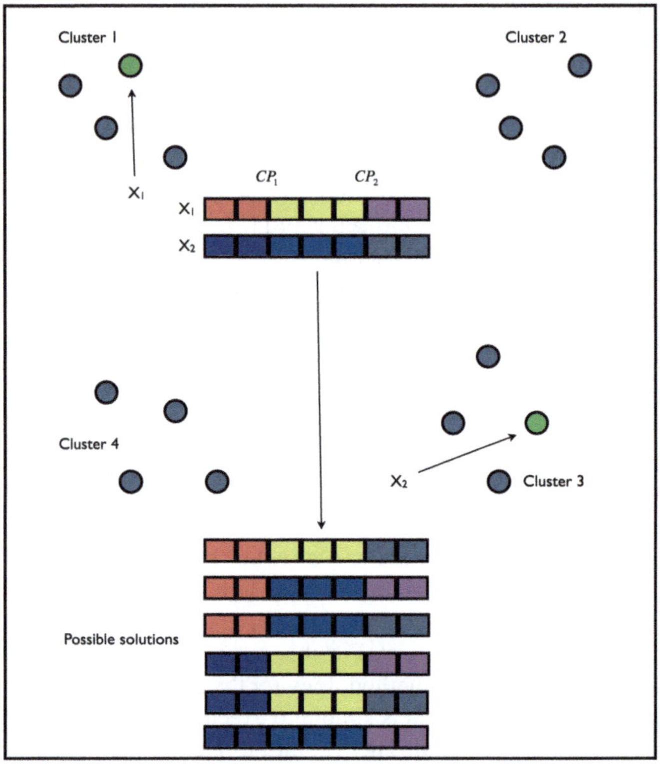

Fig. 15.15 GA representation

Table 15.6 Possible solutions from crossover

Permutation	Solution
$\{1,1,2\}$	$\{2,5,4,3,6,5\}$
$\{1,2,1\}$	$\{2,5,1,2,1,6\}$
$\{1,2,2\}$	$\{2,5,1,2,6,5\}$
$\{2,1,1\}$	$\{3,4,4,3,1,6\}$
$\{2,1,2\}$	$\{3,4,4,3,6,5\}$
$\{2,2,1\}$	$\{3,4,1,2,1,6\}$

isolated in a replicated array $x_{repl} = \{x_j, x_{j+n}, .., x\}$. The second routine is to find which values are missing from the solutions given as $x_{mis} = \{1,..,n\} \cap \{x_1, x_2, .., x_n\}$.

Since, the replicated array contains a number of sequences of replicated solutions, randomly one solution in each sequence is labelled as feasible and repatriated back into the main solution. This leaves the replicated array containing only infeasible values.

Randomly each value is selected from the missing array and inserted in the position of a replicated value in the replicated array $x_{mis} \overset{random}{\longrightarrow} x_{repl}$.

Finally, the replicated array is reinserted in the solution array with all values now feasible $x_{repl} \longrightarrow x$.

An illustrative example is given in Table 15.7.

Table 15.7 Illustrative example of repairment.

Routine	Rand	x	x_{repl}	x_{mis}
Replicated values		$\{1,3,4,3,4,$ $10,6,7,1,1\}$	$(1,1,1^*)$ $(4^*,4)$	
Missing value				$\{2,8,9\}$
Feasible solution	$\{3,1\}$	$\{*,3,4,3,*,$ $10,6,7,*,1\}$	$(1,1,1^*)$ $(4^*,4)$	
Repair solution	$\{2,3,1\}$ $\{3,1,2\}$		$\{\underset{3}{1},\underset{1}{1},\underset{2}{4}\}$	$\{\underset{2}{2},\underset{3}{8},\underset{1}{9}\}$
Final solution		$\{8,3,4,5,9,$ $10,6,7,2,1\}$		

15.4.2 *Differential Evolution Algorithm*

Differential Evolution (DE) [33], is the second heuristic selected to be used in conjunction with the clustered population. DE uses a vector perturbation methodology for crossover.

Each solution is visualized as a vector in search space. A new vector is created by the combination of four unique vectors. A schematic of DE applied to clustering is given in Figure 15.16.

There are ten working strategies for DE, but the one selected for implementation is the DE/rand/2/bin represented as in (15.6):

$$U_{i,G+1} = x_{best,G} + F \cdot (x_{j,r_1,G} - x_{j,r_2,G} - x_{j,r_3,G} - x_{j,r_4,G}) \tag{15.6}$$

This strategy was selected since it maps to the four unique clusters in the *SP*. The best solution is selected from the entire *SP* based on fitness value. Then, each random solution is selected from each distinct cluster. Again the selected values are checked for opposing *spread*. If the spread is identical, then a second round of selection occurs. A schematic is given in Figure 15.17.

$$
\begin{array}{l}
1.\ \text{Input: } D, G_{\max}, NP \geq 4, F \in (0,1+), CR \in [0,1], \text{ and initial bounds:} x^{(lo)}, x^{(hi)}. \\
\left\{\begin{array}{l}
2.\ \text{While } G < G_{\max} \\
\forall i \leq NP \left\{\begin{array}{l}
3.\ \text{Mutate and recombine:} \\
\qquad 3.1 \quad r_1, r_2, r_3, r_4 \in \{1,2,\ldots, P_{size}\}, \\
\qquad\qquad \text{randomly selected from each cluster} \\
\qquad 3.2 \quad j_{rand} \in \{1,2,\ldots,D\}, \text{ randomly selected once each } i \\
3.3\ \ \forall j \leq D, u_{j,i,G+1} = \begin{cases} x_{best,G} + F \cdot (x_{j,r_1,G} - x_{j,r_2,G} - x_{j,r_3,G} - x_{j,r_4,G}) \\ \qquad \text{if } (rand_j[0,1] < CR \vee j = j_{rand}) \\ x_{j,i,G} \quad \text{otherwise} \end{cases} \\
4.\ \text{Select Criteria}
\end{array}\right. \\
G = G + 1
\end{array}\right.
\end{array}
$$

Fig. 15.16 DE selection

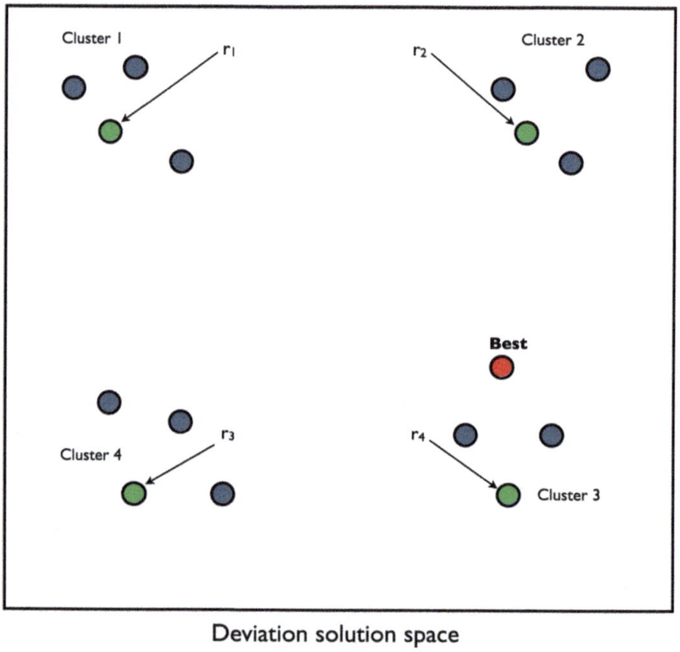

Fig. 15.17 DE selection

The selection of the cluster is random, so r_1 can be selected from any cluster with no preference. These values are subtracted given as $x_{j,r_1,G} - x_{j,r_2,G} - x_{j,r_3,G} - x_{j,r_4,G}$. The resulting value is multiplied by the scaling factor F and added to the best solution as given in Figure 15.18.

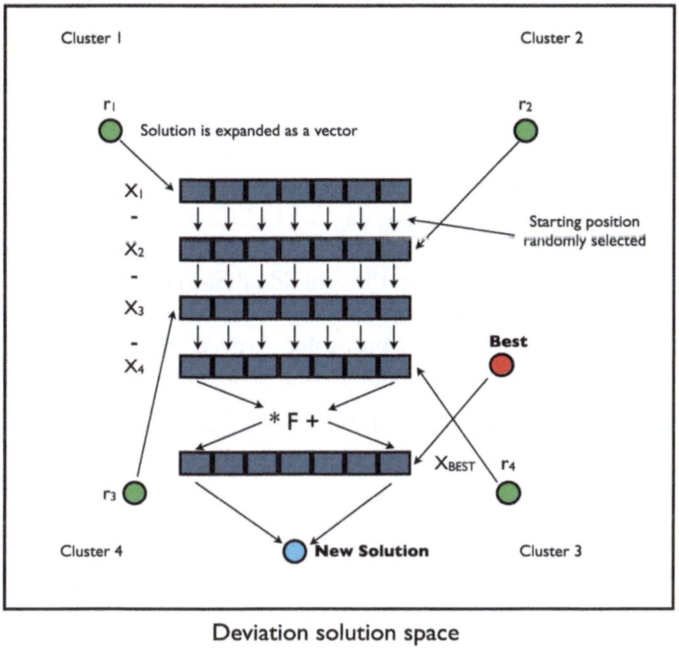

Fig. 15.18 DE crossover

The resulting value is only accepted in the new solution if a generated random number is below the given threshold provided by the controlling parameter of *CR*. This procedure provides added stochasticity to the heuristic.

15.4.3 Self Organizing Migrating Algorithm

The third utilized heuristic is SOMA [39], which is based on the competitive-cooperative behavior of intelligent creatures solving a common problem.

In SOMA, individual solutions reside in the optimized model's hyperspace, looking for the best solution. It can be said, that this kind of behavior of intelligent individuals allows SOMA to realize very successful searches.

Because SOMA uses the philosophy of competition and cooperation, the variants of SOMA are called strategies. They differ in the way as to how the individuals affect all others. The best operating strategy is called 'AllToAll' and consists of the following steps:

1. *Definition of parameters*. Before execution, the SOMA parameters (PathLength, Step, PRT, Migrations see Table 10) are defined.

2. *Creating of population.* The population *SP* is created and subdivided into clusters.
3. *Migration loop.*
   a. Each individual is evaluated by the cost function
   b. For each individual the PRT Vector is created.
   c. All individuals, perform their run towards the randomly selected solution in the opposing cluster according to (15.8). Each solution is selected from individual cluster piecewise. The movement consists of jumps determined by the Step parameter until the individual reaches the final position given by the PathLength parameter. For each step, the cost function for the actual position is evaluated and the best value is saved. Then, the individual returns to the position, where it found the best-cost value on its trajectory.

The schematic of SOMA with clustered population is given in Figure 15.19.

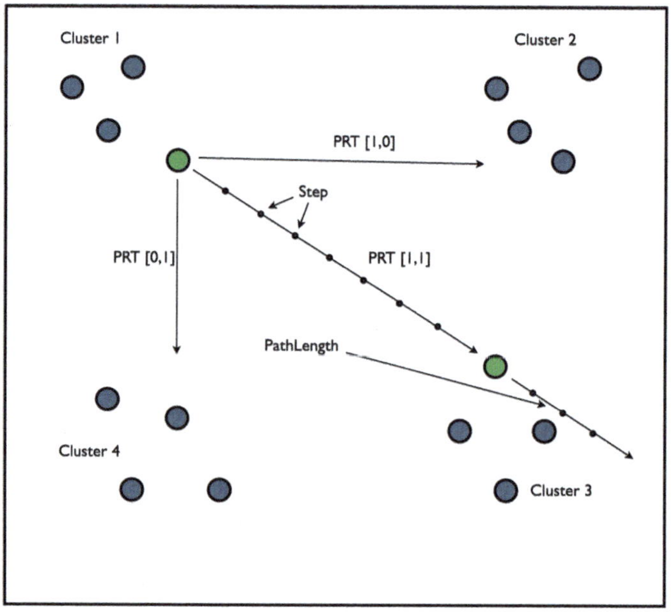

Fig. 15.19 SOMA migration utilizing clustered population

SOMA, like other evolutionary algorithms, is controlled by a number of parameters, which are predefined. They are presented in Table 15.8.

Table 15.8 SOMA parameters

Name	Range	Type
PathLength	(1.1 − 3)	Control
StepSize	(0.11 − PathLength)	Control
PRT	(0 − 1)	Control

15.4.3.1 Mutation

Mutation, the random perturbation of individuals, is applied differently in SOMA compared with other ES strategies. SOMA uses a parameter called PRT to achieve perturbation. It is defined in the range [0, 1] and is used to create a perturbation vector (PRT Vector) as shown in (15.7):

$$\begin{aligned} &if \quad rnd_j < PRT \quad \text{then} \quad PRTVector_j = 1 \\ &else \quad 0, \quad j = 1,..,n \end{aligned} \tag{15.7}$$

The novelty of this approach is that in its canonical form, the PRT Vector is created before an individual starts its journey over the search space. The PRT Vector defines the final movement of an active individual in search space.

The randomly generated binary perturbation vector controls the allowed dimensions for an individual. If an element of the perturbation vector is set to zero, then the individual is not allowed to change its position in the corresponding dimension.

15.4.3.2 Crossover

In standard ES, the crossover operator usually creates new individuals based on information from the previous generation. Geometrically speaking, new positions are selected from an N-dimensional hyper-plane. In SOMA, which is based on the simulation of cooperative behavior of intelligent beings, sequences of new positions in the N-dimensional hyperplane are generated. The movement of an individual is thus given as follows:

$$\mathbf{r} = \mathbf{r}_0 + \mathbf{m}t PRTVector \tag{15.8}$$

where:

- $\mathbf{r}$: new candidate solution
- $\mathbf{r}_0$: original individual
- $\mathbf{m}$: difference between leader and start position of individual
- t : $\in$ [0 , Path length]
- $PRTVector$: control vector for perturbation

It can be observed from (15.8) that the PRT vector causes an individual to move toward the leading individual (the one with the best fitness) in *N-k* dimensional space. If all *N* elements of the PRT vector are set to 1, then the search process is carried out in an *N* dimensional hyperplane (*i.e.* on a *N+1* fitness landscape). If some elements of the PRT vector are set to 0 then the second terms on the right−hand side of (15.8) equals 0. This means those parameters of an individual that are related to 0 in the PRT vector are not changed during the search. The number of frozen parameters, *k,* is simply the number of dimensions that are not taking part in the actual search process. Therefore, the search process takes place in an *N-k* dimensional subspace.

For each individual, once the final placement is obtained, the values are re-converted into integer format. SOMA conversion is different from that used for DE. The values are simply rounded to the nearest integer and repaired using the repairment procedure. This process was developed and selected during experimentation.

15.5 General Template

Collating all the piecewise explanation, a general generic template is now described.

1. *Initialize*: Assign the problem size n, population size P_{size}, sub population sizes SP_{struct} SP_{rand}, and the control parameters of C_A and C_E .
2. *Generate:* Randomly create SP_{rand}, half the size of P_{size}, and then structurally create SP_{struct}. These two form the basis of the population.
3. *Calculate:* Calculate the *deviation* and *spread* of each solution in the population. Taking the *deviation* values, configure the population into four clusters. The minimal separation value between the clusters is assigned as C_A. Taking the entire *SP*, the standard deviation of the *fitness* is computed. This is labelled as the C_E.
4. *Generation/Migration*
    a. Taking each *SP* in turn, the selected heuristic of GA, DE or SOMA is applied to the population.
    b. The new solution is calculated for its *deviation* and *spread.*
    c. Using the selection criteria, the solution is placed within the cluster corresponding to its deviation. If replicated solutions exist, then it is discarded. Selection is based on *fitness* and the move of the C_A and C_E.
5. *Re-calculation:* The SP is re-calculated for its cluster boundaries.
6. *Dynamic clustering:* If the value of C_A has deceased, then the boundary solutions are reconfigured. The C_E value is calculated for the new population.

The generic template is given in Figure 15.20

1. Input: $n, P_{size}, SP_{struct}, SP_{rand}, C_A \in (0.1, 1+), C_E, Gen$

2. Initialize: $SP_{rand} = \begin{cases} \forall i \le P_{size}/2 \wedge \forall j \le n : x_{i,j,G=0} = rand_j[0,1]\left(x_j^{(hi)} - x_j^{(lo)}\right) \\ i = \left\{1,2,.., P_{size}/2\right\}, j = \{1,2,...,n\}, G = 0, rand_j[0,1] \in [0,1] \end{cases}$

$$SP_{struct} = \begin{cases} k = P_{size}/2 ; z \ni \max\left(\sum_{i=1}^{z}(1+2!+...+z!)\right) \le k; x_{ascend} = \left\{x^{(lo)},..,x^{(hi)}\right\} \\ x_{descend} = \left\{x^{(hi)},..,x^{(lo)}\right\} \\ \forall i \le z, \left(x_{ascend}, x_{descend}\right) \subseteq i \xrightarrow{permutate} append\left(SP_{struct}\right) \end{cases}$$

3. Calculate $\begin{cases} \text{Deviation } \delta = \left(\dfrac{\sum_{j=1}^{n-1}\left|x_j - x_{j+1}\right|}{n}\right) : x_j \in \{x_1, x_2, .., x_n\} \\ \text{Spread } \partial = \begin{cases} +1 & if\left(x_{j-1} - x_j\right) \ge 1 \\ -1 & if\left(x_{j-1} - x_j\right) \ge 1 \end{cases} \\ C_A = \left(\delta_1, \delta_2, .., \delta_{k/5}\right) \overset{C_A}{\leftrightarrow} \left(\delta_{(k/5)+1}, \delta_{(k/5)+2}, .., \delta_{2(k/5)}\right) \overset{C_A}{\leftrightarrow} ... \overset{C_A}{\leftrightarrow} \left(\delta_{4(k/5)+1}, \delta_{4(k/5)+2}, .., \delta_k\right) \\ C_E = std\left(f(x_i)\right) : x_i \in \left\{x_1, x_2, ..., x_{P_{size}}\right\} \end{cases}$

$\begin{cases}$ 4. While $G < G_{\max}$ for each SP

$\forall i \le P_{size}$:

- 5. Mutate and recombine:

  $$u_{i,G+1} \overset{DE}{\leftarrow} \{x_1, x_2, ..., x_i\}$$

  5.1 Calculate δ and ∂ of $u_{i,G+1}$
- 6. Select

  $$x_{i,G+1} = \begin{cases} u_{i,G+1} & \text{if } f\left(u_{i,G+1}\right) \le f\left(x_{best}\right) \\ u_{i,G+1} & \text{if } > C_A \\ x_{i,G} & \text{otherwise} \end{cases}$$
- 7. Calculate C_A, C_E
- 8. Dynamic clustering

$G = G + 1$

Fig. 15.20 General Template

15.6 Quadratic Assignment Problem

QAP is an important problem in theory and practice. Formally, given n facilities and n locations, two n x n matrices $A = [a_{ij}]$ and $B = [b_{rs}]$, where a_{ij} is the distance between locations i and j and b_{rs} is the flow between facilities r and s, the QAP can be stated as follows:

$$\min_{\psi \varepsilon S(n)} \sum_{i=1}^{n} \sum_{j=1}^{n} b_{ij} a_{\psi_i \psi_j} \tag{15.9}$$

where S(n) is the set of all permutations (corresponding to the assignments) of the set of integers $\{1,\ldots,n\}$, and the ψ_i gives the location of facility i in the current solution $\psi \varepsilon S(n)$. Here $b_{ij} a_{\psi_i \psi_j}$ describes the cost distribution of simultaneously assigning facility i to location ψ_j and facility j to location ψ_i.

The term *quadratic* stems from the formulation of the QAP as an integer optimization problem with a quadratic objective function. Let x_{ij} be a binary variable which takes value 1 if facility i is assigned to location j and 0 otherwise. Then the problem can be formulated as:

$$\min \sum_{i=1}^{n} \sum_{j=1}^{n} \sum_{l=1}^{n} \sum_{k=1}^{n} a_{ij} b_{kl} x_{ik} x_{jl} \tag{15.10}$$

subject to the constraints

$$\sum_{i=1}^{n} x_{ij} = 1, \sum_{j=1}^{n} x_{ij} = 1, x \varepsilon \{0,1\} \tag{15.11}$$

According to [36], the QAP instances found in QAPLIB can be classified into four classes;

- **Unstructured, randomly generated instances:** Instances with the distance and flow matrix entries generated randomly according to an uniform distribution. The *taixxa* is an example of these instances, which are considered the most difficult to solve (we note that $x \equiv$ integer number).
- **Unstructured instances:** Instances with the grid matrix defined as the Manhattan distance between grid points on a n_1 x n_2 grid and with random flows.
- **Real-life instances:** 'Real-life' instances from practical application of the QAP. Amongst them are the layout problem of the hospital (*kra30x*)and instances corresponding to the layout of the typewriter keyboards (*bur26x*). The real-life instances have in common that the flow matrices have (in contrast to the previously mentioned randomly generated instances) many zero entries and the entries are not uniformly distributed.
- **Real-life like instances:** Because the real life like instances are mainly of small size, [36] proposed the *taixxb* instances in such a way that they resemble the distribution found in real life problems.

In order to differentiate different classes of QAP the flow dominance *fd* is used. It is defined as the coefficient of the flow matrix entries multiplied by the factor of 100 and is represented as:

$$fd(B) = 100 \cdot \frac{\sigma}{\mu} \tag{15.12}$$

where

$$\mu = \frac{1}{n^2} \cdot \sum_{i=1}^{n} \sum_{j=1}^{n} b_{ij}$$

and

$$\sigma = \sqrt{\frac{1}{n^2 - 1} \cdot \sum_{i=1}^{n} \sum_{j=1}^{n} (b_{ij} - \mu)^2}$$

The general description is that unstructured randomly generated problems with a uniform distribution have a *fd* of less than 1.2 while real life structured problems have a *fd* larger than 1.2.

15.7 Results

This section presents the results obtained from the three different sets of experimentations conducted. Each experiment was repeated 10 times with the same control values. The presented results are the best solutions obtained from these ten simulation on each instant.

All experimentation was conducted on an parallel array of 16 X-Serves with a total of 64 Quad Xeon processors all running on Grid Mathematica platform.

15.7.1 Genetic Algorithm Results

The first set of results is from Genetic Algorithms. The operational parameters of GA is given in Table 15.9.

Table 15.9 GA operational values

Parameter	**Value**
Strategy	2 Point Crossover
Mutation	Single
Population size	500 - 1000
Generations	500 - 1000

The generic and clustered GA results for the irregular problems is presented in Table 15.10.

Table 15.10 Clustered GA Irregular QAP comparison

Instant	fd	n	Optimal	GA	GA_{clust}
bur26a	2.75	26	5246670	1.64	**1.25**
bur26b	2.75	26	3817852	1.95	**1.34**
bur26c	2.29	26	5426795	1.75	**1.56**
bur26d	2.29	26	3821225	1.24	**1.21**
bur26e	2.55	26	5386879	1.52	**1.32**
bur26f	2.55	26	3782044	1.62	**1.56**
bur26g	2.84	26	10117172	1.53	**1.42**
bur26h	2.84	26	7098658	1.65	**1.54**
chr25a	4.15	26	3796	2.3	**1.56**
els19	5.16	19	17212548	0.94	**0.91**
kra30a	1.46	30	88900	1.23	**1.12**
kra30b	1.46	30	91420	1.64	**1.34**
tai20b	3.24	20	122455319	1.58	**1.21**
tai25b	3.03	25	344355646	1.61	**0.94**
tai30b	3.18	30	637117113	2.19	**1.24**
tai35b	3.05	35	283315445	2.32	**0.85**
tai40b	3.13	40	637250948	2.54	**1.12**
tai50b	3.1	50	458821517	2.75	**1.24**
tai60b	3.15	60	608215054	2.68	**1.52**
tai80b	3.21	80	818415043	3.11	**1.95**

The results of the regular problems in given in Table 15.11.

The results clearly demonstrate that using clustering improves the results of generic GA. Even though the results obtained for GA are not as competitive for the QAP instances, the main idea of this research of clustering of the population to improve the performance of metaheuristics is validated.

15.7.2 Differential Evolution Results

The second experiment is conducted with Differential Evolution algorithm. Extensive experimentation was conducted with both the regular and irregular QAP problems. Comparison is done with the DE heuristic without clustering [11].

The operational parameters of DE are given in Table 15.12.

The first part of the results is on the irregular QAP instances. The results are presented in Table 15.13. The columns represent the name of the problem, its flow

Table 15.11 Clustered GA Regular QAP comparison

Instant	fd	n	Optimal	GA	GA_{clust}
nug20	0.99	20	2570	0.98	**0.85**
nug30	1.09	30	6124	0.84	**0.82**
sko42	1.06	42	15812	0.95	**0.84**
sko49	1.07	49	23386	1.12	**0.93**
sko56	1.09	56	34458	1.35	**0.94**
sko64	1.07	64	48498	1.68	**1.23**
sko72	1.06	72	66256	2.52	**1.54**
sko81	1.05	81	90998	3.21	**2.15**
tai20a	0.61	20	703482	0.98	**0.52**
tai25a	0.6	25	1167256	**0.68**	**0.68**
tai30a	0.59	30	1818146	1.02	**0.95**
tai35a	0.58	35	2422002	1.32	**0.98**
tai40a	0.6	40	3139370	1.54	**1.22**
tai50a	0.6	50	4941410	1.62	**1.31**
tai60a	0.6	60	7208572	2.13	**1.98**
tai80a	0.59	80	13557864	3.21	**2.35**
wil50	0.64	50	48816	1.89	**0.98**

Table 15.12 DE operational values

Parameter	Value
Strategy	DE/rand/2/bin
CR	0.9
F	0.3
Population	500 - 1000
Generation	500 - 1000

dominance, problem size, optimal reported value, DE result and DE with clustering result.

Comparing the results of DE and DE_{clust}, it is easy to see that DE_{clust} performs better than DE. Of the 8 bur*xx* instances, the optimal result is obtained for all instances. On the kra*xx* and tai*xx* instances, DE_{clust} outperforms DE marginally.

The second part of the results is on the regular QAP instances as given in Table 15.14.

DE_{clust} outperforms DE in regular QAP instances. It manages to find 10 optimal instances out of the 16 tested. Of the remaining 6, DE_{clust} obtains close to 0.01% to the optimal.

Table 15.13 Clustered DE Irregular QAP comparison

Instant	fd	n	Optimal	DE	DE_{clust}
bur26a	2.75	26	5246670	0.006	**0**
bur26b	2.75	26	3817852	0.0002	**0**
bur26c	2.29	26	5426795	0.00005	**0**
bur26d	2.29	26	3821225	0.0001	**0**
bur26e	2.55	26	5386879	0.0002	**0**
bur26f	2.55	26	3782044	0.000001	**0**
bur26g	2.84	26	10117172	0.0001	**0**
bur26h	2.84	26	7098658	0.0001	**0**
chr25a	4.15	26	3796	0.227	**0.07**
els19	5.16	19	17212548	0.0007	**0**
kra30a	1.46	30	88900	0.0328	**0.024**
kra30b	1.46	30	91420	0.0253	**0.015**
tai20b	3.24	20	122455319	0.0059	**0**
tai25b	3.03	25	344355646	0.003	**0**
tai30b	3.18	30	637117113	0.0239	**0**
tai35b	3.05	35	283315445	0.0101	**0.002**
tai40b	3.13	40	637250948	0.027	**0**
tai50b	3.1	50	458821517	0.001	**0**
tai60b	3.15	60	608215054	0.0144	**0.012**
tai80b	3.21	80	818415043	0.0287	**0.014**

Table 15.14 Clustered DE Regular QAP comparison

Instant	fd	n	Optimal	DE	DE_{clust}
nug20	0.99	20	2570	0.018	**0**
nug30	1.09	30	6124	0.005	**0**
sko42	1.06	42	15812	0.009	**0**
sko49	1.07	49	23386	0.009	**0**
sko56	1.09	56	34458	0.012	**0**
sko64	1.07	64	48498	0.013	**0.006**
sko72	1.06	72	66256	0.011	**0.007**
sko81	1.05	81	90998	0.011	**0.01**
tai20a	0.61	20	703482	0.037	**0**
tai25a	0.6	25	1167256	0.026	**0**
tai30a	0.59	30	1818146	0.018	**0**
tai35a	0.58	35	2422002	0.038	**0**
tai40a	0.6	40	3139370	0.032	**0.019**
tai50a	0.6	50	4941410	0.033	**0.026**
tai60a	0.6	60	7208572	0.037	**0.012**
tai80a	0.59	80	13557864	0.031	**0.021**
wil50	0.64	50	48816	0.004	**0**

15.7.3 Self Organizing Migration Algorithm Results

The third and final experiment was conducted with SOMA. The operational parameters of SOMA is given in Table 15.15.

Table 15.15 SOMA operational values

Parameter	Value
Strategy	All-to-All
Step Size	0.21
PathLength	3
Population	500 - 1000
Migration	500 - 1000

The results are compared with those of SOMA without clustering of [12] and is given in Table 15.16.

Table 15.16 Clustered SOMA Irregular QAP comparison

Instant	fd	n	Optimal	SOMA	$SOMA_{clust}$
bur26a	2.75	26	5246670	**0**	**0**
bur26b	2.75	26	3817852	**0**	**0**
bur26c	2.29	26	5426795	**0**	**0**
bur26d	2.29	26	3821225	**0**	**0**
bur26e	2.55	26	5386879	**0**	**0**
bur26f	2.55	26	3782044	0.03	**0.01**
bur26g	2.84	26	10117172	**0**	**0**
bur26h	2.84	26	7098658	**0**	**0**
chr25a	4.15	26	3796	0.129	**0.10**
els19	5.16	19	17212548	**0**	**0**
kra30a	1.46	30	88900	**0.002**	**0.002**
kra30b	1.46	30	91420	0.03	**0.027**
tai20b	3.24	20	122455319	0.004	**0**
tai25b	3.03	25	344355646	**0**	**0**
tai30b	3.18	30	637117113	0.043	**0**
tai35b	3.05	35	283315445	**0**	**0**
tai40b	3.13	40	637250948	0.02	**0**
tai50b	3.1	50	458821517	0.2	**0.2**
tai60b	3.15	60	608215054	0.5	**0.2**
tai80b	3.21	80	818415043	0.8	**0.4**

The results of clustered SOMA with regular problems is given in Table 15.17.

Table 15.17 Clustered SOMA Regular QAP comparison

Instant	fd	n	Optimal	SOMA	$SOMA_{clust}$
nug20	0.99	20	2570	**0**	**0**
nug30	1.09	30	6124	0.02	**0**
sko42	1.06	42	15812	0.01	**0**
sko49	1.07	49	23386	0.005	**0**
sko56	1.09	56	34458	0.01	**0**
sko64	1.07	64	48498	0.06	**0.02**
sko72	1.06	72	66256	0.2	**0.04**
sko81	1.05	81	90998	0.35	**0.05**
tai20a	0.61	20	703482	**0**	**0**
tai25a	0.6	25	1167256	**0**	**0**
tai30a	0.59	30	1818146	0.01	**0**
tai35a	0.58	35	2422002	0.03	**0**
tai40a	0.6	40	3139370	0.623	**0.58**
tai50a	0.6	50	4941410	0.645	**0.42**
tai60a	0.6	60	7208572	0.62	**0.62**
tai80a	0.59	80	13557864	1.05	**0.95**
wil50	0.64	50	48816	**0**	**0**

15.8 Analysis

Comparison of the obtained results is done with some published heuristics. The first comparison is done with the irregular QAP instances. The two best performing results of DE_{clust} and $SOMA_{clust}$ is compared with the Improved Hybrid Genetic Algorithm of [20] shown as GA_1 and the highly refereed Ant Colony approach of [14] given as HAS in Table 15.18.

The best performing algorithm is DE_{clust} which obtains the best comparative result in 17 out of 20 problem instances. $SOMA_{clust}$ obtains the best results in 13 instances and HAS in 12 instances. The hybrid Genetic Algorithm approach however is able to find the optimal result in the two instances that it is applied, where the other heuristics are not so effective. For the larger size problems, DE_{clust} proves to be a better optimizer.

The second set of comparison is done with the regular QAP instances. Comparison of the clustered SOMA and DE is done with the GA (GA_1) approach of [20], greedy GA (GA_{Greedy}) of [1], GA (GA_2) of [13], Simulated Annealing algorithm (TB2M) of [3], Robust Tabu Search (RTS) of [36], Combined Simulated Annealing and Tabu Search (IA-SA-TS) of [25] and Ant Colony (HAS) of [14]. The results are given in Table 15.19.

Table 15.18 Irregular QAP comparison

Instant	**fd**	**n**	**Optimal**	GA_1	**HAS**	DE_{clust}	$SOMA_{clust}$
bur26a	2.75	26	5246670	-	**0**	**0**	**0**
bur26b	2.75	26	3817852	-	**0**	**0**	**0**
bur26c	2.29	26	5426795	-	**0**	**0**	**0**
bur26d	2.29	26	3821225	-	**0**	**0**	**0**
bur26e	2.55	26	5386879	-	**0**	**0**	**0**
bur26f	2.55	26	3782044	-	**0**	**0**	0.01
bur26g	2.84	26	10117172	-	**0**	**0**	**0**
bur26h	2.84	26	7098658	-	**0**	**0**	**0**
chr25a	4.15	26	3796	-	3.082	**0.07**	0.10
els19	5.16	19	17212548	-	**0**	**0**	**0**
kra30a	1.46	30	88900	**0**	0.629	0.024	0.002
kra30b	1.46	30	91420	**0**	0.071	0.015	0.027
tai20b	3.24	20	122455319	-	0.091	**0**	**0**
tai25b	3.03	25	344355646	-	**0**	**0**	**0**
tai30b	3.18	30	637117113	-	**0**	**0**	**0**
tai35b	3.05	35	283315445	-	0.025	0.002	**0**
tai40b	3.13	40	637250948	-	**0**	**0**	**0**
tai50b	3.1	50	458821517	-	0.192	**0**	0.2
tai60b	3.15	60	608215054	-	0.048	**0.012**	0.2
tai80b	3.21	80	818415043	-	0.667	**0.014**	0.4

Table 15.19 Regular QAP comparison

Instant	**fd**	**n**	**Optimal**	GA_1	GA_{Greedy}	GA_2	**TB2M**	**RTS**	**IA-SA-TS**	**HAS**	DE_{clust}	$SOMA_{clust}$
nug20	0.99	20	2570	-	-	-	-	-	-	**0**	**0**	**0**
nug30	1.09	30	6124	**0**	0.07	**0**	0.94	0.73	0.52	0.098	**0**	**0**
sko42	1.06	42	15812	**0**	0.250	**0**	0.66	1.03	0.46	0.076	**0**	**0**
sko49	1.07	49	23386	0.038	0.210	0.009	0.67	0.54	0.46	0.141	**0**	**0**
sko56	1.09	56	34458	**0**	0.02	0.001	0.66	0.53	0.50	0.101	**0**	**0**
sko64	1.07	64	48498	**0**	0.22	**0**	0.57	0.93	0.45	0.504	0.006	0.02
sko72	1.06	72	66256	0.042	0.29	0.014	0.60	0.52	0.48	0.702	**0.007**	0.04
sko81	1.05	81	90998	0.067	0.2	0.014	0.46	0.41	0.40	0.493	**0.01**	0.05
tai20a	0.61	20	703482	-	-	-	-	-	-	0.675	**0**	**0**
tai25a	0.6	25	1167256	-	-	-	-	-	-	1.189	**0**	**0**
tai30a	0.59	30	1818146	-	-	-	-	-	-	1.311	**0**	**0**
tai35a	0.58	35	2422002	-	-	-	-	-	-	1.762	**0**	**0**
tai40a	0.6	40	3139370	-	-	-	-	-	-	1.989	**0.019**	0.58
tai50a	0.6	50	4941410	-	-	-	-	-	-	2.8	**0.026**	0.42
tai60a	0.6	60	7208572	-	-	-	-	-	-	0.313	**0.012**	0.62
tai80a	0.59	80	13557864	-	-	-	-	-	-	1.108	**0.021**	0.95
wil50	0.64	50	48816	0.028	0.07	0.002	0.25	0.55	0.16	0.061	**0**	**0**

As with the irregular problem, DE_{clust} is the best performing algorithm. It manages to find the best value in 16 out of 17 instances, of which 10 are optimal values. $SOMA_{clust}$ is the second best heuristic with 10 best solutions, all of which are optimal values of those particular problems.

In terms of population dynamics, consider the initial population clustering of a sample population of "bur26a" instance as given in Figure 15.21.

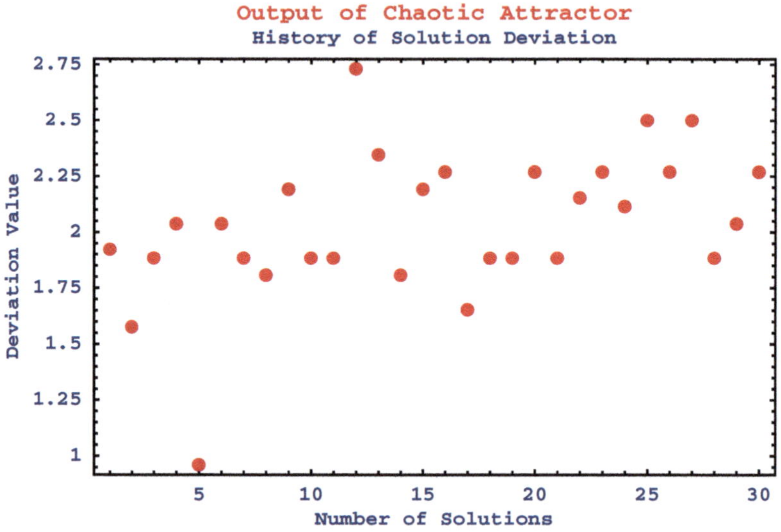

Fig. 15.21 Initial Population Clustering

The final population clustering is given in Figure 15.22.

The deviation of the solutions is from 1 - 2.75 in the initial population and 5 - 10 in the final population. This shows a *drift* of the solutions in the deviation space. Another point of interest is that the solutions are still diversified in their structure. The solutions within the clusters have converged, however the overall diversity is maintained within the population. This opens more opportunity to obtain better solutions in next generations.

The *spread* of the solutions in given in Figure 15.23.

The *Chaotic Edge* C_E of the population throughout the population generation (in this case, 200 generations) is given in Figure 15.24.

A general decline of the spread of the clusters and fitness values is seen. This is typical for a minimizing function.

The final graph of the best individual is seen in Figure 15.25.

A direct correlation is seen between the graphs of Chaotic Edge and Best Individual. The Edge is a prelude to a shift in solution space. A shift generally signifies a region of new solutions, and possibility of further improvement.

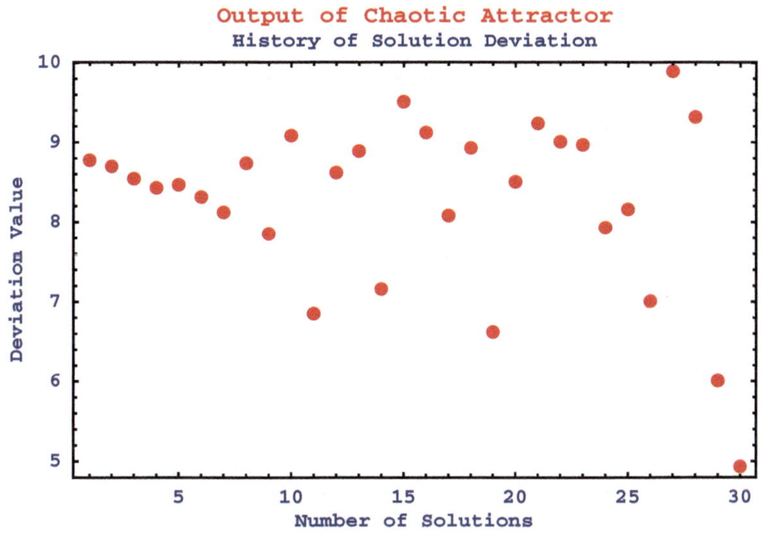

Fig. 15.22 Initial Population Clustering

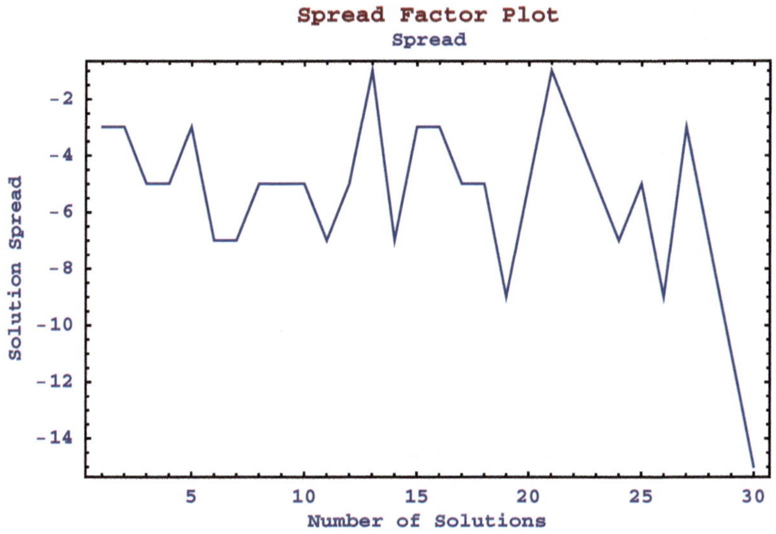

Fig. 15.23 Solution Spread

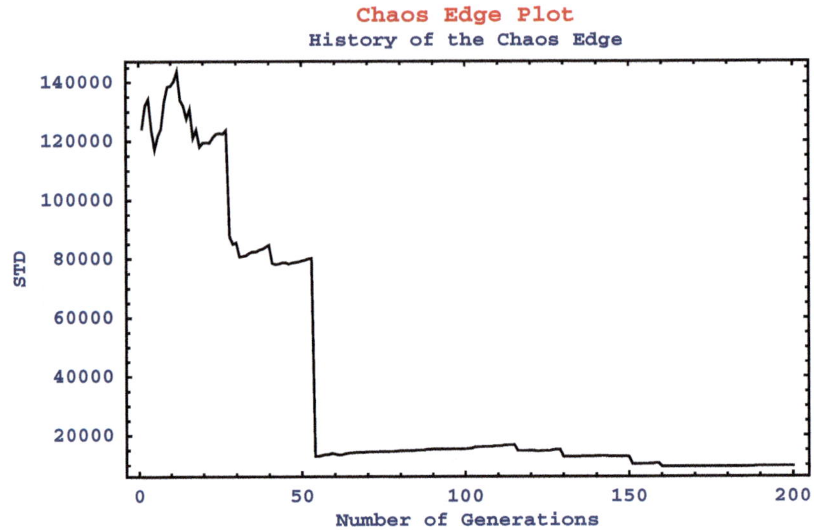

Fig. 15.24 Chaotic Edge

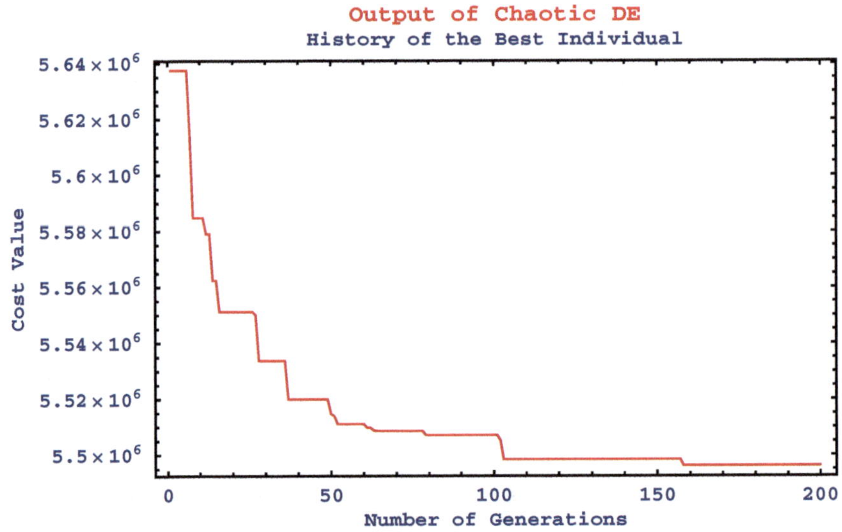

Fig. 15.25 Best Individual

15.9 Conclusion

Chaotic principles and attributes in respect to stagnation in evolutionary algorithms is the underlying principle of this research. An approach of bypassing local optima and creating a viable and diversified population is presented. This population relies on the two principles of chaos, Attractors and Edges. Attractors forms basin of solutions where solutions converge, where as an Edge is the limit along which feasible and better solutions can exist, taking in terms the information currently held by the population.

A dynamic population is devised which can be utilized by any heuristic. This population is embedded on three different heuristics of GA, DE and SOMA. Experimentation is first done with the canonical heuristics and then with clustered heuristics. A marked improvement is observed in the clustered results, which validate the approach of dynamic clustering of the population. Comparison is also done with published heuristics with very good results.

Acknowledgements. The following two grants are acknowledged for the financial support for this research.

1. Grant Agency of the Czech Republic **GARC 102/09/1680**
2. Grant of the Czech Ministry of Education **MSM 7088352102**

References

1. Ahuja, R., Orlin, J., Tiwari, A.: A descent genetic algorithm for the quadratic assignment problem. Comput. Oper. Res. 27, 917–934 (2000)
2. Aihara, K., Takabe, T., Toyoda, M.: Chaotic Neural Networks. Phys. Lett. A 6, 333–340 (1990)
3. Boelte, A., Thonemann, U.: Optimizing simulated annealing schedules with genetic programming. Eur. J. Oper. Res. 92, 402–416 (1996)
4. Burkard, R., Rendl, F.: A thermodynamically motivated simulation procedure for combinatorial optimisation problems. Eur. J. Oper. Res. 17, 169–174 (1994)
5. Chen, L., Kazuyuki, A.: Chaotic simulated annealing by a neural network model with transient chaos. Neural Networks 6(8), 915–930 (1995)
6. Connolly, D.: An improved annealing scheme for the QAP. Eur. J. Oper. Res. 46, 93–100 (1990)
7. Davendra, D.: Differential Evolution Algorithm for Flow Shop Scheduling, Bachelor Degree Thesis, University of the South Pacific (2001)
8. Davendra, D.: Hybrid Differential Evolution Algorithm for Discrete Domain Problems. Master Degree Thesis, University of the South Pacific (2003)
9. Davendra, D., Onwubolu, G.: Flow Shop Scheduling using Enhanced Differential Evolution. In: Proceeding of the 21st European Conference on Modelling and Simulation, Prague, Czech Republic, June 4-5, pp. 259–264 (2007)
10. Davendra, D., Onwubolu, G.: Enhanced Differential Evolution hybrid Scatter Search for Discrete Optimisation. In: Proceeding of the IEEE Congress on Evolutionary Computation, Singapore, September 25-28, pp. 1156–1162 (2007)
11. Davendra, D., Onwubolu, G.: Forward Backward Transformation. In: Onwubolu, G., Davendra, D. (eds.) Differential Evolution: A Handbook for Permutation-Based Combinatorial Optimization, pp. 35–80. Springer, Germany (2009)
12. Davendra, D., Zelinka, I.: Optimization of Quadratic Assignment Problem using Self-Organinsing Migrating Algorithm. Comput. Informat. 28, 169–180 (2009)

13. Drezne, Z.: A new genetic algorithm for the quadratic assignment problem. INFORMS Journal on Computing 115, 320–330 (2003)
14. Gambardella, L., Thaillard, E., Dorigo, M.: Ant Colonies for the Quadratic Assignment Problem. Int. J. Oper. Res. 50, 167–176 (1999)
15. Gleick, J.: Chaos: Making a New Science, Vintage, USA (1987)
16. Hochbam, D.: Approximation Algorithms for NP - Hard Problems. PWS Publishing Company, USA (1997)
17. Holland, J.: Adaptation in Natural and Artificial Systems. University of Michigan Press, Ann Arbor (1975)
18. Ikeguchi, T., Horio, Y.: Chaos for avoiding local minima A. Mutual Connection Neural Network Dynamics (White Paper)
19. Ishi, S., Sato, M.: Chaotic potts spin model for combinatorial optimization problems. Neural Networks 10, 941–963 (1997)
20. Ji, P., Yongzhong, W., Haozhao, L.: A solution method for the Quadratic Assignment Problem (QAP). In: Proceeding of the Sixth International Symposium on Operations Research and Its Applications (ISORA 2006), Xinjiang, China, August 8-12, pp. 106–117 (2006)
21. Koopmans, T., Beckmann, M.: Assignment problems and the location of economic activities. Econometrica 25, 53–76 (1957)
22. Lawler, E., Lensta, J., Rinnooy, K., Shmoys, D.: Sequencing and scheduling: algorithms and complexity. In: Graves, S., Rinnooy, K., Zipkin, P. (eds.) Logistics of Production and Inventory, pp. 445–522. North Holland, Amsterdam (1995)
23. Lin, F., Kao, C., Hsu: Applying the genetic approach to simulated annealing in solving NP- hard problems. IEEE Trans. Syst. Man Cybern. B Cybern. 23, 1752–1767 (1993)
24. May, R.: Stability and Complexity in Model Ecosystems. Princeton University Press, Princeton (2001)
25. Misevicius, A.: An Improved Hybrid Optimization algorithm for the Quadratic Assignment Problem. Mathematical Modelling and Analysis 9(2), 149–168 (2004)
26. Nozawa, H.: Chaos 2. Physics D 2, 377 (1992)
27. Onwubolu, G.: Optimisation using Differential Evolution Algorithm. Technical Report TR-2001-05, IAS (October 2001)
28. Onwubolu, G.: Emerging Optimisation Techniques in Production Planning and Control. Imperial Collage Press, London (2002)
29. Onwubolu, G., Clerc, M.: Optimal path for automated drilling operations by a new heuristic approach using particle swamp optimisation. Int. J. Prod. Res. 42(3), 473–491 (2004)
30. Onwubolu, G., Davendra, D.: Scheduling flow shops using differential evolution algorithm. Eur. J. Oper. Res. 171, 674–679 (2006)
31. Operations Reserach Library, `http://people.brunel.ac.uk/~mastjjb/jeb/info.htm` (Cited September 13, 2008)
32. Pinedo, M.: Scheduling: theory, algorithms and systems. Prentice Hall, Inc., New Jersey (1995)
33. Price, K.: An introduction to differential evolution. In: Corne, D., Dorigo, M., Glover, F. (eds.) New Ideas in Optimisation, pp. 79–108. McGraw Hill, International, UK (1999)
34. Price, K., Storn, R.: Differential evolution (2001), `http://www.ICSI.Berkeley.edu/~storn/code.html` (Cited September 10, 2008)

35. Sahni, S., Gonzalez, T.: P-complete approximation problems. J. ACM 23, 555–565 (1976)
36. Taillard, E.: Robust taboo search for the quadratic assignment problem. Parallel Comput. 17, 443–455 (1991)
37. Taillard, E.: Benchmarks for basic scheduling problems. Eur. J. Oper. Res. 64, 278–285 (1993)
38. Yamada, T., Aihara, K.: Nonlinear Neurodynamics and Combinatorial Optimization in Chaotic Neural Networks. J. Intell. Fuzzy Sys. 1(5), 53–68 (1997)
39. Zelinka, I.: Soma - Self Organizing Migrating Algorithm. In: Onwubolu, G., Babu, B. (eds.) New Optimization Techniques in Engineering. Springer, Germany (2004)

Chapter 16
Frontiers

Ivan Zelinka and Sergej Celikovsky

This book presents and discusses the interdisciplinary scientific field between deterministic chaos and evolutionary techniques. As demonstrated in the previous chapters, this research is very promising. In this chapter, we would like to offer a few exciting and realistic ideas and opinions for possible future directions of research and development on chaos and evolutionary techniques.

Let us first take a overview on the historical evolution of both involved disciplines.

Footprints of deterministic chaos as well as that of the evolutionary theory can be traced back to the 19th century. The first signs of chaos were discovered by the famous French mathematician Henri Poincaré when he studied the well-known three-body problem of the celestial mechanics. On the other hand, the revolutionary and yet perhaps also controversial evolutionary theory by Charles Darwin from the Great Britain generated another impulse in the human history. It is very interesting to note that independently of Darwin, the basic laws of the genetic inheritance had been defined and experimentally verified by Gregor Johann Mendel, the augustinian priest and scientist who lived in Brno, on the territory of the present Czech Republic. The sad story is that Mendel's letters about his discovery written to many scientific societies were discovered several decades thereafter, and remained unopened in the libraries! In the 19th century, both chaos theory and evolution theory were mainly of academic interest. On the contrary, in the 20th and 21st centuries, deterministic

Ivan Zelinka
Tomas Bata University in Zlin, Faculty of Applied Informatics, Nad Stranemi 4511, Zlin 76001, Czech Republic
and
VSB-TUO, Faculty of Electrical Engineering and Computer Science, 17. listopadu 15, 708 33 Ostrava-Poruba, Czech Republic
e-mail: zelinka@fai.utb.cz

Sergej Celikovsky
Control Theory Department, Institute of Information Theory and Automation, Academy of Sciences of Czech Republic, Pod Vodarenskou vezi 4, 182 08, Praha 8, Czech Republic
e-mail: celikovs@utia.cas.cz

I. Zelinka et al. (Eds.): Evolutionary Algorithms and Chaotic Systems, SCI 267, pp. 519–521.
springerlink.com

chaos and evolutionary theory brought up a growing number of real-world applications as well as new theory developments, especially after the 1950's.

The reason why deterministic chaos has become an area of engineering interest stems from the fact that many engineering applications involve nonlinear dynamical systems, which possibly generate chaotic behavior. In the past, when appropriate computational techniques that could be used to simulate or even solve such a problems were not available, these kinds of behavior had been either ignored or replaced by linear approximate models. Thanks to the modern powerful computational technologies and techniques (workstations, high performance computing, cloud computing, etc.), solving hard problems involving chaotic systems is no longer a problem. In fact, one has become capable of getting better results leading to more precise engineering outcome.

Deterministic chaos has been successfully applied to such areas as secure data encryption, oscillators synchronization, random-like number generators, system identification and reconstruction, and many others areas from systems engineering and information processing. On the other hand, the evolutionary theory, based on the principle of genetic inheritance, has also been successfully developed and applied to solve complicated and complex problems (see Chapter 2 for more details). The power of evolutionary techniques is evidenced by the fact that when being properly used, it is able to solve many hard (or "unsolvable") problems so as to obtain at least acceptable, sometimes even optimal solutions. Some typical examples are listed in Chapter 1, including the traveling salesman by Ant Colony Optimization (ACO) [3] and [4], aircraft engine improvement by Genetic Algorithms (GA), fingerprint identification [2] by GA, and many others. Applications can also be found within various fields like chemical engineering, mechanical engineering, electronics, aircraft design, logistics, manufacturing, and so on.

Evolutionary algorithms have been "transformed" into the so-called evolutionary hardware, which is currently a quite promising area of intensive research, with potential applications in robotics, defense technology and space technologies. It can be expected that evolutionary hardware will be utilized in the near future, in industrial applications as well as physical systems modeling and prediction (one possible realization has been recorded in quantum physics, [5]). Evolutionary algorithms, genetic programming and genetic programming-like techniques might also be used for engineering design of chaotic systems according to user-defined specifications, as discussed in Chapter 11. Another progress can be expected in evolutionary design of algorithms, as initiated in [1]). In this research, different versions of differential evolution were successfully synthesized by other evolutionary algorithms.

From the reports of this book (e.g., in chapter introductions and experimental results), it is foreseeable that the mutual interactions of evolutionary algorithms and chaotic dynamics are vital and valuable. Evolutionary algorithms can be used to easily handle very complex problems of informatics or to control very difficult engineering devices where chaotic dynamics play an important role (as a proper solution bypassing an "obstacle"), while chaotic behavior can also be observed in the dynamics of the evolutionary algorithms (see Chapter 15), so its "control" or elimination can significantly improve the performance of some such algorithms.

Lets summarize a few basic findings and facts. The first is that evolutionary algorithms are capable of getting solutions (at least acceptable from an engineering point of view) for hard problems whose complexity imposes so many possible solutions that there is no computer (even futuristic ones) that can verify all such solutions to find the best one (see Chapter 2). The second is that inside chaotic dynamics there is an infinite number of unstable trajectories, so it is a control engineering problem that can be used for stabilization and control, especially when the observed system has a truly black-box model (i.e., no mathematical knowledge is present to the designer). The third is that there exist some physical limits based on quantum mechanics (Chapter 2). These limits create troubles, which cannot be overcome by any existing or even hypothetical computer and thus give us computational limitations and restrictions. Therefore, one can foresee that the future of the computational techniques will, at least partially, be based on mutual fusion of evolutionary theory and chaos theory, in order to "find a shortcut" to get feasible solutions of extremely complex problems. Application of such interdisciplinary research can be expected in such fields like nanotechnology, complex networks (e.g., social networks and the Internet), automatic algorithms design, evolutionary hardware, etc. It is clear that if this indeed takes place, then the theoretical foundations will, in turn, be significantly enriched, especially in the areas of algorithm theory, computational biology, aerospace physics, complex networks, and complexity theory, and many others.

Even though the last paragraph presents the prognosis of the future impact of only the overlapping between chaos and evolution, based on the materials presented in this book, we are fairly confident that such a prognosis will eventually become a reality, which may actually happen very soon.

References

1. Oplatkova, Z.: Metaevolution - synthesis of evolutionary algorithms by means of symbolic regression, Ph.D. thesis, TBU Zlin (2007)
2. Hany, H.A., Tao, Y.: Fingerprint registration using genetic algorithms. In: 3rd IEEE Symposium on Application-Specific Systems and Software Engineering Technology (ASSET 2000), p. 148 (2000)
3. Stützle, T., Hoos, H.: The Max-Min Ant System and Local Search for the Travelling Salesman Problem. In: Bäck, T., Michalewicz, Z., Yao, X. (eds.) IEEE International Conference on Evolutionary Computation, Piscataway, pp. 309–314. IEEE Press, Los Alamitos (1997)
4. Gambardella, L.M., Dorigo, M.: Ant-Q: A Reinforcement Learning Approach to the Traveling Salesman Problem. In: Prieditis, A., Russell, S. (eds.) Proceedings of ML 1995, Twelfth International Conference on Machine Learning, Tahoe City, CA, pp. 252–260. Morgan Kaufmann, San Francisco (1995)
5. Bartels, R.A., Murnane, M.M., Kapteyn, H.C., Christov, I., Rabitz, H.: Learning from learning algorithms: Application to attosecond dynamics of high-harmonic generation. Phys. Rev. A 70, 043404 (2004)

Printing and Binding: Stürtz GmbH, Würzburg

GPSR Compliance

The European Union's (EU) General Product Safety Regulation (GPSR) is a set of rules that requires consumer products to be safe and our obligations to ensure this.

If you have any concerns about our products, you can contact us on ProductSafety@springernature.com

In case Publisher is established outside the EU, the EU authorized representative is:

Springer Nature Customer Service Center GmbH
Europaplatz 3
69115 Heidelberg, Germany

Batch number: 09699112

Printed by Printforce, the Netherlands